T0223477

Lecture Notes in Computer Science 9009

Commenced Publication in 1973
Founding and Former Series Editors:
Gerhard Goos, Juris Hartmanis, and Jan van Leeuwen

Editorial Board

David Hutchison
  Lancaster University, Lancaster, UK
Takeo Kanade
  Carnegie Mellon University, Pittsburgh, PA, USA
Josef Kittler
  University of Surrey, Guildford, UK
Jon M. Kleinberg
  Cornell University, Ithaca, NY, USA
Friedemann Mattern
  ETH Zurich, Zürich, Switzerland
John C. Mitchell
  Stanford University, Stanford, CA, USA
Moni Naor
  Weizmann Institute of Science, Rehovot, Israel
C. Pandu Rangan
  Indian Institute of Technology, Madras, India
Bernhard Steffen
  TU Dortmund University, Dortmund, Germany
Demetri Terzopoulos
  University of California, Los Angeles, CA, USA
Doug Tygar
  University of California, Berkeley, CA, USA
Gerhard Weikum
  Max Planck Institute for Informatics, Saarbrücken, Germany

More information about this series at http://www.springer.com/series/7412

C.V. Jawahar · Shiguang Shan (Eds.)

Computer Vision – ACCV 2014 Workshops

Singapore, Singapore, November 1–2, 2014
Revised Selected Papers, Part II

 Springer

Editors
C.V. Jawahar
Center for Visual Information Technology
International Institute of Information
  Technology
Hyderabad
India

Shiguang Shan
Institute of Computing Technology
Chinese Academy of Sciences
Beijing
China

ISSN 0302-9743   ISSN 1611-3349 (electronic)
Lecture Notes in Computer Science
ISBN 978-3-319-16630-8   ISBN 978-3-319-16631-5 (eBook)
DOI 10.1007/978-3-319-16631-5

Library of Congress Control Number: 2015934895

LNCS Sublibrary: SL6 – Image Processing, Computer Vision, Pattern Recognition, and Graphics

Springer Cham Heidelberg New York Dordrecht London
© Springer International Publishing Switzerland 2015
This work is subject to copyright. All rights are reserved by the Publisher, whether the whole or part of the material is concerned, specifically the rights of translation, reprinting, reuse of illustrations, recitation, broadcasting, reproduction on microfilms or in any other physical way, and transmission or information storage and retrieval, electronic adaptation, computer software, or by similar or dissimilar methodology now known or hereafter developed.
The use of general descriptive names, registered names, trademarks, service marks, etc. in this publication does not imply, even in the absence of a specific statement, that such names are exempt from the relevant protective laws and regulations and therefore free for general use.
The publisher, the authors and the editors are safe to assume that the advice and information in this book are believed to be true and accurate at the date of publication. Neither the publisher nor the authors or the editors give a warranty, express or implied, with respect to the material contained herein or for any errors or omissions that may have been made.

Printed on acid-free paper

Springer International Publishing AG Switzerland is part of Springer Science+Business Media
(www.springer.com)

Preface

The three-volume set of LNCS contains the carefully reviewed and selected papers presented at the 15 workshops that were held in conjunction with the 12th Asian Conference on Computer Vision, ACCV 2014, in Singapore, during November 1–2, 2014. These workshops were carefully selected from a large number of proposals received from almost all the continents.

This series contains 153 papers selected from 307 papers submitted to all the 15 workshops as listed below. A list of organizers for each of these workshops is provided separately.

1. Human Gait and Action Analysis in the Wild: Challenges and Applications
2. The Second International Workshop on Big Data in 3D Computer Vision
3. Deep Learning on Visual Data
4. Workshop on Scene Understanding for Autonomous Systems
5. RoLoD: Robust Local Descriptors for Computer Vision
6. Emerging Topics In Image Restoration and Enhancement
7. The First International Workshop on Robust Reading
8. The Second Workshop on User-Centred Computer Vision
9. International Workshop on Video Segmentation in Computer Vision
10. My Car Has Eyes - Intelligent Vehicles with Vision Technology
11. Feature and Similarity Learning for Computer Vision
12. The Third ACCV Workshop on e-Heritage
13. The Third International Workshop on Intelligent Mobile and Egocentric Vision
14. Computer Vision for Affective Computing
15. Workshop on Human Identification for Surveillance

Workshops in conjunction with ACCV have been emerging as a forum to present focused and current research in specific areas of interest within the broad scope of ACCV. This year, the workshops covered diverse research topics including both conventional ones (such as robust local descriptor) and newly emerging ones (such as deep feature learning). Besides direct submissions to the workshops, submissions rejected by the main conference were provided the opportunity to co-submit to the workshops, following the policy of previous ACCVs.

We would like to thank many people for their efforts in making this publication possible. General Chairs, Publication Chairs, and Local Organizing Chairs helped a lot in smoothly organizing the workshops and coming out with this proceedings. Reviewers of the individual workshops did an excellent job of selecting quality papers for the final presentation. They deserve credit for the excellent quality of the papers in this proceedings.

It is our pleasure to place these volumes in front of you.

November 2014

C.V. Jawahar
Shiguang Shan

Organization

ACCV 2014 Workshop Organizers

1. Human Gait and Action Analysis in the Wild: Challenges and Applications

Mark Nixon	University of Southampton, UK
Liang Wang	Chinese Academy of Sciences, China
Jian Zhang	University of Technology, Sydney, Australia
Qiang Wu	University of Technology, Sydney, Australia
Zhaoxiang Zhang	Beihang University, China
Yasushi Makihara	Osaka University, Japan

2. Second International Workshop on Big Data in 3D Computer Vision

Jian Zhang	University of Technology, Sydney, Australia; National ICT Australia, Australia
Mohammed Bennamoun	University of Western Australia, Australia
Fatih Porikli	NICTA, Australia
Ping Tan	National University of Singapore, Singapore
Hongdong Li	Australian National University, Australia
Lixin Fan	Nokia Research Centre, Finland
Qiang Wu	University of Technology, Sydney, Australia

3. Deep Learning on Visual Data

Wanli Ouyang	The Chinese University of Hong Kong, China
Xiaogang Wang	The Chinese University of Hong Kong, China
Kai Yu	Baidu, China
Quoc Le	Google, USA
Shuicheng Yan	National University of Singapore, Singapore

4. Workshop on Scene Understanding for Autonomous Systems (SUAS)

Sebastian Ramos	CVC, Universitat Autònoma de Barcelona, Spain
Raquel Urtasun	University of Toronto, Canada
Antonio Torralba	Massachusetts Institute of Technology, USA

Nick Barnes   NICTA and Australian National University,
                   Australia
Markus Enzweiler   Daimler AG, Germany
David Vazquez   CVC, Universitat Autònoma de Barcelona, Spain
Antonio M. Lopez   CVC, Universitat Autònoma de Barcelona, Spain

5. RoLoD: Robust Local Descriptors for Computer Vision

Jie Chen   CMV, University of Oulu, Finland
Zhen Lei   NLPR, Chinese Academy of Sciences, China
Li Liu   VIP, University of Waterloo, Canada
Guoying Zhao   CMV, University of Oulu, Finland
Matti Pietikäinen CMV, University of Oulu, Finland

6. Emerging Topics In Image Restoration and Enhancement

Zhe Hu   University of California Merced, USA
Oliver Cossairt   Northwestern University, USA
Yu-Wing Tai   KAIST, Korea
Sunghyun Cho   Samsung Electronics, Korea
Chih-Yuan Yang   University of California Merced, USA
Robby Tan   SIM University, Singapore

7. First International Workshop on Robust Reading

Masakazu Iwamura   Osaka Prefecture University, Japan
Dimosthenis Karatzas   CVC, Universitat Autònoma de Barcelona, Spain
Faisal Shafait   University of Western Australia, Australia
Pramod Sankar Kompalli Xerox Research India, India

8. Second Workshop on User-Centred Computer Vision (UCCV 2014)

Gregor Miller   University of British Columbia, Canada
Darren Cosker   University of Bath, UK
Kenji Mase   Nagoya University, Japan

9. International Workshop on Video Segmentation in Computer Vision

Michael Ying Yang   Leibniz University Hannover, Germany
Jason Corso   University of Michigan, Ann Arbor, USA

10. My Car Has Eyes - Intelligent Vehicles with Vision Technology

Xue Mei   Toyota Research Institute North America,
                   Ann Arbor, USA
Andreas Geiger   Max Planck Institute for Intelligent Systems,
                   Germany
Michael James   Toyota Research Institute North America,
                   Ann Arbor, USA
Yi-Ping Hung   National Taiwan University, Taiwan
Fatih Porikli   Australian National University, Australia
Danil Prokhorov   Toyota Research Institute North America,
                   Ann Arbor, USA

11. Feature and Similarity Learning for Computer Vision

Jiwen Lu   Advanced Digital Sciences Center, Singapore
Shenghua Gao   ShanghaiTech University, China
Gang Wang   Nanyang Technological University, Singapore
Weihong Deng   Beijing University of Posts and
                   Telecommunications, China

12. Third ACCV Workshop on e-Heritage

Takeshi Oishi   University of Tokyo, Japan
Ioannis Pitas   Aristotle University of Thessaloniki, Greece
Bo Zheng   University of Tokyo, Japan
Manjunath Joshi   DA-IICT, Gandhinagar, India
Anupama Mallik   Indian Institute of Technology, Delhi, India

13. Third International Workshop on Intelligent Mobile and Egocentric Vision (IMEV2014)

Chu-Song Chen   Academia Sinica, Taiwan, China
Mohan Kankanhalli   National University of Singapore, Singapore
Shang-Hong Lai   National Tsing Hua University, Taiwan, China
Joo Hwee Lim   Institute for Infocomm Research, Singapore
Vijay Chandrasekhar   Institute for Infocomm Research, Singapore
Liyuan Li   Institute for Infocomm Research, Singapore
Yu-Chiang Frank Wang   Academia Sinica, Taiwan, China
Shuicheng Yan   National University of Singapore, Singapore

14. Computer Vision for Affective Computing (CV4AC)

Abhinav Dhall   University of Canberra/Australian National
                   University, Australia

Roland Goecke   University of Canberra/Australian National
                 University, Australia
Nicu Sebe   University of Trento, Italy

15. Workshop on Human Identification for Surveillance (HIS)

Tao Xiang   Queen Mary University of London, UK
Nalini K. Ratha   IBM Research, USA
Venu Govindaraju   University at Buffalo, USA
Meina Kan   Chinese Academy of Sciences, China
Wei-Shi Zheng   Sun Yat-sen University, China
Marco Cristani   University of Verona, Italy

Contents – Part II

Second Workshop on User-Centred Computer Vision (UCCV 2014)

International Workshop on Video Segmentation in Computer Vision

My Car Has Eyes: Intelligent Vehicle with Vision Technology

Third ACCV Workshop on E-Heritage

Workshop on Computer Vision for Affective Computing (CV4AC)

Emerging Topics on Image Restoration and Enhancement

Multi-view Image Restoration from Plenoptic Raw Images

Shan Xu[1]([✉]), Zhi-Liang Zhou[2], and Nicholas Devaney[1]

[1] School of Physics, National University of Ireland, Galway, Republic of Ireland
s.xu1@nuigalway.ie
[2] Academy of Opto-electronics, Chinese Academy of Sciences, Beijing, China

Abstract. We present a reconstruction algorithm that can restore the captured 4D light field from a portable plenoptic camera without the need for calibration images. An efficient and robust estimator is proposed to accurately detect the centers of microlens images. Based on that estimator, parameters that model the centers of microlens array images are obtained by solving a global optimization problem. To further enhance the quality of reconstructed multi-view images, a novel 4D demosaicing algorithm based on kernel regression is also proposed. Our experimental results show that it outperforms state of the art algorithms.

1 Introduction

Plenoptic cameras, also known as light field cameras, are capable of capturing the radiance of light. In fact, the principle of the plenoptic camera was proposed more than a hundred years ago [1]. Thanks to the recent advances in optical fabrication and computational power, plenoptic cameras are already commercially available as a consumer commodity. There are several types of plenoptic cameras [2–5]. In this paper we focus on restoring the light field from the first consumer light field camera, the Lytro [6]. The light rays inside the camera are characterized by two planes, the exit pupil and the plane of the microlens array, which is known as two plane parametrization of 4D light field [7,8]. Each microlens image is an image of the exit pupil viewing at different angles on the sensor plane. However, in such a spatially multiplexing device, the price to pay is the significant loss of spatial resolution. Having the 4D light field enables both novel photographic and scientific applications such as refocusing [9], changing perspective, depth estimation [10,11] and measuring the particle's velocity in 3D [12,13]. Evidently, these applications all rely on high quality 4D light field reconstruction.

The recent growing interest in light field imaging has resulted in several papers addressing the calibration and reconstruction of the light field from a microlens-based light field camera. Donald et al. [14] proposed a decoding, calibration and rectification pipeline. Cho et al. [15] introduced a learning based interpolation algorithm to restore high quality light field images. Yunsu et al. [16] proposed a line feature based geometric calibration method for a microlens-based light field camera. All these approaches mentioned above require a uniform illuminated image as a calibration reference image. One exception is Juliet's [9]

© Springer International Publishing Switzerland 2015
C.V. Jawahar and S. Shan (Eds.): ACCV 2014 Workshops, Part II, LNCS 9009, pp. 3–15, 2015.
DOI: 10.1007/978-3-319-16631-5_1

recent work, which proposed to use dark vignetting points as a metric to find the spatial translation of the microlens array with respect to the center of image sensor (Fig. 1).

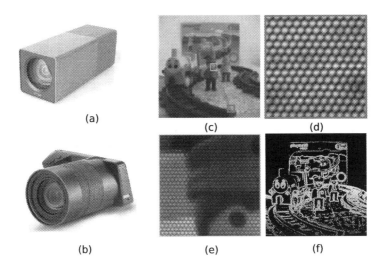

Fig. 1. (a), (b) The first and second generation Lytro cameras. (c) The microlens array raw image. (d), (e) The light field raw image (after demosaicing) with close-up views. (f) The depth estimation result from the light field raw image.

Most traditional digital color cameras use a single image sensor with a Color Filter Array (CFA) [17] on top of the senor to capture Red, Blue and Green color information. This is also a spatial-multiplexing method which gains multi-color information at the expense of losing spatial resolution. A typical light field color camera is also equipped with such a CFA sensor. The process of recovering the full color information in a single pixel is called demosaicing. Although demosaicing is a well explored topic in the image processing community, only some work has discussed demosaicing for a plentopic camera. Todor [18] proposed a demosaicing algorithm after refocusing which can reduce the color artifacts of the refocused image. Recently, Xiang et al. [19] proposed a learning based algorithm which considers the correlations between angular and spatial information. However the algorithm they proposed requires nearly an hour processing time with PC equipped with an Intel $i3 - 4130$ CPU.

In this paper, we present an efficient and robust processing pipeline that can restore the light field a.k.a the multi-view image array from natural light field images which doesn't need calibration images. We formulate estimating the parameters of microlens image center grid as an energy minimization problem. We also propose a novel light field demosaicing algorithm which is based on a 4D kernel regression that has been widely used in computer vision for the purpose of de-noising, interpolation and super-resolution [20]. It is tedious to process the light field raw image taken from different cameras or even with different optical settings which all require corresponding calibration images. As our light field reconstruction algorithm is calibration file free, it simplifies the processing complexity and reduces the file storage space. Our dataset and source code are available on-line[1].

2   The Grid Model of Plenoptic Images

In this section, we derive the relation between the ideal and the practical microlens image centers which can be described by an affine transformation. In this paper, we focus on the light field camera with a microlens array placed at the focus plane of the main lens [2].

Applying a pinhole camera model to the microlenses, the center of the main lens is projected to the sensor by a microlens as shown in Fig. 2(a). The microlens center (x_i, y_i) and its corresponding microlens image center (x'_i, y'_i) has the following geometric relation,

$$\begin{pmatrix} x'_i \\ y'_i \end{pmatrix} = \frac{Z'}{Z} \begin{pmatrix} x_i \\ y_i \end{pmatrix} \tag{1}$$

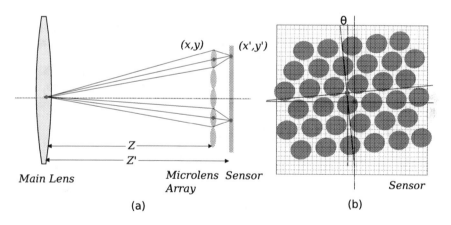

(a)   (b)

Fig. 2. (a) The main lens center is projected to sensor plane. (b) The installation errors include a rotation angle θ and a translation offset $(\Delta x, \Delta y)$. The physical center of microlens array is highlighted in red and the physical center of sensor is highlighted in blue (Color figure online).

Ideally, the microlens array has perfect regular arrangement such as a square or hexagonal grid. Nevertheless, the manufacturing error and installation error can be observed in the raw images. The lateral skew parameters σ_1, σ_2, the rotation angle θ and the translation parameters T_x, T_y are considered as the

---

[1] https://sites.google.com/site/lfmulview.

main sources of position displacement. If we use s to substitute $\frac{Z'}{Z}$ in Eq. 1, and approximate $sin\theta = \epsilon, cos\theta = 1$ as the rotation is tiny, we obtain,

$$\begin{pmatrix} x' \\ y' \\ 1 \end{pmatrix} = s \begin{pmatrix} 1 & \epsilon & 0 \\ -\epsilon & 1 & 0 \\ 0 & 0 & 1 \end{pmatrix} \cdot \begin{pmatrix} 1 & \sigma_1 & T_x \\ \sigma_2 & 1 & T_y \\ 0 & 0 & 1 \end{pmatrix} \cdot \begin{pmatrix} x \\ y \\ 1 \end{pmatrix} = s \begin{pmatrix} \sigma_2 - \epsilon & \sigma_2\epsilon + 1 & Tx \\ \sigma_1 - \epsilon & \sigma_1\epsilon + 1 & Ty \\ 0 & 0 & 1 \end{pmatrix} \cdot \begin{pmatrix} x \\ y \\ 1 \end{pmatrix},$$
$$(2)$$

Equation (2) shows that the relation between the microlens center (x_i, y_i) and its image center (x'_i, y'_i) can be expressed by an affine transform with six parameters. The above derivation explains why an affine transform matrix is preferred as a good transformation model to be used for estimating the centers of the microlens array.

3 Multi-view Images Restoration Algorithm

In decoding the 4D light field a.k.a extract the multi-view image array from a 2D raw image, the center of each microlens image is regarded as the origin of the embedded 2D data. To accurately detect the position of each microlens image is the fundamental step in restoring a high quality light field. We first introduce a robust local centroiding estimator which is insensitive to the content and shape of microlens image compared to conventional centroiding estimators. Next, we formulate estimating the grid parameters problem in terms of energy minimization. We break our brute-force search algorithm into three steps to reduce the processing time. In the last section, a 4D demosaicing algorithm is exploited to improve the quality of the reconstructed images.

3.1 Local Centroiding Estimator

In Cho et al's. and Dansereau et al.'s papers [14,15], the center of each microlens image is determined either by convolving a 2D filter or performing an eroding operation from the uniform illumination light field raw image. However, the limitation of previous methods is that the image needs to be uniform and in a circular shape. In practice, some microlens images are distorted by vignetting effect [21] and the Bayer filter makes the image less uniform. In contrast, we measure the individual microlens image centers by examining the dark pixels among the microlens images. Concretely, for either a square or hexagonal microlens array, there are dark gaps between microlens images. For example, as shown in Fig. 3, the position of the darkest spots of a hexagonal grid with respect to the center of a microlens are $p_0 = (R, \frac{R}{\sqrt{3}})$, $p_1 = (0, \frac{2R}{\sqrt{3}})$, $p_2 = (-R, \frac{R}{\sqrt{3}})$, $p_3 = (-R, -\frac{R}{\sqrt{3}})$, $p_4 = (0, -\frac{2R}{\sqrt{3}})$, $p_5 = (R, -\frac{R}{\sqrt{3}})$, where R is the radius of the hexagonal grid. For an arbitrary pixel at position $x = [x, y]^T$ of the light field raw image I, the summation of the six special surrounding pixels denoted by $P(x)$ is used for detecting the center of the microlens image. To achieve sub-pixel accuracy, we up-sample the image by a factor of 8 with cubic interpolation. Additionally, to reduce the effect of dark current, a Gaussian filter is applied before the up-sampling. The local centroiding estimator is defined as a score map $\mathcal{P}(\boldsymbol{x})$,

$$\mathcal{P}(\boldsymbol{x}) = \boldsymbol{\Sigma}_{i=0}^{5}(G_{\sigma} * I_{\uparrow 8})(\boldsymbol{x} + \boldsymbol{p_i}) \tag{3}$$

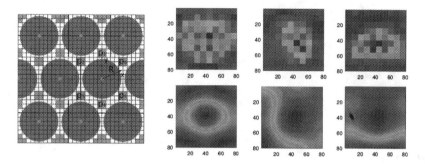

Fig. 3. Left: The hexagonal grid microlens array image. The dark gap pixels are labeled in pink. Right: top row are the microlens images with different shapes. Bottom row are our local estimator score maps $\mathcal{P}$ (Color figure online).

If the light field raw image is uniformly illuminated, $P(\boldsymbol{x})$ reaches a minimum only when $\boldsymbol{x}$ is at the center of a microlens image. The nice property of this operator is that it constantly produces minimum when $\boldsymbol{x}$ is the center of a microlens image regardless of its content. Notice that, if there are some under-exposed pixels inside the surrounding microlens images, then the multiple minimum points may exist. The center point $\boldsymbol{x}_{center}$ belongs to the minimum points set,

$$\boldsymbol{x}_{center} \in \{\boldsymbol{x}_i | \mathcal{P}(\boldsymbol{x}_i) = \mathcal{P}_{min}, i = 0, ...N\} \tag{4}$$

Evidently, our local estimator is not able to find all the microlens image centers from a natural light field raw image individually. In the next section, instead of using the local minimum to identify the individual microlens image center, we propose a global optimization scheme to estimate the global grid parameters.

3.2 Global Grid Parameters Optimization

For a natural scene, it is impractical to accurately measure the centers of those microlens images that are under-exposed or over-exposed. As a consequence, estimating the transformation matrix [22] by the method of minimizing Euclidean distance between the ideal and real point set is not applicable to this problem. Our approach first generates an up-sampled centroiding score map $\mathcal{P}$ based on our local estimators. Then we can use the summation of all pixels on the grid as a metric to measure how well is the grid fitted to the centers of the microlens array. As shown in Fig. 4, only a best fitted grid model will produce the global minimum.

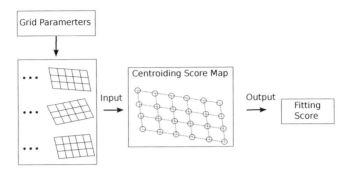

Fig. 4. Grid Fitting. Only when the grid parameters are all optimum, the fitting score reaches minimum, as highlighted in red color (Color figure online).

Thus we formulate it as a global optimization problem. The cost function $\mathcal{F}$ is defined as,

$$\mathcal{F}(s, \sigma_1, \sigma_2, \epsilon, Tx, Ty) = \Sigma_{j=1}^{M}\Sigma_{i=1}^{N}\boldsymbol{P}(\boldsymbol{T}(s, \sigma_1, \sigma_2, \epsilon, Tx, Ty) \cdot \boldsymbol{x}_{ji}) \qquad (5)$$

where $\boldsymbol{x}_{ij} = [x_{ij}, y_{ij}]^T$ is the spatial position of ijth microlens center and $\mathcal{T}$ is the homogeneous affine transformation for which the matrix form is given in Eq. 1. The cost function $\mathcal{F}$ reaches a global minimum only when the grid parameters are accurately estimated.

In our experiment, several numerical optimization methods have been applied to solve this problem. For example, the NelderMead algorithm [23] has fast convergence rates but occasionally gets stuck at a local minimum. The simulated annealing algorithm [24] as a probabilistic method guarantees the solution is a global minimum but the rate of convergence is very slow. Also tuning the parameters such as the cooling factor can be troublesome. Considering there are only small affine transformation between practical and ideal microlens image centers, we perform a coarse-to-fine brute-force searching scheme. The perfect microlens center grid $\{\boldsymbol{x}_{ji}|i = 0 \cdots N, j = 0 \cdots N\}$ is used as the initial condition and is constructed based on the geometric arrangement of the microlens array. We assume the physical separation between microlenses d and pixel size l are known parameters. For a hexagonal grid microlens array, we have,

$$\boldsymbol{x}_{ji} = \begin{cases} (i \cdot \frac{d}{l}, j \cdot \frac{\sqrt{3}d}{l}) & \text{for j is odd} \\ (i \cdot \frac{d}{l}, j \cdot \frac{\sqrt{3}d}{l}) - (\frac{d}{2}, \frac{\sqrt{3}d}{2}) & \text{for j is even} \end{cases} \qquad (6)$$

To speed up the searching, we set reasonable boundary constraints for each parameter. The spatial translation T_x and T_y are in the range of $[-\frac{d}{2l}, \frac{d}{2l}]$. We also assume the rotation and skew angle is within ± 0.1 degree. To search the parameter in 6 dimensions will be time consuming. We divide it into three steps: we first search T_x and T_y then $s, \sigma_1, \sigma_2, \epsilon$ and finally refine T_x and T_y. Each step includes several searches with different resolution as illustrated in Fig. 5(a). For each search, the optimal solution from the previous search is used as the searching center, and the searching range is narrowed down by one half. Figure 5(b)

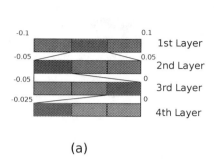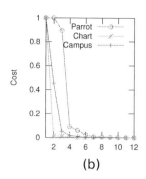

(a)   (b)

Fig. 5. (a) Sketch of our coarse-to-fine searching algorithm. The sub-region highlighted in dark color is the optimal parameter at the current layer's resolution. (b) The cost function converges within 8 iterations. Three scenes were captured with different Lytro cameras (Color figure online).

shows that with different scenes and cameras, our proposed algorithm has fast convergence. We summarize the algorithm as follows in Table 1.

As mentioned above, for a natural light field image, some parts of microlens images or entire microlens images might be under-exposed and this might influence the accuracy of our proposed algorithm. However, our experiment shows that the under-exposure effect only has minor impact on the estimation accuracy. We compare the microlens image centers estimated from the white uniform illumination scene and a natural scene with the same optical settings. The largest error is within half a pixel and it occurs only when there are large under-exposed regions (Fig. 6).

3.3   4D Light Field Demosaicing

Applying a conventional 2D demosaicing algorithm to the light field raw images produces noticeable color artifacts. The reason is that pixels on the boundary of microlens image are interpolated with the pixels from adjacent microlens images which are not their 4D neighbors. Intuitively, in contrast to 2D demosaicing, 4D demosaicing should result in better quality if the coherence of both angular and spatial information is considered. In order to infer the interest pixel value from the structure of its 4D neighbors, we use the first order 4D kernel regression method. Concretely, borrowing the notation from [20], the local 4D function $f(\boldsymbol{x})$, $\boldsymbol{x} \in \mathbb{R}^4$ at a given sample point $\boldsymbol{x}_i$, can be expressed by Taylor expansion,

$$f(\boldsymbol{x}_i) = f(\boldsymbol{x}) + (\boldsymbol{x} - \boldsymbol{x}_i)^T \nabla f(\boldsymbol{x}) + \cdots \qquad (7)$$

where $\nabla f(\boldsymbol{x}) = [\frac{\partial f(\boldsymbol{x})}{\partial x_0}, \frac{\partial f(\boldsymbol{x})}{\partial x_1}, \frac{\partial f(\boldsymbol{x})}{\partial x_2}, \frac{\partial f(\boldsymbol{x})}{\partial x_3}]^T$

Equation (7) can be converted to a linear filtering formula,

Input   : Centroiding score map $\mathcal{P}$
Output   : Optimum parameters $s_0, \sigma_{10}, \sigma_{20}, \epsilon_0, Tx_0, Ty_0$

Processing:
Step 0. Parameter initialization $s_0, \sigma_{10}, \sigma_{20}, \epsilon_0, Tx_0, Ty_0$.
Step 1. 2D search to find optimum Tx and Ty.
for $k \leftarrow 0$ **to** K **do**
 for $j \leftarrow -N$ **to** N **do**
  for $i \leftarrow -N$ **to** N **do**
   $Tx_i = Tx_0 + \delta x \cdot i$
   $Ty_i = Ty_0 + \delta x \cdot j$
   Update $\mathcal{F}$
   if $\mathcal{F} < \mathcal{F}_{min}$ **then** $\mathcal{F}_{min} \leftarrow \mathcal{F}$, $T_x = Tx_i, T_y = Ty_j$;
   ;
  end
 end
 Scale down the searching range to $[-\frac{N}{K^k}, \frac{N}{K^k}]$.
end
Step 2. 4D search for finding optimum $s, \sigma_1, \sigma_2, \epsilon$.
Update $s, \sigma_1, \sigma_2, \epsilon$ similar to Step 1.
Step 3. Refine optimum Tx, Ty similar to Step 1.

Algorithm 1. Brute-Force coarse-to-fine Searching

Table 1. Grid modeling prediction error

	ISO chart	Campus	Parrot	Toy	Flower
$l2-\textbf{norm}$	0.014	0.333	0.246	0.151	0.490

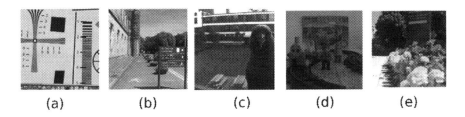

(a)   (b)   (c)   (d)   (e)

Fig. 6. Test scene. (a) ISO chart. (b) Campus. (c) Parrot. (d) Toy. (e) Flower.

$$f(\boldsymbol{x_i}) \approx \beta_0 + \boldsymbol{\beta}_1^T (\boldsymbol{x} - \boldsymbol{x_i}) \tag{8}$$

where $\beta_0 = f(\boldsymbol{x})$, $\boldsymbol{\beta}_1 = [\frac{\partial f(\boldsymbol{x})}{\partial x_0}, \frac{\partial f(\boldsymbol{x})}{\partial x_1}, \frac{\partial f(\boldsymbol{x})}{\partial x_2}, \frac{\partial f(\boldsymbol{x})}{\partial x_3}]^T$,

Therefore a 4D light field demosaicing problem is the estimation of an unknown pixel $\boldsymbol{x}$ from a measured irregularly sampled data set $\{\boldsymbol{x_i} \in \mathbb{R}^4 | i = 1, \dots, N\}$. The solution is a weighted least squares problem, in the form,

$$\hat{\boldsymbol{b}} = \underset{\boldsymbol{b}}{\mathrm{argmin}}(\boldsymbol{y} - \boldsymbol{X}\boldsymbol{b})^T \boldsymbol{K}(\boldsymbol{y} - \boldsymbol{X}\boldsymbol{b}) \tag{9}$$

where $\boldsymbol{y} = [f_1, f_2, \ldots, f_N]^T$, $\boldsymbol{b} = [\beta_0, \boldsymbol{\beta}_1^T]$, $\boldsymbol{X} = \begin{bmatrix} 1 & \boldsymbol{x} - \boldsymbol{x}_1 \\ 1 & \boldsymbol{x} - \boldsymbol{x}_2 \\ \vdots & \vdots \\ 1 & \boldsymbol{x} - \boldsymbol{x}_N \end{bmatrix}$

$$\boldsymbol{K} = diag[K(\boldsymbol{x} - \boldsymbol{x}_0), K(\boldsymbol{x} - \boldsymbol{x}_1), K(\boldsymbol{x} - \boldsymbol{x}_2), \cdots, K(\boldsymbol{x} - \boldsymbol{x}_{N-1})]$$

The detailed derivation of the above formulas in N-dimensions can be found in [20]. We use a Gaussian function as the Kernel function and only the pixels which are within a distance of 2 pixels are included in the sample data set in each dimension. In the experimental results section, we compare the demosaicing result with our proposed algorithm, 4D-quad-linear interpolation method and 2D demosaicing method.

3.4   Mult-view Reconstruction Pipeline

A plentoptic camera processing pipeline is shown in Fig. 7. Note that our processing pipeline only requires light field raw images.

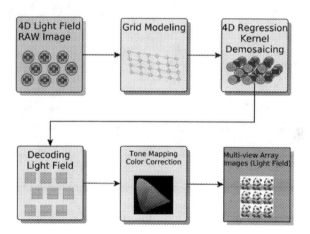

Fig. 7. Our proposed plentoptic camera processing pipeline

4   Experimental Result

Our experiment is based on the first commercially available consumer light field camera, the Lytro [6]. It has approximately 360 by 380 microlenses. There are around 10 by 10 pixels under each individual microlens. The resolution of the image sensor is $3{,}280 \times 3{,}280$ pixels. To avoid aliasing of boundary pixels of the microlens image, we extract a 9 by 9 multi-view array. Our light field reconstruction algorithm is implemented in C++. It takes around 1 min to build the grid model and 8 min to extract the whole multiview array images with an Intel $i3 - 4130$ CPU.

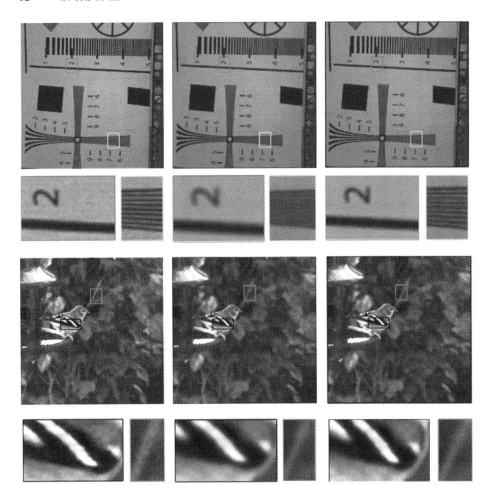

Fig. 8. Demosaicinge examples using real world examples. Left column: 2D demosaicing. Central column: 4D quad-llinear demosaicing. Right column: 4D kernel regression demosaicing.

To verify our 4D demosaicing algorithm, in Fig. 8 we compare our method with traditional 2D demosaicing and 4D quad-linear interpolation method. The result from 2D demosaicing looks sharper but also contains much more color artifacts than our result. The result from 4D demosaicing has the fewest color artifacts, but it is too blurry as each pixel is interpolated with surrounding pixels with weight proportional to the distance without considering the underlying data structure.

In Fig. 9 we also compare our reconstructed multiview image with Dansereau et al.'s [14] and Cho et al.'s [15] results. They reconstructed the images using both light field raw image and calibration image, but we only process the light field raw images. We didn't compare Cho et al.'s result after dictionary learning

Fig. 9. Top row: left is Don et al.'s [14] result, right is our result. Bottom row: left is Cho et al.'s [15] result, right is our result. The image is the central view cropped from the reconstructed multiview light field image.

as our purpose is to reconstruct the light field image with the single raw image. From the comparison, our results produce less artifacts and are less noisy than both their results.

5 Conclusion

In this paper, we have presented a simple and efficient method that is able to reconstruct the light field from the natural light field raw image without the need for reference images. To accurately extract 4D light field data from a 2D light field raw image, the parameters of the grid model of the microlens array are optimized by solving a global optimization problem. We describe our detailed implementation of coarse-to-fine brute force search. We also demonstrate that the content inside the microlens image has only minor impact on the accuracy of the grid model. For the purpose of further improving the quality of the reconstructed light field, a 4D demosaicing algorithm is introduced. In our further work, we plan to include vignetting correction and geometry distortion correction into our light field processing pipeline.

References

1. Gabriel, L.: La photographie intégrale. Comptes-Rendus, Académie des Sciences **146**, 446–551 (1908)
2. Ng, R., Levoy, M., Brédif, M., Duval, G., Horowitz, M., Hanrahan, P.: Light Field Photography with a Hand-Held Plenoptic Camera. Technical report (2005)
3. Lumsdaine, A., Georgiev, T.: Full resolution lightfield rendering. Technical report, Adobe (2008)
4. Veeraraghavan, A., Raskar, R., Agrawal, A., Mohan, A., Tumblin, J.: Dappled photography: Mask enhanced cameras for heterodyned light fields and coded aperture refocusing. In: SIGGRAPH, vol. 26 (2007)
5. Liang, C.K., Lin, T.H., Wong, B.Y., Liu, C., Chen, H.: Programmable aperture photography: multiplexed light field acquisition. In: SIGGRAPH, vol. 27, pp. 55:1–55:10 (2008)
6. Georgiev, T., Yu, Z., Lumsdaine, A., Goma, S.: Lytro camera technology: Theory, algorithms, performance analysis. In: MCP, SPIE (2013)
7. Levoy, M., Hanrahan, P.: Light field rendering. In: SIGGRAPH, pp. 31–42 (1996)
8. Gortler, S.J., Grzeszczuk, R., Szeliski, R., Cohen, M.F.: The lumigraph. In: SIGGRAPH, pp. 43–54 (1996)
9. Fiss, J., Curless, B., Szeliski, R.: Refocusing plenoptic images using depth-adaptive splatting (2014)
10. Tao, M.W., Hadap, S., Malik, J., Ramamoorthi, R.: Depth from combining defocus and correspondence using light-field cameras (2013)
11. Yu, Z., Guo, X., Ling, H., Lumsdaine, A., Yu, J.: Line assisted light field triangulation and stereo matching. In: ICCV. IEEE (2013)
12. Lynch, K., Fahringer, T., Thurow, B.: Three-dimensional particle image velocimetry using a plenoptic camera. 50th AIAA Aerospace Sciences Meeting including the New Horizons Forum and Aerospace Exposition (2012)
13. Garbe, C.S., Voss, B., Stapf, J.: Plenoptic particle streak velocimetry (ppsv): 3d3c fluid flow measurement from light fields with a single plenoptic camera. In: 16th International Symposium on Applications of Laser Techniques to Fluid Mechanics (2012)
14. Dansereau, D.G., Pizarro, O., Williams, S.B.: Decoding, calibration and rectification for lenselet-based plenoptic cameras. In: CVPR. IEEE (2013)

15. Cho, D., Lee, M., Kim, S., Tai, Y.W.: Modeling the calibration pipeline of the lytro camera for high quality light-field image reconstruction. In: ICCV. IEEE (2013)
16. Yunsu Bok, H.G.J., Kweon, I.S.: Geometric calibration of micro-lens-based light-field cameras using line features. In: ICCV. IEEE (2014)
17. Bayer, B.E.: Color image array (1976)
18. Georgiev, T.: An analysis of color demosaicing in plenoptic cameras. In: Proceedings of the 2012 IEEE Conference on Computer Vision and Pattern Recognition (CVPR), CVPR 2012, pp. 901–908. IEEE Computer Society, Washington, DC (2012)
19. Xiang Huang, O.C.: Dictionary learning based color demosaicing for plenoptic cameras. In: ICCV. IEEE (2013)
20. Takeda, H., Farsiu, S., Milanfar, P.: Kernel regression for image processing and reconstruction. IEEE Trans. Image Process. **16**, 349–366 (2007)
21. Xu, S., Devaney, N.: Vignetting modeling and correction for a microlens-based light field camera. In: IMVIP (2014)
22. Sabater, N., Drazic, V., Seifi, M., Sandri, G., Perez, P.: Light-field demultiplexing and disparity estimation (2014)
23. Nelder, J.A., Mead, R.: A simplex method for function minimization. Comput. J. **7**, 308–313 (1965)
24. Kirkpatrick, S., Gelatt, C.D., Vecchi, M.P.: Optimization by simulated annealing. Science **220**, 671–680 (1983)

On the Choice of Tensor Estimation for Corner Detection, Optical Flow and Denoising

Freddie Åström[1,2]($\boxtimes$) and Michael Felsberg[1,2]

[1] Computer Vision Laboratory, Linköping University, Linköping, Sweden
freddie.astrom@liu.se
[2] Center for Medical Image Science and Visualization (CMIV),
Linköping University, Linköping, Sweden

Abstract. Many image processing methods such as corner detection, optical flow and iterative enhancement make use of image tensors. Generally, these tensors are estimated using the structure tensor. In this work we show that the gradient energy tensor can be used as an alternative to the structure tensor in several cases. We apply the gradient energy tensor to common image problem applications such as corner detection, optical flow and image enhancement. Our experimental results suggest that the gradient energy tensor enables real-time tensor-based image enhancement using the graphical processing unit (GPU) and we obtain 40 % increase of frame rate without loss of image quality.

1 Introduction

A drawback of many current state of the art image processing methods is the high computation requirements necessary to achieve high-quality results. As a consequence, the computational constraints limit the methods applicability as useful tools in real-time applications, implying that processing-pipelines operate on suboptimal image data. Specifically, the structure tensor introduced by Förstner and Gülch [1] and Bigün and Granlund [2] is an integral part of many image processing applications, such as corner detection [3,4], optical flow [5] and tensor-based image denoising [6].

In this paper we propose to replace the structure tensor with an alternative tensor, the gradient energy tensor [7] that does not (necessarily) require a post-convolution of its tensor-components to form a rank-2 tensor. The principal difference to the structure tensor is that the gradient energy tensor use higher-order derivatives to capture the orientation in a neighbourhood. In Fig. 1 we have used the tensors for corner detection and dense optical flow and as visualized the two tensors produce very similar results. As a major contribution of this work we present an adaptive tensor-based image denoising method implemented using Nvidia CUDA programming language. Our approach is superior to those based on the structure tensor with regards to computational performance, without loss of image quality.

The structure tensor is defined as the outer product of the image gradient directions and in the case of two-dimensional images, the tensor is at most of

© Springer International Publishing Switzerland 2015
C.V. Jawahar and S. Shan (Eds.): ACCV 2014 Workshops, Part II, LNCS 9009, pp. 16–30, 2015.
DOI: 10.1007/978-3-319-16631-5_2

Corner Detection   Optical flow
Good Features to Track [4]   *Dense Lucas-Kanade* [5]

Fig. 1. An illustration that the Gradient Energy Tensor performs comparably to the Structure tensor in the two applications of corner detection and optical flow estimation (sequence Teddy frames 10 and 11 [9]). The gradient energy tensor does ***not*** (necessarily) require a post-convolution of its tensor-components, which is the case in the case of corner detection. See text for details.

rank-2 and determines the local energy distribution of a neighbourhood. In practice, to enforce robustness to noise and to form a rank-2 tensor, the structure tensor components are averaged using a post-smoothing of the tensor-components. Furthermore, without the post-convolution operation the tenor is of rank-1 and thus it cannot be used to describe corners and junctions in the image structure. The averaged structure tensor can also be viewed as a second-moment matrix, which estimates the local variance of the image data [8].

The gradient energy tensor is particularly suitable when considering graphical processing units (GPU). A GPU is a high-performance graphics card and is designed for massive parallelization of data-processing tasks. The GPU architecture is most suitable for problems with high spatial locality in the image plane and is therefore very fitting for the solution of partial differential equations (PDE). In this work we present a novel tensor-based PDE, which uses the gradient energy tensor for image enhancement and utilize the GPU architecture to enable real-time image enhancement.

The standard approach to image filtering using PDE's is defined using the structure tensor, a rank-2 tensor which is transformed such that it describes the direction parallel to the image structure. This approach does not allow for real-time computations of high-resolution colour images since the tensor is defined using several convolutions of the tensor components. In contrast, the gradient energy tensor does not requires any convolution to form a rank-2 tensor but yet performs equally well, or better in denoising, when considering the resulting peak signal to noise ratio (PSNR) and structural similarity index (SSIM) [10].

1.1 Contributions

In this work we present a lesser known tensor: the gradient energy tensor as an alternative to the commonly used structure tensor. Our main contribution is a novel PDE-based denoising scheme which utilize the gradient energy tensor to drive an image enhancement process. In Sect. 3.1 we adopt the Good Features to Track [4] framework to our tensor formulation and show comparable performance to the structure tensor using a repeatability test for different viewpoint angles. We also demonstrate that it is possible to solve the Lucas-Kanade [5] optical flow formulation using the gradient energy tensor. In Sect. 3.2 we illustrate the approach on sequences from the Middleburry optical flow dataset [9].

Finally, in Sect. 4 we do an exhaustive evaluation on high-resolution colour images and describe the GPU implementaion using Nvidias CUDA programming language. We show that by using the gradient energy tensor we significantly boost the achieved frames per second (fps) compared to the structure tensor to enable real-time image denoising.

2 Estimating Directional Information

Many image processing algorithms contain an estimate of local orientation as an integral part of the methods. Often the directional information is computed using the so called structure tensor [1,2]. The tensor is the outer product of the image gradients whereafter the components are averaged in a local neighbourhood. If the averaging operator, $w(x)$ is circular (i.e. Gaussian) then the structure tensor is isotropic in homogeneous regions. Below, we define the structure tensor and the gradient energy tensor which we show in the experimental part outperforms the structure tensor with regards to computation efficiency without compromising the accuracy of the final result.

2.1 Structure Tensor

The structure tensor, $T \in \mathbb{R}^{2 \times 2}$, is symmetric and positive semi-definite [1,2]. The tensor is defined as the outer product of the image gradients followed by post-convolutions, one for each component, i.e.

$$T(u_x, u_y) = \begin{pmatrix} \int w(\xi) u_x(x - \xi)^2 d\xi & \int w(\xi) u_x(x - \xi) u_y(x - \xi) d\xi \\ \int w(\xi) u_x(x - \xi) u_y(x - \xi) d\xi & \int w(\xi) u_y(x - \xi)^2 d\xi \end{pmatrix}. \quad (1)$$

The effect of the post-convolution operator is that T will have two non-zero eigenvalues, thus the operator can be used to estimate the orientation of image structures. In the case the convolution is given by the identity i.e. $w(x) = 1$ for $x = 0$ and $w(x) = 0$ otherwise, then the eigenvalues are given by the trace of T and 0.

2.2   Gradient Energy Tensor

The gradient energy tensor (GET) was first introduced by Felsberg and Köthe [7]. Let $Hu = \nabla\nabla^t u$ be the Hessian and $\nabla\Delta u = \nabla\nabla^t\nabla u$, then GET is

$$GET(\nabla u) = HuHu - \frac{1}{2}\Big(\nabla u[\nabla\Delta u]^t + [\nabla\Delta u]\nabla^t u\Big). \tag{2}$$

In contrast to the structure tensor (1), the gradient energy tensor (2) does *not* (necessarily) require a convolution operator to form a rank 2 tensor. The response from the two tensors are illustrated in Fig. 2. The presence of second and third-order derivatives in GET does makes it more sensitive to noise than the structure tensor, however, it allows us to capture orientation of structures not possible to detect using the structure tensor.

The second difference to the structure tensor is that the gradient energy tensor is not necessarily positive semi-definite. In applications where it is required to have a positive semi-definite tensor it is straightforward to define the eigenvalues of GET to be positive. That is, simply compute the eigenvaluedecomposition of GET and use the alternative definition

$$GET^+(\nabla u) = vv^t|\mu_1| + ww^t|\mu_2| \tag{3}$$

where v, w are the eigenvectors and μ_1, μ_2 eigenvalues of GET respectively. From (3) the tensor's orientation information is made explicit, the eigenvectors describe the local orientation of the neighbourhood and the eigenvalues describe the magnitude.

Since the positivity of the tensor is reflected in the sign of its eigenvalues the presence of negative eigenvalues can be determined on beforehand if the condition tr $(HuHu) - \nabla^t u\nabla\Delta u \geq \sqrt{l}$ is not satisfied where $l = (\text{tr } GET)^2 - 4\det(GET) \geq 0$. Since GET is symmetric it has real eigenvalues. Thus by its eigenvalue decomposition it is sufficient to show that tr $GET \geq \sqrt{l}$ in order for GET to be positive semi-definite. l is necessarily positive since $l = (a-c)^2 + 4b^2 \geq 0$ where a, b and c are the components $GET(\nabla u) = \begin{pmatrix} a & b \\ b & c \end{pmatrix}$ and

$$a = u_{xx}^2 + u_{xy}^2 - u_x(u_{xxx} + u_{xyy}) \tag{4}$$

$$b = u_{xx}u_{xy} + u_{yx}u_{yy} - \frac{1}{2}(u_x(u_{yxx} + u_{yyy}) + u_y(u_{xxx} + u_{xyy})) \tag{5}$$

$$c = u_{yy}^2 + u_{yx}^2 - u_y(u_{yxx} + u_{yyy}). \tag{6}$$

An analysis of the positivity of the 1-dimensional energy operator was done in [11]. In the remainder of the paper we apply the GET for corner detection, more specifically the good features to track approach, and GET^+ is used to compute the Lucas-Kanade optical flow and tensor-based image denoising.

2.3   Eigendecomposition of 2×2 Tensors

In the previous section we have shown that both the structure and gradient energy tensor can be used to compute the local orientation. By factorizing the

Structure Tensor (1)   *Gradient Energy Tensor* (2)

Fig. 2. Illustration of the resulting tensor-fields. The structure tensor require additional post-smoothing to form a rank-2 tensor compared to the gradient energy tensor which is defined without post-smoothing. Note that the size of the ellipses has been scaled for improved visualization.

tensors into their eigendecomposition the directional change and its magnitude is made explicit. Specifically, a matrix $S \in \mathbb{R}^{2 \times 2}$ can be decomposed into its eigenvalues $\mu_{1,2}$ and eigenvectors v, w representation such that

$$S = vv^t \mu_1 + ww^t \mu_2 \tag{7}$$

where v, w are two orthonormal vectors. The eigenvectors describe the orientation and the eigenvalues the magnitude of the directional change in a neighbourhood. The eigenvalues can be computed by solving the characteristic polynomial $\det(S - \mu I) = 0$ and the solution is given by

$$\mu_{1,2} = \frac{1}{2} \left(\operatorname{tr} S \pm \sqrt{(\operatorname{tr} S)^2 - 4 \det S} \right). \tag{8}$$

For the applications presented in this work we are not required to compute the explicit eigendecompistion (7), rather we are primarily interested in the eigenvalues. Thus, by observing that $vv^t + ww^t = I \iff ww^t = I - vv^t$, then S can be expressed as [12],

$$S = (\mu_1 - \mu_2)vv^t + I\mu_2 \tag{9}$$

and we compute the eigenvector-product vv^t as

$$vv^t = \frac{1}{(\mu_1 - \mu_2)}(S - I\mu_2). \tag{10}$$

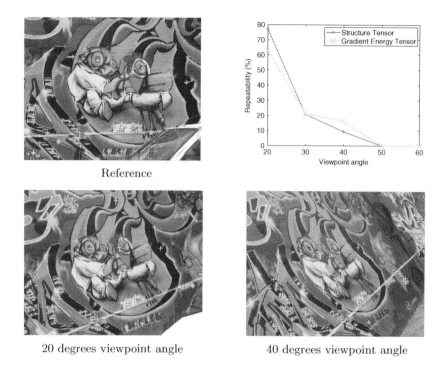

Reference

20 degrees viewpoint angle   40 degrees viewpoint angle

Fig. 3. Examples of corner detection for the structure tensor (with post-smoothing) and the gradient energy tensor (without post-smoothing) using Good Features to Track. Both tensors detect corners accurately but not always the same corners.

3   GET Corner Detection and Optical Flow

3.1   Corner Detection

The first application we consider is corner detection using Good Features to Track [4]. The problem of corner detection is part of many image processing pipelines such as interest point detection and sparse optical flow. The Good Features to Track framework detects corners by considering the eigenvalues of the structure tensor. If the structure tensor has two non-zero eigenvalues $\mu_{1,2}$, both larger than some threshold μ then the orientation tensor is necessarily invertible, *i.e.* the tensor is of rank 2. If $\min(\mu_1, \mu_2) > \mu$ where μ is often set to a fraction of the largest minimum eigenvalue then the neighbourhood contains a corner point.

In this work we set $\mu = 0.01$ and Fig. 3 shows the 128 strongest detected corners for the structure tensor and the gradient energy tensor. We use a Gaussian kernel with standard deviation 1 for post-smoothing of the structure tensor components. We also computed the repeatability measure [13] at 40 % overlap (see Fig. 3) for the *viewpoint* dataset where the viewpoint angle has been changed from 20-60 degrees from the reference image [14]. The repeatability

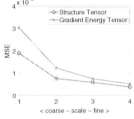

Structure Tensor   *Gradient Energy Tensor*

Fig. 4. Optical flow estimated from frames 7 and 8 of the Schefflera sequence in the Middleburry optical flow dataset [9]. Top row illustrate the vector field. On the bottom we show the obtained flow fields where the direction is colour coded and intensity corresponds to the magnitude. The graph to the right show the mean squared error between the warped image $J(x + d)$ and the reference image $I(x)$ in (11).

measure is similar for the two tensors. Note that the gradient energy tensor, in this example, does not contain a post-smoothing of the tensor components.

3.2   Optical Flow

Our second application is to apply the gradient energy tensor to the original Lucas-Kanade optical flow formulation [5] to compute a dense motion field. By minimizing the below energy functional the structure tensor appear as part of the minimizer

$$E(u) = \int_{\Omega} [J(x + d) - I(x)]^2 w(x) \, dx \tag{11}$$

where J and I are two images of size Ω with some unknown displacement vector d. The minimizer of (11) is obtained by solving (see [15])

$$\left[\int_{\Omega} [\nabla J(x) \nabla^t J(x)] w(x) \, dx \right] d = \left[\int_{\Omega} [(J(x) - I(x)) \nabla J(x)] w(x) \, dx \right]. \tag{12}$$

The bracket in the left hand side of (12) is the structure tensor, T in (1). Figure 4 shows the results when we solve (12) with the structure tensor and when we interchanged the structure tensor with the gradient energy tensor with positive

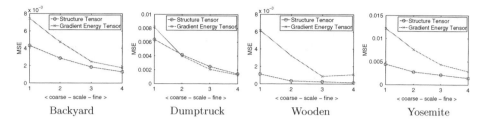

Fig. 5. Mean squared error (MSE) graphs between the first and second image after warping the first image with the estimated motion field. The sequences are part of the Middlebury dataset [9]. The estimate of the motion field in the Wooden sequence for the gradient energy tensor diverge on the finest scale, but in the other sequences the MSE yield comparable final displacement estimate.

eigenvalues, *i.e.* GET^+ in (3). Due to the large displacement between the image pairs we are required to have a post-convolution of the GET^+ components similarly to the structure tensor in order to capture the motion.

For both tensors we solve the normal Eq. (12) explicitly using the pseudo-inverse on multiple scales. The solution we obtain at a coarse scale is propagated to a finer scale and after each scale we apply a median-filter to the estimated motion field [16]. Furthermore, we found that the magnitude scaling of the gradient energy tensor eigenvalues resulted in a poor motion estimation. Therefore, we scale the eigenvalues of the gradient energy tensor using the factor $\sigma i/I$ where $i \in I = \{1, 2, 3, 4\}$ are the scales, $i = 1$ is the fines scale, and σ is the standard deviation of the Gaussian filter $w(x)$. As for the structure tensor, the selection of σ is dependent on the absolute motion present within the frames, *i.e.* if the displacement is large then a larger σ is required. In Fig. 5 we show the mean squared error between $J(x + d)$ and $I(x)$ for some additional sequences from the Middlebury optical flow dataset [9].

The optical flow formulation (11) does (obviously) not give state of the art results, however the approach illustrates that the gradient energy tensor is a possible alternative to the structure tensor. We expect that many interesting further results can be derived from this approach.

4   Iterative Tensor-Based PDE Denoising

Image enhancement methods based on partial differential equations (PDE) are often considered to be too computationally expensive for practical applications. The main bulk of computations is the iterative update scheme and the calculation of the post-convolution of the structure tensor components [6].

It is in this application that the gradient energy tensor really excels over the structure tensor. In this section we implement the proposed iterative denoising scheme on the GPU. We show how to utilize the highly parallelizable nature of iterative PDE-based denoising schemes and benefit from the locality of the gradient energy tensor. The implementation is done using Nvidias CUDA programming language with OpenGL support, the GPU we use is the GTX 670 with 4 Gb on card memory and the workstation is equipped with an Intel Xeon CPU at 3 GHz and 8 Gb of installed RAM memory. Even though the hardware specification is in the middle segment of the consumer-market we show that by using the gradient energy tensor, the proposed iterative tensor-based PDE denoising scheme can reach near maximum PSNR at 60 frames per second (fps) for a three channel colour 1280×720 pixel image (HD720p). This is a significant improvement over the structure tensor running at 30 fps while achieving similar peak signal to noise ration (PSNR) and structural similarity [11] (SSIM) values.

4.1   The Proposed Filtering Scheme

The standard formulation of tensor-based anisotropic diffusion [6] using the structure tensor is given in the PDE below with Neumann boundary condition

$$\begin{cases} u - u^0 - \beta \operatorname{div}\left(D(\nabla u)\nabla u\right) = 0 \text{ in } \Omega \\ \qquad\qquad \boldsymbol{n} \cdot \nabla u = 0 \text{ on } \partial\Omega \end{cases} \tag{13}$$

where β is a stepsize parameter which controls the smoothness of the solution u that minmizes the PDE. In (13), $D(\nabla u)$ is the *diffusion tensor* computed as

$$D(T(\nabla u)) = vv^t g(\mu_1) + ww^t g(\mu_2) \tag{14}$$

where v, w and $\mu_{1,2}$ are the eigenvectors and eigenvalues of the structure tensor T in (1). g is the diffusivity function and here we set it as $g(s) = \exp(-s/k)$ where k is the edge-stopping parameter determining the adaptivity to the image structure. Instead of using the diffusion tensor in the PDE (13) we propose to use the gradient energy tensor with positive eigenvalues, GET^+ in (3), as the tensor controlling the orientation estimate of the image structures, *i.e.* we define the following PDE

$$\begin{cases} u - u^0 - \beta \operatorname{div}\left(D^+(\nabla u)\nabla u\right) = 0 \text{ in } \Omega \\ \qquad\qquad \boldsymbol{n} \cdot \nabla u = 0 \text{ on } \partial\Omega \end{cases} . \tag{15}$$

The computation of the eigenvalues are done according (9), *i.e.*

$$D^+(GET^+(\nabla u)) = \left(\frac{g(\lambda_1) - g(\lambda_2)}{\lambda_1 - \lambda_2} \right)(GET^+ - I\lambda_2) + Ig(\lambda_2) \tag{16}$$

where $\lambda_{1,2}$ are the eigenvalues of GET^+ computed according to (8) where we set $S = GET^+$. In practice we compute (14) using (16) with $S = T$.

   In order to solve the PDEs (13) and (15), we use a forward Euler iterative scheme with finite differences to approximate the image derivatives. For a discussion on the numerical stability of iterative scheme see [17].

4.2   Implementation Details

A GPU implementation is about how to efficiently utilize the parallelism of the graphics card architecture. Using CUDA terminology, the parallelism is achieved

Table 1. Algorithm pseudo-code for the main CUDA kernels. Left: Standard approach to adaptive image diffusion using the structure tensor. Right: adaptive image diffusion using the gradient energy tensor. The convolutions are separable Gaussian functions of size 5×5 with standard deviation of 1.

Anisotropic Diffusion [6]

```
for  i=0 to maxint  do  {
  convolution_rows()
  convolution_cols()
  compute_structure_tensor()
  filter_update()
}
```

```
compute_structure_tensor() {
  gradient_products(a,b,c)
  convolution_rows(a)
  convolution_cols(a)
  convolution_rows(b)
  convolution_cols(b)
  convolution_rows(c)
  convolution_cols(c)
  remapp_eigenvalues(a,b,c)
}
```

Gradient Energy Diffusion

```
for  i=0 to maxint  do  {
  convolution_rows()
  convolution_cols()
  compute_energy_tensor()
  filter_update()
}
```

```
compute_energy_tensor() {
  a = (4)
  b = (5)
  c = (6)
  remapp_eigenvalues(a,b,c)
}
```

by dividing the image data into blocks, each block contains the threads that are to be executed in parallel on a streaming multiprocessor. Today's GPU architectures offer many memory types (global, texture, shared, local ...) and our implementation is focused on utilizing the high-performance shared memory. We achieve this by coalescing memory access when streaming data from global memory to shared memory. We considered using texture memory due to its automatic handling of Neumann boundary conditions and memory-access optimization for localized reads, however texture memory is read-only and we require dynamic updates of intermediate results. Also, shared memory is on-chip, and therefore read-access requires less clock-cycles than the texture memory. These differences in memory-latency have a significant impact on runtime performance, for example in convolutions [18].

Since we are interested in processing images with three colour channels, we simplify our implementation by defining a container describing the three colour channels red, green and blue as well as the alpha channel (required for visualization using OpenGL). The image data is read and written using 24-bits but during the filter process we use single precision float. There are primarily three steps involved in the iterative filtering scheme, pre-filtering for regularization of the image derivatives, orientation estimation and filter update. The steps are illustrated in Table 1 and highlights the primary difference between the two implementations: the computation of the structure tensor requires three full (separable) convolutions of the image data whereas the energy tensor does not

(a) Convolution rows	(b) Convolution columns	(c) Gradient energy tensor

Fig. 6. We use tiles of size 16×16 (blue regions) with padding of 2 pixels (red region) for the convolution and computation of the gradient energy tensor. We use corresponding tile layouts for computing the image gradient and filter update (Color figure online).

Fig. 7. Colour test images and the corresponding image sizes in pixels that are used in the evaluation.

require any post-convolution of the tensor-components, which is key to the gain in computation speed. Figure 6 shows the memory layout of the shared memory that we use in the CUDA kernels shown in Table 1. We use tile sizes of 16×16 and note that each entry in the tile consists of four entries *i.e.* red, green, blue and alpha channels. This approach is convenient since it both simplifies the code and facilitates efficient memory access. The padding of the tiles (the red region in Fig. 6) is done with two pixels in the case of the separable convolution since we use a Gaussian filter of size 5×5 with standard deviation of 1 for smoothing the image and tensor-components. The coefficients of the Gaussian filters is set using constant memory. The shared memory associated with the gradient energy tensor require a padding of 2 pixels in horizontal and vertical direction since the support of the third order finite difference derivative is 5 pixels. Lastly, since we only require first order diagonal derivatives, it is sufficient to read the closest corner-pixel into the shared memory, further simplifying the global memory access pattern.

In order to measure the resulting computation times, we are required to compute the resulting execution speed of the filtering methods. We do this by using the standard sdkStartTimer() and sdkStopTimer() available in CUDA. We have chosen to include the OpenGL rendering in the timing of the filter performance as shown in Table 2 since it more accurately reflects the expected real-time capabilities of video denoising where each frame in a video sequence would be visualized.

Table 2. Main function of the tensor-based image denoising method. The timer values reported in Sect. 4.3 are timed including the OpenGL rendering.

```
void display () {
  sdkStartTimer(&timer);
  unsigned int *dResult;
  // Initialize OpenGL for visualization
  diffusionFilterRGBA(dResult, ...);
  // Map dResult to OpenGL resources and draw image on display
  sdkStopTimer(&timer);
}
```

4.3 Results

The focus of our evaluation is to show that the gradient energy tensor does not compromise the denoising quality compared to the structure tensor while achieving a faster runtime. Our measures include the peak signal to noise ratio (PSNR) and the structure similarity index [10] (SSIM) for the image quality. With regards to the total runtime (measured as shown in Table 2) we report achieved frames per seconds (fps) for each method. For each of the measures a higher value is better than a lower value.

Figure 7 show the colour test image *pippin_Florida0002.bmp* from the McGill colour image database [19] with the original resolution 2560×1920 pixels. The image was downsampled (using bicubic interpolation) to 1920×1080 (HD1080) and 1280×720 (HD720), two common image resolutions in high-definition (HD) video. We corrupt each image with 20 standard deviations of normal distributed noise. The filtering scheme is iterated 10 times with a fixed update step of size 0.20 (we refer to [17] for a discussion on convergence results for iterated solutions of PDEs). The diffusivity constant (see g in (16)) is set to $k/10$ where we compute $k = (e^1 - 1)/(e^1 - 2)\sigma^2$ [20], this yields $k/10 = 0.0015$ when $\sigma = 20/255$ and each colour channel is quantized using an 8-bit representation.

Figure 8 show the obtained PSNR (a) and SSIM (b) values compared to the iteration number, as expected the two methods are comparable for the obtained error measures. The best error values are in agreement between the two methods but obtained at different iterations numbers, it is not a discrepancy in method performance but an issue of parameter tuning. In Fig. 8 (c) and (d) we show the fps that we obtain for each iteration. After four iterations, in (c), the filter using the gradient energy tensor is stable at 60 fps whereas the standard diffusion scheme using the structure tensor has dropped to 30 fps for the smallest image resolution. In (d) we show the tradeoff between PSNR and obtained fps for *both* tensors: a higher fps result in a lower PSNR value. Note that the fps count is *independent* of the image content. Future work will include a more comprehensive study of the GET orientation estimation compared to the structure tensor for other noise levels and image types than considered in this work.

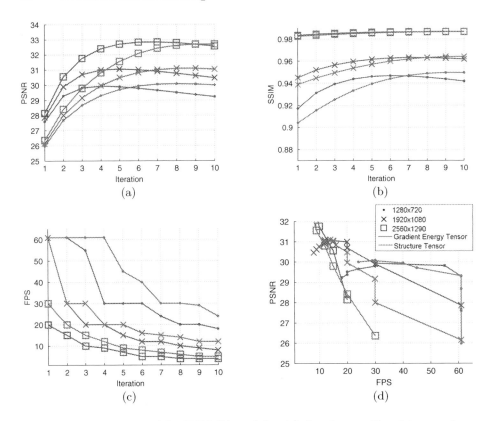

Fig. 8. Obtained PSNR (a), SSIM (b) and fps (c) for the considered image sizes. The PSNR and SSIM values are similar for the two tensors but the obtained fps is significantly higher for the gradient energy tensor.

In Fig. 9 we illustrate an example of the final denoised result, the zoomed images were cropped from the 2560×1920 resolution image and the original image was corrupted by 20 standard deviations of normal distributed additive noise. The images illustrate the denoised result after 10 iterations and note that the gradient energy tensor fps-count is 25 % higher than the structure tensor but with indiscernible image quality and PSNR (cmp. Fig. 8 (a) and (c)). In Table 3 we give the ratio of the gradient energy tensor error measures over the structure tensor error measures averaged over the iterations, *i.e.* if the ratio is identical to 1.0 there is no difference in method performance, if the value is larger than 1.0 then the ratio indicate that the gradient energy tensor performs better. The SSIM value difference is less than 10^{-3} whereas the PSNR shows a 3.2 % difference in the favour of the structure tensor, however considering the final fps-count the gradient energy tensor is up to 40 % more computationally efficient for the largest image size.

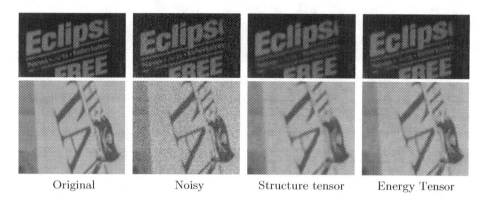

Original	Noisy	Structure tensor	Energy Tensor

Fig. 9. Two patches from the 2560 × 1920 resolution image and the corresponding denoised images at iteration 10. The applied noise was 20 standard deviation of normal distributed additive noise. Note that the difference in visual appearance is not discernible but the fps count is improved by 25 % with the gradient energy tensor.

Table 3. Ratios are computed as the measure obtained by the gradient energy tensor divided by the measure obtained by the structure tensor and averaged over iterations. A ratio of 1.0 show that the performance is identical. The SSIM and PSNR ratios are nearly identical but at up to 40 % higher fps values in the largest image resolution, the PSNR show a marginal loss in image quality for the gradient energy tensor.

Ratio	PSNR	SSIM	fps
1280 × 720	0.991	0.995	1.35
1920 × 1080	0.982	0.997	1.27
2560 × 1920	0.968	0.999	1.40

5 Conclusion

In this work we have presented the gradient energy tensor as an alternative to the structure tensor for local orientation. We have considered three applications: corner detection, optical flow and adaptive image enhancement. Due to the absence of a post-convolution of the gradient energy tensor components the tensor is highly suitable for efficient implementation on the GPU, and we showed that the tensor yield significant improvement in obtained frames per second compared to the structure tensor without compromising PSNR and SSIM error values.

Acknowledgement. This research has received funding from the Swedish Research Council through grants for the projects Visualization-adaptive Iterative Denoising of Images (VIDI) and Extended Target Tracking (ETT), within the Linnaeus environment CADICS and the excellence network ELLIIT.

References

1. Förstner, W., Gülch, E.: A fast operator for detection and precise location of distinct points, corners and centres of circular features. In: ISPRS Intercommission, Workshop, Interlaken, pp. 149–155 (1987)
2. Bigun, J., Granlund, G.H.: Optimal orientation detection of linear symmetry. In: Proceedings of the IEEE First ICCV, pp. 433–438 (1987)
3. Harris, C., Stephens, M.: A combined corner and edge detector. In: Proceedings of Fourth Alvey Vision Conference, pp. 147–151 (1988)
4. Shi, J., Tomasi, C.: Good features to track. In: CVPR 1994, pp. 593–600 (1994)
5. Lucas, B.D., Kanade, T.: An iterative image registration technique with an application to stereo vision. In: Proceedings of the 7th International Joint Conference on Artificial Intelligence - Volume 2. IJCAI 1981, pp. 674–679. Morgan Kaufmann Publishers Inc., San Francisco (1981)
6. Weickert, J.: Anisotropic Diffusion in Image Processing. Teubner-Verlag, Stuttgart (1998)
7. Felsberg, M., Köthe, U.: GET: the connection between monogenic scale-space and gaussian derivatives. In: Kimmel, R., Sochen, N.A., Weickert, J. (eds.) Scale-Space 2005. LNCS, vol. 3459, pp. 192–203. Springer, Heidelberg (2005)
8. Lindeberg, T.: Scale-Space Theory in Computer Vision. Kluwer international series in engineering and computer science: Robotics: Vision, manipulation and sensors. Springer, New York (1993)
9. Baker, S., Scharstein, D., Lewis, J.P., Roth, S., Black, M.J., Szeliski, R.: A Database and Evaluation Methodology for Optical Flow. Int. J. Comput. Vis. **92**, 1–31 (2011)
10. Wang, Z., Bovik, A.C., Sheikh, H.R., Simoncelli, E.P.: Image quality assessment: from error visibility to structural similarity. IEEE Trans. Image Process. **13**, 600–612 (2004)
11. Bovik, A.C., Maragos, P.: Conditions for positivity of an energy operator. IEEE Trans. Signal Process. **42**, 469–471 (1994)
12. Granlund, G.H., Knutsson, H.: Signal processing for computer vision. Kluwer, New York (1995)
13. Mikolajczyk, K., Schmid, C.: A performance evaluation of local descriptors. IEEE Trans.Pattern Anal. Mach. Intell. **27**, 1615–1630 (2005)
14. Mikolajczyk, K.: Implementation (2014). http://www.robots.ox.ac.uk/~vgg/research/affine
15. Tomasi, C., Kanade., T.: Detection and Tracking of Point Features. Technical report, Carnegie Mellon University Technical Report CMU-CS-91-132 (1991)
16. Sun, D., Roth, S., Black, M.J.: Secrets of optical flow estimation and their principles. In: CVPR2010, pp. 2432–2439 (2010)
17. Scherzer, O., Weickert, J.: Relations Between Regularization and Diffusion Filtering. J. Math. Imag. Vis. **12**, 43–63 (2000)
18. Podlozhnyuk, V.: Image convolution with CUDA, NVIDIA Corporation white paper, v1.0 (2007)
19. Olmos, A., Kingdom, F.A.A.: A biologically inspired algorithm for the recovery of shading and reflectance images. Percept. **33**, 1463–1473 (2004)
20. Felsberg, M.: Autocorrelation-Driven Diffusion Filtering. IEEE Trans. Image Process. **20**, 1797–1806 (2011)

Feature-Preserving Image Restoration from Adaptive Triangular Meshes

Ke Liu, Ming Xu, and Zeyun Yu[⊠]

University of Wisconsin Milwaukee, Milwaukee, WI 53211, USA
{keliu,yuz}@uwm.edu

Abstract. The triangulation of images has become an active research area in recent years for its compressive representation and ease of image processing and visualization. However, little work has been done on how to faithfully recover image intensities from a triangulated mesh of an image, a process also known as image restoration or decoding from meshes. The existing methods such as linear interpolation, least-square interpolation, or interpolation based on radial basis functions (RBFs) work to some extent, but often yield blurred features (edges, corners, etc.). The main reason for this problem is due to the isotropically-defined Euclidean distance that is taken into consideration in these methods, without considering the anisotropicity of feature intensities in an image. Moreover, most existing methods use intensities defined at mesh nodes whose intensities are often ambiguously defined on or near image edges (or feature boundaries). In the current paper, a new method of restoring an image from its triangulation representation is proposed, by utilizing anisotropic radial basis functions (ARBFs). This method considers not only the geometrical (Euclidean) distances but also the local feature orientations (anisotropic intensities). Additionally, this method is based on the intensities of mesh faces instead of mesh nodes and thus provides a more robust restoration. The two strategies together guarantee excellent feature-preserving restoration of an image with arbitrary super-resolutions from its triangulation representation, as demonstrated by various experiments provided in the paper.

Keywords: Image restoration · Image triangulation · Radial basis function · Anisotropic radial basis function · Local structure tensor

1 Introduction

Modern imaging technologies often digitize an image into a uniform array of pixels (or voxels in 3D). With uniformly sampling, the sampling density is inevitably too high in regions where intensities change slowly and too low in regions whose intensities change rapidly. Despite the ease of use in both hardware and software developments, uniformly-digitized images often pose challenges in data storage and transmission, as well as image processing, especially in 3D medical images that have been consistently and significantly grown in size in recent years. Evolving from previously commonly-used uniform sampling, non-uniform sampling

© Springer International Publishing Switzerland 2015
C.V. Jawahar and S. Shan (Eds.): ACCV 2014 Workshops, Part II, LNCS 9009, pp. 31–46, 2015.
DOI: 10.1007/978-3-319-16631-5_3

and adaptive mesh triangulation of an image has become an active research area in image processing. Image triangulation involves partitioning an image into a collection of non-overlapping small triangles called mesh elements (faces or triangles). This procedure often serves as an image coding method, meaning that an image in pixels is compressed by using a number of "super-pixels". This method is a compact way to represent images for effective data storage and transmission, and also an efficient way to process and visualize images, especially for 3D images where the number of voxels can be extremely large. In addition, the resulting mesh edges are expected to be well aligned with image featured (edges or corners) in order to maintain a faithful restoration of the original image. Mesh modeling of an image has many applications like image compression [5,7,15,17], motion tracking and compensation [6,22,27,28,38,43], image processing by geometric manipulation [19], medical image processing [34], feature detection [14], pattern recognition [29], computer vision [32], restoration [8], tomographic reconstruction [9], interpolation [36,37], image/video coding [1,2,23,25,30,44], video modeling [13] and image retargeting [21].

A common procedure of image triangulation consists of two steps: (1) generating mesh nodes (vertices) by choosing a set of sampling points defined in the image domain, and (2) connecting these mesh nodes by Delaunay triangulation [16]. Delaunay triangulation is a geometric operator and can avoid long and thin triangles that often lead to poor approximations. The selection of sampling nodes, however, is data-dependent, where the connectivity of the triangulation depends on the data set, based on which the mesh nodes are generated. Depending on how to generate mesh nodes, there are two categories of the image triangulation. The first one places mesh nodes inside the image features but near both sides of feature edges. So the triangulated images of this category show double-layer vertices at both sides of feature edges. The second category places mesh nodes directly at the feature edges, thus there are only single-layer vertices defined right on feature edges. Yang et al. [45] employed Floyd-Steinberg error-diffusion (ED) algorithm to place mesh nodes so that their spatial density varies according to the local image content. As a result, the triangulated images fall into category I. Adams [3] employed greedy-point removal (GPR) and error-diffusion scheme together to achieve meshes of quality comparable to the original GPR scheme but at a much lower computational and memory complexities. With the conjunction of smoothing operators, this method produces image triangulation of category I. Adams also proposed a framework in [4] for mesh generation by fixing various degrees of freedom available within that framework. This method performs extremely well and produces meshes of higher quality than the GPR method, and is considered as a method of category I as well. By contrast, Li et al. [26] proposed a modified version of Rippa [31] and Garland-Heckbert (GH) [20] frameworks which can generate single-layer mesh nodes on edges, and this framework generates triangulated images of category II. Another method of this category was proposed by Tu et al. [39], based on constrained Delaunay triangulations. In this method, the approximating function is not required to be continuous everywhere but with discontinuities being permitted across constrained edges of triangles in triangulation.

Both categories of image triangulation generated by the methods mentioned above have their advantages and disadvantages. For the first category (double-layer vertices), the quality of image restoration is usually better because all vertices are well defined on images features thus the intensities of pixels during image restoration will not be affected by edges. As a result, the edges in the recovered images are sharp and the Peak Signal-to-Noise Ratio (PSNR) is usually larger. While the restoration quality of methods in category I is high enough for subjective quality testing, the two layers must be very close to each other in order to have well-defined and sharp image edges. A consequence of this is that the resulting meshes always contain lots of thin and long triangles between the two layers, which could cause large approximation errors when the meshes are to be used for numerical analysis (like finite element analysis). Additionally, in many applications, the direct communication between different materials should be maintained, meaning that no "cushion" layer between materials should be introduced in the meshes. Moreover, using two layers of mesh vertices usually has more storage cost, resulting more memory space to be used and more time to transmission it over the network. Methods of category II avoid the small triangles and also the "cushion" layer problem, thus the mesh quality is usually better if proper steps are taken. However, the vertices are defined on feature edges, where the nodal intensities are ambiguously defined. That is, the intensity of an edge pixel will be contributed by both sides of the image edge. As will be shown in the experiments, the restored images often suffer from blurred and distorted feature edges if not properly addressed.

Because our proposed method is a single step in image processing, the same mesh will be used for numerical analysis in future. Because the obvious limitation of methods of category I, we are more interested in a method that lies in the second category (single-layer approaches). However, in order to address the blurring and distortion problems often seen in existing approaches in this category, we propose a method based on the radial basis function (RBF) interpolation with the following improvements: (1) rather than considering only the Euclidean distances between vertices, our method also takes into consideration the image local orientations, yielding an anisotropic radial basis function (ARBF). (2) our method does not use intensities of vertices directly, but instead we utilize the intensities of triangles to eliminate the uncertainty of nodal intensities on feature edges.

Our proposed method provides a new approach to restore image from triangular mesh. Because triangular mesh representations of images have much fewer nodes (sampling rates are often as low as 5 % – 6 %) defined, storage, transmission, and image operations like smoothing, sharpening, etc. will be much faster at the mesh domain instead of pixel domain. Also, our proposed method can be used to visual verification for image operations done in mesh domain.

The remainder of this paper is organized as the following. Section 2.1 briefly summarizes our mesh generation method. Section 2.2 introduces image restoration using traditional RBF interpolation. The proposed image restoration is presented in Sect. 2.3. Section 2.4 shows the detailed algorithm of our proposed method. Finally, Sect. 3 shows the experimental results and discussions. Section 4 concludes this paper.

2   Methods

While mesh generation from images is not the main focus of the current paper, we will first give a brief summary of this step just for completion of the present work. The traditional (isotropic) radial basis function (RBF) interpolation is then introduced, followed by the proposed anisotropic RBF-based interpolation for image restoration from meshes. The detail of the implementation algorithm is given below as well.

2.1   Adaptive Mesh Generation from Images

A series of algorithms are used to generate high quality, feature-sensitive, and adaptive meshes from a given image. Firstly, three kinds of the sample points (namely, Canny's points, halftoning points, and uniform points) are generated. Secondly, a triangular mesh is generated from these points by using constrained Delaunay triangulation. The Canny's edge detector is employed to guarantee that important image features are preserved in the meshes. A halftoning-based sampling strategy is adopted to provide feature-sensitive and adaptive point distributions in the image domain. Finally, a Delaunay-triangulation is used to generate initial quality triangulation of the image. These steps are briefly summarized below.

Canny Sample Points. Image edges are important features in an image and need to be preserved in the obtained meshes. Canny edge detector is a well-known method to deal with boundary extraction. In this paper, we use Canny edge detector to generate the initial Canny edge points and they are strictly attached to the boundary of the features of the image. However, the initial Canny edge points are too dense to yield quality meshes if all these edges are used as mesh nodes. In our method, we take the curvature information of every Canny's edge point into account and use the Principal Component Analysis (PCA) to determine the sampling density. The PCA method can detect the overall attribute of the neighbors of a certain size by a statistical way. After the PCA sampling, tiny features and features with high curvature have dense sample points and big features or features with straight lines have sparse sampled points.

Halftoning Sample Points. The edge points generated by the Canny edge detector described above can only capture pixels on or near the image edges. In order to have a decent initial mesh, one has to scatter some more points in the non-edge regions of the image. We adopt the halftoning sample points based on the approach described in [45]. This method generates the sample points using the second derivatives of an image, where most of the sample points are placed near the image features (edges).

Uniform Sample Points. Although the halftoning sample points can cover most non-edge regions of the image, it is possible that no point (either Canny or halftoning) is found in regions of almost constant intensities. We therefore generate some points uniformly to cover the rest of the images where the first two types of sample points are not located. A point (x, y) is said to be a valid uniform sample point if no Canny's or halftoning points are found in its neighborhood in a fixed distance.

Mesh Generation via Constrained Delaunay Triangulation. The sample points found above are used to generate our triangular mesh for a given image by using the Delaunay triangulation. We employed a popular open source software *Triangle* [35] for Delaunay triangulation. In order to guarantee the obtained meshes being well aligned with image edge features, we provide to *Triangle* with a set of line segments as additional constraints formed by connecting the Canny's sample points along the detected Canny's edges. With all the described strategies combined, we can generate high quality, feature-sensitive, and adaptive meshes from a given grayscale image. Some meshing examples will be shown in the result section below.

2.2   Review of Radial Basis Function (RBF) Interpolation

The traditional radial basis function interpolation is given by

$$f(\mathbf{x}) = \sum_{i=1}^{N} w_i \phi(\|\mathbf{x} - \mathbf{x}_i\|) \tag{1}$$

where the interpolated function $f(\mathbf{x})$ is represented as a weighted sum of N radial basis functions $\phi(\cdot)$, each centered differently at $\mathbf{x}_i$ and weighted by w_i. Let $f_j = f(\mathbf{x}_j)$. By given conditions $f_j = \sum_{i=1}^{N} w_i \phi(\|\mathbf{x}_j - \mathbf{x}_i\|)$, the weights w_i can be solved by

$$\begin{bmatrix} \phi_{11} & \cdots & \phi_{1N} \\ \vdots & \ddots & \vdots \\ \phi_{N1} & \cdots & \phi_{NN} \end{bmatrix} \begin{bmatrix} w_1 \\ \vdots \\ w_N \end{bmatrix} = \begin{bmatrix} f_1 \\ \vdots \\ f_N \end{bmatrix} \tag{2}$$

where $\phi_{ji} = \phi(\|\mathbf{x}_j - \mathbf{x}_i\|)$. Once the unknown weights w_i are solved, the image intensity at an arbitrary pixel can be calculated by

$$s(\mathbf{x}) = \sum_{i=1}^{N} w_i \phi(\|\mathbf{x} - \mathbf{x}_i\|) \tag{3}$$

In the traditional RBF method, the distance between the point $\mathbf{x} \in \mathbb{R}^d$ and center $\mathbf{x}_i \in \mathbb{R}^d$ is measured by Euclidean distance. Let $r = \|\mathbf{x} - \mathbf{x}_i\|$, commonly used radial basis functions include:

$$Gaussian : \phi(r) = e^{-(cr)^2}$$

$$Multiquadric\ (MQ) : \phi(r) = \sqrt{r^2 + c^2}$$

$$Inverse\ Multiquadric\ (IMQ) : \phi(r) = \frac{1}{\sqrt{r^2 + c^2}}$$

$$Thin\ Plate\ Spline\ (TPS) : \phi(r) = r^2 ln(r),$$

where c is a shape parameter. The shape parameter plays a major role in improving the accuracy of numerical solutions. In general, the optimal shape parameter depends on the densities, distributions and function values at the nodes. However, it is difficult to assign different shape parameters for each local domain. Thus, choosing shape parameters has been an active topic in approximation theory [42]. Interested readers can refer to [18,24,33,40,41] for more details.

One question about restoring image from triangular meshes is: what intensities should be used, intensities on vertices or intensities on faces? In the mesh generation approach described above, many vertices are located on image edges. These vertices are good to capture image gradients (or orientations) but not for image intensities because there is an ambiguity in assigning intensity to a node defined on an edge, as illustrated for blue nodes in Fig. 1(a). Obviously, a very small change (or error) on the location of blues nodes would make a big interpolation difference if the mesh vertices are used as the nodal values in RBFs. A better way is to use face centers as the nodal values for RBF interpolations, which can eliminate the ambiguity and is less sensitive to mesh errors. Figure 1(b) shows this idea, where the face centers are more robust to the location changes of mesh vertices. Results of vertex-based RBF interpolation and triangle-based RBF interpolation can be found in Fig. 3(c) and (d) in Sect. 3.

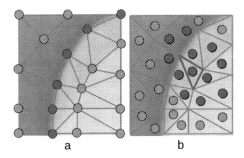

a   b

Fig. 1. Example of interpolation. (a) Interpolation by vertices. Green dots are vertices defined on feature. Blue dots are vertices defined on feature edge. (b) Interpolation by faces. Green dots are face centers. Blue dots are face centers used for interpolation of the intensities of pixels enclosed by the blue triangle(Color figure online).

Although using face centers performs better than the vertex-based RBF interpolation, the traditional RBF is isotropic in the sense that only the geometrical distance information is considered, which often causes blurring and distortion artifacts as can be seen in Fig. 3(d). To capture the anisotropicity of the image features, the direction of image edges has to be considered as well. Otherwise, nodes across feature edges may have strong influence on the pixel being interpolated. Figure 2(a) shows the cause of the blurred edge problem. $\mathbf{x}$ is the pixel whose intensity we want to find out. The intensities on nodes $\mathbf{x}_1$ and $\mathbf{x}_2$ are two of the neighbors used for interpolation. The weights of them are determined only by the Euclidean distance to $\mathbf{x}$ based on the definition of traditional RBF. However, $\mathbf{x}_1$ is on the other side of the feature edge, so it should have much less influence on $\mathbf{x}$ than $\mathbf{x}_2$. The isotropic RBF has a hyper-spherical support domain which cannot satisfy this data-dependent requirement. Thus the intensity on $\mathbf{x}$ is blurred by $\mathbf{x}_1$. By contrast, Fig. 2(b) shows the anisotropic RBF (ARBF) interpolation. The support domain of ARBF is a hyper-ellipsoid. By choosing proper shape parameter, the support domain could rule out the interfering node $\mathbf{x}_1$, or give insignificant weight to node $\mathbf{x}_1$. Thus the blurring effect will be eliminated and sharp features can be well retained. In the following subsections we will elaborate on the detail of designing anisotropic radial basis functions for image restoration.

2.3   Anisotropic Radial Basis Function (ARBF) Interpolation

The main difference between the isotropic and anisotropic RBFs is the definition of distance metrics used. As in [12], the anisotropic RBF is defined as follows:

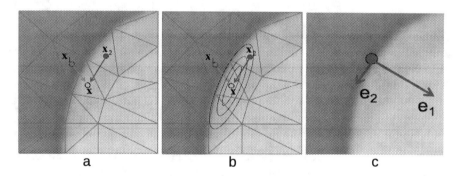

Fig. 2. Interpolation schemes. (a) Isotropic RBF interpolation. (b) Anisotropic RBF interpolation. (c) Eigenvectors on an edge pixel. $\mathbf{e}_1$ shows the normal direction. $\mathbf{e}_2$ shows the tangent direction.

Definition 1. *Given N distinct points $X = \{\boldsymbol{x}_j \in \mathbb{R}^d\}_{j=1,\ldots,N}$ and a $d \times d$ positive definite matrix $\boldsymbol{T}$, the anisotropic radial basis function associated with a radial basis function $\Phi_j(\cdot) = \phi(\|\cdot - \boldsymbol{x}_j\|_2)$ is defined by*

$$\Phi_{T,j}(\cdot) := \phi(\|\cdot - \boldsymbol{x}_j\|_T), \tag{4}$$

where $\|\boldsymbol{x}\|_T = \boldsymbol{x}^T \boldsymbol{T} \boldsymbol{x}$.

The support domain of ARBF is hyper-ellipsoid instead of a hyper-sphere in traditional RBF. Its center is $\mathbf{x}_j$, associated with the quadratic form $(\mathbf{x} - \mathbf{x}_j)^T(\mathbf{x} - \mathbf{x}_j)$. Interested readers can refer to [10,11] for more details of ARBF.

To construct the metric $\mathbf{T}$, we use the image structure tensor

$$G_\sigma * \begin{bmatrix} I_x^2 & I_x I_y \\ I_x I_y & I_y^2 \end{bmatrix}$$

where G_σ is the Gaussian smooth operator, and $\begin{bmatrix} I_x \\ I_y \end{bmatrix}$ is the image gradient at a pixel. Two eigenvectors $\mathbf{e}_1$ and $\mathbf{e}_2$ are the normal and tangent directions of the edge, respectively, as shown in Fig. 2(c). The corresponding eigenvalues are λ_1 and λ_2. The anisotropic metric is defined by

$$\mathbf{T} = \begin{bmatrix} \mathbf{e}_1 \mathbf{e}_2 \end{bmatrix} \begin{bmatrix} \lambda_1 & 0 \\ 0 & \lambda_2 \end{bmatrix} \begin{bmatrix} \mathbf{e}_1^T \\ \mathbf{e}_2^T \end{bmatrix} \tag{5}$$

Similar to the isotropic RBF but with a modified distance metric, the ARBF image interpolation problem becomes

$$s'(\mathbf{x}) = \sum_{i=1}^{N} w_i' \phi(\|\mathbf{x} - \mathbf{x}_i\|_{\mathbf{T}}) \tag{6}$$

Please note that the matrix in equation (2) should also be updated accordingly with the new distance metric $\mathbf{T}$. Therefore, the new set of weights w_i' would be different from the weights w_i in the isotropic RBF interpolation.

2.4   Algorithms

The following algorithm shows the steps of the proposed approach for image restoration from triangular meshes. The major step is the ARBF interpolation which comprises two sub-steps. First, the weight coefficients are solved by using the new distance metric $\mathbf{T}$. As stated in Sect. 2.2, this is done by taking intensities at triangle centers. Then the weights are applied to Eq. (6) to restore the intensity at each pixel.

Algorithm: Image Reconstruction

```
ImageReconstruction()
{
    // read image mesh
    loadMesh();

    // calculate intensities for each triangle center
    calculateTriangleCenters();

    // find the neighboring vertices and triangles for each node
    findNeighbors();
```

```
    // rescale eigenvalues in equation (5)
    calculateEigenvalues();

    // calculate metric T in equation (5)
    computeMetrics();

    // do the ARBF interpolation in equation (6)
    ARBFInterpolation();

    // output result
    printResult();
}

ARBFInterpolation()
{
    for (every triangle centers)
        solveCoefficients();
    end

    for (every triangles)
        for (every pixel in current triangle)
            applyCoefficientsToInterpolation();
        end
    end
}
```

3 Results and Discussion

Numerous experiments have been conducted on publicly available images by using the proposed approaches and the image restoration results are all promising. Due to the space limit, we will only consider the well-known "Lena" image and three medical images of different sizes.

Figure 3(a) is the original Lena image of size 256×256 pixels. Figure 3(b) is the result of assigning a constant intensity to all pixels in a mesh triangle (so-called piecewise interpolation). As we can see, this result shows heavy mosaic effect. Figure 3(c) is the result of iso-RBF interpolation using intensities on vertices. As previously stated on Sect. 2.3, the ambiguity of intensities on vertices blurred the result. Figure 3(d) is the result of iso-RBF interpolation using intensities on triangle centers. In this case, there is no ambiguity of intensities. So the result is much better comparing to Fig. 3(c). However, the feature edges are still blurred and some distortions are clearly seen because of the lack of directional information used in isotropic RBF. Figure 3(e) is the result of ARBF interpolation using intensities on triangle centers with multi-quadrics (MQ) basis function. The result is much better thanks to a modified distance metric that

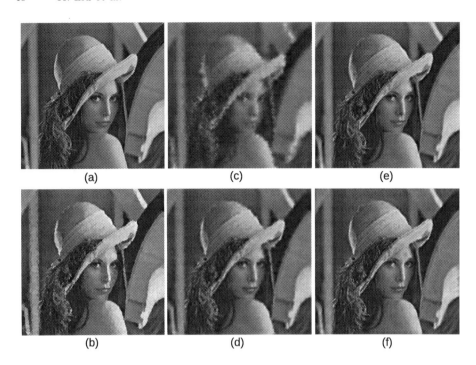

Fig. 3. Summary of restoration of Lena. (a) Original Lena image. (b) Result of piece-wise interpolation. (c) Result of vertex-based iso-RBF interpolation. (d) Result of triangle-based iso-RBF interpolation. (e) Result of triangle-based ARBF interpolation using MQ basis. (f) Result of triangle-based ARBF interpolation using IMQ basis.

incorporates both geometric distances and data-dependent feature orientations. Figure 3(f) is similar to Fig. 3(e), except that the basis function is inverse multi-quadrics (IMQ). We have also tested other basis functions like Gaussian and Thin-Plate-Spline (TPS). However, it is hard to find a proper shape parameter to get a reasonable result for Gaussian, and the TPS interpolation doesn't converge.

In Fig. 4, more details of the Lena experiment are shown. (a) is the original Lena image, the same as Fig. 3(a). (b) is the mesh generated by the method outlined in Sect. 2.1.(c) is the recovered image, which is the same as Fig. 3(e). To visually see the generated mesh and compare the difference between the original and restored images, Fig. 4(d)–(f) are the zoomed-in views of (a)–(c), respectively. As the results show, the mesh quality is high enough for subsequent numerical analysis and the recovered image is very close to the original one. As a matter of fact, the restored image looks smoother due to the smooth radial basis functions used, and the sharp edge features are well preserved. Figure 5 shows the original brain MRI, its generated mesh, and the result of ARBF interpolation using intensities on triangle centers with the MQ basis function. The zoomed-in views show the quality of mesh and restoration as well. Figure 6 shows another MRI experiment of breast. Figure 7 shows a CT-scanning experiment. From all

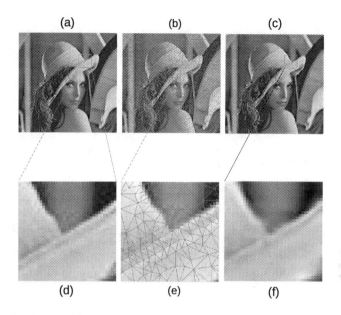

Fig. 4. Details of Lena. (a) Original Lena image. (b) Generated mesh of (a). (c) Result of triangle-based ARBF interpolation using MQ basis. (d)–(f) are zoomed-in views of (a)–(c), respectively.

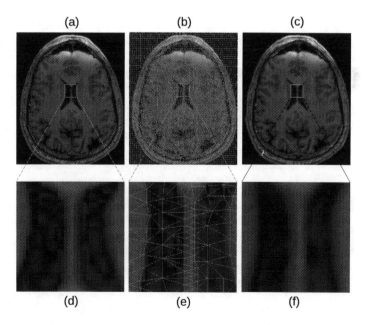

Fig. 5. Details of brain MRI. (a) Original brain MRI. (b) Generated mesh of (a). (c) Result of triangle-based ARBF interpolation using MQ basis. (d)–(f) are zoomed-in views of (a)–(c), respectively.

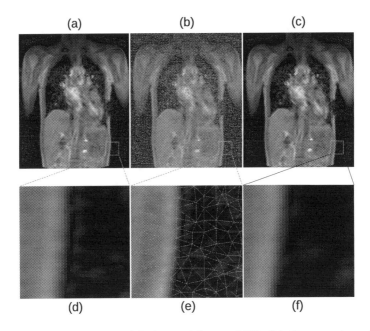

Fig. 6. Details of breast MRI. (a) Original breast MRI. (b) Generated mesh of (a). (c) Result of triangle-based ARBF interpolation using MQ basis. (d)–(f) are zoomed-in views of (a)–(c), respectively.

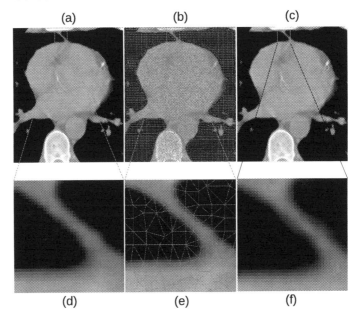

Fig. 7. Details of CT-scanned image of heart. (a) Original heart image. (b) Generated mesh of (a). (c) Result of triangle-based ARBF interpolation using MQ basis. (d)–(f) are zoomed-in views of (a)–(c), respectively.

Table 1. Summary of the Lena image (Fig. 3).

Lena (size is 256 × 256, compression ratio is 6 %)	PSNR (db)	Shape Parameter
Piecewise Interpolation	22.9703	0.5
Triangle-based ISO-RBF Interpolation	26.7367	0.5
Triangle-based ARBF Interpolation (MQ)	28.2088	0.5
Triangle-based ARBF Interpolation (IMQ)	27.1836	1.8

Table 2. Summary of the three medical images (Figs. 5, 6 and 7).

Data	Size	Compression Ratio	PSNR (db)	Shape Parameter	Time (s)
Brain	285 × 341	6 %	15.7058	0.5	1.71
Breast	512 × 512	5 %	11.8763	0.5	8.70
Heart	356 × 396	5 %	10.5208	0.5	2.73

these examples, one can see the effectiveness of the proposed aproaches for image mesh generation and feature-preserving restoration.

To give a quantitative evaluation of the restored images, we use the widely-used peak signal-to-noise ratio (PSNR) as defined below:

$$PSNR = 20 * \log_{10} \left(\frac{255}{RMSE} \right), \tag{7}$$

where

$$RMSE = \sqrt{ \frac{\sum_{i=0}^{M-1} \sum_{j=0}^{N-1} [O(i,j) - I(i,j)]^2}{M * N} },$$

where M and N are the dimensions of the image. $O(i,j)$ is the original intensity at pixel (i,j) and $I(i,j)$ is the interpolated intensity at (i,j).

Table 1 gives a summary of the Lena image using different restoration approaches. The compression ratio in the table means the ratio of the number of vertices in the mesh vs. the number of pixels in the original image. As we can see, the restored image with the anisotropic RBF interpolation gives the best PSNR score. Table 2 summarizes the other three data sets, where the running time of image restoration for each case is included and measured on a PC with 1.8 GHz CPU and 2 GB RAM. The proposed algorithms were implemented in C programming and will be released to the public.

4   Conclusions

The present paper describes a nonlinear interpolation method by using anisotropic radial basis functions and structure tensor driven metrics. Using the proposed methods, an original image can be stored and processed in the mesh format with some nice advantages including less storage requirement, faster transmission speed, and more efficient image processing due to the significantly reduced number of mesh nodes as opposed to the number of pixels in the original image. The generated meshes, after some post-processing such as mesh-based segmentation, can be readily used for further numerical analysis. The present image restoration algorithm provides an effective way to restore the image with an arbitrary super-resolution from a mesh representation, serving as a decoding algorithm for the mesh-based image coding technique. The anisotropic RBF algorithm can be used as a de-blurring process as well with sharp features well preserved in the images.

As the image restoration algorithm shows, the time complexity of the function ARBFInterpolation() is $O(m \times n)$, where m is the number of triangles and n is the number of pixels inside of a triangle. In case of 3D images or very large 2D images, the running time could be very expensive. One of our further investigations would be the parallel implementation of the proposed algorithm using GPU programming. Fortunately the present method is very straightforward to parallelize to accelerate the computations. Additionally we are also interested in the mesh-based image segmentation by using the adaptive meshes generated from the original images, and in applying the segmented meshes to image-based numerical analysis.

References

1. Adams, M.: An efficient progressive coding method for arbitrarily-sampled image data. IEEE Signal Process. Lett. **15**, 629–632 (2008)
2. Adams, M.: Progressive lossy-to-lossless coding of arbitrarily-sampled image data using the modified scattered data coding method. In: Proceedings of IEEE International Conference on Acoustics, Speech and Signal Processing, Taipei, Taiwan, pp. 1017–1020 (2009)
3. Adams, M.: A flexible content-adaptive mesh-generation strategy for image representation. IEEE Trans. Image Process. **20**, 2414–2427 (2011)
4. Adams, M.: A highly-effective incremental/decremental Delaunay mesh-generation strategy for image representation. Signal Process. **93**, 749–764 (2013)
5. Aizawa, K., Huang, T.: Model-based image coding: advanced video coding techniques for very low bit-rate applications. Proc. IEEE **83**, 259–271 (1995)
6. Altunbasak, Y., Tekalp, A.: Closed-form connectivity-preserving solutions for motion compensation using 2-d meshes. IEEE Trans. Image Process. **6**(533), 1255–1269 (1997)
7. Benoit-Cattin, H., Joachimsmann, P., Planat, A., Valette, S., Baskurt, A., Prost, R.: Active mesh texture coding based on warping and DCT. In: IEEE International Conference on Image Processing, Kobe, Japan (1999)
8. Brankov, J., Yang, Y., Galatsanos, N.: Image restoration using content-adaptive mesh modeling. In: Proceedings of IEEE International Conference on Image Processing, vol. 2, pp. 997–1000 (2003)
9. Brankov, J., Yang, Y., Wernick, M.: Tomographic image reconstruction based on a content-adaptive mesh model. IEEE Trans. Med. Imaging **23**, 202–212 (2004)
10. Casciola, G., Lazzaro, D., Montefusco, L., Morigi, S.: Shape preserving surface reconstruction using locally anisotropic RBF interpolants. Comput. Math. Appl. **51**, 1185–1198 (2006)

11. Casciola, G., Montefusco, L., Morigi, S.: The regularizing properties of anisotropic radial basis functions. Appl. Math. Comput. **190**, 1050–1062 (2007)
12. Casciola, G., Montefusco, L., Morigi, S.: Edge-driven image interpolation using adaptive anisotropic radial basis functions. J. Math. Imaging Vis. **36**, 125–139 (2010)
13. Chen, J., Paris, S., Wang, J., Matusik, W., Cohen, M., Durand, F.: The video mesh: A data structure for image-based three-dimensional video editing. In: IEEE International Conference on Computational Photography (ICCP), Pittsburgh, PA, USA, pp. 1–8 (2011)
14. Coleman, S., Scotney, B., Herron, M.: Image feature detection on content-based meshes. In: Proceedings of IEEE International Conference on Image Processing, vol. 1, pp. 844–847 (2002)
15. Davoine, F., Antonini, M., Chassery, J., Barlaud, M.: Fractal image compression based on Delaunay triangulation and vector quantization. IEEE Trans. Image Process. **5**, 338–346 (1996)
16. Delaunay, B.: Sur la sphere vide. Classe des Science Mathematics et Naturelle **7**, 793–800 (1934)
17. Demaret, L., Robert, G., Laurent, N., Buisson, A.: Scalable image coder mixing DCT and triangular meshes. In: IEEE International Conference on Image Processing, Vancouver, BC, Canada, vol. 3, pp. 849–852 (2000)
18. Divo, E., Kassab, A.: An efficient localized RBF meshless method for fluid flow and conjugate hear transfer. ASME J. Heat Transfer **129**, 124–136 (2007)
19. Garcia, M., Vintimilla, B.: Acceleration of filtering and enhancement operations through geometric processing of gray-level images. In: IEEE International Conference on Image Processing, Vancouver, BC, Canada, vol. 1, pp. 97–100 (2000)
20. Garland, M., Heckbert, P.: Fast polygonal approximation of terrains and height fields. Technical Report CMU-CS-95-181, School of Computer Science, Carnegie Mellon University, Pittsburgh, PA, USA (1995)
21. Guo, Y., Liu, F., Shi, J., Zhou, Z., Gleicher, M.: Image retargeting using mesh parametrization. IEEE Trans. Multimedia **11**, 856–867 (2009)
22. Hsu, P., Liu, K., Chen, T.: A low bit-rate video codec based on two-dimensional mesh motion compensation with adaptive interpolation. IEEE Trans. Circuits Syst. Video Technol. **11**, 111–117 (2001)
23. Hung, K., Chang, C.: New irregular sampling coding method for transmitting images progressively. In: IEEE Proceedings of Vision, Image and Signal Processing, vol. 150, pp. 44–50 (2003)
24. Kosec, G., Sarler, B.: Local RBF collocation method for Darcy flow. Comput. Model. Eng. Sci. **25**, 197–208 (2008)
25. Lechat, P., Sanson, H., Labelle, L.: Image approximation by minimization of a geometric distance applied to a 3-D finite elements based model. In: Proceedings of IEEE International Conference on Image Processing, vol. 2, pp. 724–727 (1997)
26. Li, P., Adams, M.: A tuned mesh-generation strategy for image representation based on data-dependent triangulation. IEEE Trans. Image Process. **22**, 2004–2018 (2013)
27. Marquant, G., Pateux, S., Labit, C.: Mesh and "crack lines": application to object-based motion estimation and higher scalability. In: IEEE International Conference on Image Processing, Vancouver, BC, Canada, vol. 2, pp. 554–557 (2000)
28. Nosratinia, A.: New kernels for fast mesh-based motion estimation. IEEE Trans. Circuits Syst. Video Technol. **11**, 40–51 (2001)
29. Petrou, M., Piroddi, R., Talebpour, A.: Texture recognition from sparsely and irregularly sampled data. Comput. Vis. Image Underst. **102**, 95–104 (2006)

30. Ramponi, G., Carrato, S.: An adaptive irregular sampling algorithm and its application to image coding. Image Vis. Comput. **19**, 451–460 (2001)
31. Rippa, S.: Adaptive approximation by piecewise linear polynomials on triangulations of subsets of scattered data. SIAM J. Sci. Stat. Comput. **13**, 1123–1141 (1992)
32. Sarkis, M., Diepold, K.: A fast solution to the approximation of 3-D scattered point data from stereo images using triangular meshes. In: Proceedings of IEEE-RAS International Conference on Humanoid Robots, Pittsburgh, PA, USA, pp. 235–241 (2007)
33. Sarler, B., Vertnik, R.: Meshfree explicit local radial basis function collocation method for diffusion problems. Comput. Math. Appl. **51**, 1269–1282 (2006)
34. Singh, A., Terzopoulos, D., Goldgof, D.: Deformable models in medical image analysis. IEEE Computer Society Press (1998)
35. Shewchuk, J.: Triangle: A two-dimensional quality mesh generator and Delaunay triangulator (2005). http://www.cs.cmu.edu/quake/triangle.html
36. Su, D., Willis, P.: Demosaicing of color images using pixel level data-dependent triangulation. In: Proceedings of Theory and Practice of Computer Graphics, pp. 16–23 (2003)
37. Su, D., Willis, P.: Image interpolation by pixel-level data-dependent triangulation. Comput. Graph. Forum **23**, 189–201 (2004)
38. Toklu, C., Tekalp, A., Erdem, A.: Semi-automatic video object segmentation in the presence of occlusion. IEEE Trans. Circuits Syst. Video Technol. **10**, 624–629 (2000)
39. Tu, X., Adams, M.: Improved mesh models of images through the explicit representation of discontinuities. Can. J. Electr. Comput. Eng. **36**, 78–86 (2013)
40. Vertnik, R., Sarler, B.: Meshless local radial basis function collocation method for convective-diffusive solid-liquid phase change problems. Int. J. Numer. Meth. Heat Fluid Flow **16**, 617–640 (2006)
41. Vertnik, R., Sarler, B.: Solution of incompressible turbulent flow by a mesh-free method. Comput. Model. Eng. Sci. **44**, 65–95 (2009)
42. Wang, J., Liu, G.: On the optimal shape parameters of radial basis functions used for 2-d meshless methods. Comput. Methods Appl. Mech. Eng. **191**, 2611–2630 (2002)
43. Wang, Y., Lee, O.: Active mesh - a feature seeking and tracking image sequence representation scheme. IEEE Trans. Image Process. **3**, 610–624 (1994)
44. Wang, Y., Lee, O., Vetro, A.: Use of 2-D deformable mesh structures for video coding, part II-the analysis problem and a region-based coder employing an active mesh representation. IEEE Trans. Circuits Syst. Video Technol. **6**, 647–659 (1996)
45. Yang, Y., Miles, N., Jovan, G.: A fast approach for accurate content-adaptive mesh generation. IEEE Trans. Image Process. **12**, 866–881 (2003)

Image Enhancement by Gradient Distribution Specification

Yuanhao Gong[1,2] and Ivo F. Sbalzarini[1,2(✉)]

[1] MOSAIC Group, Faculty of Computer Science, TU Dresden, Dresden, Germany
[2] MOSAIC Group, Center for Systems Biology Dresden (CSBD), Max Planck
Institute of Molecular Cell Biology and Genetics, Dresden, Germany
{gong,ivos}@mpi-cbg.de

Abstract. We propose to use gradient distribution specification for image enhancement. The specified gradient distribution is learned from natural-scene image datasets. This enhances image quality based on two facts: First, the specified distribution is independent of image content. Second, the distance between the learned distribution and the empirical distribution of a given image correlates with subjectively perceived image quality. Based on those two facts, remapping an image such that the distribution of its gradients (and therefore also Laplacians) matches the specified distribution is expected to improve the quality of that image. We call this process "image naturalization". Our experiments confirm that naturalized images are more appealing to visual perception. Moreover, "naturalness" can be used as a measure of image quality when ground-truth is unknown.

1 Introduction

Image enhancement plays a fundamental role in image processing. The usual means for image enhancement is histogram equalization or one of its variants. However, the intensity histogram greatly varies with image contents, so that there is no simple mathematical model for it. In the absence of such a model, histogram equalization assumes a uniform prior distribution. This assumption is not usually valid, causing histogram equalization to fail in many cases.

This problem is circumvented when replacing the intensity histogram with the gradient histogram, which is remarkably invariant across natural-scene images [1–3]. This fact has previously been exploited for image denoising [1], motion deblurring [2,3], and image restoration [4]. In this paper, we propose to use gradient distribution specification for image enhancement in a novel process we call "image naturalization".

The concept of image naturalization is illustrated in Fig. 1. Given a learned prior distribution, the quality of images can be improved by gradient remapping such that the new gradient and Laplace fields satisfy the prior. The result image can be reconstructed by solving a single Poisson equation.

© Springer International Publishing Switzerland 2015
C.V. Jawahar and S. Shan (Eds.): ACCV 2014 Workshops, Part II, LNCS 9009, pp. 47–62, 2015.
DOI: 10.1007/978-3-319-16631-5_4

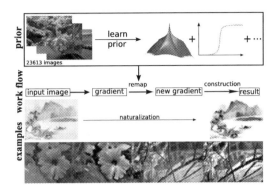

Fig. 1. Illustration of image naturalization.

1.1   First-Order Prior

It is well known that the gradient distributions of natural-scenes image have a heavy tail in log-scale [2–4]. Traditionally, such gradient distributions are modeled as generalized Laplace distributions:

$$\log(p(x)) = -k\|x\|^\alpha + \beta, \tag{1}$$

where k, α, and β are parameters, x is the gradient, and p is the gradient probability distribution. This model includes several frequently used priors, including total variation (TV) ($\alpha = 1$) [5] and Hyper-Laplacian ($\alpha = 0.6$) [3].

   This model, however, has several drawbacks. First, as shown in this paper, this model does not fit the data well. Second, all previous works treat k, α, and β as independent variables, violating the normalization of the distribution ($\int_{-\infty}^{+\infty} p\,dx \neq 1$). Third, the generalized Laplace model is computational expensive to evaluate. Here, we present a new model that overcomes all of these issues.

1.2   Second-Order Prior

In addition to first-order (i.e., gradient) statistics, second-order statistics such as mean curvature (MC) [6] and Gaussian curvature (GC) [7,8] can also be imposed as priors. Different orders of priors can be combined, e.g., in *Fields of Experts* models [9].

   However, higher-order statistics may not be necessary in all cases. In Fig. 2, the correlation between $I(\boldsymbol{x})$ and $I(\boldsymbol{x} + \boldsymbol{r})$ is shown for different orders d of derivatives:

$$Corr(d, r) := \text{correlation}(\nabla^d I(\boldsymbol{x}), \nabla^d I(\boldsymbol{x} + \boldsymbol{r})), \tag{2}$$

where $\boldsymbol{r} = (r, 0)$. The correlation reduces significantly for larger d. This explains why first- and second-order priors are so powerful for image processing, but higher-order derivatives do not necessarily improve the result. One reason is that the discrete image may not be higher-order differentiable. For image-processing

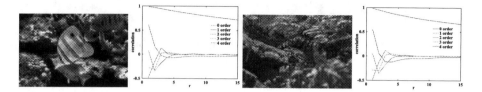

Fig. 2. $Corr(d, r)$ rapidly reduces for higher-order derivatives.

tasks, second order has repeatedly been shown to be enough [8]. Therefore, we only consider derivatives up to the order of two.

Among all second order operators, the Laplace operator is the most attractive one since it leads to a Poisson equation, as does the gradient prior. Therefore, both priors can be imposed simultaneously by solving a single Poisson equation, for which efficient solvers are readily available.

1.3 Equivalence Between Gradient Field and Original Image

The gradient field of an image is equivalent to the original image plus a single point constraint. The original image can hence be exactly reconstructed from its gradient field [10–12]. An excellent review about signal processing in the gradient domain can be found in Ref. [11]. Some recent works in this area are described in Refs. [13,14]. More details are given here in Sect. 4.2.

1.4 Our Contributions

- We provide here new parametric models for the gradient and Laplace distributions of images. Instead of modeling the probability density functions (PDF) in log-scale, we model the cumulative distribution functions (CDF). We show that the resulting models are more accurate and computationally more efficient than previous models.
- We show that the distance between the gradient/Laplace distributions of a given image and the learned prior distributions is correlated with image quality. Therefore, imposing these priors is expected to improve image quality.
- We provide an algorithm for image enhancement by gradient/Laplace distribution specification through a remapping function. A nonlinear remapping function can be approximated by a linear function, which has only one scalar variable.
- The proposed image enhancement process is parameter-free, which avoids manual parameter tuning.
- First- and second-order priors are imposed simultaneously by solving a single Poisson equation.
- We provide a simple scalar number that measures how close the gradient distribution of an image is to the prior. This number can be used to evaluate image quality in cases where ground-truth is unknown.

2   The Naturalization Prior

The naturalization prior proposed here is a linear combination of a gradient distribution prior and the consistent Laplace operator prior, i.e., the one corresponding to the divergence of the gradient. We learn these priors from a large dataset of natural-scene images. We then study the variability of the data around the priors. We provide novel parametric models for both priors and an efficient algorithm for imposing them on any given image. Finally, we define an "image naturalness factor" based on our models.

We use seven datasets of natural-scene images as shown in Table 1. Each image $I(x, y)$ was converted to 8-bit gray-scale. The gradient field $\boldsymbol{G}$ is defined as

$$\boldsymbol{G}(x, y) = (\nabla_x I(x, y), \nabla_y I(x, y)), \tag{3}$$

where we use the first-order finite-difference approximations: $\nabla_x I = I(x+1, y) - I(x, y)$ and $\nabla_y I = I(x, y + 1) - I(x, y)$. We use homogeneous Dirichlet boundary conditions at the image borders. Due to the use of 8-bit gray-scale images, possible gradients are in the integer domain $[-255, 255] \times [-255, 255]$, where we can easily construct the two-dimensional histogram of $\boldsymbol{G}$. We use G^x and G^y to denote the respective components of $\boldsymbol{G}$.

Table 1. Natural-scene image datasets: sources and sizes.

Dataset	1[a]	2[b]	3[c]	4 [d]	5[e]	6[f]	7[g]	Total
#images	1005	1000	5063	832	1491	6033	8189	23613

[a]http://www.vision.ee.ethz.ch/showroom/zubud/.
[b]http://see.xidian.edu.cn/faculty/wsdong/Data/Flickr_Images.rar.
[c]http://www.robots.ox.ac.uk/~vgg/data/oxbuildings/.
[d]http://www.comp.leeds.ac.uk/scs6jwks/dataset/leedsbutterfly/.
[e]http://lear.inrialpes.fr/~jegou/data.php.
[f]http://www.vision.caltech.edu/visipedia/CUB-200.html.
[g]http://www.robots.ox.ac.uk/~vgg/data/flowers/102/index.html.

The Laplace field L is defined as

$$L(x, y) = \Delta I(x, y), \tag{4}$$

where Δ is the Laplace operator, which is discretized using the second-order 5-point finite-difference stencil. Possible values are in the integer domain $[-1020, 1020]$.

In order to turn the histograms into probability distributions, we divide all bins by the total number of pixels mn in the image where m and n are the numbers of pixels along the x and y axes of the image. After aggregating data from all images in the dataset, we further normalize the histogram by the total number of images in the dataset. This yields the average distributions p_1^{pr} and p_2^{pr} of the gradient and Laplace operators. For color images, the priors are learned and applied separately for each color channel, performing all operations channel-wise.

2.1   Variability

We analyze how closely the natural-scene images in the training dataset match the average gradient and Laplace priors learned from them. The first row of Fig. 3 show histograms of the Root Mean Square (RMS) distances (left), Hellinger distance (middle), and Kullback Leibler divergence (right) between each image's individual distribution and the prior. The second row of Fig. 3 shows the RMS distribution for the Laplace operator for grayscale (left) and color (right) images, respectively.

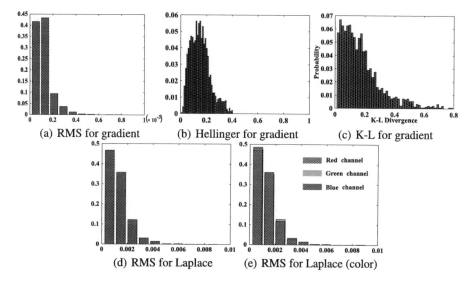

Fig. 3. Histograms of different distances between the priors and the gradient (top row) and Laplace (bottom row) distributions of individual training images.

2.2   Gradient Distribution Models

Considering G^x and G^y individually, both satisfy the heavy-tail characteristic in log-scale, which can be modeled as a hyper-Laplacian distribution [3]. Those traditional one-dimensional models fit $p(G^x)$ in log-scale, which is why $\int_{-\infty}^{+\infty} p(G^x)\mathrm{d}G^x = 1$ cannot be guaranteed. Unlike previous works, we hence propose to do the modeling on the CDF instead.

The CDF of the gradient is defined as:

$$C(\boldsymbol{G}) = \int_{-255}^{G^x} \int_{-255}^{G^y} P((u,v))\,\mathrm{d}u\mathrm{d}v. \qquad (5)$$

Observing the step property of $C(\boldsymbol{G})$, we propose to approximate the CDF with the parametric model:

$$\widetilde{C}(\boldsymbol{G}) = \left(\frac{\mathrm{atan}(T_1 G^x)}{\pi} + \frac{1}{2} \right) \left(\frac{\mathrm{atan}(T_1 G^y)}{\pi} + \frac{1}{2} \right). \qquad (6)$$

The choice of the atan function is motivated by the Student-T distribution or Cauchy distribution. This model has only one parameter to be fitted: T_1. The fitting results are shown in Table 2.

Table 2. Fits of the parametric 2D CDF model to the image datasets.

Image set	1	2	3	4	5	6	7	all
T_1	0.37	0.26	0.38	0.35	0.56	0.37	0.7	0.46
SSE	20.71	23.11	19.08	23.7	22.94	19.64	22.97	18.75
R-square	0.9995	0.9995	0.9996	0.9995	0.9995	0.9996	0.9995	0.9996

The corresponding marginal model for the distribution of G^x (analogously G^y) is:

$$\log(P(G^x)) = \log\left(\frac{T_1}{\pi}\right) - \log\left(1 + (T_1 G^x)^2\right). \tag{7}$$

We compare our model with other models in Fig. 4 and Table 3 using the optimal parameters for each model. In all cases, the present model describes the data best.

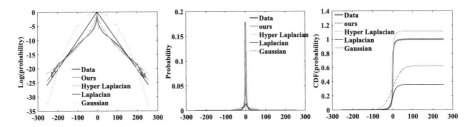

Fig. 4. Comparison of models in log scale, linear scale, and CDF.

Table 3. Comparison of our model (Eq. 7, top row) with generalized Laplace models (Eq. 1) for $\alpha = 0.6$, 1, 2 (rows two to four).

Image set	1	2	3	4	5	6	7
SSE	40.5	43.0	67.7	23.8	34.1	37.9	25.4
R^2	0.99	0.99	0.99	0.99	0.99	0.99	0.99
SSE	576	301	537	45.4	389	70.5	250
R^2	0.92	0.93	0.91	0.98	0.96	0.98	0.97
$SSE \times 10^{-3}$	1.86	3.01	3.02	3.95	2.34	3.90	3.95
R^2	0.74	0.30	0.52	0.13	0.81	0.10	0.57
$SSE \times 10^{-4}$	0.83	1.02	1.10	1.24	1.32	1.23	1.64
R^2	−0.12	−1.3	−0.72	−2.6	−0.046	−2.5	−0.75

In Fig. 5, different images and their corresponding CDFs are shown. For the blurred image (Gaussian blur, $\sigma = 3$), the frequency of small gradients is increased. For the noisy image (10 % Gaussian noise), the frequency of large gradients is increased. For the super-resolution (SR) image (upsampling factor 9), the frequency of small gradients is increased. For the bilateral filter ($w = 5$, $\sigma_s = 3$, $\sigma_c = 0.1$) and the guided filter ($r = 10$, $\epsilon = 0.01$), the frequency of small gradients is increased.

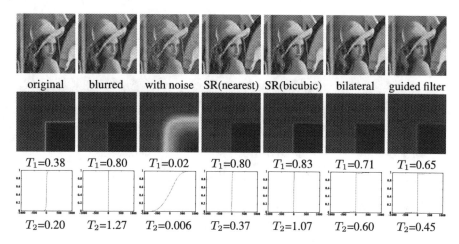

Fig. 5. Row 1: images; Row 2: their gradient CDFs; Row 3: their Laplace CDFs

2.3 Laplace Distribution Models

Also for the distribution of the Laplace operator response, we use the CDF to model the statistics:

$$L(t) = \int_{-\infty}^{t} p(\Delta I(\boldsymbol{x})) \mathrm{d}\Delta I(\boldsymbol{x}). \tag{8}$$

For the Laplace CDF, we propose the parametric model:

$$\widetilde{L}(t) = \frac{\mathrm{atan}(T_2 t)}{\pi} + \frac{1}{2}, \tag{9}$$

where T_2 is the only free parameter. The Laplace CDFs for different images are shown in Fig. 5. The figure also shows that the steps in the CDF indicate edge preservation in the spatial domain.

We test the present parametric model by adding Gaussian noise with $\sigma = [0.02 : 0.02 : 0.8]$ to the Lena image and computing the difference $\widetilde{L} - L$. The result is shown in Fig. 6. Even the maximum difference is small compared to the absolute value of L.

2.4 Naturalness Factor and Image Naturalization

For a given image, it is easy to fit T_1 and T_2 using the above parametric distribution models. The corresponding values for the priors learned from the

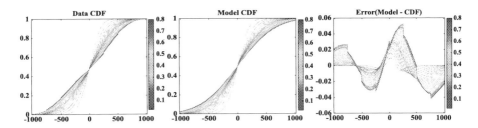

Fig. 6. Left: L, middle: $\widetilde{L}$, right: $\widetilde{L} - L$. The noise level σ is indicated by color (Color figure online).

natural-scene dataset are: $T_1^{\mathrm{pr}} = 0.38$ and $T_2^{\mathrm{pr}} = 0.14$. Comparing the values of an image to these expected ones from natural-scene images tells how close the image is to a natural-scene one.

For any image I, the **naturalness factor** N_f is defined as

$$N_f = (1 - \theta)\frac{T_1}{T_1^{\mathrm{pr}}} + \theta\frac{T_2}{T_2^{\mathrm{pr}}}, \tag{10}$$

where $\theta \in [0, 1]$ is a weight parameter. The naturalness factor N_f^c of a color image is defined separately for each color channel c. The **naturalized image** I_n is generated from I such that $T_i \approx T_i^{\mathrm{pr}}$ ($i \in \{1, 2\}$). This process is called **image naturalization**.

In our C++ implementation, we use the ternary search algorithm to find T_1 and T_2. On a 2011 MacBook Pro, this code achieves 290 Mpixel/s for 8-bit three-channel color images. Implementations in Matlab, Java, and C++ are available from MOSAIC Group web site mosaic.mpi-cbg.de.

3   Correlation with Image Quality

We show that the naturalization prior is correlated with subjectively perceived image quality. Therefore, image naturalization is expected to improve image quality. We test both the non-parametric priors and the above parametric models on the standard image quality assessment dataset LIVE [15], containing 779 images with five different types of distortions (degradations). The subjective image quality score (DMOS, difference mean opinion score) for each image is provided by LIVE. Our objective score is the average Hellinger distance (HD) between the ground truth image and distorted image, defined as:

$$score = \frac{1}{2}\mathrm{HD}(p(\nabla I^{\mathrm{true}}), p(\nabla I^{\mathrm{distort}})) + \frac{1}{2}\mathrm{HD}(p(\Delta I^{\mathrm{true}}), p(\Delta I^{\mathrm{distort}})). \tag{11}$$

We also define a score from the naturalization factor N_f. We compute N_f for both the ground-truth image and the distorted image and use the absolute difference as an objective quality score:

$$score_{N_f} = |N_f^{\text{true}} - N_f^{\text{distort}}|. \tag{12}$$

In the case where ground truth is unknown or unavailable, the score is defined with respect to the naturalization prior:

$$score_{pr} = \frac{1}{2}\text{HD}(p_1^{\text{pr}}, p(\nabla I^{\text{distort}})) + \frac{1}{2}\text{HD}(p_2^{\text{pr}}, p(\Delta I^{\text{distort}})). \tag{13}$$

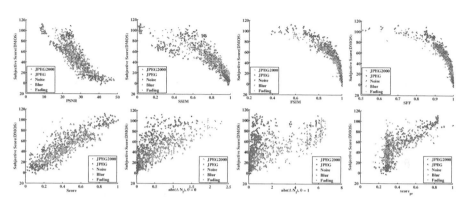

Fig. 7. Correlation of different objective image quality measures with subjectively perceived image quality (DMOS) on the LIVE benchmark dataset. First row: PSNR, SSIM, FSIM, SFF; second row: $score$, $score_{N_f}$ for $\theta = 0$ and 1, $score_{pr}$.

Table 4. Pearson, Spearman, and Kendall correlation coefficients between DMOS and the different image quality measures on the LIVE benchmark dataset.

	PSNR	SSIM	FSIM	SFF	$score$	$N_f(\theta = 0)$	$N_f(\theta = 1)$	$score_{pr}$
PCC	−0.8585	−0.8252	−0.8586	−0.8126	0.8687	0.6692	0.5199	0.761
SCC	−0.8756	−0.9104	−0.9634	−0.9649	0.8630	0.7118	0.5817	0.706
KCC	−0.6865	−0.7311	−0.8337	−0.8365	0.6745	0.5123	0.4134	0.522

The correlations between DMOS and all of these scores are shown in Fig. 7 and Table 4. Correlations are reported as Pearson linear correlation coefficients (PCC), Spearman rank-order correlation coefficients (SCC), and Kendall rank-order correlation coefficients (KCC). We use the HD in our score because it outperforms all other tested distances (L_2, L_1, cos, χ^2) for these correlation coefficients. We compare the present score with the state-of-the-art image quality assessment methods SSIM [15], FSIM [16], and SFF [17]. Among all approaches, our nonparametric method shows the best linearity between $score$ and DMOS, which is preferable for image enhancement tasks. Our result is comparable with PSNR, but does not require knowing the noise level in the image, or any other geometric information about the ground truth.

Taken together, these results confirm that imposing the gradient and Laplace distribution priors is expected to improve image quality, therefore making then good candidates for image enhancement priors.

4   Image Naturalization Algorithm

The image naturalization process consists in remapping the gradient field of the image to satisfy the prior distributions, followed by reconstructing the naturalized output image from the remapped field.

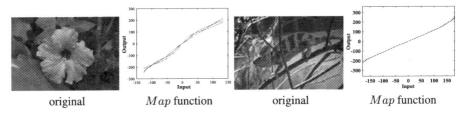

original   Map function   original   Map function

Fig. 8. Two example images and their nonlinear *Map* functions (color indicates the different color channels) (Color figure online).

4.1   Gradient Distribution Specification

Let *Map* be a function that maps the gradient field $\boldsymbol{G}$ to a new gradient field $\boldsymbol{G}_n$, which satisfies the naturalization prior:

$$\boldsymbol{G}_n = Map(\boldsymbol{G}), \quad s.t. \quad p(\boldsymbol{G}_n) = p_1^{\mathrm{pr}}, \quad p(\nabla \cdot \boldsymbol{G}_n) = p_2^{\mathrm{pr}}. \tag{14}$$

In general, *Map* can be non-parametric and nonlinear. More specifically, we use here modified exact histogram specification [18] as the *Map* function. Two example images and their nonlinear non-parametric *Map* functions are shown in Fig. 8.

4.2   Image Reconstruction

We reconstruct the naturalized image I_n from the remapped gradient field by solving the variational model:

$$\min \left\{ \int_{x \in \Omega} \|\nabla I_n - \boldsymbol{G}_n\|_2^2 \, \mathrm{d}\boldsymbol{x} \right\}, \tag{15}$$

which leads to the Poisson equation

$$\Delta I_n = \nabla \cdot \boldsymbol{G}_n. \tag{16}$$

This equation can be solved efficiently by FFT-based algorithms or wavelet solvers. A short summary of commonly available Poisson solvers is shown in Table 5.

Reconstructing an image from its gradient field is accurate and computationally efficient. An example is shown in Fig. 9. The original image (Fig. 9a) is an 8-bit grayscale image. The absolute pixel-wise RMS of the reconstruction without remapping (Fig. 9b) is 2.0 while the average intensity value is 105. The size of the image is 1881×2400 pixels and the reconstruction took 3.5 s using our MATLAB implementation of the wavelet Poisson solver on a 2 GHz Intel i7 processor. The naturalized image after nonlinear remapping is shown in Fig. 9c.

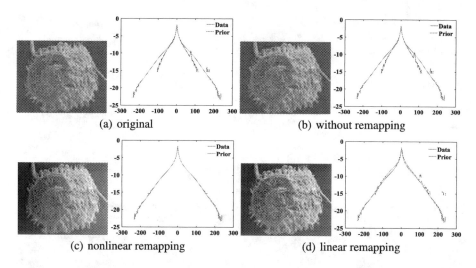

(a) original   (b) without remapping

(c) nonlinear remapping   (d) linear remapping

Fig. 9. Comparison of different types of remapping. The reconstruction errors (RMS) are 2.0, 33, and 23, respectively. The corresponding gradient distributions after remapping are shown in comparison with the naturalization prior next to the image.

Table 5. Summary of Poisson solvers (we implemented FFT and wavelet solvers).

Solver	Cholesky[h]	Jacobi	Gauss-Seidel	SOR
Type	Direct	Iterative	Iterative	Iterative
	$(mn)^3$	$(mn)^2$	$(mn)^2$	$(mn)^{3/2}$
Solver	Cholesky[i]	FFT	Multigrid	Wavelet
Type	direct	direct	iterative	direct
	$(mn)^{3/2}$	$(mn)log(mn)$	(mn)	(mn)

[h]dense Cholesky decomposition.
[i]sparse Cholesky decomposition.

4.3   Linear Approximation of *Map*

Our one-parameter model results in a linear approximation of the remapping function *Map*:

$$G_n = N_f\,G, \tag{17}$$

which is equivalent to scaling the original image:

$$I_n = N_f I. \tag{18}$$

Simply scaling the image with N_f significantly accelerates the naturalization process while still avoiding halo artifacts in the result. An example of a naturalized image after linear remapping is shown in Fig. 9d.

(a) original   (b) nonlinear map   (c) linear map   (d) HE   (e) CLAHE

(f) Retinex (t=4)   (g) Retinex (t=5)   (h) Retinex (t=6)   (i) Retinex (t=8)   (j) Retinex (t=10)

(k) original   (l) AM [19]   (m) DT [20]   (n) GF [21]

(o) L0 [22]   (p) RTV [23]   (q) WLS [14]   (r) Naturalization

Fig. 10. Comparison of image enhancement methods: nonlinear and linear naturalization, HE (Histogram Equalization), CLAHE (Contrast Limited Adaptive Histogram Equalization), Retinex (for different parameters), AM (Adaptive Manifold), DT (Domain Transform), GF (Guided Filter), L0 norm, RTV (Relative Total Variation), WLS (Weighted Least Square), and Naturalization.

5   Experiments

In the past decades, image smoothing and sharpening have received a lot of attention. However, there is to date no objective measure to decide if an image should be smoothed or sharpened. N_f can be used to automatically determine whether an image should be smoothed or sharpened. It is clear from Eq. 17 that an image gets smoothed for $N_f < 1$ and sharpened for $N_f > 1$. Two examples are shown in Fig. 10 and compared with other methods using their respective default parameters. More results and comparisons are shown in Fig. 11.

5.1   Objective Image Quality Evaluation

In addition to image naturalization, the prior can be used for objective evaluation of results from various algorithms, such as deblurring, image editing, image synthesis, etc. Significant progress has recently been made in image deblurring,

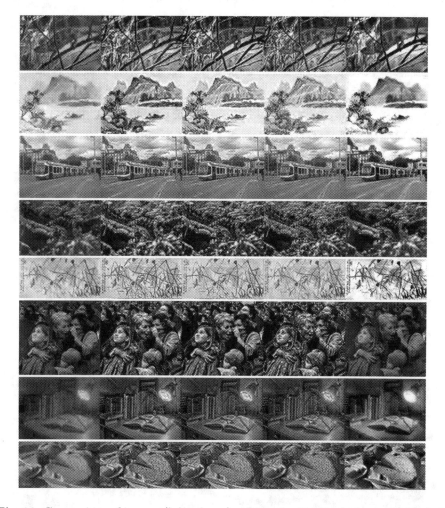

Fig. 11. Comparison of images (left column) enhanced with AM (column 2), GF (column 3), RTV (column 4), and naturalization ($\theta = 0.5$, column 5).

High Dynamic Range (HDR) compression, Poisson image editing, etc. In all these applications, the ground truth is unknown. The naturalness factor can provide an objective evaluation of the results. To the best of our knowledge, this is the first proposal of an objective measure to evaluate such processing results.

In Fig. 12, the first two rows show deblurring results (original, [24–26]); the third row considers object adding and removing by Poisson editing (original, add, original, remove); the fourth row shows HDR results ([27]); the fifth row scene rendering results (realistic image synthesis); the sixth row gives an evaluation of a panorama-stitched image. It is obvious that visually pleasing results satisfy the naturalization prior better.

(0.88, 0.99, 1.25)   (0.51, 0.64, 0.80)   (0.58, 0.69, 0.96)   (0.74, 0.90, 1.24)
(1.22, 1.32, 1.44)   (0.45, 0.53, 0.66)   (0.57, 0.63, 0.82)   (0.63, 0.74, 1.11)

(1.45, 1.45, 1.42)   (0.88, 0.91, 0.87)   (0.91, 0.91, 0.90)   (1.40, 1.41, 1.41)
(2.02, 2.05, 1.91)   (0.87, 0.89, 0.83)   (0.87, 0.87, 0.85)   (1.64, 1.64, 1.60)

(1.29, 1.34, 1.32)   (1.31, 1.37, 1.35)   (0.80, 0.81, 0.73)   (0.96, 0.98, 0.87)
(1.32, 1.37, 1.31)   (1.41, 1.47, 1.41)   (1.01, 1.01, 0.92)   (1.23, 1.25, 1.12)

(1.66, 1.70, 1.69)   (1.41, 1.44, 1.49)   (1.85, 1.93, 2.12)   (1.58, 1.67, 1.62)
(2.88, 2.95, 2.68)   (1.82, 1.85, 1.82)   (2.29, 2.50, 2.89)   (1.85, 1.92, 1.72)

(1.01, 0.90, 0.99)   (1.06, 1.06, 1.03)   (1.12, 1.15, 1.21)   (1.63, 1.68, 1.54)
(1.12, 1.06, 1.10)   (1.00, 1.01, 0.99)   (1.22, 1.25, 1.25)   (2.91, 3.16, 2.64)

(1.18, 1.18, 1.21), (1.44, 1.44, 1.44)

Fig. 12. Objective evaluation of image-processing results: N_f for each color channel (in RGB order). Better-quality results show values closer to 1.

6   Conclusion

We have presented novel parametric models for the gradient and Laplace distributions of natural-scene images. We have shown that the models are more accurate than previous models and are easy to compute. We have further shown that prior models learned from natural-scene images are correlated with image quality. Therefore, remapping an image to satisfy the learned gradient prior yields a quality-improved image. We call this novel image enhancement method "image naturalization". The remapping function has a linear approximation, which further simplifies the algorithm. The corresponding linear factor is called the "naturalness factor" and can be used to evaluate the quality of image-processing result when ground truth is unavailable.

References

1. Zhu, S., Mumford, D.: Prior learning and Gibbs reaction-diffusion. IEEE Trans. Pattern Anal. Mach. Intell. (PAMI) **19**, 1236–1250 (1997)
2. Shan, Q., Jia, J., Agarwala, A.: High-quality motion deblurring from a single image. ACM Trans. Graph. **27**, 73 (2008)
3. Krishnan, D., Fergus, R.: Fast image deconvolution using hyper-Laplacian priors. Adv. Neural Inform. Proc. Sys. **22**, 1–9 (2009)
4. Cho, T.S., Zitnick, C.L., Joshi, N., Kang, S.B., Szeliski, R., Freeman, W.T.: Image restoration by matching gradient distributions. IEEE Trans. Pattern Anal. Mach. Intell. (PAMI) **34**, 683–694 (2012)
5. Rudin, L.I., Osher, S., Fatemi, E.: Nonlinear total variation based noise removal algorithms. Physica D **60**, 259–268 (1992)
6. El-Fallah, A.I., Ford, G.E.: Mean curvature evolution and surface area scaling in image filtering. IEEE Trans. Image Proc. **6**, 750–753 (1997)
7. Lee, S.H., Seo, J.K.: Noise removal with Gauss curvature-driven diffusion. IEEE Trans. Image Proc. **14**, 904–909 (2005)
8. Gong, Y., Sbalzarini, I.F.: Local weighted Gaussian curvature for image processing. In: International Conference on Image Processing (ICIP), pp. 534–538 (2013)
9. Roth, S., Black, M.J.: Fields of experts. IJCV **82**, 205–229 (2009)
10. Fattal, R., Lischinski, D., Werman, M.: Gradient domain high dynamic range compression. ACM Trans. Graph. **21**, 249–256 (2002)
11. Agrawal, A., Raskar, R.: Gradient domain manipulation techniques in vision and graphics. In: ICCV Short Course (2007)
12. Gong, Y., Paul, G., Sbalzarini, I.F.: Coupled signed-distance functions for implicit surface reconstruction. In: IEEE International Symposium on Biomedical Imaging (ISBI), pp. 1000–1003 (2012)
13. Fattal, R.: Edge-avoiding wavelets and their applications. ACM Trans. Graph. **28**, 1–10 (2009)
14. Farbman, Z., Fattal, R., Lischinski, D., Szeliski, R.: Edge-preserving decompositions for multi-scale tone and detail manipulation. ACM Trans. Graph. **27**, 67:1–67:10 (2008)
15. Wang, Z., Bovik, A., Sheikh, H., Simoncelli, E.: Image quality assessment: from error visibility to structural similarity. IEEE Trans. Image Process. **13**, 600–612 (2004)

16. Zhang, L., Zhang, L., Mou, X., Zhang, D.: Fsim: a feature similarity index for image quality assessment. IEEE Trans. Image Process. **20**, 2378–2386 (2011)
17. Chang, H.W., Yang, H., Gan, Y., Wang, M.H.: Sparse feature fidelity for perceptual image quality assessment. IEEE Trans. Image Process. **22**, 4007–4018 (2013)
18. Coltuc, D., Bolon, P., Chassery, J.M.: Exact histogram specification. IEEE Trans. Image Process. **15**, 1143–1152 (2006)
19. Gastal, E.S.L., Oliveira, M.M.: Adaptive manifolds for real-time high-dimensional filtering. ACM TOG **31**, 33:1–33:13 (2012)
20. Gastal, E.S.L., Oliveira, M.M.: Domain transform for edge-aware image and video processing. ACM TOG **30**, 69:1–69:12 (2011). Proceedings of SIGGRAPH 2011
21. He, K., Sun, J., Tang, X.: Guided image filtering. In: Daniilidis, K., Maragos, P., Paragios, N. (eds.) ECCV 2010, Part I. LNCS, vol. 6311, pp. 1–14. Springer, Heidelberg (2010)
22. Xu, L., Lu, C., Xu, Y., Jia, J.: Image smoothing via L0 gradient minimization. ACM Trans. Graph. **30**(6), 147 (2011)
23. Xu, L., Yan, Q., Xia, Y., Jia, J.: Structure extraction from texture via relative total variation. ACM Trans. Graph. **31**, 139:1–139:10 (2012)
24. Hirsch, M., Schuler, C.J., Harmeling, S., Schölkopf, B.: Fast removal of non-uniform camera shake. In: ICCV, pp. 463–470 (2011)
25. Gupta, A., Joshi, N., Lawrence Zitnick, C., Cohen, M., Curless, B.: Single image deblurring using motion density functions. In: Daniilidis, K., Maragos, P., Paragios, N. (eds.) ECCV 2010, Part I. LNCS, vol. 6311, pp. 171–184. Springer, Heidelberg (2010)
26. Xu, L., Zheng, S., Jia, J.: Unnatural l0 sparse representation for natural image deblurring. In: CVPR, CVPR 2013, pp. 1107–1114. IEEE Computer Society, Washington, DC (2013)
27. Farbman, Z., Fattal, R., Lischinski, D.: Convolution pyramids. ACM Trans. Graph. **30**, 175:1–175:8 (2011)

A Two-Step Image Inpainting Algorithm Using Tensor SVD

Mrinmoy Ghorai[1](✉), Sekhar Mandal[2], and Bhabatosh Chanda[1]

[1] Indian Statistical Institute, Kolkata, India
mgre04@gmail.com
[2] Indian Institute of Engineering Science and Technology,
Shibpur, Howrah, India

Abstract. In this paper, we present a novel exemplar-based image inpainting algorithm using the higher order singular value decomposition (HOSVD). The proposed method performs inpainting of the target image in two steps. At the first step, the target region is inpainted using HOSVD-based filtering of the candidate patches selected from the source region. It helps to propagate the structure and color smoothly in the target region and restrict to appear unwanted artifacts. But a smoothing effect may be visible in the texture regions due to the filtering. In the second step, we recover the texture by an efficient heuristic approach using the already inpainted image. The experimental results show the superiority of the proposed method compared to the state of the art methods.

1 Introduction

Inpainting is the process of generating information into a region(s) marked by the user of an image or video in such a way that the filled region(s) is visually plausible [1,2]. Recently, this system becomes popular and widely used in various field of image processing and computer vision liker restoration (scratch removal of old photograph) and image editing (text or object removal). Previous approaches of image inpainting can be divided roughly into two categories: (i) partial differential equation (PDE) based approach for structure propagation and (ii) exemplar-based approach for texture synthesis. Here our main concern is about the second category.

The PDE-based image inpainting technique propagates information smoothly inward the target region from the surrounding source region along the isophote directions [1]. In [3] the total variation is minimized in the processed image maintaining the fidelity with the input image. The base line idea of these types of methods is to transmit the contours smoothly into the region being inpainted. The second type of approaches is exemplar-based where the selected patch (target patch) from the target region is replaced by the most similar candidate

This work is partially supported by Department of Science and Technology, Government of India (NRDMS/11/1586/09/Phase-I/Project No. 9).

Electronic supplementary material The online version of this chapter (doi:10. 1007/978-3-319-16631-5_5) contains supplementary material, which is available to authorized users.

© Springer International Publishing Switzerland 2015
C.V. Jawahar and S. Shan (Eds.): ACCV 2014 Workshops, Part II, LNCS 9009, pp. 63–77, 2015.
DOI: 10.1007/978-3-319-16631-5_5

patch from the source region. This concept is first introduced by Efors et al. for texture synthesis [4]. Criminisi et al. [2] proposed similar method where the structure completion is emphasized over the texture synthesis. For this, they proposed an ordering of the target patches from the boundary of the target region depending on the structure strength. A synthesized patch is estimated to fill the missing pixels of the target patch. Recently, sparse representation is applied in image inpainting like the other fields of image processing and computer vision because of it's robustness [5–7]. Komodakis et al. [8] proposed coherence in the image inpainting formulation by favoring the similarity with the overlapping region of patches. Liu *et al.* [9] introduced multiscale graph cuts algorithm in image inpainting as a energy minimization problem. Meur *et al.* [10,11] proposed image inpainting based on hierarchical single frame super-resolution combining different inpaint versions of the target image for different settings of the input parameters. But still these methods have several limitations related to structure and texture completion. In this paper we adopt exemplar-based inpainting algorithm proposed in [2] as our base-line method.

1.1   Motivation

To overcome the above mentioned problems, we suggest to inpaint the target region several times sequentially in a multiresolution framework. In a particular resolution, first the candidate patches similar to the target patch are filtered using transform domain approach and combine them using loopy belief propagation to infer the target patch. When the target region is totally inpainted, it is re-inpainted because a smoothing effect may appear in the texture region due to the transformation. We have two main motivation in the step of transformation based inpainting. First, it removes the artifacts from the candidate patches if appear in some of the patches. The artifacts in the set of candidate patches corresponds to the pixels which are different from most of the patches at a particular position (see supplementary). Second it is easy to inpaint with the smooth candidate patches so that information propagates smoothly in the target region. For this, we build a 3D array of candidate patches to apply the higher order singular value decomposition (HOSVD) with hard thresholding. The advantage of this transformation is that it performs not only across the height and width of the patches but also along the third dimension, that is, along different candidate patches. It reduces the variation among the candidate patches which ultimately restricts artifacts to appear in the target region and helps to propagate structure correctly inwards the target region. Several authors suggested to combine the candidate patches in different ways, the most popular and the robust one is sparse representation. But the representation depends only the known pixels of the target patch and the corresponding pixels of the candidate patches. This may create artifacts in the unknown region. The hard thresholding in HOSVD transformation remove these artifacts if present in any of the candidate patches. Also, we suggest to apply loopy belief propagation to combine the patches and obtain an approximate solution for a global optimization problem. Due to the transformation, it may produce a smoothing effect in the texture region. To avoid this, traditional texture synthesis approach is applied on the smoothly inpainted image which is obtained from the first step. The above two steps are executed in a multiresolution framework since it is easy to inpaint in the coarser resolution of the target image. The idea of the HOSVD transformation has been applied earlier for dynamic texture synthesis [12] and denoising [13].

2   Overview of the Proposed Method

Image inpainting is a challenging task for large blob type target region with complex background and random textures. In literature, different types of exemplar-based techniques have been proposed to solve the inpainting problem. The common steps of these algorithms are patch priority computation and inferring the selected target patch combining the most similar candidate patches [7,14,15]. In this paper, we propose a new approach to combine these candidate patches using HOSVD in a multiresolution framework. We briefly describe the main steps of the proposed algorithm in the next paragraph.

The proposed method have two main steps. The first step is to inpaint the target image using HOSVD-based patch filtering. For this, a target patch is selected from the boundary of the missing region and find the candidate patches similar to the target patch. For HOSVD transformation, three 3D array (for color image) of candidate patches is built-up. Then coefficient matrix of singular values is computed applying HOSVD on the 3D array for each color channel. The singular values below some threshold in the coefficient matrix represent the variation among the patches and artifacts. These coefficients are modified by hard thresholding and the patches are reconstructed by inverse transformation [13]. Then we combine these filtered patches using loopy belief propagation to determine the unknown pixels of the target patch. This technique gives more robustness in inpainting by smoothly propagating structure and color into the unknown region. The candidate patches are selected from local as well as global source region and a histogram based similarity measure is taken to approximate the unknown part of the target patch. For local patch selection, we take relatively large neighborhood surrounding the target patch so that local as well as global consistency in the inpainted region is maintained. Beside the advantage of this transform domain technique, the hard

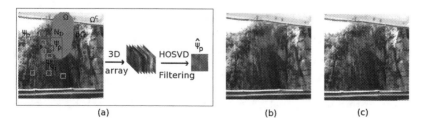

(a)   (b)   (c)

Fig. 1. Steps of HOSVD-based inpainting. (a) This step illustrates the estimation of the target patch Ψ_p using HOSVD. (b) Inpainted image after all target patch completion. (c) Image after texture recovery.

thresholding of the HOSVD may produce smoothness effect in fine texture region. To avoid it, in the second step, we recover the texture from the neighborhood of the target region in the already inpainted image. Figure 1 shows different steps of the proposed inpainting algorithm.

3 HOSVD-Based Inpainting

The proposed exemplar-based method have two core steps: (i) To select target patch on the boundary of the missing region based on some priority computation and to fill-in the target patch using HOSVD of candidate patches iteratively until the whole missing region is inpainted, and (ii) to recover appropriate texture in the smooth region reconstructed in the previous step. It is observed that the structure or edge preservation is more important than the texture synthesis since former carries major information and gives meaning to the regions in the image. Several authors [7,10,14] introduced different priority terms to choose the target patch from the structure region. Though many priority terms are available in the literature, here we employ the simplest priority measure proposed by *Criminisi et al.* [2]. For a candidate target patch Ψ_p at p, the priority term $Pr(p)$ is defined as $Pr(p) = K(p)V(p)$ where $K(p)$ is the *knowledge* term which measures the fraction of patch surrounding the pixel p is known already, and $V(p)$ is *local variation* term, which in a sense gives an idea of local structure. The proposed method with this simple priority term can produce structure better compared to the more complicated method.

3.1 HOSVD-Based Patch Completion

The main goal of the proposed method is to infer the unknown pixels of the target patch Ψ_p selected from the previous step. The very first task of this step is to select some patches similar to Ψ_p from the source region Ω^c. Then we apply patch filtering based on the HOSVD transformation to preserve the color and structure consistency in the target region. The robustness of this technique shows it's superiority to eliminate unwanted artifacts if present in some of these candidate patches. Here we also incorporate the local as well as global patch consistency.

Patch Similarity and Patch Selection: In this work our objective is to select the candidate patches in such a way that fill-in the target patch Ψ_p following the local as well as global consistency. That means after target patch completion, unwanted artifacts do not appear in the filled-in region. We consider a neighborhood window N_p centered at p and find m similar patches from N_p. But fixed m may give some patches with larger dissimilarity. Hence we find the similar patches Ψ_{q_j} of Ψ_p from N_p as

$$\mathcal{X}_p = \left\{ \Psi_{q_j} \in N_p : d_{SSD}(\Psi_p^k, \Psi_{q_j}^k) < \epsilon \text{ and } d_H(\Psi_p^k, \Psi_{q_j}) < \delta \right\} \qquad (1)$$

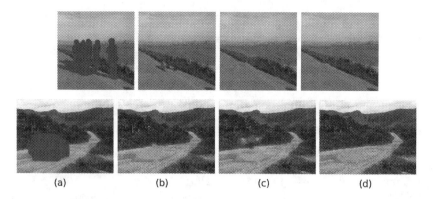

| (a) | (b) | (c) | (d) |

Fig. 2. (a) Target image with red marked target region. (b) Inpainted image using only local patches. (c) Inpainted image using only global patches. (d) Inpainted image using both local and global patches (Color figure online).

where Ψ_p^k and $\Psi_{q_j}^k$ are the known part of Ψ_p and corresponding part of Ψ_{q_j}, d_{SSD} is the sum of square difference among the patches and ϵ, δ are the threshold parameters. The histogram based dissimilarity measure d_H is defined by euclidean distance as

$$d_H(\Psi_p^k, \Psi_{q_j}) = \sqrt{\sum_{i=1}^{b} |h_p(i) - h_{q_j}(i)|^2} \qquad (2)$$

where h_p and h_{q_j} denotes the normalized histogram of the known part of the patch Ψ_p and full part of the neighbor patch Ψ_{q_j}, and b is the number of histogram bins. This measure approximates the unknown pixels of the target patch by of it's known pixels so that both the parts look similar. The histogram is computed on the intensity channel of color space.

Also some candidate patches $\mathcal{Y}_p = \{\Psi_{p_i}\}$ are obtained in the similar way as described in (1), but here the search range is the source region Ω^c instead of N_p. These candidate patches are also added to $\mathcal{X}_p$ to get a list $\mathcal{P}_p$ of local and global patches for HOSVD transformation. This idea behind the selection of candidate patches is to approximate the estimated target patch consistent with the local as well as global texture and structure. Top row of Fig. 2, shows inpainting using only local patches produce wrong texture in the target region, but when we take both the local and global patches, it works well. Similar thing happens in the bottom row also, but here local patches almost correctly recover the target region whereas global patches fail to generate proper texture. Local and global patches together, however, consistently recover the texture in the target region. Here we define $\epsilon = \lambda * n_{ch} * |\Psi_p^k|$ where λ is the factor determining the error tolerance of SSD-based dissimilarity measure, n_{ch} is the number of channel of the input image and $|\Psi_p^k|$ is the count of known pixels in Ψ_p. λ is set to 3δ where δ is the threshold of histogram-based dissimilarity measure. Since δ is a fixed parameter we may obtain an empty set of both $\mathcal{X}_p$ and $\mathcal{Y}_p$. In such a case, we decreases the priority of the current target patch Ψ_p to say half of its original priority and select a new patch based on priority. However if no such candidate patch is found we increase the value error tolerance δ by 0.1. This assumption also helps to restrict selection of target patch with large dissimilarity to both the local and the global patches.

HOSVD for Inpainting: Due to the unknown pixels of the target patch, the candidate patches selected from the local and global source region may not be similar in the unknown region. They may differ slightly from each other or some artifacts may appear in the unknown part. But here the main goal is to process the candidate patches in such a way that the unknown part of the patches looks similar to the known part. HOSVD provides the solution because it filters the candidate patches as well as takes into account the similarity among the pixels of at all locations. In the next section, we discuss about the standard SVD and it's higher order generalization HOSVD. Lastly, HOSVD is introduced in image inpainting with mentioning it's different steps.

Background: Given a matrix A of size $m \times n$, the singular value decomposition (SVD) is of the from $A = USV^T$ where U is a $m \times m$ orthonormal matrix, S is a $m \times n$ diagonal matrix of positive singular values in the descending order and V is a $n \times n$ orthonormal matrix. The columns of U and V are the eigen vectors of AA^T and A^TA respectively. The square of the singular values in S are the eigen values of AA^T (or A^TA). The HOSVD is an extension of the matrix SVD for higher order matrices [16]. Usually, matrices of order higher than 2, is called tensor. Suppose $\mathcal{A} \in R^{I_1 \times I_2 \times \dots \times I_r}$ is a tensor of order r where $I_1, I_2, \dots, I_r$ denotes the number of elements for each dimension. The r-order tensor $\mathcal{A}$ may be decomposed as

$$\mathcal{A} = \mathcal{S} \times_1 U^{(1)} \times_2 U^{(2)} \dots \times_r U^{(r)} \tag{3}$$

where $U^{(1)}, U^{(2)}, \dots, U^{(r)}$ are orthogonal matrices containing the orthonormal vectors spanning the column space of the matrix unfolding $A_{(p)}$ with $p = 1, 2, \dots, r$ and $\mathcal{S}$ is the core tensor analogous to the diagonal matrix S in the standard SVD. Note that, generally $\mathcal{S}$ is a full tensor that means it is not a diagonal matrix like S. The s-th mode tensor product $\times_s$ for a tensor $\mathcal{X} \in R^{I_1 \times I_2 \times \dots \times I_s \dots \times I_r}$ and a matrix $\mathcal{Y} \in J_s \times I_s$ may be denoted by $\mathcal{X} \times_s \mathcal{Y}$ and is a tensor $\mathcal{Z} \in R^{I_1 \times I_2 \times \dots \times J_s \dots \times I_r}$. Therefore

$$z_{i_1 i_2 \dots i_{s-1} j_s i_{s+1} \dots i_r} = \sum_{i_s} x_{i_1 i_2 \dots i_s \dots i_r} y_{j_s i_s}. \tag{4}$$

The core tensor $\mathcal{S}$ is obtained by

$$\mathcal{S} = \mathcal{A} \times_1 U^{(1)^H} \times_2 U^{(2)^H} \dots \times_r U^{(r)^H} \tag{5}$$

where H denotes the Hermitian matrix transpose operator.

There is an equivalent matrix formulation of the tensor decomposition. For this, we first define the p-mode matrix unfolding (also called matricization)

$A_{(p)} \in R^{I_p \times (I_{p+1} \times \ldots \times I_r \times I_1 \times \ldots \times I_{p-1})}$ consists of the tensor element $a_{i_1, i_2, \ldots, i_r}$ at (i_p, j) where

$$j = 1 + \sum_{l=1, l \neq p}^{r} (i_l - 1) \prod_{m=1, m \neq p}^{r} I_m \tag{6}$$

The Eq. (3) can be expressed in matrix format as

$$A_{(p)} = U^{(p)} S_{(p)} (U^{(p+1)} \otimes U^{(p+2)} \ldots U^{(r)} \otimes U^{(1)} \otimes U^{(2)} \ldots U^{(p-1)})^H \tag{7}$$

where $U^{(p)}$ is obtained from SVD of $A_{(p)}$ by

$$A_{(p)} = U^{(p)} \Sigma^{(p)} V^{(p)^H} \tag{8}$$

and the symbol $\otimes$ denotes the Kronecker product. The diagonal matrix $\Sigma^{(p)}$ is defined as

$$\Sigma^{(p)} = diag(\sigma_1^{(p)}, \sigma_2^{(p)}, \ldots, \sigma_{I_p}^{(p)}) \tag{9}$$

where $\sigma_1^{(p)}, \sigma_2^{(p)}, \ldots, \sigma_{I_p}^{(p)}$ are the Fobenius-norms of $S_{(p)}$.

The matrix formulation of Eq. (5) is

$$S_{(p)} = U^{(p)} A_{(p)} (U^{(p+1)} \otimes U^{(p+2)} \ldots U^{(r)} \otimes U^{(1)} \otimes U^{(2)} \ldots U^{(p-1)}) \tag{10}$$

Patch fill-in using HOSVD: Now we will discuss how HOSVD is applied in the proposed inpainting method. Given a target patch Ψ_p of size $m \times n$, we find K number of candidate patches $\mathcal{P}_p$ (see the previous section). So the size of the tensor is defined by $I1 = m$, $I2 = n$ and $I3 = K$. We first build a 3D array $\mathcal{A}$ using the candidate patches. Since we deal with color images, actually one 3D array is taken for each individual channel and same scheme is followed for each of the arrays. The HOSVD-based patch synthesis method consists of following steps: (1) Unfolding of $\mathcal{A}$ to $A_{(1)}, A_{(2)}, A_{(3)}$ and decomposition of $A_{(p)}$ using standard SVD to obtain $U^{(p)}, S_{(p)}$ for $p = 1, 2, 3$ using Eqs. (8) and (10), (ii) manipulation of singular values in $S_{(p)}$ and reconstruct the array $\mathcal{A}$ by inverse transformation using (7), and (iii) averaging the filtered candidate patches for obtaining an estimated target patch. Usually the coefficients are manipulated (typically by hard thresholding) to obtain the filtered patch in the HOSVD-based transform domain. Here the main purpose of patch filtering is to remove unwanted artifacts from a collection of almost similar patches and obtain a smooth version of the patches preserving the edges. The basic idea behind this approach is that it is easy to inpaint the target region surrounded by smooth patches. The random textures and structure surrounding the target region may mislead in estimating the target patch since it's some part is unknown. The singular values in the coefficient matrix represent the variation among the candidate patches. To suppress these variation, we nullify the coefficients which are below the hard threshold $\sigma \sqrt{2 \log(mnK)}$. The 3D array $\hat{\mathcal{A}}$ is then reconstructed by inverting the transformed candidate patches. Since the target patch have unknown pixels and approximating them by a set of candidate patches, some unwanted artifacts may

$$\sigma = 0.01 \qquad \sigma = 0.1 \qquad \sigma = 1.0 \qquad \sigma = 10.0 \qquad \sigma = 30.0$$

Fig. 3. Left column shows the target image and other columns show the inpainted images for different values of σ.

appear in the unknown region of the target patch. The artifacts in a set of candidate patches corresponds to the pixels of patches which are very much different from most of the patches in the set. The singular values below some threshold in the coefficient matrix corresponds to the variation among the patches and also the artifacts. The HOSVD-based filtering try to remove those artifacts from the candidate patches. In Fig. 3, we show the results of inpainting for different values of σ and it is clear that higher value of σ can efficiently remove the unwanted artifacts better from the target region. In the second figure of supplementary material, we have shown the effect of patch filtering for different set of candidate patches and in some cases artifacts are removed in the filtered patches.

Finally, we estimate the target patch by combining the filtered candidate patches. In literature, K candidate patches are combined by different approaches like sparse representation [7], comprehensive framework [14]. Some authors also consider weighted averaging [10,15] because of it's computational simplicity, defined as

$$\hat{\Psi}_p = \sum_{\Psi_q \in \mathcal{P}_p} \lambda_{p,q} \Psi_q \tag{11}$$

where Ψ_q is the candidate patch after HOSVD-based filtering and

$$\lambda_{p,q} = \frac{1}{N} \exp\left(-\frac{d_{SSD}(\Psi_p^k, \Psi_q^k)}{2\eta^2}\right). \tag{12}$$

Here d_{SSD} denotes the sum of square differences, N is the normalization constant such that $\sum_{\Psi_q \in \mathcal{P}_p} \lambda_{p,q} = 1$ and η is a scaling parameter set to 10.0. This procedure of combining several candidate patches can estimate the unknown pixels in the target image. But it does not ensure to give the global optimization solution for inpainting. Also it may introduce blur ring effect on the fine texture regions (see Fig. 4(b)). To overcome these problems we incorporate loopy belief propagation which is able to produce an approximation solution of a global optimization problem.

3.2   Loppy Belief Propagation

The problem in belief propagation is to assign a label to each unknown patch Ψ_p in the target region Ω. For the large number of labels, the algorithm suffers from the high time complexity. Komodakis et al. [8] introduced priority belief propagation (PBP) where each patch in the source region is assigned by a label. In [11] the authors used loopy belief propagation (LBP) to combine multiple inpainted images. The multiple images are obtained by inpainting the target image with different patch size and rotation. But in our case the approach is somewhere difference. We assign a lable ($z \in \mathcal{Z}$) to the target patch Ψ_p from the set of already filtered candidate patches $\mathcal{P}_p$. That means each candidate patch have a label and number of labels may vary for different target patches. Markov Random Field (MRF) formalization of the objective function can be represented by a graph $G = (\nu, \varepsilon)$. The MRF nodes ν are the lattice consisting of the target patches in the unknown region Ω and the edges ε of the MRF are the 4-neighborhood system $\mathcal{N}_4$ on the lattice. Now the problem of label assigning is to assign a label $z \in \mathcal{Z}$ to each node/patch $\Psi_p \in \Omega$ so that the total energy $\mathcal{E}$ of the MRF is minimized, where

$$\mathcal{E}(z) = \sum_{p \in \nu} V_{\mathbf{s}}(z_p) + \sum_{(p,q) \in \mathcal{N}_4} V_{\mathbf{p}}(z_p, z_q) \tag{13}$$

The single node potential (also called the *label cost*) $V_1(z_p)$ represents the cost of placing $\Psi^*_{z_p} \in \mathcal{P}_p$ over the target patch Ψ_p. The formula of the above cost may be written as

$$V_{\mathbf{s}}(z_p) = \sum_{x \in \Psi_p \cap \Omega^c} \{\Psi^*_{z_p}(x) - \Psi_p(x)\}^2 \tag{14}$$

The pairwise potential cost $V_{\mathbf{p}}(z_p, z_q)$ represents the cost of placing the patches $\Psi^*_{z_p} \in \mathcal{P}_p$ and $\Psi^*_{z_q} \in \mathcal{P}_q$ over the neighbors p, q is given by

$$V_{\mathbf{p}}(z_p, z_q) = \sum_{x \in \Psi^*_{z_p}} \{\Psi^*_{z_p}(x) - \Psi^*_{z_q}(x)\}^2 \tag{15}$$

The minimization of the above objective function $\mathcal{E}$ can be estimated using loopy belief propagation [17].

3.3   Texture Recovery

In the previous step, we obtain an inpainted image with a smooth target region preserving all the structure and color details. Textures are smoothed out due to HOSVD-based patch filtering. In this section our aim is to recover the texture sharpness using neighborhood texture information of the target region in the inpainted image obtained from the previous step. This step is similar to as the basic of inpainting by priority computation and patch completion, but here the target region is fully known by the previous step. Since a real scene

Fig. 4. (a) Target image with red marked target region, (b) image inpainted by HOSVD, (c) image inpainted by texture recovery (Color figure online).

image may contain texture, structure and smooth regions, we want to recover only those regions which are smoothed, but surrounded by texture regions. The constraint is defined in terms of edge map of the inpainted image using HOSVD. For this, we take the window N_p at the pixel p in the edge image. If $N_p \cap \Omega$ does not contain any edge pixels and $N_p \cap \Omega^c$ contains sufficient edge pixels, we consider Ψ_p as a patch in the smooth target region and must have to recover the texture of this patch. Accordingly, the final estimated target patch $\hat{\Psi}_p$ is recovered by

$$\hat{\Psi}_p = \arg \max_{\Psi_q \in N_p \cap \Omega^c} \left\{ d_{SSD}(\Psi_p, \Psi_q) < \epsilon \right\} \tag{16}$$

We take $\epsilon = \lambda * n_{ch} * |\Psi_p|$ as similar to the patch selection step. Note that, here Ψ_p and Ψ_q both are fully known. Figure 4 illustrates the efficiency and necessity of this step. The inpainted image (b) using HOSVD is smooth in the snow region. The result of the texture recovery in Fig. 4(c) shows, our proposed method is robust to recover the texture.

3.4   Multiresolution Approach

Several authors incorporate multiresolution scheme in the proposed inpainting algorithm. There are a few reasons behind this consideration. It permits to capture various details like structure in different scales. It enforces to reduce time complexity and it is also easy to inpaint on the coarse version of the image [10,14,15]. The multiresolution scheme follows an recursive process in multiple scales using spatial pyramid. First, inpainting algorithm runs at the coarsest level of the pyramid and the result of this level is considered as an initialization for the finer level for further modification. Here we use 3–5 pyramid level with resolution factor 1.5.

3.5   HOSVD vs. KSVD and BM3D in Inpainting

The main advantage in choosing HOSVD is it's simplicity compare to KSVD [18] and BM3D [19]. The KSVD algorithm learns an overcomplete dictionary

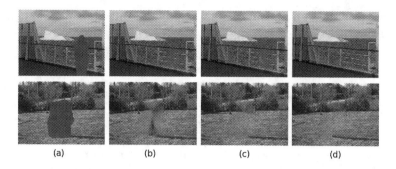

Fig. 5. (a) Target image with red marked target region. Inpainted image using (b) KSVD [18]. (c) BM3D [19]. (d) HOSVD (Color figure online).

and represent data samples by pursuit algorithm. It need some parameters which are not easy to tune, such as the number of dictionary elements, the stopping criterion for the pursuit algorithm and the trade off between data fidelity and sparsity terms. Our main aim is to jointly filter the candidate patches which is not possible by KSVD. The idea of jointly filtering multiple patches is introduced earlier in BM3D for denoising. But the algorithm is complex in some sense. The filtering of similar patches go through the 2D followed by 1D transformation in BM3D. But in HOSVD 3D filtering is not combination of such 2D and 1D filtering. BM3D have many parameters such as the choice of 2D and 1D filtering, the maximum number of similar patches, the choice of patch size depending on the noise variance, the choice of pre-filter for patch similarity in the first stage, and also the set of parameters used in Wiener filtering in the second stage. HOSVD learns spatially adaptive bases whereas BM3D uses fixed bases. HOSVD has only two parameters, the number of candidate patches and the value of sigma in the hard threshold which are common to BM3D. In [13] it is shown that HOSVD outperforms KSVD and with Wiener filtering produce

Fig. 6. (a) Target image. (b) Criminisi's [2]. (c) Komodakis's [8]. (d) Pritch's [20]. (e) He's [21]. (f) Proposed method.

better result compare to BM3D for some examples. In Fig. 5, it is shown that our HOSVD-based inpainting approach outperforms over KSVD and better than BM3D.

4   Experimental Results and Discussions

In this section, we first set the parameters used in our experiments and then test the proposed method on different types of natural images. We also compare our algorithm with the existing state-of-the-art methods based on exemplar by Criminisi [2], priority belief propagation by Komodakis [8], shift-map by Pritch [20], patch-offsets by He [21], graph cuts by Liu [9] and super-resolution by Meur [11]. We set the size of the patch to 9×9, the value of $\delta = 1.0$ and the value of $\sigma = 30.0$ in HOSVD. For local consistency, we select some candidate patches

    (a)   (b)   (c)   (d)   (e)   (f)   (g)

Fig. 7. (a) Target image. (b) Criminisi's [2]. (c) Komodakis's [8]. (d) Pritch's [20]. (e) He's [21]. (f) Liu's [9]. (g) Proposed method.

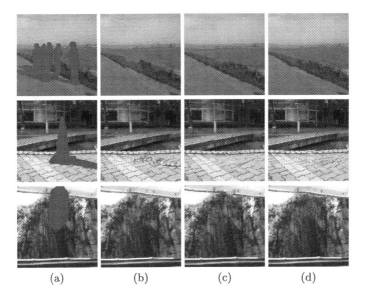

    (a)   (b)   (c)   (d)

Fig. 8. (a) Image with mask. (b) Criminisi's exemplar-based method [2]. (c) Liu's graph cuts approach [9]. (d) Proposed method.

from a restricted search window N_p (neighborhood) centering the target patch. We set the window size to $\kappa\tau$ where κ is the level of spatial pyramid ($\kappa = 1$ for the coarsest level) and τ is a fixed parameter set to 30. The MATLAB implementation of the proposed method on Intel 3.07 GHz CPU takes 65 s for the last example of Fig. 8.

Figure 6 shows that the proposed method visually outperforms Criminisi's exemplar-based [2] and Komadakis's belief propagation based [8] method, and provides comparable result to the methods proposed by Pritch *et al.* [20] based on shift-map and He *et al.* [21] based on statistics of patch offsets.

Figure 7 shows an popular example of bungee jump. The methods compared in the previous example produce wrong texture in the structure area and both

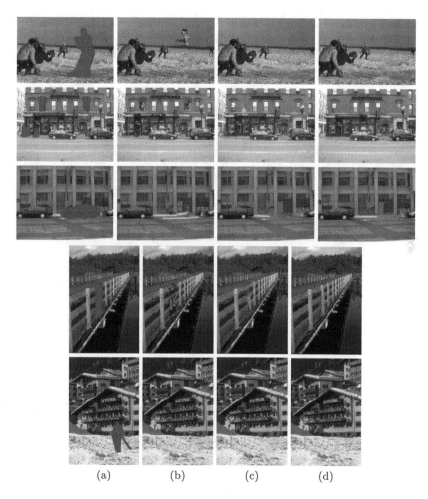

(a)   (b)   (c)   (d)

Fig. 9. (a) Image with mask. (b) Criminisi's exemplar-based method [2]. (c) Meur's super-resolution based algorithm [11]. (d) Proposed method.

(a)   (b)   (c)   (d)

Fig. 10. Some failure cases. (a, c) Target images. (b, d) Images inpaited of (a) and (c) respectively by our proposed method.

graph cuts proposed by Liu *et al.* [9] and our proposed method produce visually plausible interpolation.

In Fig. 8, we give more comparisons with Criminisi's approach and recently proposed method [9] for several natural images. From the results, it is clear that the proposed method produce better texture (second example) and structure (third example) completion compare to the other techniques.

Figure 9 illustrates the performance of the proposed method and the comparison with super-resolution based inpainting proposed by Meur *et al.* [11]. The results show that the proposed method perfectly recover the target region whereas the other methods fails in many cases to generate proper texture and structure. Note that, with simplest priority term, the proposed method produce better structure than [11] in most of the examples (zoom to see the difference).

In Fig. 10 we show some examples where our method fails to recover both texture and structure.

5   Conclusions

In this paper, we have proposed a novel inpainting algorithm using higher order singular value decomposition of candidate patches. The novelty in choosing HOSVD is that it measures the similarity among the candidate patches which robustly recover the target region with structure and color. Texture recovery step reconstruct the texture which is smoothed due to the filtering step. Experiments and comparisons show that our proposed exemplar-based algorithm can produce better results in most of the cases. In future, we plan to employ this algorithm, combined with de-blurring techniques and hue correction, to digital restoration of old heritage murals and paintings.

References

1. Bertalmio, M., Sapiro, G.: Image inpainting. In: Proceedings of the ACM SIG-GRAPH Conference on Computer Graphics, New York, USA, pp. 417–424 (2000)
2. Criminisi, A., Perez, P., Toyama, K.: Region filling and object removal by exemplar-based inpainting. IEEE Trans. Image Process. **13**, 1200–1212 (2004)
3. Chan, T., Shen, J.: Non-texture inpainting by curvature-driven diffusions. J. Vis. Commun. Image Represent. **12**, 436–449 (2001)

4. Efros, A., Leung, T.: Texture synthesis by non-parametric sampling. In: Proceedings of the IEEE International Conference on Computer Vision, vol. 2, pp. 1033–1038 (1999)
5. Fadili, M.J., Starck, J.L., Murtagh, F.: Inpainting and zooming using sparse representations. Comput. J. **52**, 64–79 (2009)
6. Shen, B., Hu, W., Zhang, Y., Zhang, Y.: Image inpainting via sparse representation. In: Proceedings of the IEEE International Conference on Acoustics, Speech and Signal Processing, pp. 697–700 (2009)
7. Xu, Z., Sun, J.: Image inpainting by patch propagation using patch sparsity. IEEE Trans. Image Process. **19**, 1153–1165 (2010)
8. Komodakis, N., Tziritas, G.: Image completion using efficient belief propagation via priority scheduling and dynamic pruning. IEEE Trans. Image Process. **16**, 2649–2661 (2007)
9. Liu, Y., Caselles, V.: Exemplar-based image inpainting using multiscale graph cuts. IEEE Trans. Image Process. **22**, 1699–1711 (2013)
10. Le Meur, O., Guillemot, C.: Super-resolution-based inpainting. In: Fitzgibbon, A., Lazebnik, S., Perona, P., Sato, Y., Schmid, C. (eds.) ECCV 2012, Part VI. LNCS, vol. 7577, pp. 554–567. Springer, Heidelberg (2012)
11. Meur, O.L., Ebdelli, M., Guillemot, C.: Heigherchical super-resolution-based inpainting. IEEE Trans. Image Process. **22**, 3779–3790 (2013)
12. Constantini, R., Sbaiz, L., Susstrunk, S.: Higher order SVD analysis for dynamic texture synthesis. IEEE Trans. Image Process. **17**, 42–52 (2008)
13. Rajwade, A., Rangarajan, A., Banerjee, A.: Image denoising using the higher order singular value decomposition. IEEE Trans. Pattern Anal. Mach. Intell. **35**, 849–862 (2013)
14. Bugeau, A., Bertalmio, M., Caselles, V.: A comprehensive framework for image inpainting. IEEE Trans. Image Process. **19**, 2634–2645 (2010)
15. Wexler, Y., Shechtman, E., Irani, M.: Space-time completion of video. IEEE Trans. Pattern Anal. Mach. Intell. **29**, 463–476 (2007)
16. Lathauwer, L.D.: Signal processing based on multilinear algebre. Ph.D. dissertation, Katholieke Universiteit Leuven, April 2013
17. Yedidia, J., Freeman, W., Weiss, Y.: Constructing free energy approximations and generalized belief propagation algorithms. IEEE Trans. Inf. Theor. **51**, 2282–2312 (2005)
18. Aharon, M., Elad, M., Bruckstein, A.: The K-SVD: an algorithm for designing of overcomplete dictionaries for sparse representation. IEEE. Trans. Signal Process. **54**, 4311–4322 (2006)
19. Dabov, K., Foi, A., Katkovnik, V., Egiazarian, K.: Image denoising by sparse 3-D transform-domain collaborative filtering. IEEE Trans. Image Process **16**, 2080–2095 (2007)
20. Pritch, Y., Kav-Venaki, E., Peleg, S.: Shift-map image editing. In: Proceedings of the IEEE International Conference on Computer Vision, pp. 151–158, September 2009
21. He, K., Sun, J.: Statistics of patch offsets for image completion. In: Fitzgibbon, A., Lazebnik, S., Perona, P., Sato, Y., Schmid, C. (eds.) ECCV 2012, Part II. LNCS, vol. 7573, pp. 16–29. Springer, Heidelberg (2012)

Image Interpolation Based on Weighted and Blended Rational Function

Yifang Liu[1,2], Yunfeng Zhang[1,2(✉)], Qiang Guo[1,2], and Caiming Zhang[1,2]

[1] School of Computer Science and Technology,
Shandong University of Finance and Economics, Jinan 250014, China
yfzhang@sdufe.edu.cn
[2] Shandong Provincial Key Laboratory of Digital Media Technology,
Jinan 250014, China

Abstract. Conventional linear interpolation methods produce interpolated images with blurred edges, while edge directed interpolation methods make enlarged images with good quality edges but with details distortion for some cases. An adaptive rational-based algorithm for the interpolation of digital images with arbitrary scaling factors is proposed. In order to remove artifacts, we construct a new interpolation model with weight and blend, which are used for preserving the clear edge and detail. The proposed model is blended by basic rational interpolation model and three rotated rational models. The weight coefficients are determined by the edge information from different scale based on point sampling. Experimental results show that the proposed method produces images with high objective quality assessment value and good visual quality.

1 Introduction

Image interpolation has a wide range of applications which aims to reconstruct a high resolution (HR) image from the low-resolution (LR) image. The most common interpolation methods are bilinear, bicubic, cubic spline algorithm, etc. [3,6]. These conventional methods are the approximation of sinc function which corresponds to ideal filtering [14]. These methods have a relatively low complexity, but suffer from several types of visual degradation around "edges".

To solve these problems, many adaptive interpolation algorithms have been developed [2,4,7–10,13,15–18]. These algorithms can be broadly divided into two categories: discrete method and continuous method. In discrete method, new edge-directed interpolation (NEDI) [8] estimates high resolution covariances form low resolution image based on the geometric duality; In [17], for a pixel to be interpolated two observation sets are defined in two orthogonal directions, and then fuse the directional interpolation results by minimum mean square-error estimate; Zhang and Wu [18] develop a soft-decision interpolation method which is able to estimate missing pixels by groups instead of by pixels. These discrete algorithms which consider more adaptive image information can improve the visual effect. However, these methods deliver not a continuous function but

© Springer International Publishing Switzerland 2015
C.V. Jawahar and S. Shan (Eds.): ACCV 2014 Workshops, Part II, LNCS 9009, pp. 78–88, 2015.
DOI: 10.1007/978-3-319-16631-5_6

a set of subpixel values which are not suitable for resampling after, for example, rotation [10], and they have a much higher computational complexity than conventional methods. Besides, these methods sometimes generate speckle noise or distortion of edges [18].

Once a digital image is converted into an interpolating continuous function, we can resample it to obtain resized and rotated images at a better precision [10]. In fact, the continuous methods create a HR image though constructing a interpolating patch. Hu and Tan [4] presents a method for preserving the contours or edges based on adaptive osculatory rational interpolation kernel function. Zhang et al. [16] constructs a fitting surface by using image data as constraints to reverse sampling process for improving fitting precision. However, these continuous methods all suffer form blurred edge in some ways. Recently, a bivariate rational interpolation with parameters based on the function values is studied in [12,19]. The rational function has a simple and explicit expression, and compared with other methods, it can keep the natural attributes of image better. Because the rational model is suitable for resizing natural image, it performs the details well and a relative clear edge. Unfortunately, the single bivariate rational model(basic model (Fig. 2a)) is asymmetric and does not meet the structures of natural image. So it can produce zigzagging artifacts around the edge regions.

In this paper, we construct an adaptive interpolation function with weight and blend based on rational function model. To reduce the zigzagging edge generated by the basic model, we rotate the basic rational model 3 times to get 4 interpolating functions. The proposed model is weighted and blended by them. The weight coefficients which contain edge information are adaptively calculated by distance, gradient and difference quotient based on point sampling, and can be used to keep the edge attributions. Experiments show that the proposed approach performs better than conventional and discrete methods in preserving edge.

This paper addresses the problem of constructing an adaptive weighted rational function such that the resized image has better precision and visual quality. We use blending model not only to maintain the image natural attribution but also to preserve the structure of image, and the adaptive weight coefficients can preserve edge information from different aspects. Furthermore, point sampling can ensure that more information in a cell can be used to determined the weight coefficients.

The paper is arranged as follows. In Sect. 2, the proposed interpolated model based on basic rational model is introduced. Section 3 shows the performance of the method.

2   Description of Proposed Method

In this section, the interpolation function with unknown weight coefficients is proposed, and the key problem is to determine the weight coefficients. Then the weight coefficients are determined by different scale edge information of local image.

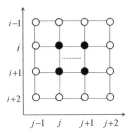

Fig. 1. The interpolation model

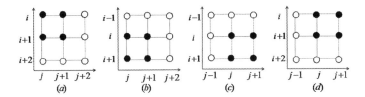

Fig. 2. The rotation model: (a) Basic model; (b), (c) and (d) are 90, 180 and 270° counterclockwise rotation of (a) respectively.

2.1 The Interpolation Model

Thought rational function has good features to maintain details, it suffers from some visual problems around edge region. The proposed model can preserve edge region well as well as detail region. Let $[i, i+1; j, j+1]$ be the interpolated region. The proposed interpolation model based on rational spline function is showed in Fig. 1. The rectangle region surrounded by 4 black points is the interpolated region. All 16 points within the interpolation model are involved in the interpolation. And Fig. 2 shows the decomposition of the proposed model. The proposed model is weighted and blended by the 4 submodels. In Fig. 2, (a) means the basic rational spline interpolation model; (b) represents the 90° counterclockwise rotation of (a) in Fig. 1 model, and the rotation center is the interpolation region; in the same way, (c) and (d) are 180° and 270° counterclockwise of (a). Figure 3 is another expression of Fig. 2, and (a), (b), (c) and (d) correspond respectively. Let the basic rational spline interpolation model (a) denotes $P_1(x, y)$, then (b), (c) and (d) are denoted $P_2(1 - y, x)$, $P_3(1 - x, 1 - y)$ and $P_4(y, 1 - x)$ respectively. All these four submodels can produce the same patch $[i, i + 1; j, j + 1]$. The proposed weighted and blended rational function $P_{i,j}(x, y)$ is expressed in Eq. 1.

$$P_{i,j}(x, y) = aP_1(x, y) + bP_2(1 - y, x) \\ + cP_3(1 - x, 1 - y) + dP_4(y, 1 - x), \tag{1}$$

where a, b, c and d are unknown weight coefficients.

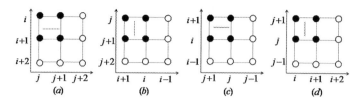

Fig. 3. The rotation model: (a) $P_1(x,y)$; (b) $P_2(1-y,x)$; (c) $P_3(1-x,1-y)$; (d) $P_4(y,1-x)$.

Now the basic rational spline function (Fig. 2(a)) is given. A bivariate rational interpolation with parameters based on the function values is constructed in [12, 19]. Let $P_{i,j}(x,y)$ be a bivariate function defined in the region $[i, i+1; j, j+1]$. Denote the pixel value by $I_{i,j}$. For any point $(x,y) \in [i, i+1; j, j+1]$, the bivariate rational interpolating function $P_{i,j}(x,y)$ can be expressed as

$$P_{i,j}(x,y) = \sum_{r=0}^{2}\sum_{s=0}^{2} \omega_r(x,\alpha_i)\omega_s(y,\beta_j)I_{i+r,j+s}, \tag{2}$$

where

$$\omega_0(t,\delta) = \frac{(1-t)^2(\delta+t)}{(1-t)\delta+t},$$

$$\omega_1(t,\delta) = \frac{t(1-t)\delta + 3t^2 - 2t^3}{(1-t)\delta+t},$$

$$\omega_2(t,\delta) = \frac{-t^2(1-t)}{(1-t)\delta+t}.$$

Considering the basic model, 9 points $I_{i+r,j+s},(r,s=0,1,2)$ are used to construct the patch $P_1(x,y)$ which crosses the 4 black points, and these 9 points have different basis functions. However, it would suffer from blurred edges. There are two main reasons. On the one hand, for a nature image, it will result some artifacts around edges because of its asymmetry; on the other hand, the function is constructed by x-direction first and then y-direction, which results the advantage on x-direction [19]. The proposed weighted and blended interpolation model can refrain from these two factors. Obviously, Fig. 1 contains 16 points and the interpolated region is located in the center of the model. And it is easy to know that the disadvantage of y-direction is eliminated through the rotation. For example, there is a horizontal direction edge marked in red as shown in Fig. 1. And Fig. 3 shows the changes of the direction of the red edge during rotation. We can see that in (a) and (c), the red edge is still horizontal, while it rotates to vertical direction in (b) and (d) models. It means that the proposed interpolation model balances the effect of different edge directions.

Then the edge information is used as constraints to determine the weight coefficients. It would not only be able to ensure good visual perception of detail areas, but also make the edge regions avoid zigzagging.

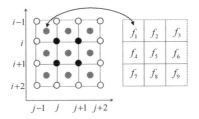

Fig. 4. Every cell can be regard as a point.

2.2   Adaptive Weights

From discussed above, we know that the weight coefficients are vital to the interpolation effect. And the weight coefficients should reflect local structure information and natural attributions of image. Following, the determination of the unknown in Eq. 1 is discussed. Adaptive interpolation means that the way the neighboring pixels influence the value of the interpolated pixel depends on local properties [11]. Thus the weight coefficients should be different when the construction of near pixels is different. In the proposed model, the 4×4 pixels region is divided into 4 overlapping subregions, and the contribution of each subregion to the intermediate patch determines the weight coefficients. The 1×1 rectangular region composed of four pixels can be seen as a basic cell. It can be seen from Fig. 4 that every subregion can be regraded as four basic cells, then the whole interpolation model contains 9 cells denoted as $f_s, (s = 1, \cdots, 9)$. If we know the relationship between the 8-connected neighbors cells and intermediate cell f_5, it is easy to determine the four weight coefficients in Eq. 1.

For all $s = 1, 2, 3, 4, 6, 7, 8, 9$, let w_s represent the impact factors between the patch f_s and f_5. In Fig. 4, intuitively, f_5, f_6, f_8 and f_9 constitute the subregion P_1, and f_5 is the interpolated region. Thus we consider the relationship between f_6, f_8, f_9 and f_5 determines the contribution of the subregion P_1 to the intermediate interpolated region. And other subregions are in similar way. Then the a, b, c and d can be expressed as follows:

$$a = \frac{w_6 + w_8 + w_9}{W}, b = \frac{w_2 + w_3 + w_6}{W}, \tag{3}$$
$$c = \frac{w_1 + w_2 + w_4}{W}, d = \frac{w_4 + w_7 + w_8}{W},$$

where $W = w_1 + w_2 + w_7 + w_9 + 2w_2 + 2w_4 + 2w_6 + 2w_8$, which means normalization of the weight coefficients.

Now we discuss how to determine w_s. It is not easy to measure the relationship between two patches, while there are more approaches to measure the relationship between two points. Therefore a cell is regarded as a point such that the relationship between two cells can be approximately replaced by two points. Usually the pixel value can be regarded as the sample value of a continuous patch [16]. The approximate point sampling value of a cell is expressed as

$$\iint_{A_{ij}} P_{i,j}(x,y)dxdy = \overline{P_{ij}} \times S_{ij},$$

where A_{ij} is a cell region, and S_{ij} is the area of A_{ij}, $P_{i,j}(x,y)$ represents function of a patch. Thus the point sampling value of patch can be approximately achieved by $\overline{P_{ij}}$. And the point values of f_1 to f_9 can be calculated. It is converted to a problem of computing the relationship between the intermediate point and the 8-connected neighbors points respectively.

In a natural image, distance, gradient and different quotient can measure the relationship between different pixels from different attributions. The *distance* measures the pixel space relationship, and the local *gradient* shows the edge information in a cell, and the *difference quotient* indicates the edge information among the whole model region. Thus, the local *gradient* and *difference quotient* can describe the edge from different scale. We focus on three factors *distance, gradient,* and *difference quotient* to determine the w_s, as shown in Eq. 4

$$w_s = F(distance, gradient, difference\ quotient). \tag{4}$$

Based on the weight expression of bilateral filter which contains distance and gray value, we construct a trilateral weighted expression. For all $s = 1, 2, 3, 4, 6, 7, 8, 9$, the Eq. 4 can be represented as

$$w_s = e^{-\frac{w_s^1}{h_1^2}} \times e^{-\frac{w_s^2}{h_2^2}} \times e^{-\frac{w_s^3}{h_3^2}}, \tag{5}$$

where w_s^1, w_s^2 and w_s^3 represent *Distance, Gradient,* and *Difference Quotient* respectively, and h_1, h_2 and h_3 are adjusting parameters. Then the unknown w_s^1, w_s^2 and w_s^3 are calculated.

First, the weight coefficients depend on the distance between each point and the intermediate point. If f_s is closer to the intermediate point, the weight coefficient will be greater. It is shown as

$$w_s^1 = (x_5 - x_s)^2 + (y_5 - y_s)^2, \tag{6}$$

where x and y are the local coordinates of these points.

Second, the weight coefficients also depend on the local gradients. The local gradient is expressed as

$$w_s^2 = |f_x'|^2 + |f_y'|^2, \tag{7}$$

where the f_x' and f_y' are the local gradient of a cell around the interpolation patch. In essence the smaller the local gradient of a pixel is, the more influence it should have on the intermediate pixel [11]. Obviously, the small scale edge information is considered due to the gradients as one of the factors in a cell.

Third, the second difference quotient is taken into account. If there is an edge along the vertical direction, f_2 and f_8 should have the closest connection to f_5. They are defined as

$$w_2^3 = |2f_5 - f_2 - f_8|^2, \tag{8}$$

w_1^3, w_3^3 and w_4^3 are defined in the same way. And $w_6^3 = w_4^3$, $w_7^3 = w_3^3$, $w_8^3 = w_2^3$, $w_9^3 = w_1^3$. This factor reflects the large scale edge information in the whole region of 4×4, and the smaller the w_s^3 is, the more effect it should have on the f_5.

All the unknown factors are calculated. Substituting Eq. 5 into Eq. 3 gets the adaptive weights. Since the quality around the edges plays an important role in the visual effect of an image, $P_{i,j}(x,y)$ should reflect the characteristics around the edges as well as possible. In Eq. 7, the local gradient which infers the local small scale edge is involved, and in Eq. 8 all the pixels in the whole window are contained to determine weight coefficients which means that the large scale edge is considered. Thus the final weight coefficients a, b, c and d are adaptive by edge.

3   Experiments

The proposed method is compared with recent interpolation algorithms: new edge-directed interpolation (NEDI) [8], soft-decision interpolation (SAI) [18], sparse mixing estimators (SME) [9] and robust soft-decision interpolation (RSAI) [5]. All experiments are performed with softwares provided by the authors of these algorithms[1]. We have used 7 images as our benchmark images (Fig. 5). We downsmaple these HR images to get the corresponding LR images. Table 1 gives the PSNRs and SSIMs generated by all algorithms for the images in Fig. 5. It can be seen that the proposed method has a highest average PSNR and SSIM among all the algorithms.

Fig. 5. Benchmark images (Color figure online).

Table 1. PSNR and SSIM results of the reconstructed HR images.

Method	NEDI		SME		SAI		RSAI		Proposed	
	PSNR	SSIM	PSNR	SSIM	PSNR	SSIM	PSNR	SSIM	PSNR	SSIM
Barbara	22.36	0.8513	23.98	0.8731	23.55	0.8638	23.37	0.8618	**24.51**	**0.8795**
Fence	19.82	0.6853	21.47	0.7314	20.82	0.7182	21.556	0.7353	**21.56**	**0.7355**
Airplane	25.69	0.8556	25.80	0.8708	26.01	0.8769	26.01	0.8762	**26.11**	**0.8770**
Lake	25.58	0.8606	26.95	0.8820	26.78	0.8838	26.98	0.8844	**27.17**	**0.8870**
Milkdrop	28.85	0.9027	29.35	0.9092	29.72	0.9176	29.78	0.9170	**30.61**	**0.9216**
Girl	29.73	0.9668	30.58	0.9562	29.49	**0.9676**	29.60	0.9598	**31.16**	0.9609
Wall	23.94	0.8812	24.72	0.8903	24.63	0.8930	24.77	**0.8947**	**25.10**	0.8916
Average	25.14	0.8576	26.12	0.8733	25.86	0.8744	26.01	0.8756	**26.60**	**0.8790**

---

[1] The source code of the proposed method is opened, please request the first author.

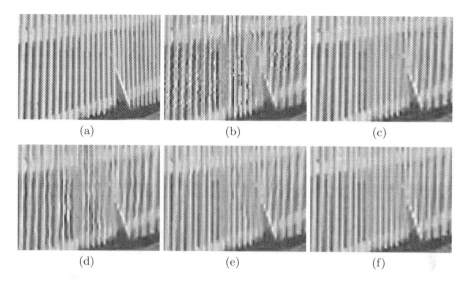

Fig. 6. Comparison on *Fence*. (a) Original image, (b) NEDI, (c) SME, (d) SAI, (e) RSAI, (f) Proposed method.

Fig. 7. Comparison on *Wall*. (a) Original image, (b) NEDI, (c) SME, (d) SAI, (e) RSAI, (f) Proposed method.

Figures 6, 7, 8 and 9 compare the interpolated images obtained by different algorithms. These images are cropped by red rectangle in Fig. 5. Figure 8 shows the edge information, and the others show details. For Fig. 8, all the algorithms perform similar results in edge region. We can see that NEDI suffers from some noisy interpolation artifacts (Figs. 7, 9, 6(b) because of the fixed interpolation window. And SAI method also suffers from noisy artifacts in Figs. 9(d) and 6(d). RSAI performs better than SAI but produces some unconnected stripes (Figs. 7(e) and 6(e)). Although SME has similar visual quality with the proposed method, the objective quality assessment value is lower than the proposed algorithm. Therefore, the proposed method can keep the edge region well, and it can perform better detail areas than other algorithms. Moreover, we also compared the proposed method with the methods in papers DFDF [17], KR [13],

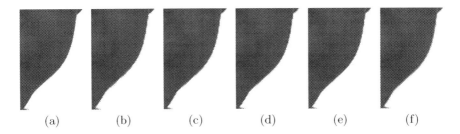

(a)   (b)   (c)   (d)   (e)   (f)

Fig. 8. Comparison on *Girl*. (a) Original image, (b) NEDI, (c) SME, (d) SAI, (e) RSAI, (f) Proposed method.

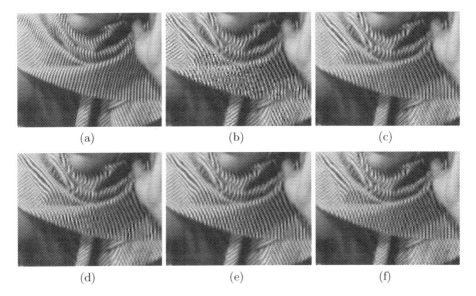

(a)   (b)   (c)

(d)   (e)   (f)

Fig. 9. Comparison on *Barbara*. (a) Original image, (b) NEDI, (c) SME, (d) SAI, (e) RSAI, (f) Proposed method.

INEDI [1], the proposed method has better vision quality and objective quality assessment value as well.

4   Conclusions

We propose an adaptive image interpolation method using rational function. The rational function is weighted and blended to remove artifacts. The edge information is used as constraints to determinate the weights adaptively. The new method has the advantage in that it can easily zoom the image into multiples. Our method can perform well on PSNRs and SSIMs. Furthermore, the proposed method produces clean edges and fine details.

Acknowledgement. This work was partially supported by Projects of International Cooperation and Exchanges NSFC (61020106001), National Natural Science Foundation of China under Grant 61373080, Grant 61202150, Grant 61373078.

References

1. Asuni, N., Giachetti, A.: Accuracy improvements and artifacts removal in edge based image interpolation. In: VISAPP, vol. 1, pp. 58–65 (2008)
2. Dong, W., Zhang, L., Lukac, R., Shi, G.: Sparse representation based image interpolation with nonlocal autoregressive modeling. IEEE Trans. Image Process. **22**, 1382–1394 (2013)
3. Hou, H., Andrews, H.: Cubic splines for image interpolation and digital filtering. IEEE Trans. Acoust. Speech Sig. Process. **26**, 508–517 (1978)
4. Hu, M., Tan, J.: Adaptive osculatory rational interpolation for image processing. J. Comput. Appl. Math. **195**, 46–53 (2006)
5. Hung, K., Siu, W.: Robust soft-decision interpolation using weighted least square. IEEE Trans. Image Process. **21**, 1061–1069 (2012)
6. Key, R.: Cubic convoluion interpolation for digital image processing. IEEE Trans. Acoust. Speech Sig. Process. **29**, 1153–1160 (1981)
7. Li, M., Nguyen, T.: Markov random field model-based edge-directed image interpolation. IEEE Trans. Image Proc. **17**, 1121–1128 (2008)
8. Li, X., Orchard, M.: New edge-directed interpolation. IEEE Trans. Image Process. **10**, 1521–1527 (2001)
9. Mallat, S., Yu, G.: Super-resolution with sparse mixing estimators. IEEE Trans. Image Process. **19**, 2889–2900 (2010)
10. Matsumoto, S., Kamada, M., Mijiddorj, R.: Adaptive image interpolation by cardinal splines in piecewise constant tension. Optim. Lett. **6**, 1265–1280 (2012)
11. Shezaf, N., Abramov-Segal, H., Sutskover, I., Ran, B.: Adaptive low complexity algorithm for image zooming at fractional scaling ratio. In: Proceeding of the 21st IEEE Convention of the Electrical and Electronic Engineers, IEEE, pp. 253–256 (2000)
12. Sun, Q., Bao, F., Zhang, Y., Duan, Q.: A bivariate rational interpolation based on scattered data on parallel lines. J. Vis. Commun. Image R. **24**, 75–80 (2013)
13. Takeda, H., Farsiu, S., Milanfar, P.: Kernel regression for image processing and reconstruction. IEEE Trans. Image Process. **16**, 349–366 (2007)

14. Thévenaz, P., Blu, T., Unser, M.: Image interpolation and resampling. In: Bankman, I. (ed.) Handbook of Medical Imaging, Processing and Analysis, pp. 392–420. Academic Press, San Diego (2000)
15. Zhang, C., Wang, J.: C-2 quartic spline surface interpolation. Sci. China F **45**, 417–432 (2002)
16. Zhang, C., Zhang, X., Li, X., Cheng, F.: Cubic surface fitting to image with edges as constraints. In: Proceedings of the IEEE International Conference on Image Processing, IEEE (2013)
17. Zhang, L., Wu, X.: An edge-guided image interpolation algorithm via directional filtering and data fusion. IEEE Trans. Image Process. **15**, 2226–2238 (2006)
18. Zhang, X., Wu, X.: Image interpolation by adaptive 2-d autoregressive modeling and soft-decision estimation. IEEE Trans. Image Process. **17**, 887–896 (2008)
19. Zhang, Y.F., Bao, F.X., Zhang, C.M., Duan, Q.: A weighted bivariate blending rational interpolation function and visualization control. J. Comput. Anal. Appl. **14**, 1303–1321 (2012)

First International Workshop
on Robust Reading (IWRR2014)

Text Localization Based on Fast Feature Pyramids and Multi-Resolution Maximally Stable Extremal Regions

Alessandro Zamberletti$^{(\boxtimes)}$, Lucia Noce, and Ignazio Gallo

Department of Theoretical and Applied Science, University of Insubria, Varese, Italy
a.zamberletti@uninsubria.it

Abstract. Text localization from scene images is a challenging task that finds application in many areas. In this work, we propose a novel hybrid text localization approach that exploits Multi-resolution Maximally Stable Extremal Regions to discard false-positive detections from the text confidence maps generated by a Fast Feature Pyramid based sliding window classifier. The use of a multi-scale approach during both feature computation and connected component extraction allows our method to identify uncommon text elements that are usually not detected by competing algorithms, while the adoption of approximated features and appropriately filtered connected components assures a low overall computational complexity of the proposed system.

1   Introduction

Text localization from scene images has recently gained attention due to its potential application in various areas.

Using the categorization criteria of Pan *et al.* [1], algorithms for text localization can be classified as either region-based [1–4] or connected component CC-based [5–9]. Region-based methods exploit local features and sliding window classifiers to identify potential regions of text and build text confidence maps, while CC-based methods are based on the observation that text characters usually show uniform characteristics and therefore appear as stable connected components within the processed images.

Both of the previously mentioned approaches have disadvantages: region-based methods need to process the image in a multi-scale manner to obtain satisfying results, this usually causes those methods to be computationally expensive as they spend most of their processing time performing feature computation at the different scales. Moreover, sliding window classifiers for text localization are prone to false-positive errors as some local regions in scene images are virtually indistinguishable from text characters [10].

Most CC-based text localization methods [5–9] identify stable connected components using Maximally Stable Extremal Regions (MSER) [11]. Even though the basic assumption of CC-based algorithms is that text characters always appear as MSER, this does not always hold true, *e.g.* almost none of the published CC-based algorithms participating in ICDAR'13 [12] competition successfully detect blurred

© Springer International Publishing Switzerland 2015
C.V. Jawahar and S. Shan (Eds.): ACCV 2014 Workshops, Part II, LNCS 9009, pp. 91–105, 2015.
DOI: 10.1007/978-3-319-16631-5_7

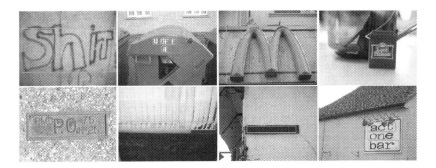

Fig. 1. Examples of uncommon and difficult text components successfully detected by the proposed method (images from ICDAR'03 and ICDAR'13).

or uncommon (graffiti, company logos, *etc.*) text characters, as those elements either do not appear as stable connected components or are discarded due to their irregular geometric properties.

In this work, we pair Fast Feature Pyramids and Aggregated Channel Features [13] with Multi-resolution Maximally Stable Extremal Regions (MR-MSER) [14] to propose a hybrid algorithm for text localization that exploits the key ideas of region-based and CC-based methods but tries to overcome some of their previously mentioned limitations.

Without losing detection accuracy, in multi-scale approaches some image features (gradients, *etc.*) can be approximated from nearby scales within the same feature pyramid, instead of being explicitly computed at every level, to reduce by 2 orders of magnitude the time required to complete the feature computation process [13]. In our method, an approximated feature based classifier trained with natural, synthetic and semi-synthetic data, is used to efficiently build text confidence maps that are subsequently refined using MR-MSER.

Throughout our experiments we prove that MR-MSER excels at extracting entire words of text from scene images as single connected components, this also holds true for words composed by uncommon and difficult character fonts. We exploit this ability to discard false-positive text regions from the text confidence maps generated by the sliding window classifier.

In our system, most of the initially extracted MR-MSER are stacked and discarded; together with the use of approximated feature, this choice assures that the proposed method maintains an acceptable computational complexity even though it employs a multi-scale approach during both feature computation and connected component extraction.

As shown by the publicly available detection results for ICDAR'13 (some examples are provided in Fig. 1), despite its simplicity, the proposed approach succeeds where competing CC-based text localization methods usually fail, and achieves good results for ICDAR's Challenge 2 Task 1.[1]

---

[1] http://dag.cvc.uab.es/icdar2013competition/?ch=2&com=results
Method: iwrr2014.

2   Related Works

2.1   Region-Based Text Localization

Among region-based text localization methods [1–4], the works that are closer to the proposed method are the ones of Pan *et al.* [1] and Wang *et al.* [4].

Pan *et al.* [1] build a text confidence map by processing images in a sliding window manner, using Waldboost and HOG features. The confidence map is used, together with other geometric features and a Multi-layer Perceptron, to compute the binary and the unary weights of a component neighbourhood graph built over a set of connected components extracted using Niblack's text binarization algorithm. CRF are used to filter out non-text components from the graph, while the remaining neighboring elements are clustered together into Minimum Spanning Trees to form text words. In our approach, we exploit a similar text confidence map to identify potential regions of text.

Wang *et al.* [4] perform end-to-end text recognition using Random Ferns and Pictorial Structures. The part of their work that is related to ours is the choice of using synthetic positive training data: roughly 1000 images are synthesized per character using 40 different fonts, adding Gaussian noise and applying random affine deformations (similarly to [2]). The classifier trained exclusively using synthetic positive data achieves the same F-score of a NN classifier trained with HOG features extracted from native data.

Another novel idea from [4] is the choice of extracting negative training samples from classes of Microsoft Research Cambridge Object Recognition Image Database (MSRC) [15]: classes like *buildings* and *countryside* resemble the background patterns of ICDAR's images and help in reducing the number of false-positive detections generated by sliding window classifiers.

2.2   CC-Based Text Localization

Most CC-based text localization methods [5–7] either exploit MSER [11] to identify potential text components that are filtered and clustered together to form words, or use the Stroke Width Transform [10] algorithm to identify connected components having low intra-stroke variance [16].

Neumann *et al.* [17] proved that Extremal Regions extracted from multiple image channels cover almost 95 % over ground-truth character annotations for ICDAR'11; however, to decrease the complexity of the system, only a suboptimal subset of those channels is used in [17].

Other works focused on maximizing the effectiveness of MSER in terms of number of text elements successfully captured as stable components, *e.g.* Li *et al.* [6] showed that blurred and low quality characters become stable when extracting MSER from images incorporating gradient magnitude and intensity channels information.

Another technique for improving the coverage of MSER is the one of Forssén and Lowe [14]: a pyramid of images is built and MSER, called MR-MSER, are extracted at multiple scales (1 scale per octave). This multi-resolution approach

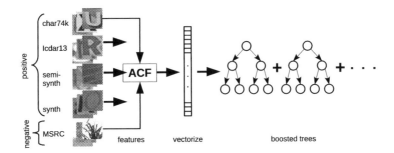

Fig. 2. Aggregated Channel Features (ACF) extracted from negative samples from MSRC and positive natural, synthetic and semi-synthetic samples from different datasets are used to train boosted depth-two decision trees.

causes some of the unstable regions in the original image to become stable at low scales in the pyramid, where the original image has lost most of its details as it has been sub-sampled and blurred multiple times with a Gaussian kernel.

In the proposed approach, we adopt the multiple image channel technique of [17] to extract words of text from scene images by computing multi-channel MR-MSER, appropriately filtering out useless regions extracted within the same pyramid's octave to keep an acceptable computational complexity.

2.3 Fast Feature Pyramids

Fast Feature Pyramids [13] revolutionized multi-scale sliding window approaches by showing that some image features (gradients, *etc.*) can be approximated from nearby scales within the same pyramid rather than being computed explicitly; since their introduction, Fast Feature Pyramids have been used in many works to build effective and efficient rigid object recognition detectors [18].

Based on the analysis of [19] on how to build the best classifier for rigid object recognition, we use a bootstrapped approximated feature based multi-scale linear classifier to perform text localization from scene images.

3   Proposed Model

The proposed approach is presented in this section: a binary classifier based on Fast Feature Pyramids and Aggregated Channel Features (Sect. 3.1) is trained using natural, synthetic and semi-synthetic data collected from multiple datasets or artificially generated (Sect. 3.2); predictions from the classifier are used to build a text confidence map in which potential regions of text are highlighted (Sect. 3.3); the text confidence map is used, together with MR-MSER (Sect. 3.4), to identify potential bounding boxes for lines of text in the processed image (Sect. 3.5).

An analysis of the computational complexity of the proposed approach and implementation details are provided in Sect. 3.6 and Table 1.

(a) source   (b) scale 1.0   (c) scale 0.5   (d) scale 0.25   (e) scale 0.12

Fig. 3. MSER extracted at different levels of the pyramid capture different details: at low scales (≤0.5), characters are merged together and words are captured as single components, this also holds true for uncommon fonts (*e.g.* "Apocalypse Now"); in some instances, difficult characters that are not detected at the original scale are correctly identified as stable connected components at lower levels in the pyramid (*e.g.* "£99", graffiti).

3.1   Text Region Detector

The first step in our pipeline is to build a text confidence map by detecting potential regions of text using a multi-scale sliding window ACF detector [13].

ACF uses Aggregated Channel Features, which are extracted by smoothing the processed image with a $[1\ 2\ 1]/4$ filter and then computing 10 different channels: normalized gradient magnitude, histogram of oriented gradients (6 orientations) and LUV. The channels are then condensed into 4×4 blocks and once again smoothed using the same approximated Gaussian kernel before being concatenated together to form single descriptors.

In our system, ACF is tuned to reach acceptable detection rates for text detection from scene images by setting the sliding window size to 32×32 pixels and the window stride to 16 pixels both in the horizontal and vertical directions. To deal with the large variation in size of text components in ICDAR datasets,

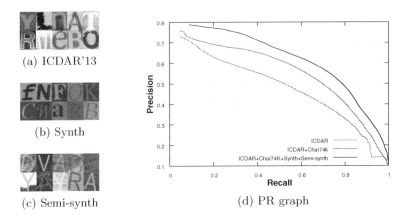

(a) ICDAR'13

(b) Synth

(c) Semi-synth

(d) PR graph

Fig. 4. Augmenting the positive training set with synthetic and semi-synthetic data increases the detection rate of the approximated feature based classifier.

we increase the size of the image pyramid by computing 1 octave above the canonical image scale. The final image pyramid goes from 2× the size of the original image I to at most 32×32 pixel and has 8 scales per octave. For each octave, 7 scales out of 8 are approximated using λ coefficients [13] inferred from 1000 random samples from the positive training set.

In our experiments, increasing the number of scales per octave or decreasing the number of approximated scales per octave did not affect the final results, on the other hand, decreasing the size of the image pyramid deeply affects the final detection rate, *e.g.* removing the highest octave while maintaining the same window size almost halves the accuracy of the classifier as tiny text components are not correctly detected. The same occurs when removing low pyramid levels or when improperly altering the size of the sliding window.

The ACF classifier is composed by 2048 quickly boosted [20] depth-two decision trees (3 stump classifiers per tree, as in Fig. 2).[2] Multiple rounds of bootstrapping are performed, at each round false-positive samples collected from the previous round are added to the negative training set, this shifts the decision boundary of the classifier and reduces false-positive detections in the subsequent round. Unlike [21], false-negative are not bootstrapped because text elements classified as background are usually identified using MR-MSER, as described in Sect. 3.5. Even when using 100000 samples and 3 rounds of bootstrapping, the classifier can be trained in less than 3 min on a Intel® Core i5 (see Table 1).

3.2   Training Data

Detection rates of linear classifiers are affected by both the quality/amount of training samples and the discriminative power of features extracted from those samples. Considering that state-of-the-art results have been obtained in rigid

---

[2] http://vision.ucsd.edu/~pdollar/toolbox/doc/index.html.

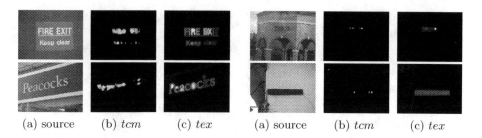

Fig. 5. True-positive regions discarded when thresholding the text confidence map *tcm* (b) are recovered in the textness map *tex* using MR-MSER (c).

object recognition by methods based on ICF and ACF [13, 18], we assume that good results may also be obtained in text detection using the same set of features when sufficient training data is collected. For this reason, we not only gather positive/negative samples from multiple datasets but we also generate additional semi-synthetic positive images by combining natural and synthetic images.

The process of extracting negative samples is straightforward: images not containing text are collected from some classes of MSRC database [15] (*benches, chairs, buildings, chimneys, kitchen utensils, miscellaneous, scenes, trees* and *windows*). In total, 1843 images containing only background components are gathered. For each image, a 4-level image pyramid (20 %, 50 %, 80 % and 100 % of the original image size) is built and 32 × 32 pixels patches are randomly extracted from all the pyramids, until a total of 50000 negative samples are gathered. Extracting negative examples at multiple scales reduces the number of false-positive detections generated at low octaves in the feature pyramid.

Gathering positive training samples is a challenging task, poor detection rates were obtained when training the ACF classifier using just the ≈5400 samples from ICDAR'13 [12] train set (see Fig. 4). To obtain acceptable performances, we augmented the set of positive training data with: ≈8000 images from the *GoodImg* class of Char74k English dataset [22], ≈6200 artificial images from the publicly available Synth dataset [4] (vertically cropped to remove neighbouring characters) and ≈30000 semi-synthetic samples obtained by combining natural background patches from MSRC images with synthetic fonts.

More in detail, semi-synthetic images are generated by placing random sized artificial characters in random positions over the images previously collected from MSRC to extract negative samples, random jitter (translation and rotation) is applied to increase the robustness of the classifier. Characters are cropped to their bounding boxes (leaving at most 5 pixels of random padding in every direction) and sub-sampled/up-sampled to 32 × 32 pixels. In order to keep an acceptable degree of contrast between the synthetic character and its surroundings, we compute the histogram of the patch on which the character is pasted and discard samples that are human unreadable (zero contrast between character and background).

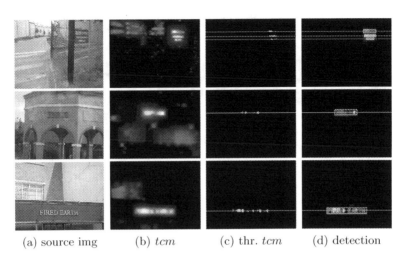

(a) source img   (b) *tcm*   (c) thr. *tcm*   (d) detection

Fig. 6. Text line formulation algorithm pipeline: the text confidence map (b) is thresholded (c) and words are identified using both the textness map *tex* and MR-MSER (d), final components are grouped together using Mean-shift.

Figure 4 shows how the positive sample sets we aggregate complement each others: samples from ICDAR train set and Char74k (Fig. 4a) contain uncommon and handmade characters that cannot be artificially generated; synthetic data from Synth (Fig. 4b) is useful to learn the shapes of artificial characters placed on plain backgrounds; semi-synthetic samples (Fig. 4c) are often placed on cluttered backgrounds and degraded due to sub-sampling/up-sampling, and thus represent a good connection point between synthetic and natural data.

3.3   Text Confidence Map

Let $\{s_0, \ldots, s_n\}$ be the scores assigned by the trained ACF classifier to each position of the sliding window in the image pyramid built for the processed image; a greedy Non-maximum Suppression (NMS) is performed to discard overlapping regions. In detail: (i) we discard regions having score lower than the average detection score $\mu(\{s_0, \ldots, s_n\})$; (ii) resize the remaining ones to half of their sizes to obtain a good separation between detected text regions, as in [4]; (iii) iterate over them by descending score and, if the region has not yet been suppressed, we suppress all the other non-suppressed regions having intersection-over-union $IoU > 0.5$ with the one currently selected; (iv) surviving regions are restored to their original sizes.

Using the suppressed regions we define a set of local text confidence maps $\{tc_0, \ldots, tc_j\}$, one for each level of the image pyramid. The final text confidence map tcm is obtained by stacking all the local confidence maps together $tcm = \frac{\sum_{i=1}^{j} tc_i}{n}$. Finally, tcm is normalized in $[0, 1]$ and thresholded at $t = 0.5$ to remove false-positive regions. True-positive regions discarded because of the threshold are recovered using MR-MSER, as explained in Sect. 3.5 (see Fig. 5).

3.4 Textness Map and MR-MSER

The text confidence map tcm is used, together with MR-MSER [14], to generate a *textness* map tex in which the value of each pixel denotes the probability it belongs to a text component in the original image I.

To extract MR-MSER, we compute 7 channels for I (RGB, HSI and ∇) and build an independent scale pyramid for each channel. MR-MSER are detected at each level of the pyramid, which has 1 octave per scale and a minimum size of 256×256 pixels; images in the pyramid are obtained by blurring and subsampling using a 6-tap Gaussian kernel with $\sigma = 1$. To reduce the final number of MR-MSER and discard the duplicate ones, at each level of each pyramid we retain only the larger MSER and filter out the smaller nested regions. This significantly decreases the final number of extracted MR-MSER: on average we discard more than 2000 regions from the $\approx$2500 initially identified.

Similarly to the text confidence map tcm, the *textness* map tex is built by iterating over the extracted MR-MSER and, for each of them, increasing the value of its pixels in the tex map by the average value of those pixels in tcm.

3.5 Text Line Formulation

The last step in a text localization framework consists in identifying the bounding boxes for words of text in the processed image; we formulate an algorithm that can be applied to different datasets without extensive tuning.

We propose a peak-based text grouping algorithm, in detail: (i) local maxima of the column-wise histogram of tex are identified, those peaks correspond to rows $\{r_0, \ldots, r_k\}$ of tex having maximum *textness* value compared to their neighbours; (ii) for each peak row r_i, connected components $\{cc_0, \ldots, cc_q\}$ intersecting r_i in the text confidence map tcm are identified; (iii) each cc_i is resized to the size of the minimum bounding box enclosing MR-MSER extracted from the image that have a pixel-wise $IoU > 0.2$ with cc_i; (iv) each resized cc_i is assigned a score computed as the average intensity of its pixels in tex, and overlapping components are suppressed (as in Sect. 3.3); (v) neighbours connected components are merged into text lines using Mean-shift, components are clustered on the basis of their centroid positions. The pipeline is summarized in Fig. 6.

In phase (iii), we reshape regions labelled by the classifier as potential text areas according to the boundaries of MR-MSER extracted from the image. As it is possible to observe from Figs. 3 and 7a, MR-MSER extracted at low levels in the scale pyramid are often able to capture entire words (instead of single characters) as at those low levels most details of the original image are lost, and this causes characters to be merged together and words to be identified as single stable regions. Exploiting the word detection ability of MR-MSER to discard noise regions from the text confidence map without worrying about losing true-positive areas is the key idea of our method.

By grouping text components just on the basis of their centroid positions (ignoring scale, orientation, *etc.*), our algorithm can capture text in any possible orientation, even though ICDAR datasets do not contain examples of non-horizontal text components. The major drawback of ignoring orientation *etc.*

Table 1. Implementation details. Times refer to a 640×480 image and ≈ 500 MR-MSER processed on a desktop Intel$^\circledR$ Core i5.

Task	Time (s)	Implementation type
Gathering p/n training data	103.40	Parallel
Training the classifier	185.58	Par. (Seq. feature computation)
Text confidence map	0.29	Parallel
Textness map	0.45	Parallel
Text line formulation	0.01	Parallel

is that the proposed line formulation method frequently aggregates lines and phrases into single bounding boxes; such behaviour is penalized by some evaluation metrics (see Sect. 4.3), and additional processing may be required to split the detected bounding boxes into words before they are passed on to OCR engines.

3.6 Implementation Details

Timings information for the proposed approach are given in Table 1: gathering positive/negative samples and training the classifier for ICDAR'13 dataset require less than 5 min on a desktop Intel$^\circledR$ Core i5 with 12 GB RAM.

On average, a 640×480 image can be fully processed in ≈ 0.75 s. The computational complexity of the method can be further reduced by decreasing the number of channels from which MR-MSER are extracted, at the cost of sacrificing the accuracy of the whole system.

Using the classifier configuration of Sect. 3.1 and the training data from Sect. 3.2, RAM consumption during training is ≈ 6 GB. On average, ≈ 500 MB of RAM are sufficient to build the *textness* map for a 640×480 image.

4 Experiments

In this section, we provide an experimental evaluation of the components described in Sect. 3: the detection rate of the ACF text region detector introduced in Sect. 3.1 is evaluated in Sect. 4.1; MR-MSER are compared with MSER at detecting entire words of text in Sect. 4.2; text localization results achieved by the proposed approach for ICDAR'03 and ICDAR'13 datasets are presented in Sect. 4.3 and compared with competing published algorithms.

4.1 Classifier and Training Data

Figure 4d shows the PR curves for multiple ACF classifiers trained using the same parameter configuration but different training samples. PR curves are computed as in [2]: the text confidence map *tcm* is thresholded multiple times to yield binary decisions at each pixel and compared pixel-wise with ground-truth annotations from ICDAR'13 test set.

Image Channels	MR-MSER		MSER	
	chars	words	chars	words
∇	0.56	0.69	0.52	0.56
RGB	0.63	0.40	0.56	0.25
HSI	0.62	0.51	0.56	0.36
HSI $\cup$ ∇	0.71	0.77	0.67	0.65
RGB $\cup$ ∇	0.72	0.73	0.68	0.61
HSI $\cup$ RGB	0.70	0.56	0.64	0.41
HSI $\cup$ RGB $\cup$ ∇	0.75	0.78	0.71	0.66

(a) MR-MSER vs. MSER

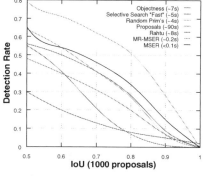

(b) Word detection accuracy

Fig. 7. Evaluation of MR-MSER for word detection: (a) MR-MSER are compared with MSER at detecting single characters and entire words, while varying the image channels from which they are extracted, as in [17]; (b) word detection accuracy evaluation and timings information for MR-MSER, MSER and object proposal methods on ICDAR'03, while varying the intersection-over-union IoU coverage tolerance.

This experiment shows how training data affects the performance of the ACF classifier: unsatisfying detection rates are obtained when training using just the samples from ICDAR'13 train set; significantly better results are obtained when combining ICDAR'13 train data with samples from Char74k; acceptable detection accuracies are achieved when augmenting the positive training set with synthetic and semi-synthetic data collected as described in Sect. 3.2.

In every experiment the training set has been kept balanced, meaning that the number of negative samples has always been equal to the number of positive samples. AUC of the PR curve for the classifier trained with natural, synthetic and semi-synthetic data is higher than the one of [2], proving the effectiveness of approximated feature based sliding window classifiers for text localization.

4.2   Word Detection via MR-MSER

The proposed method relies on the ability of MR-MSER to identify entire words of text from scene images (see Sects. 3.4 and 3.5).

In Fig. 7a, MR-MSER and MSER are compared at the task of identifying single characters (*chars*) and entire words (*words*) on ICDAR'03, while varying the image channels from which MR-MSER and MSER are extracted. In our experiment, a character/word is considered identified if there exists at least one MR-MSER/MSER, from the ones extracted, whose bounding box has an intersection-over-union $IoU > 0.5$ with the ground-truth annotation of that character/word.

Even though in this experiment we compared filtered MR-MSER (smaller nested regions are discarded as described in Sect. 3.4) with unfiltered MSER (all the regions are retained for evaluation), MR-MSER outperform MSER both at detecting single characters and entire words for all the evaluated combinations of image channels. On average, the extraction of MR-MSER requires 0.2 s using a parallel implementation (multiple scales within the same octave are extracted at the same time), while the computation of MSER requires 0.1 s per image.

In order to measure how accurately MR-MSER detect entire words of text at low blurred scales in the image pyramid, in Fig. 7b they are evaluated on ICDAR'03 while varying the IoU coverage tolerance. Since text characters and words satisfy some of the conditions analysed in [23], object proposal methods have also been added to the comparison to see whether they constitute a valid alternative to MR-MSER or MSER at detecting words of text in scene images. Results are measured as in [24]: for each algorithm, at most 1000 bounding boxes per image are selected from the ones initially extracted; Detection Rate (DR, y-axis) is the percentage of ground-truth words *covered* by those bounding boxes. A ground-truth word is *covered* if there exists at least one bounding box, among the 1000 selected, that has an $IoU > x$ with the ground-truth bounding box of that word. The value of x varies on the x-axis, by increasing x we require the identified bounding boxes to match more precisely the ground-truth data in order for a word to be considered *covered*.

The comparison is carried out as follows: (i) Objectness: among the $\approx$1850 ranked proposals generated per image, the top 1000 are selected for evaluation. MS, CC and SS cues are learnt from 50 images from ICDAR'03 training set; (ii) Selective Search: evaluated in its *fast* variant, 1000 windows are uniformly sampled from the ones initially extracted; (iii) Prims: a grid search is performed in $[0, 5]$ for color similarity, common border ratio and size, the parameters providing the best results for 1000 unique windows and $IoU > 0.5$ are used for evaluation; (iv) Proposals: evaluation is performed considering the bounding boxes surrounding the identified ranked segmentations proposals, top 1000 windows are selected from the ones initially extracted; (v) MR-MSER: extracted as described in Sect. 3.4, the bounding boxes surrounding each MR-MSER are considered for evaluation, on average, no more than 500 windows per image are generated; (vi) MSER [11]: extracted from RGB, HSI and ∇ channels, the bounding boxes surrounding 1000 unique MSER are uniformly sampled from the initial set. For references to the evaluated object proposals algorithms see [24].

MR-MSER prove their effectiveness as robust word detector from scene images by achieving higher detection accuracies throughout all the tolerance values.

4.3   Text Localization Results

In Table 2a and b, the proposed text localization approach is evaluated on ICDAR'03 and ICDAR'13 datasets.

ICDAR 2003 [28] contains a total of 509 images: 258 for training and the remaining 251 for testing. The classifier is trained using 45000 positive samples from ICDAR'03 train set, Char74k, Synth and Semi-synth and 45000 negative samples from MSRC. Precision, Recall and F-score are computed by looking for the best match between each detected bounding boxes and each ground-truth annotation [5]. This evaluation metric penalizes approaches that detect text at

Table 2. Text localization results for ICDAR's Challenge 2 Task 1.

(a) ICDAR'03. Evaluation metric: [28]

Method	Precision	Recall	F-score
Li [6]	0.79	0.64	0.71
Kim [5]	0.78	0.65	0.71
Proposed	**0.71**	**0.74**	**0.70**
TD-Mixture [16]	0.69	0.66	0.67
Yi [25]	0.73	0.67	0.66
Epshtein [10]	0.73	0.60	0.66
Li	0.62	0.65	0.63
Chen	0.60	0.60	0.58
Neumann [7]	0.59	0.55	0.57
Zhang	0.67	0.46	0.55

(b) ICDAR'13. Evaluation metric: [12]

Method	Precision	Recall	F-score
Proposed	**0.86**	**0.70**	**0.77**
Yin [9]	0.88	0.66	0.76
Neumann [26]	0.88	0.65	0.74
Bai [27]	0.79	0.68	0.73
Shi [8]	0.85	0.63	0.72
Shijian	0.75	0.69	0.72
Yang	0.70	0.65	0.67
Fabrizio	0.74	0.53	0.62
Baseline	0.61	0.35	0.44
Inkam	0.31	0.35	0.33

line level, as only *one-to-one* matches are taken into account. Since our method often captures entire phrases as single components, it generates numerous *many-to-one* detections and therefore performs slightly worse than [5,6].

ICDAR 2013 [12] contains a total of 462 images: 229 for training and 233 for testing. The classifier is trained using 50000 positive samples from ICDAR'13 train set, Char74k, Synth and Semi-synth and 50000 negative samples from MSRC. Unlike ICDAR'03, results are measured using a new evaluation metric [12], which takes into account *one-to-one*, *one-to-many* and *many-to-one* matches between ground-truth annotations and detected bounding boxes. The competition protocol penalizes methods that perform text localization at character level (*one-to-many*) but does not inflict any penalty to methods that provide text detection at line level (*many-to-one*). F-score of the proposed method is higher than competing approaches both when measured using ICDAR's evaluation metric or DetEval [29]. Complete results are available on ICDAR's web page. For references to all the evaluated algorithms see [12,28].

Using classic MSER in place of MR-MSER, F-score of the proposed method decreases by roughly 10 % on both datasets, as expected from the analysis of Sect. 4.2, where multi-channel MR-MSER covers 78 % of ICDAR's ground-truth words while MSER provides a coverage of 66 %.

Negative detection results are provided in Fig. 8, the proposed method fails when MSER extracted at multiple scales do not capture text components or when the text confidence map is noisy and text components are lost due to threshold (*e.g.* "HHH CELCON"). It is in fact possible to obtain different values of Precision/Recall by shifting the threshold value used during the text confidence map building phase described in Sect. 3.3: lower threshold values increase the Recall of the algorithm and decrease its Precision, while higher values discard more components from the text confidence map and therefore decrease the Recall of the whole system while increasing its overall Precision.

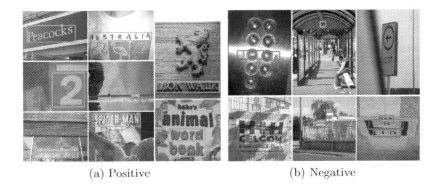

(a) Positive   (b) Negative

Fig. 8. Examples of positive and negative text localization results for ICDAR'13.

5   Conclusion

A novel method for text localization from scene images has been proposed, it exploits both the latest advancements in rigid object recognition and MR-MSER to obtain good results for text localization from scene images. In the proposed solution, stable connected components are not discarded on the basis of their geometric properties; this assures that uncommon text fonts that are typically filtered out as noise elements by competing approaches are correctly retained and identified. Thanks to the use of approximated multi-resolution features and appropriately filtered connected components extracted in a multi-scale multi-channel manner, the proposed system is computationally efficient to train and test. This enables its application to numerous problems in which execution and training times are critical factors.

References

1. Pan, Y.F., Hou, X., Liu, C.L.: Text localization in natural scene images based on conditional random field. In: Proceedings of the ICDAR (2009)
2. Coates, A., Carpenter, B., Case, C., Satheesh, S., Suresh, B., Wang, T., Wu, D.J., Ng, A.Y.: Text detection and character recognition in scene images with unsupervised feature learning. In: Proceedings of the ICDAR (2011)
3. Mishra, A., Alahari, K., Jawahar, C.: Scene text recognition using higher order language priors. In: Proceedings of the BVMC (2012)
4. Wang, K., Babenko, B., Belongie, S.: End-to-end scene text recognition. In: Proceedings of the ICCV (2011)
5. Koo, H.I., Kim, D.H.: Scene text detection via connected component clustering and non-text filtering. IEEE Trans. IP **22**, 2296–2305 (2013)
6. Li, Y., Jia, W., Shen, C., Hengel, A.: Characterness: an indicator of text in the wild. IEEE Trans. IP **23**, 1666–1677 (2014)
7. Neumann, L., Matas, J.: A method for text localization and recognition in real-world images. In: Kimmel, R., Klette, R., Sugimoto, A. (eds.) ACCV 2010, Part III. LNCS, vol. 6494, pp. 770–783. Springer, Heidelberg (2011)

8. Shi, C., Wang, C., Xiao, B., Zhang, Y., Gao, S.: Scene text detection using graph model built upon maximally stable extremal regions. Pattern Recogn. Lett. **34**, 107–116 (2013)
9. Yin, X.C., Yin, X., Huang, K.: Robust text detection in natural scene images. IEEE Trans. PAMI **36**, 970–983 (2013)
10. Epshtein, B., Ofek, E., Wexler, Y.: Detecting text in natural scenes with stroke width transform. In: Proceedings of the CVPR (2010)
11. Matas, J., Chum, O., Urban, M., Pajdla, T.: Robust wide baseline stereo from maximally stable extremal regions. In: Proceedings of the BMVC (2002)
12. Karatzas, D., Shafait, F., Uchida, S., Iwamura, M., Bigorda, L., Mestre, S., Mas, J., Mota, D., Almaz, J., Heras, L.: ICDAR 2013 robust reading competition. In: Proceedings of the ICDAR (2013)
13. Dollár, P., Appel, R., Belongie, S., Perona, P.: Fast feature pyramids for object detection. IEEE Trans. PAMI **36**, 1532–1545 (2014)
14. Forssén, P.E., Lowe, D.G.: Shape descriptors for maximally stable extremal regions. In: Proceedings of the ICCV (2007)
15. Crimisi, A.: Microsoft Research Cambridge Object Recognition Image Database (2004)
16. Yao, C., Bai, X., Liu, W., Ma, Y.: Detecting texts of arbitrary orientations in natural images. In: Proceedings of the CVPR (2010)
17. Neumann, L., Matas, J.: Real-time scene text localization and recognition. In: Proceedings of the CVPR (2012)
18. Mathias, M., Timofte, R., Benenson, R., Gool, L.V.: Traffic sign recognition: how far are we from the solution? In: Proceedings of the IJCNN (2013)
19. Benenson, R., Mathias, M., Tuytelaars, T., Gool, L.V.: Seeking the strongest rigid detector. In: Proceedings of the CVPR (2013)
20. Appeal, R., Fuchs, T., Dollár, P., Perona, P.: Quickly boosting decision trees pruning underachieving features early. In: Proceedings of the ICML (2013)
21. Villamizar, M., Andrade-Cetto, J., Sanfeliu, A., Moreno-Noguer, F.: Bootstrapping boosted random ferns for discriminative and efficient object classification. Pattern Recogn. **45**, 3141–3153 (2012)
22. de Campos, T.E., Babu, B.R., Varma, M.: Character recognition in natural images. In: Proceedings of the VISAPP (2009)
23. Alexe, B., Deselaers, T., Ferrari, V.: What is an object? In: Proceedings of the CVPR (2010)
24. Manen, S., Guillaumin, M., Gool, L.V.: Prime object proposals with randomized prims algorithm. In: Proceedings of the ICCV (2013)
25. Yi, C., Tian, Y.: Localizing text in scene images by boundary clustering, stroke segmentation, and string fragment classification. IEEE Trans. IP **21**, 4256–4268 (2012)
26. Neumann, L., Matas, J.: On combining multiple segmentations in scene text recognition. In: Proceedings of the ICDAR (2013)
27. Bai, B., Yin, F., Liu, C.L.: Scene text localization using gradient local correlation. In: Proceedings of the ICDAR (2013)
28. Lucas, S.M., Panaretos, A., Sosa, L., Tang, A., Wong, S., Young, R.: ICDAR 2003 robust reading competition. In: Proceedings of the ICDAR (2003)
29. Wolf, C., Jolion, J.M.: Object count/area graphs for the evaluation of object detection and segmentation algorithms. IJDAR **8**, 280–296 (2006)

A Hybrid Approach to Detect Texts in Natural Scenes by Integration of a Connected-Component Method and a Sliding-Window Method

Yojiro Tonouchi[1]([✉]), Kaoru Suzuki[1], and Kunio Osada[2]

[1] Corporate Research and Development Center,
Toshiba Corporation, Tokyo, Japan
yojiro.tonouchi@toshiba.co.jp
[2] IT Reseach and Development Center,
Toshiba Solutions Corporation, Tokyo, Japan

Abstract. Text detection in images of natural scenes is important for scene understanding, content-based image analysis, assistive navigation and automatic geocoding. Achieving such text detection is challenging due to complex backgrounds, non-uniform illumination, and variations in text font, size, and orientation. In this paper, we present a novel hybrid approach for detecting text robustly in natural scenes. We connect two text-detection methods in parallel structure: (1) a connected-component method and (2) a sliding-window method and outputs basically both results. The connected-component method generates text lines based on local relations of connected components. The sliding-window method consisting of a novel Hough Transform-based method generates text lines based on global structure. These two text-detection methods can output complementary results, which enables the system to detect various texts in natural scenes.

Testing with the ICDAR2013 text localization dataset shows that the proposed scheme outperforms the latest published algorithms and the parallel structure consisting of the two different methods contributes to decreasing false negatives and improves recall rate.

1 Introduction

Text detection in natural scenes has a wide range of applications, such as augmented reality devices, image retrieval, and robotic navigation. Extraction of texts from natural scenes, however, is much more difficult than reading texts from scanned materials. Images which capture natural scenes are variable and subject to the influences of background clutter, lighting conditions, shadowing at different distances as well as the different perspectives, rotations, and skews of the image itself. Moreover, texts in natural scenes can appear anywhere in the image and can have different sizes and layouts.

A great number of works deals directly with detection of text from natural images. In general, the methods for detection text can be broadly categorized

© Springer International Publishing Switzerland 2015
C.V. Jawahar and S. Shan (Eds.): ACCV 2014 Workshops, Part II, LNCS 9009, pp. 106–118, 2015.
DOI: 10.1007/978-3-319-16631-5_8

into three groups: Connected component methods, sliding-window methods and hybrid methods. Connected component methods (CC method) have recently grown more popular and have reported promising performance on the ICDAR 2013 Robust Reading Competition [3]. These methods first detect individual connected components by using local properties (color, stroke-width, etc.) and assuming that the selected property does not change much for neighboring characters. The connected components are then grouped into higher structures such as words and text lines in subsequent stages. The advantage of CC methods is that the complexity typically does not depend on the parameters of the text (range of scales, orientations, fonts) and character segmentation is typically obtained, which can be used for optical character recognition (OCR). But they are sensitive to clutter and occlusions which change connected component structure. Sliding-window methods scan the image at a number of scales and the presence of text is estimated. Generally, a feature vector extracted from each local region is fed into a classifier which estimates the likelihood of text. Then, neighboring text regions are merged to generate text lines. These methods are generally more robust to noise in the image than other methods. On the other hand, a major limitation of these methods is the high computational complexity due to the need to scan the image at several scales. Additionally, these algorithms are typically unable to detect highly slanted text. Hybrid methods integrate CC methods and sliding-window method in order to take advantages of both methods.

In this paper, we present a new hybrid system which connects a CC method and a sliding-window method in parallel to detect text lines. The CC method generates text lines by means of a neighborhood graph. It generates lines based on local relations between connected components. On the other hand, in the sliding-window method, we adopt a new extended Hough transform-based text line generator. The extended Hough transform uses not only position information but also size information of text region in generation of text lines. It generates lines based on global structure that a text line consists of characters which have similar sizes and lie on a nearly straight line. This system outputs both text lines which two methods generate except in the case of overlapped regions. It can detect various text patterns in natural scenes by using these two types of results, which complement one another.

The rest of this paper is organized as follows. Previous work is presented in Sect. 2 and our proposed algorithm is described in detail in Sect. 3. Experimental results are presented in Sect. 4, and we conclude this paper in Sect. 5.

2   Previous Work

The majority of recently published methods for text detection are CC methods [2,8,12,13]. The methods differ in their approach to individual character detection, which can be based on edge detection [2,12], character energy calculation [13] or extreme region (ER) detection [8]. Although these methods pay close attention to individual character detection, final segmentation (extractor of connected components) is done at a low level using only local features.

Sliding-window Methods [1,6] use a window which is moved over the image, and the presence of texts is estimated on the basis of local image features. While these methods are generally more robust to noise in the image than other methods, their computational complexity is high because of the need to search with many windows of different sizes, aspect ratios and, potentially, rotations. Additionally, support for slanted or perspective distorted text is limited, and sliding-window methods do not always provide text segmentation information accurate enough to be used for character recognition.

A hybrid method was presented by Pan et al. [10]. It consists of a text region detector and a connected components extractor based on an image local binarization in series. It makes use of size information obtained by the region detector in the image local binarization process and filter out non-text region. It connects the region detector with the connected components extractor directly to decrease false positives in the connected component extraction process.

3   Algorithm Description

As mentioned above, both CC methods and sliding-window methods have weak points. Neither alone detects texts universally well. Figures 1 and 2 show some results of our two text detection methods. These figures show that the CC method and the sliding-window method output complementary results.

Hybrid methods which integrate CC methods and sliding-window method are likely to take advantages of both methods. Pan et al. [10] connect them in serial. But we connect them in parallel with an aim to decrease false negatives in text detection. In this paper, we present a hybrid system consisting of two text detection methods (a CC method and a sliding-window method) and a method of integrating the results. The overall scheme of our algorithm is summarized in Fig. 3. It proceeds by these steps:

1. CC-based text detection (CC method).
2. Sliding window-based text detection (sliding-window method).
3. Integration of text lines.

3.1   CC-based Text Detection

Our CC-based text string detection method has four steps:

1. Image binarization.
2. Connected component extraction.
3. Connected component verification.
4. Text-line generation from connected components.

Image Binarization. Niblack's binarization algorithm [9] is adopted. The formula is defined as

Fig. 1. Some texts in natural scenes which can be detected by the CC method **(left)** but not the sliding-window method **(right)**. Rectangles indicate detected texts.

Fig. 2. Some texts in natural scenes which cannot be detected by the CC method **(left)** but can the sliding-window method **(right)**. Rectangles indicate detected texts.

$$Niblack(x) = \begin{cases} black, & \textbf{if} gray(x) < \mu(x) - k \cdot \sigma(x); \\ white, & \textbf{if} gray(x) > \mu(x) + k \cdot \sigma(x); \\ gray, & \textbf{other}, \end{cases} \qquad (1)$$

where $\mu(x)$ and $\sigma(x)$ are the mean and standard deviation of intensity within a constant-radius window centered on the pixel x; k is a smoothing term and

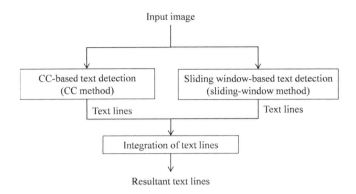

Fig. 3. Flowchart of our approach.

gray() is gray-scale transform from color image. For a binarized image, black or white values are extracted as text candidate while gray values are not considered further.

Connected Component Extraction. In binarized images, the method connects adjacent pixels which have same value (black or white) and similar stroke width in order to extract connected components. Stroke width can be calculated by the stroke width transform (SWT) [2].

Connected Component Verification. To eliminate non-character connected components, the method rejects connected components which satisfy $N/L < T_{sh}$. The quantity N/L is sharpness and T_{sh} is a threshold parameter. Sharpness is the ratio of N, the number of pixels in the connected component, to L, the number of pixels where gradient magnitudes exceed another threshold parameter T_{gm}. The gradient magnitudes are extracted at each pixel by $\sqrt{v^2 + h^2}$ where v and h are vertical and horizontal gradient respectively, which are calculated by an edge detector such as Sobel filter.

Text Line Generation from Connected Components. The text line generation has two steps. First, a neighborhood graph, in which nodes are connected components, is constructed, with two connected components sharing an edge if and only if they have similar color, brightness, position, size, and stroke width. Second, connected components are grouped according to whether they lie on a nearly straight line by searching each node and its attached edges in the neighborhood graph by a bottom-up approach. All pairs of connected components in the neighborhood graph are checked to see whether the pair can be grouped in order of increasing distance. To be grouped, a pair of connected components must satisfy any one of the following three conditions.

Case 1: Neither connected components belongs to a group.

Case 2: One of the connected components belongs to a group and the other connected component does not but lies on the line of the group which the first belong to.

Case 3: The connected components belong to different groups, and each lies on the line of the other's group.

Figure 4 is an example of text line generation from connected components.

3.2   Sliding Window-Based Text Detection

Our sliding windows-based text detection method has two steps:

1. Character candidate detection.
2. Generation of Text line from character candidates.

Character Candidate Detection. First, the method detects character candidates. We use the detector proposed by Kozakaya et al. [5], which they used to detect the faces of cats. We trained a character dictionary with character samples and used this dictionary for detecting characters.

The system cascades (i) joint Haar-like features with AdaBoost [7] and (ii) co-occurrence histograms of oriented gradients (CoHOG) descriptors with a linear support vector machine (SVM) classifiern [11]. The joint Haar-like feature improves discriminative performance by considering co-occurrence of multiple Haar-like features. It can be calculated very quickly due to the integral image technique. A strong classifier is learned by stagewise selection of the joint Haar-like features according to the AdaBoost algorithm. The CoHOG descriptor is based on gradient histogram. Gradient-based features are widely used in human detection and object detection. The CoHOG descriptor is calculated from histograms of paired gradient orientations, which we call a co-occurrence matrix. The co-occurrence matrices with the various orientation pairs are able to capture precise local shape information at multiple scales, and then the extracted CoHOG descriptors are evaluated with a linear classifier obtained by linear SVMs. The combination of these two distinct classifiers enables fast and accurate character detection as, Fig. 5 shows.

Generation of Text Line from Character Candidates. Second, the method generates text line from the character candidates. We use the extended Hough transform proposed by Kohno et al. [4] to detect text lines which consist of an array of similar size character candidates. The normal Hough transform detects line consisting of pixels (Fig. 6). The parameter space is two-dimensional, with parameters ρ and θ determining the line:

$$\rho = x \cos \theta + y \sin \theta. \tag{2}$$

Kohno et al. [4] extends the Hough transform to generate text lines, which consist of characters and have width (Fig. 7). In the extended Hough transform,

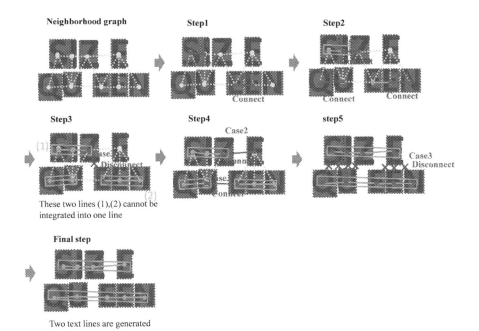

Fig. 4. An example of text line generation from connected components. (**Upper left**) Yellow broken lines represent edges of the neighborhood graph. (**From upper middle to lower right**) Text line generation proceeds in steps. Yellow broken lines represent potential groupings. Red solid lines represent current accepted groupings and red broken lines represent rejected groupings. Orange solid lines represent groupings which had been accepted and orange rectangles represent the resultant text lines.

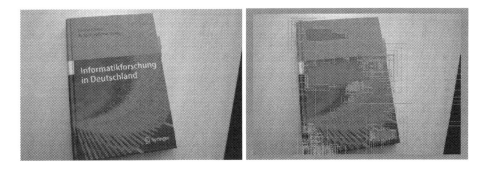

Fig. 5. An example of character candidate detection. (**Left**) Original image, (**Right**) Character detection results by (i) joint Haar-like features with AdaBoost and (ii) CoHOG descriptors with a linear SVM. Gray rectangles represent regions detected by (i) but rejected by (ii). Red rectangles represent regions detected by (i) and accepted by (ii).

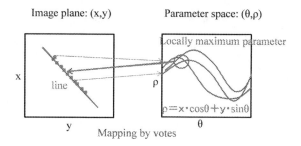

Fig. 6. Normal Hough transform. **(Green point)** pixels which are to be voted.

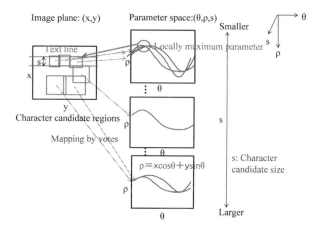

Fig. 7. The extended Hough transform. **(Green rectangles)** character candidate regions which are voted.

the parameter space is three dimensional consisting of the previous parameters (ρ and θ) and the new parameter s. s represents text-line width as obtained from the character candidate's height. In the extended Hough transform, character candidate regions (x, y, s) are voted for, instead of pixels. Here, (x, y) is the center coordinate of a character candidate. The method can detect text lines consisting of character candidates which have similar sizes and lie on a nearly straight line. It can detect not only text line position parameter (ρ, θ) but also line width s. Figure 8 shows an example of detection by the extended Hough transform. This example shows that it can generate neighboring text lines which have different region and different sizes even if character candidate regions overlap each other. We summarize differences between the two Hough transforms in Table 1.

3.3   Integration of Two Text String Detection

The proposed method integrates results from two types of text string candidates. We will briefly remind readers of the results' characteristics here. Figure 9 shows results from each type of text detection method. The CC method tends to output

Fig. 8. Text detected by the extended Hough transform. (**Left**) Character candidate regions, (**Right**) text lines generated by the extended Hough transform.

Table 1. Differences between the normal Hough transform and the extended Hough transform.

	The normal Hough	The extended Hough
Voting unit	Pixel (dot)	Region (character candidate)
Voting value	x, y	x, y, s
Parameters space	ρ, θ	ρ, θ, s

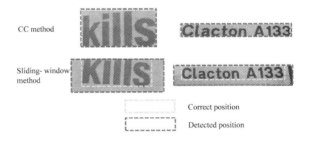

Fig. 9. Positions detected by CC method (**upper**) and sliding-window method (**Lower**). The CC method gives rectangular results. The sliding-window method gives distortion-corrected quadrilaterals. Yellow broken rectangles represent correct text positions. Red broken rectangles represent detected text positions.

candidates which are more accurately positioned than those detected by the sliding-window method, and so we give higher weight to results from the CC method.

The integration of results from the two methods is performed in the following three cases as shown Fig. 10.

Case 1: Isolation. All regions are accepted.

Case 2: Inclusion. Outer regions are accepted. Inner regions are rejected.

Case 3: Overlap. Regions generated by the CC method are accepted. Regions generated by the sliding-window method are rejected.

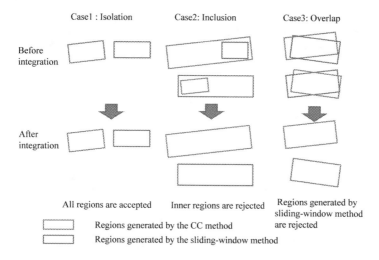

Fig. 10. Integration of two text string detection. Green rectangles represent regions generated by the CC method. Blue rectangles represent regions generated by the sliding-window method. Black rectangles represent integrated regions.

This has the obvious effect of prioritizing results from the CC method. Figure 11 shows an example initial candidates and the integrated result. This text string integration procedure can be summarized as follows.

$L := \emptyset$
for all $t_i^c \in M, t_j^c \in N$
    if $t_i^c \supset t_j^s$ **then** $L := L \cup \{t_i^c\}$
    elseif $t_i^c \subset t_j^s$ **then** $L := L \cup \{t_j^s\}$
    elseif $\text{region}(t_i^c \cap t_j^s) >$ Threshold **then** $L := L \cup \{t_i^c\}$
    else $L := L \cup \{t_i^c, t_j^s\}$
endfor
return L

In this procedure, M represents results from the CC method, N represents results from the sliding-window method and L is resultant text lines.

4   Experiments

We evaluate the performance of the proposed method on a public dataset:The ICDAR 2013 Robust Reading competition test dataset [3]. It contains 1095 words in 233 images. We use 61 images which we collected as training dataset for the character candidate detector. Under the ICDAR 2013 competition evaluation scheme [3], the method achieves the recall of 67.85 %, precision of 86.80 % and an F-score corresponding to 76.17 % in text localization This F-score is better than that of the winner of the ICDAR 2013 Robust Reading competition

(i)CC method   (iii)Integration   (ii)Sliding-window method

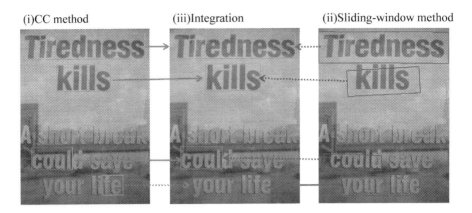

Fig. 11. An example of the integration method. **(Left)** Results of the CC method. **(Right)** Results of the sliding-window method. **(Middle)** Results of the integration. Rectangles are detected text regions, colored by method.

Table 2. The text detection results.

Method	Recall (%)	Precision (%)	F-score
CC alone	61.52	81.81	70.23
Sliding window alone	58.54	85.19	69.39
Integration (Proposed)	67.85	86.80	76.17
Winner of ICDAR2013	66.45	88.47	75.89

$(75.89\%)^1$. Table 2 show the performances of our three methods: (1) our CC method alone, (2) our sliding window method alone and (3) our total system (the proposed method) and the winner of ICDAR2013 Robust Reading competition. These results show that the integration process improves recall rate.

Figure 12 shows some examples of text detection by our CC method (left), our sliding-window method (center), and the proposed method (right).

5   Conclusion

We presented a hybrid text detection algorithm which is effective on natural scene images. We built two text detection methods: a CC method and a sliding-window method.

In the sliding-window method, we adopt the extend Hough transform to generate text lines based on global text line structure that a text line consists of characters which have similar sizes and lie on a nearly straight line.

---

[1] In the ICDAR 2013 Robust Reading competition on-line web-page [3], two methods (anon2014, SWT) reaches 77.03% and 76.80% (F-score). These two methods exceeds the winner of ICDAR2013.

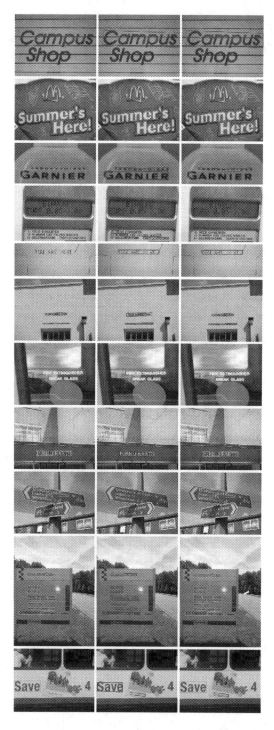

Fig. 12. Examples of text detected from the ICDAR2013 dataset. **(left)** CC method only, **(middle)** sliding-window method only and **(right)** integration method (Proposed)

By connecting it to the CC method based on local relations between connected components in parallel, the method produces fewer false negatives and improves recall rate. Extensive testing on the ICDAR2013 test dataset shows that the proposed scheme outperforms the latest published algorithms.

We plan to combine the text detection algorithm with a text recognition algorithm to build a complete text detection and recognition system. Text recognition would also help to reduce false positives and improve the precision rate.

References

1. Chen, X., Yuille, A.L.: Detecting and reading text in natural scenes. In: CVPR (2004)
2. Epshtein, B., Ofek, E., Wexler, Y.: Detecting text in natural scenes with stroke width transform. In: CVPR (2010)
3. ICDAR 2013 Robust Reading Competition (2013). http://dag.cvc.uab.es/icdar 2013competition
4. Kohno, Y., Aoki, Y., Hamamura, T., Irie, B.: A method for license plate recognition using relative location of extracted numbers (in Japanese). In: Vision Engineering Workshop I-37 (2007)
5. Kozakaya, T., Ito, S., Kubota, S., Yamaguchi, O.: Cat face detection with two heterogeneous features. In: IEEE ICIP (2009)
6. Lee, J.J., Lee, P.H., Lee, S.W., Yuille, A.L., Koch, C.: AdaBoost for text detection in natural scene. In: ICDAR (2011)
7. Mita, T., Kaneko, T., Hori, O.: Joint haar-like features for face detection. In: ICCV (2005)
8. Neumann, L., Matas, J.: Real-time scene text localization and recognition. In: CVPR (2012)
9. Niblack, W.: An Introduction to Digital Image Processing, pp. 115–116. Prentice-Hall, Englewood Cliffs (1986)
10. Pan, Y.F., Hou, X., Liu, C.L.: A hybrid approach to detect and localize texts in natural scene images. IEEE Trans. Image Process. **20**(3), 800–813 (2011)
11. Watanabe, T., Ito, S., Yokoi, K.: Co-occurrence histograms of oriented gradients for pedestrian detection. In: Wada, T., Huang, F., Lin, S. (eds.) PSIVT 2009. LNCS, vol. 5414, pp. 37–47. Springer, Heidelberg (2009)
12. Yao, C., Bai, X., Liu, W., Ma, Y., Tu, Z.: Detecting texts of arbitrary orientations in natural images. In: CVPR (2012)
13. Zhang, J., Kasturi, R.: Character energy and link energy-based text extraction in scene images. In: Kimmel, R., Klette, R., Sugimoto, A. (eds.) ACCV 2010, Part II. LNCS, vol. 6493, pp. 308–320. Springer, Heidelberg (2011)

Robust Text Segmentation in Low Quality Images via Adaptive Stroke Width Estimation and Stroke Based Superpixel Grouping

Anna Zhu, Guoyou Wang$^{(\boxtimes)}$, and Yangbo Dong

State Key Lab for Multispectral Information Processing Technology,
School of Automation, Huazhong University of Science and Technology,
Wuhan, China
gywang_2004@126.com

Abstract. Text segmentation is an important step in the process of character recognition. In literature, there have been numerous methods that work very well in practical applications. However, when an image includes strong noise or surface reflection distraction, accurate text segmentation still faces many challenges. Observing that the stroke width of text is stable and significantly different from that of reflective regions generally, we present a novel method for text segmentation using adaptive stroke width estimation and simple linear iterative clustering superpixel (SLIC-superpixel) region growing in this paper. It consists of four following steps: The first is to normalize image intensity to overcome the influence of gray changes. The second utilizes the intensity consistency to compute normalized stroke width (NSW) map. The third is to estimate the optimal stroke width through searching for the peak value of the histogram of normalized stroke width, the text polarity is also determined. Finally, we propose a local region growing method for text extraction using SLIC-superpixel. Unlike current existing methods of computing stroke width, such as gray level jump on a horizontal scan line and gradient-based SWT methods, the proposed method is based on the statistics of stroke width in the whole image. Hence the stroke width estimation is not only invariant in scale and rotation, but also more robust to surface reflection and noise than that of those methods based only on the pairs of sudden changes of intensity or gradient maps. Experiments with many real images, such as laser marking detonator codes, notice signatures and vehicle license plates, *etc.*, have shown that the proposed algorithm can work well in noised images and also achieve comparable performance with current state-of-the-art method on text segmentation from low quality images.

1 Introduction

The OCR systems update rapidly nowadays and most of current commercial OCR engines work well on high quality documents. However, the recognition precision is greatly influenced by integrity of the segmented characters. OCR systems perform poorly on noisy and poor quality documents due to the lack of

© Springer International Publishing Switzerland 2015
C.V. Jawahar and S. Shan (Eds.): ACCV 2014 Workshops, Part II, LNCS 9009, pp. 119–133, 2015.
DOI: 10.1007/978-3-319-16631-5_9

valid image segmentation under various variations, such as non-uniform illumination, surface reflection, perspective distortions, and background outliers, etc.
As a part of text analysis in practical applications, text segmentation in low
quality images, also named text binarization, still remains to be an important
research topic.

Up to now, there already has been a lot of research on text segmentation
reported in the literature and can be mainly classified into two categories: statistical thresholding methods and machine learning methods. In the first category, it has two subclasses: global and local. The global ones employ information
on the whole of image. The well known global method is presented by Otsu [1]
which is to acquire the optimal segmenting threshold of an entire image by the
between-class variance maximization. However, when applied to an image where
the distraction has similar grayscale to the text, it will produce bad results due
to the inseparability in gray. The local thresholding methods use features only
in a locally region and the locally segmented results constitute the final segment
result. Among locally adaptive methods, Niblack's method [2] has been proved
to give the best performance in those images of spatial grayscale changing slowly.
However, it might fail when the text is embedded in complex backgrounds or
text polarity is initially unknown. [3,4] presented some modified versions of the
Niblack's method to improve binarized results. Satish's edge based local method
[5] uses the gray value of edge pixels to decide the threshold in a window and
performs well in license plate recognition (LPR) application, but they still fail
to distinguishing text and distraction with similar intensity. The same defect
is shared with the current used Maximally Stable Extremal Regions (MSER)
method [6] which is often applied to text detection and achieves great success.

Recently, various machine learning methods have been popularly proposed
in segmentation [7–9]. The basic idea of machine learning is to extract accurate
invariant descriptions for specific target recognition by some kind of pattern
clustering or classification algorithm. However, these methods also face different
problems, like time consuming, insensitive to affine intensity changes, parameter
selection, etc.

As stroke width of texts is not only stable but also significantly different from
that of reflective regions, the stroke width can be used to detect text by grouping
neighboring pixels with similar stroke width into connected components as letter
candidates. The algorithm described in [10] scans an image horizontally, looking
for pairs of sudden changes of intensity. Then the regions between changes of
intensity are examined for color constancy and stroke width. Surviving regions
are grouped within a vertical window of size w and if enough regions are found,
a stroke can be presented. The limitations of this method include a number of
parameters, such as finding vertical window size w tuned to the scale of the text,
inability to deal with arbitrary directions of strokes. Moreover, the method [11]
also uses the idea of stroke width similarity for detecting text overlays in video
sequences. The limitations of the method include the need for integration over
scales and orientations of the filter, and the inherent attenuation to horizontal
texts again. Another definition of stroke relates to remote sensing images [12].

In this case, the algorithm defines a stroke as a contiguous part of an image that forms a band of a nearly constant width and describes a local image operator (called as stroke width transform-SWT) that computes the width of the most likely stroke pixels by using gradient maps. This method has been evaluated on natural scene text images with variant fonts and languages and achieves a desirable performance. However, it is sensitive to noise and changing intensity due to the usage of parallel edges.

Unlike the methods mentioned above based on finding pairs of sudden changes of intensity or pairs of edge pixels with roughly opposite directions, we propose a robust algorithm that extracts normalized stroke width (NSW) based on the image intensity similarity in a surrounding window for text extraction in low contrast images. Then a local region growing method based on superpixel level is applied on the NSW map to recall missed text regions using intensity similarity. This paper focuses only on how text can be separated accurately so as to increase recognition results after the text-line is located. This method is insensitive to noise and changing intensity due to its locally statistical intensity similarity. Experimental results on real images captured by camera, such as detonator laser marking code, notice signal and vehicle license plate, show that the proposed algorithm can perform more robust in poor quality images than the existing methods.

The rest of the paper is structured as follows. Firstly, we introduce the proposed method in Sect. 2. In Sect. 3, we analyze the parameters. Experiments and results are presented in Sect. 4 and conclusions are drawn in Sect. 5.

2   The Proposed Algorithm

Since human visual system needs only the brightness channel to recognize character shapes [13], we firstly convert the image from color to gray scale, then to further split the text from the gray image.

2.1   Image Intensity Normalization

In many text images, the text is often obscured due to local shadows or surface reflections. To remove this disturbance, we apply to each pixel of the original image $I(x, y)$ a contrast normalization procedure $I(x, y) \leftarrow 0.5 + (I(x, y)-m)/(3\sigma)$. m and σ are the local mean and standard deviation computed by a doubly binomial weight window of width $2H + 1$, where H is the height of the text image.

2.2   Pixel-Based Stroke Width Estimation

Considering that most of texts are composed of many lines with constant width, so the stroke width is a significant feature of the text. Inspired by [14] for line detection, we define the intensity consistency. As shown in Fig. 1, $P(x, y)$ locates in a surrounding window with size of $r \times r$ centered in pixel $C(x_o, y_o)$. Then the similarity between $P(x, y)$ and $C(x_o, y_o)$ can be described as

$$\text{con}(x, y, x_o, y_o) = \begin{cases} 1 & \text{if } |I(x, y) - I(x_o, y_o)| \le t \\ 0 & \text{if } |I(x, y) - I(x_o, y_o)| > t \end{cases} \tag{1}$$

where t is the threshold of intensity consistency comparison, $I(x, y)$ and $I(x_o, y_o)$ is intensity of the pixel C and P, respectively.

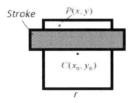

Fig. 1. The surrounding window.

Based on the intensity consistency between center pixel and other pixels in the window, the normalized stroke width (NSW) is calculated as:

$$w(x_o, y_o) = \frac{\sum_{(x,y) \in W_{r \times r}} \text{con}(x, y, x_o, y_o)}{r^2} \tag{2}$$

where r^2 is the number of pixels in the window, the stroke width of each pixel can be calculated by $w(x_o, y_o, t) \times r$. In order to detect the stroke structure of text, we will discuss the NSW as follows:

(1) If $C(x_o, y_o)$ locates in an uniform background or wider stroke with width exceeding r, $w(x_o, y_o) = 1$.
(2) When $C(x_o, y_o)$ moves onto the boundary of backgrounds adjacent to stroke, if stroke width is larger than half of r, $w(x_o, y_o) = 0.5$, else $w(x_o, y_o) > 0.5$.
(3) When $C(x_o, y_o)$ moves on the boundary of stroke adjacent to backgrounds, if stroke width is larger than half of r, $w(x_o, y_o) = 0.5$, else $w(x_o, y_o) < 0.5$.
(4) When $C(x_o, y_o)$ moves to other locations on the stroke, if stroke width is larger than half of r, $w(x_o, y_o) > 0.5$, else $w(x_o, y_o) < 0.5$.

From the above mentioned, we can see that when $C(x_o, y_o)$ belongs to backgrounds, $w(x_o, y_o) \ge 0.5$. On the contrary, when $C(x_o, y_o)$ locate on stroke, if the stroke width is less than half of r, $w(x_o, y_o) \le 0.5$, else $w(x_o, y_o) \ge 0.5$. Hence we set the maximum width $w_{max} = 0.5$, the stroke widths are computed by

$$S(x_o, y_o) = \begin{cases} w(x_o, y_o) & \text{if } w(x_o, y_o) < w_{max} \\ 0 & \text{otherwise} \end{cases} \tag{3}$$

where $S(x_o, y_o)$ is normalized stroke image, in which the width is less than 0.5. If we set w_{max} equal to different ranges, those width-specified lines can also be

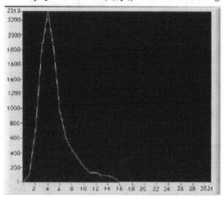

(a) Detonator laser marking code image with size 250×50

(b)NSW map: points with $w(x_0, y_0)$ <0.5 label as foreground

(c) Histogram of NSW (stroke width=5)

Fig. 2. NSW and histogram of stroke width in low quality images.

captured. The threshold t and window size r will be discussed in next section. The NSW map is presented in Fig. 2(b) where the non-zero pixels' values are binary to 1 for show.

In order to estimate the stroke width of text in an image that includes strong noise and reflection interference, we compute the histogram of the normalized stroke map, and then estimate the stroke width of text by searching for the peak value of the histogram. Figure 2 gives the histogram statistics of NSW by our proposed algorithm.

2.3   Text Polarity Determination

Among the foreground points, majority of points locate on stroke, only a small amount of them belong to backgrounds or stains, so we define the 0–1 NSW map as coarse binary image. To remove non-text parts and extract text stroke accurately, we firstly determine text polarity. Firstly, we compute the mean value of middle white pixels WP and outer contour pixels CP of NSW map in the gray scale image correspondingly. WP and CP are defined as follow:

$$WP = \{(x,y) \,|\, NSW(x,y) = 1, NSW(x \pm 1, y) = 1 \text{ and } NSW(x, y \pm 1) = 1\}$$

$$CP = (x,y) \,|\, NSW(x,y) = 0, NSW(x + (or-)1, y) = 0 \text{ and } NSW(x + (or-)2, y) = 1 \text{ or } NSW(x, y + (or-)1) = 0 \text{ and } NSW(x, y + (or-)2) = 1$$

Fig. 3. Background region removing.

After we obtain the number and position of $\boldsymbol{WP}$ and $\boldsymbol{CP}$, the mean value of the two pixel sets in the gray scale image are computed. The ratio is defined as:

$$R = \frac{m_{WP}}{m_{CP}} \tag{4}$$

The text polarity is bright when $\boldsymbol{R} > 1$ and dark when $\boldsymbol{R} < 1$. $\boldsymbol{R} = 1$ is an abnormal situation illustrating that the text region and background is inseparable. This specific situation has never happened in our practical experiment so we ignore it.

2.4   Text Segmentation Refinement

Since many text pixels in the intersection are missed and background regions may also be false extracted, the NSW map can only be regarded as coarse segmentation result. In this section, we address how to recall the missed text pixels and filter out the false extracted regions. Since we get the text polarity, a simple thresholding method can be applied to remove the background pixels. If a global threshold is chosen, it should be the normalized image mean m0. We set the white pixels, in which its value is less than m0, to 0. Shown in Fig. 3, in the renewed NSW map, the background region in the intersection of '4' is removed and the same to character '9'.

For recalling missed text parts, the searching work is performed superpixel by superpixel instead of pixel by pixel. We select simple linear iterative clustering (SLIC) superpixels [15] for text image segmentation. It's a local clustering of pixels in a constrained region and the distance measure enforces the compactness and regularity of its shape. We use gray level clustering instead of the Lab space online. Here we introduce the SLIC-superpixels in brief.

The simple linear iterative clustering (SLIC) performs in a given color space and x, y coordinates. This algorithm needs the input of desired number of superpixels firstly. Assume the number is $\boldsymbol{N}$, the image size is $\boldsymbol{w} \times \boldsymbol{h}$, the approximate size of each initial superpixel is $(\boldsymbol{w} \times \boldsymbol{h})/\boldsymbol{N}$ and the interval of neighbored superpixel center is $\boldsymbol{d} = \sqrt{(\boldsymbol{w} \times \boldsymbol{h})/\boldsymbol{N}}$. With the given number, size and interval, we can get the center of each superpixel in the image roughly. Then new center of each superpixel is computed as the average vector of all the pixels belonging to the cluster. The process is repeated iteratively until convergence. Instead of using simple Euclidean norm, we define a distance measure $\mathbf{D}$ as follows:

$$\mathbf{D}_i = \|I - I_i\|$$
$$\mathbf{D}_{xy} = \sqrt{(x - x_i)^2 + (y - y_i)^2}$$
$$\mathbf{D} = \mathbf{D}_i + \frac{m}{d}\mathbf{D}_{xy} \tag{5}$$

where m is a compactness constant. The distance is computed within a $2\mathbf{d} \times 2\mathbf{d}$ region around each superpixel center on the xy plane. This measure, weighting the two distances: $\mathbf{D}_i$ and $\mathbf{D}_{xy}$, offers a good balance between gray value similarity and spatial proximity. We set the superpixel number $\mathbf{N} = (\mathbf{w} \times \mathbf{h})/(\mathbf{w}_s - 1)^2$ and this parameter will be discussed in the next section. $\mathbf{w}_s$ is the obtained stroke width. The result of performing SLIC to the example detonator image is shown in Fig. 4. Initially, we search the corresponding region on the renewed NSW map for each superpixel. If the space ratio of white pixels to black pixels are greater than 0.8, this superpixel is identified on the text stroke and defined as text superpixel. For other text superpixels' retrieval, we use a local region growing method [16] while the algorithm is performed on superpixel level instead of pixel level. For each non-identified superpixel, if one of its neighbors is the text superpixel, we compute the following intensity difference of two superpixels:

$$\triangle I = d_m + \frac{1}{2}d_a + \frac{1}{4}d_n \tag{6}$$

Definition 1. (Average intensity difference)

$$d_a = \left| \frac{\sum_{g_i \in r_i} g_i}{n_i} - \frac{\sum_{g_j \in r_j} g_j}{n_j} \right|$$

where g_i, n_i denotes the intensity and number of pixels in superpixel r_i. This expression is used to compute the average intensity difference between two neighboring superpixels.

Definition 2. (Middle intensity difference)

$$d_m = |g_{mi} - g_{mj}|$$

where g_{mi} is the middle value of intensity in superpixel r_i.

Definition 3. (Neighbored intensity difference)

$$d_n = \left| \frac{\sum_{g_i \in r_{\text{neighbori}}} g_i}{n_{\text{neighbori}}} - \frac{\sum_{g_{\text{neighborj}} \in r_{\text{neighborj}}} g_j}{n_j} \right|$$

where $n_{\text{neighbori}}$, $n_{\text{neighborj}}$ are number of adjacent pixels in node r_i and r_j, $r_{\text{neighbori}}$, $r_{\text{neighborj}}$ are adjacent regions in two neighbored superpixels. d_n represents the average adjacent pixels intensity difference. If two neighboring superpixels both locate on the stroke, this value nearly equal to 0 for the continuity

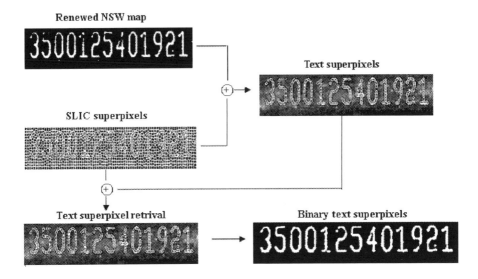

Fig. 4. Flowchart of text superpixel extraction.

of character stroke. However, if one locates on stroke and the other one on background, the value of d_n is just as large as deviation in edge detection.

If the intensity difference $\triangle I < T$, the non-identified superpixel is labeled as text pixel. If $\triangle I \geq T$, the superpixel is labeled as checked superpixel. T is the intensity difference threshold which will be discussed in the next section. Then a repeat work is done on the non-checked superpixels until no new text superpixel is found. The final extraction result is shown in Fig. 4.

3   Parameter Analysis

Four important parameters in our framework are discussed in this section. All of them are selected adaptively demonstrating the robustness of our proposed method.

3.1   Intensity Consistency Threshold t

Considering that human visual response relates to local background luminance, the threshold t of intensity consistency comparison can be calculated as follows:

$$t = k \times mean(I)$$

where k is constant coefficient and it is set to 0.8 in our experiment. $mean()$ is average operator. It's a global threshold which will result in many text pixels missing due to the uneven illumination. Even if we use block strategy to select t locally, the result will be not improved much for a local growing method is implemented sequentially which will recall most lost strokes.

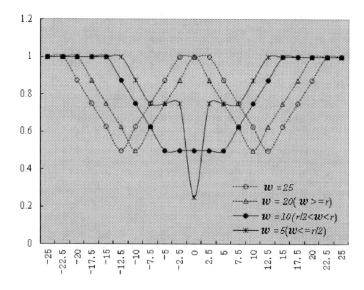

Fig. 5. Relationship of w and r.

3.2   Surrounding Window Size r

This section discusses the relationship of text stroke width w and instantiated surrounding window size r. Example is shown below with $r = 20$. The x axel is the position offset of text stroke and center of surrounding window. y axis is the normalized stroke width estimation.

From Fig. 5, we get that if $w < r/2$, the normalized stroke width will be less than 0.5 when the center of surrounding window is on the stroke. This feature can be used to extract text stroke. Therefore, we set $r > 2w$. w is estimated by simply applying Canny edge detection and finding the middle value of all parallel horizontal and vertical edges' width.

3.3   Initial Superpixels Number N

If N is smaller than a certain value, the segmentation may be blurred. Most of the superpixels located on the stroke contain too much background pixels. We call it coarse segmentation. The coarse segmentation cannot distinguish between background and character in superpixel level in the next step, for both of information may coexist in a single superpixel. If N is too much greater than the proper value, the segment result owns low under-segment error and high boundary recall [15]. However, it is not profitable in time taken for the next step whose complexity is O($NlogN$). We consider it as over segmentation. If each superpixel in the image contains only one kind of pixels-background pixels or character pixels, it is fitting segmentation. In this situation, the stroke decomposed to less superpixels and each superpixel owns unambiguous information. In Fig. 6(b), three different sizes of superpixels are designed for the synthesized stroke and

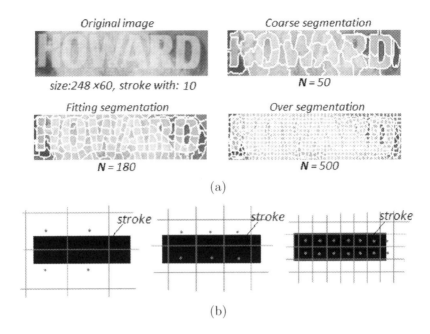

Fig. 6. Relationship between stroke width and superpixel numbers.

the red points represent the centers of seed superpixels around the stroke. We can find that the center may not fall in the stroke area if the interval d is greater than stroke width w_s in the left. In the middle is the situation $w_s/2 < d < w_s$, only one center locates on the stroke in the column array corresponding to horizontal stroke. If $w_s/2$ as shown in the right, more than one center of superpixels will fall in the stroke area. The three cases correspond to the three different segments in Fig. 6(a). The fitting segment is the proper one and satisfies the defined condition. We choose number $N = (\boldsymbol{w} \times \boldsymbol{h})/(w_s - 1)^2$.

3.4   Intensity Difference Threshold T

As we know, the response of the eye to changes in the intensity of illumination is known to be nonlinear. The visual phenomenon depicted in Fig. 7 is the profile of the Just Noticeable Difference (JND) [17] in psychophysics concept.

JND is the minimum amount by which stimulus intensity must be changed in order to produce a noticeable variation in a sensory experience. Over a wide range of intensities, the expression is $\frac{\triangle I}{I} = \xi$.

$\triangle I/I$ is called Weber fraction, where the original intensity is I, $\triangle I$ is the addition amount required for the difference to be perceived (the JND), and ξ is a constant called the Weber constant. When out of this range, the JND is markedly changed with variant intensity of I. By experimentation, we chose $\xi = 0.05$ for all applications in this paper. Then the intensity difference threshold $T = \xi \times I$. I represents superpixel owning lower intensity value between two comparable superpixels.

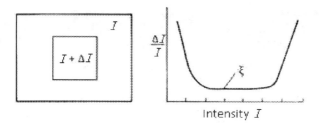

Fig. 7. Contrast sensitivity measurements.

4   Experimental Result

In this section we compare the performance of our collected dataset with that of other classical text extraction methods published in the literature in their ability of binarization.

4.1   Data Collection

Since our method process images in low contrast with uniform background and text, the benchmark datasets are unsuited for our evaluation. Therefore, we select a collected dataset which consists of four different kinds of images. Those images are screenshot detected from detonator laser marking code, license plates, part of ICDAR 2011 scene images and synthetic images as shown in Fig. 8.

The collected dataset has 4 types and totally 498 images including 212 detonator laser marking code images (T 1), 164 license plates images (T 2), 84 natural scene images (T 3) and 28 synthetic images (T 4). Those images suffer from different degradations, like uneven lighting, curved or/and shiny surfaces, low contrast to the backgrounds, different text fonts and sizes, different resolutions etc. As many as 4732 characters are contained in the test image. All candidate images are processed as black characters on white background for shown. Figure 9 presents some examples of the binary results.

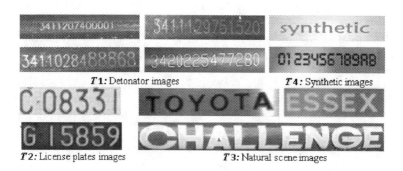

Fig. 8. Four types images in our datasets.

Fig. 9. Some examples of the segmented text.

4.2   Result Comparison

We use two simple evaluation methods to verify the binarization performance [18] and OCR performance: the character extraction rate (CER) and the character recognition rate (CRR). They are defined as:

$$CER = N_{segment}/N \qquad CRR = N_{recognize}/N$$

where $N_{segment}$ is the number of characters completely extracted without lost strokes or connecting to background, $N_{recognize}$ is the number of characters correctly recognized by commercial OCR software ABBYY Fine Reader 9.0, and N is the total number of the characters.

Besides, we define some rules to judge whether the extracted character can be counted as correctly recognized characters for CER: 1. An character which connects with large size background connected components is not counted. 2. An character with part missing should not be counted (missing size larger than 20 % of the character).

To highlight the effectiveness of our proposition, we compare our proposed method with seven classical and competing binarization techniques:

– Otsu's global thresholding method derived from discriminated analysis [1].
– A local version of Otsu's method, where a sliding window is shifted across the image and the threshold is calculated from the histogram of the gray values of the window.
– Satish's edge based local method [5] which detects the edges in image and then passes a threshold surface through the edge pixels in a window locally.
– Niblack's method [2] which calculates a threshold surface $T = m + k \times s$ using the mean m and the variance s of the gray values by shifting a rectangular window across the image with parameter $k = -0.2$.
– X. Chen's local thresholding method [4]. It's a variant algorithm of Niblack's that selects window size adaptively.
– MSER based method [6]. It is widely used in text detection using the gray level stability of text region and background regions.
– MRF method [9]. It belongs to high level image processing method and the test includes training part.

Fig. 10. Binarization results of different methods. (a) Global Otsu. (b) A local windowed version of Otsu. (c) Edge based local detection method. (d) Niblack's method. (e) X. Chen's local thresholding method. (f) MSER based method. (g) MRF based method. (h) Our proposed method.

Figure 10 shows three images taken from the collected dataset and results of the five different binarization techniques and our proposed method. As can be seen in Fig. 10, Otsu's method suffers from the known disadvantages of global thresholding methods, *i.e.*. imprecise separation of text and background in case of blur or high variations of the gray values. The windowed version improves the segment precision, but creates additional non-text structures. As a result of the fact that the shift window may calculate the threshold even in areas where only background exists. The disadvantage of edge based method is that it's very sensitive to noise and depends much on the quality of edge detection. Niblack's method has the same problem with local OTSU method, which segments the characters very well, but also suffers from noise in the zones without text. X. Chen's method, which is considered as the most successful binarization method recently, improves Niblack's method and performs better meaning while increases the computational time severely. MSER based method performs well for natural scene images, license plates images and synthetic images except detonator images. This arises from the uneven gray scale of text and low contrast in the detonator images. The MRF method is time consuming which takes 47 s on average to produce the final binary results but our method only takes 12 s on average with the same system. Besides, our proposed method combines text

Table 1. Performance for different binarization methods.

Bin. method	CER					CRR				
	(T 1)	(T 2)	(T 3)	(T 4)	Total	(T 1)	(T 2)	(T 3)	(T 4)	Total
Global OTSU	72.2%	92.6%	84.4%	76.5%	79.4%	69.5%	90.4%	63.5%	75.2%	74.9%
Local OTSU	81.6%	94.1%	79.5%	74.4%	83.8%	74.7%	90.2%	60.6%	70.2%	75.3%
Edge based	92.3%	93.5%	91.6%	92.9%	92.7%	90.3%	91.1%	86.4%	90.5%	90.2%
Niblack's method	85.2%	97.4%	85.0%	93.1%	89.0%	80.8%	96.3%	72.0%	90.5%	86.8%
X. Chen's method	94.8%	98.6%	92.7%	97.6%	96.4%	93.5%	96.9%	84.4%	96.4%	93.6%
MSER method	90.6%	92.0%	94.0%	98.3%	93.8%	86.5%	95.8%	87.2%	97.0%	91.4%
MRF method	97.4%	96.0%	94.0%	98.1%	96.8%	95.8%	94.5%	83.9%	97.8%	95.2%
Our method	97.3%	96.0%	92.2%	96.1%	96.2%	96.0%	95.2%	84.2%	96.0%	95.7%

width and intensity similarity in superpixel level and can not only extracts text with the same stroke width but also can automatically filter out background spots that is smaller than text width. An evaluation of these binarization techniques using OCR are given in the Table 1.

As we can see, our method is comparable with the best binary algorithm that depicts in X. Chen's method. Although its binary performance is better from a human view, the OCR system does not give a better recognition result than our method for recognize text from low quality images.

5   Conclusion

In this paper, we present a new algorithm to segment text from low quality images. We consider that the text consists of wide lines and employ not only the intensity information but also the structure of the text stroke. Besides, a new text stroke width computation method is presented by using intensity similarity in a surrounding window. Utilizing the stroke width information, we propose a superpixel by superpixel growing method based on the text stoke width to recall missing text stokes. The presented algorithm is robust due to that all of the parameters are selected adaptively. In this paper, we collect our datasets including four types of low quality images suffering from different degradations. To demonstrate the effectiveness of our approach, a commercial OCR system is used to evaluate the precision performance and we achieve comparable result with those best binary algorithms.

The limitation of our method is that the it cannot be used in text images with various stroke width. And for complex background text segmentation, it proves to be no superior. Therefore, we will search for more general solution for text segmentation with more complex background in future work.

References

1. Otsu, N.: A threshold selection method from gray-level histograms. Automatica **11**, 23–27 (1975)
2. Trier, O.D., Jain, A.K.: Goal-directed evaluation of binarization methods. IEEE Trans. Pattern Anal. Mach. Intell. **17**, 1191–1201 (1995)

3. Wolf, C., Jolion, J.M.: Extraction and recognition of artificial text in multimedia documents. Formal Pattern Anal. Appl. **6**, 309–326 (2004)
4. Chen, X., Yuille, A.L.: Detecting and reading text in natural scenes. In: Proceedings of the 2004 IEEE Computer Society Conference on Computer Vision and Pattern Recognition, CVPR 2004, vol. 2, pp. II–366. IEEE (2004)
5. Satish, M., Lajish, V., Kopparapu, S.K.: Edge assisted fast binarization scheme for improved vehicle license plate recognition. In: 2011 National Conference on Communications (NCC), pp. 1–5. IEEE (2011)
6. Yin, X., Huang, K., Hao, H.: Robust text detection in natural scene images (2013)
7. Mancas-Thillou, C., Gosselin, B.: Color text extraction with selective metric-based clustering. Comput. Vis. Image Underst. **107**, 97–107 (2007)
8. Li, J., Tian, Y., Huang, T., Gao, W.: Multi-polarity text segmentation using graph theory. In: 15th IEEE International Conference on Image Processing, ICIP 2008, pp. 3008–3011. IEEE (2008)
9. Mishra, A., Alahari, K., Jawahar, C.: An MRF model for binarization of natural scene text. In: 2011 International Conference on Document Analysis and Recognition (ICDAR), pp. 11–16. IEEE (2011)
10. Subramanian, K., Natarajan, P., Decerbo, M., Castañòn, D.: Character-stroke detection for text-localization and extraction. In: Ninth International Conference on Document Analysis and Recognition, ICDAR 2007, vol. 1, pp. 33–37. IEEE (2007)
11. Jung, C., Liu, Q., Kim, J.: A stroke filter and its application to text localization. Pattern Recogn. Lett. **30**, 114–122 (2009)
12. Epshtein, B., Ofek, E., Wexler, Y.: Detecting text in natural scenes with stroke width transform. In: 2010 IEEE Conference on Computer Vision and Pattern Recognition (CVPR), pp. 2963–2970. IEEE (2010)
13. Fairchild, M.D.: Color Appearance Models. John Wiley & Sons, New York (2013)
14. Liu, L., Zhang, D., You, J.: Detecting wide lines using isotropic nonlinear filtering. IEEE Trans. Image Process. **16**, 1584–1595 (2007)
15. Achanta, R., Shaji, A., Smith, K., Lucchi, A., Fua, P., Susstrunk, S.: Slic superpixels compared to state-of-the-art superpixel methods. IEEE Trans. Pattern Anal. Mach. Intell. **34**, 2274–2282 (2012)
16. Liu, Q., Jung, C., Moon, Y.: Text segmentation based on stroke filter. In: Proceedings of the 14th Annual ACM international Conference on Multimedia, pp. 129–132. ACM (2006)
17. Shen, J.: On the foundations of vision modeling: I. Weber's law and weberized tv restoration. Physica D: Nonlinear Phenom. **175**, 241–251 (2003)
18. Shi, C., Xiao, B., Wang, C., Zhang, Y.: Adaptive graph cut based binarization of video text images. In: 2012 10th IAPR International Workshop on Document Analysis Systems (DAS), pp. 58–62. IEEE (2012)

Efficient Character Skew Rectification in Scene Text Images

Michal Bušta[(✉)], Tomáš Drtina, David Helekal, Lukáš Neumann,
and Jiří Matas

Department of Cybernetics, Centre for Machine Perception,
Czech Technical University, Prague, Czech Republic
bustam@fel.cvut.cz

Abstract. We present an efficient method for character skew rectification in scene text images. The method is based on a novel skew estimators, which exploit intuitive glyph properties and which can be efficiently computed in a linear time. The estimators are evaluated on a synthetically generated data (including Latin, Cyrillic, Greek, Runic scripts) and real scene text images, where the skew rectification by the proposed method improves the accuracy of a state-of-the-art scene text recognition pipeline.

1 Introduction

Scene text detection and recognition is an open problem with many interesting applications. When compared to traditional printed document OCR (which is considered to be a solved problem), scene text recognition faces additional challenges such as complex backgrounds, variations of font face, size, texture and image distortions. The distortions may include perspective distortion, blur distortion, motion blur or a combination of any of the above.

In this paper, we focus on the problem of text skew estimation and subsequential rectification, in order to improve text recognition accuracy. We refer to *skew* as the angle of individual letters to the text direction, whereas we refer to *rotation* as orientation of a word baseline relative to the horizontal direction (see Fig. 1).

We propose five novel character skew estimators, which exploit intuitive glyph properties and which can be efficiently computed in a linear time. The estimators are first evaluated on a synthetically generated data and then an existing TextSpotter pipeline [1,2] is adapted to evaluate the impact of character skew rectification to the scene text recognition accuracy.

The rest of the paper is structured as follows. In Sect. 2 previous work is presented, in Sect. 3 the problem of scene text geometry is introduced, in Sect. 4 the proposed method is described and in Sect. 5 the experimental validation is given. The paper is concluded in Sect. 6.

© Springer International Publishing Switzerland 2015
C.V. Jawahar and S. Shan (Eds.): ACCV 2014 Workshops, Part II, LNCS 9009, pp. 134–146, 2015.
DOI: 10.1007/978-3-319-16631-5_10

Fig. 1. Character skew and rotation.

2   Related Work

The most common way to estimate text skew is using the projection profiles [3,4]. In these methods, text skew is iterated in small increments and for each value a vertical projection profile is calculated (see Fig. 2). The projection profile looks like a series of peaks and valleys, which correspond to spaces between characters; for the correct skew the peaks of the vertical projection profile are going to be the highest. This can be computationally expressed by a number of measures, such the entropy of the histogram, standard deviation variation or sum of squares. The major drawback of these methods is their computational complexity, since the interval of possible skews has to be iterated in small increments (this may even prohibit the use of such methods on devices with lower performance, such as mobile phones).

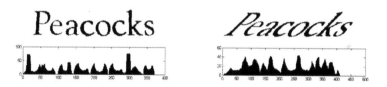

Fig. 2. In previous work, text skew is estimated by iterating all skew values and selecting an optimal value using vertical projection profiles.

Another group of methods focuses on skew estimation for whole documents. The methods are based on the Hough transform [5,6], distance transform [7], gradient Directions [8] and Focused Nearest Neighbor Clustering (FNNC) of interest points [8]. The main limitation of these methods is the fact that they require a whole document with a lot of text in a binarized form to estimate the skew, which is often unusable for scene text where binarization (segmentation) may not be as accurate and which usually consists of short text snippets, sometimes even with different skew or orientation.

3   Scene Text Geometry

Assuming that scene text typically appears on planar surfaces (see Fig. 3), the text transformation in an image can be modelled in the standard pinhole camera model as a homography.

Fig. 3. Scene text typically appears on planar surfaces.

Let us recall that a homography **H** can be decomposed as

$$\mathbf{H} = \mathbf{EASP} = \begin{pmatrix} s\cos(\theta) & s\sin(\theta) & t_x \\ -s\sin(\theta) & s\cos(\theta) & t_y \\ 0 & 0 & 1 \end{pmatrix} \begin{pmatrix} 1/b & 0 & 0 \\ 0 & 1 & 0 \\ 0 & 0 & 1 \end{pmatrix} \begin{pmatrix} 1 & -\tan(\alpha) & 0 \\ 0 & 1 & 0 \\ 0 & 0 & 1 \end{pmatrix} \begin{pmatrix} 1 & 0 & 0 \\ 0 & 1 & 0 \\ l_x & l_y & 1 \end{pmatrix} \quad (1)$$

where **E** is a 2D rigid transformation, **A** is an anisotropic scaling transformation, **S** is a skew transformation and **P** is a projective transformation.

A homography has 8 degrees of freedom: rotation θ, isotropic scaling factor s, skew α, anisotropic scaling factor b, translation factors t_x, t_y, and perspective shortening l_x and l_y.

For character recognition, the scaling and translation factors s, b, t_x and t_y are not important since in the OCR stage character regions are typically normalized to a fixed size (this applies to both training and recognition time).

Exploiting the fact that in a typical character detection pipeline text lines are formed before the character recognition stage (see Fig. 4), the direction of the text (i.e. the text bottom line direction) can be used as an estimate for the rotation θ.

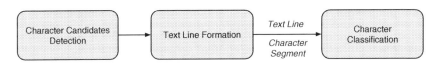

Fig. 4. In typical scene text recognition pipelines the text direction is known before the character recognition stage.

If we assume that the effect of perspective shortening is not significant, the only remaining parameter required for text rectification is the skew α. The skew α of a text in an image depends on the position of the plane in the scene, but also on the topographical parameters of the font (e.g. a text can be written in italics). The skew α may therefore be different for different words on the same plane and it is beneficial to estimate skew for each word individually.

In this paper, we focus on estimating the skew α from word segmentations (which may contain only few characters). Applying a reverse transformation to rectify the text skew (and rotation) improves OCR accuracy and helps to significantly reduce the size of the training set as fonts with different slanting (e.g. italics) don't have to be included.

Fig. 5. Rotation rectification is not sufficient for scene text images as text is typically perspectively distorted. Scene text image (left). Rotation θ rectified (top right). Rotation θ and skew α rectified (bottom right).

Let us also note that omitting skew estimation can lead to problems, since text in scene images is often perspectively distorted and rectifying only rotation leaves the character segmentations slanted (see Fig. 5).

4   Skew Estimation

In this paper, we propose five novel skew estimators for character segmentation or its polygonal approximation (the Ramer-Douglas-Peucker algorithm [9] was used to calculate the boundary approximation).

The estimators are based on intuitive character glyphs features (e.g. character symmetricity, dominant stroke, etc.), are easy to implement and have a low computational cost (most of them are linear in number of character glyph points).

4.1   Vertical Dominant (VD)

Assumption 1. *The dominant tangent direction of an undistorted character is vertical.*

In other words, the dominant tangent direction corresponds to the skew α (see Fig. 6). To find the dominant tangent direction, each edge of the polygonal approximation contributes in its direction with a weight proportional to its length to a Parzen-window density estimate.

$$P(\alpha) = \frac{1}{N} \sum_{i=0}^{N} \frac{l_i}{\delta\sqrt{2\pi}} \exp\left(-\frac{1}{2}\left(\frac{\alpha - \alpha_i}{\delta}\right)^2\right) \qquad (2)$$

where N is the number of edges in approximated contour, l_i is the edge length, α_i is the edge angle and δ ($\delta = 3°$) is the standard deviation of the Gaussian PDF (smoothing parameter).

The skew estimate $\hat{\alpha}$ then corresponds to the maximum of the density estimate

$$\hat{\alpha} = \arg\max(P(\alpha)) \qquad (3)$$

where the maximal value of $P(\alpha)$ gives a confidence estimate.

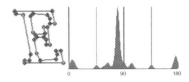

Fig. 6. Vertical Dominant (VD). Character polygonal approximation, edges with the dominant tangent direction highlighted (left). Weighted Parzen-window density estimate $P(\alpha)$ (right).

4.2 Vertical Dominant on Convex Hull (VC)

It is expected (and verified on the training dataset), that the Vertical Dominant (VD) estimator will perform poorly on characters such as S, 7, where the dominant tangent angle direction is not vertical. However, if the dominant tangent direction is found on a convex hull instead of the segmentation polygonal approximation, a more accurate estimate can be obtained for some characters (see Fig. 7).

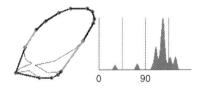

Fig. 7. Vertical Dominant on Convex Hull (VC). Character convex hull, edges with the dominant tangent direction highlighted (left). Weighted Parzen-window density estimate $P(\alpha)$ (right).

4.3 Longest Edge (LE)

Assumption 2. *The longest edge of a character is usually vertical.*

The skew estimate $\hat{\alpha}$ is the direction of the longest edge in the polygonal approximation (see Fig. 8).

Fig. 8. Longest Edge (LE). Character polygonal approximation and the longest edge highlighted.

4.4 Thinnest Profile (TP)

Assumption 3. *The angle of the thinnest profile of a character corresponds to the skew angle.*

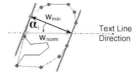

Fig. 9. Thinnest Profile (TP). Character polygonal approximation, its convex hull and the pair of parallel lines with the shortest distance. The thinnest profile w_{min} is normalized using the angle α_i between the thinnest profile and the text line direction

The thinnest profile is found by the Rotating Calipers algorithm [10] on the convex hull of the character polygonal approximation. The width of the profile is normalized using the angle of the measured profile $w_{norm} = w_{min}/\cos\alpha_i$ where α_i is the angle of current profile (see Fig. 9). The Rotating Calipers algorithm exploits an analogy of a rotating spring-loaded vernier caliper - every time one blade of the caliper lies flat against an edge of the polygon, the algorithm forms an antipodal pair with the point or edge touching the opposite blade. The complete "rotation" of the caliper around a convex polygon finds all antipodal pairs and is carried out in $O(N)$ time (where N is the number of convex hull contour points). In our implementation, Sklansky's algorithm [11] is used to form the convex hull from the polygonal approximation, whose complexity is linear in the number of contour points.

4.5 Symmetric Glyph (SG)

Assumption 4. *The top and the bottom parts of a character are vertically symmetric.*

First, the topmost and the bottom-most points are found as the pixels with longest distance from the center of bounding box of letter glyph in the direction perpendicular to the text line direction. Next, the leftmost and the rightmost

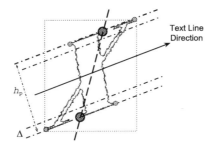

Fig. 10. Symmetric Glyph (SG). The most left and the most right points (green), the top and bottom centers (red) (Color figure online).

pixels are found in the band of the height Δ, which is placed around the top-most (respectively the bottom-most point) and which is parallel to the text line direction. Finally, the top (respectively the bottom) center is found as the center of the line passing trough the most left and the most right points on the top (the bottom) of the character - see Fig. 10. Estimated skew $\hat{\alpha}$ is then angle of the line connecting the top and bottom center points.

In our experiments, we set $\Delta = 0.08h$, where h is the height of the character.

We have observed that if the process of skew estimation using the Symmetric Glyph property is repeated (i.e. the region is rectified using the estimated skew $\hat{\alpha}$ and a new estimate is created for the rectified region), the estimation accuracy is improved. Therefore, in our experiments, the process of estimating and rectification is repeated until convergence.

5   Experiments

A testing set of 10 randomly selected fonts was first created to tune the parameters of the skew estimators. The optimal values of each paramater have been selected by a simple grid-search.

5.1   Synthetic Characters

In the first experiment, characters from 5 different scripts (Latin, Cyrillic, Georgian, Greek and Runic) and 50 random fonts for each script (the selected fonts were not present in the training set) were distorted with a random skew α_{gt}. We then compared the difference of the skew estimate $\hat{\alpha}$ of each skew estimator with the ground truth skew α_{gt}. If the estimated skew differs less than $3°$ we consider the estimate as correct (see Table 1). The standard deviation of the angle difference is measured as well. At last but not least, skew estimation error was broken down by ground truth skew value to see how well small or large distortions are estimated by each estimator (see Fig. 11).

It can be observed that the most informative is the Thinnest Profile (TP) estimate, which has the highest accuracy and amongst all scripts. On contrary,

Table 1. Skew estimation accuracy a and standard deviation σ on synthetic characters.

	Cyrillic		Georgian		Greek		Latin		Runic		All	
	$a[\%]$	$\sigma[°]$	$a[\%]$	$\sigma[°]$	$a[\%]$	$\sigma[°]$	$a[\%]$	$\sigma[°]$	$a[\%]$	$\sigma[°]$	$a[\%]$	$\sigma[°]$
VD	73.5	10.8	69.1	12.3	67.8	12.7	64.6	14.0	79.7	14.3	68.7	12.7
VC	70.2	9.5	54.5	12.8	61.7	10.8	64.9	8.6	59.2	13.4	66.1	9.5
LE	67.2	12.9	62.0	16.6	58.5	14.9	57.8	15.6	78.7	14.5	61.7	14.5
TP	74.1	5.9	67.2	6.3	68.3	6.5	68.8	6.0	68.3	8.5	70.7	6.1
SG	51.0	11.1	40.1	10.4	61.2	8.0	58.6	8.1	62.7	8.5	56.1	9.4

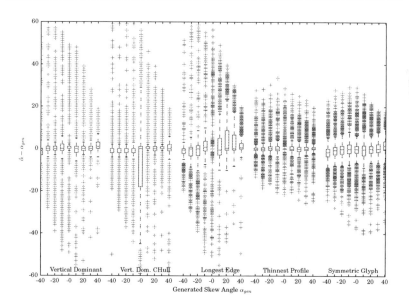

Fig. 11. Skew estimation error as a function of the ground truth skew.

the Symmetric Glyph (SG) estimate performs the worst on the Cyrillic and Georgian scripts, but is the third best on the Greek script (which suggests that Greek glyphs are more symmetric) and it has the second most consistent estimates for both small and large skews.

In the second experiment, we evaluate how much the estimators are complementary to each other. In Fig. 12 we show the density plot of skew estimation error for each pair. From the plot it can be observed that the Vertical Dominant on Convex Hull (VC), the Symmetric Glyph (SG) and the Thinnest Profile (TP) estimators complement each other.

5.2   Synthetic Text

In this experiment, synthetic images where generated for a random subset of dictionary words using a font randomly selected from the Google Fonts directory

Table 2. Skew estimation accuracy a and standard deviation σ of the combined estimator on synthetic text, dependent on the text length

	2 characters		3 characters		4 characters	
	$a[\%]$	$\sigma[^\circ]$	$a[\%]$	$\sigma[^\circ]$	$a[\%]$	$\sigma[^\circ]$
VD+VC+SG+TP	75.9	6.4	90.5	3.6	95.4	1.9

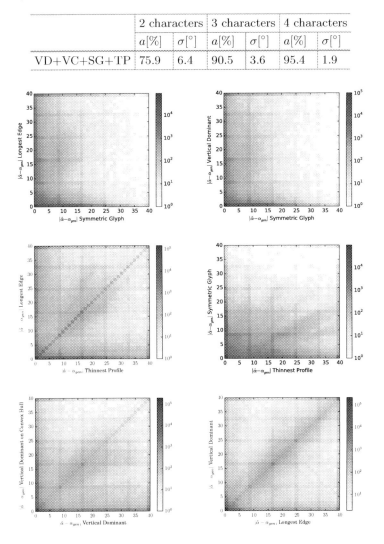

Fig. 12. Skew estimation error $|\hat{\alpha} - \alpha_{\text{gt}}|$ for each estimator pair.

(which contains approximately 1200 fonts). For each randomly generated word, a random skew in the interval $\langle -40^\circ, 40^\circ \rangle$ was applied. The skew estimate is again considered as correct if it differs less than 3° from the ground truth.

In order to benefit from the complementarity of estimators (see Sect. 5.1), the 4 most complementary estimators (Vertical Dominant, Vertical Dominant on Convex Hull, Thinnest Profile and Symmetric Glyph) were combined into a single estimator. Each estimator votes for its estimated skew $\hat{\alpha}$ with a trained probability, and the resulting skew is obtained as the highest peak in a histogram calculated over all word characters.

It can be observed that the combined estimator predicts skew correctly for 75.9 % of words which consist of 2 characters only; the accuracy then grows to 95.4 % for 4-character words (see Table 2 for estimation accuracy, Fig. 13 for examples of correctly estimated text skew).

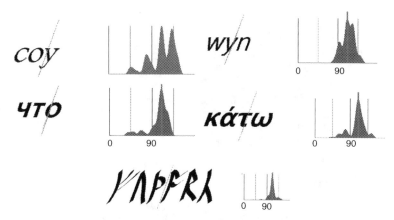

Fig. 13. Samples of a correct skew estimation in the synthetic text dataset. The ground truth skew α_{gt} marked green, the estimated skew $\hat{\alpha}$ marked red (Color figure online).

The estimation typically fails for short words which contain many characters that consist of slanted strokes in an undistorted view (such as v, w or y), which create false peaks in the skew histogram (see Fig. 14). However, we can see from the histogram that the confidence of such estimates is relatively low and therefore for practical applications for such low confidence deskewing of the word can be skipped.

Fig. 14. Samples of an incorrect skew estimation in the synthetic text dataset. The ground truth skew α_{gt} marked green, the estimated skew $\hat{\alpha}$ marked red (Color figure online).

5.3 ICDAR 2013 Dataset

In the last experiment, the proposed method was incorporated into the TextSpotter pipeline [1,2]. In order to evaluate the impact of skew rectification in real-world images, we compared the number of correctly recognized words in the

Fig. 15. Samples from the ICDAR 2013 Dataset where the text recognition accuracy was improved. Source image (left), original word segmentation (top right) and rectified word segmentation (bottom right).

ICDAR 2013 dataset [12] (a word is correctly recognized if all its characters are recognized correctly, using case-sensitive comparison).

It can be observed (see Table 3) that the skew rectification using the proposed method improved the recognition accuracy by 21 words, which is relatively significant, given the fact that most of the text in the dataset is horizontal (examples of the words with improved text recognition are shown in the Fig. 15). Note that the localization recall of TextSpoter is only 65.38 %, therefore the recognition recall of 49.95 % implies that 75 % of the words that were detected were actually recognized correctly.

Table 3. Text recognition accuracy of the TextSpotter pipeline on the ICDAR 2013 dataset with and without skew rectifcation

	correctly recognized words
Rotation rectification	526 (48.03 %)
Rotation and skew rectification	547 (49.95 %)
Localization recall	*65.38 %*

6   Conclusions

We have presented five novel character skew estimators, which exploit intuitive glyph properties and which can be efficiently computed in linear time. On synthetically generated data the estimators were able to accurately estimate skew for 5 different scripts in different fonts even for short snippets of text. On real-world images of the ICDAR 2013 dataset the proposed method improved the text recognition accuracy of the state-of-the-art TextSpotter pipeline [1,2].

The future work includes learning a custom voting function for the combined estimator and handling of scene text on non-planar surfaces.

Acknowledgment. The authors were supported by the EU project MASELTOV (FP7-ICT-288587) and by the Czech Science Foundation Project GACR P103/12/G084. Lukáš would also like to acknowledge the support of the Google PhD Fellowship in Computer Vision and the Google Research Award.

References

1. Neumann, L., Matas, J.: On combining multiple segmentations in scene text recognition. In: 2013 12th International Conference on Document Analysis and Recognition (ICDAR), pp. 523–527. IEEE, California (2013)
2. Neumann, L., Matas, J.: Text localization in real-world images using efficiently pruned exhaustive search. In: 2011 International Conference on Document Analysis and Recognition (ICDAR), pp. 687–691. IEEE Computer Society Conference Publishing Services (2011). (IEEE Computer Society Offices 2001 L Street N.W. Suite 700 Washington, DC 20036–4928, United States)
3. Baird, H.S.: Document Image Analysis. IEEE Computer Society Press, Los Alamitos (1995)
4. Papandreou, A., Gatos, B.: A novel skew detection technique based on vertical projections. In: 2011 International Conference on Document Analysis and Recognition (ICDAR), pp. 384–388 (2011)
5. Epshtein, B.: Determining document skew using inter-line spaces. In: 2011 International Conference on Document Analysis and Recognition (ICDAR), pp. 27–31 (2011)
6. Srihari, S., Govindaraju, V.: Analysis of textual images using the hough transform. Mach. Vis. Appl. **2**, 141–153 (1989)

7. Bar-Yosef, I., Hagbi, N., Kedem, K., Dinstein, I.: Fast and accurate skew estimation based on distance transform. In: The Eighth IAPR International Workshop on Document Analysis Systems, DAS 2008, pp. 402–407 (2008)
8. Sun, C., Si, D.: Skew and slant correction for document images using gradient direction. In: Proceedings of the Fourth International Conference on Document Analysis and Recognition, vol. 1, 142–146 (1997)
9. Ramer, U.: An iterative procedure for the polygonal approximation of plane curves. Comput. Graph. Image Process. 1, 244–256 (1972)
10. Toussaint, G.: Solving geometric problems with the rotating calipers (1983)
11. Sklansky, J.: Finding the convex hull of a simple polygon. Pattern Recogn. Lett. 1, 79–83 (1982)
12. Shahab, A., Shafait, F., Dengel, A.: Icdar 2011 robust reading competition challenge 2: Reading text in scene images. In: 2011 International Conference on Document Analysis and Recognition (ICDAR), pp. 1491–1496 (2011)

Performance Improvement of Dot-Matrix Character Recognition by Variation Model Based Learning

Koji Endo, Wataru Ohyama$^{(\boxtimes)}$, Tetsushi Wakabayashi, and Fumitaka Kimura

Graduate School of Engineering, Mie University, 1577 Kurimamachiya-cho,
Tsu-shi, Mie 5148507, Japan
{endo,ohyama}@hi.info.mie-u.ac.jp

Abstract. This paper describes an effective learning technique for optical dot-matrix characters recognition. Automatic reading system for dot-matrix character is promising for reduction of cost and labor required for quality control of products. Although dot-matrix characters are constructed by specific dot patterns, variation of character appearance due to three-dimensional rotation of printing surface, bleeding of ink and missing parts of character is not negligible. The appearance variation causes degradation of recognition accuracy. The authors propose a technique improving accuracy and robustness of dot-matrix character recognition against such variation, using variation model based learning. The variation model based learning generates training samples containing four types of appearance variation and trains a Modified Quadratic Discriminant Function (MQDF) classifier using generated samples. The effectiveness of the proposed learning technique is empirically evaluated with a dataset which contains 38 classes (2030 character samples) captured from actual products by standard digital cameras. The recognition accuracy has been improved from 78.37 % to 98.52 % by introducing the variation model based learning.

1 Introduction

Dot-matrix characters are widely used for clarifying important information of a product such as consumption or appreciation expiration dates. The dot-matrix characters must be printed directly on the products in order to make both consumers and producers being able to read information about the products. Automatic reading system for the dot-matrix characters is promising for reduction of cost and labor required for quality control of products.

Figure 1 shows examples of actual camera-captured dot-matrix characters. As implied by the figure, recognition of dot-matrix characters contains several types of difference from standard character recognition. Since a dot-matrix character is constructed by multiple dots which are observed as multiple separated connected components in recognition process, a preprocessing which connects these dots is required to handle a character as one connected component. Although dot-matrix characters are constructed by specific dot patterns, variation of character

© Springer International Publishing Switzerland 2015
C.V. Jawahar and S. Shan (Eds.): ACCV 2014 Workshops, Part II, LNCS 9009, pp. 147–156, 2015.
DOI: 10.1007/978-3-319-16631-5_11

appearance due to three-dimensional rotation of printing surface, diffusing or squeeze out of ink and lack of dots in a printing pattern is not negligible.

For accurate recognition of these dot-matrix characters, several attempts have been proposed [1–5]. These methods are mainly divided into two groups, i.e. preprocessing-based and training-based methods. The preprocessing-based methods employ several ad hoc preprocessing techniques such as blob-connection, slant and rotation correction to restrict appearance variations of captured dot-matrix characters [3–5]. However, dot-matrix characters used in actual production scene contain a large amount of variation in matrix font patterns, dot size, printing quality and degradation. Construction of an universal preprocessing technique is quite difficult. Training-based methods employ classification models which are trained to capture possible appearance variation of dot-matrix characters. Artificial neural networks were employed as classifiers in literatures [1,2].

It is also known that there are many undocumented dot-matrix recognition products for factory automation. Many of them simplify the recognition task by restriction of appearance variation of character using controlled environment for character image capturing. And it is also relatively easy to recognize controlled dot-matrix characters using actual training dataset consists of the same capturing environment. If an universal recognition technique which is applicable uncontrolled or less-controlled environment, it should have certain amount of contributions for industrial scene.

In this paper, the authors propose a technique improving accuracy and robustness of dot-matrix character recognition against such variation, using variation model based learning. The variation model based learning generates training samples containing four types of appearance variation and trains a Modified Quadratic Discriminant Function (MQDF) classifier using generated samples. The effectiveness of the proposed learning technique is empirically evaluated with a dataset which contains 38 classes (2030 character samples) captured from actual products.

The paper is organized as follows. In Sect. 2, generation models for training dataset is described. Classification process on this research is shown in Sect. 3. Section 4 provides information about evaluation experiments and results. Finally, conclusions of the paper is given by Sect. 5.

2   Variation Model Based Learning

Training dataset which correctly reflect a model of data generation is necessary for a classifier obtaining high recognition performance. Since obtaining a priori generation model is generally difficult, an approach where the generation model is estimated from given or sampled training data is employed. Although the size of training data does not guarantee the accuracy of estimated model, it is expected that a large size of training data contributes improving statistical accuracy of the model.

Small size of training data sometimes cause insufficient accuracy of estimated model. The generative learning, which artificially generates training dataset from

Fig. 1. Examples of actual camera-captured dot-matrix characters

Fig. 2. Variation of actual dot-matrix font face

assumptions or a priori knowledge of data generation process, has been proposed as a promising solution for similar situation [6].

The variation model based learning proposed in this paper also generates large scale training data containing four possible appearance variations of dot-matrix characters to recognize. In this section, we describe data generation process for these appearance variations.

2.1   Multiple Dot-Matrix Font Faces

Figure 2 shows examples of actual dot-matrix characters in the same class. As shown in the figure, multiple dot-matrix font faces are employed even they are same characters. The proposed method employs two types matrices, i.e. 5 by 7 and 5 by 5 which are widely used by actual products. Figure 3(a) and (b) shows generated 5 by 7 and 5 by 5 standard dot-matrix characters, respectively.

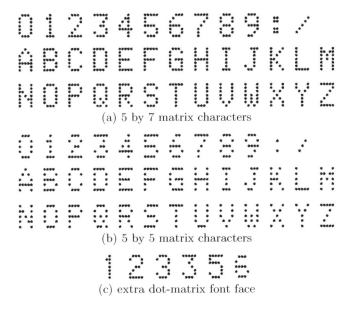

(a) 5 by 7 matrix characters

(b) 5 by 5 matrix characters

(c) extra dot-matrix font face

Fig. 3. Generated multiple dot-matrix font face

Some character class consists of multiple font face to improve readability by human. For instance, '3' contains multiple font face as shown by Fig. 2. To correctly recognize these characters, the proposed method also generate extra font faces addition in the standard characters. Specifically, one extra font face is added for each of '1', '2', '5' and '6', and two are added for '3'. Figure 3(c) shows actual extra font faces.

2.2   Three Dimensional Rotation

A major problem of camera-based character recognition is three-dimensional rotation of captured characters. Occurred situation in dot-matrix character recognition is that characters printed on rounded surface easily change their appearance by spacial relationship between the camera and the product. To handle these appearance variation, the proposed method generates rotated dot-matrix characters and add them to the training dataset. Figure 4 illustrates generation process of three dimensional rotation characters.

2.3   Size of Dots

Dots constructing characters are easily diffused in printing. Diffusing dots also causes large appearance variation. To handle such variation, the proposed method generates the training dot-matrix characters with multiple diameter d. Figure 5 shows examples of dot-matrix character 'A' generated with different d.

(a) three dimensional (b) x-axis rotation (c) y-axis rotation (d) z-axis rotation
rotation

Fig. 4. Generation of variation model by three dimensional rotation

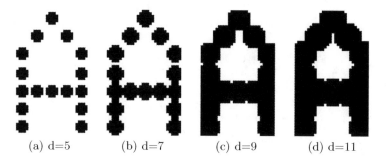

(a) d=5   (b) d=7   (c) d=9   (d) d=11

Fig. 5. Appearance variation by difference of diameter of dots

2.4   Missing Dots

The dots constructing characters sometimes disappear due to printing error or
capturing environment. When even one dot disappears, the dot-matrix character changes significantly in appearance. To recognize characters containing
missing dots, the method generates characters of which one dot is deleted as
shown by Fig. 6. Additionally, deleting one dot from a character results ambiguous appearance between two different character class. These ambiguous characters are excluded from training dataset. Figure 7 shows examples of excluded
dot-matrix patterns.

3   Classification

3.1   Gradient Feature Extraction

In this research, the gradient feature vector [7] are used for classification. The
gradient feature vector is composed of directional histogram of gradient of the
input character image. In this section, we summarize the gradient feature extraction. The gradient feature extraction is performed as in the following steps:

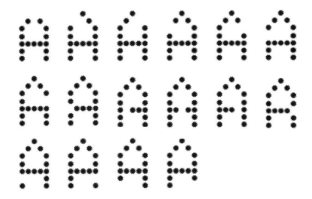

Fig. 6. Examples of generated character of which one constructing dot is deleted

Step 1: A 2×2 mean filtering is applied 4 times on the input image.

Step 2: The gray-scale image obtained in Step 1 is normalized so that the mean gray scale values lie in the range of -1 to $+1$, with a mean value of 0.

Step 3: Roberts filter is applied to the image to obtain gradient image. The direction of the gradient is initially quantized into $4D$ directions and the strength of the gradient is accumulated for each of the quantized direction. The strength and the direction of the gradient are defined by (1) and (2) respectively.

$$f(x, y) = \sqrt{(\Delta u)^2 + (\Delta v)^2}, \tag{1}$$

$$\theta(x, y) = \tan^{-1} \frac{\Delta v}{\Delta u}, \tag{2}$$

$$\Delta u = g(x + 1, y + 1) - g(x, y),$$

$$\Delta v = g(x + 1, y) - g(x, y + 1).$$

Step 4: The enclosing rectangular of the input character is divided into $2m - 1$ square blocks in height and width respectively and the histogram of $(2m - 1)^2 \times 4D$ dimension is extracted.

Step 5: Directional histogram of $(2m - 1) \times (2m - 1)$ blocks and $4D$ directions are down sampled into $m \times m$ blocks and D directions using Gaussian filters, and the histogram of $m^2 \times D$ dimension is obtained.

In this research, m and D are set as 6 and 8, respectively, so we obtained $6^2 \times 8 = 288$ dimensional original feature vector.

3.2 MQDF Classfier

The MQDF [8] classifier was employed. The MQDF is expressed by:

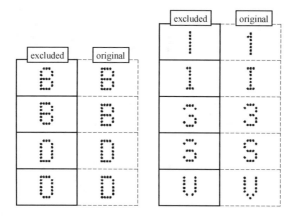

Fig. 7. Excluded dot-matrix characters due to ambiguous appearance

$$g(X) = (N + N_0 + n - 1) \ln[1 + \frac{1}{N_0\sigma^2}[||X - M||^2$$

$$- \sum_{i=1}^{k} \frac{\lambda_i}{\lambda_i + \frac{N_0}{N}\sigma^2} \{\Phi_i^T (X - M)\}^2]]$$

$$+ \sum_{i=1}^{k} \ln\left(\lambda_i + \frac{N_0}{N}\sigma^2\right) - 2\ln P(\omega), \qquad (3)$$

where, X denotes a n-dimensional gradient feature vector of input dot-matrix character, M is mean vector of training samples, λ_i and Φ_i are i-th eigenvalue and corresponding eigenvector of covariance matrix of training samples, respectively. k is a parameter which denotes the number of eigenvectors used for classification. N and $P(\omega)$ denote the number of training data and a priori probability of class ω. σ^2 is variance assuming spherical a priori distribution of X and determined by mean of all eigenvalues in all classes.

The parameter k which denotes the number of eigenvector used for classification is determined by prior experiment with verification dataset, i.e. a sub-set of experimental dataset. N_0 is defined by:

$$N_0 = \frac{\alpha}{1 - \alpha}N, \qquad (4)$$

where, the parameter α $(0 < \alpha < 1)$ is also determined by prior experiment.

4   Experiments and Results

To confirm the effectiveness of the proposed variation model based learning, we conducted evaluation experiments.

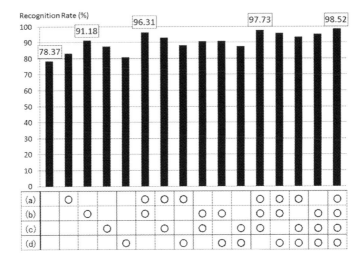

Fig. 8. Improvement of recognition performance by variation model based learning

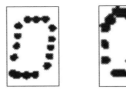

Fig. 9. Examples of failure character recognition due to significant degradation of dot-matrix patterns

4.1   Evaluation Dataset

The evaluation dataset employed in this research consists of 2030 dot-matrix character images of 38 character classes. Each character image is captured from actual industrial, medical and food products by digital cameras and extracted and binarized manually. As shown in Fig. 1, the evaluation dataset contains appearance variations due to distortion, blur, multiple font-face and three-dimensional rotation. Since the character classes '0 (numeral)' and 'O (alphabetical)' have same appearance, they are handled as a same character class.

4.2   Experimental Results

Figure 8 shows recognition performances by combinations of training data generation methods. (a)–(d) in the table below bar chart denote:

(a) multiple dot-matrix font face,
(b) three dimensional rotation,
(c) multiple size of dots,
(d) missing dots,

respectively. Circles in the table mean that a corresponding variation model is employed for data generation.

The original recognition accuracy of 78.37 % was improved by introducing generated training data and the highest recognition accuracy of 98.52 % has been obtained when all of four variation models were employed for training data generation. This results imply that training data generation reflecting appearance variation is effective for performance improvement even when no actual captured character images are available in the training dataset.

Failure recognition samples are shown in Fig. 9. Since significant degradation of characters were not included in the training dataset, the MQDF classifier could not handle these degraded characters.

5   Conclusions

In this paper, the authors propose a technique improving accuracy and robustness of dot-matrix character recognition against such variation, using variation model based learning. The variation model based learning generates training samples containing four type of appearance variation and trains a MQDF classifier using generated samples. The effectiveness of the proposed training data generation was confirmed by the experiments using actual camera-captured dot-matrix characters. The MQDF classifier trained generated dataset, which did not contain any actual captured data, successfully recognize dot-matrix characters against appearance variations happened in usual industrial scene.

Future study topics include (1) further performance improvement by optimization of data generation conditions, (2) integration of the proposed recognition method and a dot-matrix character detection method and (3) introducing failure printing detection.

Acknowledgement. A part of this research is supported by OMRON Corporation.

References

1. Yanikoglu, B.A.: Pitch-based segmentation and recognition of dot-matrix text. Int. J. Doc. Anal. Recogn. **3**, 34–39 (2000)
2. Namane, A., Soubari, E.H., Meyrueis, P.: Degraded dot matrix character recognition using CSM-based feature extraction. In: Proceedings of the 10th ACM Symposium on Document Engineering, pp. 207–210. ACM (2010)
3. Grafmüller, M., Beyerer, J.: Segmentation of printed gray scale dot matrix characters. In: Proceedings of 14th World Multi-conference on Systemics, Cybernetics and Informatics WMSCI, vol. 2, pp. 87–91 (2010)
4. Du, Y., Ai, H., Lao, S.: Dot text detection based on fast points. In: 2011 International Conference on Document Analysis and Recognition (ICDAR), pp.435–439. IEEE (2011)
5. Mohammad, K., Agaian, S.: Practical recognition system for text printed on clear reflected material. ISRN Mach. Vis. **2012**, 1–16 (2012)

6. Ishida, H., Yanadume, S., Takahashi, T., Ide, I., Mekada, Y., Murase, H.: Recognition of low-resolution characters by a generative learning method. In: Proceedings of CBDAR, pp. 45–51 (2005)
7. Shi, M., Fujisawa, Y., Wakabayashi, T., Kimura, F.: Handwritten numeral recognition using gradient and curvature of gray scale image. Pattern Recogn. **35**, 2051–2059 (2002)
8. Kimura, F., Wakabayashi, T., Tsuruoka, S., Miyake, Y.: Improvement of handwritten Japanese character recognition using weighted direction code histogram. Pattern Recogn. **30**, 1329–1337 (1997)

Scene Text Recognition: No Country for Old Men?

Lluís Gómez$^{(\boxtimes)}$ and Dimosthenis Karatzas

Computer Vision Center, Universitat Autònoma de Barcelona, Barcelona, Spain
{lgomez,dimos}@cvc.uab.es

Abstract. It is a generally accepted fact that Off-the-shelf OCR engines do not perform well in unconstrained scenarios like natural scene imagery, where text appears among the clutter of the scene. However, recent research demonstrates that a conventional shape-based OCR engine would be able to produce competitive results in the end-to-end scene text recognition task when provided with a conveniently preprocessed image. In this paper we confirm this finding with a set of experiments where two off-the-shelf OCR engines are combined with an open implementation of a state-of-the-art scene text detection framework. The obtained results demonstrate that in such pipeline, conventional OCR solutions still perform competitively compared to other solutions specifically designed for scene text recognition.

1 Introduction

The computer vision research community has dedicated a significant research effort on text extraction systems for text in natural scene images over the last decade. As a result, scene text extraction methods have evolved substantially and their accuracy has increased drastically in recent years [6], see the evolution of detection algorithms in the ICDAR competitions in Fig. 1. However, the problem is still far from being considered solved: note that the winner methods in the last ICDAR competition achieve only 66 and 74 % recall in the tasks of text localization and text segmentation respectively, while the best scoring method in cropped word recognition achieved a 83 % recognition rate. The main difficulties of the problem stem from the extremely high variability of scene text in terms of scale, rotation, location, physical appearance, and typeface design. Moreover, text in scene images may suffer from several degradations like blur, illumination artifacts, or may appear in extremely low quality.

Most of the published methods on Robust Reading systems focus on specific problems of the end-to-end pipeline, i.e. detection [3,15], extraction [5], and recognition [14] are traditionally treated as separate problems. This is quite natural if one takes into account that earlier works on text detection were hardly able to do detection with acceptable rates, see e.g. the 38 % f-score of the winner method in the first ICDAR competition [7], to furthermore think on approaching the full end-to-end problem. More recently, encouraged by the increased accuracy rates in detection and segmentation, end-to-end recognition methods seem viable to produce decent results.

ⓒ Springer International Publishing Switzerland 2015
C.V. Jawahar and S. Shan (Eds.): ACCV 2014 Workshops, Part II, LNCS 9009, pp. 157–168, 2015.
DOI: 10.1007/978-3-319-16631-5_12

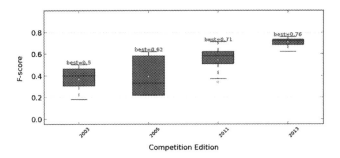

Fig. 1. Evolution of text detection f-score results of the participant methods in the ICDAR Robust Reading competitions. Notice that although dataset may have changed among different editions, results are still reasonably comparable.

In this paper we are interested in end-to-end unconstrained scene text recognition, i.e. in methods that do not make use of small fixed lexicons for the recognition. The firsts works approaching this complex task were proposed more than ten years ago [2], but papers reporting results in public benchmarks date from just few years ago [9,20]. Nowadays there exist reliable applications based on such technology that are already in the hands of millions[1] [1]. However, such real world applications are limited to very well delivered conditions, e.g. horizontally oriented and well focussed bilevel text, and often rely on large-scale data center infrastructure. Thus, there is still room for research on more robust and efficient methods.

A typical experiment frequently repeated to demonstrate the need of specific techniques for scene text detection and recognition is to attempt to process a raw scene image with a conventional OCR engine. This normally produces a bunch of garbage on the recognition output. Obviously this is not the task for which OCR engines have been designed and the recognition may be much better if we provide it with a pixel level segmentation of the text. Figure 2 show the output of the open source Tesseract[2] [17] OCR engine for a raw scene image and for its binarized text mask obtained with a scene text extraction method.

Through this paper we experimentally demonstrate that in such pipeline off-the-shelf OCR engines yield competitive results to state-of-the-art solutions specifically designed for scene text recognition.

While being this primarily an engineering work, we think there are two important aspects of it that can be of broad interest from a research perspective: one is about the question of whether pixel-level text segmentation is really useful for the final recognition, the other is about how stronger language models affect the final results.

---

[1] Word Lens and the Google Translate service are examples of a real applications of end-to-end scene text detection and recognition that have acquired market-level maturity.

[2] http://code.google.com/p/tesseract-ocr/.

In the following sections we make a comprehensive review of related work, describe the set-up of our end-to-end pipeline, and show the results of our experiments. Finally we close the paper with a valuable discussion.

2   Related Work

Table 1 summarize several aspects of language models used in existing end-to-end approaches for unconstrained recognition, and a more detailed description of such methods is provided afterwards. Table 1 compares the use of character/word n-grams (as well as their sizes), over-segmentation, and dictionaries (here the "Soft" keyword means the method allow Out-Of-Dictionary word recognition, while "Hard" means only "In-Dictionary" words are recognized).

Table 1. Summary of different language model aspects of end-to-end scene text recognition methods.

Method	Dictionary	Char n-gram	Word n-gram	Over-segmentation
Neumann *et al.* [9–12]	Soft (10k)	bi-gram	No	Yes
Wang *et al.* [21]	Hard (hunspell)	No	No	Yes
Neumann *et al.* [13]	No	3-gram	No	Yes
Yao *et al.* [22]	Soft (100k)	bi-gram	No	No
Bissacco *et al.* [1]	Soft (100k)	8-gram	4-gram	Yes

In [9] Neumann and Matas propose the classification of Maximally Stable Extremal Regions (MSERs) as characters and non-characters, and then the grouping of character candidates into text lines with multiple (mutually exclusive) hypotheses. For the recognition each MSER region mask is normalized and resized to a fixed size in order to extract a feature vector based on pixel directions along the chain-code of its perimeter. Character recognition is done with a SVM classifier trained with synthetic examples. Ambiguous recognition of upper-case

Fig. 2. Off-the-shelf OCR engine recognition accuracy may increase to acceptable rates if we provide pixel level text segmentation instead of raw pixels.

and lower-case variants of certain letters (e.g. "C" and "c") are tackled as a single class, and then differentiated using a typographic model that also serves to split the line into words. Finally, a combination of bi-gram and dictionary-based language model scores each text line hypothesis individually and the most probable hypothesis is selected. The authors further extended the text detection part of their method in several ways [10,11], increasing their end-to-end recognition results. In [12] they add a new inference layer to the recognition framework, where the best sequence selection is posed as an optimal path problem, solved by a standard dynamic programming algorithm, allowing the efficient processing of even more segmentation hypotheses.

Wang *et al.* [21] propose the use of Convolutional Neural Networks together with unsupervised feature learning to train a text detector and a character recognizer. The responses of the detector in a multi-scale sliding window, with Non-Maximal Suppression (NMS), give rise to text lines hypotheses. Word segmentation and recognition is then performed jointly for each text line using beam search. For every possible word the character recognizer is applied with an horizontal sliding window, giving a score matrix that (after NMS) can be used to compute an alignment score for all words in a small given lexicon. Words with low recognition score are pruned as being "non-text". Since the method is only able to recognize words in a small lexicon provided, in order to perform a more unconstrained recognition the authors make use of an off-the-shelf spell checking software to generate the lexicon given the raw recognition sequences.

In a very related work to the one presented in this paper Milyaev et al. [8] demonstrate that off-the-shelf OCR engines can still perform well on the scene text recognition task as long as appropriate image binarization is applied to input images. For this, they evaluate 12 existing binarization methods and propose a new one using graph cuts. Their binarization method is combined with an AdaBoost classifier trained with simple features for character/non-character classification. And the components accepted by the classifier are used to generate a graph by connecting pairs of regions that fulfill a set of heuristic rules on their distance and color similarity. Text lines obtained in such way are then split into words and passed to a commercial OCR engine[3] for recognition.

In [13] Neumann and Matas propose the detection of constant width strokes by convolving the gradient image with a set of bar filters at different orientations and scales. Assuming that characters consist in a limited number of such strokes a set of candidate bounding-boxes is created by the union of bounding boxes of 1 to 5 nearest strokes. Characters are detected and recognized by matching the stroke patterns with an Approximate Nearest Neighbor classifier trained with synthetic data. Each candidate bounding box is labelled with a set of character labels or rejected by the classifier. The non rejected regions are then agglomerated into text lines and the word recognition is posed as an optimal sequence search by maximizing an objective function that combines the probabilities of the character classifier, the probability of the inter-character spacing difference

---

[3] OCR Omnipage Professional, available at http://www.nuance.com/.

of each triplet, the probability of regions relative positioning, and the characters adjacency probability given by a 3-gram language model.

Yao et al. [22] propose an arbitrary oriented scene text detection and recognition method that extracts connected components in the Stroke Width Transform (SWT) domain [3]. Component analysis filters out non-text components using a Random Forest classifier trained with novel rotation invariant features. This component level classifier performs both text detection and recognition. Remaining character candidates are then grouped into text lines using the algorithm proposed by Yin et al. [23], and text lines are split into words using the method in [3]. Finally, the authors propose a modified dictionary search method, based on the Levenshtein edit distance but relaxing the cost of the edit operation between very similar classes, to correct errors in individual character recognition using a large-lexicon dictionary[4]. To cope with out-of-dictionary words and numbers, n-gram based correction [18] is used if the distance with closest dictionary word is under a certain threshold.

Bissacco et al. [1] propose a large-scale end-to-end method using the conventional multistage approach to text extraction. In order to achieve a high recall text detection the authors propose to combine the outputs of three different detection methods: a boosted cascade of Haar wavelets [19], a graph cuts based method similar to [15], and a novel approach based on anisotropic Gaussian filters. After splitting text regions into text lines a combination of two over-segmentation methods is applied, providing a set of possible segmentation points for each text line. Then beam search is used to maximize a score function among all possible segmentations in a given text line. The score function is the average per-character log-likelihood of the text line under the character classifier and the language model. The character classifier is a deep neural network trained on HOG features over a training set consisting of around 8 million examples. The output layer of the network is a softmax over 99 character classes and a noise (non-text) class. At test time this classifier evaluates all segmentation combinations selected by the beam search. The language model used in the score function to be optimized by the beam search is a compact character-level ngram model (8-gram). Once the beam search has found a solution the second level language model, a much larger distributed word-level ngram (4-gram), is used to rerank.

As a conclusion of the state of the art review we can see that the majority of reviewed methods make use of similar language models, based on a dictionary of frequent words and a character n-gram (usually a bi-gram). A much stronger language model has been proposed by Bissacco et al. [1]. On the other hand, while the system in [1] makes use of three different detection methods combined with an independent recognition module, in other cases both detection and recognition are treated together, e.g. using the same features for detection and recognition [22] or using multiple character detections as an over-segmentation cue for the recognition [10], giving rise to more compact and efficient methods.

---

[4] Word list is provided by the Microsoft Web N-Gram Service (http://webngram. research.microsoft.com/info/) with top 100k frequently searched words on the Bing search engine.

3   End-to-End Pipeline

In order to ensure that our results are reproducible we adopt a publicly available implementation of a well known algorithm for text detection. Concretely, the used OpenCV text module[5] implements, among others, the Class Specific Extremal Regions (CSER) algorithm initially proposed by Lukás Neumann & Jiri Matas [11], and the Exhaustive Search algorithm of the same authors [10].

The main idea behind Class-specific Extremal Regions is similar to the MSER in that suitable Extremal Regions (ERs) are selected from the whole component tree of the image. However, this technique differs from MSER in that selection of suitable ERs is done by a sequential classifier trained for character detection, i.e. dropping the stability requirement of MSERs and selecting class-specific (not necessarily stable) regions.

On the other hand, the Exhaustive Search algorithm was proposed in [10] for grouping ERs corresponding to character candidates into candidate text lines (for horizontally aligned text). The algorithm models a verification function that efficiently evaluates all possible ER sequences. The algorithm incorporates a feedback loop that allows to recover errors in the initial ER selection, by searching among the discarded ERs those that fit well with the detected text lines hypotheses. This feedback loop proves to be particularly important when doing the final recognition.

At this point it is fair to notice that the text detection implementation used in our experiments differs in several ways from the original work in [11]. Particularly, we work in a single channel projection (gray level image), i.e. we do not use different color channels, thus relying in a simplified pipeline that allows for faster computation at the expense of lower recall. Moreover, in our case the geometric normalization (if needed) and word splitting process is left to the OCR engine.

3.1   Text Recognition

The output of the described text detection algorithm is a set of text line hypotheses corresponding to groups of ERs. Those groups of ERs give rise to a pixel level segmentation of the detected text elements that can be directly fed to the OCR engine. Apart from that we further evaluate in our experiments the recognition performance by feeding three alternative inputs to the OCR: the gray scale cropped box of the detected lines, their Otsu's binarizations, and another binary image obtained with simple thresholding by setting as foreground all pixels with an intensity value in the range $(min(I_{ER_l}), max(I_{ER_l}))$, where I_{ER_l} is the set of intensity values of pixels corresponding to extremal regions of a given text line l.

We feed the results of the detection/segmentation pipeline to two well known off-the-shelf OCR engines: the open source project Tesseract[6] [17], and the commercial software ABBYY Fine Reader SDK[7]. The set-up for both OCR engines

---

[5] http://docs.opencv.org/trunk/modules/text/doc/erfilter.html.

[6] http://code.google.com/p/tesseract-ocr/.

[7] http://finereader.abbyy.com/.

is minimal: we set the recognition language to English and specify a closed list of characters (52 letters and 10 digits), we also set the OCR to interpret the input as a single text line, apart from that we use the default parameters.

The recognition output is filtered with a simple post-processing junk filter in order to eliminate garbage recognition, i.e. sequences of identical characters like "IIii" that may appear as a result of trying to recognize repetitive patterns in the scene. Concretely we discard the words in which more than half of their characters are recognized as one of "i", "l", or "I".

4   Experiments

We conduct all our experiments on the ICDAR2011 dataset in order to compare with all other methods that provide end-to-end results. The ICDAR2011 test set consists of 255 scene images with different sizes. The ground truth is defined at the word level using axis oriented bounding boxes and their corresponding text annotations. The evaluation protocol considers a valid recognition when the estimated word bounding box has at least 80 % recall with a ground-truth word and all its characters are recognized correctly (case sensitive).

Table 2 show the obtained results using the different end-to-end variants described before. Overall performance of *ABBYY* OCR engine is better than the obtained with *Tesseract*, and in both engines the better results are obtained with the CSER segmentation. *Tesseract* results with the Otsu thresholded image and the raw grey scale bounding boxes are exactly the same, which seems to indicate that the internal binarization method in *Tesseract* may be equivalent to the Otsu thresholding.

Table 2. End-to-end recognition evaluation in the ICDAR 2011 dataset comparing different segmentation approaches.

Method	Precision	Recall	F-score
CSER + ABBYY	59.2	39.3	47.2
CSER + Raw bbox + ABBYY	67.1	35.8	46.7
CSER + $(min(I_{ER_l}), max(I_{ER_l}))$ + ABBYY	65.9	36.1	46.6
CSER + Otsu + ABBYY	65.3	35.2	45.8
CSER + Tesseract	52.9	32.4	40.2
CSER + $(min(I_{ER_l}), max(I_{ER_l}))$ + Tesseract	60.4	29.3	39.5
CSER + Otsu + Tesseract	63.2	28.1	38.9
CSER + Raw bbox + Tesseract	63.2	28.1	38.9

Table 3 shows the comparison of the best obtained results of the two evaluated OCR engines with current state of the art. In both cases our results outperform [11] in total f-score while, as stated before, our detection pipeline is a simplified implementation of that method. However, it is important to notice that our recall is in general lower that in the other methods, while it is our precision what makes the difference in f-score. A similar behaviour can be seen for the method of Milyaev *et al.* [8], which also uses a commercial OCR engine for the recognition. Such higher precision rates indicate that in general off-the-shelf OCR engines are doing very good in rejecting false detections.

Notice that Table 3 does not include the method in [1] because end-to-end results are not available. However, it would be expected to be in the top of the table if we take into account their cropped word recognition rates and their high recall detection strategy.

Table 3. End-to-end results in the ICDAR 2011 dataset comparing our proposed pipeline with other state-of-the-art methods. Methods marked with an asterisk evaluate on a slightly different dataset (ICDAR 2003).

Method	Precision	Recall	F-score
Milyaev *et al.*[8]	66.0	46.0	54.0
CSER + ABBYY	59.2	39.3	47.2
Yao *et al.* [22]	49.2	44.0	45.4
Neumann and Matas [13]	44.8	45.4	45.2
CSER + Tesseract	52.9	32.4	40.2
Neumann and Matas [12]	37.8	39.4	38.6
Neumann and Matas [11]	37.1	37.2	36.5
Wang *et al.* [21] *	54.0	30.0	38.0

It is indicative at this point comparing the recognition performance with the results of the detection module alone. For this we calculate precision and recall of line-level detection, i.e. we count a given detection bounding box as true positive if it overlaps more than 80 % with a ground truth bounding box, irrespective of the recognition output. The detection precision and recall obtained in this way are 56.7 and 73.5 % respectively. Notice that this detection rates are not comparable with the standard ICDAR evaluation framework for text localization, but just an indicative value in order to assess the effects of the recognition module in the final results.

Figures 3, 4, and 5 show qualitative results of the **CSER+Tesseract** pipeline. The end-to-end system recognizes words correctly in a variety of different situations, including difficult cases, e.g. where text appears blurred, or with non standard fonts. Common misspelling mistakes are, most of the time, due to missed diacritics, rare font types, and/or poor segmentation of some characters. Also in many cases the OCR performs poorly because the text extraction algorithm is not able to produce a good segmentation, e.g. in challenging situations or in cases where characters are broken in several strokes.

As part of our experiments we have also implemented a similar system to the **CSER+Tesseract** pipeline, but using MSER to be less computationally expensive, that reaches real-time end-to-end recognition with similar accuracy in a 640 × 480 video stream on a standard PC. For this experiment the whole pipeline is set exactly as described in Sect. 3 but replacing the CSER character detector by the standard MSER algorithm. Table 4 show the speedup of this implementation compared to **CSER+Tesseract**, and the accuracies of both pipelines in the ICDAR2011 dataset. Obviously the required time to process a single image is not constant and depends, among other, on the number of detected text lines that are fed to the OCR engine. This makes difficult to measure the time performance in a real-time video stream, as an indicative result we provide the average processing time in a 640 × 480-resized version

Fig. 3. Qualitative results on well recognized text using the CSER+Tesseract pipeline.

Fig. 4. Common misspelling mistakes using the CSER+Tesseract pipeline are due to missed diacritics, rare font types, and/or poor segmentation of some characters.

Fig. 5. Common errors when segmentation is particularly challenging or characters are broken in several strokes.

of the ICDAR2011 dataset, which is 177 milliseconds. This processing time can be further reduced by using multi-threaded OCR workers, reaching an average frame-rate of $8fps$. in the ICDAR resized dataset with a 2.3 GHz quad-core i7 processor.

Table 4. End-to-end recognition performance in the ICDAR 2011 dataset comparing different region extraction approaches.

Method	Precision	Recall	F-score	Avg. time (ms.)
CSER + Tesseract	52.9	32.4	40.2	851.0
MSER + Tesseract	48.1	30.7	37.5	420.7

Finally, we have done a set of experiments in order to evaluate the specific impact of Tesseract's language model in the final recognition results of the **CSER+Tesseract** pipeline. The language model of the Tesseract engine has several components including an efficient segmentation search, a heuristic word rating algorithm, and a word bigram model. The main component is the set of different word dictionaries that are used for word rating in combination with different heuristic scores, like script/case/char-type consistency, and with the individual ratings of the character classifier. We have evaluated the **CSER+Tesseract** pipeline deactivating any dictionary-based correction. As can be seen in Table 5 the impact of the dictionary-based correction component accounts around 9 % of total f-score of the system.

Table 5. End-to-end recognition accuracy comparison in the ICDAR 2011 dataset by deactivating Tesseract's dictionaries.

Method	Precision	Recall	F-score
CSER + Tesseract	52.9	32.4	40.2
CSER + Tesseract (No-dict)	43.7	24.6	31.5

5   Discussion

We have evaluated two off-the-shelf OCR engines in combination with a scene text extraction method for the task of end-to-end text recognition in natural scenes, demonstrating that in such a pipeline conventional OCR solutions still perform competitively compared to other solutions specifically designed for scene text recognition. Moreover, our results are based in a simplified implementation of the text detection pipeline in [11], thus it is to be expected that a complete implementation of the original work would eventually improve the obtained results. Our findings, inline with other works reporting similar experiments [8], call to a discussion on the underlying factors of such results.

It is well known that complex language models are very useful in OCR applications where the character classifier has high error rates [18], for example in languages such as Arabic or Hindi. And this is certainly the case in scene text recognition, where state of the art methods are usually bounded to character recognition accuracies around 80 %.

Although document analysis from scanned documents in English language is not the same case, and thus traditional OCR for such task may rely much more in the character classifier confidence, it is still true that off-the-shelf engines, like the ones used in our experiments, have more developed language models than the usually found in scene text recognition methods.

As conclusion we can say that stronger language models, as found in off-the-shelf OCR engines or in the photoOCR system in [1], can make an important difference in the final recognition accuracy of end-to-end methods, and may be further investigated in the future.

On the other hand, regarding the usefulness of pixel level segmentation methods for scene text recognition we conclude that such techniques are nowadays able to provide state of the art accuracies for end-to-end recognition. It is true however that scene text is not always binarizable and that in such cases other techniques must be employed. But current segmentation methods combined with existing OCR technologies may produce optimal results in many cases, by taking advantage of more than 40 years of research and development in automated reading systems [4], e.g. all the accumulated knowledge of shape-based classifiers, and the state of the art in language modelling for OCR.

Acknowledgement. This project was supported by the Spanish project TIN2011-24631 the fellowship RYC-2009-05031, and the Catalan government scholarship 2013 FI1126. The authors want to thanks also Google Inc. for the support received through the GSoC project, as well as the OpenCV community, specially to Stefano Fabri and Vadim Pisarevsky, for their help in the implementation of the scene text detection module evaluated in this paper.

References

1. Bissacco, A., Cummins, M., Netzer, Y., Neven, H.: Photoocr: reading text in uncontrolled conditions. In: International Conference on Computer Vision (ICCV) (2013)
2. Chen, X., Yuille, A.L.: Detecting and reading text in natural scenes. In: Computer Vision and Pattern Recognition (CVPR) (2004)
3. Epshtein, B., Ofek, E., Wexler, Y.: Detecting text in natural scenes with stroke width transform. In: Computer Vision and Pattern Recognition (CVPR) (2010)
4. Fujisawa, H.: Forty years of research in character and document recognition an industrial perspective. Pattern Recogn. **41**, 2435–2446 (2008)
5. Gomez, L., Karatzas, D.: Multi-script text extraction from natural scenes. In: International Conference on Document Analysis and Recognition (ICDAR) (2013)
6. Karatzas, D., Shafait, F., Uchida, S., Iwamura, M., Gomez, L., Robles, S., Mas, J., Fernandez, D., Almazan, J., de las Heras, L.P.: ICDAR 2013 robust reading competition. In: International Conference on Document Analysis and Recognition (ICDAR) (2013)

7. Lucas, S.M., Panaretos, A., Sosa, L., Tang, A., Wong, S., Young, R.: ICDAR 2003 robust reading competitions. In: International Conference on Document Analysis and Recognition (ICDAR) (2003)

8. Milyaev, S., Barinova, O., Novikova, T., Kohli, P., Lempitsky, V.: Image binarization for end-to-end text understanding in natural images. In: International Conference on Document Analysis and Recognition (ICDAR) (2013)

9. Neumann, L., Matas, J.: A method for text localization and detection. In: Assian Conference on Computer Vision (ACCV) (2010)

10. Neumann, L., Matas, J.: Text localization in real-world images using efficiently pruned exhaustive search. In: International Conference on Document Analysis and Recognition (ICDAR) (2011)

11. Neumann, L., Matas, J.: Real-time scene text localization and recognition. In: Computer Vision and Pattern Recognition (CVPR) (2012)

12. Neumann, L., Matas, J.: On combining multiple segmentations in scene text recognition. In: International Conference on Document Analysis and Recognition (ICDAR) (2013)

13. Neumann, L., Matas, J.: Scene text localization and recognition with oriented stroke detection. In: International Conference on Computer Vision (ICCV) (2013)

14. Novikova, T., Barinova, O., Kohli, P., Lempitsky, V.: Large-lexicon attribute-consistent text recognition in natural images. In: Fitzgibbon, A., Lazebnik, S., Perona, P., Sato, Y., Schmid, C. (eds.) ECCV 2012. LNCS, vol. 7577, pp. 752–765. Springer, Heidelberg (2012)

15. Pan, Y.F., Hou, X., Liu, C.L.: Text localization in natural scene images based on conditional random field. In: International Conference on Document Analysis and Recognition (ICDAR) (2009)

16. Shahab, A., Shafait, F., Dengel, A.: ICDAR 2011 robust reading competition challenge 2: reading text in scene images. In: International Conference on Document Analysis and Recognition (ICDAR) (2011)

17. Smith, R.: An overview of the tesseract OCR engine. In: International Conference on Document Analysis and Recognition (ICDAR) (2007)

18. Smith, R.: Limits on the application of frequency-based language models to OCR. In: International Conference on Document Analysis and Recognition (ICDAR) (2011)

19. Viola, P., Jones, M.: Rapid object detection using a boosted cascade of simple features. In: Computer Vision and Pattern Recognition (CVPR) (2001)

20. Wang, K., Babenko, B., Belongie, S.: End-to-end scene text recognition. In: ICCV (2011)

21. Wang, T., Wu, D.J., Coates, A., Ng, A.Y.: End-to-end text recognition with convolutional neural networks. In: International Conference on Pattern Recognition (ICPR) (2012)

22. Yao, C., Bai, X., Liu, W.: A unified framework for multi-oriented text detection and recognition. In: IEEE Transactions on Image Processing (TIP) (2014)

23. Yin, X.C., Yin, X., Huang, K., Hao, H.W.: Robust text detection in natural scene images. In: IEEE Transactions on Pattern Analysis and Machine Intelligence (TPAMI) (2013)

A Machine Learning Approach to Hypothesis Decoding in Scene Text Recognition

Jindřich Libovický[1]([⊠]), Lukáš Neumann[2], Pavel Pecina[1], and Jiří Matas[2]

[1] Institute of Formal and Applied Linguistics, Charles University in Prague,
Praha 1, Czech Republic
libovicky@ufal.mff.cuni.cz
[2] Centre for Machine Perception, Czech Technical University in Prague,
Praha 6, Czech Republic

Abstract. Scene Text Recognition (STR) is a task of localizing and transcribing textual information captured in real-word images. With its increasing accuracy, it becomes a new source of textual data for standard Natural Language Processing tasks and poses new problems because of the specific nature of Scene Text. In this paper, we learn a string hypotheses decoding procedure in an STR pipeline using structured prediction methods that proved to be useful in automatic Speech Recognition and Machine Translation. The model allow to employ a wide range of typographical and language features into the decoding process. The proposed method is evaluated on a standard dataset and improves both character and word recognition performance over the baseline.

1 Introduction

Scene Text Recognition (STR) is a computer vision task which aims to automatically localize all text areas in an image and to recognize (transcribe) their textual content. The problem has been receiving significant attention of the scientific community since the textual information is heavily present in real-world images with a large application potential. However, only manually assigned metadata is commonly available for image retrieval or content analysis. Manual annotation is costly or infeasible given the steadily rising data volumes.

STR also poses new problems for Natural Language Processing (NLP) as text in real-world images often consists of very few words or snippets without other textual context (see Fig. 1a). Even a plain transcription can be quite difficult and the current state-of-the-art STR methods achieve character accuracy of only about 70 % on standard datasets [1]. Additionally, interpretation of Scene Text can heavily depend on visual clues not present in the textual information itself (e.g., a meaning of the stand-alone word "visa" is completely different when written on a direction sign at an embassy and when written on a credit card), so novel joint techniques integrating computer vision and NLP are appropriate for such situations.

In this work, we integrate typographical and language features into the state-of-the-art end-to-end STR pipeline [2] and improve its accuracy by using structured machine learning approach inspired by hypotheses decoding in automatic

© Springer International Publishing Switzerland 2015
C.V. Jawahar and S. Shan (Eds.): ACCV 2014 Workshops, Part II, LNCS 9009, pp. 169–180, 2015.
DOI: 10.1007/978-3-319-16631-5_13

speech recognition [3,4] and machine translation [5,6]. The proposed method is evaluated on the ICDAR 2013 dataset, a standard benchmark for STR evaluation [1].

The rest of this paper is structured as follows. Section 2 summarizes previous work in this field with focus on the hypotheses decoding algorithms. In Sect. 3, we describe and formalize the decoding problem we aim to solve in this paper. Section 4 presents the features we use in the models and methods we use for training the models. Section 5 describes the process of preparing the training data. The method is evaluated in Sect. 6 and the results discussed in Sect. 7. Section 8 concludes the work.

2   Related Work

For a comprehensive survey of text detection and localization in STR we refer the reader to work of Zhang et al. [7].

One criterion for categorization of STR methods is the requirement for prior manual text localization. The methods of Mishra et al. [8] and Novikova et al. [9] omit the localization phase and require a human annotator to first "cut out" all words and then recognize (transcribe) text of each of the manually cropped word images. The requirement for manual text localization makes the methods impractical for processing of larger datasets.

Another criterion is the requirement to know the set words that may appear in an image. The method of Wang et al. [10] is given a *lexicon* (a list of 50 to 500 words) for each processed image and aims at localizing one or more of the lexicon words in the image. The method achieves high precision in text recognition, but its applicability is limited by the requirement of having a fixed lexicon for each image prior to the recognition and therefore it cannot be used to acquire new textual data.

TextSpotter [2,11] is to our knowledge the only lexicon-free method for STR which operates in an end-to-end setup, i.e., it implements both text localization and recognition and requires no manual annotation or lexicon to transcribe text in images. It first detects image regions corresponding to individual characters, joins them into text line hypotheses, and then recognizes the characters using an Optical Character Recognition (OCR) classifier.

The final stage of STR is usually string decoding, which disambiguates character recognition in the context of the entire strings. The hypotheses typically form a graph with vertices representing the character regions (there may be either one vertex with distribution over multiple characters or one vertex per character) and edges between the regions which follow each other in the transcription. The decoding itself can be approached in various ways.

Mishra et al. [8] use the Conditional Random Fields (CRFs) to decide between different segmentations. Each bounding box is represented only by its first-best character recognition and is assigned a unary feature – its classification score. Potentially neighboring character hypotheses are connected by binary factors encoding language model score. The feature weights are estimated empirically.

Roy et al. [12] first detect a line on which characters lie. Then a Hidden Markov Model (HMM) is used to decode the string from a set observed by sliding windows. This approach allows both multiple segmentation and multiple hypotheses per window, however by having most of the windows emitting empty output they loose possibility to use a language model score for adjacent characters.

A similar approach to the one presented in this paper is used by Shi et al. [13]. After a text area is detected, it is segmented into character bounding boxes. For each bounding box, a classifier produces several hypotheses which are then decoded (disambiguated) using a linear chain CRF. However, in this case the character segmentation is decided beforehand, and multiple transcription hypotheses a segment hypotheses are allowed.

An end-to-end STR pipeline, PhotoOCR [14], uses a machine-translation-like beam search to explore all possible paths with pruning over the least probable. This approach is able to deal with both multiple hypotheses and multiple segmentations. In another end-to-end pipeline [15], similarly to our work, the structured perceptron is used for optimization of the weights in a dynamic programming decoding.

In case of recognition of words from a prior lexicon, the decoding can be constructed such that it could only produce the words listed in the lexicon. Novikova et al. [9] employs a trie-shaped weighted finite state automaton, Wang et al. [10] use dynamic programing for searching areas that may correspond to words from a dictionary. This produces a list of candidate locations which are filtered using a binary classifier.

Field [16] recently came with two different approaches. One approach is string decoding realized as a parse of a probabilistic context free grammar for English syllables. In the second approach, she first identifies identical characters which necessarily must have the same transcription. The string is decoded using a HMM with a character bigram model for transition probabilities. The inference is done by integer linear programming which allows to ensures the previously detected character' identities.

3   Problem Description

In TextSpotter, each character appearing in an image may be detected several times (each detection corresponds to a different image segmentation) and each such detection may be assigned one or more character labels (transcription hypotheses) by the OCR module. Such an approach is beneficial, because it allows to keep multiple hypotheses for character segmentation and its labels for a later stage of the pipeline, where the final decision can be made by exploiting context of the character in the word (we define a *word* simply as a sequence of characters).

In this paper, the space of all word transcription hypotheses is modelled as a directed acyclic graph, where each vertex corresponds to a character segmentation with a character label (assigned by the OCR module). Two vertices are

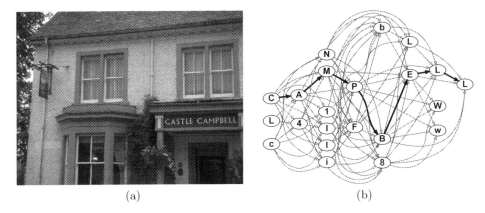

(a)   (b)

Fig. 1. Scene text in a real-world image (a). Graph representing transcription hypotheses for the word "CAMPBELL" (the path representing the ground truth transcription is bolded) (b)

connected by an oriented edge iff the character associated with the first vertex can immediately follow the character associated with the second vertex (see Fig. 1b). Therefore, each path unambiguously induces one possible transcription of the whole word. Finding the final word transcription can then be formalized as finding the maximum weighted path where each edge is assigned a weight indicating how likely the two characters follow each other in the character sequence of the word, i.e., we globally optimize values of local cost functions. For solving this problem we need to learn a function that estimates edge weights given a set of features that are assigned to each edge.

Formally, we define a transcription hypotheses graph $G = (V, E, s, t, l, \phi)$ where (V, E) is a Directed Acyclic Graph (DAG), $s \in V$ is the start node with no incoming edges, $t \in V$ is the target node with no outgoing edges, $l : V \setminus \{s, t\} \rightarrow$ $[\text{a} - \text{z} \text{A} - \text{Z}]$ is label assignment for the vertices, and $\phi : E \rightarrow \mathbb{R}^m$ is a feature function that assigns an m-dimensional real-valued feature vector to each edge. We denote the set of all the hypotheses graphs as $\mathcal{X}$.

We want find a path $\mathbf{y} = (y_1 = s, y_2, \ldots, y_{k-1}, y_k = t)$ from the start node to the target node such that concatenation of labels of the vertices on the path, $l(y_2), \ldots, l(y_{k-1})$, form ground. For that purpose we want to learn a function $f : \mathbb{R}^m \rightarrow \mathbb{R}$ assigning each edge a number based on its feature vector such that the ground truth string will be the maximum weighted path $\mathbf{y}$ from s to t.

For each transcription hypothesis graph, we also define a second-order transcription hypothesis graph whose vertices represent edges in the original graphs (i.e., character bigrams) and edges between are pairs of adjacent edges in the original graph (i.e., represent character trigrams). This enable us to use features that describe properties of triplets of potentially adjacent characters. The size of such graphs is quadratic in size of the original graphs.

4   Proposed Method

The proposed method extends TextSpotter[1] by using additional features and employing machine learning for parameter training and decoding.

4.1   Features

TextSpotter employs a linear combination of four feature functions to infer edge weights (segmentation threshold compatibility, OCR confidence of the second character, confidence of the second character fitting the inferred text line and a heuristic language model score – character bigram probability modified by hand-crafted rules).

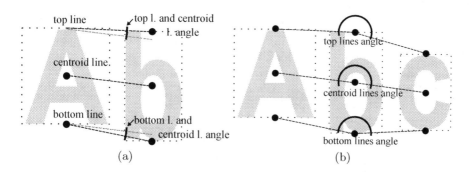

Fig. 2. Typographical features used in the first-order model (a) and the second-order (b) model.

Our first-order (bigram) model extends the method by adding the following 20 typographical and language features:

– ratios of width, height, and area of the character on the edge,
– mutual angles of the top line, bottom line, and centroid line (see Fig. 2a),
– conditional character bigram probability, and
– binary features coding character patterns (two digits, lower case + uppercase letter, both the same case, lower to upper case).

The second-order (trigram) model then employs the following 9 additional features capturing trigram properties:

– adjacent top lines, bottom lines, and centroid lines angles (see Fig. 2b),
– adjacent spaces ratio,
– conditional character trigram probability, and
– 4 binary features coding character patterns (digits only, Xxx, xxx|XXX, xXX|
  xXx|xxX|XxX, where X stand for an upper-case character and x for a lower-case
  letter).

---

[1] We used the current version of TextSpotter available at http://www.textspotter.org.

All the ratios were computed as absolute values of the difference of logarithms. All features are standardized to have a zero mean and unit variance on the training data.

4.2   Model

In TextSpotter, the model parameters (linear combination coefficients) are tuned by a simple grid search. An alternative is to treat the problem as standard classification and train a classifier to predict for each edge how likely it is to lie on the ground truth path. Such classification is called local – the edges are scored independently from each other – those lying on the ground truth paths are used as positive training examples and the others as negative ones (we resampled the training data to have the same proportion of positive and negative examples). We examined several standard machine-learning algorithms implemented in WEKA [17]: Logistic Regression, Support Vector Machine (SVM) with various kernels, Random Forest, and multilayer Perceptron with various hidden layer configurations.

In local classification, the information about the final output is ignored during training. The constraints for the output structure (a graph path) apply in decoding (testing) phase only. A more appropriate solution is structured prediction where the same decoding algorithm is used during training, but the classification is global and allows the parameters to be optimized also with respect to the inference algorithm. To employ the structured prediction we need to formulate the problem as finding a structure (in this case a path) which is maximal with respect to a function that is a dot product of a weight vector and a feature function of the structure, here the sum of the feature values along the path. This means:

$$\hat{\mathbf{y}} = \underset{\mathbf{y} \in \mathcal{Y}_{\mathbf{x}}}{\operatorname{argmax}} \, \mathbf{w}^T \Psi(\mathbf{x}, \mathbf{y}) = \mathbf{w}^T \sum_{e \in \mathbf{y}} \phi(e)$$

where $\mathbf{w}$ is a learned m-dimensional weight vector. While training a prediction model we want to estimate the weight vector $\mathbf{w}$ where $\mathbf{w}$ is an m-dimensional weight vector optimized to maximize the number of training instances correctly classified.

We examined two state-of-the-art techniques for the weights optimization: the structured perceptron [18] and structured SVM approximated by the cutting plane algorithm [19].

The structured perceptron [18] is a simple modification of the standard perceptron algorithm. The weight vector is iteratively updated by the difference of the feature vector of the currently estimated solution and the ground truth solution.

The structured SVM algorithm aims to optimize the weight vector such that the dot product is an upper bound estimate of a loss function, i.e., unlike the perceptron, it only distinguishes between partially and entirely incorrect solutions. A quadratic programming formulation capturing this requirement would demand exponentially many conditions for each of the training instances and its computation would be intractable. For this reason, we use an iterative approximate algorithm which finds the most violated conditions in each iteration and add them as constraints to the quadratic programming problem.

Based on the results of our preliminary experiments, we also added score from the best performing (the Random Forest) local edge classifier to the to feature vector for the structured learning. It aims to combine an advantage of capturing non-linear relations between the features in the local classifier and the knowledge of the inference algorithm during the learning procedure.

5 Data

The ICDAR 2013 dataset [1] was used to generate the training data for our experiments. The training set consists of 229 real-world images with 849 words in total. For each word, the ground truth annotation is given in the form of a bounding box coordinates and a reference transcription. For our purposes, we ignored all punctuation marks because its annotation is inconsistent across the dataset. The training data was generated by running the initial stages of the TextSpotter pipeline on each image to generate word transcription hypotheses graphs and then in each graph, the ground truth path representing the correct transcription is selected (if it exists) by matching graph vertices to the ground truth annotation of the word. A path is matched with the ground truth annotation iff

– the sequence of the character labels along the path corresponds to the reference transcription or its prefix of a length at least four, and
– all character centroids lie in the bounding box.

This process produced a total of 1607 graphs (including multiple graphs for the same word because the TextSpotter pipeline processes the same image several times using different visual transformations of the original image). We were able to match a subset of 812 graphs with the ground truth, out of which a random sample of 568 graphs was used to train the model and the remaining 244 graphs were used for intrinsic evaluation (see Sect. 6).

6 Evaluation

In the first experiment, we evaluate how well the model learns to find the ground truth path. A straightforward measure for the correctness of finding the longest path in a graph could be the Hamming loss (number of incorrectly used edges), or precision and recall of correctly selected edges. However, this measures cannot be compared between the graphs with bigram and trigram edges. Therefore, we only use the string based metrics, which were averaged over all graphs in the test set:

– Levenshtein (edit) distance $\bar{d}$ of the output string and the ground truth string;
– relative Levenshtein distance $\bar{d}_r$ of the output and ground truth strings, i.e., the edit distance divided by the length of the ground truth string, saying how likely is a character to be incorrect; and
– full string accuracy $\bar{a}$ – proportion of correctly selected paths.

Fig. 3. Samples from the ICDAR 2013 dataset. Note the improvements of the proposed method over TextSpotter [2].

The results of the intrinsic evaluation is provided in Table 1. The evaluated methods brought substantial improvement over the baseline method. The best results were achieved by the Random Forest classifier [20] trained locally on the edges and with a structurally trained model that used the Random Forest output as a feature. Bringing more information to the model by using higher order graphs improved only the performance of the local linear regression and structured SVM classifier. The second-order graphs contain quadratically more edges which adds complexity to the learning algorithm, which may be a reason why it lead to a better result with large margin training but worse with Perceptron.

In the second experiment, we used the standard metrics (see Table 2) on the ICDAR 2013 dataset to evaluate the effect of replacing the hypothesis decoding method on the performance of the STR pipeline.

The ICDAR dataset is commonly used to evaluate performance of STR methods, but to our knowledge the only method to report performance in the

Table 1. Results of the intrinsic evaluation (Levenshtein distance $\bar{d}$, relative Leven-shtein distance $\bar{d}_r$, full string accuracy $\bar{a}$).

Model	Order	$\bar{d}$	$\bar{d}_r$	$\bar{a}$
TextSpotter	–	.647	.134	.685
Logistic Regression	1st	.701	.127	.561
SVM, Gauss. kernel	1st	.480	.094	.737
Multilayer Perceptron	1st	.471	.100	.754
Random Forest	1st	**.332**	**.068**	.807
Structured Perceptron	1st	.463	.092	.738
Structured SVM	1st	.439	.082	.750
Str. Perc + Rand. for	1st	.377	.080	**.816**
Str. SVM + Rand. for	1st	.377	.080	**.816**
Logistic Regression	2nd	.660	.121	.631
SVM, Gauss. kernel	2nd	.599	.118	.657
Multilayer Perceptron	2nd	.598	.115	.676
Random Forest	2nd	.398	.075	.779
Structured Perceptron	2nd	.488	.104	.701
Structured SVM	2nd	.402	.077	.775
Str. Perc + Rand. for	2nd	.504	.101	.725
Str. SVM + Rand. for	2nd	.398	.077	.779

end-to-end setup is TextSpotter [2] as other methods either focus solely on text localization or on cropped word recognition. The recent ICDAR 2013 Robust Reading competition [1] also only listed participants in these two limited setups.

The *text localization* measures how well the method is able to localize text areas in an image, the *character* and *word retrieval* measure the proportion of correctly transcribed characters (respectively words) in the dataset, when a character (a word) is considered to be correctly transcribed when it is localized correctly and the textual content is identical (using case-sensitive comparison) with the ground truth label. Most of the models substantially improved the recognition precision (the best result was achieved by the structured SVM), while the recall is only marginally worse. The combined model achieved very good results in the character recognition both in precision and recall, however it lowered the whole word recognition performance.

7   Discussion

The intrinsic evaluation showed a significant improvement over the baseline method, however the biggest effect on the whole STR pipeline is in improving precision of character recognition (8 percentage points improvement for

Table 2. Results of the pipeline evaluation (precision P, recall R, and F_1 measure).

Model	Order	Text localization			Character retrieval			Word retrieval		
		P	R	F_1	P	R	F_1	P	R	F_1
TextSpotter	–	.828	**.629**	**.715**	.786	.625	.696	.421	**.368**	.392
Logistic Regression	1st	.831	.600	.697	.769	.593	.670	.340	.340	.316
SVM, Gauss. kernel	1st	.822	.605	.697	.780	.601	.679	.387	.387	.359
Multilayer Perceptron	1st	.814	.574	.673	.797	.596	.682	.394	.394	.330
Random Forest	1st	.809	.577	.673	.810	.606	.693	.414	.345	.376
Structured Perceptron	1st	**.842**	.617	.710	.800	.626	.703	.424	.360	.389
Structured SVM	1st	.794	.585	.673	.821	.614	.703	**.425**	.364	**.393**
Str. Perc + Rand. for	1st	.833	.605	.701	**.885**	**.689**	**.775**	.404	.344	.372
Str. SVM + Rand. for	1st	.833	.605	.701	**.885**	**.689**	**.775**	.404	.344	.372
Logistic Regression	2nd	.843	.606	.705	.784	.604	.682	.387	.329	.356
SVM, Gauss. kernel	2nd	.808	.589	.681	.788	.604	.684	.403	.347	.373
Multilayer Perceptron	2nd	.829	.578	.681	.798	.590	.679	.387	.327	.355
Random Forest	2nd	.804	.570	.667	.808	.602	.690	.425	.353	.386
Structured Perceptron	2nd	.836	. 613	.707	.808	.624	.704	.425	.359	.389
Structured SVM	2nd	.802	.590	.680	.812	.617	.701	.418	.357	.385
Str. Perc + Rand. for	2nd	.810	.605	.694	.808	.625	.705	.410	.359	.383
Str. SVM + Rand. for	2nd	.820	.599	.692	.818	.626	.709	.404	.360	.381

the combined model). This can be attributed to a better language modelling where knowledge of character bigram (trigram) statistics can help to distinguish between similar characters, as seen in Fig. 3. The character case pattern features also prevented the differently cased letters to appear within words. Nevertheless, the precision of the word recognition is improved only slightly, because of the relative strictness of the evaluation protocol (a word is considered as correctly recognized only when all its characters match the ground truth, using case-sensitive comparison).

The recall in all three STR metrics (localization, character and word retrieval) remains virtually unchanged compared to the baseline method, which is desired because an improved hypotheses decoding method cannot contribute to detect more characters in the earlier stages of the pipeline but it could incorrectly reject true characters.

The results of classifiers trained on the edges locally show that the classifiers with non-linear decision boundaries performed much better than the linear ones. The gain from the non-linearity is comparable from the gain of using the inference algorithm during the learning. This suggests that even better results could be achieved using a structured prediction method utilizing a non-linear decision boundary.

Using the trigram edges did not lead to any significant improvement in the decoding accuracy. We hypothesize that the additional information from the trigram features did not outweigh the added complexity of finding a path in quadratically bigger graph. However, choosing more informative character trigram features can lead to a different trade-off between the decoding complexity and the informativeness of the features.

8 Conclusions

We proposed a method for hypothesis decoding in a Scene Text Recognition pipeline. Our approach is based on structured prediction and allows to exploit a larger number of features. When plugged-in to an end-to-end STR system together with additional typographical and language features proposed in this work, it achieves a state-of-the-art precision for character and word recognition on the standard ICDAR 2013 dataset and brings a substantial improvement on the character level.

Except the already mentioned non-linear structured prediction methods, another improvements in the word decoding could be achieved by using some global features evaluating the produced words. Additional rescoring of a list word of hypotheses using a syllable based language model or a corpus based spell checking may be a way how to increase also the whole word recognition scores.

A natural follow-up is connecting the recognized words into longer logical segments and thus enabling to use the STR output as an input to Machine Translation and Information Retrieval. The knowledge of which words belong together and how they follow each other can provide more informative input for further processing than just bags of words and moreover it can provide further accuracy improvements to STR methods.

Acknowledgements. This research has been funded by the Czech Science Foundation (grant number P103/12/G084). Lukáš would also like to acknowledge the Google PhD Fellowship in Computer Vision and the Google Research Award.

References

1. Karatzas, D., Shafait, F., Uchida, S., Iwamura, M., Mestre, S.R., Mas, J., Mota, D.F., Almazan, J.A., de las Heras, L.P., et al.: ICDAR 2013 robust reading competition. In: 2013 12th International Conference on Document Analysis and Recognition (ICDAR), pp. 1484–1493. IEEE (2013)
2. Neumann, L., Matas, J.: On combining multiple segmentations in scene text recognition. In: 2013 12th International Conference on Document Analysis and Recognition (ICDAR), pp. 523–527. IEEE (2013)
3. Ghoshal, A., Jansche, M., Khudanpur, S., Riley, M., Ulinski, M.: Web-derived pronunciations. In: IEEE International Conference on Acoustics, Speech and Signal Processing, 2009. ICASSP 2009, pp. 4289–4292. IEEE (2009)
4. Bilmes, J.A.: Graphical models and automatic speech recognition. In: Johnson, M., Khudanpur, S.P., Ostendorf, M., Rosenfeld, R. (eds.) Mathematical Foundations of Speech and Language Processing, pp. 191–245. Springer, New York (2004)

5. Daumé III, H., Langford, J., Marcu, D.: Search-based structured prediction. Mach. Learn. **75**, 297–325 (2009)
6. Koehn, P., Hoang, H., Birch, A., Callison-Burch, C., Federico, M., Bertoldi, N., Cowan, B., Shen, W., Moran, C., Zens, R., et al.: Moses: open source toolkit for statistical machine translation. In: Proceedings of the 45th Annual Meeting of the ACL on Interactive Poster and Demonstration Sessions, pp. 177–180. Association for Computational Linguistics (2007)
7. Zhang, H., Zhao, K., Song, Y.Z., Guo, J.: Text extraction from natural scene image: a survey. Neurocomputing **122**, 310–323 (2013)
8. Mishra, A., Alahari, K., Jawahar, C.: Top-down and bottom-up cues for scene text recognition. In: 2012 IEEE Conference on Computer Vision and Pattern Recognition (CVPR), pp. 2687–2694. IEEE (2012)
9. Novikova, T., Barinova, O., Kohli, P., Lempitsky, V.: Large-lexicon attribute-consistent text recognition in natural images. In: Fitzgibbon, A., Lazebnik, S., Perona, P., Sato, Y., Schmid, C. (eds.) ECCV 2012, Part VI. LNCS, vol. 7577, pp. 752–765. Springer, Heidelberg (2012)
10. Wang, K., Babenko, B., Belongie, S.: End-to-end scene text recognition. In: 2011 IEEE International Conference on Computer Vision (ICCV), pp. 1457–1464. IEEE (2011)
11. Neumann, L., Matas, J.: Real-time scene text localization and recognition. In: 2012 IEEE Conference on Computer Vision and Pattern Recognition (CVPR), CA, USA, pp. 3538–3545. IEEE (2012)
12. Roy, S., Roy, P.P., Shivakumara, P., Louloudis, G., Tan, C.L., Pal, U.: HMM-based multi oriented text recognition in natural scene image. In: 2013 2nd IAPR Asian Conference on Pattern Recognition (ACPR), pp. 288–292. IEEE (2013)
13. Shi, C., Wang, C., Xiao, B., Zhang, Y., Gao, S., Zhang, Z.: Scene text recognition using part-based tree-structured character detection. In: 2013 IEEE Conference on Computer Vision and Pattern Recognition (CVPR), pp. 2961–2968. IEEE (2013)
14. Bissacco, A., Cummins, M., Netzer, Y., Neven, H.: PhotoOCR: reading text in uncontrolled conditions. In: 2013 IEEE International Conference on Computer Vision (ICCV), pp. 785–792. IEEE (2013)
15. Weinman, J., Butler, Z., Knoll, D., Feild, J.: Toward integrated scene text reading. IEEE Trans. Pattern Anal. Mach. Intell. **36**, 375–387 (2014)
16. Feild, J.: Improving text recognition in images of natural scenes. Ph.D. thesis, University Massachusetts Amherst (2014)
17. Hall, M., Frank, E., Holmes, G., Pfahringer, B., Reutemann, P., Witten, I.H.: The WEKA data mining software: an update. ACM SIGKDD Explor. Newslett. **11**, 10–18 (2009)
18. Collins, M.: Discriminative training methods for hidden Markov models: theory and experiments with perceptron algorithms. In: Proceedings of the ACL-02 Conference on Empirical Methods in Natural Language Processing, vol. 10, pp. 1–8. Association for Computational Linguistics (2002)
19. Joachims, T., Finley, T., Yu, C.N.J.: Cutting-plane training of structural SVMs. Mach. Learn. **77**, 27–59 (2009)
20. Svetnik, V., Liaw, A., Tong, C., Culberson, J.C., Sheridan, R.P., Feuston, B.P.: Random forest: a classification and regression tool for compound classification and QSAR modeling. J. Chem. Inf. Comput. Sci. **43**, 1947–1958 (2003)

Perspective Scene Text Recognition with Feature Compression and Ranking

Yu Zhou[1], Shuang Liu[1], Yongzheng Zhang[1(✉)], Yipeng Wang[1], and Weiyao Lin[2]

[1] Institute of Information Engineering, Chinese Academy of Sciences, Beijing 100093, China
{zhouyu,zhangyongzheng}@iie.ac.cn
[2] School of Electronic, Information and Electrical Engineering, Shanghai Jiao Tong University, Shanghai 200240, China

Abstract. In this paper we propose a novel character representation for scene text recognition. In order to recognize each individual character, we first adopt a bag-of-words approach, in which the rotation-invariant circular Fourier-HOG features are densely extracted from an individual character and compressed into middle level features. Then we train a set of two-class linear Support Vector Machines in a one-vs-all schema to rank the compressed features by their contributions to the classification. Based on the ranking result we further select and keep those top rated features to build a compact and discriminative codebook. By using densely extracted features that are rotation-invariant and efficient, our method is capable of recognizing perspective texts of arbitrary orientations, and can be combined with the existing word recognition methods. Experimental results demonstrates that our method is highly efficient and achieves state-of-the-art performance on several benchmark datasets.

1 Introduction

Nowadays with the widespread availability of low cost devices equipped with cameras, lots of natural scene images that contain text information are generated. Texts in images contain much semantic information, which could be used to build useful applications such as street sign interpretation, content based image retrieval and product recognition.

Unlike traditional document optical character recognition (OCR) [1], which has achieved sufficient accuracy for practical application, recognizing text in uncontrolled environments is still a challenge. This is because texts in uncontrolled environments often suffer from low resolution, blur, non-uniform illumination and cluttered background. Especially, many images captured by handheld devices may suffer from perspective distortion, which is still an open problem in the computer vision community.

Y. Zhou and S. Liu are contributed equally.

© Springer International Publishing Switzerland 2015
C.V. Jawahar and S. Shan (Eds.): ACCV 2014 Workshops, Part II, LNCS 9009, pp. 181–195, 2015.
DOI: 10.1007/978-3-319-16631-5_14

Many text detection methods, such as [2–5] already provides character level segmentations. Thus, we are mainly concerned with segmented scene text. Most existing scene text recognition methods are focused on recognizing frontal scene text, while the task of recognizing perspective text of arbitrary orientations is still not well addressed. Some methods use text rectification to deal with perspective distortion with the assumption that the shape of the text can be extracted with high accuracy and the text is in a straight line [6,7]. This approach however, might not work well because of the aforementioned interference factors. Besides, methods that rely on rigid features, e.g., histogram of oriented gradient(HOG) [8], achieved satisfying performance on frontal text recognition datasets. However it cannot be used for recognizing perspective text directly [9,10]. To solve this problem, one can train a classifier with character examples of all possible poses. However, this approach is not realistic since it is expensive to collect and label examples. Also, to describe texts in such a way would make a model too complex and computationally inefficient. Therefore, it is important to develop new methods to effectively represent and recognize characters in natural scene images.

Usually, to discriminate among a fairly large number of classes (e.g., 62 character classes), low level features have to be extracted densely to provide enough information. These low level features can be directly used in a bag-of-words approach [11,12]. However non-text regions often introduce strong noise that are not helpful. These features have to be processed equally when searching the visual word codebook, and this can be very costly. Besides, since there are only a few distinctive prototypes for each character class, considering all of these low level features individually is inappropriate. To address the above problem, we observe that characters can be reduced to a few reoccurring shape prototypes. For example, intuitively, from a rotation-invariant point of view "o" is made entirely of curve strokes, and "N" is made of straight strokes and two identical turns. These patterns are obvious and prevalent regardless of languages. Also, it can be expected that the features of non-symmetrical characters such as "R" and "G" will have more variation than those of symmetrical ones. If these patterns can be discovered we can build a more compact and discriminative middle level representation with them. Although, a rotation-invariant representation would cause some confusions in individual character recognition (e.g., between "N" and "Z"), language models can be designed to effectively correct these confusions by incorporating lexicons.

Therefore, in this paper, we propose to extract rotation-invariant features densely and compressing them into compact and discriminative middle level features. These features are automatically learned and independent to language. We first extract circular Fourier-HOG (CHOG) [13] densely from character images, which serves as the underlying low level rotation invariant feature. Due to the nature of scene text, it is the rotation that causes the most signification variance when the viewpoint changes. Although CHOG is only invariant to in-plane rotation, when trained with a multi-scale bag of words scheme on sufficient samples from different viewpoints, it can tolerate perspective distortion to some extent. Next, to compress the CHOG features we use k-means to partition them into a number of prototypes depending on the variation. Then, we rank these prototypes and use the top rated features to build a compact visual word codebook. Finally we retrain a character classifier with the new codebook. In this way the size of our codebook and the computational workload of our method is significantly decreased. To evaluate the effectiveness of our method, we incorporate our method into the classic PLEX word recognition pipeline [14] and compare our method to the state-of-the-art methods on scene character recognition and scene word recognition. Our method outperforms the state-of-the-art methods on perspective text recognition while being orders of magnitude faster, and its performance is competitive on frontal text recognition.

The remaining of this paper is organized as follows: in Sect. 2 we discuss related work. We describe our method in detail in Sect. 3, followed by the implementation details and the experimental results of our method. Finally we conclude our paper in Sect. 6.

2   Related Work

A variety of methods have been developed for scene text detection and recognition. Maximally stable extremal region [1,6,15], stroke width transform [3] and HOG [14] have been successfully applied in scene text detection. The outputs of the detection algorithm are usually the bounding boxes of either characters or words. HOG templates have also been used to match character instances in test images with training examples [16]. Shi et al. [17] proposed to use manually designed deformable part-based model to represent characters. Most of these methods are mainly concerned with only frontal text.

Using rotation invariant feature such as scale invariant feature transform (SIFT) to describe characters has been proved successful. Phan [12] proposed to use dense SIFT instead of normal SIFT to describe individual characters. With the original SIFT, the descriptors are only extracted at sparse interest points. Since scene characters suffer from deformations such as blurring and uneven illumination, the number of detected interest points is not sufficient. The dense SIFT defined in the literature was designed for scene classification, which does not require rotation invariance [18]. An extraction scheme that fixes the position and size but allows the orientation of the interest points to vary was devised in their work, which provides rotation invariance. The work of Phan et al. provides helpful insights into perspective scene text recognition, and it also introduces two datasets that are used to benchmark perspective recognition performance.

SIFT aligns a local coordinate system to the dominant gradient direction at each detected interest point, which relies on the assumption that such a dominant gradient orientation is available. SIFT does not work well for arbitrary positions or dense feature computation, and it is a main source of error in dense image alignment [19]. Most recent text recognition approaches skip this step and use the non-invariant dense HOG features with a sliding window classifier.

CHOG, on the other hand, offers a well defined rotation behavior by representing circular HOG on Fourier domain. The window function of CHOG is

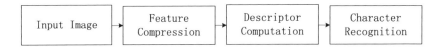

Fig. 1. Flow diagram of our method

isotropic, unlike rectangle spatial window, the descriptor rotates with respect to rotation of its underlying data without leading to any discrete binning artifacts in the histogram. Also, there is no need for interpolation. CHOG can be computed both densely and efficiently, while still being highly discriminative like HOG with rectangle spatial window. Therefore, we use CHOG as the low level building block of our method.

3   Proposed Method

An overview of our method is shown in Fig. 2. The flow diagram of our method is illustrated in Fig. 1. In the following section we describe the procedure for each step.

3.1   Character Representation

In order to sufficiently describe character images, local descriptor of the images has to be computed densely. First, for both training and testing we resize all character images to 48×48. Next, to capture the multi-scale structure in a

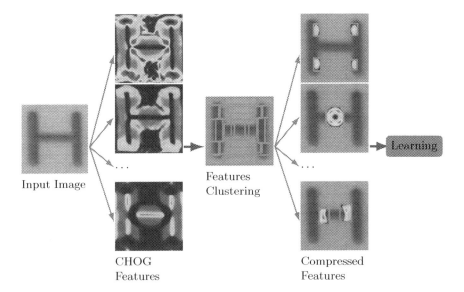

Fig. 2. Overview of our method. For a single character, the densely extracted rotation-invariant features are compressed and ranked by their discriminativeness, which are used to retrain the final character classifier.

pixel's surrounding, Gaussian window functions of different sizes are used. The radial profile of the circles is nested and Gaussian smoothed ($e^{\frac{-(r-d)^2}{2\sigma^2}}, \sigma \in \mathbb{R}$), to ensure that the corresponding CHOG descriptors are neither suffering from discretization effects nor from small deformations. A local CHOG at position $x \in \mathbb{R}^2$ is computed by collecting all magnitudes of gradients within the window function w contributing to orientation $n \in \mathbb{R}^2, \|n\| = 1$ according to the continuous distribution function

$$\mathrm{CHOG}\{f\}_w(\mathbf{x}, \mathbf{n}) = \int_{\mathbf{r} \in \mathbb{R}^2} \|\mathbf{g}(\mathbf{r})\| \delta_{\mathbf{n}}(\hat{\mathbf{g}}(\mathbf{r})) w(\mathbf{x} - \mathbf{r}) d\mathbf{r} \tag{1}$$

where $\mathbf{g} : \mathbb{R}^2 \rightarrow \mathbb{R}^2, g = \nabla f$ is the gradient field of the image f, $S1$ denotes the unit-circle. $\hat{\mathbf{g}} = \mathbf{g}/\|\mathbf{g}\|, \hat{\mathbf{g}} := \mathbb{R}^2 \rightarrow S_1$ the gradient orientation field and $\mathbf{n} \in S_1$ is the current direction histogram entry taken into account. $\delta_{\mathbf{n}} : S_1 \rightarrow \mathbb{R}$ denotes the Dirac delta function on the circle that selects those gradients out of $\mathbf{g}$ with orientation $\mathbf{n}$. Next, the CHOG features are represented in terms of the orthogonal (periodic) circular Fourier basis functions. Thanks to the rotation preserving characteristic, CHOG features rotates smoothly with respect to the underlying image data. We refer the interested readers to [13].

In our experiments specifically, we use more nested circles for larger scale structure, and less nested circles for smaller scale structure. Using more nested circles will result in longer and more detailed descriptors and vice versa. The rationale behind this is that structures of different scales are supposed to carry different amount of information, thus it would make sense to use descriptors of different detail to describe structures of different scale. From our understanding of the description in [12], the standard SIFT implementation was used. The length of the SIFT feature vector is 128 for all key points of different sizes. Using long feature vector to describe small scale image will inevitable introduce more noise and unnecessarily slow down the recognition. Therefore, the features we extracted are supposed to have the advantage of reduced complexity and higher quality [20].

3.2   Feature Compression

The final descriptor of a single character is a histogram computed by assigning its visual words to their nearest visual codes. The variation of the low level features of any particular character is different but limited. Most of the densely extracted low level features vary from their neighbors slightly, except for those around sharp gradient adjustment. Non-text regions and smooth regions often do not contribute to the classification. Yet to find the most similar visual word in the codebook the distance has to be computed for all these features nonetheless, which is a huge waste of computation time. The image resolution of the character segmented from the scene image is generally very small. Therefore, as suggested in [12], keypoint detection methods fail to produce enough high quality keypoints for meaningful classification. Moreover, although the cost of clustering is non-trivial, we can compress the densely extracted features to a compact subset, in which all the features are distinctive from one another. We can both benefit from the discriminative power of densely extracted features and compute the descriptor (histogram) faster. As a result, the combined computational workload is still significantly smaller than directly using densely extracted feature.

Of course, if somehow we could use only the discriminative and distinctive features to build the histogram, we could avoid the unnecessary computation. We propose to densely extract CHOG features on every pixel, then use k-means to compress these features while explaining a sufficient percentage of variation necessary for effective recognition. Here we denote the "percentage explained" as the ratio of the sum of the standard deviation of each clusters to the standard deviation of the whole image. Although the sum of the variation of the features is readily available, we cannot directly determine how many clusters to use. Through empirical analysis on the training samples, we learn the boundary of the number of clusters that is needed to explain a required percentage. We also learn the relationship between the number of clusters and the percentage of variation explained with linear regression. At runtime, we start with the number of clusters predicted by the linear regression model and run k-means for a few iterations, then increase the number of clusters and restart until the required percentage of variation is explained or the number of clusters reaches the upper boundary. The resulting centroids of the clusters are compressed, stable and more importantly distinctive from one another. The computation overhead of k-means clustering is easily compensated because the computation workload using the compressed features instead of the original densely computed raw features is significantly reduced. The outline is outlined in Procedure 1. Finally, to exploit the spatial characteristic for each compressed feature vector, we add another dimension by appending the distance between the centroid corresponding to the compressed feature and the center of the character to the end. This is before the global vocabulary construction. In a rotation invariant setting it is difficult to differentiate between characters like "U","C". Doing so has virtually no cost, we can both keep the rotation invariance and build a more discriminative vocabulary, since in general "U" are more slender than "C".

3.3   Feature Selection

To build a compact and discriminative codebook, we need to rank and filter out those visual words that contribute weakly to the classification. First we perform k-means clustering on the compressed features of all the training samples to generate an initial codebook. We then use this codebook to compute the initial histograms of each training samples. These histograms are used to train a set of two-class linear support vector machines (SVMs) in a one-vs-all schema, whose weights are used to rank the visual words by their relevance. We keep the top rated ones of each character class and build a new codebook. We then compute new histograms while discarding features that are not found in the new codebook. Finally we train a multi-class radial basis function (RBF) kernel SVM with the new histograms as our character classifier. Multi-class SVM has a well defined probabilistic output, which can be useful for further natural language processing. Ranking the feature by their relevance helps us to focus on discriminative features and discard unnecessary features.

Procedure 1. Feature Compression

Input: Raw features, required percentage, variation predictor
Output: Compressed features, percentage explained
 1: **Variation Computation:** Compute the standard deviation of the raw features.
 2: **Cluster Initialization:** Predict the number of clusters needed to explain the required percentage of variation.
 3: **Feature Clustering:** Start clustering and compute percentage of variation explained when it converges.
 4: **Reiteration:** Increase the number of clusters and restart until required percentage is explained or the number of clusters exceeds the upper boundary, then repeat step 1.

The one-vs-all schema is used to rank the features. SVMs classify data via finding a separating hyperplane with the maximal margin between two classes. Given a set of the visual word histograms $\mathbf{x}_i \in \mathbb{R}, i = 1, \ldots, l$ and character classes $y_i \in \{1, \ldots, 62\}, i = 1, \ldots, l$, we use a one-vs-all schema for each individual character class to train a set linear SVMs that solve the following unconstrained optimization problem:

$$\min_{\mathbf{w},b} \quad \frac{1}{2}\mathbf{w}^T\mathbf{w} + C \sum_{1}^{l} \xi(\mathbf{w}, b; \mathbf{x}_i, y_i) \tag{2}$$

where $\xi(\mathbf{w}, b; \mathbf{x}_i, yi)$ is a loss function, and $C \geq 0$ is a penalty parameter on the training error. In this paper specifically we use L2-loss linear SVMs. For testing instance $\mathbf{x}$, the decision function is

$$f(\mathbf{x}) = sgn(\mathbf{w}^T \phi(\mathbf{x}) + b) \tag{3}$$

where the mapping function is $\phi(x) = x$. After the set of linear SVMs are obtained for each character class, the weights $\mathbf{w} \in \mathbb{R}$ in Eq. 2 can be used to decide the relevance of each feature [21]. Because the absolute value of the weights $\mathbf{w}$ in linear SVMs indicates the importance of a particular feature in the decision function Eq. 3. The features can be ranked by sorting the corresponding weights [22]. We only keep the top $\mathbf{K}$ ranked features. This procedure is outlined in Procedure 2. Some of the top ranked features sorted by character class is shown in Fig. 3, note that the nested window functions are not displayed because they are hard to visualize, thus the bounding box contains merely the center positions of the CHOG features, which capture the underlying surrounding structures.

4   Implementation Detail

In our implementation we densely extract CHOG features of three different scales on every pixels. By nesting window functions of different sizes we extracted

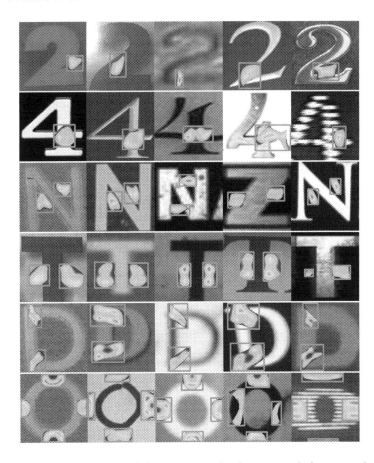

Fig. 3. Top rated compressed feature examples from several character classes

feature vectors of different lengths, which describe the surrounding area with different radius and detail. The window functions are given by $w1 := \{d = 0, \sigma = 4\}$, $w1 := \{d = 0, \sigma = 4\}$; $w2 := \{d = 4, \sigma = 6\}$ and $w1 := \{d = 0, \sigma = 4\}$; $w2 := \{d = 4, \sigma = 6\}$; $w3 := \{d = 10, \sigma = 12\}$, where σ indicates the outer radius and d indicates the inner radius. To compare the computational cost, we used the average dimension of the feature vectors. The average length of the extracted CHOG feature vectors is 60. On average the extraction procedure for each image takes 200 ms.

In terms of feature compression we obtain the best result in experiments when roughly 95 % of the variation of individual character image is explained. A linear regression model is trained to describe the relationship between the percentage of variation explained and the number of clusters used in k-means. At runtime this model is used to determine the initial number of clusters we use on the testing images. After a few initial iterations we increase the number of clusters and restart until the number of clusters reaches the upper boundary or the required

---

Procedure 2. Feature Ranking

Input: Training samples, visual word codebook
Output: Top ranked features and retrained classifier
 1: **Intial Train:** Train a set of two-class L2-loss linear SVMs for each individual character class using one-vs-all schema, using grid search to find best penalty parameters C on the visual word codebook.
 2: **Feature Ranking:** Rank the visual words according to the absolute values of weights in the set of SVMs for each individual character class.
 3: **Feature Selection:** Truncate the visual word codebook, keeping only the top K ones in each individual character class.
 4: **Model Retrain:** Retrain a multi-class SVM to obtain the final classifier.

---

percentage of variation is explained. In our implementation on average, this procedure needs 10 iterations with 20 clusters on 48×48 feature vectors whose average length is 60. Supposing on a visual word codebook of size **N** the workload for finding the most similar **K** prototype with **M** dimension is $\mathbf{N} \times \mathbf{M} \times \mathbf{K}$, this procedure is equivalent to a workload of $10 \times 20 \times 60 \times (48 \times 48) = 6912000$, which takes 100 ms on average.

For feature ranking, the relationship between the number of features and the accuracy of the classification is illustrated in Fig. 4. Take the visual word codebook used for SVT-Perspective for example. As illustrated in Fig. 4, the best performance was achieved when the size of the vocabulary learned from compressed CHOG features is 250. The computation workload defined above equals to $1 \times 250 \times 60 \times 20 = 300000$. The combined computation workload of our method is only 13.0 % of the state-of-the-art perspective character recognition method described in [12], which is equivalent to $1 \times 3000 \times 128 \times (12 \times 12) = 55296000$. Consequently, on SVT-Perspective (with the original lexicons), the average processing time of our method is 3.5 s while the method in [12] is 38.6 s.

5 Experimental Results

Proposed method has been evaluated on several benchmark datasets. Results on these datasets are compared to the state-of-the-art methods. For character recognition our method was evaluated and compared to other methods on ICDAR-Char and SVT-Char. ICDAR-Word and SVT-Word are used for frontal word recognition, while SVT-Perspective-Word and MSRA-TD-500 are used for the recognition of perspective word.

The PLEX framework [14] requires a lexicon for each word image, which is not provided in ICDAR-Word by default. For fairness we used the same lexicons and lexicon construction method in [14]. Following previous work [10,12,17,23] on this dataset we skipped the words with less than 3 characters, and those with non-alphanumeric characters, and then constructed additional lexicons denoted as *Full* which contain all the combined ground truth words.

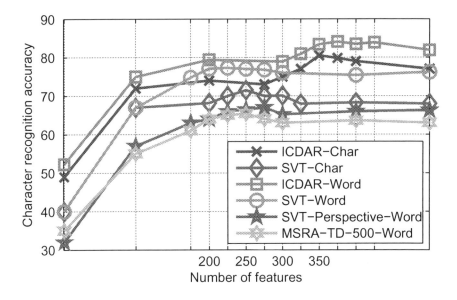

Fig. 4. Recognition accuracy of our method with different vocabulary size

5.1   Classifier Training

Our classifier was trained only on frontal examples. To train the character classifier we harvested labeled character images from ICDAR-Char, SVT-Char, IIIT 5K-Word and Chars74K. ICDAR-Char is a character level subset of the ICDAR 2003 Robust Word Recognition Competition [24] dataset. Similar to ICDAR, the SVT dataset [14,25] also contains both word level annotations on full images and character level annotations. IIIT 5K-Word [23] contains word level and character level annotation of cropped word images only. Chars74K [26] contains only cropped characters. Our character detector in the PLEX framework was also trained on these datasets. Leave-one-out cross validation was performed on these four datasets.

Theoretically, the performance of our method on perspective scene text datasets would not degrade as much as those methods that only focused on frontal text recognition, because CHOG is inherently more suited for dense computation compared with SIFT and our representation is more robust. The experiments in Sect. 5.4 demonstrated the effectiveness of our method.

5.2   Character Recognition

We evaluated the character recognition performance of our method on both ICDAR-Char and SVT-Char. Table 1 lists the performance on these two datasets. Some characters in SVT suffer from perspective distortion, which makes SVT different from ICDAR in that the characters in ICDAR are mainly frontal. This difference is reflected by the recognition accuracy difference between two datasets.

Our method achieved the best result on SVT-Char. Without context, it is hard to tell the difference between some characters such as "W" and "M", "q" and "b", etc. Nevertheless, the performance of our method on ICDAR-Char is better than that of [12] and is only slightly worse than that of [27]. Although the performance gap can be partially explained by the aforementioned reason, it suggests that neural network based methods are still better suited for frontal character recognition tasks accuracy-wise. These were the two main causes for the performance gap between our methods and [27] on frontal word recognition.

Table 1. Character recognition accuracy (in%). Only our method and [12] aims to solve perspective character recognition, others focus only on frontal character recognition.

Method	ICDAR-Char	SVT-Char
Proposed	80.5	**71.5**
FineReader 9.0 [28]	21.0	11.7
K. Wang (PLEX) [14]	64.0	N.A
Mishra [9]	N.A	61.9
Coates [29]	81.7	N.A
T. Wang [27]	**83.9**	N.A
Yi [30]	76.0	N.A
Phan [12]	75.6	67.0

5.3   Frontal Word Recognition

Although we focused on perspective scene text recognition, for a comprehensive comparison we still evaluated the word recognition performance of our method on frontal text datasets. The word level ICDAR-Word subset of ICDAR was used to evaluate a variety of scene text recognition methods in cropped images. Most of the word images in ICDAR-Word are frontal, whereas the SVT-Word dataset contains more perspectively distorted words. The word level SVT-Word subset of SVT was also used in the evaluation. Table 2 lists the word recognition results. Some recognition examples are shown in Fig. 5. We achieved the state-of-the-art performance on SVT-Word and outperformed several other methods on ICDAR-Word. This result accords with our analysis in Sect. 5.2.

5.4   Perspective Word Recognition

We used the words with English characters and digits in SVT-Perspective-Word and MSRA-500 to evaluate our method on perspective word recognition. The recognition accuracy is listed in Table 3. The performance degradation between frontal and perspective text is also measured in Table 4. The smaller gap between the performance on SVT-Word and SVT-Perspective-Word indicates that our method is more robust against rotation and perspective distortion.

Table 2. Frontal word recognition accuracy (in%).

Method	ICDAR-Word	SVT-Word
Proposed	84.1	**77.3**
FineReader 9.0 [28]	56.0	36.0
K. Wang et al. [14]	76.0	57.0
Mishra et al. [9]	81.8	73.3
Mishra et al. [23]	80.3	73.6
T. Wang et al. [27]	**90.0**	70.0
Phan et al. [12]	82.2	73.3

Table 3. Recognition accuracy on perspective text (in%).

Method	SVT-Perspective Word	SVT-Perspective Word (Full)	MSRA-TD500-Word (Full)
Proposed	**67.0**	**45.7**	**65.4**
FineReader 9.0 [28]	16.9	9.7	23.2
K. Wang et al. [14]	40.5	26.1	44.5
Mishra et al. [9]	45.7	25.7	27.8
T. Wang et al. [27]	40.2	32.4	20.8
Phan et al. [12]	62.3	42.2	58.4

Table 4. Degradation in performance between frontal texts and perspective texts taken from [12] (in%).

Method	SVT- Word	SVT-Perspective Word	% Change
Proposed	**77.3**	**67.0**	-13.3%
FineReader 9.0 [28]	35.0	16.9	-51.7
K. Wang et al. [14]	57.0	40.5	-28.9
Mishra et al. [9]	73.3	45.7	-37.7
T. Wang et al. [27]	70.0	40.2	-42.6
Phan et al. [12]	73.7	62.3	-15.5

In addition to leave-one-out cross validation, we also excluded SVT-Char from the training set when SVT-Perspective-Word is being tested. In this way we eliminated the possibility of rigged result because the samples in SVT are very similar to those in SVT-Perspective. The fact that our method suffered the least from the perspective distortion revealed that even it was only trained on frontal examples it generalizes well on perspective texts of arbitrary orientations. Some example recognition results are shown in Fig. 6.

Fig. 5. Example recognition results on frontal text

Fig. 6. Example recognition results on perspective text

6   Conclusion

In this paper, we adopt a bag-of-words approach to recognize scene character. We propose to use CHOG as the building block, which has desirable properties such as rotation invariance and efficient dense computation. We compress the densely extracted features to summarize the over-complete raw features to essential subsets, which greatly reduced the computation workload. Finally, we rank the compressed features to build a compact and discriminative visual word codebook. Our method achieves the state-of-the-art performance on perspective scene text recognition, and its performance on frontal scene text recognition is competitive with many state-of-the-art methods. Moreover, for perspective scene text recognition our method is many times faster than [12], which previously was state-of-the-art method.

Acknowledgment. This paper is partially supported by National Natural Science Foundation of China under Contract nos. 61303170, 61402472 and 61471235, and also supported by the National High Technology Research and Development Program of China (863 programs)under Contract nos. 2013AA014703 and 2012AA012803.

References

1. Neumann, L., Matas, J.: Real-time scene text localization and recognition. In: CVPR (2012)
2. Neumann, L., Jiri, M.: Real-time scene text localization and recognition. In: CVPR (2012)
3. Epshtein, B., Ofek, E., Wexler, Y.: Detecting text in natural scenes with stroke width transform. In: CVPR (2010)
4. Chen, H., Tsai, S.S., Schroth, G., Chen, D.M., Grzeszczuk, R., Girod, B.: Robust text detection in natural images with edge-enhanced maximally stable extremal regions. In: ICIP (2011)
5. Huang, W., Lin, Z., Yang, J., Wang, J.: Text localization in natural images using stroke feature transform and text covariance descriptors. In: ICCV (2013)
6. Neumann, L., Matas, J.: A method for text localization and recognition in real-world images. In: Kimmel, R., Klette, R., Sugimoto, A. (eds.) ACCV 2010, Part III. LNCS, vol. 6494, pp. 770–783. Springer, Heidelberg (2011)
7. Dance, C.R.: Perspective estimation for document images. In: Electronic Imaging (2002)
8. Dalal, N., Triggs, B.: Histograms of oriented gradients for human detection. In: CVPR (2005)
9. Mishra, A., Alahari, K., Jawahar, C.: Top-down and bottom-up cues for scene text recognition. In: CVPR (2012)
10. Novikova, T., Barinova, O., Kohli, P., Lempitsky, V.: Large-lexicon attribute-consistent text recognition in natural images. In: Fitzgibbon, A., Lazebnik, S., Perona, P., Sato, Y., Schmid, C. (eds.) ECCV 2012, Part VI. LNCS, vol. 7577, pp. 752–765. Springer, Heidelberg (2012)
11. Csurka, G., Dance, C.R., Fan, L., Willamowski, J., Bray, C.: Visual categorization with bags of keypoints. In: ECCV (2004)
12. Phan, T.Q., Shivakumara, P., Tian, S., Tan, C.L.: Recognizing text with perspective distortion in natural scenes. In: ICCV (2013)
13. Skibbe, H., Reisert, M.: Circular fourier-hog features for rotation invariant object detection in biomedical images. In: ISBI (2012)
14. Wang, K., Babenko, B., Belongie, S.: End-to-end scene text recognition. In: ICCV (2011)
15. Chen, H., Tsai, S.S., Schroth, G., Chen, D.M., Grzeszczuk, R., Girod, B.: Robust text detection in natural images with edge-enhanced maximally s table extremal regions. In: ICIP (2011)
16. Weinman, J.J., Learned-Miller, E., Hanson, A.R.: Scene text recognition using similarity and a lexicon with sparse belief propagation. IEEE TPAMI $31(10)$, 1733–1746 (2009)
17. Shi, C., Wang, C., Xiao, B., Zhang, Y., Gao, S., Zhang, Z.: Scene text recognition using part-based tree-structured character detection. In: CVPR (2013)
18. Bosch, A., Zisserman, A., Muñoz, X.: Scene classification via pLSA. In: Leonardis, A., Bischof, H., Pinz, A. (eds.) ECCV 2006. LNCS, vol. 3954, pp. 517–530. Springer, Heidelberg (2006)
19. Lin, W.Y., Liu, L., Matsushita, Y., Low, K.L., Liu, S.: Aligning images in the wild. In: CVPR (2012)
20. Napoleon, D., Pavalakodi, S.: A new method for dimensionality reduction using k-means clustering algorithm for high dimensional data set. Int. J. Comput. Appl. (2011)

21. Guyon, I., Weston, J., Barnhill, S., Vapnik, V.: Gene selection for cancer classification using support vector machines. Mach. Learn. **46**, 389–422 (2002)
22. Chang, Y.W., Lin, C.J.: Feature ranking using linear svm. In: JMLR (2008)
23. Mishra, A., Alahari, K., Jawahar, C.: Scene text recognition using higher order language priors. In: BMVC (2012)
24. Sosa, L.P., Lucas, S.M., Panaretos, A., Sosa, L., Tang, A., Wong, S., Young, R.: ICDAR 2003 robust reading competitions. In: ICDAR (2003)
25. Wang, K., Belongie, S.: Word spotting in the wild. In: Daniilidis, K., Maragos, P., Paragios, N. (eds.) ECCV 2010, Part I. LNCS, vol. 6311, pp. 591–604. Springer, Heidelberg (2010)
26. de Campos, T., Babu, B.R., Varma, M.: Character recognition in natural images. In: VISAPP (2009)
27. Wang, T., Wu, D.J., Coates, A., Ng, A.Y.: End-to-end text recognition with convolutional neural networks. In: ICPR (2012)
28. ABBYY FineReader Professional 9.0 (2008). http://www.abbyy.com/
29. Coates, A., Carpenter, B., Satheesh, S., Suresh, B., Wang, T., Wu, D.J., Ng, A.Y.: Text detection and character recognition in scene images with unsupervised feature learning. In: ICDAR (2011)
30. Yi, C., Yang, X., Tian, Y.: Feature representations for scene text character recognition: A comparative study. In: ICDAR (2013)

Second Workshop on User-Centred Computer Vision (UCCV 2014)

3D Interaction Through a Real-Time Gesture Search Engine

Shahrouz Yousefi$^{(\boxtimes)}$ and Haibo Li

KTH Royal Institute of Technology, 100 44 Stockholm, Sweden
shahrouz@kth.se

Abstract. 3D gesture recognition and tracking are highly desired features of interaction design in future mobile and smart environments. Specifically, in virtual/augmented reality applications, intuitive interaction with the physical space seems unavoidable and 3D gestural interaction might be the most effective alternative for the current input facilities such as touchscreens. In this paper, we introduce a novel solution for real-time 3D gesture-based interaction by finding the best match from an extremely large gesture database. This database includes the images of various articulated hand gestures with the annotated 3D position/orientation parameters of the hand joints. Our unique matching algorithm is based on the hierarchical scoring of the low-level edge-orientation features between the query frames and database and retrieving the best match. Once the best match is found from the database in each moment, the pre-recorded 3D motion parameters can instantly be used for natural interaction. The proposed bare-hand interaction technology performs in real-time with high accuracy using an ordinary camera.

1 Introduction

Currently, people interact with the digital devices through the track pads and touchscreen displays. The latest technology offers single or multi-touch gestural interaction on 2D touchscreens. Although this technology has solved many limitations in human mobile device interaction, the recent trend reveals that people always prefer to have intuitive experiences with their digital devices. For instance, popularity of the Microsoft Kinect can demonstrate the idea that people enjoy experiences that give them the freedom to act like they would in the real world. However, when we discuss the next generation of digital devices such as AR glasses and smart watches we should also consider the next generation of interaction facilities. The important point is to select a suitable space and develop a technology for effective and intuitive interaction. An effective solution for natural interaction is to extend the interaction space from 2D surface to real 3D space [1,2]. For this reason, vision-based 3D gestural interaction might be hired to facilitate a wide range of applications where using physical hand gestures are unavoidable. Specifically, in future wearable devices such as Google Glass, 3D gestural interaction with augmented environments might be extremely useful. Therefore, developing an efficient and robust interaction technology seems to

© Springer International Publishing Switzerland 2015
C.V. Jawahar and S. Shan (Eds.): ACCV 2014 Workshops, Part II, LNCS 9009, pp. 199–213, 2015.
DOI: 10.1007/978-3-319-16631-5_15

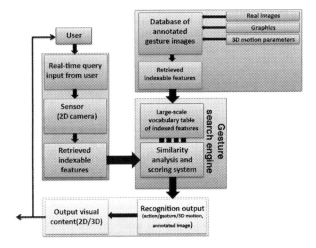

Fig. 1. System overview of the real-time gesture retrieval system. For each query image, the best corresponding match with the tagged motion information will be retrieved through the gesture search engine.

be a need for the near future. From technical perspective, due to the complexity, diversity and flexibility of the hand poses and movements, recognition, tracking and 3D motion analysis are challenging tasks to perform on hand gestures. In order to handle these difficulties, we decided to shift the complexity from classical pattern recognition problem to large-scale gesture retrieval system. Due to the possibility of forming a large-scale image database, the new problem is to find the best match for the query among the whole database. In fact, for a query image or video, representing a unique hand gesture with specific position and orientation of the joints, the challenging task is to retrieve the most similar image from the database that represent the same gesture with maximum similarity in position and orientation. Our matching method is based on the scoring of the database images with respect to the similarity of the low-level edge-orientation features to the query frame. By forming an advanced indexing system in an extremely large lookup table, the scoring system performs the search step and the best out-put result will be retrieved efficiently. Since in the offline step we annotate the database images with the corresponding global and local position/orientation of the joints, after the retrieval step, the motion parameters might be immediately used to facilitate the interaction between user and device in various applications (see Fig. 1).

2   Related Work

Designing a robust gesture detection system, using a single camera, independent of lighting conditions or camera quality is still a challenging issue in the field of computer vision. A common method for gesture detection is marker-based app-roach. Most of the augmented reality applications are based on marked gloves for accurate and reliable fingertip tracking, [3,4]. However, in marker-based methods users have to wear special inconvenient markers. Moreover, some strategies rely on object segmentation by means of shape or temperature, [5–7]. Robust finger detection and tracking could be gained by using a simple threshold on the infrared images. Despite the robustness, thermal-based approaches require expensive infrared cameras which are not provided to most devices. Many gesture tracking systems are based on new depth sensors such as Kinect, but due to the size and power limitations they are only available for stationary systems [8]. In addition, feature-based algorithms for gesture tracking have been employed in various applications, [7,9]. Model-based approaches are also being used in this area, [10,11].

Generally, all these techniques are computationally expensive, which is not suitable for our purposes. Another set of methods for hand tracking are based on color segmentation in appropriate color space, [5,12]. Color-based techniques are always sensitive to lighting conditions that degrades the quality of recognition and tracking. Other approaches such as template matching and contour-based methods often work for specific hand gestures, [13]. In new smartphones and tablets, accelerometer-based approaches recognize hand gesture motions by using the device's acceleration sensor, [14,15]. Reference [16], use visual color markers for detecting the fingertips to facilitate the gesture-based interaction in augmented reality applications on mobile phones. References [17,18], perform marker-less visual fingertip detection, based on the color analysis and computer vision techniques for manipulating the applications in human device interaction. Reference [19], perform HMM to recognize different dynamic hand gesture motions. Reference [20], use visual marker or shape recognition to augment and track the virtual objects and graphical models in augmented reality environments.

Unfortunately, most of the computer vision algorithms perform quite complex computations for detection and recognition of objects or patterns. For this reason we should find a totally innovative way to integrate the existing solutions with the minimum level of complexity and maximum efficiency. Another important point to mention is that the current technology is mostly limited to gesture detection and global motion tracking not real 3D motion analysis, while in many cases 3D parameters such as position and orientation of the hand joints might be used for manipulation in different applications. Therefore, besides the gesture recognition system we need to retrieve the 3D motion parameters of the hand joints (27 degrees of freedom for one hand). In our innovative solution we treat this issue as a large-scale retrieval problem. In fact, this is the main reason behind choosing very low-level features for efficient detection and tracking system. During the recent years, interesting works have been done on the large-scale image search topic. References [21,22], perform the sketch-based image search based on the indexed oriented chamfer matching and bag-of-features descriptors, respectively. Reference [23], introduces the matching based on distribution of oriented patches. The major problem with image search systems is that although you might receive interesting results in the first top matches but you also might find irrelevant results. Since our plan is to use the retrieval system for designing a real-time interaction scenario, we expect to achieve around 100 % correct detection and accurate 3D motion retrieval. In this work, we demonstrate that how our contribution leads to the effective and efficient 3D gesture recognition and tracking that can be applied to various applications.

3   System Description

As briefly explained before, our recognition and tracking system is based on the low-level edge-orientation features that can be achieved by hierarchical scoring of the similarity between the query and database images. Since hand gestures do not provide complex textured patterns, they are not suitable enough for detecting stable features such as SIFT or SURF. On the other hand, for robustness and efficiency of the detection and tracking, we cannot rely on color-based or shape-based approaches. These are the main reasons behind the selection of edge-based scoring system. As a result, the proposed method works independent of lighting conditions, variety of users, and different environments.

3.1   Pre-processing on the Database

Our database contains a large set of different hand gestures with all the potential variations in rotation, positioning, scaling, and deformations. Besides the matching between the query input and database, one important feature that we aim to achieve is to retrieve the 3D motion parameters from the query image. Since query inputs do not contain any pose information, the best solution is to associate the motion parameters of the query to the best retrieved match from the database. For this reason, we need to annotate the database images with their ground-truth motion parameters, P_{D_i}, and O_{D_i}. In the following we explain how the pre-processing on the database is performed.

Annotation of Global Position/Orientation to the Database: During the process of providing the database, one way to measure the corresponding motion parameters of the hand gesture is to attach the motion sensor to the user's hand and synchronize the image frames with the measured parameters. Another approach is to use computer vision techniques and estimate the parameters from the database itself. Since we could capture extremely clear hand gestures with a uniform background in the database, we could apply the second approach to estimate the global position of the gestures in each frame. As sample hand gestures are shown in Fig. 3, by using the common methods such as computing the area, bounding box, ellipse fitting, etc., we can estimate the position and scale of the user's gesture. On the other hand, to estimate the orientation of the user's gesture in x, y, and z directions, we apply *Active Motion Capture* technique [24, 25]. In active motion capture, during the process of making database, we mount the vision sensor on the user's hand to accurately measure and report the motion parameters in each captured frame. The vision sensor captures and tracks the stable SIFT features from the environment. Next, we find feature point correspondences by matching feature points between consecutive frames. Then the fundamental matrix for each image pair is computed using robust iterative RANSAC algorithm. Due to the fact that the matching part might be degraded by noise, the RANSAC algorithm is used to detect and remove the wrong matches(outliers) and improve the performance. Running RANSAC algorithm, the candidate fundamental matrix is computed based on the 8-point algorithm. The fundamental matrix F is the 3×3 matrix that satisfies the epipolar constraint:

$$x_i^{'T} F x_i = 0 \tag{1}$$

where x_i and $x_i^{'}$ are a set of image point correspondences. Each point correspondence provides one linear equation in the entries of F. Since F is defined up to a scale factor, it can be computed from 8 point correspondences. If the intrinsic parameters of the cameras are known, as they are in our case, the cameras are said to be calibrated. In this case a new matrix E can be introduced by equation:

$$E = K^{'T} F K \tag{2}$$

where the matrix E is called the essential matrix, $K^{'}$ and K are 3×3 upper triangular calibration matrices holding intrinsic parameters of the cameras for two views. Once the essential matrix is known, the relative translation and rotation matrices, t and R can be recovered. Let the singular value decomposition of the essential matrix be:

$$E \sim U diag(1,1,0) V^T \tag{3}$$

where U and V are chosen such that $\det(U) > 0$ and $\det(V) > 0$ ($\sim$ denotes equality up to scale). If we define the matrix D as:

$$D \equiv \begin{bmatrix} 0 & 1 & 0 \\ -1 & 0 & 0 \\ 0 & 0 & 1 \end{bmatrix} \tag{4}$$

Then $t \sim t_u \equiv \begin{bmatrix} u_{13} & u_{23} & u_{33} \end{bmatrix}^T$ and R is equal to $R_a \equiv UDV^T$ or $R_b \equiv UD^TV^T$. If we assume that the first camera matrix is $[I \mid 0]$ and $t \in [0,1]$, there are then 4 possible configurations for second camera matrix: $P_1 \equiv [R_a \mid t_u]$, $P_2 \equiv [R_a \mid -t_u]$, $P_3 \equiv [R_b \mid t_u]$ and $P_4 \equiv [R_b \mid -t_u]$. One of these solutions corresponds to the right configuration. In order to determine the true solution, one point is reconstructed using one of four possible configurations. If the reconstructed point is in front of both cameras, the solution corresponds to the right configuration.

Once the right configuration is obtained, the relative rotation between two consecutive frames are computed and can be tagged to the corresponding captured database image (Fig. 2).

Annotation of Local Joint Motions to the Database: In order to annotate the local motion of the hand joints to the database we have used a semi-automatic system. In this system we manually mark the fingertips and all the hand joints including the finger joints and wrist in each and every frame of the database. Afterwards, our system automatically stores the exact position of the marked points according to the image coordinates and generates the connection between the joints in form of a skeletal model. The joints information and hand model can be used after the retrieval step (see Fig. 3, right).

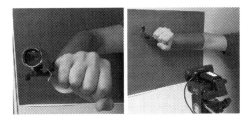

Fig. 2. Active motion capture setup for tagging the rotation parameters to the database images. The hand-mounted camera captures the global 3D rotation parameters and the static camera records the database frames. Both cameras are synchronized to automatically assign the real-time motion parameters to database frames.

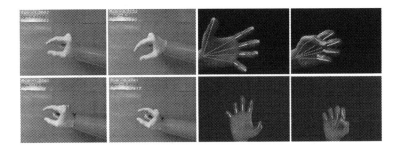

Fig. 3. Left: real-time measurement of the global orientation of the hand gestures in the database images using Active Motion Capture system. R_x, R_y, R_z represent the rotation of the hand gesture around 3D axes in degrees. Right: semi-automatic annotation of the joint positions and skeletal model to the database images.

Defining and Filling the Edge-Orientation Table: Suppose that all the database images, D_{1-k}, are normalized and resized to $m \times n$ pixels and their corresponding edge images, ED_{1-k}, are computed by common edge detection methods such as Canny edge detection algorithm. Therefore, in each binary edge image, any single edge pixel can be represented by its row and column position. Moreover, it is possible to compute the orientation of the edge pixels, α_e, from the gradient of the image in x and y directions: $\alpha_e = atan(d_y/d_x)$. In order to simplify the problem, as it is demonstrated in Fig. 4(top left), we divide the space to eight angular intervals, where the direction of each edge pixel belongs to the one of these intervals. As a result, each single edge pixel will be represented by its position and angle: (x_e, y_e, α_e). In order to make a global structure for edge-orientation features we need to form a large table to represent all the possible cases that each edge-orientation pixel might happen. If we consider the whole edge database with respect to the position and orientation of the edges, (x_e, y_e, α_e), a large vector with size $m \times n \times n_\alpha$, can define all the possibilities, where m and n are number of rows and columns in normalized database images and n_α is the number of angle intervals. For instance, for 320×240 images and 8 angle intervals we will have a vector with length, 614400. After we formed this structure, each (x_i, y_j, α_l) block should be filled with the indices

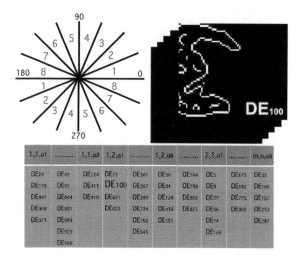

Fig. 4. Top-left: associated angle intervals for edge pixels; Top-right: sample database edge image. The corresponding positions-orientation block to each single edge pixel will be marked with the index of the database image in the edge-orientation table.

of all database images that have edge at the same row, i, and column, j, with similar orientation interval, l. Figure 4 shows how the edge-orientation table is filled with database images.

3.2 Query Processing and Matching:

The first step in the retrieval and matching process is edge detection. This process is the same as edge detection in the database processing but the result will be totally different, because for the query gesture we expect to have large number of edges from the background and other irrelevant objects. In the following we explain how the scoring system works.

Direct Scoring: Assume that each query edge image, QE_i, contains a set of edge points that can be represented by the row-column positions and specific directions. Basically, during the first step of scoring process, for all single query edge pixels, $QE_i \mid_{(x_u,y_v)}$, similarity function to the database images at that specific position is computed as:

$$Sim(QE_i, DE_j) \mid_{(x_u,y_v)} = \begin{cases} 1 \; if \left\{ QE_i \mid_{(x_u,y_v)} \neq 0 \right\} \\ \quad \wedge \left\{ DE_j \mid_{(x_u,y_v)} \neq 0 \right\} \\ \quad \wedge \left\{ (\alpha_i \cong \alpha_j) \mid_{(x_u,y_v)} \right\} \\ \\ 0 \; otherwise \end{cases} \tag{5}$$

If this condition is satisfied for the edge pixel in the query image and the corresponding database images, the first level of scoring starts and all the database images that have an edge with similar direction at that specific coordinate

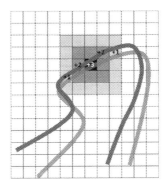

Fig. 5. The scoring process for a single edge pixel is depicted. Red and green patterns represent the database and query, respectively. Here, for the pixel marked with black, the associated scores for the red pattern with respect to the neighbor scoring are shown. The scores will be accumulated for the index of the corresponding database image. The same process will be done for all the edge pixels in the query pattern in comparison with all the database images (Colour figure online).

receive +3 points in the scoring table. Similarly, for all the edge pixels in the query image the same process is performed and corresponding database images receive their +3 points. Here, we need to clarify an important issue that might be considered during the scoring system. The first step of scoring system satisfies our need where two edge patterns from the query and database images exactly cover each other, whereas in most real cases two similar patterns are extremely close to each other in position but there is not a large overlap between them (as demonstrated in Fig. 5). For these cases that regularly happen, we introduce the first and second level neighbor scoring. A very probable case is when two extremely similar patterns do not overlap but fall on the neighboring pixels of each other. In order to consider these cases, besides the first step scoring, for any single pixel we also check the first level 8 neighboring and second level 16 neighboring pixels in the database images. All the database images that have edge with similar direction in the first level and second level neighbors receive +2 and +1 points respectively. In short, scoring system is performed for all the edge pixels in the query with respect to the similarity to the database images in three levels with different weights. Finally, the accumulated scores of each database image is calculated and normalized and the maximum scores are selected as first level top matches. The process of scoring for a single edge pixel is depicted in Fig. 5.

Reverse Scoring: In order to find the closest matches among the first level top matches, the reverse comparison system is required. Reverse scoring means that besides finding the similarity of the query gesture to the database images $(Sim(Q_i, D))$, the reverse similarity of the selected top database images to the query gesture should be computed. In fact, direct scoring system only retrieves the best matches based on the similarity of the query to them. This similarity

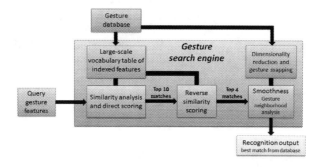

Fig. 6. Gesture search engine blocks in detail.

might have happened due to the noisy parts of the query gestures. For instance, edge-orientation features of the background of the query image might be similar to a gesture database image. This similarity might cause the wrong detection. Therefore, similarity of the selected top database images to the query should be analyzed as well. Since database images are noise-free (plain background), similarity of the selected top matches to the query is a more accurate criterion.

Combination of the direct and reverse similarity functions will result in a much higher accuracy in finding the closest match from the database. The final scoring function will be computed as: $S = [Sim(Q_i, D) \times Sim(D, Q_i)]^{0.5}$. The highest values of this function returns the best top matches from the database images for the given query gesture. In this work best top ten matches are selected in direct similarity. In reverse similarity analysis, best four database images of the previous step are selected. Afterwards, the smoothness process is performed to estimate the closest motion parameters for the query gesture image (see Fig. 6).

Another additional step in a sequence of gestural interaction is the smoothness of the gesture search. Smoothness means that the retrieved best matches in a video sequence should represent a smooth motion. Basically, this process is performed in the following steps.

Weighting the Second Level Top Matches: In order to increase the accuracy of the 3D motion estimation, after the reverse scoring, we retrieve the tagged parameters from the four top matches and estimate the query motion parameters based on the weighted sum of them as follows:

$$P_Q = aP_{Dm_1} + bP_{Dm_2} + cP_{Dm_3} + dP_{Dm_4} \tag{6}$$

$$O_Q = aO_{Dm_1} + bO_{Dm_2} + cO_{Dm_3} + dO_{Dm_4} \tag{7}$$

Note that P and O represent the x-y-z tagged position and orientation, respectively. Q and Dm_i represent the query and $i - th$ best database match. Mostly, in the experiments, a, b, c, and d are set to 0.4, 0.3, 0.2, and 0.1. At this step the best motion parameters can be estimated for the first query in a video sequence.

Dimensionality Reduction for Motion Path Analysis: In order to perform a smooth retrieval, we analyze the database images in high dimensional space to

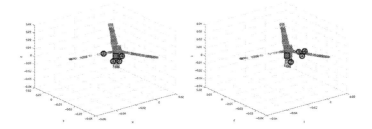

Fig. 7. Left: 3D motion estimation based on the top matches and neighborhood analysis. Red square indicates the best match from the previous frame. Numbered circles show the location of the top four matches for the current query frame. Based on the proposed algorithm, number 3 from the left plot and number 2 from the right plot should be ignored in the computations (Colour figure online).

detect the motion paths. Motion paths indicate that which gestures are closer to each other and fall in the same neighborhood in high dimension. The algorithm searches the motion paths to check which of these top matches is closer to the best found match for the previous frame. Therefore, if some of the selected top matches are not in the neighborhood area of the previous match, they should not affect the final selection and consequently the estimated 3D motion. For this reason, from the second query frame, the neighborhood analysis is performed and the irrelevant matches will be out from weighting the motion parameters.

For dimensionality reduction and gesture mapping different algorithms have been tested. The best achieved results that properly mapped the database images to visually distinguishable patterns are performed by Laplacian method. As demonstrated in Fig. 7, database images are automatically mapped to four branches. The direction of each branch shows the position of the hand gestures towards the four corners of the image frame. Clearly higher density of the points in the central part is due to the availability of the database images around the center area of the image frames. By using this pattern, from the second query matching, we can remove the noisy results. For instance, if one of the top four matches is out of the neighborhood of the previous match, it will be removed and weighing will be applied on the rest of the selected matches (see Fig. 7-left).

Another important point to mention is that if for any reason, the final top matches for the query frame are wrong (mainly due to the direct scoring), for the next frame the neighborhood analysis should not be considered. Otherwise the wrong detections significantly affect the estimated motion parameters. Therefore, if majority of the top four matches of the current frame are not from the neighborhood area of the previous match, they should be considered as a reference for estimating the 3D motion parameters and minority should be ignored from the computations (see Fig. 7-right).

Motion Averaging: Suppose that for the query images Q_{k-n}-Q_k ($k > n$), best database matches are selected. In order to smooth the retrieved motion in a sequence, the averaging method is considered. Thus, for the $k + 1$th query

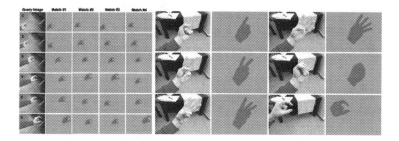

Fig. 8. Left: experimental results on a sample query video sequence of Grab gesture. The retrieved top four matches are shown on the samples. Right: different hand gestures and the corresponding best matches from the database.

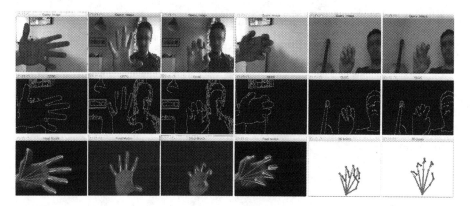

Fig. 9. First row: sample query frames from real-time video. Second row: detected edge from the query frames. Third row: corresponding best matches from the database with the annotated joint information.

image, position and orientation can be computed based on the estimated position/orientation of the n previous frames as follows:

$$P_{Q_{k+1}} = \frac{1}{n} \sum_{i=k-n+1}^{k} P_{Q_i} \qquad (8)$$

$$O_{Q_{k+1}} = \frac{1}{n} \sum_{i=k-n+1}^{k} O_{Q_i} \qquad (9)$$

Here, P_Q and O_Q represent the estimated position and orientation for the query images, respectively. Position/orientation include all 3D information (translation and rotation parameters with respect to x, y, and z axes). Therefore, motion parameters of each query image will be estimated by averaging the motion parameters of the certain number of previous image frames. According to the experiments, for $3 \leq n \leq 5$, averaging can be performed properly. For instance, if $n = 3$, motion averaging starts from the 4th query frame. 3D position and orientation of the 4th query frame will be estimated by the three previous frames and so on.

4   Experimental Results

The process of making database images and tagging the corresponding rotation parameters are implemented in C++. We synchronized two web-cams, one mounted on the user's hand to capture the hand motion and a static one to record the images for the database. Since the whole process is performed in real-time, the 3D hand motion parameters will be immediately tagged, as a separate text file, to each frame captured by the static camera. In order to provide extremely clear images for database, we covered user's arm and camera with similar paper to the background color. With some adjustments in the color intensity we could finally provide clear database images containing the user's gesture with a plain black or green background.

The matching experiments are conducted on different gesture databases. First, we provided the database with the specific hand gesture from a single user including all the variations in positioning, orientation and scaling (about 1500 images). During the second step, we extended the database to more than 3000 images of different dynamic hand gestures using one to five fingers and similarly including all the position/orientation variations. Finally, we added extra images to the database including the indoor and outdoor scenes, objects, etc. to test the robustness of the algorithm (totally more than 6000 images) (Figs. 8 and 9).

Our early experiments were conducted on a 2.93 GHz Core2Due PC. During the test step we used query gesture images from totally different environments with different backgrounds and lighting conditions. All the database images were organized in different scales for the experiments (320×240, 160×120, 80×60 and few tests on 40×30). Obviously, the processing time is the feature that changes over the tests on different scales. We could reach reasonable processing time on the largest gesture database with size 320×240. The performance seems to satisfy the image-based retrieval at this level. Other important criteria to consider are the robustness in gesture recognition and accuracy in retrieving the 3D parameters. Our algorithm works with around 100 % accuracy rate in recognition of the same gesture as the query even in the low-resolution case where we reduced the size of the database images to 80×60. We could achieve quite promising results in retrieving the 3D motion parameters up to the database images of size 160×120. In general the optimal point to achieve the best performance with respect to accuracy and efficiency is the test on gesture database with about 3000 entries with the image size of 320×240. We implemented the latest version of our system in Xcode environment on a Macbook Pro using the embedded camera. With this system we could easily achieve the real-time processing. The details about the performance of the system are depicted in Table 1.

As discussed before, direct scoring, reverse scoring, weighting the top matches, and finally the motion averaging are the main four steps in estimation of the

Table 1. Performance of the system with respect to the database size, image size, efficiency and accuracy.

Database size	Image size	Proc. time sec.	Reco. rate	3D accuracy 1–5
6000(*ges. + oth.*)	320 × 240	≈0.064	≈100 %	4
6000(*ges. + oth.*)	160 × 120	≈0.049	≈100 %	3
6000(*ges. + oth.*)	80 × 60	≈0.029	≈100 %	3
3000(*ges.*)	320 × 240	≈0.048	≈100 %	4
3000(*ges.*)	160 × 120	≈0.036	≈100 %	4
3000(*ges.*)	80 × 60	≈0.027	≈100 %	3
1500(*grabges.*)	320 × 240	≈0.033	≈100 %	4
1500(*grabges.*)	160 × 120	≈0.025	≈100 %	4
1500(*grabges.*)	80 × 60	≈0.016	≈100 %	3

best motion information for the query image. During the direct scoring step top ten matches will be selected. Although many of these ten matches might be close enough to the query frame, but for accuracy reasons the best matches should represent the closest entries of the database to the query frame. Therefore, reverse scoring refines the top four from the previous step. Extending the reverse scoring to more entries can improve the final results but due to the efficiency reasons (reverse scoring substantially increases the processing time), this step is limited to ten top matches. Afterwards, we retrieve the annotated parameters from the first four top matches and estimate the query motion parameters based on the weighted sum of them. In cases that some entries are ignored due to the neighborhood analysis, weights will be allocated to the rest of the top matches. In general, reverse scoring and weighting system significantly improve the smoothness of the motion in a video sequence and remove the noisy results. In the final step, motion averaging is applied to enhance the fluctuations in the sequence of retrieved motion. Since the idea behind this work is to facilitate the future human device interaction in various applications, we should concentrate on effective hand gestures that might be useful in a wide range of applications. Based on the related works, the most effective hand gestures in 3D application scenarios are the family of Grab gesture [1,2] (including all dynamic deformations and variations) which is widely used in 3D manipulation, pick and place, and controlling in augmented/virtual reality environments. For this reason, these gestures are considered in most of our experiments while other hand gestures show the similar performance in the tests.

5   Conclusion and Future Work

In this work we proposed a novel solution for high degrees of freedom gesture recognition, tracking and 3D motion retrieval based on the gesture search engine. The proposed algorithm has successfully passed the inventive step and has been filed as a patent application (US Patent pending) in January 2014. This method might be used in real-time gestural interaction with stationary or hand-held devices in a wide range of applications where gesture tracking and 3D manipulation are useful. Currently, we are implementing this technology on mobile devices to improve the quality of interaction in future applications. Here, an important point to mention is how to choose a reasonable size for the gesture database. Obviously, diversity and spatial resolution of the hand gestures are two main factors that directly affect the database size. In general, as discussed in [10], the hand motion has 27 degrees of freedom. Due to the correlation of the joint angles the dimension might significantly be reduced by applying dimensionality reduction techniques. In the current implementation, the vocabulary table can represent all possible indexable features that might occur (length of the search table is fixed). This indicates that the complexity of the processing does not depend on the size of the database and the current defined structures can handle substantially larger databases. Another important point is how to store the database. In fact, we do not need to store the database images. Instead, we only store the corresponding motion parameters and the search table. Size of the search table for database images of 320×240 is 614,400. According to our estimation each word in the search table will be marked with less than 100 entries of the database. Therefore, considering 2 bytes for storing each index of the database images in the search table we need around 100 MB of memory to store the whole search table. Obviously this amount of memory can be handled on any device. However, capturing, organizing and annotation of an extremely large database require substantial efforts which will be considered in the future work. Clearly, if we only target the gesture recognition and tracking, several thousand images are enough, but if we seek for high resolution 3D motion estimation we should increase the database size.

References

1. Yousefi, S.: 3D Photo Browsing for Future Mobile Devices. In: Proceedings of the ACMMM12, Nara, Japan, 29 October–2 November, 2012
2. Yousefi, S.: Enabling media technologies for mobile photo browsing. Licentiate thesis. Umea University, Department of Applied Physics and Electronics. Umea, Sweden (2012). ISBN 978-91-7459-426-3
3. Dorfmueller-Ulhaas, K., Schmalstieg, D.: Finger tracking for interaction in augmented environments. In: 2nd ACM/IEEE Int'l Symposium on Augmented Reality (2001)
4. Maggioni, C.: A novel gestural input device for virtual reality. In: Virtual Reality Annual International Symposium, pp. 118–124. IEEE (1993)
5. Hardenberg, C.V., Berard, F.: Bare-hand human-computer interaction. In: Proceedings of the 2001 Workshop on Perceptive User Interfaces. ACM International Conference Proceeding Series, Orlando, Florida, vol. 15, pp. 1–8 (2001)
6. Iwai, D., Sato, K.: Heat sensation in image creation with thermal vision. In: ACM SIGCHI International Conference on Advances in Computer Entertainment Technology (2005)
7. Kolsch, M., Turk, M.: Fast 2D hand tracking with flocks of features and multicue integration. In: Proceedings of the Computer Vision and Pattern Recognition Workshop (2004)

8. Ren, Z., Meng, J., Yuan, J., Zhang, Z.: Robust hand gesture recognition with kinect sensor. In: ACM Multimedia, pp. 759–760 (2011)
9. Erol, A., Bebis, G., Nicolescu, M., Boyle, R., Twombly, X.: Vision-based hand pose estimation: a review. Comput. Vis. Image Underst. **108**, 52–73 (2007)
10. Stenger, B., Thayananthan, A., Torr, P., Cipolla, R.: Model-based hand tracking using a hierarchical Bayesian filter. IEEE Trans. Pattern Anal. Mach. Intell. **28**(9), 1372–1384 (2006)
11. Yang, R., Sarkar, S.: Gesture recognition using hidden markov models from fragmented observations. In: Proceedings of the IEEE Computer Society Conference on Computer Vision and Pattern Recognition (2006)
12. Bencheikh, M., Bouzenada, M., Batouche, M.: A new method of finger tracking applied to the magic board. In: Conference on Industrial Technology (2004)
13. Zhou, H., Ruan, Q.: Finger countour tracking based on model. In: Conference on Computers, Comunications, Control and Power Engineering, p. 503 (2002)
14. Arce, F., Valdez, J.: Accelerometer-based hand gesture recognition using artificial neural networks. In: Castillo, O., Kacprzyk, J., Pedrycz, W. (eds.) Soft Computing for Intelligent Control and Mobile Robotics. SCI, pp. 67–77. Springer, Heidelberg (2011)
15. Choi, J., Song, K., Lee, S.: Enabling a gesture-based numeric input on mobile phones. In: IEEE International Conference on Consumer Electronics (ICCE), pp. 151–152 (2011)
16. Hrst, W., Wezel, C.: Gesture-based interaction via finger tracking for mobile augmented reality. Multimedia Tools Appl. **62**, 1–26 (2012)
17. Baldauf, M., Zambanini, S., Fröhlich, P., Reichl, P.: Markerless visual fingertip detection for natural mobile device interaction. Mobile HCI, pp. 539–544 (2011)
18. Lee, D., Lee, S.: Vision-based finger action recognition by angle detection and contour analysis. ETRI J. **33**(3), 415–422 (2011)
19. Hannuksela, J., Barnard, M., Sangi, P., Heikkil, J.: Camera-Based Motion Recognition for Mobile Interaction. ISRN Signal Processing (2011)
20. Hagbi, N., Bergig, O., El-Sana, J., Billinghurst, M.: Shape recognition and pose estimation for mobile augmented reality. In: 8th IEEE International Symposium on Mixed and Augmented Reality (ISMAR 2009), IEEE Computer Press (2009)
21. Cao, Y., Wang, C., Zhang, L., Zhang, L.: Edgel index for large-scale sketch-based image search. In: Proceedings of the IEEE Conference on Computer Vision and Pattern Recognition (CVPR), Colorado Springs, USA, pp. 761-768 (2011). ISBN 978-1-4577-0394-2
22. Eitz, M., Hildebrand, K., Boubekeur, T., Alexa, M.: Sketch-based image retrieval: benchmark and bag-of-features descriptors. IEEE Trans. Vis. Comput. Graphics **17**(11), 1624–1636 (2011)
23. Ikizler, N., Duygulu, P.: Human action recognition using distribution of oriented rectangular patches. In: Workshop on Human Motion, pp. 271–284 (2007)
24. Kondori, F.A., Liu, L.: 3D Active human motion estimation for biomedical applications. In: World congress on Medical Physics and Biomedical Engineering (WC2012). Beijing, China (2012)
25. Kondori, F.A.: Human motion analysis for creating immersive experience. Licentiate thesis, Department of Applied Physics and Electronics, Umea University, Sweden (2012). ISBN 9789174594164

Debugging Object Tracking Results by a Recommender System with Correction Propagation

Mingzhong Li and Zhaozheng Yin[(✉)]

Department of Computer Science,
Missouri University of Science and Technology, Rolla, USA
yinz@mst.edu

Abstract. Achieving error-free object tracking is almost impossible for state-of-the-art tracking algorithms in challenging scenarios such as tracking a large amount of cells over months in microscopy image sequences. Meanwhile, manually debugging (verifying and correcting) tracking results object-by-object and frame-by-frame in thousands of frames is too tedious. In this paper, we propose a novel scheme to debug automated object tracking results with humans in the loop. Tracking data that are highly erroneous are recommended to annotators based on their debugging histories. Since an error found by an annotator may have many analogous errors in the tracking data and the error can also affect its nearby data, we propose a correction propagation scheme to propagate corrections from all human annotators to unchecked data, which efficiently reduces human efforts and accelerates the convergence to high tracking accuracy. Our proposed approach is evaluated on three challenging datasets. The quantitative evaluation and comparison validate that the recommender system with correction propagation is effective and efficient to help humans debug tracking results.

1 Introduction

Automated visual object tracking is very useful to monitor objects over a long period and analyze their behavior. Multi-Hypothesis Tracking (MHT) [10] and Joint Probabilistic Data Association Filters (JPDAF) [4] are two representative examples for multi-object tracking. To reduce the computational cost, tracklet stitching [5] is proposed: first reliable tracklets are generated which are fragments of tracks formed by conservative grouping of detection responses, then the tracklets are connected by the Hungarian algorithm [6]. Bonneau et al. [1] proposes a tracklet linking method in which a minimal path among tracklets is obtained by using dynamic programming in order to track quantum dots in a living cell. Zhang et al. [13] proposes a minimum-cost flow network to resolve the global data association of multiple objects over time.

In real world applications such as uncovering hidden patterns of a complex biological process, high quality object tracking algorithms are required to accurately track bio-specimens over a long period. But, due to the numerous challenges in biomedical data such as appearance similarity, heavy occlusion and

© Springer International Publishing Switzerland 2015
C.V. Jawahar and S. Shan (Eds.): ACCV 2014 Workshops, Part II, LNCS 9009, pp. 214–228, 2015.
DOI: 10.1007/978-3-319-16631-5_16

clutter, it is extremely difficult to achieve perfect tracking performance without any error. To pursue solid scientific discovery and error-free health diagnosis, biologists and doctors are willing to exchange a small amount of their human efforts to check the automated tracking results manually. Hence, it is worth to consider how to incorporate human efforts to debug (verify and correct) the tracking results, which leads to the following three questions:

(1) Manually checking each object's trajectory frame by frame is very costly for human labors, which we cannot afford. How to find out which tracking data are error-prone thus they are worth to be checked by human?

(2) Checking tracking data on specimens captured over months with thousands of frames is too tedious for an individual. How can we integrate crowdsourcing to check the data collectively?

(3) There might be analogy between different error nodes in the tracking data, and the error can also affect its nearby data. How can we propagate the costly human correction to other unchecked data and automatically correct similar errors such that human burden is alleviated and the convergence to the best tracking accuracy is accelerated?

Recommender systems [2,8,9,11,14] are capable of using historical data of a user to infer her/his preference on items and then predicting other items that the user might like. Websites such as Google.com, Amazon.com, Ebay.com, etc. have widely equipped their search engines with specialized recommender systems to help their customers find their preferred commodities. Particularly, content-based recommender systems analyze a set of documents and/or descriptions of items previously rated by a user, and build a model to predict the user's interest based on the features of the object ratings [8,9]. How to construct a proper user profile by collecting data representing the user's preference, is the key of content-based recommender systems.

In this paper, we assume no object tracking algorithm can achieve perfect tracking performance in challenging scenarios. Instead of aiming at developing object detection and tracking algorithms, our focus is to investigate how to debug existing object tracking results with humans in the loop. The main contributions of this paper include:

(1) we propose a novel recommender system to assist multiple human annotators to debug tracking data collectively. Tracking data with high error likelihood are recommended to each individual annotator based on their debugging histories. The verification and correction made by annotators are collected for subsequent correction propagation and user profile updating procedures;

(2) we propose a correction propagation scheme, which propagates the corrected information to other track data affected by the corrected data, based on the verification and corrections made by multi-annotators.

The paper is organized as below. In Sect. 2, we describe a basic data-association method for multi-object tracking. In Sect. 3, we present the recommender system to debug tracking data with multi-annotators in the loop. In Sect. 4, we introduce how to propagate human corrections to other unchecked tracking data to accelerate the debugging process. Experimental results are presented in Sect. 5.

2   Multi-object Tracking

We formulate the multi-object tracking problem in the framework of "tracking-by-detection". First, detected objects in individual frames are considered as *nodes* and they are connected frame-by-frame into short reliable trajectories (a.k.a, *tracklets*). Second, these short tracklets are linked into longer and longer tracklets gradually by a sequential procedure (*fine-to-coarse tracklet association*). Finally, detection-related and tracklet-related features are generated for every node of every tracklet, which are used in the recommender system and correction propagation.

2.1   Tracklet Generation

Every detected object candidate in a frame is represented as a *node* with corresponding features such as color distribution, gradient histogram, object shape, location, etc. We denote $\mathbf{f}(n_i^t) = [\mathbf{f}_1(n_i^t), \ldots, \mathbf{f}_K(n_i^t)]$ as the vector of K features of node i in frame t. The dissimilarity cost between a pair of nodes in two consecutive frames is computed as

$$c(n_i^t, n_j^{t-1}) = \begin{cases} \frac{1}{K} \sum_{k=1}^{K} \frac{\left\| \mathbf{f}_k(n_i^t) - \mathbf{f}_k(n_j^{t-1}) \right\|}{\Delta_k}, & \text{if } \left\| \mathbf{f}_k(n_i^t) - \mathbf{f}_k(n_j^{t-1}) \right\| \leq \Delta_k, \forall k \in [1, K] \\ \infty, & \text{otherwise} \end{cases}$$

$$(1)$$

where $\| \cdot \|$ is the L_2 norm and Δ_k is the normalization factor of the kth feature. For example, when $\mathbf{f}_k$ is the location feature, Δ_k controls the spatial gating region (i.e., the size of local neighborhood to search a node's correspondence between consecutive frames).

Given I nodes in frame t and J nodes in frame $t-1$, a cost matrix $\mathbf{C} = [c(n_i^t, n_j^{t-1})]$ with size I-by-J is generated and we apply the Hungarian algorithm [6] onto it to solve the linear assignment problem (i.e., corresponding nodes between frames t and $t-1$ are connected). After sequentially performing the Hungarian algorithm between consecutive frames, *tracklets* are generated for a given video (e.g., Fig. 1(a)). Note that, (1) we use small gating regions in the frame-by-frame assignment, which generates tracklets with less errors but also causes short broken tracklets when objects move beyond the gating regions; (2) the Hungarian algorithm solves the 1-to-1 bipartite assignment problem but it can not solve the 2-to-1 or 1-to-2 assignment problem when there exists object merging or division, which causes broken or wrong connections among tracklets; (3) it is usually difficult to have perfect detection results for every frame, hence false positives and miss detections will cause broken or wrong connections among the tracklets. In the following two subsections, we describe how to gradually link the short tracklets into longer object trajectories.

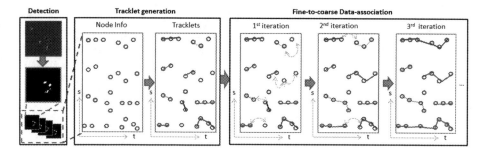

Fig. 1. Multi-object tracking.

2.2 Fine-to-Coarse Tracklet Association

We denote the ith tracklet $\mathbf{T}_i$ by its node set, $\mathbf{T}_i = \{n_i^{s_i}, n_i^{s_i+1}, \ldots, n_i^{e_i}\}$, where $n_i^{s_i}$ and $n_i^{e_i}$ represent the nodes in the start and end frame s_i and e_i, respectively. Five types of hypotheses are considered when associating tracklets:

(1) Translation (1-to-1): the head of tracklet $\mathbf{T}_j$ is associated with the tail of tracklet $\mathbf{T}_i$ with the cost:

$$c(\mathbf{T}_i \rightarrow \mathbf{T}_j) = \begin{cases} \frac{1}{K+1} \sum_{k=1}^{K+1} \frac{\left\| \mathbf{f}_k^+(n_i^{e_i}) - \mathbf{f}_k^+(n_j^{s_j}) \right\|}{\Delta_k}, \\ \quad \text{if} \left\| \mathbf{f}_k^+(n_i^{e_i}) - \mathbf{f}_k^+(n_j^{s_j}) \right\| \leq \Delta_k, \forall k \in [1, K+1], \\ \infty, \text{otherwise} \end{cases} \tag{2}$$

where $\mathbf{f}^+(n_i^{e_i}) = [\mathbf{f}(n_i^{e_i}), \theta(n_i^{e_i})]$ and $\mathbf{f}^+(n_j^{s_j}) = [\mathbf{f}(n_j^{s_j}), \theta(n_j^{s_j})]$ are the augmented feature vectors for the end and start nodes of tracklets $\mathbf{T}_i$ and $\mathbf{T}_j$, respectively. $\theta(\cdot)$ denotes a node's trajectory orientation in the tracklet.

(2) Division (1-to-2): the tail of a tracklet is associated with the heads of two tracklets with the cost:

$$c(\mathbf{T}_i \rightarrow (\mathbf{T}_{j_1}, \mathbf{T}_{j_2})) = c(\mathbf{T}_i \rightarrow \mathbf{T}_{j_1}) + c(\mathbf{T}_i \rightarrow \mathbf{T}_{j_2}) + c(n_{j_1}^{s_{j_1}}, n_{j_2}^{s_{j_2}}) \tag{3}$$

(3) Merging (2-to-1): the tails of two tracklets are associated with the head of a tracklet with the cost:

$$c((\mathbf{T}_{i_1}, \mathbf{T}_{i_2}) \rightarrow \mathbf{T}_j) = c(\mathbf{T}_{i_1} \rightarrow \mathbf{T}_j) + c(\mathbf{T}_{i_2} \rightarrow \mathbf{T}_j) + c(n_{i_1}^{e_{i_1}}, n_{i_2}^{e_{i_2}}) \tag{4}$$

(4) Disappearing (1-to-0): the tail of a tracklet is not linked to any other tracklet with the cost:

$$c(\mathbf{T}_i \rightarrow \phi) = \begin{cases} \frac{d^{(t)}(n_i^{e_i}, e)}{\Delta t}, & \text{if } d^{(t)}(n_i^{e_i}, e) \leq \Delta t, \\ \frac{d^{(s)}(n_i^{e_i})}{\Delta s}, & \text{if } d^{(s)}(n_i^{e_i}) \leq \Delta s, \\ \eta, & \text{otherwise.} \end{cases} \tag{5}$$

where $d^{(t)}(n_i^{e_i}, e)$ denotes the temporal distance from the ending node of $\mathbf{T}_i$ to the last frame. $d^{(s)}(n_i^{e_i})$ denotes the spatial distance from the ending node of

$\mathbf{T}_i$ to the image boundary. During object tracking, three scenarios cause the disappearing cases: (i) objects at the end of a video will disappear; (ii) objects close to the image boundary may move out of the view field; and (iii) every object may be missed by the detection or occluded by other objects/background so it is associated with a constant cost η (we choose η as the maximum of all non-infinity $c(\cdot \to \cdot)$).

(5) Appearing (0-to-1): similar to 1-to-0 case we define the cost for 0-to-1 hypothesis as:

$$c(\phi \to \mathbf{T}_i) = \begin{cases} \frac{d^{(t)}(n_i^{s_i}, s)}{\Delta t}, & \text{if } d^{(t)}(n_i^{s_i}, s) \le \Delta t, \\ \frac{d^{(s)}(n_i^{s_i})}{\Delta s}, & \text{if } d^{(s)}(n_i^{s_i}) \le \Delta s, \\ \eta, & \text{otherwise.} \end{cases} \quad (6)$$

Denoting the number of tracklets in a video as N and the number of all possible hypotheses among the N tracklets as M, we catenate the costs of all hypotheses into a M-by-1 vector $\mathbf{c}$ and define a constraint matrix $\mathbf{Q}$ of size M-by-$2N$. For example, if the hth hypothesis is $\mathbf{T}_i \to (\mathbf{T}_{j_1}, \mathbf{T}_{j_2})$ involving tracklets i, j_1 and j_2, then $\mathbf{Q}(h,i) = 1$, $\mathbf{Q}(h, N + j_1) = 1$, $\mathbf{Q}(h, N + j_2) = 1$ and all other elements of the hth row of $\mathbf{Q}$ are zero. The tracklet association is obtained by solving the following Linear Integer Programing (LIP) problem:

$$\arg\min_{\mathbf{x}} \mathbf{c}^T \mathbf{x}, \quad s.t. \ \mathbf{Q}^T \mathbf{x} = \mathbf{1} \quad (7)$$

where $\mathbf{x}$ is an M-by-1 binary vector and $\mathbf{x}_h = 1$ indicates that the hth hypothesis is selected to be true in the optimal solution. The objective function $(\mathbf{c}^T \mathbf{x})$ aims to find an optimal $\mathbf{x}$ to minimize the total cost of selected hypotheses. The constraint $(\mathbf{Q}^T \mathbf{x} = \mathbf{1})$ ensures that one tracklet is only associated at most once on its head and tail.

Rather than solving the global tracklet association only once with a large gating region which may introduce a significant amount of errors during association, we gradually increase the gating regions (Δ_k) and iteratively solve the corresponding LIP problem, thus the short tracklets are linked into longer and longer ones in a fine-to-coarse manner with less errors (e.g., Fig. 1(b)).

2.3   Features for Nodes in Tracklets

In Table 1, we list the features used for nodes in tracklets. Given node n_k^t of tracklet $\mathbf{T}_k$ in frame t, it has features related to both object detection $(\mathbf{f}(n_k^t))$ and tracklet association.

where $|\cdot|$ denotes the cardinality of a set and $\delta(\mathbf{T_k}, \mathbf{T_j})$ is an indicator function $(\delta(\mathbf{T_k}, \mathbf{T_j}) = 1$ when $\mathbf{T_j}$'s head is within the gating region of the tail of $\mathbf{T_k}$; $\delta(\mathbf{T_k}, \mathbf{T_j}) = 0$, otherwise).

For example, in Fig. 2(a), if node a is linked to node x by data association, $c_s(a)$ and $c_s(a)$ will be the cost of associate the tracklet of a to the tracklet of x by Eq. 2. If the gating region has been increased twice before a and x are linked, $c_g(a)$ and $c_g(x)$ will be 2. $l(a)$ is the length of the shortest tracklet within a's gating region, which is the length of the tracklet with starting node y, hence

Table 1. Features for nodes in tracklets.

$c_s(n_k^t)$	The cost of the hypothesis involving n_k^t
$c_g(n_k^t)$	Number of times the gating region has been increased
$l(n_k^t)$	Length of the shortest tracklet among $\{\mathbf{T}_i : \delta(\mathbf{T}_k, \mathbf{T}_i) \neq 0, i \neq k\}$
$c_t(n_k^t)$	$|\{\mathbf{T}_j : \delta(\mathbf{T}_k, \mathbf{T}_j) \neq 0, j \neq k\}|$
$c_h(n_k^t)$	$|\{\mathbf{T}_i : \delta(\mathbf{T}_i, \mathbf{T}_k) \neq 0, i \neq k\}|$

$l(a) = 2$. Similarly, $l(b) = 3$ and $l(c) = 3$. In the relation graph (Fig. 2(b)) of Fig. 2(a), $c_t(\cdot)$ and $c_h(\cdot)$ compute the degrees of corresponding nodes, thus $c_t(a) = 2$, $c_t(b) = 2$ and $c_t(c) = 1$; $c_h(x) = 2$, $c_h(y) = 1$ and $c_h(z) = 2$.

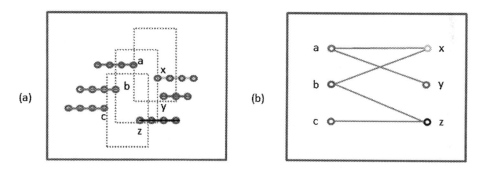

Fig. 2. (a) Tracklets and gating regions (blue dotted windows); (b) Relation graph (Color figure online).

$c_s(n_k^t)$ stores the latest association cost involving the node n_k^t. The higher $c_s(n_k^t)$ is, the more unreliable the association happened on node n_k^t is. The motivation of considering $c_g(n_k^t)$ as one of the features of node n_k^t is that we want to evaluate at which stage node n_k^t is associated to the longest possible trajectory at last. If the stage is high, which means the gating region has been increased many times, the association on node n_k^t is more likely to be a mistake.

Very short tracklets near node n_k^t are highly possible to be false positives and associating them with node n_k^t causes errors, thus we consider $l(n_k^t)$ as one feature. Similar reasons lead us to consider $c_t(n_k^t)$ and $c_h(n_k^t)$: when there are more association possibilities around node n_k^t, the association on node n_k^t may be more erroneous.

All these features are combined with the object-detection-related features into a feature vector $\mathbf{F}(n_k^t) = [\mathbf{f}(n_k^t), c_s(n_k^t), c_g(n_k^t), l(n_k^t), c_t(n_k^t), c_h(n_k^t)]$ to describe node n_k^t. The association-related parts $(c_s(n_k^t), c_g(n_k^t), l(n_k^t), c_t(n_k^t), c_h(n_k^t))$ are updated only when an association hypothesis involving n_k^t is within the optimal solution of the LIP problem in Eq. 7. Details of updating the association-related node features are summarized in Algorithm 1 below.

Algorithm 1. Node Feature Updating in Fine-to-Coarse Association

---

Input : $Tracklets$: $\{\mathbf{T}_i\}$; gating region increasing rate: α;
Initialization : $\forall n_k^t$, $c_s(n_k^t) \leftarrow 0, c_g(n_k^t) \leftarrow 0$, $l(n_k^t) \leftarrow 0, c_t(n_k^t) \leftarrow 0, c_h(n_k^t) \leftarrow 0$,
$\beta \leftarrow [\Delta_1, \dots, \Delta_{K+1}]$;
Repeat
    Solve the LIP problem in Eq. 7;
    for any selected association hypothesis linking tracklets $\mathbf{T_p}$ with $\mathbf{T_q}$
        $c_s(n_q^{s_q}) \leftarrow c(\mathbf{T}_p, \mathbf{T}_q), c_s(n_p^{e_p}) \leftarrow c(\mathbf{T}_p, \mathbf{T}_q);$//the current hypothesis' cost
        $c_g(n_p^{e_p}) + +, c_g(n_q^{s_q}) + +;$//the times of gating regions being increased
        compute $l(n_p^{e_p})$, $l(n_q^{s_q});$//the length of the shortest tracklet nearby
        $c_t(n_p^{e_p}) \leftarrow |\{\mathbf{T_j} : \delta(\mathbf{T_p}, \mathbf{T_j}) \neq 0, j \neq p\}|;$
        $c_h(n_q^{s_q}) \leftarrow |\{\mathbf{T_i} : \delta(\mathbf{T_i}, \mathbf{T_q}) \neq 0, i \neq q\}|;$
    end for
    $\beta \leftarrow \beta + \alpha \cdot \beta$//increase the gating regions
    update $\{\mathbf{T}_i\}$ with the optimization result;
Until no change happens to the association.

---

3   Recommender System

The key idea of content-based recommender system is to estimates the profile parameters of users, $\{\theta^{(u)}, u = 1, \dots, U\}$, using the available feature vectors of targets $\{\mathbf{x}^{(i)}, i = 1, \dots, n_x\}$

$$\theta^{(u)} : \arg\min_{\theta^{(u)}} \sum_{i:r(i,u)=1} \left(\theta^{(u)^T} \mathbf{x}^{(i)} - \mathbf{y}^{(i,u)} \right)^2 + \lambda \|\theta^{(u)}\| \tag{8}$$

where U and n_x denote the number of users and targets, respectively. $r(i,u) = 1$ if user u has recommendation ($\mathbf{y}^{(i,u)}$) on target i. λ is the coefficient for the regularization term. For any user u, we learn a parameter vector $\theta^{(u)}$ representing the user's preference. Given any new target j with feature $\mathbf{x}^{(j)}$, we predict user u's recommendation on target j as $\theta^{(u)^T} \mathbf{x}^{(j)}$.

In our system the users are annotators who can verify tracking results and correct corresponding errors, and the targets are a large pool of nodes from all linked tracklets with features. Figure 3 shows the workflow of our recommender system and correction propagation:

First, each of the U users selects a small portion of nodes from the large node pool independently and classifies them into *positive* nodes (nodes with tracking errors) and *negative* nodes (nodes without any tracking error). All other nodes unselected by any user are transferred to the uncertain node pool. Note that this manual initialization step only needs to be done once.

Second, each user's profile parameters are learned from the positive and negative node sets.

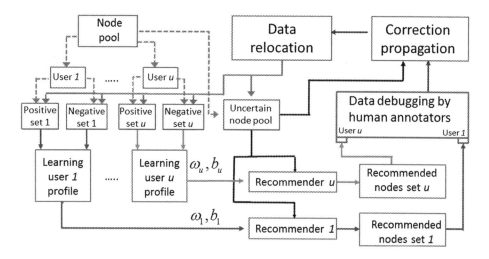

Fig. 3. Workflow of our recommender system.

Third, the recommendation on every node in the uncertain node pool by every user is computed by the user's profile and node's feature vector.

Fourth, top-ranked recommendations are sent to users for verification and correction.

Fifth, the corrections made by human are automatically propagated to other uncertain nodes and their feature vectors are updated accordingly.

Finally, the nodes in uncertain node pool after correction propagation are either relocated to positive/negative sets of users for updating users' profiles or still remain in the uncertain pool if not affected by the correction propagation. The process is iterated until the uncertain node pool is empty.

Different from the least square cost in Eq. 3 for recommender learning, we use a linear SVM to learn the profile parameters of any user u

$$\arg \min_{\mathbf{w}_u, b_u, \lambda} \left\{ \sum_i \lambda_i [y^{(i,u)} (\mathbf{w}_u{}^T \psi(\mathbf{F}^{(i,u)}) + b_u) - 1] + \frac{1}{2} \|\mathbf{w}_u\| \right\} \quad (9)$$

where $\mathbf{F}^{(i,u)}$ is the feature vector of node i in user u's positive/negative training node sets. $\mathbf{w}_u$ and b_u define the maximum-margin hyperplane that classifies $\mathbf{F}^{(i,u)}$ according to its class labels $\mathbf{y}^{(i,u)}$. $\psi(\cdot)$ is the kernel function (linear in this paper).

After learning the profile of user u, recommendation on any node j in the uncertain node pool by user u (i.e., verify the node or not) can be computed by the linear SVM score:

$$\mathbf{w}_u{}^T \psi(\mathbf{F}^{(j,u)}) + b_u \quad (10)$$

A small set of nodes from the uncertain node pool which have suspiciously high scores (i.e., high probability of tracking errors) are recommended to user u for verification and correction. For example, top 20 nodes are recommended to a user in each iteration. The profiles of different users are learnt independently hence different sets of nodes are recommended for different users to verify and correct, but their corrections can be collected together for efficient correction propagation discussed below.

4   Correction Propagation

In tracking data, neighboring nodes affect each other. In order to accelerate the debugging efficiency, we propose a correction propagation approach to spread out the correction information of corrected nodes to their neighboring nodes in the uncertain node pool.

First of all, a graph-based Propagation Set Detection (PSD) algorithm is proposed. In the graph G, each vertex is a node in the node pool and edge exists between two nodes if and only if there is an association hypothesis involving both of the nodes in the data-association step (e.g., the relation graph in Fig. 2(b)). Given users' corrections involving nodes $\mathbf{R}$, we detect the propagation set by the algorithm below:

Algorithm 2. Propagation Set Detection

Input : Graph $\mathbf{G}$, correction set $\mathbf{R}$
Output : node set $\mathbf{V}_{PSD}$;
Initialization : queue $\mathbf{Q} \leftarrow \mathbf{R}$; node set $\mathbf{V}_{PSD} \leftarrow \mathbf{R}$;
Repeat
    $t \leftarrow \mathbf{Q}.dequeue;$//get the first element of the queue
    for all edges of node t in $\mathbf{G}$, e;
        $v \leftarrow \mathbf{G}.adjacentVertex(t,e)$;
        if $v \notin \mathbf{V}_{PSD}$, **then** add v to $\mathbf{V}_{PSD}$, enqueue v onto $\mathbf{Q}$;
        end if;
    end for;
Until $\mathbf{Q}$ is empty
Return $\mathbf{V}_{PSD}$;

By implementing this PSD algorithm, we find all the nodes influenced by the current corrections in tracking data. We denote the affected nodes and corrected nodes as set V_{PSD}. Using the human corrections as hard constraints, we perform the LIP problem in Eq. 7 on V_{PSD}, which updates the tracklet association and corresponding node features, i.e., automatically propagates correction information to nodes close to corrected nodes. We use the gating regions to control how far the correction can be propagated in the local neighborhood. While the recommender system is run iteratively, this correction propagation performs like the Butterfly Effect and sweeps gradually over the entire node pool.

After correction propagation, data relocation is performed before we move to the next iteration:

(1) top μ nodes with high scores are recommended for a user to check (e.g., $\mu = 20$) . After human verification and correction, the top μ nodes are separated

into two subsets: nodes with errors are assigned to Positive Set and nodes without errors are assigned to Negative Set;

(2) the nodes in V_{PSD} which have low scores after correction propagation are moved to Negative Set. Those with high scores are moved to Positive Set and the rest remains in the uncertain node pool;

(3) rated nodes by the recommender system with low scores are assigned to Negative Set, only if they are not in the propagation set $\mathbf{V}_{PSD}$ found by **PSD** algorithm.

Our iterative recommender system with correction propagation is summarized in Algorithm 3:

Algorithm 3. Iterative Recommender System with Correction Propagation

---

Input : node set **V** of all nodes and their features;
Initialization : **Uncertain Node Pool=V**; **Temporary Set P**=∅;
Positive Set← Pick μ nodes with errors in tracking data;
Negative Set ← Pick μ nodes without errors in tracking data;
Repeat
    Update **RecommenderProfiles** using Eq. 9;
    Compute the scores of nodes in the **Uncertain Node Pool** by Eq. 10;
    for all node v ∈ **V**
        if node score of v > ω;
            if v is one of the top μ nodes, **then** recommend v for human check;
                if v is verified as a node with errors, **then** add v to **Positive Set**;
                else add v to **Negative Set**;
                end if;
            else add v to **P**;
            end if;
        else if node score of v < $\omega/2$;
            add v to **Negative Set**;
        end if;
        Uncertain Node Pool ← **Uncertain Node Pool** − v;
    end for;
    Find $\mathbf{V}_{\mathbf{Propagation}}$ using **Algorithm 2**;
    Implement data-association algorithm within $\mathbf{V}_{\mathbf{Propagation}}$ where human corrections are added as additional hard constraints;
    Add **P** − **P** ∩ $\mathbf{V}_{\mathbf{Propagation}}$ to **Uncertain Node Pool**;
until **Uncertain Node Pool** is empty.

---

5   Experimental Results

5.1   Datasets

To test our proposed recommender system, we perform experiments on three different biomedical image sequences. Specifications of these datasets are summarized in Table 2, while some sample images are shown in Fig. 4. In Datasets 1 and 2 which are downloaded from [15], the main challenges are the frequent occurring of cell merging and division (causing many tracklets in a small local neighborhood), false positives in detection (causing distractions and wrong associations) and miss detections (causing broken tracklets). In dataset 3 obtained from [7], the main challenges are false positive distractions due to low image contrast, fast motion blurring and object camouflaging.

Table 2. Specifications of datasets.

	#images	Object type	#Objects per frame	Image size
Set1	400	Stem cells	20–100	1392×1040
Set2	380	Stem cells	100–400	1392×1040
Set3	10000	Fruit flies	52	848×480

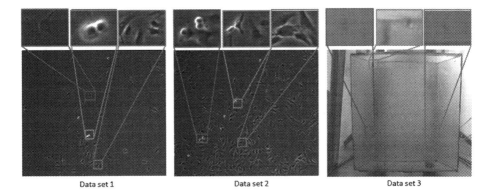

Fig. 4. Examples of 3 datasets.

5.2   Quantitative Evaluation

To evaluate how well our recommender system and correction propagation can assist human annotators on debugging object tracking results, we use two metrics:

(1) efficiency: how fast will the number of uncertain nodes reduce so less human effort is needed to verify and correct nodes?

(2) effectiveness: what percentage of false nodes (nodes with tracking errors) is detected by the recommender system (i.e., how effective can our system guide human annotators towards false nodes)? Figure 5(a) shows the number of nodes remaining in the uncertain node pool at different iterations for the 3 datasets. We observe that it decreases drastically when using our recommender system and

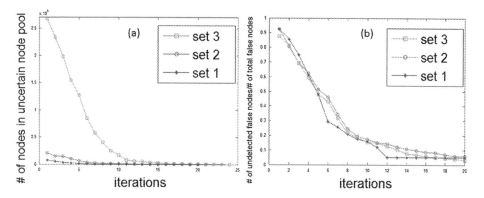

Fig. 5. (a) Number of nodes in the uncertain node pool; (b): # of undetected false nodes/# of total false nodes.

correction propagation to assist human annotators, which proves that our system is efficient with less human effort to debug object tracking results. Figure 5(b) shows "# of undetected false nodes/# of total false nodes" at different iterations on the 3 datasets. It is noticeable that the percentage curves drop quickly, and within 20 iterations the number of undetected false nodes falls below 5 % out of the total false nodes, which proves that our recommender system can effectively guide human annotators to find questionable tracking results for correction.

5.3   Quantitative Comparison

In order to demonstrate the effect of our recommender system and correction propagation, we compare the following four approaches:

(1) **Random selection without propagation.** Nodes in the uncertain node pool are randomly selected by humans to verify and correct. Human corrections are not propagated to neighboring nodes.
(2) **Random selection with propagation.** Nodes in the uncertain node pool are randomly selected by humans to verify and correct. Human corrections are propagated to neighboring nodes.
(3) **Recommendation without propagation.** Nodes in the uncertain node pool are recommended by our system for humans to verify and correct. Human corrections are not propagated to neighboring nodes.
(4) **Recommendation with propagation.** Nodes in the uncertain node pool are recommended by our system for humans to verify and correct. Human corrections are propagated to neighboring nodes. As shown in Fig. 6, **Random selection without propagation** is very inefficient for human to debug the object tracking results since the human randomly verifies uncertain nodes without any guidance and no further usage is applied to human correction in each iteration. With correction propagation applied to the random selection, **Random selection with propagation** decreases the number of uncertain nodes

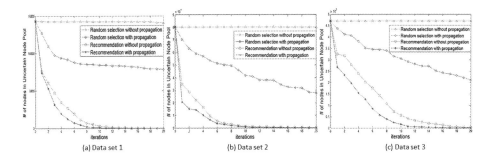

Fig. 6. Uncertain node pool shrinking rates of 4 different approaches on 3 datasets.

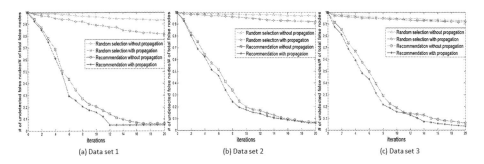

Fig. 7. (# of undetected false nodes/# of total false nodes) of 4 different approaches.

faster, but the human annotators still have no clue on what nodes should be checked. **Recommendation without propagation** adds recommendation to humans for debugging tracking results, which reduces the number of uncertain nodes further faster. Finally, when **Recommendation with propagation** is applied together, the uncertain node pool shrinks the fastest, which means that it takes much less time and human labeling costs.

From Fig. 6, we observe that both our recommender system and correction propagation can efficiently reduce the size of the uncertain node pool. We evaluate the effectiveness of the four approaches in terms of *"# of undetected false nodes/# of total false nodes"*. In Fig. 7, our recommender system with correction propagation performs beyond the other 3 approaches in helping humans detect nodes with tracking errors. Another observation from Fig. 7 is that random selection has much lower effectiveness than recommender system since it has no guidance on which nodes should be verified and corrected.

5.4   Qualitative Examples

In Fig. 8 we present some examples to show how our recommender system helps humans find similar false tracking data. In the left box, a cell is detected as multiple fragments and similar cases are found by the recommender system. In the right box, figure shows the wrong ID associations due to nearby distractors.

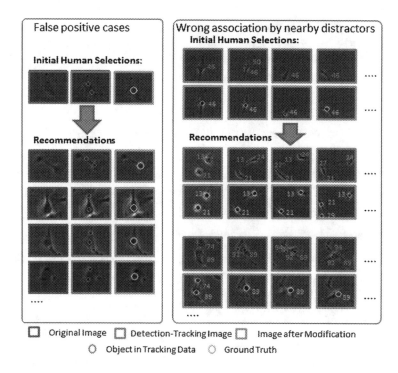

Fig. 8. Examples of recommended nodes for human verification and correction.

In our recommender system, all of these tracking errors can be represented by their nodes' feature vectors. Initially, human selects some false nodes and makes corresponding corrections, then the recommender system search similar false nodes in the uncertain node pool and let human to verify and correct them.

6   Conclusion

In this paper, a novel iterative recommender system is proposed to help humans debug tracking errors in data generated by various tracking algorithms. Instead of object-by-object or frame-by-frame checking, human annotators only need to debug a sparse set of nodes recommended by the recommender system in each iteration. Multiple human debuggers work on the tracking data independently and their debugging results are collected together. Since every correction made on one node will result in a chain reaction involving other neighboring nodes, we propagate all the corrections in the neighborhood, which ensures the tracking consistency and speeds up the debugging process. Each annotator's profile parameters on the recommender system is updated based on their new debugged nodes. The process is iterated until the uncertain node pool is empty. We tested our approach on three sets of biomedical image sequences. The results show that our recommender system with correction propagation can efficiently and effectively guide human annotators to debug tracking data.

Acknowledgement. This research was supported by NSF EPSCoR grant IIA-1355406 and NSF CAREER award IIS-1351049, University of Missouri Research Board, ISC and CBSE centers at Missouri University of Science and Technology.

References

1. Bonneau, S., et al.: Single quantum dot tracking based on perceptual grouping using minimal paths in a spatiotemporal volume. IEEE Trans. Image Process. **14**(9), 1384–1395 (2005)
2. Burke, R.: Knowledge-based recommender systems. Encycl. Libr. Inf. Syst. **69**(Supplement 32), 175–186 (2000)
3. Cortes, C., et al.: Support-vector networks. Mach. Learn. **20**, 273–297 (1995). Springer
4. Fortmann, T., et al.: Sonar tracking of multiple targets using joint probabilistic data association. IEEE J. Ocean. Eng. **8**(3), 173–184 (1983)
5. Huang, C., Wu, B., Nevatia, R.: Robust object tracking by hierarchical association of detection responses. In: Forsyth, D., Torr, P., Zisserman, A. (eds.) ECCV 2008, Part II. LNCS, vol. 5303, pp. 788–801. Springer, Heidelberg (2008)
6. Kuhn, H.: The Hungarian Method for the assignment problem. Nav. Res. Logist. Q. **2**, 83–97 (1955)
7. Li, M., et al.: Track fast-moving tiny flies by adaptive LBP feature and cascaded data association. In: ICIP (2013)
8. Lops, P., et al.: Content-based recommender systems: state of the art and trends. In: Ricci, F., Rokach, L., Shapira, B., Kantor, P.B. (eds.) Recommender Systems Handbook. Springer, New York (2011)
9. Park, D., et al.: A literature review and classification of recommender systems research. Expert Syst. Appl. **39**, 10059–10072 (2012)
10. Reid, D.: An algorithm for tracking multiple targets. IEEE Trans. Autom. Control **24**(6), 843–854 (1979)
11. Resnick, P., et al.: Recommender systems. Commun. ACM **40**, 56–58 (1997)
12. Yilmaz, A., et al.: Object tracking: a survey. ACM Comput. Surv. **38**(4), 1–45 (2006). Article 13
13. Zhang, L., et al.: Global data association for multi-object tracking using network flows. In: CVPR (2008)
14. Zhang, T., Iyengar, Y.: Recommender systems using linear classifiers. J. Mach. Learn. Res. **2**, 313–334 (2002)
15. http://www.celltracking.ri.cmu.edu/downloads.html

An Abstraction for Correspondence Search Using Task-Based Controls

Gregor Miller[✉] and Sidney Fels

Human Communication Technologies Laboratory,
University of British Columbia, Vancouver, BC V6T 1Z4, Canada
gregor@ece.ubc.ca

Abstract. The correspondence problem (finding matching regions in images) is a fundamental task in computer vision. While the concept is simple, the complexity of feature detectors and descriptors has increased as they provide more efficient and higher quality correspondences. This complexity is a barrier to developers or system designers who wish to use computer vision correspondence techniques within their applications. We have designed a novel abstraction layer which uses a task-based description (covering the conditions of the problem) to allow a user to communicate their requirements for the correspondence search. This is mainly based on the idea of *variances* which describe how sets of images vary in blur, intensity, angle, etc. Our framework interprets the description and chooses from a set of algorithms those that satisfy the description. Our proof-of-concept implementation demonstrates the link between the description set by the user and the result returned. The abstraction is also at a high enough level to hide implementation and device details, allowing the simple use of hardware acceleration.

1 Introduction

Computer vision has recently seen a rise in production of real-world applications, such as on mobile device's cameras (for stitching and face detection) and real-time pose estimation for games and other gesture-based interfaces. However most of these are developed by experts in computer vision, in dedicated teams designing reliable units of software to perform these tasks. The application of these techniques could be increased if they were easier to use and deploy, however there has not been much work published on how to provide access to these sophisticated techniques to developers; effective use of these methods requires extensive knowledge of how the algorithms work and how their parameters affect the results, expertise beyond the scope of mainstream developers or system designers (termed *users* for the rest of the paper).

The contribution of this paper is a task-based description applied as an abstraction to the correspondence problem in computer vision, to hide the details of specific methods and their configuration. The abstraction may be employed by users to describe the type of vision problem they are trying to solve, and our

© Springer International Publishing Switzerland 2015
C.V. Jawahar and S. Shan (Eds.): ACCV 2014 Workshops, Part II, LNCS 9009, pp. 229–242, 2015.
DOI: 10.1007/978-3-319-16631-5_17

novel interpreter uses the description to select an appropriate algorithm (with parameters derived from the user's description).

There are numerous benefits of a higher-level abstraction for computer vision: users can focus on their application/system's main task, without focussing on algorithms and parameter tuning; subsequent improvements in techniques for particular problems can be incorporated later with re-implementation; various back-ends can be supported, allowing specific methods to be employed for different requirements e.g. low power for mobile or high speed and accuracy using servers; hardware-acceleration can be used seamlessly, optionally in coordination with CPU; and finally, computer vision expertise can be more readily adopted by researchers in other disciplines, such as HCI and graphics.

If any abstraction is used to access vision methods, hardware and software developers of the underlying mechanisms are free to continually optimise and add new algorithms. This idea has been applied successfully in many other fields, notably graphics with OpenGL, and is the main goal of OpenVL [1] for computer vision. OpenVL is an abstraction framework which hides algorithmic detail and provides developers with access to sophisticated vision methods, such as segmentation [2], human body pose estimation [3] and face detection [4]. The work presented here applies a similar methodology to construct a task-based description of correspondence search at a low enough level to maintain flexibility but high enough for mainstream developers to apply successfully, within the OpenVL framework.

Various technology companies have also recognised the need for a solution to this problem, although most have focussed on hardware acceleration and not higher-level abstraction. A working group at Khronos (a standards body) are developing a low-level hardware abstraction layer called OpenVX to accelerate vision methods[1]; this layer would sit beneath libraries such as OpenCV [5] in order to accelerate existing library calls (much like projects such as OpenVIDIA[2]). Unfortunately OpenCV presents algorithms directly to the developer which requires that they have significant expertise in computer vision - otherwise they are not able to take full advantage of the library. Our proposed abstraction would act as an additional higher-level layer to hide the details of correspondence algorithms and hardware acceleration from developers and allow them to focus on developing applications.

The correspondence problem is a fundamental challenge in computer vision with many robust solutions for a given set of narrow conditions. It is important in many real-world applications such as image stitching, super-resolution, image stabilisation, camera calibration, object detection and 3D reconstruction. To provide flexible approaches to these applications to non-expert users, we must first abstract the complexity to a higher, more intuitive level.

---

[1] http://www.khronos.org/vision.
[2] http://openvidia.sourceforge.net.

2   Related Work

Previous attempts on simpler access have generally been in the development of vision or image processing frameworks which present lists of algorithms; the contributions in general have been how the algorithms are presented. Developments in artificial intelligence were used in an attempt to automate the vision pipeline [6–8]. Others provided higher-level access to vision algorithms through object-oriented methodologies for accessibility and reusability reasons, such as in the Image Understanding Environment project (IUE) [9]. Some have attempted to solve specific problems, such as the OpenTL framework [10] which tries to unify efforts on tracking in real-world scenarios In general, the algorithm categorisation and direct access of these approaches requires users to have expert knowledge of vision methods.

Vision applications can be created by using a data flow structure to connect components, using visual programming interfaces such as the Khoros software development environment [11] and Apple's Quartz Composer[3]. These contain components such as colour conversion, feature extraction, spatial filtering, statistics and signal generation, among others. Declarative programming languages have also been used to provide vision functionality in small, usable units, e.g. ShapeLogic[4] or FVision [12], although they are limited in scope due to the difficulty of combining logic systems with computer vision. While these methods provide a simpler method to access and apply methods, there is no abstraction above the algorithmic level, and so users of these frameworks must have a sophisticated knowledge of vision to apply them effectively. As a more graphically intuitive method to overcome the computer vision usability problems, the RADIUS project [13] employed user-manipulated geometric scene models to help guide the choice of algorithm. The level of abstraction provides good usability although power, breadth and flexibility are reduced.

There are many open vision libraries that provide common vision functionality, such as OpenCV [5], Mathworks Vision Toolbox[5] and Gandalf[6]. These libraries often provide utilities such as camera capture or image conversion as well as suites of algorithms, which has previously been shown to lessen the effectiveness on application [14]. All of these software frameworks and libraries provide vision components and algorithms without any context of how and when they should be applied, and so often require expert vision knowledge for effective use. For example, many feature detectors/descriptors are provided by OpenCV but with no indication of under what conditions each works most effectively. Our goal with this paper is to outline a higher-level abstraction for access to these methods through an intuitive task-based interface.

---

[3] https://developer.apple.com/technologies/mac/graphics-and-animation.html.
[4] http://www.shapelogic.org.
[5] http://www.mathworks.com/products/computer-vision.
[6] http://gandalf-library.sourceforge.net.

3   Task-Based Correspondence

The primary aim of our contribution is to provide non-experts in computer vision with intuitive access to sophisticated feature matching techniques. We define a description model for users to specify what the problem is they wish to solve, instead of the current method of defining *how* to solve it. The description is used by our framework to select appropriate feature descriptors, configure their parameters and execute them to return the required result.

3.1   Abstraction Through Description

We are using a simple definition of the correspondence problem: among a set of images, find regions whose structure is similar, based on a required strength threshold. The central idea of the abstraction for correspondence is *variances*. Assuming two regions $R_1 \in I_1, R_2 \in I_2$ have identical structure (where I_1, I_2 are images, not necessarily different), the variances describe how the two regions differ. There are many possible variances which generally indicate a difference in appearance (such as intensity) or a distortion of the structure (such as blur or tilt).

Note that our definition covers regions, and not points, despite the result often being a central point in the region. The result from our framework is in the form of regions or central points, as decided by the requirements of the user. Note also that our definition does not define variances as between *images*, but regions. This is due to the fact that there may be matches occurring within a single image, or the set of images may have variations in the property e.g. selective focus. In the future, we will be introducing variances also as requirements, to allow users to specify that matches which have a particular variance (e.g. blur) are not needed.

Each variance is associated with an expected quantity of how much it varies. Currently it is challenging to be highly specific with these, as the descriptors have not been evaluated to this level of detail. Therefore we employ a simple scale from *None* to *High* to allow the user to indicate the level of variance expected.

Each variance is intended to be orthogonal to the others within their description space, to avoid overlap in the description and to encourage completeness. Our eventual goal is to create a unified space for vision descriptions, to apply to all problems, which can be interpreted into algorithms and parameters to provide the user with a solution. The descriptions should be kept as small as possible while maintaining the largest possible coverage of the description space, to help minimise the complexity as the description language is extended.

The other main component of the description from the variances is the *constraints*. These essentially control the execution of the methods and the search space for correspondences. The search space can be defined as *Set* or as *Image*. In *Set*, the search for matches occurs over 'the set' of all regions taken from all images, and each region gets N matches from the required quantity. For *Image*, the search occurs across images only i.e. a region in one image can only be matched to regions in other images, and N matches will be returned per image.

The search operation can be augmented with a qualifier of where to search: *All*, the default, uses all regions or images; *Source* overrides the default search and instructs the framework to only return matches from regions in the same image as the to-be-matched region; *Exact* allows the images to be used to be specified, or a number to be specified to constrain the number of best matches to a subset of images. The final constraints for the correspondence problem are *Quantity* (how many matches to return per region, either within the *Set* or from the *Images*) and *Strength*, the main (simple) threshold to control the reliability of the returned match.

3.2   Algorithm Selection

The interpreter is the second layer in our framework, and receives the description from the user passed through the interface (e.g. API). It is responsible for choosing the appropriate feature matching algorithms based on the described variances, defining the parameters of each algorithm based on the input description, and post-processing constraints which can't be defined as a parameter (e.g. finding the top N matches in K images).

The addition of an algorithm to the framework is accomplished through a 'plug-in' system, defined using an internal interface. Each algorithm must implement this interface; the interpreter then uses it to provide the algorithm with the input images and the full user-defined description. The algorithm returns matches in the interface-defined representation, so that all algorithms return the same type to the user.

The process of adding a new algorithm to the framework is as follows:

– The problem conditions (for correspondence, this is the variances) under which the algorithm is designed to return reliable results, defined using expected tolerance to variances (i.e. how invariant the method is). The set of conditions defines a larges dimensional volume in which algorithms occupy sub-volumes.
– The input must be converted to the format used by the algorithm.
– After execution the results must be checked for compliance with the constraints, and flagged for processing in the interpreter if needed.
– The match results must be converted into the global framework's format for presentation to the user (as regions or points).

It is extremely challenging to define the conditions for the algorithm based on a higher-level description: there may be many ranges under which it works well; it may perform best under certain optimal circumstances, but perform well enough under other conditions; it may not work as well as other algorithms when condition volumes intersect, and so a ranking system may be needed.

There is also the problem of how to define the operating conditions: the creator of the algorithm could define them (or an expert in the area), or we could use a standardised dataset within an evaluation framework to generate them. We could also use machine learning techniques to automatically determine the conditions based on the description. In our framework, we have defined the conditions (as 'experts') using evaluations of feature descriptors from the literature; our resources for the condition definitions are presented in Sect. 4.2.

The correspondence interpreter in our framework will select all algorithms which match the user-defined description and constraints, to provide as many matches as possible. The interface allows the user to specify the efficiency required, which can tune the number of methods chosen down. This is also based on the level of parallel processing available on the host machine and allowed by the user.

4 Evaluation of the Task Description

Our framework is implemented in C++, with three separate layers to allow for simple replacement of any one layer. The description layer sits on top and acts as the interface between the user and the lower layers. It is through this that the user provides a description of the correspondence problem. The description is passed to the second layer, the interpreter, a thin layer which chooses which algorithms are appropriate given the description, and then configures each algorithm's parameters. The interpreter can also define any necessary pre- or post-processing operations (such as noise removal or image scaling). The lowest layer of the three is where the algorithms sit. For this work we have defined a condition set for five feature descriptor algorithms for matching image regions. Each algorithm is registered with the interpretation layer along with its optimal operating problem conditions (defined in Table 1). The conditions are defined using the variance components of the description. Given a lack of available specific evidence for the performance of each feature descriptor, their capabilities are approximated using a four-scale *None*, *Low*, *Medium* and *High*.

4.1 Parameters

The correspondence description provides the following set of variances: position, size, orientation, blur, intensity and tilt distortion. All are measured on the scale mentioned previously. Unfortunately the use of such a scale is not consistent across all variances, and so each must be documented individually.

- Position: Indicates the expected level of variation in the region's position in image coordinates. This can encode problem conditions such as the quantity of expected overlap among images. This is usually used purely to reduce the search space, and is not used in algorithm selection.
- Size: *None* indicates identical size, *High* means changes of 100 % or more. Sizes are measured in units of image width (based on source region).
- Orientation: 2D rotation up to 180° (either direction).
- Blur: It is challenging to determine a scale of blur which is easily understood, and to avoid measurements based on pixels (since they are not directly relevant to the vision/matching problem). In future we will need a more sophisticated

description for blur, taking into account multiple blur kernels and parameters. For now, since this type of variation is not accounted for within feature descriptors, a simpler description will do. The scale we use starts from zero blur (*None*, i.e. circle of confusion has a radius lower than the current sampling density); *Low* blur is defined as a gaussian blur with $\sigma = 1.0$ and a kernel size of 0.5 % image width; *High* blur uses the same definition but with a kernel size of 2 % image width.

- Intensity: This is also challenging - we treat it as an absolute measurement, so *Low* is approximately 10 % difference, *Medium* is 30 % and *High* is above 50 %.
- Tilt Distortion: This represents the central viewing angle difference between views (and could also account partly for lens distortions). The variance is measured on a scale from 0° to 45° (*None* to *High*).

Table 1. The mapping from description to algorithm within our interpreter, using the variances between images. The methods satisfy any permutation of their conditions.

Algorithm	Size	Orientation	Blur	Intensity	Tilt distortion
SIFT [15]	High	High	High	Medium	Medium
SURF [16]	Medium	Low	High	High	Low
ORB [17]	Low	High	Low	Low	None
MSER [18]	None	None	None	Low	Medium
FREAK [19]	Medium	None	High	High	Medium

4.2 Feature Matching Algorithms

The algorithms chosen to sit in the lowest layer of our framework are shown in Table 1. Each algorithm registers itself with the interpreter using the variance capabilities indicated in the table. The set of mappings from description to algorithm (defined in the Table) are quite expansive, and could be more specific with a more direct evaluation of capabilities. In the cases where a user supplies a description which is not represented by our interpreter's algorithms, we can perform a 'best effort' and provide the closest match, or provide an informative error.

The mapping from variances to algorithms were inferred using multiple sources: size, orientation, blur and intensity for SIFT, SURF and ORB[7]; ORB comparison to SIFT and SURF for orientation [17]; size, orientation, intensity and tilt distortion for SIFT and SURF [20]; orientation, blur, intensity and affine transforms for SIFT and SURF [21]; size and tilt distortion for SIFT and MSER [22]; orientation, blur and tilt distortion for SIFT and MSER [23]; size, blur,

---

[7] http://computer-vision-talks.com/articles/2011-08-19-feature-descriptor-compari son-report/.

intensity and tilt distortion for MSER [24]; and finally, size, blur, intensity and tilt distortion for SIFT, SURF and FREAK [19].

The conditions chosen for each algorithm are examples, and not meant to be definitive. Choosing the conditions is a difficult problem, and perhaps the best solution is an evaluation of each algorithm under all known conditions on a ground-truth dataset. We can then assign the conditions from this evaluation or use a learning-based approach to bypass the literal conditions completely - although this would require a differently designed interpreter.

By default, the interpreter executes as many of the algorithms as it can, based on how well they match the user's description. This will provide the highest number of matches, but is also slower or uses more resources. The user can override this behaviour by prioritising efficiency over quantity. The user may also choose to switch implementations from the CPU-based to the GPU-based algorithms (which use SIFT and SURF only): this is seamless and the actual algorithms included are not exposed to the user. This allows the user to take advantage of the parallel hardware and provide a faster computation. The interface does not change for multiple implementations, ensuring that the code written for one device type will work on any other device type.

The strength constraint does not use the variances, even though some algorithms may have a higher strength than others. Instead it is used to modify the parameters of the algorithms, such as the number of nearest neighbours to require for a match or the minimum distance between feature vectors.

4.3   Results

In Figs. 1 and 2 we illustrate four different descriptions (using variances) which describe their pair of images. For each example three results are provided, at High, Medium and Low Strengths. Figure 3 shows results for the four previous examples but without providing any description. The images used for this demonstration were taken from a feature detection evaluation site[8] and a feature evaluation paper [21].

The first example in Fig. 1 is of two images containing a lot of detail (church exterior). The variances defined lead to SIFT being chosen by the interpreter to find the correspondences. The results are generally good at all strengths (likely due to the high detail), with some errors apparent at Low strength. The second example in Fig. 1 has images with a significant level of structure, with the two images showing the same building at two focal lengths. The description supplied here invoked SIFT, SURF and FREAK and provided excellent correspondences, with some errors at Low strength.

The first example in Fig. 2(a) contains two images with a large size variation and low-to-medium in all other variances. The interpreter executed SIFT to accommodate the size and tilt variation. At a High strength the results are sparse but accurate, and as the strength is lowered more matches are provided but many noticeable errors occur. This is due to the more extreme difference in images.

---

[8] http://lear.inrialpes.fr/people/mikolajczyk/Database/det_eval.html.

Example (a) - Variances:
Position(Medium); Size(None); Orientation(Low);
Blur(None); Intensity(Low); Tilt Distortion(Medium).

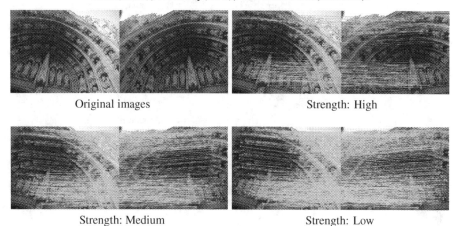

Original images   Strength: High

Strength: Medium   Strength: Low

Example (b) - Variances:
Position(Medium); Size(Medium); Orientation(None);
Blur(None); Intensity(None); Tilt Distortion(Low).

Original images   Strength: High

Strength: Medium   Strength: Low

Fig. 1. The original images for each example are shown along with the results at three strength levels, with the stated variances above each set of images. Example (a) illustrates when almost all variances are set, and the returned matches at all three strengths with good results. Example (b) also demonstrate good results, but with noticeable mismatches at lower strengths.

Example (a) - Variances:
Position(Medium); Size(High); Orientation(Low);
Blur(Low); Intensity(Low); Tilt Distortion(Low).

Original images   Strength: High

Strength: Medium   Strength: Low

Example (b) - Variances:
Position(None); Size(None); Orientation(None);
Blur(High); Intensity(None); Tilt Distortion(None).

Original images   Strength: High

Strength: Medium   Strength: Low

Fig. 2. The original images for each example are shown along with the results at three strength levels, with the stated variances above each set of images. It is noticeable in Example (a) that in more extreme cases like this, the strength can have a noticeable effect on quality of returned match. This is also seen in Fig. 1(a), indicating that perhaps strength plays a more important role in size differentials. In Example (b), the level of blur is so high that it inhibits feature extraction, even at the lowest strength (although the results obtained at a low strength are quite accurate). This may indicate that we should provide a finer level of control over the strength for the user.

Examples with all variances set to *None*

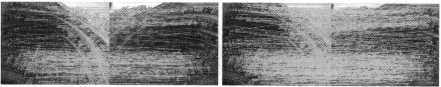

Strength: High   Strength: Medium

Strength: High   Strength: Medium

Strength: High   Strength: Medium

Strength: High   Strength: Medium

Fig. 3. The examples here demonstrate the result when you set all variances to *None*, and so all feature algorithms are executed. At a high strength, this provides reasonable results, with only the occasional error (such as in the book example). At medium strength it is clear that there are many more errors (and many more matches) than in the case where the specific problem was described. In all cases, this took significantly longer to compute, since all algorithms were used. Given the higher quality of correspondence and the decrease in time taken, this demonstrates the improvement which can be made with knowledge of the problem conditions in addition to the reduced level of expertise required to use our framework.

The second example illustrates a single image matched with a blurred version of itself. The description translated to the execution of SIFT, SURF and FREAK, providing a few correspondences. The results seem to be as good when the strength is lowered, which may indicate we should provide more control over these cases to the user (to allow for a variation of strength or method when blur is a significant variation).

To see how well the framework performs in the absence of any information, we tested on all previous examples and gave the interpreter a description with all variances set to None. All methods are executed in this case. The results are generally acceptable for High Strengths, with only a few mistakes on the more extreme examples. There are significantly more errors in the medium case, along with a higher number of returned correspondences. The time taken was much longer in all cases due to executing all algorithms.

5   Conclusion

We have presented a novel abstraction for the correspondence problem which provides users with the ability to target specific conditions for the correspondence problem. This allows them to find the most efficient and accurate algorithm(s) without requiring a high level of expertise in feature detectors and descriptors. We demonstrated the results produced when providing our framework with a description of variances, and illustrated how the matches were of higher quality than not providing any information.

We intend to expand the variances, possibly to multiple levels of description satisfaction: for example, SURF can do well enough on tilt distortion but might need to be controlled via strength by the user; alternatively ORB is generally faster than the others, but not as accurate in certain conditions - this could be an override control for the user.

Acknowledgements. We would like to gratefully acknowledge the support of the *Natural Sciences and Engineering Research Council of Canada (NSERC)* and the *Canadian Graphics, Animation and New Media Network of Centres of Excellence (GRAND NCE)*.

References

1. Miller, G., Fels, S.: OpenVL: a task-based abstraction for developer-friendly computer vision. In: Proceedings of the 13th IEEE Workshop on the Applications of Computer Vision (WACV), WVM 2013, pp. 288–295. IEEE (2013)
2. Miller, G., Jang, D., Fels, S.: Developer-friendly segmentation using OpenVL, a high-level task-based abstraction. In: Proceedings of the 1st IEEE Workshop on User-Centred Computer Vision (UCCV), WVM 2013, pp. 31–36. IEEE, New York City (2013)
3. Oleinikov, G., Miller, G., Little, J.J., Fels, S.: Task-based control of articulated human pose detection for openvl. In: Proceedings of the 14th IEEE Winter Conference on Applications of Computer Vision, WACV 2014, pp. 682–689. IEEE, New York City (2014)

4. Jang, D., Miller, G., Fels, S., Oldridge, S.: User oriented language model for face detection. In: Proceedings of the 1st Workshop on Person-Oriented Vision (POV), WVM 2011, pp. 21–26. IEEE, New York City (2011)
5. Bradski, G., Kaehler, A.: Learning OpenCV: Computer Vision with the OpenCV Library, 1st edn. O'Reilly Media, Inc., Sebastopol (2008)
6. Matsuyama, T., Hwang, V.: SIGMA: a framework for image understanding integration of bottom-up and top-down analyses. In: Proceedings of the 9th International Joint Conference on Artificial Intelligence, vol. 2, pp. 908–915. Morgan Kaufmann Publishers Inc. (1985)
7. Kohl, C., Mundy, J.: The development of the image understanding environment. In: Proceedings of the Conference on Computer Vision and Pattern Recognition, CVPR 1994, pp. 443–447. IEEE Computer Society Press, Los Alamitos (1994)
8. Clouard, R., Elmoataz, A., Porquet, C., Revenu, M.: Borg: a knowledge-based system for automatic generation of image processing programs. Trans. Pattern Anal. Mach. Intell. 21, 128–144 (1999)
9. Mundy, J.: The image understanding environment program. IEEE Expert Intell. Syst. Appl. 10, 64–73 (1995)
10. Panin, G.: Model-based Visual Tracking: The OpenTL Framework, 1st edn. Wiley, Chichester (2011)
11. Konstantinides, K., Rasure, J.R.: The Khoros software development environment for image and signal processing. IEEE Trans. Image Process. 3, 243–252 (1994)
12. Peterson, J., Hudak, P., Reid, A., Hager, G.D.: FVision: a declarative language for visual tracking. In: Ramakrishnan, I.V. (ed.) PADL 2001. LNCS, vol. 1990, pp. 304–321. Springer, Heidelberg (2001)
13. Firschein, O., Strat, T.M.: RADIUS: Image Understanding For Imagery Intelligence, 1st edn. Morgan Kaufmann, San Francisco (1997)
14. Makarenko, A., Brooks, A., Kaupp, T.: On the benefits of making robotic software frameworks thin. In: Proceedings of the Workshop on Measures and Procedures for the Evaluation of Robot Architectures and Middleware, IROS 2007. IEEE, New York City (2007). Invited Presentation
15. Lowe, D.G.: Distinctive image features from scale-invariant keypoints. Int. J. Comput. Vis. 60, 91 (2004)
16. Bay, H., Ess, A., Tuytelaars, T., Van Gool, L.: Speeded-up robust features (surf). Comput. Vis. Image Underst. 110, 346–359 (2008)
17. Rublee, E., Rabaud, V., Konolige, K., Bradski, G.: Orb: An efficient alternative to sift or surf. In: Proceedings of the 2011 International Conference on Computer Vision, ICCV 2011, pp. 2564–2571. IEEE Computer Society, Washington, DC (2011)
18. Forssén, P.E., Lowe, D.: Shape descriptors for maximally stable extremal regions. In: IEEE International Conference on Computer Vision, vol. CFP07198-CDR. IEEE Computer Society, Rio de Janeiro (2007)
19. Ortiz, R.: Freak: Fast retina keypoint. In: Proceedings of the 2012 IEEE Conference on Computer Vision and Pattern Recognition (CVPR), CVPR 2012, pp. 510–517. IEEE Computer Society, Washington, DC (2012)
20. Bauer, J., Sünderhauf, N., Protzel, P.: Comparing several implementations of two recently published feature detectors. Intell. Auton. Veh. 6, 143–148 (2007)
21. Juan, L., Gwon, O.: A comparison of SIFT, PCA-SIFT and SURF. Int. J. of Image Process. 3, 143–152 (2009)
22. Yu, G., Morel, J.M.: A fully affine invariant image comparison method. In: IEEE International Conference on Acoustics, Speech and Signal Processing, 2009, ICASSP 2009, pp. 1597–1600 (2009)

23. Morel, J.M., Yu, G.: Asift: a new framework for fully affine invariant image comparison. SIAM J. Img. Sci. **2**, 438–469 (2009)
24. Mikolajczyk, K., Tuytelaars, T., Schmid, C., Zisserman, A., Matas, J., Schaffalitzky, F., Kadir, T., Gool, L.V.: A comparison of affine region detectors. Int. J. Comput. Vis. **65**, 43–72 (2005)

Interactive Shadow Editing from Single Images

Han Gong$^{(\boxtimes)}$ and Darren Cosker

Department of Computer Science, University of Bath, Bath, UK
hg299@cs.bath.ac.uk

Abstract. We present a system for interactive shadow editing from single images which includes the manipulations of shape, distribution, sharpness and darkness of shadows according to the features of existing shadows. We first obtain a shadow-free image, shadow boundary and its registered sparse shadow scales using an existing shadow removal method. The modifiable features of the shadow are synthesised from the sparse shadow scales. According to the user-specified shadow-shape and its attributes, our system generates a new shadow matte and composites it into the original image, while also allowing editing of existing shadows. We share our executable for open comparison in community.

1 Introduction

Shadows are ubiquitous in natural images. Advanced shadow editing, which may include the manipulations of different shadow properties, is often desirable for Graphical artists. A shadow generally consists of two parts which are the umbra and penumbra. The umbra is the darkest region of shadow with constant illumination while the penumbra, i.e. shadow boundary, has transitioning illumination between the fully dark and lit area. Shadow effects can be represented as a multiplicative scale field, i.e. a matte. An image $\mathcal{I}_c$ can be represented as a Hadamard product of a shadow scale field $\mathcal{S}_c$ and a shadow-free image $\mathcal{I}_c$ as follows:

$$\mathcal{I}_c = \mathcal{I}_c \circ \mathcal{S}_c \qquad (1)$$

where c is a channel of RGB colour space. The lit area's scale are 1 and the other areas' scales are between 0 and 1.

In this paper, we propose a system for interactive shadow editing from single images which synthesises the features of existing shadow and preserves the naturalness of newly generated shadows. Our system provides friendly and flexible user-controls for defining the shape, darkness, softness, and colour of existing or new shadows. Furthermore, it does not require users to manually analyse and adjust the shadow properties in an image, as the manipulatable shadow properties are automatically synthesised by our system. Compared with the artificial shadows generated by Computer Graphics rendering, our shadow modification is based on the existing real shadows and our newly modified and added shadows can be highly consistent with the original shadows in a scene. The potential usages of this tool include but are not limited to: (1) moving existing objects and their shadows from one image to another; and (2) artistic modification of shadow.

© Springer International Publishing Switzerland 2015
C.V. Jawahar and S. Shan (Eds.): ACCV 2014 Workshops, Part II, LNCS 9009, pp. 243–252, 2015.
DOI: 10.1007/978-3-319-16631-5_18

This paper focuses on a shadow editing framework that utilises the output information from an existing shadow extraction method. Based on a state-of-the-art shadow removal dataset [1], we have both visually and quantitatively tested our system on various shadow scenes, which includes scenes with strong texture background, broken, soft and coloured shadows.

2   Related Work

Naive shadow editing using existing image manipulation software, e.g. direct change of brightness or blurring for shadow boundaries, requires a considerable amount of manual adjustment to align the appearances of modified shadow to the original shadow. This alteration requires delicate editing steps and unavoidably results in unnatural artefacts around shadow boundaries. These features of shadow editing are not available in current image manipulation software, e.g. Photoshop and GIMP.

Detection and removal of shadow are the prerequisites for extracting the features of a shadow. Approaches to shadow removal from single images can be categorised as either automatic [2–4] or user-aided [1,5–10]. The difference between these is whether the shadow detection process is automatic or user-guided. Automatic shadow removal methods generally rely on intrinsic image decomposition [2,3] or shadow feature learning [4]. User-aided methods generally achieve more accurate and reliable shadow detection results at the practical cost of varying degrees of user input, such as quad map [5], shadow boundary [6,8], and sample of intensity [1,9,10]. As for the removal phase, some of them [2–5,7] adopt pixel-wise optimisation while others [1,6,8–11] rely on the analysis of intensity profiles passing through shadow boundary.

Recent shadow removal work [5,7,8] presents some basic examples of shadow editing including complete removal, duplication, distortion and sharpness adjustment of the original shadow. The applications of this work only apply simple image manipulation to the original shadow matte and do not provide a manipulatable model for arbitrary shadow modification. A shadow editing system is first proposed in [11] which requires users to specify some boundary points of shadow. Their shadow edge model is manipulatable and supports controls for sharpness, darkness and shape of shadows. However, users are only allowed to move the specified boundary points of shadow which limits its range of amendable shapes and are not allowed to add new shadow segments. In summary, natural shadow editing for arbitrary shapes and properties is still unexplored.

2.1   Contributions

Given our review of state-of-the-art approaches, we propose the following two contributions:

(1) **A model for shadow editing from single images.** We propose a model that analyses the features of an existing shadow in an image and provides parameters for users to edit the extracted shadow and add new shadows.

Unlike previous work (e.g. [11]), our shadow editing model is fast, highly flexible and provides many more controls of shadow properties. The model is also universal and compatible with several shadow removal methods.

(2) **Easy user interaction design for shadow editing.** An easy interface is proposed for users to freely and quickly define the shape of shadow and control various properties of shadow.

To summarise, we believe our contributions are important in this area due to our significant improvements in extracting and utilising controllable properties from existing shadow and the ease in which a user may interact with a shadow scene.

3   Interactive Shadow Editing Model

In this section, we overview our algorithm first in brief then expend on technical details for each step. Our algorithm consists of three steps (see Fig. 2):

(1) **Pre-processing** (Sect. 3.1). A shadow-free image, shadow boundary information, and shadow scale attenuation profiles are first obtained. This information is converted into a field of shadow softness and a synthesised intensity attenuation profile.

(2) **Synthesis of shadow matte** (Sect. 3.2). Based on the pre-processed information, a shadow model with various tunable parameters is generated. A modified shadow matte is synthesised using a distance transform.

(3) **Composition of shadow** (Sect. 3.3). The modified shadow is seamlessly composited into the original shadow image (Fig. 1).

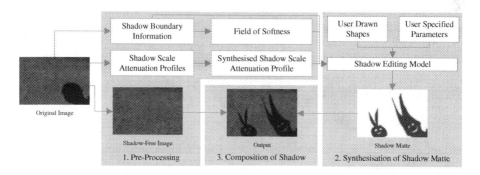

Fig. 1. Our shadow editing pipeline. The 3 main blocks in this chart correspond to the 3 main steps of our shadow editing system.

3.1   Pre-processing

Pre-processing provides a shadow-free image, shadow boundaries, a field of shadow softness and a synthesised shadow scale attenuation profile required by our shadow editing model.

Shadow Extraction. Shadow extraction is the first step for shadow editing which provides necessary information for realistic shadow re-creation. To extract shadow from a single image, a state-of-the-art shadow removal method [1] is applied. This shadow removal method gives a shadow-free image, a shadow mask, and many shadow scale attenuation profiles perpendicular to the shadow boundary. Some other shadow removal methods [6,8–11] rely on intensity profile analysis are also compatible to our system.

Synthesisation of Shadow Scale Attenuation. Shadow creation is simplified by synthesising intensity attenuation. All extracted shadow scale profiles are re-scaled to a unique length (length of the longest profile by default). A synthesised shadow scale attenuation profile is then generated by profile-wise averaging all re-scaled shadow scale profiles. The averaging process cancels texture noise. To accelerate the computation for variable penumbra widths, a look-up table like profile is pre-computed. The synthesised shadow scale attenuation profile is evenly re-sampled to a large number (1000 in our implementation) of data sites using a piecewise cubic Hermite polynomial [12]. Shadow scales on wide penumbra can thus be queried by finding the closest scale from the large number of data sites.

Generation of Softness Field. When the shape of shadow is changed from the original shadow image, it is unknown how soft the penumbra of the newly created parts should be. This is solved by generating a field of shadow softness from known penumbra widths of original shadow boundary points. The problem is equalised to in-painting an image with known pixel values (shadow scale profile lengths) at original shadow boundary points and unknown values for the remaining of pixels. A spring-metaphor based in-painting method [13] is adopted to smoothly interpolate and extrapolate the unknown values.

3.2   Synthesisation of Shadow Matte

Our goal is to generate a shadow matte according to user specified shadow properties. Given a shadow mask, whether the original or not, its crisp shadow boundaries can be located using a method for boundary tracing from a binary image [14]. The resulting boundaries should exclude the boundaries of the image border. A typical solution to generate a shadow matte is to re-generate the sparse scales of sampling lines perpendicular to shadow boundaries and in-paint for the other unknown shadow scales in image like the algorithm to form a dense shadow scale field from sparse scales described in [1]. However, this can be computationally costly for interactive performance as the in-painting process is comparatively slow and every modification of shadow revokes the in-painting process. Instead, a linear time Euclidean distance transform [15] is used to generate the shadow matte quickly. The procedure for the generation of shadow mattes is described in Algorithm 1 where χ is a function computes Euclidean distance transform [15] (its first output D refers to a matrix of Euclidean distance and its second output L refers to a label index matrix of closest boundary point), ϕ is a function which

---

Algorithm 1. Generation of Shadow Matte

---

    input   : point set of shadow boundary $\mathbf{B}$, field of shadow softness F,
                synthesised shadow scale attenuation profile A, shadow mask N
    output: shadow matte S

1 Initialise M as a zero matrix in the size of original image;
2 Initialise S as a matrix of ones in the size of original image;
3 $M(\mathbf{B}) \leftarrow 1$;   `/* mark shadow boundary points */`
4 $(D, L) \leftarrow \chi(M)$;   `/* perform a Euclidean distance transform */`
5 **foreach** *point* $p \in \mathbf{B}$ **do**
6      $E \leftarrow \{x | L(x) = \phi(p)\}$;   `/* find pixels closet to current point */`
        `/* divide the point set E into lit and shadow parts        */`
7      $E_l \leftarrow \{x | x \in E \cap N(x)\}$;   `/* lit part */`
8      $E_s \leftarrow \{x | x \in E \cap x \notin E_l\}$;   `/* shadow part */`
        `/* perform distance conversion                              */`
9      $R \leftarrow D(E)$;
10     $R(E_l) = \max(0.5 + D(E_l)/F(p), 1)$;
11     $R(E_s) = \min(0.5 - D(E_s)/F(p), 0)$;
12     $r \leftarrow 1000$; `/* number of cached intensity attenuation data sites */`
13   **foreach** RGB *colour channel* $c \in \{1, 2, 3\}$ **do** convert distance to shadow
        scale
14     $\mid$ $S(E, c) \leftarrow A(\max(\kappa(rR), 1), c)$;
15   **end**
16 **end**

---

returns the index of a pixel in image, κ is a function that truncates each value of a set of data.

    Algorithm 1 already handles variable shadow shape. To provide more controls for softness, darkness, and colour of shadow, more parameters are added to enable tuning:

Softness of Shadow. A parameter of global softness adjustment is provided. This is achieved by simply multiplying the field of softness F in Algorithm 1 with a scaling factor v_s as follows:

$$F_n = \max(Fv_s, 2) \tag{2}$$

where F_n is the modified field of softness and an additional saturation operation is applied to ensure the width of penumbra is at least 2 pixel wide.

Darkness of Shadow. A parameter for darkness adjustment is provided. This is achieved by modifying the synthesised shadow scale attenuation profile A in Algorithm 1 using a scaling factor V_d which controls the variation of shadow scale attenuation. The profile A is modified as follows:

$$\begin{cases} m(c) = \mu(A(1 \ldots r, c)) \\ A_n(1 \ldots r, c) = \max(v_d(A(1 \ldots r, c) - m(c)) + m(c), 0) \end{cases} \tag{3}$$

where A_n is the updated profile, c is the RGB channel index, r is the number of pre-computed data sites in Algorithm 1, μ is a function which computes the mean of a set of values. An additional saturation operation is applied to ensure the shadow scale is non-negative.

Colour of Shadow. Three parameters of colour adjustment are provided. Each of them controls the intensity strength of its RGB colour channel. The 3 parameters are represented as a 1-by-3 vector V_k. The adjustment is done by further modifying Eq. 3 as follows:

$$\begin{cases} m(c) = \mu(A(1\ldots r, c)) \\ o_k(c) = A(1, c) \\ d_k(c) = V_k(c) - o_k(c) \\ q_k(c) = (1 - V_k(c))/(1 - o_k(c)) \\ A_n(1..r, c) = \max(q_k(c)v_d(A(1..r, c) - m(c)) + m(c) + d_k(c)/2, 0) \end{cases} \quad (4)$$

where o_k is an intensity vector representing the original colour of the shadow, d_k and q_k are the error and ratio between original and adjusted colours of shadow respectively which change the variation and mean of shadow scale attenuation profile A.

After the adjustment, A_n may contain shadow scale values greater than 1, it is thus normalised by dividing all data sites by its lit end value as follows:

$$A_n(1..r, c) = A_n(1..r, c)/A_n(r, c) \quad (5)$$

3.3   Composition of Shadow

According the Eq. 1, the edited shadow can be composited using the Hadamard product of the edited shadow matte and the shadow-free image. This approach of composition ensures that the original properties of the surface background, e.g. texture and reflectance, are preserved.

4   User Interaction

In this section, we describe our user interaction and its related algorithm. Figure 2 shows the prototype of our user interface. This interface can be divided into a drawing section and a configuration section.

4.1   Drawing Section for Shape Modification

In previous work [11], users are required to specify multiple sparse control points defining shadow boundary. This user interaction requires multiple delicate mouse clicks to ensure accuracy of shadow modification. After the initial specification, users can drag the previously defined boundary points to change the shape of shadow. However, its modifiable shapes are limited by the initial boundary points

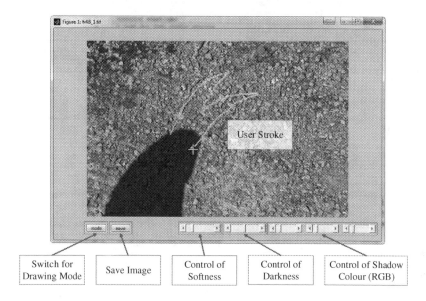

Fig. 2. Graphic user interface.

which also disallows users to add new segments of shadow. The interaction can be cumbersome for complex shape editing as users have to move multiple boundary points. To alleviate this burden as well as to increase the freedom of editing, we introduce an interaction to modify the shape of shadow by simply adding and subtracting user drawn shapes.

At start-up, our system detects, removes shadow and loads the original shadow according to the original shadow shape. Users then draw the boundary of new shape for addition or subtraction. The system automatically connects the two ends of a boundary curve using a straight line when the boundary is not closed. The drawing actions have two types:

Shadow Addition. Shadow addition is equivalent to adding or sometimes merging a shape to the original shadow shape. This is done by applying a logical *or* image operation as follows:

$$N_n = N \vee N_a \tag{6}$$

where N refers to the shadow mask in Algorithm 1, N_a is the additional shape drawn by user, N_n is the updated shadow mask.

Shadow Subtraction. Similar to Eq. 6, shadow subtraction is done by applying two logical image operations as follows:

$$N_n = N \wedge \neg N_a \tag{7}$$

As our shadow matte generation in Algorithm 3 already handles arbitrary shadow masks, users can draw any complicated shapes rather than changing a single existing shape in previous work [11].

4.2   Configuration Section for Other Controls

The other non-drawing controls are placed in the configuration section. The "mode" toggle button switches drawing mode between addition and subtraction of shape. The parameters for tuning shadow editing model are controlled by scrollbars.

5   Evaluation

In this section, we demonstrate our shadow editing system with examples of different types of shadow and show the quantitative evaluation result of our shadow reconstruction based on a state-of-the-art shadow removal dataset [1]. Our executables and data are made available to the community.

5.1   Demonstration of Shadow Editing

The shadow editing steps for various modification and shadow scenes are show in Fig. 3. Please also see our supplementary material for the video demonstration of these examples. Our algorithm is implemented in MATLAB script (non-MEX) and it gives interactive performance on a 2.4 Ghz machine. Figure 3 also provides

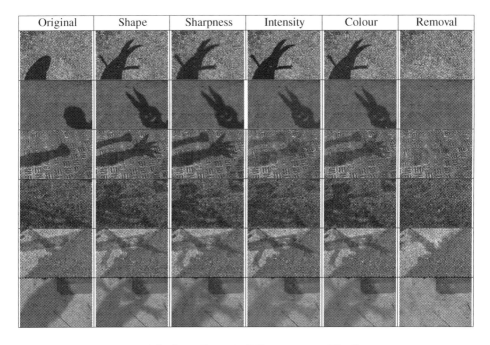

Fig. 3. Demonstration of shadow editing in different scenes: The first two rows are examples of easy scenes. The remaining rows are examples for scenes with strong texture background, broken shadow, coloured shadow, and soft shadow respectively.

Table 1. Error of shadow reconstruction according to 4 attributes. The intensities of image are normalised and the errors shown in this table are in percent (multiplied by 100). The "E" and "E*" indicate the error score where all pixels in the image are used, and just shadow area pixels respectively. For each score of each attribute, the images with other predominant attributes (degree = 3) are not used. Hence, test cases have a strong single bias towards one of the attributes. "Other" refers to a set of shadow cases showing no markedly predominant attributes (degree = 1). "M" refers to the average score for each category. Standard deviations are shown in brackets.

	Texture				Soft				Broken				Colour				Other
	1	2	3	M	1	2	3	M	1	2	3	M	1	2	3	M	
E	1(0)	1(1)	3(1)	2(1)	1(0)	1(1)	3(1)	2(1)	1(0)	2(1)	3(1)	2(1)	1(1)	1(0)	2(1)	2(1)	2(1)
E*	0(0)	0(0)	0(0)	0(0)	0(0)	0(0)	0(0)	0(0)	0(0)	0(0)	0(0)	0(0)	0(0)	0(0)	1(1)	0(0)	0(0)

a qualitative evaluation of the shadow removal and editing algorithms. The real shadows are modified seamlessly in the images of different scenes. This qualitative demonstration reflects the compatibility of our shadow editing tool. The shadow editing process only changes the illumination and the background features, e.g. texture and reflectance, are preserved.

5.2   Quantitative Evaluation of Shadow Reconstruction

Although our system can produce new shadows perceptually similar to the original shadow, there may still be minor errors introduced in shadow composition phase. To evaluate any error and judge the quantitative performance of our system, we reconstruct the original shadow image using our shadow editing model with the extract shadow information and compute the Root Mean Square Error (RMSE) between the generated image and the original image. The evaluation is based on a state-of-the-art shadow removal dataset [1] containing 214 categorised shadow images in variable scenes. As shown in Table 1, our system produces negligible errors of shadow reconstruction and has similar accuracies for all 4 categories. The accuracy of reconstruction is often not crucial for shadow editing but can be important for photo forging.

6   Conclusion

We have presented a shadow editing system with a user-friendly interface. It enables users to freely and quickly modify various properties of existing shadows in images which include shape, darkness, softness, and colour. We have demonstrated and quantitatively evaluated our shadow editing system. Future work includes: (1) the evaluation of the interface with actual users; (2) making use of estimated light sources and geometry from images to provide intelligent suggestions for users to create plausible shadows.

Acknowledgements. This work was funded by the China Scholarship Council and the University of Bath.

References

1. Gong, H., Cosker, D.: Interactive shadow removal and ground truth for variable scene categories. In: British Machine Vision Conference (BMVC) (2014)
2. Finlayson, G.D., Drew, M.S., Lu, C.: Entropy minimization for shadow removal. Int. J. Comput. Vision **85**, 35–57 (2009)
3. Yang, Q., Tan, K.H., Ahuja, N.: Shadow removal using bilateral filtering. IEEE Trans. Image Process. **21**, 4361–4368 (2012)
4. Guo, R., Dai, Q., Hoiem, D.: Paired regions for shadow detection and removal. IEEE Trans. Pattern Anal. Mach. Intell. **35**, 2956–2967 (2012)
5. Wu, T.P., Tang, C.K., Brown, M.S., Shum, H.Y.: Natural shadow matting. ACM Trans. Graph. (TOG) **26**, 8 (2007)
6. Liu, F., Gleicher, M.: Texture-consistent shadow removal. In: Forsyth, D., Torr, P., Zisserman, A. (eds.) ECCV 2008, Part IV. LNCS, vol. 5305, pp. 437–450. Springer, Heidelberg (2008)
7. Shor, Y., Lischinski, D.: The shadow meets the mask: pyramid-based shadow removal. Comput. Graph. Forum **27**, 577–586 (2008)
8. Su, Y.F., Chen, H.H.: A three-stage approach to shadow field estimation from partial boundary information. IEEE Trans. Image Process. **19**, 2749–2760 (2010)
9. Arbel, E., Hel-Or, H.: Shadow removal using intensity surfaces and texture anchor points. IEEE Trans. Pattern Anal. Mach. Intell. **33**, 1202–1216 (2011)
10. Gong, H., Cosker, D., Li, C., Brown, M.: User-aided single image shadow removal. In: Proceeding IEEE International Conference on Multimedia and Expo (ICME), vol.1, p. 1 (2013)
11. Mohan, A., Tumblin, J., Choudhury, P.: Editing soft shadows in a digital photograph. IEEE Comput. Graphics Appl. **27**, 23–31 (2007)
12. Fritsch, F.N., Carlson, R.E.: Monotone piecewise cubic interpolation. SIAM J. Numer. Anal. **17**, 238–246 (1980)
13. Bertalmio, M., Sapiro, G., Caselles, V., Ballester, V.: Image inpainting. In: Proceedings of the 27th Annual Conference on Computer Graphics and Interactive Techniques, SIGGRAPH 2000, pp. 417–424 (2000)
14. Schalkoff, R.J.: Digital image processing and computer vision. Wiley, New York (1989)
15. Maurer Jr., C.R., Qi, R., Raghavan, V.: A linear time algorithm for computing exact euclidean distance transforms of binary images in arbitrary dimensions. IEEE Trans. Pattern Anal. Mach. Intell. **25**, 265–270 (2003)

Hand Part Classification Using Single Depth Images

Myoung-Kyu Sohn$^{(\boxtimes)}$, Dong-Ju Kim, and Hyunduk Kim

Department of IT Convergence, Daegu Gyeongbuk Institute of Science
and Technology (DGIST), Daegu, South Korea
smk@dgist.ac.kr

Abstract. Hand pose recognition has received increasing attention as an area of HCI. Recently with the spreading of many low cost 3D camera, researches for understanding more natural gestures have been studied. In this paper we present a method for hand part classification and joint estimation from a single depth image. We apply random decision forests (RDF) for hand part classification. Foreground pixels in the hand image are estimated by RDF, which is called per-pixel classification. Then hand joints are estimated based on the classified hand parts. We suggest robust feature extraction method for per-pixel classification, which enhances the accuracy of hand part classification. Depth images and label images synthesized by 3D hand mesh model are used for algorithm verification. Finally we apply our algorithm to the real depth image from conventional 3D camera and show the experiment result.

1 Introduction

Vision-based gesture recognition is one of the possible solutions for HCI (Human Computer Interface) as it provides natural interaction between people and all kind of devices. There have accordingly been many studies on gesture recognition techniques [1,2]. Hand gesture is emerging topic, which enables interactions between human and computer more naturally. As only hand part of the human body is needed for analysis of gesture, hand gesture recognition is more efficient in some case in contrast to human activity recognition based on full body movement. It can simply be categorized three problems for recognizing hand gesture, which are hand detection, pose estimation and gesture classification.

Hand gesture recognition research has often focused on techniques to detect hand on the frame from regular RGB camera. Light condition, cluttered background, skin color et al. make it difficult to find the hand from the RGB image. With the commercial release of depth camera such as Kinect [3] and Xtion, the segmentation process are much more simplified through depth information. Thus, techniques for recognizing hand pose using segmented hand depth information have been developed [4–6].

Hand pose estimation generally can be categorized into two groups: shape-based, 3D model-based. Shape-based approaches generally match the shape features of the detected hand to a shape features from a predefined hand shape

© Springer International Publishing Switzerland 2015
C.V. Jawahar and S. Shan (Eds.): ACCV 2014 Workshops, Part II, LNCS 9009, pp. 253–261, 2015.
DOI: 10.1007/978-3-319-16631-5_19

database. Doliotis et al. [7] extracts contour of the hand which implies the shape and the boundary of the hand. After normalizing the features for translation invariance, the depth similarity between two images was measured to classify hand pose. Liu et al. [8] use depth images acquired by a time-of-flight camera for hand gesture recognition. The authors detect hand by depth difference between foreground and background and conduct Chamfer distance matching technique to measure shape similarity. Ren et al. [9] propose a Finger-Earth Mover's distance to measure the dissimilarities between hand shapes. Suryanarayan et al. [10] propose scale and rotation invariant hand pose recognition techniques. The authors suggest a volumetric shape descriptor which implies 2D image data with depth information.

In 3D model-based approaches, the pose estimation is an optimization problem that minimizes the difference between 3D hand models in database and a detected hand. Oikonomidis et al. [11] propose a generative single hypothesis model-based pose estimation method. They apply particle swarm optimization for 3D hand pose recovery. In [12], the authors present model-based tracking with an articulated hand model and recognize the pose using an unscented Kalman filter.

With the help of depth information, many researches have been carried out to classify hand pose with less effort in detecting hand. However, most researches focus on above approaches such as shape-base pose estimation or 3D model based matching algorithm. Skeleton-based approach can be used for pose estimation but has been still challenging task. Shotton et al. [13] present state of the art work in pose estimation by skeleton configuration. In their pioneering work, they use the random decision forests(RDF) [14] to classify human's body part for each pixel. Then, a local clustering method is applied to find a center of the each body part. In [15], the authors has applied the approach by Shotton et al. to hand image instead of body image and they show the feasibility of the skeleton-based hand pose estimation.

In this paper, we adopt the idea of human pose estimation by Shotton et al. in estimating hand skeleton. We suggest more robust feature extraction method to enhance the performance of the recognition system and compare the simulation results.

2   Data

For classifier to work well in practice, it is important that training data covers variety of hand pose encountered in real life. A solution is to collect a large number real data. Gathering large amount of real data is not easy problem, many research in this field has often focused on the techniques to overcome the lack of training data. However, Shotton et al. shows that synthesized large, varied dataset outperforms in their evaluation.

To classify each of hand part, 3D hand mesh models created by a modeling tool is used for synthesizing various hand pose data. The hand model has 21 different parts which have joints with their center. Depth image and label image are generated from the mesh model and the label image indicates the different label as different color. Every finger has four parts except thumb with three part. Palm has relatively large area in hand so we divided it into two parts, upper palm and lower palm. We build five datasets with different hand gesture. Ten mesh models with different character are used for rendering depth image and label image. Size of 150 K images with different character and different pose are synthesized by the 3D mesh model and automating script. Figure 1 shows the rendered depth image and label image respectively.

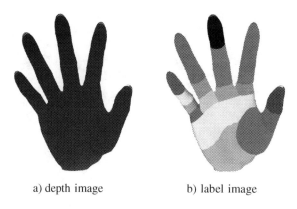

a) depth image   b) label image

Fig. 1. Synthesized image from 3D mesh model

3   Hand Part Classification and Joint Estimation

Random decision forests are used for classification of hand part. We employ per-pixel classification to recognize hand part from a image. Training is a procedure of deciding good parameters of each node for splitting data into child nodes with high information gain. Every node split the data with Eq. 1.

$$Q_L(\phi) = \{(I,p)|f_\theta(I,p) > \tau\}$$
$$Q_R(\phi) = \{(I,p)|f_\theta(I,p) \leq \tau\} \tag{1}$$

Feature responses of input data are calculated using randomly selected candidate parameters $\theta = (u, v, \tau)$ at every node. From these feature response f_θ, information gain $G(\phi)$ is obtained. In training phase we choose splitting parameters which generate maximum information gain. The gain is computed by Eq. 2.

$$G(\phi) = H(Q) - \sum_{S \in \{L,R\}} \frac{|Q_S(\phi)|}{Q} H(Q_S(\phi))$$
$$\phi^* = argmaxG(\phi) \tag{2}$$

where H is Shannon entropy, information gain is estimated from distribution of the each labels in the all input data of each node. ϕ is set of candidate parameter and ϕ^* is the selected split parameter from tree learning which maximize the information gain at node.

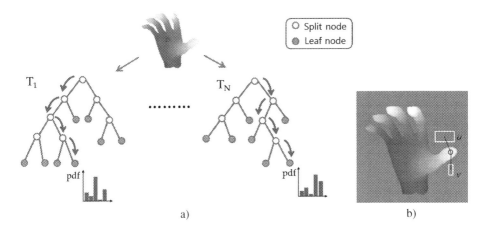

Fig. 2. (a) Random decision forests. Every split node learns a split function in a training phase and test data traces the tree by split function. The leaf node has a probability density of class label. (b) Feature extraction for a selected pixel position. Parameter u, v are random sized offset patch

In classification, input data goes to the leaf node by the split function at each node. Each input data has the learned posterior $p_t(c|I, p)$ at the leaf node in each tree. The posterior distributions of each tree in the forests are averaged for the final classification.

Feature of each pixel has a large impact on overall system and performance [16]. We suggest random offset patches from pixel x instead of using offset pixel. While the offset pixel was used as a feature extraction in the state-of-the-art algorithm in Shotton et al.

$$f_\theta(I, p) = \frac{1}{k} \sum_{i=1}^{k} d_I(p + \frac{u(i)}{d_I^0}) - \frac{1}{l} \sum_{j=1}^{k} d_I(p + \frac{v(j)}{d_I^0}) \qquad (3)$$

The element values of u and v are (x, y) values within the random sized offset patch. The center of the offset patch is a offset value from a position x. The size of patch is also normalized according to the depth d_I^0, which is averaged depth value of the hand. Feature value calculated by difference value between depth values by two patches, which are mean depth value of each patch. The normalized size of patch and the normalized depth value ensure 3D translation invariant. Figure 2(b) Shows two random offset patches for the feature extraction from a position.

Since each pixel is classified with its posterior probability, joints of hand part can be estimated using this information. The global 3D centers of mass for each part can be used as a simple method. In this method the outliers largely degrade the global estimate. Instead of using global center, we apply mean shift for finding local mode of the joint.

4 Experiments

To evaluate the proposed algorithm we have built five datasets and tried our algorithm to each datasets. Each dataset consists of consecutive hand pose frames from one pose to another pose. Each hand gestures are folding thumb, folding two fingers, folding tree fingers, fold four fingers and folding fiver fingers respectively. Training data has large effect on the performance of the recognition system. It is commonly known that containing good coverage of the variation such as scale, hand size and shape and camera pose in training data result in good performance in generalization issue. Depth and scale translation invariance are considered explicitly in the features of the random decision forests. For the hand size and shape invariance, we have used ten different hand characters including man, woman, child, etc. We have changed the camera pose using the automating script to handle the rotation invariance. Each dataset consists of approximately 30 K images. Figure 3 shows the example of different hand pose with different camera pose and the example of dataset-5 which has gesture images of folding five fingers.

Random forests with ten trees have been used for training. Input data consists of 2500 random pixels from each images of the dataset. Pixels are selected randomly in the foreground of the image with 150×150 size. 2000 candidate features (θ) and 20 candidate threshold (τ) are used for training trees. We conduct 5×2-fold cross validation for each dataset. We report the average per-pixel accuracy for per-pixel classification and the average precision for joint estimation.

Feature. We have compared our feature to simple pixel depth comparison approach suggested by Shotton et al. In Fig. 4(a) we show the average per-pixel classification accuracy for each dataset and our approach outperforms the Shotton algorithm for all datasets. As mentioned, using the normalized average of the depth between two different patches instead of using depth between two pixels gives more reliable feature value for the split function in the forests. Feature by offset patch makes the classifier to be more robust to variations of hand pose and shape.

Maximum Offset Patch. We show how maximum offset patch size affects accuracy of the system. Width and height of the patch for feature is selected randomly from a range between 1 to maximum offset patch size. In this experiment we fixed the length of maximum offset u, v to 60 pixel meter. Figure 4(b) show the result for each dataset. Accuracy increases as maximum offset patch size increases in all dataset. But the accuracy tends to decrease from maximum

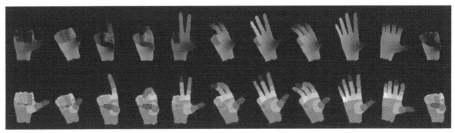

a) Examples from all the datasets

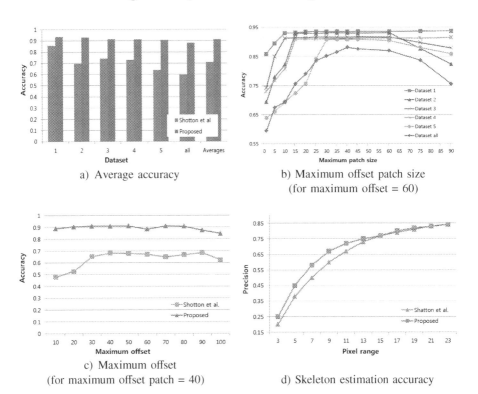

b) Examples from dataset-5 (images captured from five finger folding gesture)

Fig. 3. Examples of test and training dataset

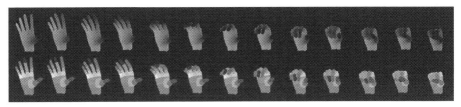

a) Average accuracy

b) Maximum offset patch size
(for maximum offset = 60)

c) Maximum offset
(for maximum offset patch = 40)

d) Skeleton estimation accuracy

Fig. 4. Accuracy results of the experiment and comparison

patch size of 60. It is likely caused by that the offset patches tend to locate out of the image bound. The maximum offset patch having a size of 1 is same as offset pixel used in Shotton method. The dataset-5 is relatively harder gesture than other datasets and the dataset-1 is relatively easier gesture than other gesture. The result also shows that the harder gesture dataset has the lowest accuracy. We show maximum offset patch size affects more on the harder dataset. The best performance is found at 40 for maximum offset patch size.

Maximum Offset. Offset is randomly selected from 0 to maximum offset value in Eq. 3. Shotton show the range of offset has also a large effect on accuracy, and as the maximum offset is increased, the classifier is able to use more spatial context to make its decision. Fig. 4(c) depicts the accuracy according to the offset value. This offset value has less effect on the accuracy in the proposed method, while the accuracy largely decreases in lower than 30 of maximum offset value in offset pixel method.

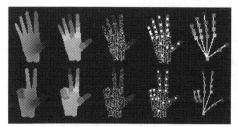 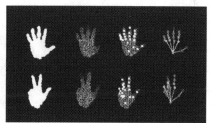

(a) result from synthetic data	(b) result from real data

(a) result from synthetic data
(Depth, Ground truth label, estimated label,
joint proposal and skeleton respectively)

(b) result from real data
(Depth, estimated label, joint proposal
and skeleton respectively)

Fig. 5. Hand part classification and skeleton estimation result

Joint Estimation. The ground truth of joint position is calculated using global 3D centers of mass for each hand part of the ground truth label image. We compare this value to the estimated joint position of the estimated label image. Joint occluded in the ground truth label image is excluded in precision calculation. Precision on joint estimation are shown in Fig. 4(d). Since our approach has higher accuracy in per-pixel classification, precision of joint estimation also higher accuracy as compared to Shotton et al. We applied the algorithm to the real depth data captured by conventional 3D depth camera which has 320×240 resolution. The final result images of synthetic and real data are shown in Fig. 5

5 Conclusion

In this paper we present a per-pixel classification for hand part and joint estimation. We synthesize depth image and ground truth label image to learn the random decision forests. Then we verify the algorithm using synthetic depth image and real depth image. Using the suggested depth comparison feature extraction method, we show the classification accuracy outperforms compared to the state-of-the-art method. To apply this algorithm to the various applications, various hand pose data should be trained. As a future work, we plan to construct more various hand pose dataset and improve random decision forests to classify the various hand pose.

Acknowledgement. This work was supported by the DGIST R&D Program of the Ministry of Education, Science and Technology of Korea (14-IT-03). It was also supported by Ministry of Culture, Sports and Tourism (MCST) and Korea Creative Content Agency (KOCCA) in the Culture Technology (CT) Research & Development Program (Immersive Game Contents CT Co-Research Center).

References

1. Rautaray, S.S., Agrawal, A.: Vision based hand gesture recognition for human computer interaction: a survey. Artif. Intell. Rev. 1–54 (2012)
2. Alon, J., Athitsos, V., Yuan, Q., Sclaroff, S.: A unified framework for gesture recognition and spatiotemporal gesture segmentation. IEEE Trans. Pattern Anal. Mach. Intell. **31**, 1685–1699 (2009)
3. Microsoft: Kinect camera. http://www.xbox.com/en-us/kinect
4. Hackenberg, G., McCall, R., Broll, W.: Lightweight palm and finger tracking for real-time 3d gesture control. In: 2011 IEEE Virtual Reality Conference (VR), pp. 19–26. IEEE (2011)
5. Doliotis, P., Stefan, A., McMurrough, C., Eckhard, D., Athitsos, V.: Comparing gesture recognition accuracy using color and depth information. In: Proceedings of the 4th International Conference on PErvasive Technologies Related to Assistive Environments, p. 20. ACM (2011)
6. Tara, R., Santosa, P., Adji, T.: Hand segmentation from depth image using anthropometric approach in natural interface development. Int. J. Sci. Eng. Res. **3**, 1–4 (2012)
7. Doliotis, P., Athitsos, V., Kosmopoulos, D., Perantonis, S.: Hand shape and 3D pose estimation using depth data from a single cluttered frame. In: Bebis, G., et al. (eds.) ISVC 2012, Part I. LNCS, vol. 7431, pp. 148–158. Springer, Heidelberg (2012)
8. Liu, X., Fujimura, K.: Hand gesture recognition using depth data. In: Proceedings of Sixth IEEE International Conference on Automatic Face and Gesture Recognition, pp. 529–534. IEEE (2004)
9. Ren, Z., Yuan, J., Zhang, Z.: Robust hand gesture recognition based on finger-earth mover's distance with a commodity depth camera. In: Proceedings of the 19th ACM International Conference on Multimedia, pp. 1093–1096. ACM (2011)
10. Suryanarayan, P., Subramanian, A., Mandalapu, D.: Dynamic hand pose recognition using depth data. In: 2010 20th International Conference on Pattern Recognition (ICPR), pp. 3105–3108. IEEE (2010)
11. Oikonomidis, I., Kyriazis, N., Argyros, A.A.: Efficient model-based 3d tracking of hand articulations using kinect. In: BMVC, vol. 1, p. 3 (2011)
12. Stenger, B., Mendonça, P.R., Cipolla, R.: Model-based hand tracking using an unscented kalman filter. In: BMVC, vol. 1, pp. 63–72 (2001)

13. Shotton, J., Sharp, T., Kipman, A., Fitzgibbon, A., Finocchio, M., Blake, A., Cook, M., Moore, R.: Real-time human pose recognition in parts from single depth images. Commun. ACM **56**, 116–124 (2013)
14. Breiman, L.: Random forests. Mach. Learn. **45**, 5–32 (2001)
15. Keskin, C., Kıraç, F., Kara, Y.E., Akarun, L.: Real time hand pose estimation using depth sensors. In: Fossati, A., Gall, J., Grabner, H., Ren, X., Konolige, K. (eds.) Consumer Depth Cameras for Computer Vision, pp. 119–137. Springer, London (2013)
16. Lepetit, V., Lagger, P., Fua, P.: Randomized trees for real-time keypoint recognition. In: IEEE Computer Society Conference on Computer Vision and Pattern Recognition, CVPR 2005, vol. 2, pp. 775–781. IEEE (2005)

Human Tracking Using a Far-Infrared Sensor Array and a Thermo-Spatial Sensitive Histogram

Takashi Hosono[1]([⊠]), Tomokazu Takahashi[2] , Daisuke Deguchi[3], Ichiro Ide[1], Hiroshi Murase[1], Tomoyoshi Aizawa[4], and Masato Kawade[4]

[1] Graduate School of Information Science, Nagoya University, Nagoya, Japan
hosonot@murase.m.is.nagoya-u.ac.jp,
{ide,murase}@is.nagoya-u.ac.jp
[2] Faculty of Economics and Information,
Gifu Shotoku Gakuen University, Gifu, Japan
ttakahashi@gifu.shotoku.ac.jp
[3] Information and Communications Headquarters, Nagoya University, Nagoya, Japan
ddeguchi@nagoya-u.jp
[4] Corporate R&D, OMRON Corporation, Kyoto, Japan
{aizawa,kawade}@ari.ncl.omron.co.jp

Abstract. We propose a human body tracking method using a far-infrared sensor array. A far-infrared sensor array captures the spatial distribution of temperature as a low-resolution image. Since it is difficult to identify a person from the low-resolution thermal image, we can avoid privacy issues. Therefore, it is expected to be applied for the analysis of human behaviors in various places. However, it is difficult to accurately track humans because of the lack of information sufficient to describe the feature of the target human body in the low-resolution thermal image. In order to solve this problem, we propose a thermo-spatial sensitive histogram suitable to represent the target in the low-resolution thermal image. Unlike the conventional histograms, in case of the thermo-spatial sensitive histogram, a voting value is weighted depending on the distance to the target's position and the difference from the target's temperature. This histogram allows the accurate tracking by representing the target with multiple histograms and reducing the influence of the background pixels. Based on this histogram, the proposed method tracks humans robustly to occlusions, pose variations, and background clutters. We demonstrate the effectiveness of the method through an experiment using various image sequences.

1 Introduction

In order to analyze human behavior, visible-light cameras are widely adopted for vision systems. However, they may cause uncomfortableness to the users because it could be easy to identify a person from the captured images. No matter how useful vision systems are, it is often difficult to install them in locations where users refuse them, e.g. toilets, bedrooms, and offices. Considering this problem, using systems based on sensitive floor [1,2] and far-infrared sensors network [3,4] could allow human tracking avoiding privacy issues. The former uses pressure

© Springer International Publishing Switzerland 2015
C.V. Jawahar and S. Shan (Eds.): ACCV 2014 Workshops, Part II, LNCS 9009, pp. 262–274, 2015.
DOI: 10.1007/978-3-319-16631-5_20

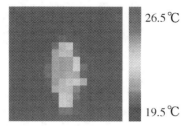

26.5 ℃

19.5 ℃

(a) Visible-light image.   (b) Low-resolution thermal image.

Fig. 1. Example of an output of a 16×16 far-infrared sensor array. It is very diffi-cult to identify the person from (b). Note that, this far-infrared sensor array outputs temperature lower than the actual body temperature. (a) is captured by a visible light camera put on the same position as the far-infrared sensor array.

sensitive floors that can detect human footsteps. The latter uses far-infrared sensors that are distributed and connected to each other by network. However, these solutions are not suitable for interactive systems such as gesture recognition because they cannot capture human body shapes.

A far-infrared sensor array is a device composed of a small number of far-infrared sensors, that is expected to be used for interactive systems [5]. Figure 1 shows an example of an output obtained from a 16×16 far-infrared sensor array. As we can see from this figure, it represents the spatial distribution of temperature as a low-resolution image. Since the image only shows the rough shape of an object, unlike visible-light cameras, far-infrared sensor arrays would cause less uncomfortableness to users because they are not easily identified. Accordingly, they can be installed in places where it is not preferable to install visible-light cameras due to privacy issues. In addition, far-infrared sensor arrays can capture the rough shape of a human body. Thus, they can also be used in an interactive system such as that for gesture recognition [6]. For the above reasons, there is a demand for a human behavior analysis system that uses a far-infrared sensor array. Among important technologies required to realize the human behavior analysis, in this paper, we propose a human body tracking method using a far-infrared sensor array.

There are some researches related to human body tracking using a far-infrared sensor array. Takahata *et al.* developed a method which localizes heat sources using a far-infrared sensor array [7]. This method simply selects a sensor which outputs the highest temperature to localize a heat source. Thus, it does not consider occlusions nor the effect of other heat sources. Meanwhile, various tracking algorithms which use visible-light cameras are developed until today. In general, tracking algorithms aim to extract and describe the feature of the target object efficiently. The feature points extraction approach is widely used in many tracking algorithms [8–10]. This approach extracts feature points from the target objects. This allows fast tracking with high accuracy by using only points effective for tracking the target object. However, in case of far-infrared sensor arrays, feature points extraction is difficult because the output thermal image is in very low-resolution. The intensity histograms approach is also widely used to represent a target object. This feature is simple and robust to pose variation. In this approach, multiple local histograms [11] that are made from multiple regions in the target object is used in the state-of-the-art methods [12–14] because a single histogram could not preserve spatial information. These methods are robust to occlusion and can also be computed fast by using the integral histogram [15]. However, if the method using multiple local histograms is applied to the far-infrared sensor array output, each region becomes very small, e.g. 2×2 pixels. For the above reason, a feature that could represent the target object in low-resolution thermal images is needed.

The proposed method introduces a thermo-spatial sensitive histogram algorithm as a feature that represents the target object in a low-resolution thermal image. Unlike the conventional histograms, in case of the thermo-spatial sensitive histogram, a voting value is weighted depending on the distance to the target's position and the difference from the target's temperature. The proposed method constructs the histograms for multiple positions in the target human body region without reducing the number of pixels to be considered as in multiple local histograms [11]. In addition, this histogram reduces the influence of the background pixels by focusing to a particular temperature. Thus, this histogram can effectively represent the target human body in a low-resolution thermal image. Using this histogram, the proposed method can track humans robustly to occlusions, pose variations, and background clutters.

The remainder of this paper is organized as follows: Sect. 2 describes the thermo-spatial sensitive histogram. Next, Sect. 3 describes the human tracking algorithm using the thermo-spatial sensitive histogram. Then, results of an experiment to confirm the effectiveness of the proposed method are reported in Sect. 4. Finally, we conclude this paper in Sect. 5.

2   Thermo-Spatial Sensitive Histogram

Conventionally, when a histogram $\mathbf{H}$ is made from the whole image, the value of bin b $(b = 1, \ldots, B)$ is defined as:

$$\mathbf{H}(b) = \sum_{i=1}^{N} Q(T_{\boldsymbol{x}_i}, b), \tag{1}$$

where N is the number of pixels, $\boldsymbol{x}_i$ is the location of the i-th pixel, B is the total number of bins, and $Q(T_{\boldsymbol{x}_i}, b)$ returns 1 if the intensity value $T_{\boldsymbol{x}_i}$ belongs to bin b, otherwise returns 0. In general, this histogram is created from the region of the target object. In case of tracking, histogram largely changes if occlusions occur because it describes the object with only one histogram.

To resolve this issue, He et $al.$ developed the locality sensitive histogram [16] that preserves spatial information. In this paper, we call this histogram "spatial sensitive histogram" to distinguish it from other histograms introduced later.

Let $\mathbf{H}_{\boldsymbol{x}_p}$ denote the spatial sensitive histogram computed at a target pixel $\boldsymbol{x}_p$. It can be written as:

$$\mathbf{H}_{\boldsymbol{x}_p}(b) = \sum_{i=1}^{N} \alpha_1^{||\boldsymbol{x}_p - \boldsymbol{x}_i||} \cdot Q(T_{\boldsymbol{x}_i}, b), \tag{2}$$

where $0 < \alpha_1 < 1$ is a parameter that controls the influence of the distance between $\boldsymbol{x}_i$ and $\boldsymbol{x}_p$. In this way, this histogram can preserve the spatial information. By using multiple histograms constructed while moving $\boldsymbol{x}_p$ in the target region, tracking robust to occlusion can be achieved.

Spatial sensitive histogram can partially reduce the influence of the background pixels. In tracking, reducing the influence of the background pixels is very important because those around the target object change frame by frame, and degrades the tracking accuracy. In order to reduce the influence of the background further, we propose the "thermal sensitive histogram". In case of visible-light cameras, human bodies are observed as a set of pixels with various intensity values due to the variety of their appearance. Meanwhile, a far-infrared sensor array captures human bodies as similar temperature because it captures the object's temperature. Therefore, this histogram is constructed by voting values weighted by the distance to the target object's temperature. Let $\mathbf{H}^{T_m}$ denote the thermal sensitive histogram computed when the target object's temperature is T_m. It can be written as:

$$\mathbf{H}^{T_m}(b) = \sum_{i=1}^{N} \alpha_2^{|T_m - T_{\boldsymbol{x}_i}|} \cdot Q(T_{\boldsymbol{x}_i}, b), \tag{3}$$

where T_m is the target object's temperature calculated by selecting the median of pixel values in the bounding box surrounding the human body, and $0 < \alpha_2 < 1$ is a parameter that controls the influence of the distance to the target object's temperature. In general, the temperature of the region that is occluded by an object is observed lower than that of the human body, and that of humans who are in front of the target human are observed higher due to the characteristic of the sensor. Thus, this histogram can reduce the influence of background pixels. However, it could not preserve the spatial information like in the conventional histogram. Therefore, we propose a "thermo-spatial sensitive histogram" that combines the above two histograms. The value of bin b of a thermo-spatial sensitive histogram $\mathbf{H}_{\boldsymbol{x}_p}^{T_m}$ made at pixel $\boldsymbol{x}_p$ is defined as:

$$\mathbf{H}_{\boldsymbol{x}_p}^{T_m}(b) = \sum_{i=1}^{N} \alpha_1^{||\boldsymbol{x}_p - \boldsymbol{x}_i||} \cdot \alpha_2^{|T_m - T_{\boldsymbol{x}_i}|} \cdot Q(T_{\boldsymbol{x}_i}, b), \tag{4}$$

The proposed method constructs multiple histograms while moving $\boldsymbol{x}_p$ in the target region like the method that uses the spatial sensitive histogram [16].

3   Human Tracking by a Far-Infrared Sensor Array

We assume that a target human body region in the initial frame is given manually as a rectangular region. First, multiple thermo-spatial sensitive histograms

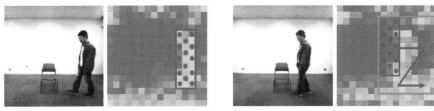

(a) Template frame.　　　　　　　　(b) Current frame.

Fig. 2. Tracking using thermo-spatial sensitive histograms. (a) shows the template thermal image and a visible-light image taken at the same time. (b) is an example of the current frame of the thermal image and a visible-light image taken at the same time. The red rectangles represent the template and candidate regions. The red points indicate the sampling points to make the thermo-spatial sensitive histograms. The orange rectangle represents the search region (Color figure online).

are made at a number of sampling points in the target region in the initial frame as template histograms. Next, in the succeeding frame, the region that maximizes the similarity to the template is searched by a sliding window approach. Then, the template histograms are updated to adopt the change of the target's appearance. Details of each step are described below.

3.1 Tracking Using Thermo-Spatial Sensitive Histograms

The proposed method makes the thermo-spatial sensitive histograms for each sampling point in the target region. This means the proposed method represents the target object with multiple histograms. The size of the target region and the relative position of the sampling points to the region are fixed, to compare each histogram to the template histogram. These histograms represent the target object region with multiple overlapping regions.

First, the temperature of the target human body represented as T_m in Eq. (4) is calculated as the median of pixel values in the region. In this way, it is possible to select the temperature of the human body without being influenced by background pixels in the region. To find a target human body region in the current frame, we define the similarity between the template and candidate regions as:

$$S(R_t, R_c) = \sum_{i=1}^{N} I(\mathbf{H}_{\boldsymbol{x}_i}^{T_m}, \mathbf{H}_{\boldsymbol{x}_i'}^{T_m}), \tag{5}$$

where R_t and R_c represent the template and candidate regions, respectively, N is the number of sampling points in the region, $\mathbf{H}_{\boldsymbol{x}_i}^{T_m}$ is the i-th histogram in the template and $\mathbf{H}_{\boldsymbol{x}_i'}^{T_m}$ is the i-th histogram in the candidate. To measure the similarity between the histograms, we use the histogram intersection function [17] defined as:

$$I(\mathbf{H}_{\boldsymbol{x}_i}^{T_m}, \mathbf{H}_{\boldsymbol{x}_i'}^{T_m}) = \sum_{b=1}^{B} \min(\mathbf{H}_{\boldsymbol{x}_i}^{T_m}(b), \mathbf{H}_{\boldsymbol{x}_i'}^{T_m}(b)). \tag{6}$$

(a) Template frame.   (b) Current frame.

Fig. 3. Online template update. The red points in (a) indicate sampling points to make thermo-spatial sensitive histograms, and the orange points in (b) indicate the sampling points of the updated histograms that had stable similarities to the template. In this case, template histograms are updated only on histograms made from the orange points (Color figure online).

Thus, a candidate region R_c that maximizes the similarity $S(R_t, R_c)$ is determined as the human body region in the current frame. Note that candidate regions are restricted to the area around the tracking result of the previous frame to reduce computational cost. Figure 2 shows an example of the tracking procedure, where the red rectangles in (a) and (b) represent the template and candidate regions, respectively, the red points are the sampling points where the thermo-spatial sensitive histograms are made and the orange rectangle in (b) is the area restricted to find the target region in the current frame.

3.2 Online Template Update

In general, the target human's appearance continues to change. Therefore, the template histograms are required to be updated online to achieve stable tracking. The proposed method updates each histogram individually to decrease the effect of posture changes and occlusions. This online template update is performed by:

$$\mathbf{H}_{\boldsymbol{x}_i}^{T_m} = \mathbf{H}_{\boldsymbol{x}_i'}^{T_m} \quad \text{if } F_1 \cdot M < I(\mathbf{H}_{\boldsymbol{x}_i}^{T_m}, \mathbf{H}_{\boldsymbol{x}_i'}^{T_m}) < F_2 \cdot M, \tag{7}$$

where $\mathbf{H}_{\boldsymbol{x}_i'}^{T_m}$ is the histogram after the tracking, M is the median of all histogram similarities between the template and tracking results, and F_1, F_2 are parameters that control the ease of updating. A template histogram is updated only when the histogram similarity is stable by using this algorithm. Thus, this algorithm can avoid being affected by posture changes and occlusions. Figure 3 illustrates the online template update algorithm, where the red rectangles in (a) and (b) represent the template region and the tracking result, respectively, the red points represent the sampling points for the template, and the orange points in (b) are the sampling points whose histograms were updated because the similarity was stable in the current frame. In Fig. 3(b), since the lower body of the person is hidden, the similarity decreases. Therefore, the method updates the template only for the sampling points in the upper body that has stable similarities.

Fig. 4. Far-infrared sensor array used in the experiment. 16 × 16 far infrared sensors are integrated in one chip.

4   Experiment

To confirm the effectiveness of the proposed method, we conducted an experiment by using actual data. Improvement in accuracy of the proposed method was evaluated by comparing it with two comparative methods. To conduct the experiment, we created a dataset using a 16 × 16 far-infrared sensor array. Figure 4 shows the far-infrared sensor array used in the experiment. We describe below the dataset and the experimental conditions, then discuss the results from the experiment.

4.1   Creation of a Dataset

We captured various scenes using the far-infrared sensor array whose frame rate was 10 fps and temperature resolution was 0.1 degrees. The dataset consisted of 13 sequences composed of 1,729 frames in total. This dataset contains sequences that have many posture changes, occlusions, and background clutters. In case of thermal images, the background clutters were caused by various heat sources including humans, PC displays, and high room temperature. In addition, we prepared not only sequences capturing a human as a target, but also complex sequences captured in a room with other heat sources including other humans, in order to confirm that the proposed method can be used in actual situations. Figure 5 shows samples of the 13 sequences in the dataset, where the red rectangles indicate the ground-truth. For each sequence, we captured the whole body of the target human in the first frame and determined the target region manually to construct the initial template histograms. Visible-light image sequences were also captured at the same time in order to provide ground-truth regions frame-by-frame manually. Note that only the thermal image sequences were input to the program. The size of ground-truth regions were fixed in each sequence.

4.2   Experimental Conditions

To confirm the effectiveness of using the proposed thermo-spatial sensitive histogram, we compared it with two methods using different types of histograms. One was the spatial sensitive histogram [16] that was used in the state-of-the-art

(a) Sitting on a chair 1.

(b) Sitting on a chair 2.

(c) Extending arms.

(d) Occlusion by human 1.

(e) Occlusion by human 2.

(f) Occlusion by whiteboard 1.

(g) Occlusion by whiteboard 2.

(h) Occlusion by locker 1.

(i) Occlusion by locker 2.

(j) Occlusion by doll and table.

(k) Passing in front of human.

(l) Walking in a room 1.

(m) Walking in a room 2.

Fig. 5. Examples of sequences from the dataset used in the experiment. Visible-light images were taken at the same time to provide ground-truth regions. Red rectangles in the thermal images represent the ground-truth regions (Color figure online).

algorithm of tracking with a visible-light camera. Another was the thermal sensitive histogram. All three methods used the same tracking algorithm described in Sect. 3. However, the comparative method that used the thermal sensitive histogram tracked with only a single histogram in its template because the same histogram was constructed for all the sampling points. In our implementation, both parameters α_1 and α_2 in Eqs. (2), (3), and (4) were set to 0.5, F_1, F_2 in Eq. 7 were set to 0.98, 1.02, respectively. In addition, when making multiple histograms from a region, sample points were set at an interval of 1 pixel, and each histogram bin was set at an interval of 1 degree. Furthermore, the candidate regions were restricted within ± 2 pixels in horizontal and vertical directions from the previous tracking result when the human body region was searched in the current frame.

We used two types of criteria to evaluate each method. One was the success rate that is the percentage of the number of frames where the target regions were successfully tracked. We determined that tracking was successful if the tracking result satisfied the PASCAL criterion [18] defined as:

$$\frac{|B^i \cap G^i|}{|B^i \cup G^i|} \geq 0.5, \tag{8}$$

where B^i denotes the region of the tracking result in frame i, and G^i denotes the ground-truth region in frame i. Another evaluation criterion was the center location error that is the L_2 distance of center locations between the tracking result and the ground-truth. By these evaluation criteria, we evaluated the tracking accuracy both discretely and continuously.

4.3 Results and Discussion

Tables 1 and 2 show the tracking performance for the 13 sequences. As shown in this table, the proposed method using the thermo-spatial sensitive histogram provided the best performance for almost all sequences. In the success rate criteria, the average score improved approximately 20 % compared to the comparative methods. Furthermore, in the center location error criteria, the average score improved approximately 0.5 pixels. This may seem a small improvement, but the shift of one pixel in the 16×16 pixels template image is large. Actually, in this experimental setting, the error of 0.5 pixels was equivalent to approximately 0.1 m. From the above, the validity of the proposed method was confirmed quantitatively.

Figure 6 shows the screenshots of the tracking results, where the red, brown and pink rectangles represent the tracking result by the thermo-spatial sensitive histogram, the spatial sensitive histogram, and the thermal sensitive histogram, respectively. Figure 6(a), (b), and (c) show the results of sequences including posture changes. As shown in these figures, the proposed method using the thermo-spatial sensitive histogram can capture the human body stably because putting weight on pixels that had values near human temperature decreased the influence of background pixels which had lower values than the human temperature. Figure 6(d) to (j) show the results of sequences including occlusions.

Table 1. The success rates [%] of the 13 sequences. The highest value among the methods for each sequence is shown in bold. The total number of frames is 1,729.

Sequence	Thermo-Spatial (proposed)	Spatial	Thermal
Sitting on a chair 1	**85**	53	66
Sitting on a chair 2	**82**	12	81
Extending arms	**57**	10	46
Occlusion by human 1	34	**35**	9
Occlusion by human 2	**59**	45	41
Occlusion by whiteboard 1	72	**81**	73
Occlusion by whiteboard 2	**100**	99	81
Occlusion by locker 1	**89**	81	43
Occlusion by locker 2	**88**	48	12
Occlusion by doll and table	**75**	61	26
Passing in front of human	10	19	**35**
Walking in a room 1	**93**	86	76
Walking in a room 2	**70**	51	47
Average	**70**	52	49

Table 2. The average center location errors [pixels]. The smallest error among the methods for each sequence is shown in bold. The total number of frames is 1,729.

Sequence	Thermo-Spatial (proposed)	Spatial	Thermal
Sitting on a chair 1	**0.7**	1.2	0.9
Sitting on a chair 2	**0.7**	1.9	0.9
Extending arms	**1.3**	3.0	1.9
Occlusion by human 1	**2.1**	2.2	3.7
Occlusion by human 2	**1.4**	2.1	2.5
Occlusion by whiteboard 1	1.1	**0.7**	1.0
Occlusion by whiteboard 2	1.1	**1.0**	1.6
Occlusion by locker 1	**0.6**	0.8	2.2
Occlusion by locker 2	**1.0**	1.7	3.5
Occlusion by doll and table	**1.1**	1.2	2.6
Passing in front of human	2.9	2.9	**1.8**
Walking in a room 1	**0.6**	0.7	1.2
Walking in a room 2	**1.3**	**1.3**	1.5
Average	**1.2**	1.6	1.9

(a) Sitting on a chair 1.

(b) Sitting on a chair 2.

(c) Extending arms.

(d) Occlusion by human 1.

(e) Occlusion by human 2.

(f) Occlusion by whiteboard 1.

(g) Occlusion by whiteboard 2.

(h) Occlusion by locker 1.

(i) Occlusion by locker 2.

(j) Occlusion by doll and table.

(k) Passing in front of human.

(l) Walking in a room 1.

(m) Walking in a room 2.

Fig. 6. Example of tracking results. The red, brown, and pink rectangles represent the tracking results by using thermo-spatial, spatial, and thermal sensitive histograms, respectively.

As shown in these figures, the proposed method can capture the human body accurately despite the occlusions. Especially in Fig. 6(e) and (i), the effect that reduces the influences of other heat sources in constructing the thermo-spatial sensitive histograms was effective. Furthermore, it can be seen that the proposed method can capture the human body stably under situations with background clutters in Fig. 6(l) and (m). From the above, it was demonstrated that the proposed human tracking using thermo-spatial sensitive histograms was robust to posture changes, occlusions, and background clutters. However, as shown in Fig. 6(k), the proposed method failed to track in the sequence where the target human passed in front of another human. In this case, the proposed method tracked the other human mistakenly because there was only a small difference in temperature between the target human and the other human. This problem could be solved by improving the histogram feature to distinguish more detailed differences. In addition, improving the tracking algorithm will also be needed.

5   Conclusion

In this paper, we proposed a human tracking method using a far-infrared sensor array. The proposed method uses the thermo-spatial sensitive histogram that is suitable to represent the human body in low-resolution thermal images. This histogram can reduce the influence of background pixels. Since the histogram can preserve the spatial information, it is possible to describe the feature of the target human body in detail by making multiple histograms for each sampling point in the target region. The experimental results show that the proposed method performed human tracking more robust to posture changes, occlusions, and background clutters than the state-of-the-art tracking method for visual images.

As future work, we will develop a tracking method which can track a human under a situation that several humans cross. In addition, we will develop a method which can also deal with scale change due to various distances between the human and the far-infrared sensor array. Furthermore, we will investigate interactive applications such as gesture recognition that uses far-infrared sensor arrays.

Acknowledgment. Parts of this research were supported by MEXT, Grant-in-Aid for Scientific Research.

References

1. Sousa, M., Techmer, A., Steinhage, A., Lauterbach, C., Lukowicz, P.: Human tracking and identification using a sensitive floor and wearable accelerometers. In: Proceedings of the 11th IEEE International Conference on Pervasive Computing and Communications, pp. 166–171 (2013)
2. Steinhage, A., Lauterbach, C.: Monitoring movement behavior by means of a large area proximity sensor array in the floor. In: Proceedings of the 2nd Workshop on Behaviour Monitoring and Interpretation, pp. 15–27 (2008)

3. Hao, Q., Brady, D., Guenther, B.D., Burchett, J., Shankar, M., Feller, S.: Human tracking with wireless distributed pyroelectric sensors. IEEE Sens. J. **6**, 1683–1696 (2006)
4. Zappi, P., Farella, E., Benini, L.: Tracking motion direction and distance with pyroelectric IR sensors. IEEE Sens. J. **10**, 1486–1494 (2010)
5. Ohira, M., Koyama, Y., Aita, F., Sasaki, S., Oba, M., Takahata, T., Shimoyama, I., Kimata, M.: Micro mirror arrays for improved sensitivity of thermopile infrared sensors. In: Proceedings of the 24th IEEE International Conference on Micro Electro Mechanical Systems, pp. 708–711 (2011)
6. Wojtczuk, P., Armitage, A., Binnie, T., Chamberlain, T.: PIR sensor array for hand motion recognition. In: Proceedings of the 2nd International Conference on Sensor Device Technologies and Applications, pp. 99–102 (2011)
7. Takahata, A., Shimada, Y., Yoshioka, F., Yoshida, M., Kimata, M., Ota, T.: Infrared position sensitive detector (IRPSD). In: Infrared Technology and Applications XXXIV, Proceedings of the SPIE, vol. 6940, pp. 694031-1–694031-11 (2008)
8. Baker, S., Matthews, I.: Lucas-Kanade 20 years on: a unifying framework. Int. J. Comput. Vis. **56**, 221–255 (2004)
9. Song, D., Zhao, B., Tang, L.: A tracking algorithm based on SIFT and Kalman filter. In: Proceedings of the 2012 International Conference on Computer Application and System Modeling, pp. 1563–1566 (2012)
10. Yan, Y., Wang, J., Li, C., Wu, Z.: Object tracking using SIFT features in a particle filter. In: Proceedings of the 3rd IEEE International Conference on Communication Software and Networks, pp. 384–388 (2011)
11. Adam, A., Rivlin, E., Shimshoni, I.: Robust fragments-based tracking using the integral histogram. In: Proceedings of the 2006 IEEE Computer Society Conference on Computer Vision and Pattern Recognition, vol. 1, pp. 798–805 (2006)
12. Cehovin, L., Kristan, M., Leonardis, A.: An adaptive coupled-layer visual model for robust visual tracking. In: Proceedings of the 2011 IEEE International Conference on Computer Vision, pp. 1363–1370 (2011)
13. Kwon, J., Lee, K.M.: Tracking of a non-rigid object via patch-based dynamic appearance modeling and adaptive basin hopping monte carlo sampling. In: Proceedings of the 2009 IEEE Computer Society Conference on Computer Vision and Pattern Recognition, pp. 1208–1215 (2009)
14. Shahed Nejhum, S.M., Ho, J., Yang, M.H.: Visual tracking with histograms and articulating blocks. In: Proceedings of the 2008 IEEE Computer Society Conference on Computer Vision and Pattern Recognition, pp. 1–8 (2008)
15. Porikli, F.: Integral histogram: A fast way to extract histograms in Cartesian spaces. In: Proceedings of the 2005 IEEE Computer Society Conference on Computer Vision and Pattern Recognition, vol. 1, pp. 829–836 (2005)
16. He, S., Yang, Q., Lau, R.W.H., Wang, J., Yang, M.: Visual tracking via locality sensitive histograms. In: Proceedings of the 2011 IEEE Computer Society Conference on Computer Vision and Pattern Recognition, pp. 2427–2434 (2013)
17. Barla, A., Odone, F., Verri, A.: Histogram intersection kernel for image classification. In: Proceedings of the 2003 IEEE International Conference on Image Processing, vol. 3, pp. 513–516 (2003)
18. Everingham, M., Van Gool, L., Williams, C., Winn, J., Zisserman, A.: The PASCAL visual object classes (VOC) challenge. Int. J. Comput. Vis. **88**, 303–338 (2010)

Feature Point Tracking Algorithm Evaluation for Augmented Reality in Handheld Devices

Amila Perera[✉], Akila Pemasiri, Sameera Wijayarathna,
Chameera Wijebandara, and Chandana Gamage

Department of Computer Science and Engineering,
University of Moratuwa, Moratuwa, Sri Lanka
amila.10@cse.mrt.ac.lk

Abstract. In augmented reality applications for handheld devices, accuracy and speed of the tracking algorithm are two of the most critical parameters to achieve realism. This paper presents a comprehensive framework to evaluate feature tracking algorithms on these two parameters. While there is a substantial body of knowledge on these aspects, a novel feature introduced in this paper is the use of error associated with the estimated directional movement in performance measurements to improve the evaluation framework. The work described in this paper is a comparative evaluation of nine widely used feature point tracking algorithms using the developed measurement framework and the results are interpreted based on the characteristics of the algorithms as well as the characteristics of test image sequences.

1 Introduction

In the field of computer vision, feature point tracking is the standard method most often used to extract motion information in an image sequence. There are many applications of feature point tracking in robotics, human computer interaction, surveillance and activity monitoring. In this study we have focused on evaluation of feature point tracking algorithms, which can be used to estimate the 3D pose of a calibrated 2D camera in augmented reality applications. This is important, for example, when we want to project a 3D object in a scene captured on a 2D camera. In this evaluation we have considered algorithms error rate, speed, robustness to unexpected changes of image properties and suitability for augmented reality applications.

The well-known feature point tracking algorithms require the image sequence to be of good quality. Algorithms use varying sets of assumptions (for example brightness consistency assumption in Kanade Lucas Tomasi (KLT feature tracker). The processing power consumed by different algorithms vary widely. In all of these algorithms, occurrence of a non-admissible event in an image sequence adds an element of uncertainty that affects the robustness of the output.

In this study, nine feature point tracking algorithms have been evaluated and compared. The objective of this study is to understand the performance

© Springer International Publishing Switzerland 2015
C.V. Jawahar and S. Shan (Eds.): ACCV 2014 Workshops, Part II, LNCS 9009, pp. 275–288, 2015.
DOI: 10.1007/978-3-319-16631-5_21

and robustness of the algorithms under the criteria of their capability to operate on handheld mobile platforms with built-in cameras. The images captured using such devices may possess a lower image quality and may be subjected to involuntary device movements.

This paper is organized as follows. Section 2 includes the details of the algorithms used and Sect. 3 describes the data set used for the evaluation. Section 4 focuses on the evaluation methods and Sect. 5 describes the results.

2   Tracking Algorithms

Feature trackers can be categorized to two main types as Trackers using Optical Flow Algorithms and Trackers using Feature Descriptor Matching Algorithms.

For both of these types, feature points are required to be identified as an initial step. Typically, feature points allow them to be distinctively identified among other points.

Feature descriptors that store locality information of points are used in feature descriptor matching algorithms. Speeded Up Robust Features (SURF) [6], Scale Invariant Feature Transform(SIFT) [3], Shi-Tomasi Corner Detector [7], Binary Robust Invariant Scalable Keypoints (BRISK) [8], and Oriented FAST and Rotated BRIEF (ORB) [4] are feature detectors that are used in the implementation of algorithms that are compared in this work. Also, the feature detectors SURF, SIFT, BRISK and ORB have feature description capabilities in addition to detection. Furthermore, Fast Retina Keypoint (FREAK) [1] feature descriptor has also been used in this comparison. Trackers using optical flow algorithms and trackers using feature descriptor matching algorithms with different types of feature detectors and descriptors have been evaluated. Descriptions of different types of algorithms along with the names that are being referred in this paper are shown in Table 1.

It should be noted that when tracking feature points using a combination of a detector and a descriptor, certain combinations (for example, SIFT and

Table 1. Description of trackers.

Tracker	Type of tracker	Feature detector	Feature descriptor
BRISK	Feature matching	BRISK	BRISK
FARNEB	Dense optical flow, using Gunnar Farnebacks algorithm	Shi-Tomasi	-
FREAK	Feature matching	SURF	FREAK
KLT	Lukas Kanade Optical flow	Shi-Tomasi	-
ORB	Feature matching	ORB	ORB
SIFT	Feature matching	SIFT	SIFT
SIFTSURF	Feature matching	SIFT	SURF
SURFSIFT	Feature matching	SURF	SIFT
SURF	Feature matching	SURF	SURF

FREAK) are not used if an improved version of an algorithm is used in another combination (for example, SURF and FREAK where SURF is an improved version of SIFT [6]). Furthermore, we have selected to evaluate only the currently known best performing algorithm in a family of algorithms (for example, Harris Corner detector [5] is not evaluated in favour of Shi-Tomasi Corner detector).

3   Evaluated Dataset

In order to select the best tracking algorithm for augmented reality applications in handheld devices, it was decided to use a dataset that include all the challenging factors in the expected context. The challenges considered were motion-blur, illumination variation, scale variation, camera rotation, occlusion, shaky camera

Table 2. Description of dataset.

Video	Main challenging factors	Number of Frames
Resolution : 1280*720		
1	Illumination variation, Scale variation	540
2	Autofocus, Motion blur, Very slow camera movement	469
3	Motion blur, High speed movements in all direction	290
4	Camera rotation, Shaky camera	273
5	Camera rotation, Autofocus, Motion blur	436
6	Camera rotation and translation, Autofocus, Motion blur	428
7	Camera rotation and translation, Ellipse shape camera path	700
8	Autofocus, Movement in all directions, Change in background texture	656
9	Illumination change, Autofocus	488
Resolution : 1920*1080		
10	Illumination change, Autofocus, Motion blur, Specular reflection	142
11	Background texture change	526
12	Specular reflection, Very slow camera movements, Camera rotation	393
13	Scale variation	548
14	Camera rotation, translation, Specular reflection	604
15	Camera rotation and translation, Scale variation, Shaky camera movements	526
16	Camera rotation and translation, Scale variation, Shaky camera movements	450
Resolution : 640*480		
17	Camera rotation and translation, Scale variation, Shaky camera movements	502

and changes of the focal point. The occurrence of these features in the dataset is elaborated in the Table 2.

All the videos used as the dataset was captured using several mobile devices to ensure that the quality and the characteristics mimic live data which is expected to be received by an operational system.

4   Evaluation Method

The evaluation requires a set of feature points in the image sequence with known positions. This is called the ground truth. To obtain the ground truth few points were tracked throughout the image sequence manually. These points were selected after carefully observing the whole image sequence for features that can be clearly distinguished by the human eye (for example, the marked point in Fig. 1). The manual tracking of selected points was done by more than one tester and the averages of the tracked points were taken as the ground truth in order to reduce the human error.

Fig. 1. Example of a manually tracked point.

The execution of implemented tracking algorithms with prepared video input gives a sequence of homography matrices as output. From these, for each frame in each image sequence, a perspective transformation is applied by using the first frame of the image sequence as the reference. Let ground truth points of the reference frame be P'_{g0}, P'_{g1}, P'_{g2} and P'_{g3}. Then, using the homography matrix corresponding to a frame, the approximate positions of P_{g0}, P_{g1}, P_{g2} and P_{g3} in that frame is computed as shown in Eq. 1.

$$< P'_{g0}, P'_{g1}, P'_{g2}, P'_{g3} >= H < P_{g0}, P_{g1}, P_{g2}, P_{g3} > \tag{1}$$

P'_{g0}, P'_{g1}, P'_{g2} and P'_{g3} are the estimated points in n^{th} frames which are corresponding to the ground truth points of that frame and H is the homography matrix.

Accuracy and performance are the two aspects which are considered in this paper. In order to measure the accuracy, the estimates taken are (1) the distance between projected points and the ground truth, which is the distance error and (2) the optical flow deviation angle, which is the directional error.

4.1 The Distance Error

Taking the point P_1 in Fig. 2 as a point of the ground truth and P_2 in (b) of Fig. 2 as the corresponding tracked point by the algorithm the Euclidean distance between P_1 and P_2 was measured using Eq. 2. There d stands for the Euclidean distance between P_1 and P_2, x_1, x_2, y_1 and y_2 are the corresponding x, y coordinates of the points.

$$d = \sqrt{(x_1 - x_2)^2 + (y_1 - y_2)^2} \tag{2}$$

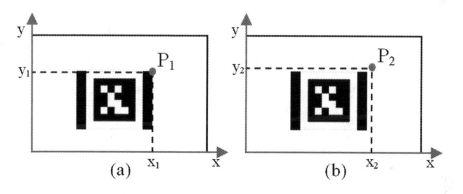

Fig. 2. Distance between the tracking points.

4.2 The Directional Error

In Fig. 3(a) P_1 is a ground truth point in the nth frame and in Fig. 3(c), P_2 is a point that is being tracked by the algorithm, corresponding to P_1 in Fig. 3(a). P'_1 and P'_2 are the points in the $(n+1)^{th}$ frame and those are corresponding to P_1 and P_2 in Fig. 3(a) and (c). Taking the vector between P_1 and P'_1 as U' and the vector between P_2 and P'_2 as V' where

$$V' = (v_1, v_2) \tag{3}$$

$$U' = (u_1, u_2) \tag{4}$$

The cosine of the angle between U' and V' can be obtained by Eq. 5.

$$cos(\alpha) = \frac{u_1.v_1 + u_2.v_2}{\sqrt{(u_1^2+u_2^2)*(v_1^2+v_2^2)}} \tag{5}$$

This angle is an indication of the directional movements.

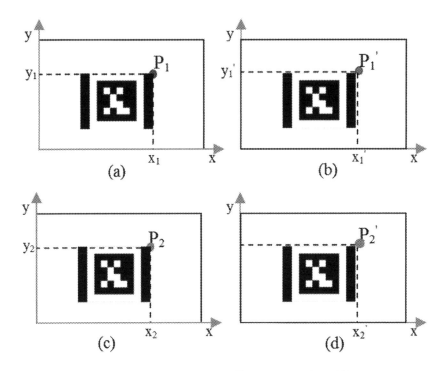

Fig. 3. The transition between n^{th} frame and $(n+1)^{th}$ frame.

4.3 Speed

In order to obtain an estimation of the performance of each algorithm the number of frames processed for a single unit of time was taken, using the same operating platform conditions.

4.4 Aggregated Score

For a complete comparison all three measurements need to be aggregated into one score that will give the final ranking. The approach used in this work is to standardize the scores of each criterion and doing a weighted summation (Eq. 6).

$$FinalScore_T = \frac{\beta * Z_{S,T} + \frac{\gamma}{Z_{d,T}} + \frac{\delta}{Z_{\alpha,T}}}{\beta + \gamma + \delta} \tag{6}$$

In Eq. 6 β, γ and δ are the weights used according to the required priority for the speed, the distance error and for the directional error.

$$S = \frac{\sum_{i=1}^{v} \sum_{j=1}^{f_i} \frac{1}{t_{i,j}}}{\sum_{i=1}^{v} \sum_{j=1}^{f_i} 1} \tag{7}$$

$$E_d = \frac{\sum_{i=1}^{v} \sum_{j=1}^{f_i} \sum_{k=1}^{p_{i,j}} d_{i,j,k}}{\sum_{i=1}^{v} \sum_{j=1}^{f_i} \sum_{k=1}^{p_{i,j}} 1} \tag{8}$$

$$E_\alpha = \frac{\sum_{i=1}^{v} \sum_{j=1}^{f_i} \sum_{k=1}^{p_{i,j}} \alpha_{i,j,k}}{\sum_{i=1}^{v} \sum_{j=1}^{f_i} \sum_{k=1}^{p_{i,j}} 1} \tag{9}$$

where v is the size of dataset, f is the number of frames in i^{th} video and $t_{i,j}$ is the time taken to process the jth frame of ith video and $p_{i,j}$ is the number of points marked as groundtruth in j^{th} frame of ith video. d is the distance error (Eq. 2) and α is the directional error (Eq. 5).

In Eq. 6 $Z_{S,T}$ is the standardized value of S in Eq. 7 for the selected tracker T, $Z_{d,T}$ is the standardized value of E_d in Eq. 8 and $Z_{\alpha,T}$ is the standardized value of E_α in Eq. 9.

5   Results

In analyzing the results, it was necessary to identify the characteristics of the image sequences. For this purpose, the amount of approximated blur presented in a video and the average contrast and color depth of each image in a video were taken in to account.

To obtain approximated blur measurement, the reciprocal of number of edges in the scene was considered. The number of edges was obtained using Canny Edge Detector [2].

To identify the illumination change, the three channel RGB image sequences were converted in to grey scale. For each image the average of pixel values were taken as it reflects the average intensity of the image.

For each image the standard deviation of pixel values were taken as the measurement of color depth. The results of the analysis are summarized below. In the figures related to interpretation of results the legend which is followed is depicted in Fig. 4. The area shown by a red rectangle is the focused area in each graph.

5.1   Tracker BRISK

This tracker exhibited poor performance in the presence of blur caused by motion as well as in the presence of blur caused by autofocus. This is due to BRISK descriptors storing different values for the same feature in the presence of blur. This behavior of the tracker on Video 2 is illustrated by Fig. 5.

5.2   Tracker FARNEB

This tracker indicates a gradual increase in error with time (Figs. 6 and 7).

Sudden differences in image intensities are not handled well (Fig. 8). As this is an optical flow algorithm this tracker cannot handle fast rotations of the

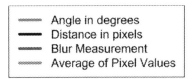

Fig. 4. Legend used in graphs.

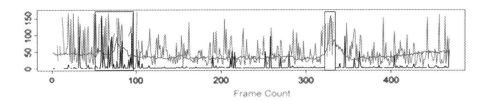

Fig. 5. Distance, angle and blur measure of Video 2 —Tracker BRISK.

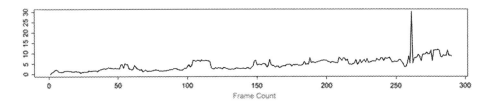

Fig. 6. Distance measure of Video 3 — Tracker FARNEB.

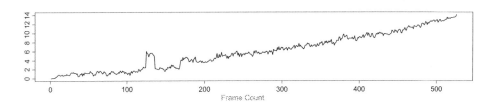

Fig. 7. Distance measure of Video 11 — Tracker FARNEB.

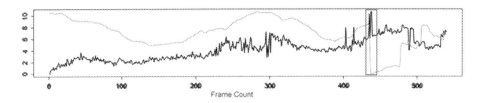

Fig. 8. Distance measure and average of pixel values of Video 1 — Tracker FARNEB.

Fig. 9. Camera rotation in Video 12.

Fig. 10. Points tracked in Video 12 –Tracker FARNEB.

camera. Figure 9 illustrates four frames from the image sequence of Video 12 which contains camera rotation along with the tracked points.

It can be observed that moving closer to a tracked feature is handled well if the feature gives the ability to identify the magnitude of flow in two orthogonal directions. Features points that resemble corners are most suited for this scenario. Figure 10 illustrates three frames from the image sequence of Video 14 taken at difference distances along with the tracked point. From that it can be identified that corners are the best tracked points. Poorly tracked points are marked with colored rectangles.

5.3   Tracker FREAK

This tracker is not capable of handling blur caused by autofocus or fast motion of the camera (Fig. 11). As observed, orientation changes and scale variations of the features are handled well by tracker FREAK.

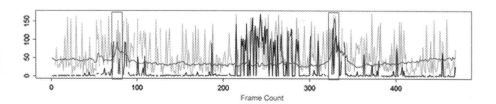

Fig. 11. Distance, angle and blur measure of Video 2 – Tracker FREAK.

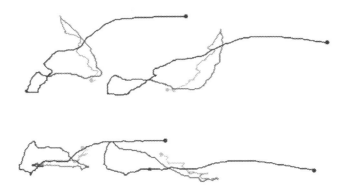

Fig. 12. Trajectories of ground truth points in Video 13.

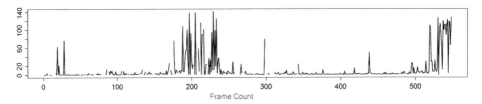

Fig. 13. Distance measure of Video 13 — Tracker FREAK.

Figure 12 illustrates the trajectories of ground truth points in Video 13. Starting points are indicated in red and end points are indicated in blue. It can be seen that the video has been subjected to scale variation. The distance error related to that is illustrated in Fig. 13.

5.4   Tracker KLT

This tracker depicts a gradual increasing error proportionate to the time (Fig. 14). In Tracker KLT, sudden difference in image intensity of image sequence disrupts the tracking process (Fig. 15). Blurring caused by autofocus or fast motion does not have an effect on this tracker (Fig. 16). This tracker as an optical flow algorithm which incorporates a simple motion model cannot handle rotation.

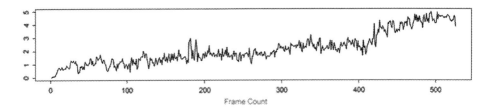

Fig. 14. Distance measure of Video 11 – Tracker KLT.

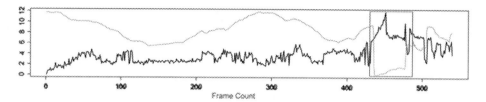

Fig. 15. Distance measure and average of pixel values of Video 1 – Tracker KLT.

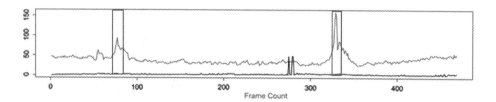

Fig. 16. Distance measure and blur measurement of Video 2 – Tracker KLT.

5.5   Tracker ORB

Tracker ORB has the capability to handle sudden intensity changes (Fig. 17). Accuracy of tracking decreases with scaling of the features being tracked. Motions blur and blur caused by autofocus does not affect the tracking performance (Fig. 18) and this tracker exhibits a considerable accuracy even at the time where the orientation changes of the features exist.

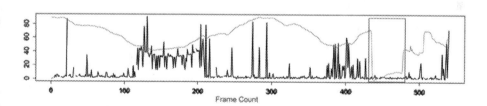

Fig. 17. Distance measure and average of pixel values of Video 1 – Tracker ORB.

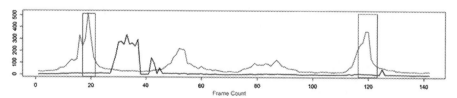

Fig. 18. Distance measure and blur measurement of Video 10 – Tracker ORB.

5.6   Tracker SIFT, Tracker SIFTSURF, Tracker SURFSIFT, Tracker SURF

These trackers use different combinations of SIFT and SURF feature detector and descriptor. The performance of trackers that use SURF as the descriptor show poor performance relative to the trackers that use SIFT descriptor.

6   Conclusion

The overall evaluations of the algorithm's consider the algorithms accuracy and speed. For accuracy two measures are used which are the distance error and the directional error. For speed the number of frames processed per second is used.

By considering only the criterion of error distance it can be concluded that Lucas Kanade optical flow algorithm (Tracker KLT) demonstrates the best performance consistently. But it should be noted that this algorithm does not handle rotation accurately due to the motion model utilized in the algorithm. Both optical flow algorithms have shown low distance error in comparison to feature matching algorithms. Among the feature matching algorithms Tracker ORB algorithm and the Tracker SURFSIFT algorithm shows consistent performance which is in the range of the optical flow algorithms. It can be concluded that optical flow algorithms can process translation and give consistent tracking but fails on rotation. Feature matching algorithms can handle rotation but fails on motion blur and give inconsistent tracking output.

When the evaluated algorithms are listed in ascending order according to the number of frames processed per second, it can be concluded that Tracker KLT is the fastest algorithm. It can be observed that the margin between Tracker KLT (Tracker 4) and the rest of the algorithms is very high. Most of the feature matching algorithms have very low speed scores except for the algorithm which uses BRISK detector and descriptor. The dense optical flow algorithm is also in the range of feature matching algorithms. This is due to the fact that KLT is a sparse optical flow algorithm and only considers a few number of feature points compared to the feature matching algorithms and dense optical flow algorithm.

According to the directional error, it can be concluded that optical flow algorithms which takes into account the motion details found within two consecutive images to estimate the succeeding position of a point will give more accurate output in contrast to feature matching algorithms which only use the locality information stored in the feature vectors. The dense optical flow algorithm has the best score for this criterion.

Modern mobile devices has the ability to capture high resolution images typically having camera specifications in range 4MP to 20MP or more. But two devices with the same camera specification can give different qualities in the same conditions. It is clear that the higher resolutions give accurate results but at the cost of speed. This is due to the fact that mobile platforms have restrictions on the number of calculations per second and memory available. Some of the interesting issues found was due to software features such as auto focusing and auto white balancing. Auto focusing created blur in the image sequence while

AWB created intensity changes which were undetectable to normal viewing but disrupted Tracker performance.

Normally a user using a mobile device for AR will be restricted to certain types of movements. Movements that include translation, circular motion around an estimated center (example: Walking around an augmented object) and scaling (Moving towards and backwards). Pure rotations can be expected to be less frequent as it is not a normal viewing movement as it does not change the perspective of viewing. Thus it would be much profitable to have a fast and accurate tracker for the more prevailing movements and have another algorithm to handle rotations supporting the main algorithm in use. According to the results it can be concluded that sparse optical flow algorithms are better equipped for handling translation at high speeds whereas feature matching algorithms take much more time to process a frame but better at handling rotations (orientation changes). This is due to the fact that sparse optical flow algorithms consider the information within a constrained area around selected keypoints and use a simplified motion model which allows for fast processing. Feature matching uses the information from the complete frame and takes much more time for processing as the complexity of calculations are much higher but this allows to find correspondence between images regardless of the motion/scaling that occurred by not having restrictions on the motion model.

Our context of augmented reality requires both speed and accuracy for a satisfactory outcome. When the scores are calculated (Eq. 6) with the values $\beta = 0.3$, $\gamma = 0.4$ and $\delta = 0.3$ the final score of each tracker is in Table 3.

Table 3. Final score of trackers.

Tracker	Final Score
KLT	2.27411145
FARNEB	0.231517254
ORB	0.142172204
FREAK	−0.326357422
SURFSIFT	−0.371670024
SIFT	−0.388266766
BRISK	−0.461264852
SURF	−0.560889913
SIFTSURF	−0.672624851

7   Future Work

It must be noted that the implementation of these algorithms are basic.Using different optimization techniques the performance of these algorithms can be improved. Furthermore there are many combinations of descriptors, detectors and optical flow algorithms that have not been evaluated in this study. Evaluating these optimizations and new algorithms can be done using the same method proposed in this paper.

References

1. Alahi, A., Ortiz, R., Vandergheynst, P.: FREAK: fast retina keypoint. In: IEEE Conference on Computer Vision and Pattern Recognition (2012)
2. Canny, J.: A computational approach to edge detection. IEEE Trans. Pattern Anal. Mach. Intell. **8**, 679–698 (1986)
3. David, G.L.: Distinctive image features from scale-invariant keypoints. Int. J. Comput. Vis. **60**, 91–110 (2004)
4. Ethan, R., Vincent, R., Gary, R.B.: ORB: an efficient alternative to SIFT or SURF. In: International Conference on Computer Vision, pp. 2564–2571 (2011)
5. Harris, C., Stephens, M.: A combined corner and edge detector. In: Proceedings of the 4th Alvey Vision Conference, pp. 147–151 (1988)
6. Bay, H., Tuytelaars, T., Van Gool, L.: SURF: speeded up robust features. In: Leonardis, A., Bischof, H., Pinz, A. (eds.) ECCV 2006. LNCS, vol. 3951. Springer, Heidelberg (2006)
7. Shi, J., Tomasi, C.: Good features to track. In: 9th IEEE Conference on Computer Vision and Pattern Recognition (1994)
8. Stefan, L., Margarita, C., Roland, S.: BRISK: binary robust invariant scalable keypoints. In: International Conference on Computer Vision, pp. 2548–2555 (2011)

Colour Matching Between Stereo Pairs of Images

Stephen Willey[1,2]([⊠]), Phil Willis[1], Jeff Clifford[2], and Ted Waine[2]

[1] University of Bath, Bath, UK
S.Willey@bath.ac.uk
[2] Double Negative, London, UK

Abstract. This paper outlines the process of colour matching stereo pairs of images using a disparity map as input along with the original images. We describe the functionality of a plugin developed for Nuke, which we call EyeMatch, that performs this automatic colour matching under conditions set by the user. The user is presented with various parameters to fine tune the matching process, but no prior knowledge of any of the underlying techniques is necessary. Results are produced quickly, allowing a trial-and-error based approach to fine tuning these parameters, and results are sufficiently accurate to be used in the post-production pipeline.

1 Introduction

3D movies have become much more prevalent in the last few years, almost becoming the standard format for any major release, with five out of the top ten grossing movies of 2012[1], and eight of the top ten for 2013[2], being released in the format. There has, however, been a backlash against the format which is due, in no small part, to the many examples of poorly executed, or seemingly unnecessary, 2D to 3D conversions that offer an uncomfortable viewing experience that can easily ruin an otherwise perfectly good movie.

The beam splitter rig is the most commonly used technique in production for capturing native 3D footage. This involves having one camera facing forward and the other facing perpendicular to it with a semi-reflective plane between them which will allow half of the light to reach one camera and half to reach the other. There are some major problems associated with this approach to stereo image capture, as can be seen in the results (Sect. 5). Firstly, the alignment of the images can be incorrect and inconsistent. This is due to the fact that the beamsplitter rig is rather large and any movement of it will result in each camera moving independently ever so slightly, which is enough to cause big problems in the post-production pipeline. Secondly, the use of two separate cameras and the introduction of a semi-reflective screen causes some noticeable colour differences between the images.

---

[1] Box Office Mojo 2012 Worldwide Grosses: http://www.boxofficemojo.com/yearly/chart/?view2=worldwide&yr=2012.

[2] Box Office Mojo 2013 Worldwide Grosses: http://www.boxofficemojo.com/yearly/chart/?view2=worldwide&yr=2013.

© Springer International Publishing Switzerland 2015
C.V. Jawahar and S. Shan (Eds.): ACCV 2014 Workshops, Part II, LNCS 9009, pp. 289–298, 2015.
DOI: 10.1007/978-3-319-16631-5_22

These colour differences are caused by multiple factors, the three main ones being polarisation, hardware, and angle. The semi-reflective screen introduced by the beamsplitter rig naturally causes some polarisation errors due to the fact that it selectively allows some light to pass through while reflecting the rest. If there is an area of a scene that is reflecting light at a particular polarisation, rather than a broad range of polarisations as is typical, it is possible for it to appear as a highlight in just one of the images. Another problem inherent in using two cameras is that, no matter how much care is taken to configure them exactly the same, there is always the possibility of a slight difference in hardware translating to some slight difference in the image recorded. Finally, the fact that each camera is recording the scene from a slightly different angle means that there is the chance of colour differences that would be perfectly natural in the real world, but would cause problems in a stereo projection (e.g. iridescence).

In this paper we discuss the use of a plugin, EyeMatch, which we have created for The Foundry's digital compositing software Nuke[3], that attempts to correct the colour differences between stereo pairs of images. EyeMatch is implemented as a Nuke node in order to fit seamlessly into the post-production pipeline, since many artists are already familiar with Nuke and use it on a daily basis. This also has allowed us to take advantage of many of Nuke's existing features.

2   Related Work

The problem presented could be solved in many different ways. The most basic solution would be to take the colour palette of one image and apply it to the other through, for example, histogram reshaping [5]. Another option would be to first segment the image, then perform a more specialised colour match between each element as defined by the chosen segmentation algorithm. Alternatively, the desired result could be achieved on a per pixel basis through techniques such as optical flow [1]. The main factors to be considered in choosing an appropriate solution are speed, accuracy and scalability.

There are many methods for adjusting the whole colour palette of an image using another image as a reference [3,4] but it is clear in our case that this will not be enough, and a more sophisticated approach is necessary to make more localised adjustments rather than just making a global change.

HaCohen et al. [6] compute dense correspondences in order to calculate the necessary colour change to bring about some regularity within a collection of images. The actual changes made are applied globally to each image, however, in order to avoid many of the complications brought about by making local alterations. Due to the nature of the issues we face with the stereo problem, it will be necessary to make many local changes and a global solution will not be sufficient.

When dealing with colour adjustments over video sequences, rather than still images, it is important to apply some temporal smoothing, as highlighted

---

[3] The Foundry: Nuke 8.0. (2013) Software available at http://www.thefoundry.co.uk/products/nuke-product-family/.

by Bonneel et al. [7]. In this paper we see that sudden changes of scene content, such as a new character walking into frame, can have a big effect on the colour adjustment algorithm and need to be accounted for.

Reinhard et al. [8] introduce the idea of converting between colour spaces in order to aid the transfer process. Due to observing that all three channels often require modification in RGB space, it is suggested that using one with less of a correlation between channels would be advantageous.

3 Motivation

3D films are popular among audiences because of the immersive experience they offer. They are popular with studios and content creators because they can command a higher premium for viewing. Due to this popularity, it is important to do it right. We aim to minimise the amount of work the viewer has to do to reconcile a stereo image by ensuring an accurate colour match between views. In this way, we hope to create a more pleasant experience for moviegoers.

If the colours are not quite matched up, or the alignment isn't quite right, then the viewer is going to have to work harder to make sense of the images in front of them and that can lead to fatigue and headaches [2]. There are many other factors that can lead to these problems, such as the projectors in a particular cinema not being set up correctly, leading to problems such as ghosting, but these issues are outside of our control, and so we focus on those that can be fixed during the post-production process.

4 Method

Having investigated the existing methods employed to alter colours within images, it became apparent that a hybrid of several suggestions would be necessary. A global colour palette shift [5] would not be appropriate for this problem due to the fact that large occluded areas are likely to exist between images and this will skew the palette of each image. Therefore, our primary focus has been on creating a colour matcher which will work on local areas in order to match colours in a more accurate and more reliable way.

Our method relies on an accurate disparity map to work properly, and for this we have some in-house tools based on the Lucas-Kanade method of optical flow. There are other methods, including commercial solutions such as Nuke's Ocula toolset, for disparity map generation but the important thing is that a reasonably accurate and reliable disparity map is available for the EyeMatch algorithm.

Due to the fact that Nuke is used heavily in the pipeline at Double Negative, EyeMatch was created as a node that makes use of many existing features in the core Nuke toolset. For example, the two images (or sequences) are taken, along with the disparity map data, as a single input and immediately processed using several Nuke plugins, such as "Blur", to reduce the effects of noise, and "iDistort", which uses the disparity map to shift the pixels in one view to the position of their corresponding pixels in the other, before being fed into the core matching algorithm.

4.1   Algorithm Breakdown

This algorithm takes as its input the left eye view, the right eye view, and the disparity map describing the relationship between the pixels in each. Figures 1 and 2 show the x and y disparity information for our example scene, while Figs. 3 and 4 show the input images themselves. The first step is to shift the pixels of one image by the values in the disparity map, so that both views are aligned, then break the images up according to the grid dimensions provided by the user.

Outliers are dealt with by the following process: The mean and variance are calculated for each colour channel and any values lying more than one standard deviation from the mean are changed to the mean value, this practically eliminates noise and produces a smoother result overall. For each colour channel of each pixel, values are then arranged in order from lowest to highest. Using the disparity information, the corresponding pixels are then found in the second view and the same process of ordering and removing outliers is completed.

Rather than comparing each pixel to its exact counterpart according to the disparity information, pixels are matched based on their position in their respective ordered lists. Again, this creates a smoother and more robust result since each pixel difference is minimised and it is less likely that there will be any extreme adjustment calculations. Once each colour value for each pixel in the grid square to be adjusted are divided by their counterpart in the grid square to be matched, the resulting values can then be reordered to their initial positions to give the necessary colour change for each individual pixel.

Once this is completed for every section of the image, we are presented with a colour difference map describing the exact per-pixel changes required to match the colours of one view to the other. This colour difference map can be use to match the colours by multiplying each pixel with its corresponding pixel in the view to be adjusted.

Due to the fact that the image is gridded before processing, at this stage there will often be some very distinct lines at the borders of these grid squares. The simple solution to this is to blur across these boundaries in order to create a smoother result. From an artist's point of view, it is better to have a smooth result overall even if this comes at the expense of accuracy to some parts of the image. This is the final stage for a single pair of images, but for a sequence we must finish with some temporal smoothing to avoid flickering on playback. This is accomplished with a standard temporal median filter provided by Nuke which produces the results we are after.

4.2   User Defined Parameters

The user interface shown in Fig. 5 shows the parameters that can be specified by the user in order to adapt the algorithm for different situations, the features of which are described below. Once a user changes one of the parameters, it is

Fig. 1. Disparity map showing x-disparity values from left to right view

Fig. 2. Disparity map showing y-disparity values from left to right view

Fig. 3. Left view input image from a beamsplitter rig

Fig. 4. Right view input image from a beamsplitter rig

applied immediately. Depending on the resolution of the images, the updated results will be displayed in the Nuke viewer in a matter of seconds.

Grid Size. The grid size can be specified, and does not need to be defined as square since height and width can be adjusted independently. This is important since there is always a tradeoff between desired smoothness and required accuracy. A smaller grid size will result in a greater accuracy (as long as a good quality disparity map is provided as input) but may produce a less robust result as a smaller grid square will mean less pixels being looked at when deciding what constitutes an outlier. This means there is more chance of some noise sneaking through. Conversely, a larger grid size will result in a smoother result but could result in some errors around edges, especially those with large occlusions.

Blur. The user is also presented with two blur options. The first of these is for blurring the input disparity map, which can help when the disparity map is of a lower quality. This will have a negative impact on precision but is very helpful for smoothing out noise. Secondly, the user may specify the blur of the output. Some blurring is always necessary in order to lessen the obvious boundaries between grid squares, but sometimes more will be needed or less would be desired in order to get the necessary result. The user is also given the choice of a median

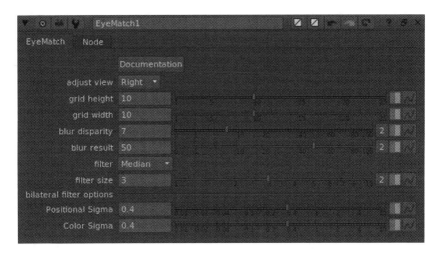

Fig. 5. The user interface showing default values as requested by users during development. Smaller values result in more locally specific changes, but require a more accurate disparity map; larger values will create a smoother result, at the expense of accuracy.

filter, which is appropriate for most tasks, or a slightly more sophisticated, but much slower, bilateral filter.

Fig. 6. By viewing the left and right views in alternating grid squares of the same image, it is easy to see the colour differences present throughout (Colour figure online).

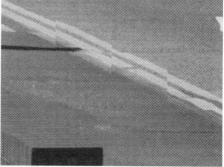

Fig. 7. A closer look at a portion of the scene with a fairly uniform colour, with alternating squares taken from the left and right view, highlights the difference between views (Colour figure online).

Fig. 8. Another close-up view, with alternating squares taken from the left and right views, further shows the colour differences that must be corrected in another portion of the scene (Colour figure online).

5   Results

We have produced some encouraging results with this algorithm and even made extensive use of it in the post production pipeline. There are some areas for improvement, but also some areas where the output produced is better than the leading commercial alternatives. The simplicity of this algorithm is its greatest

Fig. 9. Output right view produced using default values for the EyeMatch plugin, as shown in Fig. 5.

Fig. 10. Once the images have been run through our plugin, the colours are much more closely matched. This image shows the original left view in alternating squares with the colour corrected right view (Colour figure online).

strength and allows it to work robustly for large areas of a sequence, making it useful in spite of the errors which can be found in other areas of the output.

In order to better show the problem at hand, and to demonstrate the efficacy of our plugin, we have created checkerboard images with alternating squares from

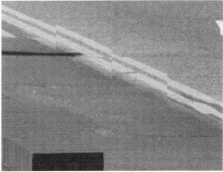

Fig. 11. The right view matches so closely that only by looking at the lamppost on the right is it clear that this image shows alternating squares from the original left view and the colour corrected right view (Colour figure online).

Fig. 12. The road portion of the scene also shows a very close colour match when viewed as alternating squares of the original left view and the colour corrected right view (Colour figure online).

Fig. 13. Incorrect colour matching at the extreme edge of the image, caused by a lack of corresponding information in the left view (Colour figure online).

Fig. 14. Incorrect colour matching due to the fact that the car occludes the wall too much, leaving EyeMatch without enough information to work with (Colour figure online).

the left and right view. Figure 6 highlights the general problem well, showing not only the existence of colour differences between views, but also its non-uniform nature - some areas are already well matched. A closer look at some areas of the image (Figs. 7 and 8) further shows the extent that colour differences can reach.

Following the use of the EyeMatch node on default settings (as shown in Fig. 5), the result shown in Fig. 9 is produced. This colour matched right view is shown in alternating grid squares with the original left view in Fig. 10. Another look at the close up sections shown earlier reveals the accuracy of the result (Figs. 11 and 12).

6 Conclusion

We have presented an approach to correct the problem of colour mismatch when recording images on separate cameras and from slightly different viewpoints. Users can adjust parameters according to their requirements for a particular scene and see the results quickly, allowing them to select the best options on a trial-and-error basis if necessary. The user does not require any knowledge of any of the underlying techniques in order to use this tool, and acceptable results can be obtained quickly, and with very little effort.

7 Limitations and Future Work

This technique has been used in production and proved effective for many areas of images, though it is not usable as a complete package solution. It has been reported by artists to be particularly good for matching skin tones, which is an area some other solutions seem to struggle with. The main problem with this solution, however, is that it does not currently handle occlusions in any way and so when presented with particularly noticeable areas of occlusion it will produce poor results, as can be seen in Figs. 13 and 14. Potential ways to handle these occlusion errors include looking to the surrounding areas to interpolate a sensible value, or to look further along the sequence to see if the occluded area becomes visible. This temporal solution raises more issues, however, such as the chance of a lighting change between frames.

The EyeMatch node relies heavily on having a good disparity map as input. Therefore future work, aside from better occlusion handling, will involve improving optical flow algorithms and introducing some occlusion detection at an early stage in order to better handle them further down the pipeline.

Acknowledgments. We would like to thank Engineering and Physical Sciences Research Council for funding this research. Many thanks to Darren Cosker, Paul Hogbin, Oliver James and Alexandros Gouvatsos for useful feedback and guidance.

References

1. Lucas, B.D., Kanade, T., et al.: An iterative image registration technique with an application to stereo vision. In: IJCAI 81, pp. 674–679 (1981)
2. Shibata, T., Kim, J., Hoffman, D.M., Banks, M.S.: The zone of comfort: predicting visual discomfort with stereo displays. J. Vis. **11**, 11 (2011)
3. Pitié, F., Kokaram, A.C., Dahyot, R.: Automated colour grading using colour distribution transfer. Comput. Vis. Image Underst. **107**, 123–137 (2007)
4. Grundland, M., Dodgson, N.A: Color histogram specification by histogram warping. Electron. Imaging Int. Soc. Opt. Photonics, 610–621(2005)
5. Pouli, T., Reinhard, E.: Progressive histogram reshaping for creative color transfer and tone reproduction. In: Proceedings of the 8th International Symposium on Non-Photorealistic Animation and Rendering, pp. 81–90 (2010)
6. HaCohen, Y., Shechtman, E., Goldman, D.B., Lischinski, D.: Optimizing color consistency in photo collections. ACM Trans. Graph. **32**, 38 (2013)
7. Bonneel, N., Sunkavalli, K., Paris, S., Pfister, H.: Example-based video color grading. ACM Trans. Graph. **32**, 2 (2013)
8. Reinhard, E., Adhikhmin, M., Gooch, B., Shirley, P.: Color transfer between images. IEEE Comput. Graphics Appl. **21**, 34–41 (2001)

User Directed Multi-view-stereo

Yotam Doron[1][✉], Neill D.F. Campbell[1], Jonathan Starck[2], and Jan Kautz[1,3]

[1] University College London, London, UK
y.doron@cs.ucl.ac.uk
[2] The Foundry, London, UK
[3] NVIDIA Research, Guildford, UK

Abstract. Depth reconstruction from video footage and image collections is a fundamental part of many modelling and image-based rendering applications. However real-world scenes often contain limited texture information, repeated elements and other ambiguities which remain challenging for fully automatic algorithms. This paper presents a technique that combines intuitive user constraints with dense multi-view stereo reconstruction. By providing annotations in the form of simple paint strokes, a user can guide a multi-view stereo algorithm and avoid common failure cases. We show how smoothness, discontinuity and depth ordering constraints can be incorporated directly into a variational optimization framework for multi-view stereo. Our method avoids the need for heuristic approaches that edit a depth-map in a sequential process, and avoids requiring the user to accurately segment object boundaries or to directly model geometry. We show how with a small amount of intuitive input, a user may create improved depth maps in challenging cases for multi-view-stereo.

1   Introduction

Multi-view-stereo (MVS) aims to reconstruct the dense geometry of a scene from a set of calibrated images. A large number of MVS methods reconstruct geometry in the form of a *depth map* that assigns a distance value to each pixel in a given input image. Depth maps are a useful representation of geometry with applications in image-based rendering, 3D modeling and augmented reality, and can be merged to create more complete models of a scene [1].

Depth reconstruction is a mature area of research, with methods that perform well on scenes that obey certain assumptions. The first assumption is that the scene geometry has a similar local appearance across different views, and that surfaces are sufficiently textured to distinguish between correct and incorrect geometry under projection. The second main assumption is that depth varies smoothly in regions of low image texture and that discontinuities in depth coincide with strong image edges. These are encoded in the data and smoothness terms common to almost all methods, with data terms based on photo-consistency scores and smoothness achieved by global optimization [2] or local filtering [3].

© Springer International Publishing Switzerland 2015
C.V. Jawahar and S. Shan (Eds.): ACCV 2014 Workshops, Part II, LNCS 9009, pp. 299–313, 2015.
DOI: 10.1007/978-3-319-16631-5_23

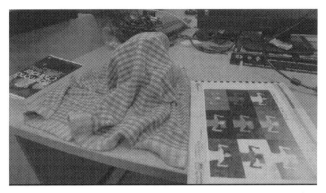

(a) Source input image (neighboring views not shown)

(b) Depth map from standard method [2]

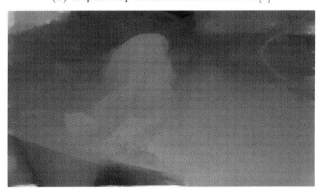

(c) Depth map after interactive editing with user-in-the-loop

Fig. 1. Overview of the advantage of our user-assisted multi-view stereo. Standard variational approaches to depth map estimation often have shortcomings when presented with real world scenes, for example the specular highlights on the brochure (a) result in holes in the depth map (b). In our approach, we provide the user with interactive tools (in the form of brush strokes) to correct for such short-comings

In reality, these assumptions are only partially valid, as many scenes contain objects with large areas of low-texture, specular reflections and colour discontinuities that are unrelated to changes in geometry (Fig. 1).

As a result, fully automatic methods have trouble reconstructing general scenes. Instead, we propose to enable the user to provide intuitive constraints to aid the reconstruction process. In our method we allow the user to apply depth smoothness, depth discontinuity and depth ordering constraints, all of which are specified by the user as simple scribbles. We incorporate the user's constraints directly into a variational optimization framework that simultaneously optimizes for both a regularized depth map derived from on a photo-consistency cost volume as well as the constraints.

Our work shows how user input can help correct depth reconstruction. In this paper we make the following contributions:

- We introduce a framework to incorporate user constraints, motivating our choice of algorithm by the need for interactive refinement.
- We define a set of simple edit operations to correct common failure cases in multi-view-stereo. Our edit operations are intuitive, do not require the user to draw accurate object boundaries and do not rely on fixed sequential operations such as superpixel segmentation.
- We demonstrate how user constraints may be directly incorporated into a state of the art variational depth reconstruction algorithm and show how this can be used to correct depth reconstruction.
- We provide a clear derivation of our solver from the energy formulation.

2   Related Work

User-guided monocular shape and depth recovery. In scenarios such as 2D-3D conversion, a user may want to recover geometry given their knowledge of the image contents. Several techniques exist to interpolate empirical depth or disparity values painted onto monocular image or video data, while others use relative constraints on depth [4]. Guttmann *et al.* [5] train a classifier on user-scribbled disparity values provided on the first and last frames of a shot, using the results to constrain a coarse-to-fine quadratic optimization. Brosch *et al.* [6] also require users to paint disparity values onto the first and last frame of a sequence. Their technique attempts to maintain consistent depth over time by combining the scribbles with the results of a video over-segmentation algorithm and optical flow. In our approach, rather than directly painting depth values, the user indirectly improves the depth computed by multi-view-stereo by providing constraints.

The Depth Director system of Ward *et al.* [7] uses superpixel data, along with sparse structure-from-motion information to assign depth to video frames. The system provides a segmentation tool, allowing users to adjust depth variation and orientation within a region. Additionally, users may choose a template shape, such as a car, to fit to a selected region. Liao *et al.* [8] ask users to provide relative depth input (ordering and equality) at keyframes. The system also uses depth from SFM features where available and applies a temporal perspective depth correction stage based on optical flow. Their user-study suggests that it is more intuitive for users to specify the relative depth constraints than to paint colour-coded empirical depth values.

High level user interactions have been applied to intrinsic image decomposition and geometric modelling. Two recent methods [9,10] include in their optimization user-provided local constraints: constant-reflectance, constant-illumination and fixed-illumination. In single-image modelling, Zhang et al. [11] optimize for a mesh satisfying user-provided normal, positional and curvature constraints. Toeppe et al. [12] propose a variational framework for single-view single-object modelling using a ballooning energy formulation. The recent work of Chen et al. [13] demonstrates modelling of complex objects with simple inputs, using symmetry constraints and automatic alignment to image structures.

User-guided stereo. The recent system of Zhang et al. [14] aims to improve the quality of disparity maps generated from a stereo video sequence by allowing a range of user edits. The user corrects disparity at keyframes indirectly with several interactive tools which alter unary and pair-wise cost terms in an label-based optimization framework. The system relies on a GrabCut-based object selection tool [15] to segment objects. The user may fit a parametric model, using sparse feature matches within a selected object, a disparity alignment tool, a smoothness brush and a discontinuity brush. The paper also presents a method for propagation of edited disparity through the sequence, with the edited disparity map and disparity costs used soft constraints in non-keyframes. This approach is related to ours in that the user edits also influence energy terms in an optimization, however our constraints do not require the user to first accurately segment object boundaries, and we target the multi-view-stero case rather than (binocular) stereo video.

Photogrammetry. There has been extensive work on image-based modelling using photogrammetry [16], with recent methods [17] providing high quality results from inaccurate user input. In the VideoTrace system [18], a user interactively creates a model from video footage. The interactions are assisted by pre-computed sparse stereo and superpixel information and the user may define symmetries to complete geometry in areas not visible in the footage.

3   Our Approach

The input to our system is a sequence of images $\{I_1 \cdots I_N\}$ for which the camera motion has been calculated using, for example, structure-from-motion with an off-the-shelf tool. This provides us with a projection matrix P_i defining the camera pose, and intrinsic parameters such as focal length, at each image (or frame of a video sequence).

Given this calibrated image sequence, the goal of our method is to estimate a dense depth map for one or more of the images. Without loss of generality, we shall discuss calculating the depth for a single image at a time; this image shall be referred to as the *source* or *reference* image and denoted I_s. For the source frame I_s, a subset of adjacent frames are used to automatically reconstruct the source frame depth D_s.

In this section we provide details on our approach to interactive depth reconstruction. We begin by providing details of how to build a photo-consistency cost volume that encodes appearance constraints from neighboring views. We then describe the basic energy model that uses this cost volume, along with regularisation enforcing smoothness constraints, to define the most likely depth map. Next, we add user constraints to modify the energy function in a principled manner to improve the quality of the depth map in an interative refinement process with the user in the loop. Finally, we provide details of the optimization scheme used to solve our energy model, including the user constraints, in an efficient manner that allows for interactive editing.

3.1   Photo-Consistency

Our depth estimation process first builds a photo-consistency cost volume. The cost volume acts as a cached data term in our optimization, as in [2] and is computed before any user interactions are introduced. We test a range of candidate depths for each pixel x in the source image I_s, spaced linearly in inverse depth (disparity). Our photo-consistency error for the pixel x taking a candidate depth d is defined as

$$C(x,d) = \frac{1}{K} \sum_{I_j \in \mathcal{N}(I_s)} |I_s(x) - I_j(\pi_{s,j}(x,d))| . \tag{1}$$

Here we have used $\pi_{s,j}(x,d)$ to denote the back-projection from I_s to a depth d along a ray through x, followed by projection into image coordinates in a neighbouring frame I_j, and, $I_j(x')$ to denote the interpolated colour at pixel x' in I_j. In addition, $\mathcal{N}(I_s)$ denotes the local neighbourhood of frames around I_s, and K normalizes by the number of frames where the re-projected pixel is inside the image bounds.

3.2   Energy

Taking the lowest cost per pixel in the photo-consistency cost volume, *i.e.*

$$d(x) = \arg \min_x C(x,d), \tag{2}$$

would provide us with a very noisy estimate of the depth map; instead, [2,19], we apply spatial regularization using a variational energy formulation that combines the photo-consistency cost with a term that encourages gradient sparsity in the depth map. Formally, we define our energy as

$$E[d(x)] = \int_\Omega \lambda C(x,d(x)) + g(x) \, ||\nabla d(x)||_\epsilon \, dx \tag{3}$$

where we use the photo-consistency cost from (1) and a Total Variation (TV) prior on the gradient of the depth map, $\nabla d(x)$, and integrate over the image domain Ω. The TV term is weighted by the inhomogeneous contrast sensitive term

$$g(x) = \exp\left(-\gamma \ ||\nabla I_s||\right) \tag{4}$$

that encourages depth discontinuities to coincide with image intensity discontinuities. We use the Huber norm

$$||\mathbf{s}||_\epsilon = \begin{cases} \frac{||\mathbf{s}||^2}{2\epsilon} & \text{if } \ ||\mathbf{s}|| \le \epsilon \\ ||\mathbf{s}|| - \frac{\epsilon}{2} & \text{if } \ ||\mathbf{s}|| > \epsilon \end{cases} \tag{5}$$

on the depth map gradient. For clarity, we will drop the dependency on x in our notation such that $d(x) \mapsto d$ and $g(x) \mapsto g$.

The energy in (3) is non-convex and difficult to optimize directly. To overcome this issue we use a quadratic relaxation, similar to [20] and [2]. We introduce an auxiliary depth variable, $v(x) \mapsto v$, and approximate (3) with the auxiliary energy $E_{\text{aux}}[d, v]$ as

$$\int_\Omega g \ ||\nabla d||_\epsilon + \frac{1}{2\theta}(d - v)^2 + \lambda\, C(x, v) \ \mathrm{d}x. \tag{6}$$

We observe that as $\theta \to 0$ we will have $v \to d$ and thus (6) $\to$ (3).

By decoupling the regularization from the data term, we obtain two sub-problems. Fixing $v = v'$, we have a problem $\min_d E_{\text{aux}}[d, v']$ that is convex in d as

$$\min_d \int_\Omega g \ ||\nabla d||_\epsilon + \frac{1}{2\theta}(d - v')^2 \ \mathrm{d}x \tag{7}$$

and by fixing $d = d'$ we have a problem $\min_v E_{\text{aux}}[d', v]$ that can be solved point-wise for v as

$$\min_v \int_\Omega \frac{1}{2\theta}(d - v)^2 + \lambda\, C(x, v) \ \mathrm{d}x. \tag{8}$$

We will show how to solve this alternation optimization in Sect. 3.4 using a primal-dual saddle point technique; before this, we provide details of how to extend this standard energy model to include user constraints.

3.3   Including User Constraints

We now describe how to extend the basic energy model to include the user in the reconstruction process (user-in-the-loop). This takes to form of the user providing brush strokes from a toolbox of three constraints targeted against specific shortcomings of the standard variational approaches. We now describe each of these three tools in further detail.

Smoothness. There are regions where the photo-consistency values in the cost volume may be noisy or incorrect, *e.g.* the Lambertian assumption fails in the prescence of specular highlights, or image intensity discontinuities may encourage artifical depth discontinuities. In such regions we can smooth the solution from neighboring regions containing the correct depth by downweighting the photo-consistency term and relying on the gradient regularizer to fill in a smooth surface.

We maintain a brush bitmap $B_{sm}(x) \in [0,1]$ and use it to modulate the weight on the data term in (3); we allow λ to vary across the image and set it to $\lambda = \alpha(1 - B_{sm})$, again dropping the explicit dependency on x for clarity. We used feathered brush strokes to ensure a smooth transistion. We note that we must modify the sub-problem in (7) since the quadratic relaxation means that the depth is still influenced by v. We therefore include λ in the coupling term as

$$\min_d \int_\Omega g \, ||\nabla d||_\epsilon + \frac{\lambda}{2\theta} (d - v')^2 \, \mathrm{d}x \tag{9}$$

such that when $\lambda \to 0$ the data term is decoupled and the regularization will take over and smooth the resulting depth map.

Boundary Discontinuities. There are the opposite cases where the solution will be too smooth in regions where there should be a discontinuity in the depth; for example, foreground and background objects with similar coloring may result in a low contrast intensity edge disguising a true discontinuity in depth. To tackle this problem we provide a second brush bitmap $B_{dc}(x) \in [0,1]$ that increases the contrast sensitive edge term of (4) as

$$g(x) = \exp\left(- \gamma \left(1 + \mu \, B_{dc} \right) \, ||\nabla I_s|| \right). \tag{10}$$

We also need to downweight the data term in these regions since the photo-consistency term can lead to the phenomenon of foreground ening [3]. Whilst previous approaches have addressed this with adaptive support weights or including view selection in the optimization [21], this requires changing or re-computing the cost volume and removes the efficiency advantages of pre-caching the photo-consistency costs. Instead, we make use of the discontinuity brush and again downweight the λ term again such that

$$\lambda = \alpha \left(1 - B_{sm} \right) \left(1 - B_{dc} \right). \tag{11}$$

Ordering Constraints. Errors in the photo-consistency volume can give rise to the an incorrect local minimum where even if smoothness and discontinuity constraints are preserved, a distinct surface may appear in the wrong layer (either too close or too far from the camera). Our third tool makes use of two brush strokes where a user can select two nearby image regions and apply the constraint that one is closer to the camera than the other. Multiple instances of such pairwise constraints can be built up as necessary given the interactive reconstruction feedback available to the user.

We will illustrate with a single pair of brush strokes; we define a foreground brush $B_{\text{fore}}(x)$ and corresponding background brush $B_{\text{back}}(x)$ with the contraint that all the foreground brush pixels $\{x_{\text{f},i}\} \in B_{\text{fore}}$ are closer to the camera than the background brush pixels $\{x_{\text{b},j}\} \in B_{\text{back}}$. We proceed by matching each foreground pixel $x_{\text{f},i}$ to the nearest pixel in the background set $x_{\text{b},m(i)}$ such that

$$m(i) = \arg\min_{j \in B_{\text{back}}} \left\| x_{\text{f},i} - x_{\text{b},j} \right\|. \tag{12}$$

This can be performed efficiently using a k-d tree. We then form a set of linear inequality constraints with a minimum threshold distance in depth t_{dist} that must separate the two layers which gives us that

$$\Phi\left[d\right] + t_{\text{dist}} \mathbf{1} < \mathbf{0} \tag{13}$$

where $\Phi\left[d\right]$ denotes

$$\begin{bmatrix} 0 \cdots -1 \cdots 1 \cdots 0 \\ \vdots \quad \vdots \quad \vdots \end{bmatrix} \begin{bmatrix} \cdot \\ d(x_{\text{f},i}) \\ \cdot \\ d(x_{\text{b},m(i)}) \\ \cdot \end{bmatrix}. \tag{14}$$

We can apply this constraint to the energy model using a set of Lagrangian multipliers $\mathbf{r} \in \mathcal{R}^{|\{x_{\text{f},i}\}|}$ to augment sub-problem (9) to

$$\min_{d} \max_{\mathbf{r}} \int_{\Omega} g \, \left\| \nabla d \right\|_{\epsilon} + \frac{\lambda}{2\theta} (d - v')^2 + \langle \mathbf{r}, (\Phi\left[d\right] + t_{\text{dist}} \mathbf{1}) \rangle \, dx \tag{15}$$

and maximizing with respect to $\mathbf{r}$ such that $\mathbf{r} \geq 0$.

3.4   Optimization

In Sect. 3.2 we described how to split our energy model into two sub-problems to be solved in alternation. We first consider the sub-problem, with the auxiliary variables v fixed, of (15); this can be solved with a primal-dual approach [22].

Auxiliary Sub-Problem. Taking (15), we first dualize the regularisation term $f(d) = g \, \left\| d \right\|_{\epsilon}$, with $g > 0$. The Legendre-Fenchel transform of $f(\cdot)$ is given by

$$f^{\star}(p) = g \max_{d} \left\{ g^{-1} \langle d, p \rangle - \left\| d \right\|_{\epsilon} \right\} \tag{16}$$

$$= \frac{\epsilon}{2g} \left\| p \right\|^2 + \delta\left(\frac{p}{g}\right) \tag{17}$$

where $\delta(\cdot)$ is the indicator function

$$\delta(p) = \begin{cases} 0 & \text{if } \|p\| \leq 1 \\ \infty & \text{otherwise} \end{cases} \tag{18}$$

and we use the scaling property

$$f(x) = a\,h(x) \implies f^{\star}(p) = a\,h^{\star}\left(\frac{p}{a}\right) \tag{19}$$

for $a > 0$. We then add a dual variable $q(x) \mapsto \mathbf{q}$ and obtain the saddle point problem $\int_{\Omega} L\,\mathrm{d}x$ as

$$\int_{\Omega} \langle \mathbf{q}, \nabla d \rangle - \frac{\epsilon}{2g}\,\|\mathbf{q}\|^2 - \delta\left(\frac{\mathbf{q}}{g}\right) + \frac{\lambda}{2\theta}(d - v)^2 + \langle \mathbf{r}, (\varPhi\,[d] + t_{\text{dist}}\mathbf{1})\rangle\,\mathrm{d}x \tag{20}$$

that we minimize with respect to d and maximise with respect to $\mathbf{q}$ and $\mathbf{r}$. Taking partial derivatives we obtain

$$\frac{\partial L}{\partial \mathbf{q}} = \nabla d - \frac{\epsilon}{g}\mathbf{q} \tag{21}$$

$$\frac{\partial L}{\partial \mathbf{r}} = (\varPhi\,[d] + t_{\text{dist}}\mathbf{1}) \tag{22}$$

$$\frac{\partial L}{\partial d} = -\nabla \cdot \mathbf{q} + \frac{\lambda}{\theta}(d - v). \tag{23}$$

We then discretize for descent in d and ascent in $\mathbf{q}$ and $\mathbf{r}$ and solve for $\mathbf{q}^{k+1}$, $\mathbf{r}^{k+1}$ and d^{k+1} as

$$\mathbf{q}^{k+1} = g\,\frac{(\mathbf{q}^k + \sigma_q \nabla d^k)}{(g + \sigma_q \epsilon)} \tag{24}$$

$$\mathbf{r}^{k+1} = \mathbf{r}^k + \sigma_r\,(\varPhi\,[d] + t_{\text{dist}}\mathbf{1}) \tag{25}$$

$$d^{k+1} = \frac{(\theta\,d^k + \sigma_d\,(\theta\,\nabla\cdot\mathbf{q}^{k+1} + \lambda v))}{(\theta + \lambda\sigma_d)} \tag{26}$$

with appropriate step sizes σ_q, σ_r and σ_d, as in [23]; the updates are shown in Algorithm 1. We note that two projection steps are required. The $\mathbf{q}$ updates require projection into the norm ball, represented by the operation $\pi[\cdot]$. The $\mathbf{r}$ updates require projection into the positive half-plane, respresented by $I^+[\cdot]$.

Cost Volume Sub-Problem. The second sub-problem, with the primal variables d fixed, of (8) may be solved using a simple point-wise search in the cost volume. As noted in [2], this search may be accelerated by keeping track of a search depth range for each pixel over subsequent iterations. We also perform a single Newton step in v in each iteration, as in [2], to obtain sub-sample accuracy within the cost volume and reduce depth quantization artefacts.

Algorithm 1. Depth optimization

$d^0(x) \leftarrow \min_{d} C(x, d(x))$

$v^0(x) \leftarrow d^0(x)$

$n \leftarrow 0$

while not converged **do**

    Projected gradient ascent for dual variable

    $\mathbf{q}^{k+1} \leftarrow g\,\pi\left[(g + \sigma_q \epsilon)^{-1} g\left\{\mathbf{q}^k + \sigma_q \nabla d^k\right\}\right]$

    Projected gradient ascent for constraint multipliers

    $\mathbf{r}^{k+1} \leftarrow I^+\left[\mathbf{r}^k + \sigma_r\left(\Phi\left[d\right] + t_{\text{dist}}\mathbf{1}\right)\right]$

    Gradient descent step for primal variable

    $d^{k+1} \leftarrow \left(1 + \frac{\lambda\sigma_d}{\theta}\right)^{-1}\left(d^k + \sigma_d\left(\nabla \cdot \mathbf{q}^{k+1} + \frac{\lambda v}{\theta}\right)\right)$

    Exhaustive point-wise search for auxiliary variable

    $v^{k+1} \leftarrow \min_{v}\left(\lambda\,C(x, v^k) + \frac{\lambda}{2\theta}(d^{k+1} - v^k)^2\right)$

    Update coupling parameter

    $\theta \leftarrow \theta(1 - \beta n)$

    $n \leftarrow n + 1$

end while

return d

Efficiency. The overall optimization process is given in Algorithm 1. We note that each update step can be performed efficiently and in parallel for each pixel. As demonstrated by the results of [2], this allows for real-time performance when implemented on the GPU. In addition, the cost volume calculation (Eq. 1) may be performed in parallel on the GPU using the texture units to perform efficient image interpolation.

4   Results

We demonstrate the results of our system, showing the effect of the different user interactions on the reconstructed depth. In this paper we show a reconstruction of the flower and lawn dataset from Zhang *et al.* [24] (in Figs. 3 and 4) and a desk scene (in Fig. 2) that features a number of violations of standard MVS assumptions. In our experiments, we select the 8 closest views to the reference frame and sample 100 depth values to construct the cost volume. For the flower and lawn scenes, we use the camera pose provided by [24] and for the desk scene we use the NUKEX camera tracker [25].

Figure 2 shows the impact of the different interactions compared to the result of the baseline variational method [2]. The desk scene features a glossy non-Lambertian surface with strong image edges that do not coincide with depth discontinuities. The resulting hole in the depth map is corrected with smoothness constraints in our method. The discontinuity edits reduce the depth smearing on the top edge of the cloth in the desk scene due to low image contrast.

Figures 3 and 4 compare our results for the flower and lawn datasets to the depth maps computed by [24] and to the results of baseline variational method [2] without any additional constraints. The user annotations are shown in the last row. In the flower and lawn datasets, the method of [24] produces smooth-looking depth maps but loses a significant amount of detail, while the baseline method [2] is able to recover more detail but suffers from edge fattening artefacts and in large areas of low texture. Our method allows the user to selectively maintain detail, for example in the leaves of the flower, while imposing smoothness in other areas of the image, such as the gravel and the bottom edge of the image. In the lawn dataset, the severe artefacts in the sky and the boundaries of the figure and bench are improved.

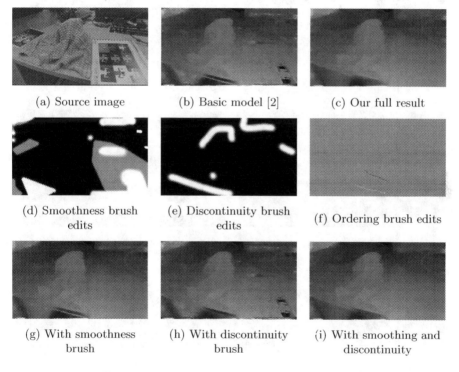

(a) Source image   (b) Basic model [2]   (c) Our full result

(d) Smoothness brush edits   (e) Discontinuity brush edits   (f) Ordering brush edits

(g) With smoothness brush   (h) With discontinuity brush   (i) With smoothing and discontinuity

Fig. 2. Our results for the desk dataset. We demonstrate that considerable improvement can be made over the basic model by adding the user to the reconstruction loop. We show our full result in (c) for comparison to the basic model result in (b). We have filled in the holes in the reconstruction and improved the quality and ordering of the discontinuities. We show the individual brush strokes in (d)–(f) and results for subsets of the brush strokes in (g)–(i).

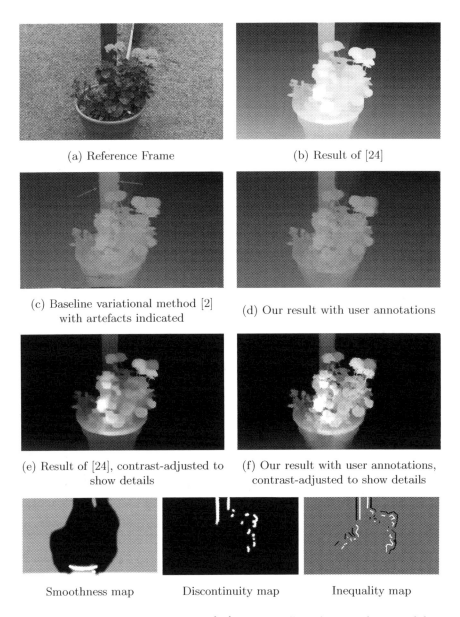

(a) Reference Frame

(b) Result of [24]

(c) Baseline variational method [2]
with artefacts indicated

(d) Our result with user annotations

(e) Result of [24], contrast-adjusted to
show details

(f) Our result with user annotations,
contrast-adjusted to show details

Smoothness map   Discontinuity map   Inequality map

Fig. 3. Results on the flower dataset of [24]. By controlling the smoothness and discontinuity terms locally, a user can obtain a smooth result while selectively maintaining fine detail, illustrated here by the detail in the leaves. The depth maps in (e) and (f) were adjusted globally with a colour-curve tool for visualisation.

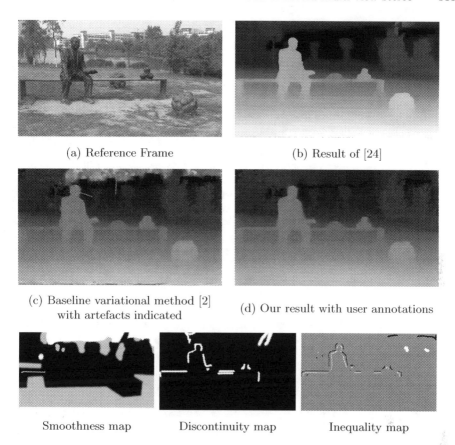

(a) Reference Frame

(b) Result of [24]

(c) Baseline variational method [2]
with artefacts indicated

(d) Our result with user annotations

Smoothness map   Discontinuity map   Inequality map

Fig. 4. Results on the lawn dataset of [24]. Both [24] and [2] fail to recover depth discontinuities in areas of low contrast between the bench and the lawn. The baseline method cannot determine depth for the textureless sky region, while [24] loses detail in the background trees and enlarges foreground objects such as the legs and sides of the bench. These problem areas are improved with user annotations.

5   Conclusion

In this paper we have shown how simple user annotations may significantly improve the quality of depth maps reconstructed from multiple images. Our method builds on state-of-the-art continuous depth recovery algorithms with the annotations modifying terms in the energy formulation, and does not require users to draw precise object boundaries or to paint absolute depth values. We believe this is particularly well-suited to mobile devices, where user input is limited to coarse strokes and image-based-rendering applications such as depth defocus are gaining popularity.

Acknowledgements. We would like to thank Anastasios Roussos and Fabio Viola for helpful discussions. This work has been supported by the EngD VEIV Centre at UCL, The Foundry and EPSRC grant EP/I031170/1.

References

1. Campbell, N.D.F., Vogiatzis, G., Hernández, C., Cipolla, R.: Using multiple hypotheses to improve depth-maps for multi-view stereo. In: Forsyth, D., Torr, P., Zisserman, A. (eds.) ECCV 2008, Part I. LNCS, vol. 5302, pp. 766–779. Springer, Heidelberg (2008)
2. Newcombe, R.A., Lovegrove, S.J., Davison, A.J.: DTAM: dense tracking and mapping in real-time. In: ICCV (2013)
3. Rhemann, C., Hosni, A., Bleyer, M., Rother, C., Gelautz, M.: Fast cost-volume filtering for visual correspondence and beyond. In: Proceedings of the 2011 IEEE Conference on Computer Vision and Pattern Recognition, CVPR 2011, pp. 3017–3024. IEEE Computer Society, Washington, DC (2011)
4. Sýkora, D., Sedlacek, D., Jinchao, S., Dingliana, J., Collins, S.: Adding depth to cartoons using sparse depth (In)equalities. Comput. Graph. Forum **29**, 615–623 (2010)
5. Guttmann, M., Wolf, L., Cohen-Or, D.: Semi-automatic stereo extraction from video footage. In: ICCV (2009)
6. Brosch, N., Rhemann, C., Gelautz, M.: Segmentation-based depth propagation in videos. In: ÖAGM/AAPR (2011)
7. Ward, B., Kang, S.B., Bennett, E.P.: Depth director: A system for adding depth to movies. IEEE Comput. Graph. Appl. **31**, 36–48 (2011)
8. Liao, M., Gao, J., Yang, R., Gong, M.: Video stereolization: Combining motion analysis with user interaction. IEEE Trans. Vis. Comput. Graph. **18**, 1079–1088 (2012)
9. Bousseau, A., Paris, S., Durand, F.: User-assisted intrinsic images. In: ACM Transactions on Graphics (TOG), vol. 28, p. 130. ACM (2009)
10. Shen, J., Yang, X., Jia, Y., Li, X.: Intrinsic images using optimization. In: 2011 IEEE Conference on Computer Vision and Pattern Recognition (CVPR), pp. 3481–3487. IEEE (2011)
11. Zhang, L.Z.L., Dugas-Phocion, G., Samson, J.S., Seitzt, S.: Single view modeling of free-form scenes. In: Proceedings of the 2001 IEEE Computer Society Conference on Computer Vision and Pattern Recognition, CVPR 2001, vol. 1 (2001)
12. Töppe, E., Oswald, M.R., Cremers, D., Rother, C.: Image-based 3d modeling via cheeger sets. In: Kimmel, R., Klette, R., Sugimoto, A. (eds.) ACCV 2010, Part I. LNCS, vol. 6492, pp. 53–64. Springer, Heidelberg (2011)
13. Chen, T., Zhu, Z., Shamir, A., Hu, S.M., Cohen-Or, D.: 3-sweep: extracting editable objects from a single photo. ACM Trans. Graph. (Proceedings of SIGGRAPH Asia 2013) **32**, Article 195 (2013)
14. Zhang, C., Price, B.: High-quality stereo video matching via user interaction and space-time propagation. In: 3DTV-Conference (2013)
15. Rother, C., Kolmogorov, V., Blake, A.: "GrabCut": Interactive foreground extraction using iterated graph cuts. In: ACM SIGGRAPH 2004 Papers, SIGGRAPH 2004, pp. 309–314. ACM, New York (2004)

16. Debevec, P.E., Taylor, C.J., Malik, J.: Modeling and rendering architecture from photographs: a hybrid geometry- and image-based approach. In: Proceedings of the 23rd Annual Conference on Computer Graphics and Interactive Techniques, SIGGRAPH 1996, pp. 11–20. ACM, New York (1996)
17. Arikan, M., Schwärzler, M., Flöry, S., Wimmer, M., Maierhofer, S.: O-snap: optimization-based snapping for modeling architecture. ACM Trans. Graph. **32**, 6:1–6:15 (2013)
18. van den Hengel, A., Dick, A., Thormählen, T., Ward, B., Torr, P.H.S.: VideoTrace: rapid interactive scene modelling from video. In: ACM Transactions on Graphics (TOG), vol. 26, p. 86. ACM (2007)
19. Chambolle, A., Pock, T.: A first-order primal-dual algorithm for convex problems with applications to imaging. J. Math. Imaging Vis. **40**, 120–145 (2011)
20. Steinbrücker, F., Pock, T., Cremers, D.: Large displacement optical flow computation withoutwarping. In: ICCV (2009)
21. Zheng, E., Dunn, E., Jojic, V., Frahm, J.M.: Patchmatch based joint view selection and depthmap estimation (2014)
22. Chambolle, A., Pock, T.: A first-order primal-dual algorithm for convex problems with applications to imaging. J. Math. Imaging Vis. **40**, 120–145 (2011)
23. Pock, T., Chambolle, A.: Diagonal preconditioning for first order primal-dual algorithms in convex optimization. In: ICCV (2011)
24. Zhang, G., Jia, J., Wong, T.T., Bao, H.: Consistent depth maps recovery from a video sequence. IEEE Trans. Pattern Anal. Mach. Intell. **31**, 974–988 (2009)
25. The Foundry: NUKEX. http://www.thefoundry.co.uk/products/nuke-product-family/nukex/ (2014)

Towards Efficient Feedback Control in Streaming Computer Vision Pipelines

Mohamed A. Helala$^{(\boxtimes)}$, Ken Q. Pu, and Faisal Z. Qureshi

Faculty of Science, University of Ontario Institute of Technology,
Oshawa, ON, Canada
{Mohamed.Helala,Ken.Pu,Faisal.Qureshi}@uoit.ca

Abstract. Stream processing is currently an active research direction in computer vision. This is due to the existence of many computer vision algorithms that can be expressed as a pipeline of operations, and the increasing demand for online systems that process image and video streams. Recently, a formal stream algebra has been proposed as an abstract framework that mathematically describes computer vision pipelines. The algebra defines a set of concurrent operators that can describe a pipeline of vision tasks, with image and video streams as operands. In this paper, we extend this algebra framework by developing a formal and abstract description of feedback control in computer vision pipelines. Feedback control allows vision pipelines to perform adaptive parameter selection, iterative optimization and performance tuning. We show how our extension can describe feedback control in the vision pipelines of two state-of-the-art techniques.

1 Introduction

Recently, there has been a rapid growth of applications capable of generating vast amounts of images and videos. Examples of such applications include online image and video sharing services (e.g., Flickr[1] and ImageNet[2]), video surveillance systems [1–4], and satellite imagery [5–7]. An emerging direction for understanding and harnessing such big visual data is *stream processing*. Stream processing represents a category of methods that process infinite sequences of data, also called data streams. In this paper, we are interested in image and video streams which we refer to as *Vision Streams*. In order to process vision streams, researchers use stream processing concepts such as pipelines to construct online computer vision algorithms [8–14]. A question that then arise is how can we formally and efficiently describe online computer vision pipelines? In database community, this question has been answered for text stream processing by developing stream algebra frameworks [15–18]. A stream algebra defines a set of abstract algebraic operators with well defined semantics that process streams as operands. These operators are used to build mathematical expressions that declaratively construct stream processing

---

[1] Flickr: https://www.flickr.com/ (*last accessed on 7 September 2014*).
[2] ImageNet: http://www.image-net.org/ (*last accessed on 7 September 2014*).

© Springer International Publishing Switzerland 2015
C.V. Jawahar and S. Shan (Eds.): ACCV 2014 Workshops, Part II, LNCS 9009, pp. 314–329, 2015.
DOI: 10.1007/978-3-319-16631-5_24

pipelines. For example, Demers *et al.* [18] proposed an algebra to express queries on event streams. Chkodrov *et al.* [15] described an implementation of a stream algebra that extends relational algebra for data streams. There are several advantages for stream algebras. For example, they can provide formal methods to resolve pipeline bottlenecks, implement dynamic reconfiguration, apply incremental evaluation, define common optimization methods, etc.

Database stream algebras are designed for structured textual streams. So, they are inapplicable for vision streams with unstructured visual content. Despite such challange, there exist some software frameworks such as OpenVL [19] and GStreamer [20], that address the efficient implementation of vision pipelines. However, these frameworks do not define a stream algebra, and lack a formal definition of vision pipelines. Recently, Helala *et al.* [21] presented a stream algebra for computer vision pipelines. This algebra revises the previous database stream algebras, and provides an abstraction for formally expressing computer vision pipelines. The algebra contains operators for both data processing and flow rate control. Two online vision algorithms have been expressed in this algebra by [21]. However, these algorithms are only for feedforward pipelines. This limits the applicability of the algebra to other online computer vision systems that use feedback control to perform tasks such as parameter tuning [22–25], and iterative optimization [14]. These tasks are widely used in online vision algorithms. For example, Sherrah [23] presented an algorithm for continuous real-time parameter tuning of a people tracking surveillance system. Supancic *et al.* [25] explored parameter tuning for long term tracking. Iterative optimization was also studied by [14] to iteratively align Flickr's photo streams.

In this paper, we are studying feedback control loops in computer vision pipelines. Specifically, we extend the stream algebra of [21] to provide an algebraic description of feedback control; here we focus on parameter tuning, and iterative optimization. Our description of feedback control is formal and abstract, which makes it reusable by several online computer vision pipelines [14,22–25]. The paper is organized as follows: Sect. 2 briefly reviews the algebra of [21], and discusses feedback control. Section 3 describes the feedback control of two state-of-the-art online computer vision algorithms. Then, we provide discussions in Sect. 4, and conclude the paper in Sect. 5.

2   Stream Algebra

The stream algebra in [21] contains three main parts: a common notation, a set of operators, and the formal semantics used to write pipeline expressions. This section gives a brief review of the algebra, and states the operators used in this paper. Finally, we discuss our algebraic extensions that provide an abstract and formal definition of feedback control in vision pipelines.

2.1   Notation

The algebra in [21] defines a data stream as an infinite sequence of data chunks with two function $\lambda x : x \rightarrow s$, and $\leftarrow s$, to write to and read from a stream s, respectively. The algebra indicates the set of all possible streams as $\mathbf{S}$. To indicate the type of the stream data, the notation $\mathbf{S}\langle T \rangle$ is used, where T signify the data type. A stream operator is defined as a mapping function $h : \mathbf{S}^m \to \mathbf{S}^n :$ $S_{\mathrm{in}}^1, \ldots, S_{\mathrm{in}}^m \to S_{\mathrm{out}}^1, \ldots, S_{\mathrm{out}}^n$, that maps m input streams to n output streams. We can define an operator by only using the following constructs as defined by [21]:

• Shared states:
  state u
  -Indicates that u is a state for subsequent loops.
• Concurrency:
  loop : *body of loop*
  -Iterates over the body forever.
  -Each loop runs in its own thread.
  -All loops of the *same* operator share the states.
  -If there are multiple concurrent loops, **loop**$_j$ indicates the j-th concurrent session.
	• Atomicity:
	  { *statements* }
	  -Executes the statements as an atomic operation.
	• Stream I/O:
	  -$x \leftarrow s$ reads from a stream s, and saves the result in x.
	  -$e \to s$ writes the expression e to stream s.
	• Attribute Access:
	  -$x.y$ accesses attribute y defined as part of variable x.

2.2 Operators

An operator can have zero or more parameters. These parameters can be simple functions, or initial values. We start by discussing the data processing operators.

Map is an operator, that *synchronously* reads from k incoming streams, applies a user-defined function $f : X_1 \times X_2 \times \ldots X_k \to Y$ on the read values, and writes the computed values to an outgoing stream. The operator is parametrized by the user-defined function. $\mathrm{MAP}(f) : \mathbf{S}\langle X_1 \rangle \times \ldots \mathbf{S}\langle X_k \rangle \to \mathbf{S}\langle Y \rangle$

$$\textbf{loop} : f(\leftarrow S_{\mathrm{in}}^1, \ldots, \leftarrow S_{\mathrm{in}}^k) \to S_{\mathrm{out}}$$

Reduce is an operator that has an internal state $u : U$. The operator reads from an incoming stream and applies a user defined function $g : U \times X \to U \times Y$. This function takes the saved state and the read value. Then, it computes a new state, and an output value. The operator updates the internal state, and writes the computed value to the output stream. The operator is parametrized by the function g, and an initial state u_0. $\mathrm{REDUCE}(g, u_0) : \mathbf{S}\langle X \rangle \to \mathbf{S}\langle Y \rangle$

$$\textbf{state}\quad u = u_0$$
$$\textbf{loop} : u, y = g(u, \leftarrow S_{\mathrm{in}})$$
$$y \to S_{\mathrm{out}}$$

Copy reads from an incoming stream, and synchronously duplicates the read value to all outgoing streams. **Copy** has no parameters. $\mathrm{COPY}() : \mathbf{S} \to \mathbf{S}^n$

$$\textbf{loop} : x \leftarrow S_{\mathrm{in}}$$
$$x \to S_{\mathrm{out}}^i \quad \text{for all } i \leq n$$

Filter has one incoming stream and two outgoing streams S_{out}^1 and S_{out}^2. It applies a used-defined predicate $\theta : X \to$ boolean, on the incoming values. It then writes values with θ true to S_{out}^1, and values with θ false to S_{out}^2. FILTER(θ) : $\mathbf{S} \to \mathbf{S}^2$

$$\textbf{loop} : x \leftarrow S_{in}$$
$$\textbf{if } \theta(x) \textbf{ then } x \to S_{out}^1 \textbf{else } x \to S_{out}^2$$

Ground ends an incoming stream. GROUND : $\mathbf{S} \to \emptyset$

$$\textbf{loop} : \leftarrow S_{in}$$

Now, we will discuss the rate controlling operators:

Latch has one incoming stream S_{in} and two outgoing streams S_{out}^1 and S_{out}^2. It reads from S_{in}, synchronously writes to S_{out}^2, and asynchronously writes to S_{out}^1. It performs the asynchronous write by saving the most-recent incoming reading, and writing it to S_{out}^1, whenever it is possible. So S_{in} and S_{out}^1 have different data rates. LATCH() : $\mathbf{S} \to \mathbf{S}^2$.

$$\textbf{state } u = \textbf{nil}$$
$$\textbf{loop}_1 : x \leftarrow S_{in} \qquad\qquad \textbf{loop}_2 : \{u \to S_{out}^1\}$$
$$\{u = x \ ; x \to S_{out}^2\}$$

Cut is similar to **Latch**, but it writes the incoming values only once to the asynchronous output stream. For the extra writes, in case S_{out}^1 has a higher data rate, **nil** is used. CUT() : $\mathbf{S} \to \mathbf{S}^2$

$$\textbf{state } u = \textbf{nil}$$
$$\textbf{loop} : x \leftarrow S_{in} \qquad\qquad \textbf{loop} : \{y = u \ ; u = \textbf{nil}\}$$
$$\{u = x \ ; x \to S_{out}^2\} \qquad\qquad y \to S_{out}^1$$

Mult has k incoming streams S_{in}^k, and one output stream S_{out}. The operator reads one value at a time from each incoming stream, forms a vector $(x_1, ..., x_k)$, and synchronously writes this vector to the outgoing stream. MULT() : $\mathbf{S}^k \to \mathbf{S}$

$$\textbf{loop} : \begin{bmatrix} \leftarrow S_{in}^1 \\ ... \\ \leftarrow S_{in}^k \end{bmatrix} \to S_{out}$$

Left-Mult is similar to **Mult**; however, it has only two incoming streams S_{in}^1 and S_{in}^2. It latches on S_{in}^2 to make the outgoing data rate depends only on the rate of S_{in}^1, and independent of S_{in}^2. LEFT-MULT : $\mathbf{S}^2 \to \mathbf{S}$

$$S^1, \ S^2 = \text{LATCH}(S_{in}^2) \ ; \ \text{GROUND}(S^2)$$
$$\textbf{loop} : \begin{bmatrix} \leftarrow S_{in}^1 \\ \leftarrow S^1 \end{bmatrix} \to S_{out}$$

Add asynchronously merges together k incoming streams S_{in}^k into one outgoing stream. ADD $: \mathbf{S}^k \rightarrow \mathbf{S}$.

$$\text{for } j \leq k$$
$$\mathbf{loop}_j : x \leftarrow S_{\text{in}}^j$$
$$x \rightarrow S_{\text{out}}$$

2.3 Feedback Control

Feedback control is an essential task in several online computer vision algorithms [14,22–25] that perform parameter tuning, or iterative optimization. These algorithms evaluate the current output results to enhance the future outputs. In this section, we extend the stream algebra of [21] by providing a formal and abstract description of feedback control in computer vision pipelines.

We assume an input stream I_1, a vision pipeline with a sequence of operators $X_1, ..., X_k$, and an output stream O. Each operator X_j has an input stream I_j, and an output stream I_{j+1}. In order to define feedback control for operator X_j, we assume a return stream R_j, a sequence of evaluation operators $E_1^j, ..., E_{k_j}^j$, and a feedback stream F_j. Given these assumptions; we describe the feedback loop in a vision pipeline as in Fig. 1. We discuss two types of feedback control, single-point and multi-point. In single-point, we only control one operator X_j. So, the pipeline has one return stream R, and one feedback stream F. In order to obtain the return stream R, we can apply either the COPY, FILTER, or CUT operator on the output stream O. This is represented by the equation,

$$R, O' \triangleq \text{COPY}()(O). \tag{1}$$

O' now represents the output. We can next apply a sequence of evaluation operators $E_1, ..., E_n$ on R to compute the feedback stream F. If the feedback loop performs parameter tuning, then F represents a stream of new parameters. We can then apply a LEFT-MULT operator to attach the new parameters to the input stream I_j, which updates the internal parameters of the pipeline operator X_j. This is represented by the equation,

$$I_j' \triangleq \text{LEFT-MULT}()(I_j, F). \tag{2}$$

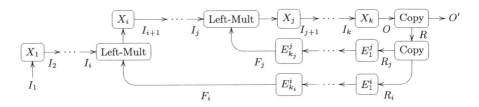

Fig. 1. An example of multi-point feedback control. Arrows show stream flow directions. Letters on arrows represent stream names. I_1, and O' indicate input and output streams, respectively.

I'_j in this case represents the input to operator X_j. Note that, the feedback loop that we just defined is synchronized with the original vision pipeline. If we replace COPY in Eq. 1 by CUT, then we have an asynchronous feedback loop that evaluates samples from the output stream O. Furthermore, in case of iterative optimization that reprocesses the output, we can replace LEFT-MULT in Eq. 2 by an ADD operator. In this case, the stream I'_j will have interleaved elements from the streams I_j and F.

In multi-point feedback control, we can control more than one operator. This is performed by applying COPY on the return stream R several times to get m duplicate streams, which we use to define the feedback loops of m operators. Figure 1 shows multi-point feedback for two operators X_i and X_j. Each operator has a distinct set of evaluation operators to construct its feedback stream.

3   Feedback Control in Computer Vision Pipelines

There is a large interest in developing online computer vision algorithms that use feedback control to perform parameter tuning or iterative optimization tasks. These tasks allow an online algorithm to adapt itself continuously to different scene contexts, or iteratively improve output results over time. In this section, we discuss two state-of-the-art algorithms [14,24] that process vision streams and apply feedback control to perform parameter tuning and iterative optimization. Without loss of generality, we will discuss how we can effectively express the feedback control of these algorithms using our algebraic extensions.

3.1   Online Adaptation of Tracking Parameters

Online tracking of moving objects is one of the fundamental problems in computer vision, and several trackers have been proposed. However, if the input video stream has an unknown scene, it becomes difficult to select a tracking algorithm. Chau et al. [24] proposed a method for online parameter tuning that can adapt a tracking algorithm to scene changes. This method works in two phases: an offline training phase, and an online control phase.

In the offline training phase, the algorithm takes as input, a set of training videos together with annotated moving objects, annotated trajectories, and a tracking algorithm with parameters. The method tracks people as the moving objects using the appearance based tracking algorithm of [26], which is controlled by six parameters. For each training video, the algorithm extracts context features from every frame. The context features are a vector of six elements that describes the density of moving objects, their occlusion level, and appearance characteristics. The algorithm then segments the video into consecutive clips, based on similarity of the context feature vectors. For each clip, the algorithm performs parameter optimization to select the best parameter values of the given tracking algorithm. Then, a clustering step is performed to cluster contexts from all training videos. Afterwards, the best tracking parameters are defined for each cluster. Finally, the context clusters together with their best parameters are stored in a database D.

In the online stage, the algorithm takes as input, a video stream $V = \{V_i | i = 0, 1, 2...\}$. For every frame $V_i \in V$, people is detected using the HOG-based detection algorithm in [27]. This generates the objects stream $B = \{B_i | i = 0, 1, 2...\}$, where every B_i represents a list of detected objects (people) in frame V_i. The algorithm then records the detected objects from every frame in a temporal window of interval $\triangle t_1$. After that, the method attaches with every frame $V_i \in V$, the recent list of recorded objects up to frame V_i. This generates the stream $U = \{U_i | i = 0, 1, 2, ...\}$, where $U_i = (V_i, B_i)$. The method also defines the parameters feedback stream $F = \{F_i | i = 0, 1, 2, ...\}$, where every F_i is a vector of tracking parameters. We will see later how this stream is generated. Every F_i is attached to U_i to generate the tracker input stream $Q = \{Q_i | i = 0, 1, 2, ...\}$, where $Q_i = (V_i, B_i, F_i)$. The tracking algorithm takes the stream Q as input and updates the tracking parameters using $F_i \in Q_i$. It also generates a list of trajectories T_i for every frame. The tracker attaches the trajectories T_i to $(V_i, B_i) \subset Q_i$, and constructs the output trajectories stream $J = \{J_i | i = 0, 1, 2, ...\}$, where $J_i = (V_i, B_i, T_i)$. The algorithm in [24], then defines a feedback control loop that takes the output stream J and records the video frames $V_i \in J_i$ in a temporal window of interval $\triangle t_2$. This defines a clip stream $C = \{C_i | i = 0, 1, 2, ...\}$. This stream is used with J to define the loopback stream $H = \{H_i | i = 0, 1, 2, ...\}$, where $H_i = (V_i, B_i, T_i, C_i)$. Now, for every H_i, the algorithm uses $(V_i, B_i, T_i) \subset H_i$ to calculate two error scores: An object interaction score s_1 to calculate the overlap of objects, and a tracking error score s_2 to measure the tracking quality. Given two thresholds Th_1 and Th_2, if $s_1 > Th_1$ and $s_2 > Th_2$, the algorithm declares an error and retrieves the best tracking parameters suitable to the context defined by the clip $C_i \in H_i$, from the database D. Otherwise, we continue using the current parameters. This generates the feedback stream F that was used previously together with the stream U to define the tracker input stream Q. This illustrates the feedback control loop of [24].

Now, we will describe the online control stage of [24] in the algebra of [21], and use our algebraic extensions to define the feedback control loop. We start by defining the following data types,

$$\texttt{Frame} : \texttt{2DImage}; \quad \texttt{Video} : \textbf{S} \langle \texttt{Frame} \rangle; \quad \texttt{Clip} : \textsc{List} \langle \texttt{Frame} \rangle$$
$$\texttt{Histogram} : \textsc{List} \langle \mathbb{R} \rangle; \quad \texttt{Object} : \mathbb{R}^8 \times \texttt{Histogram}; \quad \texttt{Params} : \mathbb{R}^6$$
$$\texttt{FrameInfo} : \texttt{Frame} \times \textsc{List} \langle \texttt{Object} \rangle; \quad \texttt{Trajectory} : \textsc{List} \langle \mathbb{R}^2 \rangle;$$
$$\texttt{TrackInput} : \texttt{Frame} \times \textsc{List} \langle \texttt{Object} \rangle \times \texttt{Params}$$
$$\texttt{TrackInfo} : \texttt{Frame} \times \textsc{List} \langle \texttt{Object} \rangle \times \textsc{List} \langle \texttt{Trajectory} \rangle$$
$$\texttt{LoopBack} : \texttt{Frame} \times \textsc{List} \langle \texttt{Object} \rangle \times \textsc{List} \langle \texttt{Trajectory} \rangle \times \texttt{Clip}$$

where a `Frame` is a 2D image, a `Video` is a stream of frames, a `Clip` is a list of frames, and a `Histogram` is a vector of values. An `Object` is a pair (a, b), where b : `Histogram`, and a : $\mathbb{R}^8$ is a vector that represents the following object features: 2D shape ratio, 2D area, color covariance in RGB, and dominant color in RGB. `Params` is a vector that represents the 6 parameters of the tracking algorithm [26]. `FrameInfo` is a pair (v, w), where v : `Frame` and w : `LIST` $\langle$`Object`$\rangle$. A `TrackInput` is a 3 elements vector (v, w, p), where

p : Params. A Trajectory is a list of 2D points. TrackInfo is a 3 elements vector (v, w, e), where e : LIST $\langle$Trajectory$\rangle$. Finally a LoopBack is a 4 elements vector (v, w, e, c), where c : Clip.

Given an input video stream $V \in$ Video, we copy V into two identical streams V_1, V_2 using a COPY operator,

$$V_1, V_2 \triangleq \text{COPY}()(V). \tag{3}$$

We will now process V_1 and return later to discuss the use of V_2. We define a function f_1 : Frame $\rightarrow$ LIST $\langle$Object$\rangle$ that detects objects from every frame $V_i \in V_1$ using the HOG-based detection algorithm of [27]. This function can be used with a MAP operator to define the objects stream B : $\mathbf{S}$ $\langle$LIST $\langle$Object$\rangle\rangle$,

$$B \triangleq \text{MAP}(f_1)(V_1). \tag{4}$$

We then define the function,

g_1 : LIST $\langle$Object$\rangle$ $\times$ LIST $\langle$Object$\rangle$ $\rightarrow$ LIST $\langle$Object$\rangle$ $\times$ LIST $\langle$Object$\rangle$

$g_1(u, x) = \{$ for all $z \in u$

if $(\text{now}() - \text{arrival-time}(z) \geq \triangle t_1)$ then

$u = u \ominus z$ //remove z from u

$u = u \oplus x$ //append x to u

return(u, u) $\}$

This function maintains a window u of the most recent objects, in a time interval $\triangle t_1$. The function starts by deleting old objects. Then, it adds the new objects x to u, and returns the updated window as output. We can use the g_1 function together with a REDUCE operator to generate the objects summary stream M : S $\langle$LIST $\langle$Object$\rangle\rangle$,

$$M \triangleq \text{REDUCE}(g_1, \text{Empty-List})(B). \tag{5}$$

Now, we will go back and use the stream V_2. Note that this stream is a copy of the input video stream V. We synchronize the stream M with the stream V_2 using a MULT operator to generate the stream U : S $\langle$FrameInfo$\rangle$,

$$U \triangleq \text{MULT}()(V_2, M). \tag{6}$$

At this step, we define the feedback stream F : S $\langle$Params$\rangle$. We will show later how we generate this stream when we discuss feedback control. We synchronize the stream F with the stream U using a LEFT-MULT operator to generate the stream Q : S $\langle$TrackInput$\rangle$,

$$Q \triangleq \text{LEFT-MULT}()(U, F). \tag{7}$$

We then define the tracking function f_2 : TrackInput $\rightarrow$ TrackInfo. This function takes the vector (v, w, p) : TrackInput, and applies the tracking algorithm on v and w using the given parameters p. The function then outputs the object trajectories which we attach to the pair (v, w), to generate the output y : TrackInfo. We can use this function together with a MAP operator to process the Q stream and generate the trajectories stream $J : S \langle \text{TrackInfo} \rangle$,

$$J \triangleq \text{MAP}(f_2)(Q). \tag{8}$$

Now, we discuss how we use our algebraic description to express the feedback control of [24]. We start by evaluating the equation, $R, J' \triangleq \text{COPY}()(J)$, to copy J into the output stream J' and the return stream R. Then, we define the function,

$g_2 : \text{Clip} \times \text{TrackInfo} \rightarrow \text{Clip} \times \text{LoopBack}$

$$
\begin{aligned}
g_2(u, x) = \{ \ &\text{for all} \ z \in u \\
&\text{if } (\text{now}() - \text{arrival-time}(z) \geq \triangle t_2) \text{ then} \\
&\quad u = u \ominus z \quad //\text{remove } z \text{ from } u \\
&\quad u = u \oplus x.v \quad //\text{append frame } x.v \text{ to } u \\
&\quad y = x \oplus u \quad //\text{append clip } u \text{ to } x \\
&\text{return}(u, y) \ \}
\end{aligned}
$$

The function g_2 maintains a clip u that has the most recent frames in a time interval $\triangle t_2$. It is similar to the function g_1, however g_2 appends the recorded clip to the input x, to form the output y : LoopBack. We use this function with a REDUCE operator to process the return stream R, and generate the loopback stream $H : S \langle \text{LoopBack} \rangle$,

$$H \triangleq \text{REDUCE}(g_2, \text{Empty-List})(R). \tag{9}$$

The algorithm in [24] defines two scoring functions to calculate the tracking errors. These functions are $f_3 : \text{FrameInfo} \rightarrow \mathbb{R}$ for object interaction score, and $f_4 : \text{TrackInfo} \rightarrow \mathbb{R}$ for tracking error score. We use these functions to define another function,

$g_3 : \text{Params} \times \text{LoopBack} \rightarrow \text{Params} \times \text{Params}$

$$
\begin{aligned}
g_3(u, x) = \{ \ &s_1 = f_3(x.v, x.w) \\
&s_2 = f_4(x.v, x.w, x.e) \\
&\text{if } (s_1 > Th_1 \text{ and } s_2 > Th_2) \text{ then} \\
&\quad p = \text{search-db}(x.c, D) \\
&\quad \text{return}(p, p) \\
&\text{return}(u, u) \ \}
\end{aligned}
$$

Note that this function starts by calculating the error scores of the current frame. If the scores s_1 and s_2 are larger than the given thresholds, we search the database D for the best parameters that meet the context of the most recent clip $x.c$, and output the new parameters. Otherwise, we return the old parameters u.

This function can be used together with a REDUCE operator to process the loopback stream H, and generate the feedback stream $F : S \langle \mathtt{Params} \rangle$,

$$F \triangleq \text{REDUCE}(g_3, \text{Initial-Params})(H). \tag{10}$$

Remember that we used the stream F together with the stream U as input to the LEFT-MULT operator in Eq. 7 to define the tracker input stream Q. In this way, we use our definition of single-point feedback control to continuously update the tracker parameters.

3.2 Iterative Optimization for Aligning Photo Streams

Recently, there has been a large interest in the analysis of Web photo streams. This interest is driven by the existence of several photo hosting websites that contain huge amounts of personal photo collections. Kim et al. [14] proposed a recent algorithm to build common storylines from Flickr's photo streams. Their algorithm takes as input, a set of n photo streams $I = \{I_k | k = 1, 2, 3..n\}$ from different Flickr users that share a common user activity. Each photo in a stream I_k stores its capture time, a spatial pyramid histogram as a visual descriptor, and a set of foreground regions that is initially empty. Every stream I_k is divided into a sequence of photo blocks $B_k = \{B_{\{k,i\}} | i = 0, 1, 2...\}$, where each block $B_{\{k,i\}}$ represents the photos taken by a user over a certain period of time $\triangle t_1$ (for example, a day). Each block also stores the earliest and latest capture times of its photos. The algorithm then iterates between two tasks, an alignment task, and a cosegmentation task.

In the alignment task, the algorithm starts by reading one block from each stream, and constructs the block-list stream $L = \{L_i | i = 0, 1, 2...\}$. Every $L_i \in L$ is a list of blocks. Then, the algorithm selects from every list L_i, a set of blocks E_i that overlap in a timeline by at least a period $\triangle t_2$ (for example, an hour). We also attach to E_i, an iteration number N, that is initially zero. This generates the overlapped blocks stream $Q = \{Q_i | i = 0, 1, 2...\}$, where $Q_i = (E_i, N_i)$. The algorithm defines a feedback stream F which is of the same type as Q. We will see later how this stream is constructed. We add the elements of the two streams Q and F together to construct the interleaved stream $P = \{P_j | j = 0, 1, 2...\}$, where $P_j = (E_j, N_j)$. Note that the stream P contains interleaved elements from the two streams Q and H. For each block b of $E_j \in P_j$, the algorithm calculates the similarity of b to other blocks in E_j. This is performed for two blocks $(b_1, b_2) \subset E_j$ by first finding for each photo $x \in b_1$, its best visually similar neighbor $y \in b_2$. This similarity is calculated using a distance function f_4 : Photo $\times$ Photo $\rightarrow \mathbb{R}$. If both x and y have foreground regions defined, f_4 returns the distance between the histograms of these regions. Otherwise, f_4 returns the distance between the spatial pyramid histograms of photos. The algorithm also defines another distance function f_5 : Time $\times$ Time $\rightarrow \mathbb{R}$, that calculates the difference between the capture times of two photos. The algorithm then measures the similarity between two blocks b_1 and b_2 using an energy function f_6 : Block $\times$ Block $\rightarrow \mathbb{R}$, that sums the similarity defined by f_4 and f_5

along the correspondent photos between b_1 and b_2. The method then maps every list of blocks $E_j \in P_j$ into a graph G_j : Graph $\langle$Block, Block $\times$ Block$\rangle$. This graph has the blocks of E_j as vertices. An edge exists between two blocks, if they are best similar to each other, in other words, they have the minimal distance to each other according to f_6. This generates the blocks-graph stream $Z = \{Z_j | j = 0, 1, 2...\}$, where $Z_j = (G_j, N_j)$. This stream defines the output of the alignment step.

The cosegmentation step takes as input, the blocks-graph stream Z. This step maps every graph $G_j \in Z_j$ into a new graph Y_j : Graph $\langle$Photo, Photo $\times$ Photo$\rangle$. This is performed by first collecting all photos from the blocks of G_j. These photos define the vertices of Y_j. Then, the algorithm adds an edge between two photos (x, y) if they are a correspondent pair of two different blocks. In addition, for every photo x in a block b, the method adds edges to its k nearest neighbors from the same block b. We append the graph Y_j to Z_j to construct the photos-graph stream $M = \{M_j | j = 0, 1, 2...\}$, where $M_j = (G_j, N_j, Y_j)$. We also increment N_j by 1. Afterwards, the algorithm in [28] defines for each photo in the vertices of $Y_j \in M_j$, a set of m foreground regions. This is performed by applying the cosegmentation technique of [28] and belief propagation between each photo and its neighbors in Y_j. If a photo already have foreground regions defined, then they are enhanced.

The algorithm of [14] iterates between the alignment task and the cosegmentation task. This is achieved by defining a feedback loop that first filters the photos-graph stream M based on the iteration number $N_j \in M_j$. If $N_j \geq N_{stop}$ then M_j is sent to the output, Otherwise, it is sent to the feedback loop. N_{stop} defines the maximum number of iterations. This defines two streams M' and R with similar type to stream M, where M' is the output stream, and R is the return stream. The stream R is then mapped to the feedback stream $F = \{F_l | l = 0, 1, 2...\}$, where $F_l = (E_l, N_l)$. E_l is the list of blocks in the vertices of $G_l \in R_l$ and N_l is the iteration number. Remember that the feedback stream F was added to the stream Q to define the interleaved stream P in the alignment step. Note also that the block photos in F have foreground regions defined. These regions enhance the matching of the feedback blocks in the alignment step. Consequently, This improves the output of the cosegmentation step, which closes the feedback loop defined by [14] for iterative optimization.

Now, we will express the iterative optimization of [24] using our algebraic extensions for feedback control. We define the following data types,

Shape : LIST $\langle \mathbb{R}^2 \rangle$;   Region : Shape $\times$ Histogram
Photo : 2DImage $\times$ Time $\times$ Histogram $\times$ LIST $\langle$Region$\rangle \times \mathbb{R}$
Block : LIST $\langle$Photo$\rangle \times$ Time^2;   BlocksInfo : LIST $\langle$Block$\rangle \times \mathbb{R}$;
BlocksGraphInfo : GRAPH $\langle$Block, Block $\times$ Block$\rangle \times \mathbb{R}$
PhotosGraphInfo : BlocksGraphInfo $\times$ GRAPH $\langle$Photo, Photo $\times$ Photo$\rangle$

where, a Shape is a list of 2D points. A Region is a 2D vector (s, h), where s : Shape, and h : Histogram is the region descriptor. We reuse the definition of Histogram from the previous section. A Photo is a vector of five variables

(a, t, h, r, nn), where a : 2DImage, t : $Time$ is the capture time, h : Histogram is a visual descriptor, r : LIST $\langle$Region$\rangle$ is a list of foreground regions (initially empty), and nn is the index of the nearest neighbor photo. A Block is a 3D vector (b, t_1, t_2), where b : LIST $\langle$Photo$\rangle$, t_1 : $Time$ is the earliest capture time of photos in b, and t_2 is the latest capture time. A BlocksInfo is a vector (w, itr), where w : LIST $\langle$Block$\rangle$, and itr : $\mathbb{R}$ indicates the iteration number. A BlocksGraphInfo is a vector (c_1, itr), where c_1 is a graph on a set of blocks. Finally, PhotosGraphInfo is a 2D vector (q, c_2), where q : BlocksGraphInfo, and c_2 is a graph on a set of photos.

For simplicity, we will consider that we have three input photo streams $I = \{I_k | k = 1..3\}$. We now define the following function,

g_4 : Block $\times$ Photo $\to$ Block $\times$ Block

$$g_4(u, x) = \{ \text{ if duration}(u) \geq \triangle t_1 \text{ then}$$
$$u' = \emptyset; y = u$$
$$\text{else}$$
$$u' = u \oplus x \quad //\text{append } x \text{ to Block } u$$
$$y = \emptyset$$
$$\text{return}(u', y) \quad \}$$

This function keeps appending the incoming photos to a block u' and setting the output y to an empty block. When the duration of the block $\triangle t = u'.t_2 - u'.t_1$ is larger than a certain interval $\triangle t_1$ (A day in [14]), the function copies the block u' to the output y and resets u' back to an empty block. We can use this function together with a REDUCE and FILTER operators to process every stream $I_k \in I$ and generate a corresponding stream B_k : $S \langle$Block$\rangle$,

$$B_k, E_k \triangleq \text{FILTER}(\lambda \, x : \, |x| \neq 0) \circ \text{REDUCE}(g_4, \text{Empty-List})(I_k). \tag{11}$$
$$\text{GROUND}()(E_k). \tag{12}$$

where the $\circ$ operator is defined by [21] as a composition operator, that supplies the output stream from the right operand as an input stream to the left operand. The FILTER operator removes the empty blocks from the output stream of the REDUCE operator. These empty blocks define the stream E_k which is ignored by a GROUND operator. We can now synchronize the three streams B_1, B_2 and B_3 using a MULT operator to construct the block-list stream L : $S \langle$LIST $\langle$Block$\rangle\rangle$,

$$L \triangleq \text{MULT}()(B_1, B_2, B_3). \tag{13}$$

We then define the function f_7 : LIST $\langle$Block$\rangle \to$ BlocksInfo. This function selects from every list $L_i \in L$, blocks that overlap in a timeline by at least a period $\triangle t_2$ (An hour in [14]). The function also attaches to the list of selected blocks, an iteration number, which is initially zero. We can use f_7 together with a MAP operator to define the stream Q : $S \langle$BlocksInfo$\rangle$,

$$Q \triangleq \text{MAP}(f_7)(L). \tag{14}$$

At this step, we define the feedback stream $F : S \langle \texttt{BlocksInfo} \rangle$. We will show later how we generate this stream when we discuss the feedback loop. We add the two streams Q and F using an ADD operator to generate the stream $P : S \langle \texttt{LIST} \langle \texttt{Block} \rangle \rangle$,

$$P \triangleq \text{ADD}()(Q, F). \tag{15}$$

Now, we define two functions $f_8 : \texttt{BlocksInfo} \rightarrow \texttt{BlocksGraphInfo}$, and $f_9 : \texttt{BlocksGraphInfo} \rightarrow \texttt{PhotosGraphInfo}$. The function f_8 maps a blocks list in $\texttt{BlocksInfo}$ into a graph of blocks, where an edge exists between two blocks if they are nearest neighbors according to the function f_6. Note that f_6 gives better similarity distance if block photos have good foreground regions. The function f_9, on the other hand, maps a graph of blocks into a graph of photos, and increments the iteration number by 1. Cosegmentation is applied between each photo and its neighbors in the photos graph to define foreground regions or enhance existing ones. We can use the two functions f_8 and f_9 together with two MAP operators to define the photos-graph stream $M : S \langle \texttt{PhotosGraphInfo} \rangle$,

$$M \triangleq \text{MAP}(f_9) \circ \text{MAP}(f_8)(P). \tag{16}$$

Now, we will discuss the feedback control of [14] to perform iterative optimization using our algebraic extensions. We start by applying a FILTER operator on the stream M to define the output stream M' and the return stream R,

$$M', R \triangleq \text{FILTER}(\lambda\, x :\ x.q.itr \geq N_{stop})(M). \tag{17}$$

We access the attribute $q : \texttt{BlocksGraphInfo}$ from each element $M_j \in M$. Then, we test if the iteration number of q reached the maximum number of iterations. If the test is true M_j is sent to M', if not, it is sent to R. We then define a function $f_{10} : \texttt{PhotosGraphInfo} \rightarrow \texttt{BlocksInfo}$ which maps every element in R into an element of type $\texttt{BlocksInfo}$. This is performed by converting the vertices of the blocks-graph attached to $\texttt{PhotosGraphInfo}$ into a list of blocks, and copying the current iteration number. The function f_{10} together with a MAP operator define the feedback stream $F : S \langle \texttt{BlocksInfo} \rangle$,

$$F \triangleq \text{MAP}(f_{10})(M_2). \tag{18}$$

Remember that we used the stream F together with the stream Q as input to the ADD operator in Eq. 15 to define the stream P. This shows how our definition of single-point feedback control describes iterative optimization in [14].

4   Discussions

The examples discussed in the previous section, demonstrate that we can effectively express feedback control in online computer vision algorithms using our abstract algebraic description presented in Sect. 2.3. In each discussed algorithm, we first express its vision pipeline by a set of algebraic equations based on the operators defined by [21]. Then, we use our abstract definition of single-point feedback control to describe both the parameter tuning and the iterative optimization tasks. For example, the algebraic description of the online tracking example [24] shows a flexible integration of our single-point feedback to tune tracking parameters. Additionally, We can easily extend this single-point feedback control to multi-point feedback control (Fig. 1). We can use this, for example, to control both the parameters of tracker and the parameters of the HOG-based people detector algorithm [27] used by [24] to detect moving objects. So, our multi-point description can scale up feedback to control several operators of a vision pipeline. Furthermore, if the vision pipeline has several output streams, we may extend our multi-point feedback to process multiple return streams. In this case, we can use rate control operators such as CUT and LATCH to replace COPY in Fig. 1, and define asynchronous feedback control that does not affect the flow rate of the feedforward pipeline. This also motivates that we may perform multi-point feedback between different computer vision pipelines, that work at separate data rates, without affecting one another.

The discussed iterative optimization example [14] suggests that our formal feedback description may be useful for other tasks such as adaptive learning and incremental evaluation. In addition, our feedback description may be used for implementing blocking resolution in vision pipelines. For example, we can make every pipeline operator attach its estimated runtime to the output stream. Then, these runtimes can be collectively monitored by feedback control to send back appropriate actions to the operators. So, our formal feedback description opens new directions for tasks such as real-time debugging, performance monitoring and bottleneck identification.

5   Conclusion

This paper develops abstract and formal methods for describing and implementing feedback control in computer vision pipelines—online vision algorithms that process images and videos (vision streams). These formal methods build upon an existing stream algebra to flexibly integrate feedback control to vision pipelines. We have demonstrated our formal methods for two state-of-the-art online computer vision algorithms that implement feedback control for parameter tuning and iterative optimization. We also discussed how our methods can scale up feedback to control different stream operators of the vision pipeline. Our work opens up new research directions to study real-time debugging, dynamic reconfiguration, bottleneck identification, adaptive learning, and performance tuning of vision pipelines.

References

1. Zhao, B., Fei-Fei, L., Xing, E.: Online detection of unusual events in videos via dynamic sparse coding. In: CVPR, Colorado Springs, pp. 3313–3320 (2011)
2. Helala, M., Pu, K., Qureshi, F.: Road boundary detection in challenging scenarios. In: AVSS, pp. 428–433 (2012)

3. Meghdadi, A., Irani, P.: Interactive exploration of surveillance video through action shot summarization and trajectory visualization. IEEE Trans. Vis. Comput. Graph. **19**, 2119–2128 (2013)

4. Yenikaya, S., Yenikaya, G., Düven, E.: Keeping the vehicle on the road: A survey on on-road lane detection systems. ACM Comput. Surv. **46**, 1–2 (2013)

5. Ozcanli, O., Dong, Y., Mundy, J., Webb, H., Hammoud, R., Victor, T.: Automatic geo-location correction of satellite imagery. In: IEEE CVPR Workshops (2014)

6. Wischounig-Strucl, D., Quartisch, M., Rinner, B.: Prioritized data transmission in airborne camera networks for wide area surveillance and image mosaicking. In: IEEE CVPR Workshops, pp. 17–24 (2011)

7. Yuping, L., Medioni, G.: Map-enhanced uav image sequence registration and synchronization of multiple image sequences. In: IEEE CVPR, Minneapolis, Minnesota, USA, pp. 1–7 (2007)

8. Ryoo, M.S.: Human activity prediction: Early recognition of ongoing activities from streaming videos. In: ICCV, Barcelona, Spain, pp. 1036–1043 (2011)

9. Lu, C., Shi, J., Jia, J.: Online robust dictionary learning. In: IEEE CVPR, pp. 415–422 (2013)

10. Xu, C., Xiong, C., Corso, J.J.: Streaming hierarchical video segmentation. In: Fitzgibbon, A., Lazebnik, S., Perona, P., Sato, Y., Schmid, C. (eds.) ECCV 2012, Part VI. LNCS, vol. 7577, pp. 626–639. Springer, Heidelberg (2012)

11. Loy, C., Hospedales, T., Xiang, T., Gong, S.: Stream-based joint exploration-exploitation active learning. In: CVPR, pp. 1560–1567 (2012)

12. Al Harbi, N., Gotoh, Y.: Spatio-temporal human body segmentation from video stream. In: Wilson, R., Hancock, E., Bors, A., Smith, W. (eds.) CAIP 2013, Part I. LNCS, vol. 8047, pp. 78–85. Springer, Heidelberg (2013)

13. Yang, J., Luo, J., Yu, J., Huang, T.: Photo stream alignment and summarization for collaborative photo collection and sharing. IEEE Trans. Multimedia **14**, 1642–1651 (2012)

14. Kim, G., Xing, E.: Jointly aligning and segmenting multiple web photo streams for the inference of collective photo storylines. In: CVPR, pp. 620–627 (2013)

15. Chkodrov, G., Ringseth, P., Tarnavski, T., Shen, A., Barga, R., Goldstein, J.: Implementation of stream algebra over class instances, Google patents (2013)

16. Broy, M., Stefanescu, G.: The algebra of stream processing functions. Theoret. Comput. Sci. **258**, 99–129 (2001)

17. Carlson, J., Lisper, B.: An event detection algebra for reactive systems. In: Proceedings of the 4th ACM International Conference on Embedded Software, pp. 147–154 (2004)

18. Demers, A., Gehrke, J., Hong, M., Riedewald, M., White, W.: A general algebra and implementation for monitoring event streams. Cornell University, Technical report (2005)

19. Shen, C., Little, J., Fels, S.: Towards OpenVL: Improving real-time performance of computer vision applications. In: Kisačanin, B., Bhattacharyya, S.S., Chai, S. (eds.) Embedded Computer Vision. Advances in Pattern Recognition, pp. 195–216. Springer, London (2009)

20. GStreamer. http://gstreamer.freedesktop.org (2014). Accessed: 26 January 2014

21. Helala, M.A., Pu, K.Q., Qureshi, F.Z.: A stream algebra for computer vision pipelines. In: IEEE CVPR Workshops (2014)

22. Kisilev, P., Freedman, D.: Parameter tuning by pairwise preferences. In: BMVC (2010)

23. Sherrah, J.: Learning to adapt: a method for automatic tuning of algorithm parameters. In: Blanc-Talon, J., Bone, D., Philips, W., Popescu, D., Scheunders, P. (eds.) ACIVS 2010, Part I. LNCS, vol. 6474, pp. 414–425. Springer, Heidelberg (2010)
24. Chau, D., Badie, J., Bremond, F., Thonnat, M.: Online tracking parameter adaptation based on evaluation. In: IEEE International Conference on AVSS, pp. 189–194 (2013)
25. III, J.S., Ramanan, D.: Self-paced learning for long-term tracking. In: CVPR, Washington, DC, USA, pp. 2379–2386 (2013)
26. Chau, D., Bremond, F., Thonnat, M.: A multi-feature tracking algorithm enabling adaptation to context variations. In: ICDP 2011, pp. 1–6 (2011)
27. Corvee, E., Bremond, F.: Body parts detection for people tracking using trees of histogram of oriented gradient descriptors. In: IEEE International Conference on AVSS, pp. 469–475 (2010)
28. Kim, G., Xing, E.: On multiple foreground cosegmentation. In: IEEE CVPR, pp. 837–844 (2012)

International Workshop on Video Segmentation in Computer Vision

Background Subtraction: Model-Sharing Strategy Based on Temporal Variation Analysis

Yufeng Chen, Kun Zhao[✉], Wenzhe Wu, and Shikai Liu

Beijing Institute of Technology, Zhongguancun Road, Haidian District,
Beijing, People's Republic of China
zhkflame@163.com

Abstract. This paper presents a new approach for moving detection in complex scenes. Different with previous methods which compare a pixel with its own model and make the model more complex, we take an iterative model-sharing strategy as the process of foreground decision. The current pixel is not only compared with its own model, but may also compared with other pixel's model which has similar temporal variation. Experiments show that the proposed approach leads to a lower false positive rate and higher precision. It has a better performance when compared with traditional approach.

1 Introduction

For many computer vision applications, detecting moving objects from a video sequence is a critical and fundamental part. A common approach to detect moving objects is background subtraction. Background subtraction have been widely used for video surveillance, traffic management, tracking and many other applications.

The purpose of a background subtraction algorithm is to distinguish moving objects from the background. A good background model should be able to achieve some desirable properties: accurate in shape detection, reliable in different light conditions, flexible to different scenarios, robust to different models of background, efficient in computation [1]. In developing a good background subtraction algorithm, there are many challenges such as dynamic scenes, illumination changes, shadows and other limitations. Researchers have been devoted to develop new methods and techniques to overcome different limitations. Different methods have been proposed to solve difference issues and challenges.

In order to deal with complex scenes, most traditional methods take the strategy of making each pixel's model contain more information. This strategy may lead to the model of each pixel too complex, and still not contain enough information because of the random movement. These methods all compare the current pixel with its own models. We take another strategy that the current pixel may compared with other pixels model which has similar temporal variation. In order to determine whether they are similar to each other, we use stability which is get by wavelet transform to quantify the pixel's variation over

© Springer International Publishing Switzerland 2015
C.V. Jawahar and S. Shan (Eds.): ACCV 2014 Workshops, Part II, LNCS 9009, pp. 333–343, 2015.
DOI: 10.1007/978-3-319-16631-5_25

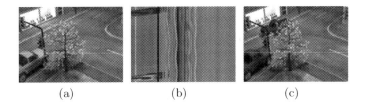

(a)   (b)   (c)

Fig. 1. Temporal variation of a chosen line

time. Figure 1 shows the temporal variation of a chosen line which is marked
as white. Figure 1(a) is the frame at time a and Fig. 1(b) is the frame at time
b. Figure 1(c) represents the variation of the chosen line from time a to b. We
can see different variations in different areas. The dynamic background such as
waving trees changed more obviously and this means their variation is less stable.

In this paper, we present a new approach which use model-sharing strategy
based on temporal variation analysis. In Sect. 2, we review the previous work
related to background subtraction. Section 3 describes our methods in detail.
Section 4 discusses experimental results including comparison with other algorithms.

2   Related Works

Numerous methods for detecting moving objects have been proposed in the past.
The most direct method to identify moving objects is using frame differences
which focus on the changes between two frames. Large changes are considered
as foreground. W^4 [2] model use the variance found in a set of background images
with the maximum and minimum intensity value and the maximum difference
between consecutive frames. A very popular method to model each pixel is to
use the Gaussian distribution. Pfinder [3] assumes that at a particular image
location the pixels over a time window are single Gaussian distributed. Mixture
of Gaussians(MoG) [4] is a widely used approach for background modeling. Here
each pixel is modeled as a mixture of weighted Gaussian distributions. Pixels
which are detected as background are used to update the Gaussian mixtures
by an update rule. These methods can deal with small or gradual changes in
the background, but if the background includes only one static scene, they may
fail when background pixels are multi-modal distributed or widely dispersed in
intensity [1]. Such methods lack the adaptability to the dynamic scenes. Lee [5],
Tuzel [6] and other researchers have improved MOG in different aspects. More
detailed reviews can be find in [7]. Kim [8] propose an approach for detecting
moving objects by building an codebook model. The method can deal with peri-
odic variations over time, but it is not able to capture complex distributions of
background values.

Over the past years, the non-parametric method has been extensively studied. A non-parametric method was proposed by Elgammal [9] to estimate the

density function for each pixel from many samples by making use of kernel density estimation technique. Wang [1] propose SACON, in which the model of each pixel include a history of the N most recent pixel values and the background model values are updated using a First-in First-out strategy. Barnich and Droogenbroeck [10] propose ViBe and this approach learn and update the back ground model of each pixel based on a random strategy. The PBAS [11] use a history of N image value as the background model as in SACON and use a similar random update rule as in ViBe. They all belong to the non-parametric method. Our method also follow a non-parametric paradigm, further more, we add the model-sharing strategy to detect moving objects iteratively.

3 Model-Sharing Strategy

3.1 Regional and Temporal Analysis

In [12], "case-by-case model sharing" is used for bidirectional analysis. A given background model is selected from a database and it is not always selected for the same pixel, moreover, it could be shared by several pixels. We also use the model sharing method, but we didn't use a database to query a suitable model. Our model-sharing strategy is based on two concepts.

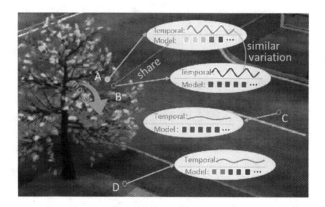

Fig. 2. Example of using model-sharing strategy

1. Different background areas have different stability, same region has similar stability. The stability is used to describe temporal variation. For instance, stable background such as buildings and roads have no change or change a little, so they have a higher stability. Dynamic background such as waving trees changed dramatically over time due to the winds, so they have a lower stability, and this lead to a higher false positives rate.

2. The movement of dynamic background is random and regional. Such as grass, trees, and rivers, they move randomly in a region. Due to the randomness, the model of a pixel in this region may not contain enough information and this could affect the accuracy of segmentation.

The two concepts will help us to find the models that could be shared. Figure 2 shows an example of using model-sharing strategy. The pixels A and B in the dynamic area, the leafs may move to their positions. In model learning process, the leaf at A, the model of A could contain the leaf's color information. If the leaf moved from A to B, because B's model didn't contain the leaf's information and this may lead to error detection. A and B are both trajectory of the leaf, if they can shared their models so that the current pixel B could share the model of A, it may reduce error detection. But we don't know which model should be shared. In order to simulate the random motion, we use the reverse speculation. Randomly and iteratively chose a pixel around B, the chosen pixel's model may include the leaf color information or not. If they have a similar temporal variation, compare B with the chosen pixel's model. If they match successfully, B should be detected as background, otherwise chose another pixel which has a similar variation to match B again until reaching the iterative boundary.

3.2   Quantify the Variation Over Time

We mentioned that different background area has different variations over time. In order to distinguish different areas for model-sharing, we need to quantify the variation in time-domain by stability.

x^t represent a pixel x at time t. First, we judge if x^t is stable by using $B(x)$. $B(x)$ is preserved by an array of k recently observed pixel values which was detected as background.

$$B(x) = \{B_1(x), B_2(x), \cdots, B_k(x) | B_i(x) \in background, 1 \leqslant i \leqslant k\} \quad (1)$$

Then, we can get different frequency coefficients $\{C_1, C_2, \cdots, C_m\}$ by taking Haar wavelet transform on $B(x)$. The coefficients could represent how much the pixel x^t change in different frequency. We assume that if all the coefficients are in a given range, $|C_i < R_{coe}|$, $1 < i < m$, then x^t should be judged as stable, otherwise it should be judged as unstable.

$$s(x^t) = \begin{cases} 1, & if \ |C_1| < R_{coe} \cap |C_2| < R_{coe} \cap \cdots |C_m| < R_{coe} \\ -1, & otherwise \end{cases} \quad (2)$$

We use S_x^t to describe the stability of the pixel x at time t.

$$S_x^t = \sum_{i=0}^{n} s(x^i) \quad (3)$$

S_x^t represent the accumulation of $s(x^i)$. We define the upper and lower bound for S_x^t, $0 < S_x^t < R_{max}$. R_{max} represent the maximum of S_x^t, so that S_x^t is in a reasonable bounds. The larger the S_x^t is, the pixel x is more likely in stable areas. Figure 3 shows the distribution of S_x^t in different scenes. Dynamic background could be distinguished effectively.

(a)   (b)

Fig. 3. Distribution of S_x^t in different scenes

3.3   Segmentation Decision

The use of model-sharing strategy will simplified the background model. In our method, the background model $M(x)$ s only initialized by the first n frames.

$$M(x) = \{M_1(x), M_2(x), \cdots, M_n(x)\} \tag{4}$$

Different from other methods, our segmentation decision is iterative for model-sharing. In the current frame, not every pixel need to share other pixels models, only the pixel detects as foreground should be concerned share others. Because if a pixel is detected as background means that its model has enough information to match the current value, so it need not to be detected again. A lower S_x^t means x is in a more dynamic area which has a higher false positives rate, so it needs more information. The number of iteration L_x is decided by S_x^t. We define $0 < L_x < R_{iter}$. Here, $L_x = 1$ imply that no other models is shared, R_{iter} represent the maximum iterations.

$$L_x = \lfloor \frac{R_{max}}{S_x^t} \rfloor \tag{5}$$

$F(x, M(x'))$ represent the pixel x is compared with the model of x', x' is a random chosen pixel around x satisfied $|S_x^t - S_{x'}^t| < R_{stab}$, R_{stab} is a given threshold. $F(x, M(x))$ represent compare x with its own model.

If the distance between the current pixel value $I(x)$ and $M_k(x')$ is smaller than a decision threshold R_d, we decide they can match each other.

$$F(x, M(x)) = \begin{cases} Background, & if \quad \#(x, M(x')) < \#_{min} \\ F(x, M(x')), & else\ if\ L_x > 1 \\ Foreground, & else \end{cases} \tag{6}$$

We use $\#(x, M(x'))$ represent the match number when x is compared with $M(x')$. If $\#(x, M(x'))$ ia larger than a given threshold $\#_{min}$ the pixel x should be detected as background. Otherwise, if $L_x > 1$, x should be detected iteratively to share other pixels model until L_x reach the boundary value.

3.4   Update Mechanism

The update mechanism is essential to adapt to changes over time, such as lighting changes, shadows and moving objects.

We use a similar random update mechanism as in ViBe. For a pixel x, $M_i(x)(i \in 1 \cdots N)$ is randomly chosen from its model $M(x)$. The value of $M_i(x)$, is replaced by the current pixel value in a random time subsampling. This mechanism allows the current value to be learned into the background model and ensures a smooth decaying lifespan for the samples stored in the background pixel models. The current value could also update the model which has be shared with x successfully. For example, when x is compared with the model of x' and detected as background, the current value could also update $M(x')$. The propagate rules is different from ViBe's that randomly chosen a neighborhood pixel to update.

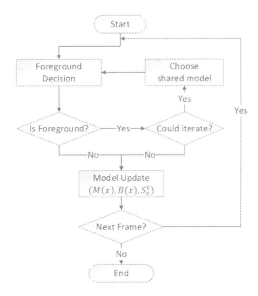

Fig. 4. The process flow of the proposed method

Figure 4 shows the process flow of the proposed method. The foreground decision is iterative until L_x reach the boundary value. If it meets the iterative condition, a random chosen model will be shared for foreground decision. $B(x)$ is updated to calculate S_x^t, S_x^t is used to choose a suitable shared model. $M(x)$ is updated to adapt to environmental changes.

4   Experimental Results

4.1   Determination of Parameters

As mentioned in previous discussions, there are some parameters in our app-roach. Since the approach is evaluated on the open database, we seek an optimal set of parameters that give the best balanced performance in different categories.

K = 8: K is the number of recent observed pixel values which was detected as background. K need to be the integer power of 2 for taking Haar wavelet transform. In order to reflect the stability of a pixel and reduce the computational complexity, K = 8 is suitable.

N = 24: N is the number of samples in background model. The higher values of N provide a better performance. In the interval between 20 and 30, it will lead to best performance. They tend to saturate for values higher than that interval.

$\#min$: The number of matched samples need to be greater than or equal to $\#min$ when the current pixel is detected as background. An optimum is found at $\#min = 2$. The same optimal value has been found in [10].

$R_d = 20$: R_d used as a threshold to compare a new pixel value with pixel samples. The same optimal value also has been found in [10].

$R_{coe} = 5$: R_{coe} is used to measure the different frequency coefficients to judge if the current pixel is stable at time t. The larger the R_{coe} is, the more likely the current pixel judged as stable.

$R_{max} = 250$: R_{max} represent the maximum of S_x^t so that S_x^t in a reasonable bounds.

$R_{iter} = 7$: R_{iter} represent the maximum iteration, The larger the R_{iter} is, the more models may be shared and need more calculations.

$R_{stab} = 30$: R_{stab} is used to measure if the pixel x and a randomly chosen pixel x' have similar temporal variation, so as to judge if x could share the model of x' or not.

4.2   Performance Evaluation

The results are evaluated by some metrics including FPR, FNR, Recall, Precision and F-measure.

$$FPR = \frac{FP}{FP + TN} \qquad FNR = \frac{FN}{TP + FN} \tag{7}$$

$$Recall = \frac{TP}{TP + FN} \qquad Precision = \frac{TP}{TP + FP} \tag{8}$$

$$F - measure = 2 \cdot \frac{Recall \cdot Precision}{Recall + Precision} \tag{9}$$

The proposed method is compared with some typical methods provided in BGSLibrary [13] on the SABS dataset [14] which includes a multitude of chal-lenges for general performance overview such as illumination changes, dynamic background and shadows.

Table 1. Performance compared with other approaches

	FPR	FNR	Recall	Precision	F-measure
MOG [15]	0.014	0.251	0.748	0.573	0.649
GMG [16]	0.047	0.117	0.882	0.323	0.473
MultiLayer [17]	0.014	0.177	0.822	0.595	0.691
ViBe [10]	0.013	0.251	0.749	0.596	0.664
PBAS [11]	0.013	**0.091**	**0.897**	0.617	0.735
Ours	**0.006**	0.276	0.753	**0.751**	**0.752**

Table 1 shows the results of different metrics when compared with other methods. It can be seen that our method has the best performance in Precision, FPR and F-measure. In other performance metrics, our methods has an average performance.

We also evaluate our approach on eleven real scenes from different categories in CDNET video dataset [18]. This dataset include scenarios with baseline, dynamic background, camera jitter intermittent object motion, shadow thermal, bad weather, low frame-rate, night videos, PTZ and turbulence.

Table 2. Performance in different scenarios

	FPR	FNR	Recall	Precision	F-measure
Baseline	0.0002	0.087	0.912	0.974	0.942
Dynamic background	0.0009	0.412	0.581	0.910	0.714
Camera jitter	0.0005	0.383	0.616	0.984	0.758
Intermittent object motion	0.0021	0.551	0.449	0.893	0.598
Shadow	0.0002	0.216	0.783	0.894	0.876
Thermal	0.0051	0.328	0.671	0.926	0.778
Bad weather	0.0001	0.283	0.716	0.998	0.834
Low framerate	0.0011	0.488	0.511	0.887	0.649
Night videos	0.0092	0.289	0.710	0.685	0.697
PTZ	0.0021	0.381	0.612	0.814	0.704
Turbulence	0.0001	0.236	0.763	0.778	0.771
Over all	0.0019	0.332	0.667	0.895	0.756

Table 2 results performance of the proposed approach in all eleven scenarios. Our method has best performance on Precision and FPR for all categories. This is because the process of foreground decision is iterative in model-sharing strategy and it can improve accuracy. Table 3 shows comparative segmentation maps in different scenarios by using different methods.

Table 3. Comparative segmentation maps in different scenarios

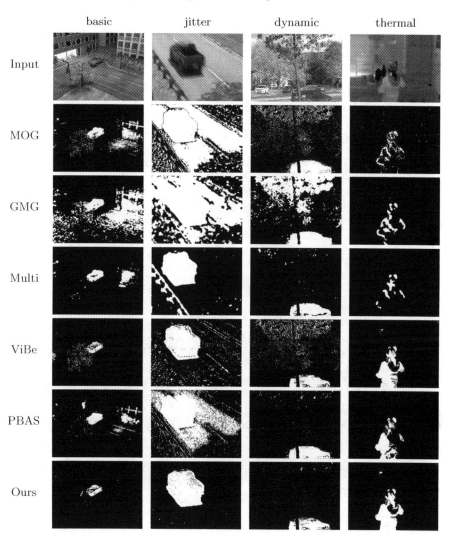

5   Conclusion

In this paper, a model-sharing strategy based on temporal variation analysis is proposed for moving detection in complex scenes. The current pixel is not only compared with its own model, but may also be compared with the model of other pixel which has similar temporal variation. This method could offset the lack of information for a pixel. Comparison with previous methods shows that the proposed approach has better performance on precision and FPR, this lead to better performance on F-measure which seem to be best for a balanced comparison.

References

1. Hanzi, W., David, S.: A consensus-based method for tracking: modelling background scenario and foreground appearance. Pattern Recogn. **40**, 1091–1105 (2007)
2. Haritaoglu, I., Harwood, D., Davis, L.S.: W 4: real-time surveillance of people and their activities. IEEE Trans. Pattern Anal. Mach. Intell. **22**, 809–830 (2000)
3. Wren, C.R., Azarbayejani, A., Darrell, T., Pentland, A.P.: Pfinder: real-time tracking of the human body. IEEE Trans. Pattern Anal. Mach. Intell. **19**, 780–785 (1997)
4. Stauffer, C., Grimson, W.E.L.: Adaptive background mixture models for real-time tracking. In: IEEE Computer Society Conference on Computer Vision and Pattern Recognition 1999, vol. 2. IEEE (1999)
5. Lee, D.S.: Effective gaussian mixture learning for video background subtraction. IEEE Trans. Pattern Anal. Mach. Intell. **27**, 827–832 (2005)
6. Tuzel, O., Porikli, F., Meer, P.: A bayesian approach to background modeling. In: IEEE Computer Society Conference on Computer Vision and Pattern Recognition-Workshops, CVPR Workshops 2005, p. 58. IEEE (2005)
7. Cristani, M., Farenzena, M., Bloisi, D., Murino, V.: Background subtraction for automated multisensor surveillance: a comprehensive review. EURASIP J. Adv. Sig. Process. **2010**, 43 (2010)
8. Kim, K., Chalidabhongse, T.H., Harwood, D., Davis, L.: Real-time foreground-background segmentation using codebook model. Real-Time Imaging **11**, 172–185 (2005)
9. Elgammal, A., Harwood, D., Davis, L.: Non-parametric model for background subtraction. In: Vernon, D. (ed.) ECCV 2000. LNCS, vol. 1843, pp. 751–767. Springer, Heidelberg (2000)
10. Olivier, B., Marc, V.D.: Vibe: a universal background subtraction algorithm for video sequences. IEEE Trans. Image Process. **20**, 1709–1724 (2011)
11. Hofmann, M., Tiefenbacher, P., Rigoll, G.: Background segmentation with feedback: the pixel-based adaptive segmenter. In: 2012 IEEE Computer Society Conference on Computer Vision and Pattern Recognition Workshops (CVPRW), pp. 38–43. IEEE (2012)
12. Shimada, A., Nagahara, H., Taniguchi, R.I.: Background modeling based on bidirectional analysis. In: 2013 IEEE Conference on Computer Vision and Pattern Recognition (CVPR), pp. 1979–1986. IEEE (2013)
13. Sobral, A.: Bgslibrary: an opencv c++ background subtraction library. In: IX Workshop de Viso Computacional (WVC 2013) (2013)
14. Brutzer, S., Hoferlin, B., Heidemann, G.: Evaluation of background subtraction techniques for video surveillance. In: 2011 IEEE Conference on Computer Vision and Pattern Recognition (CVPR), pp. 1937–1944. IEEE (2011)
15. Zivkovic, Z.: Improved adaptive gaussian mixture model for background subtraction. In: Proceedings of the 17th International Conference on Pattern Recognition, ICPR 2004, vol. 2, pp. 28–31. IEEE (2004)
16. Godbehere, A.B., Matsukawa, A., Goldberg, K.: Visual tracking of human visitors under variable-lighting conditions for a responsive audio art installation. In: American Control Conference (ACC), pp. 4305–4312. IEEE (2012)

17. Yao, J., Odobez, J.M.: Multi-layer background subtraction based on color and texture. In: IEEE Conference on Computer Vision and Pattern Recognition, CVPR 2007, pp. 1–8. IEEE (2007)

18. Goyette, N., Jodoin, P.M., Porikli, F., Konrad, J., Ishwar, P.: Changedetection.net: a new change detection benchmark dataset. In: 2012 IEEE Computer Society Conference on Computer Vision and Pattern Recognition Workshops (CVPRW), pp. 1–8. IEEE (2012)

A Fast Object Detecting-Tracking Method in Compressed Domain

Zenglei Qian, Jiuzhen Liang[✉], Zhiguo Niu, Yongcun Xu, and Qin Wu

Internet of Things Engineering College, Jiangnan University, Binhu District,
Wuxi, Jiangsu Province, China
jzliang@jiangnan.edu.cn

Abstract. The traditional pixel domain tracking algorithms are often applied to rigid objects which move slowly in simple background. But it performs very poor for non-rigid object tracking. In order to solve this problem, this paper proposes a tracking method of rapid detection in compressed domain. Convex hull formed by Self-adaptive boundary searching method and rule-based clustering are adopted for the detector in order to reduce the complexity of the algorithm. At the tracking stage, Kalman filtering is used to forecast the location of the objective. Meanwhile, as the whole process is completed in the compressed domain, it can meet the real-time requirement compared with other algorithms. And it tracks the target more precisely. The experimental results show that the proposed method has the following properties: (1) more advantages in tracking small-sized objects; (2) a better effect when track a fast moving objects; (3) faster tracking speed.

1 Introduction

It's known that long-term target tracking in a video stream is one of the most important subjects in computer vision. It has broad application in many fields such as security surveillance, intelligent transportation. The main job of tracking is to estimate the location of the target in the subsequent frames. Given a bounding box defining the object of interest in a single frame, our goal is to automatically determine the bounding box of the object or indicate that the object is not visible in succeeding frames that follows. The video stream is processed at certain frame rate and the process may run indefinitely long. Although numerous algorithms have been proposed in literature, object tracking remains a challenging problem due to target appearance change caused by pose, illumination, occlusion, and motion, etc. So far, no single tracker has been able to deal with all these factors at the same time.

There are two major groups of approaches to segment and track moving objects in a video sequence, distinguished by the domain in which they operate: pixel domain and compressed domain.

Z. Kalal et al. designed a novel framework (TLD [13]) that decomposed the long-term tracking task into three subtasks: tracking, learning, and detection. But it does not perform well in case of full out-of-plane rotation and tracks a

© Springer International Publishing Switzerland 2015
C.V. Jawahar and S. Shan (Eds.): ACCV 2014 Workshops, Part II, LNCS 9009, pp. 344–359, 2015.
DOI: 10.1007/978-3-319-16631-5_26

single object. Zhang et al. [31] proposed an effective and efficient tracking algorithm with an appearance model based on features extracted. A very sparse measurement matrix was adopted to efficiently compress features from the foreground targets and background. Wu et al. proposed a multi-scale tracking algorithm [27] which combines the random projection-based appearance model with the bootstrap filter framework. The proposed scale-invariant normalized rectangle feature was adopted to characterize a target object of different scales to be tracked.

Pixel domain approaches mentioned above have the potential for higher accuracy, but also require higher computational complexity. "Compressed-domain" approaches make use of the data from the compressed video bit stream, such as motion vectors (MVs), block coding modes, motion-compensated prediction residuals or their transform coefficients, etc. [7,14,15]. Though the lack of full pixel information often leads to relative lower accuracy. The main advantage of compressed-domain methods in practical applications is their lower computational cost [11,19,24]. Currently, The compression standard these methods commonly used is highly compressed digital video coding standard (H.264/AVC), which was formulated by the Joint Video Team ITU-T Video Coding Experts Group (VCEG) and MPEG coalition (JVT, Joint Video Team) [25,26]. It plays an extremely important role in the field of video retrieval and monitoring for its fast segmentation algorithm. Therefore, compressed-domain methods are thought to be more suitable for real-time applications [3,35], as a mean shift clustering is used in [3] to segment moving objects from MVs and partition size in H.264 bit-stream [4,5]. However, some of them are still characterized by high complexity [16,28,29]. You et al. [29] presented an algorithm to track multiple moving objects in H.264/AVC compressed video based on probabilistic spatio-temporal MB filtering and partial decoding. Their work assumes stationary background and relatively slow-moving objects.

In this paper, we propose and discuss a Fast Object Detecting-Tracking Method in Compressed Domain, firstly, the tracker estimates the objects motion between consecutive frames under the assumption that the frame-to-frame motion is limited and the object is visible. While the tracker is likely to fail and never recover if the object moves out of the camera view. Once the tracker failed to track the target, the detector will correct tracking results in time and reduce the occurrence of the lost. We will make an integration of the tracking results and detecting results, draw the best prediction result as the final output.

2   Related Work

The long-term tracking can be approached either from tracking or from detection perspectives. Tracking algorithms estimate the object motion. Trackers only require initialization. They are fast, and produce smooth trajectories. On the other hand, they accumulate errors during runtime (drift) and typically fail if the object disappears from the camera view. Research in tracking aims at developing increasingly robust trackers that track "longer". The post failure behavior is not directly addressed. Detection-based algorithms estimate the object location in every frame independently. Detectors do not drift and do not fail if the object disappears from the camera view. However, they require an offline training stage and therefore cannot be applied to unknown objects.

2.1   Detection

In recently years, Many methods have been developed for moving object detection in H.264/AVC bitstream domain. Sabirin et al. presented a graph-based algorithm [23] for detecting and tracking moving objects in H.264/AVC bitstream domain. The spatio-temporal graph first constructed represents the blocks with non-zero motion vectors or non-zero residues and their relations between two frames. A method proposed in [17] is that it uses both motion vectors and macroblock information, including macroblock data sizes, types, and partitions, to track and detect the target even when it stops. Because motion vector information is not generated when the moving object stops, use of only motion vector tracking cannot consistently track the target consistently. However, by using the addition of macroblock information, [17] demonstrate how target tracking can be maintained when the target stops. The [17] proposed method thereby improves target tracking. Khatoonabadi et al. [12] presented a method for tracking moving objects in H.264/AVC-compressed video sequences using a spatio-temporal Markov random field (STMRF) model. Sabirin et al. [23] detecting and tracking moving objects by treating the encoded blocks with non-zero motion vectors and/or non-zero residues as potential parts of objects in H.264/AVC bitstreams. Fang Yumin et al. [8] proposed a novel object segmentation approach using background estimation. To reduce the processing time for the subsequent global motion compensation, the inter-frame coding modes are applied to estimate the background region.

2.2   Tracking

Object tracking is the task of estimation of the object motion. We use Kalman filter to complete our task tracking. Kalman filter is the minimum-variance state estimator for linear dynamic systems with Gaussian noise [20]. Even if the noise is non-Gaussian, the Kalman filter is the best linear estimator. In addition, Kalman filter is the minimum-variance linear state estimator for linear dynamic systems with non-Gaussian noise [21]. For nonlinear systems it is not possible, in general, to implement the optimal state estimator in closed form, but various modifications of Kalman filter can be used to estimate the state. These modifications include the extended Kalman filter [21], the unscented Kalman filter [10], and the particle filter [6]. Meanwhile, [34] presents an extended Kalman filter adopted to estimate the current location based on the estimated parameters and the previous location estimate. In [2], the strong Kalman filter is enhanced with the improved time-variant recession matrix and applied in tracking the given video target with fast mobility. The enhanced strong Kalman filter is found to be more robust than generic Kalman filter in resisting process uncertainties to which the fast mobile target is usually subjected. To deal with the actual process of tracking moving objectsappear in occlusion and interference problems caused bycamera movement, [9] designs a moving object tracking system using the algorithm based on Camshift and Kalman filter.

3   Framework of Detection-Tracking in Compressed Domain

The aim of this section is to provide a framework for detecting-tracking moving objects system in compressed domain based on MVs (see Fig. 1 for the overall framework). As we see, the compressed bit-stream is partially decoded by the H.264/AVC decoder which generate the motion vector. Then the MV of the current frame (t-th frame) is fed into the preprocessor to get the more stable and reliable MV field. Followed that, the corresponding field is entered into the detector in which we obtained multiple motion objects regions by means of boundary searching and rule-based clustering. Under the assumption that the frame-to-frame motion is limited and the object is visible, our tracker estimates the motion of the target between consecutive frames by Kalman filter. Finally, the result which is integrated with the tracker and detector is regarded as the final output.

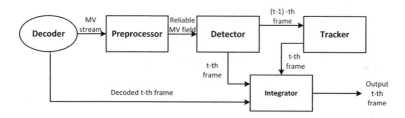

Fig. 1. CDT method framework

3.1   Preprocessor

To filter the noisy motion vectors, a spatio-temporal motion vector consistency filter (STF) is presented in paper [18]. Let $(x, y)^t$ represents the pixel coordinate (x, y) in the t-th frame, and $mv(x, y)^{t \rightarrow t_{ref}}$ represents the motion vector of pixel (x, y) in the t-th frame with respect to the reference frame t_{ref}.

Backward Iterative Cumulate. This is accomplished by dividing the cumulative motion vector by frame number, according to Motion Vector Normalization Eqs. (1) and (2).

$$N\left(MV^{t \rightarrow t_{ref}}\right) = \frac{MV^{t \rightarrow t_{ref}}}{t - t_{ref}} \tag{1}$$

$$P(i, j)^{t \rightarrow t_{ref}} = (i, j)^t$$

$$P\left(i,j\right)^{t\rightarrow(t-1)} = \left(i,j\right)^{t} + MV^{i,j}\left(t\right) = \left(\hat{i},\hat{j}\right)^{t-1} \tag{2}$$

$$P\left(i,j\right)^{t\rightarrow(t-k)} = P\left(P\left(i,j\right)^{t\rightarrow(t-k+1)}\right)^{t\rightarrow(t-k)} \quad (k > 2)$$

$N\left(MV^{t\rightarrow t_{ref}}\right)$ is the normalized motion vector. In this section, the necessity of backward iterative cumulate is demonstrated by the MV denoising theory in the time domain.

Spatial Consistency. Spatial consistency is calculated between the current MVs and neighborhood MVs in distance (see Eqs. (3) and (4)).

$$MVNCI_{amp}^{i,j} = \left|\frac{MV_{amp}^{i,j} - \frac{1}{N}\sum\limits_{(i',j')\in\theta} MV_{amp}^{i',j'}}{\frac{1}{N}\sum\limits_{(i',j')\in\theta} MV_{amp}^{i',j'}}\right| \tag{3}$$

$$MVNCI_{dir}^{i,j} = \frac{1}{N}\sum\limits_{(i',j')\in\theta}\left|MV_{dir}^{i,j} - MV_{dir}^{i',j'}\right| \tag{4}$$

$MVNCI_{amp}^{i,j}$ represents the degree of $MV_{amp}^{i,j}$ deviated from it neighborhood nonzero MV, $MV_{amp}^{i',j'}$ is the surrounding MVs, θ represents eight-neighborhood structures and N represents the number of nonzero motion vector of the neighborhood.

Temporal Consistency. Vector matching ratio is used to analyze the MVs compared with MVs of the same position in the previous frame. Temporal consistency between the consecutive frames is calculated by Eqs. (5) and (6).

$$R\left(\overrightarrow{a},\overrightarrow{b}\right) = \begin{cases} 0 & if\ \overrightarrow{a} = \overrightarrow{b} = (0,0) \\ \dfrac{\|\overrightarrow{a}-\overrightarrow{b}\|}{\|\overrightarrow{a}\|+\|\overrightarrow{b}\|} & otherwise \end{cases} \tag{5}$$

$$MVTCI\left(MV^{i,j}\left(t\right)\right) = \tag{6}$$
$$\prod_{k=1}^{n} R\left(MV\left(P\left(i,j\right)^{t\rightarrow t-k+1}\right), MV\left(P\left(i,j\right)^{t\rightarrow t-k}\right)\right)$$

3.2 Detector

The detection algorithm alone is unable to narrow the search area because the compressed stream has no color information as object feature, we propose a fast moving objects detection algorithm based on boundary searching with convex hull. The algorithm consists of three parts as follows.

Adaptive Boundary Searching. Without inner information of moving objects, we use local search by locating a moving region instead of global search, so that to reduce the time complexity. This algorithm sets a 3×3 offset template

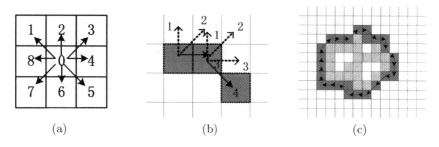

Fig. 2. Adaptive boundary blocks scanning. (a) Eight-direction template. (b) The scanning order of the eight-direction. The dotted lines and serial numbers represent the order of scanning. (c) The grey blocks represent the boundary blocks after scanning.

which is used to represent the eight-direction neighborhood block of the current search block (see Fig. 2(a)).

The zero is the current block, and 1 to 8 are searching blocks in eight-direction which saved the offset value relative to the current search block, as described in Eq. (7).

$$(x_{cur}, y_{cur}) = (x_{cur}, y_{cur}) + (x_{offset}, y_{offset}) \tag{7}$$

After the eight-direction template is defined, we define the first initial search block using the raster scan method and ensure that its searching blocks are boundary blocks by scanning blocks in clockwise direction (as shown in Fig. 2). In the current direction, the block is not a moving block, we should turn to the block in next direction and so on, until the eight searching directions is completed or found a motion block. The target in tracking is only related with its contour, getting only the convex hull points instead of all points in the target can greatly reduce the computational complexity. Graham scanning method is token in this paper (see Fig. 3).

Rule Based Clustering. After the procedure discussed above, some scattered connected regions of moving objects are got. In this section the final moving object is obtained by clustering the scattered motion subregions following some rules. The number of iterations often depends on the number of global motion vector blocks. And the consumption of the iterative time decides the segmentation algorithm performance. In this paper, clustering the motion subregions can greatly reduce the number iterations.

This paper puts forward the clustering rules contained of two elements: one is the direction and amplitude of motion vectors, the other is the distance between motion subregions. Two new distances are defined in Eqs. (8) and (9).

$$p_{dis} = \underset{1<i<n, i\neq j}{Hausdorff} \|\overrightarrow{p_i}, \overrightarrow{p_j}\| \tag{8}$$

$$p_{NCI} = \mu_1 p_{amp}^{(i)(j)} + \mu_2 p_{dir}^{(i)(j)} = \mu_1 \frac{\left|MV_{amp}^{(i)} - MV_{amp}^{(j)}\right|}{MV_{amp}^{(j)}} + \mu_2 \frac{\left|MV_{dir}^{(i)} - MV_{dir}^{(j)}\right|}{MV_{dir}^{(j)}} \tag{9}$$

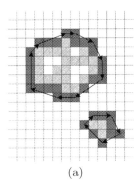 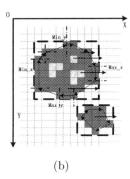

(a)   (b)

Fig. 3. In (a), the mesh blocks, grey blocks and twill blocks are the Mask after filtering in motion vector field, the grey blocks represent the boundary blocks on the boundary searching, the intermediate-blank blocks represent the missing blocks, and the mesh blocks are convex hull points. In (b), the rough borders are searching area, whose internal is used to fill points by ray method for avoiding repeated searching.

where $p_{amp}^{(i)(j)}$ and $p_{dir}^{(i)(j)}$ represent deviation in amplitude and direction respectively, $\vec{p_i}$ and $\vec{p_j}$ are the positions of block i and j, respectively. In this way, the motion subregions having common feature are clustered into a whole moving object. The constraint condition is $\mu_1 + \mu_2 = 1$.

3.3   Tracker

Object tracking is the problem of estimating the positions and other relevant information of moving objects in image sequences. Our tracker is based on Kalman filter [22], which is a classical motion prediction model and optimal recursive algorithm. The Kalman filter solves the general problem of estimating the state $x \in R^n$ of a discrete-time controlled process that is governed by the linear stochastic difference Eqs. (10) and (11).

$$X_k = AX_{k-1} + BU_{k-1} + W_{k-1} \tag{10}$$

$$Z_k = HX_k + V_k \tag{11}$$

where X_k is the system state at time k, U_k is the amount system control, A, B is system parameters, Z_k is measured values of time k, H is a parameter of the measurement system. The random variables Wk and V_k represent the process and measurement noise respectively. According to the system model, we can predict the present state based on the former state in equation, and P is denoted as the covariance in Eqs. 12 and 13.

$$\hat{X}_k = A\hat{X}_{k-1} + BU_{k-1} \tag{12}$$

$$P_k^- = AP_{k-1}A_k^T + Q \tag{13}$$

P_k^- is the covariance corresponding to $\hat{X}_k^-$, Q is the covariance of the system process. Equations 12 and 13 are the predicted results of system. Combing the predicted and measured values, we can get the following iterative (14)–(16).

$$\hat{X}_k = \hat{X}_k^- + K_k(Z_k - H\hat{X}_k^-) \tag{14}$$

$$K_k = P_k^- H^T \left(H P_k^- H^T + R\right)^{-1} \tag{15}$$

$$P_k = (I - K_k H) P_k^- \tag{16}$$

where K_k is the Kalman Gain, and of which the basic principle can be represented by (12)–(16).

4   CDT Implementation

This section describes our implementation of the CDT framework.

4.1   Initialization

In order to track the target, we need to artificially specify an object as target from the first frame, and the decoded stream should be preprocessed. At first, the first frame of the video sequence is decoded, and the initial bounding box is artificial demarcated. The real tracking target will be obtained by the detector. Spatial similarity of the bounding boxes in the detector and artificial demarcated is measured by overlap between bounding boxes, which is defined as a ratio between intersection and union.

4.2   Object Detector

The detector scans the input image by searching multiple moving objects and decides which one is the tracking target we wanted. The algorithm of boundary searching is shown in Algorithm 1.

Algorithm 1: Adaptive Boundary Searching

```
Algorithm1 (Mask, Bpoint, Tukenum)
  Data: Mvfield // MV of current frame
  MVsize = sizeof(Mvfield) // size of the current frame
  Result: Mask // moving object mask in 1 or 0,
               // 1 means object, 0 means background
        Bpoint // the boundary points of motion subregions
        Tukenum // the number of the motion subregion
  begin:
    tukenum := 0, Mask := 0 in height*width;
    for i:=1:Mvsize(Mask(i)==1), do
        if Mvfield(i) is interval in MV threshold && Mask(i)=0 then
            save startpoint with address of Mvfield(i);
            while i is not startpoint, do
```

```
            for j:=8-direction searching, do
                if Mvfield and Mask is appropriate, then
                    Mask(j):=1,Bpoint:=Mvfield(j)
                if 8-direction searching is not find block, then
                    return the original path;
        tukenum++, Bpoint:=Coverxhull(Bpoint);//get the convex
        //hull vertices remove the points inner convex hull;
end
```

In this way, the number of iteration is decreased a lot, thus the computational complexity is reduced. The *Mask* and *Bpoint* contain the scattered motion regions are the inputs of the clustering algorithm's inputs. Then the clustering algorithm is shown in Algorithm 2.

Algorithm 2: Rule Based Clustering

```
Algorithm2 (Edge, Objectnum)
  Data: Bpoint // the boundary points of motion subregions
  Tukenum // the number of the motion subregion
  Meanmv //the average of MV in x and y direction of subregion
  Result: Edge // the points of convex hull of final objects
        Objectnum // the number of the final objects
  begin:
    for i := 1:tukenum do
        for j :=i+1:tukenum do
            pdis(i,j):= compute the Hausdorff distance between
            Bpoint(i) and Bpoint(j);
            if pdis(i,j) < Tdis then
                pamp(i,j) := compute the deviation degree
                in amplitude of MeanMv;
                pdir(i,j):= compute the deviation degree
                in direction of MeanMv;
                pNCI := u1pamp(i,j)+u2pdir(i,j);
                if pNCI < TNCI then
                    Edge := Convexhull(Bpoint);
        Objcetnum:= update tukenum;
end
```

The threshold of Rule-based clustering needs to be initialized, where the value of μ_1 and μ_2 are different for different videos. If the target is moving slowly and its moving direction keeps unchanged, its moving direction can be determined by the MVs in the motion subregions. And the value of μ_2 need to be set large, generally 0.8. If the target is moving quickly and its direction changes rapidly, then the motion vector will be more messy. In that case, MV has major influence on determining the degree of deviation in magnitude, the value of μ_1 requires to be large, and generally 0.6. Figure 4 shows the segmentation performance of our detecting algorithm, it works well.

Fig. 4. The performance of segmentation in detector, these video sequence is respectively called 'foreman', 'silent', 'stefan', 'walk' and 'container' in the version of JM12.2 of H.264/AVC.

4.3   Object Tracking

The tracking component of CDT is based on Kalman Filter tracker, which represents the object by a bounding box(called tBB) and estimates its motion between consecutive frames. Here we need to consider two issues. Firstly, the input of tracker is the output of detector. When the detector fails to detect, the tracker can automatically recognize the false detection and avoid unnecessary errors. Secondly, the output of the detector depends entirely on the boundary searching and rule based clustering algorithm. Note that, the detector and the tracker propose in this paper an independent of each other. The tracker can do a smooth filtering and prediction to the detector, and the result is re-applied to the detector and correct it.

Failure Detection. The tracker can do a preliminary identification of the output of the detector. We define perdition position of the target at the i-th frame as $tBB(i)$, and the output of detector at the i-th frame is defined as $dBB(i)$. We define e is $e = tBB(i) - dBB(i)$. If the deviation e is greater than two times the average of the moving distance, we consider the output of detector is failure. the trace $tBB(i+1)$ is predicted according to $dBB(i-1)$, and so on. See Fig. 5.

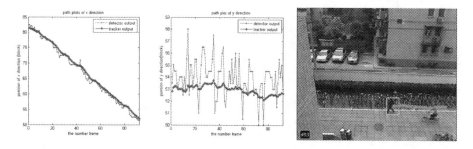

Fig. 5. The output path of detector and tracker. The first two plots are the path of x direction and y direction, the third figure is trajectory diagram of the video 'Walk'.

5   Experiments and Analysis

Our experiment is implemented in MATLAB R2011b. The system configuration is: Intel(R), Core(TM), i5-3470@3.20 GH, 4 GB memory, the OS is Microsoft Windows 8. In order to test the effectiveness of our tracking algorithm, we compare it with six algorithms, MIL [1], WMIL [30], STC [33], FCT [32], TLD [13]. The experiments compare the tracking error, error rate and frame rate. Five compressed video sequences are used in this experiment. All sequences were encoded using the H.264/AVC JM12.2 encoder as shown in Table 1. The initialized bounding box, of which including the four elements: the first two x,y coordinate of the upper left corner of the bounding box, the others coordinate of the length and width of the target object.

Table 1. Information of video sequences used in the experiments

Video sequence	Numbers of frames	Initial bounding box
Boat	180	(63,106,224,79)
Container	170	(61,40,230,71)
Stefan	90	(195,74,102,176)
Coastguard	82	(105,116,99,48)
Walk	92	(316,194,27,36)

5.1   Tracking Accuracy Experiment

From Fig. 6, one may find that the backgrounds in the test sequences a, b, d are relative simple, so other tracking algorithms learn good models from training samples which produce good tracking results. However, the third test sequence is tracking a object which moves rapidly and the background is complex and changing rapidly. In that case, the limitations of other tracking algorithms are exposed. The rapid movement leads to great changes of learning samples, which makes it impossible for other algorithms to track the object accurately. And the fifth sequence is a non-rigid object with complex background. The STC, TLD and MIL algorithms can not track the object well. In this paper, the CDT Algorithm detector is based on motion vector. And the motion vector is predicted from the inter frames, which contains part of the real motion information. So the algorithm is less sensitive to the speed of the object and background complexity.

Tracking error rate is an important index in detection tracking algorithm. In this paper, we define that the object is incorrectly tracked when the distance between the center of the tracker to the stand center of the moving object is more e_T. The value of e_T is depended on the size of the moving objects. We set $e_T = \mu D_0$, where D_0 is the size of the movement object. μ is [0,0.2]. It can be more objective to observe the changes of different tracking error rate with the error threshold for these settings. As shown in Figs. 7 and 8.

Fig. 6. Screenshots from some of the sampled tracking results (From top to bottom: *a.* Boat *b.* Container *c.* Stefan *d.* Coastguard *e.* Walk). Red is CDT, purple is MIL, green is WMIL, blue is STC, yellow is FCT, cyan is TLD (Color figure online)

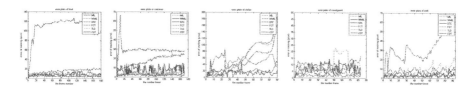

Fig. 7. Error plots for all video sequences labeled with ground truth.

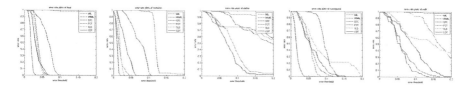

Fig. 8. Error rate plots for all video sequences.

5.2   Frame Rate Tests

In this experiment, maximum framerate, minimum, frame rate, average framerate are calculated for maximum frame rate, minimum, frame rate, average frame rate by the six algorithms. The results are shown in Table 2. From this table, one may find that both the FCT and CDT algorithms have obvious advantages in the frame rate. Especially for the CDT algorithm, it has outstanding advantages on the average frame rate and high frame rate, but not much effect on the minimum frame rate. The reason is that the core of the CDT algorithm is the detector, which takes the majority time in the whole algorithm. Boundary searching and clustering rules depend on the number of motion vectors and their dispersion, large numbers of motion vectors and dispersion increase the time consumption of the detector algorithm. For example, the container sequence, the tracking object is larger, and in the stefan sequence, there is lots of noise due to the fast movement of the object. However, as for the average frame rate, it still has good advantages.

Table 2. Frame rate comparison of six algorithms

Video sequence	Fps	CDT	MIL	WMIL	STC	FCT	TLD
Boat	Max-fps	**42.06**	10.24	24.56	28.72	**33.09**	14.56
	Mean-fps	**36.08**	5.97	20.92	21.67	**32.20**	13.31
	Min-fps	11.03	3.82	9.52	**14.28**	**18.27**	12.29
Container	Max-fps	**38.08**	8.68	24.56	27.89	33.05	17.54
	Mean-fps	**26.42**	5.76	20.12	21.49	**31.78**	13.37
	Min-fps	6.97	3.88	7.71	9.51	**18.26**	**11.96**
Stefan	Max-fps	**59.69**	9.75	23.18	32.02	**32.32**	16.62
	Mean-fps	**30.90**	7.22	21.35	28.90	**32.06**	12.95
	Min-fps	3.70	5.65	**10.57**	10.19	**18.32**	7.94
Coastguard	Max-fps	**57.33**	9.04	24.79	**36.62**	33.05	19.93
	Mean-fps	**39.43**	6.94	20.94	30.76	**31.81**	16.30
	Min-fps	**16.59**	5.53	8.99	15.60	**18.41**	11.18
Walk	Max-fps	**59.48**	9.03	24.40	**41.35**	33.89	19.88
	Mean-fps	**50.29**	6.80	20.15	**32.11**	27.68	16.08
	Min-fps	**23.28**	5.09	7.67	11.38	**18.42**	12.47

6   Conclusions

This paper presents a method of rapid detection in compressed domain and tracking target. At the tracking stage, Kalman filtering is used to forecast the location of the object. Convex hull formed by Self-adaptive boundary searching method and rule-based clustering are adopted for the detector in order to reduce the complexity of the algorithm to a large extent. Meanwhile, as the whole process is completed in the compressed domain, it can better meet the real-time requirement comparing with other algorithms. Meanwhile, it tracks the object more precisely. However, due to the fact that less features can be used in a compressed domain. The real-time is gained at the expense of the loss certain accuracy of tracking. In our experiment, more intense shaking will be observed as side effect. The reason is that the smallest unit of image information in compressed domain is a 4×4 block. In our further research, we will come up with more effective way to decrease the shaking.

References

1. Babenko, B., Yang, M.-H., Belongie, S.: Visual tracking with online multiple instance learning. In: 2009 IEEE Conference on Computer Vision and Pattern Recognition, CVPR 2009, pp. 983–990 (2009)
2. Chen, K., Zhao, X., Xu, T., Napolitano, M.R.: Enhanced strong Kalman filter applied in precise video tracking for fast mobile target. In: 2010 IEEE International Conference on Progress in Informatics and Computing (PIC), vol. 2, pp. 875–878 (2010)
3. Chen, Y.-M., Bajic, I.V.: A joint approach to global motion estimation and motion segmentation from a coarsely sampled motion vector field. IEEE Trans. Circ. Syst. Video Technol. **9**, 1316–1328 (2011)
4. Chen, Y.-M., Bajic, I.V., Saeedi, P.: Moving region segmentation from compressed video using global motion estimation and Markov random fields. IEEE Trans. Multimedia **3**, 421–431 (2011)
5. Chen, Y.-M., Bajic, I.V., Saeedi, P.: Motion segmentation in compressed video using Markov random fields. In: 2010 IEEE International Conference on Multimedia and Expo (ICME), pp. 760–765 (2010)
6. Doucet, A., De Freitas, N., Gordon, N.: Sequential Monte Carlo Methods in Practice. Springer, New York (2001)
7. Fei, W., Zhu, S.: Mean shift clustering-based moving object segmentation in the H.264 compressed domain. IET Image Process. **4**, 11–18 (2010)
8. Fang, Y., Lin, W., Chen, Z., Tsai, C., Lin, C.: A video saliency detection model in compressed domain. IEEE Trans. Circuits Syst. Video Technol. **24**(1), 27–38 (2014)
9. Huang, S., Hong, J.: Moving object tracking system based on camshift and Kalman filter. In: 2011 International Conference on Consumer Electronics, Communications and Networks (CECNet), pp. 1423–1426 (2011)
10. Julier, S.J., Uhlmann, J.K.: Unscented filtering and nonlinear estimation. Proc. IEEE **3**, 401–422 (2004)
11. Käs, C., Nicolas, H.: An approach to trajectory estimation of moving objects in the H.264 compressed domain. In: Wada, T., Huang, F., Lin, S. (eds.) PSIVT 2009. LNCS, vol. 5414, pp. 318–329. Springer, Heidelberg (2009)
12. Khatoonabadi, S.H., Bajic, I.V.: Video object tracking in the compressed domain using spatio-temporal Markov random fields. IEEE Trans. Image Process. **1**, 300–313 (2013)
13. Kalal, Z., Mikolajczyk, K., Matas, J.: Tracking-learning-detection. IEEE Trans. Pattern Anal. Mach. Intell. **34**, 1409–1422 (2012)

14. Liu, Z., Lu, Y., Zhang, Z.: Real-time spatiotemporal segmentation of video objects in the H.264 compressed domain. J. Vis. Commun. Image Represent. **3**, 375–290 (2007)

15. Liu, Z., Lu, Y., Zhang, Z.: An efficient compressed domain moving object segmentation algorithm based on motion vector field. J. Shanghai Univ. (Engl. Edn.) **12**, 221–227 (2008)

16. Liu, Z., Zhang, Z., Shen, L.: Moving object segmentation in the H.264 compressed domain. Opt. Eng. **1**, 017003 (2007)

17. Maekawa, E., Goto, S.: Examination of a tracking and detection method using compressed domain information. In: 2013 Picture Coding Symposium (PCS). Waseda University, Shinjuku-ku, Tokyo, Japan, pp. 141–144 (2013)

18. Moura, R.C., Hemerly, E.M.: A spatiotemporal motion-vector filter for object tracking on compressed video. In: 2010 Seventh IEEE International Conference on Advanced Video and Signal Based Surveillance (AVSS), pp. 427–434 (2010)

19. Mezaris, V., Kompatsiaris, I., Boulgouris, N.V., Strintzis, M.G.: Real-time compressed-domain spatiotemporal segmentation and ontologies for video indexing and retrieval. IEEE Trans. Circ. Syst. Video Technol. **14**, 606–621 (2004)

20. Rhodes, I.B.: A tutorial introduction to estimation and filtering. IEEE Trans. Autom. Control **6**, 688–706 (1971)

21. Simon, D.: Optimal State Estimation: Kalman, H Infinity, and Nonlinear Approaches. Wiley, Hoboken (2006)

22. Salmond, D.: Target tracking: introduction and Kalman tracking filters. Target Tracking: Algorithms and Applications (Ref. No. 2001/174), IEE, p. 1 (2011)

23. Sabirin, H., Kim, M.: Moving object detection and tracking using a spatio-temporal graph in H.264/AVC bitstreams for video surveillance. IEEE Trans. Multimedia **3**, 657–668 (2012)

24. Treetasanatavorn, S., Rauschenbach, U., Heuer, J., Kaup, A.: Bayesian method for motion segmentation and tracking in compressed videos. In: Kropatsch, W.G., Sablatnig, R., Hanbury, A. (eds.) DAGM 2005. LNCS, vol. 3663, pp. 277–284. Springer, Heidelberg (2005)

25. Treetasanatavorn, S., Rauschenbach, U., Heuer, J., Kaup, A.: Model based segmentation of motion fields in compressed video sequences using partition projection and relaxation. In: 2005 Visual Communications and Image Processing, p. 59600D (2005)

26. Treetasanatavorn, S., Rauschenbach, U., Heuer, J., Kaup, A.: Stochastic motion coherency analysis for motion vector field segmentation on compressed video sequences. In: 2005 Proceedings of WIAMIS, April 2005

27. Wu, Y., Jia, N., Sun, J.: Real-time multi-scale tracking based on compressive sensing. Vis. Comput. **31**(4), 471–484 (2014). Springer

28. You, W., Houari Sabirin, M.S., Kim, M.: Moving object tracking in H.264/AVC bitstream. In: Sebe, N., Liu, Y., Zhuang, Y., Huang, T.S. (eds.) MCAM 2007. LNCS, vol. 4577, pp. 483–492. Springer, Heidelberg (2007)

29. You, W., Sabirin, M.S.H., Kim, M.: Real-time detection and tracking of multiple objects with partial decoding in H.264/AVC bitstream domain. In: Proceedings of SPIE, vol. 7244, pp. 72440D–72440D-12 (2009)

30. Zhang, K., Song, H.: Real-time visual tracking via online weighted multiple instance learning. Pattern Recogn. **1**, 397–411 (2013)

31. Zhang, K., Zhang, L., Yang, M.-H.: Real-time compressive tracking. In: Fitzgibbon, A., Lazebnik, S., Perona, P., Sato, Y., Schmid, C. (eds.) ECCV 2012, Part III. LNCS, vol. 7574, pp. 864–877. Springer, Heidelberg (2012)

32. Zhang, K., Zhang, L., Yang, M.: Fast Compressive Tracking. IEEE Trans. Pattern Anal. Mach. Intell. **36**(10), 2002–2015 (2015)
33. Zhang, K., Zhang, L., Yang, M.-H., Zhang, D.: Fast Tracking via Spatio-Temporal Context Learning (2013). arXiv preprint arXiv:1311.1939
34. Zhang, L., Chew, Y.H., Wong, W.-C.: A novel angle-of-arrival assisted extended Kalman filter tracking algorithm with space-time correlation based motion parameters estimation. In: 2013 9th International Wireless Communications and Mobile Computing Conference (IWCMC), pp. 1283–1289 (2013)
35. Zeng, W., Du, J., Gao, W., Huang, Q.: Robust moving object segmentation on H.264/AVC compressed video using the block-based MRF model. Real-Time Imag. **4**, 290–299 (2005)

Automatic RoI Detection for Camera-Based Pulse-Rate Measurement

Ron van Luijtelaar[1], Wenjin Wang[2(✉)], Sander Stuijk[2], and Gerard de Haan[2]

[1] Profit Consulting, Apeldoorn, The Netherlands
[2] Electronic System Group, Electrical Engineering Department,
Eindhoven University of Technology, Eindhoven, The Netherlands
w.wang@tue.nl

Abstract. Remote photoplethysmography (rPPG) enables contactless measurement of pulse-rate by detecting pulse-induced colour changes on human skin using a regular camera. Most of existing rPPG methods exploit the subject face as the Region of Interest (RoI) for pulse-rate measurement by automatic face detection. However, face detection is a sub-optimal solution since (1) not all the subregions in a face contain the skin pixels where pulse-signal can be extracted, (2) it fails to locate the RoI in cases when the frontal face is invisible (e.g., side-view faces). In this paper, we present a novel automatic RoI detection method for camera-based pulse-rate measurement, which consists of three main steps: sub-region tracking, feature extraction, and clustering of skin regions. To evaluate the robustness of the proposed method, 36 video recordings are made of 6 subjects with different skin-types performing 6 types of head motion. Experimental results show that for the video sequences containing subjects with brighter skin-types and modest body motions, the accuracy of the pulse-rates measured by our method (94 %) is comparable to that obtained by a face detector (92 %), while the average SNR is significantly improved from 5.8 dB to 8.6 dB.

1 Introduction

Home-based healthcare monitoring is growing in popularity as modern technologies allow the medical devices to be comfortably and easily accessible. In order to assess the most basic body functions, vital physiological signals such as pulse-rate, respiration rate or blood oxygen saturation have to be measured. An extensively employed noninvasive method for pulse-rate monitoring is the Photoplethysmography (PPG) introduced in 1937 [1]. It uses contact probes, consists of a dedicated light source and photo-detector, to detect vital fluctuation of optical absorption in skin tissues. However, the contact probes that need to be attached to human body is highly inconvenient and uncomfortable to use, i.e., it cannot be used in scenario like fitness or surveillance; it is strictly prohibited in case of sensitive subjects with damaged skin or neonates. In contrast, non-contact based vital signs monitoring is preferred for its unobtrusiveness and noninvasiveness. A promising alternative is the recently introduced camera-based pulse-rate monitoring - remote photoplethysmography (rPPG) [2,3].

© Springer International Publishing Switzerland 2015
C.V. Jawahar and S. Shan (Eds.): ACCV 2014 Workshops, Part II, LNCS 9009, pp. 360–374, 2015.
DOI: 10.1007/978-3-319-16631-5_27

In the human cardiovascular system, blood pulse propagating throughout the body changes the blood volume in vessels. rPPG measures the pulse-signal by detecting the optical absorption of haemoglobin variation across the light spectrum using a regular camera. Although the recent developments in rPPG technology have demonstrated promising performance in measuring pulse-signals under ambient light conditions [4–7], a basic problem is not addressed in all these existing approaches: how to optimally initialise the Region of Interest (RoI) for pulse-rate monitoring. In previous rPPG work, the most commonly used approach for selecting the RoI is to (1) assume that the face contains essential skin pixels for pulse-signal extraction, (2) perform automatic face detection using a Viola-Jones face detector [4,6]. However, this approach is restricted to the face-similar RoI and likely to fail when the face is invisible or occluded. Other skin segmentation techniques that only based on the assumptions of human skin appearance (e.g., skin colour or texture) [8] cannot distinguish the skin regions from background scenes with skin-similar patterns.

In 2011, Schmitz [5] proposed an automatic skin detection method using the pulse-signal as a feature, since characteristics of pulse-signals can prevent the false detection of skin-similar objects in the scene. But this method is typically designed for stationary subjects, which cannot tackle with subjects' body motions. Therefore, in this paper, we aim to improve the motion robustness of automatic RoI detection for rPPG monitoring. The proposed method consists of three steps: (1) Subregion tracking: the frames in a pre-defined time interval are segmented into subregions for local pulse-signal extraction; (2) Feature extraction: a set of spatiotemporal features (e.g., pulse-signal frequency and skin chromaticity) are constructed to discriminate the skin/non-skin regions; and (3) Skin-region clustering: the subregions are classified into the skin/non-skin clusters using the feature vectors, and the skin cluster is found as the RoI for pulse-rate monitoring.

The organisation of this paper is as follows: In Sect. 2, we provide an overview of the proposed method and illustrate the steps in detail. In Sect. 3, the experiment and evaluation metric are set up to verify the proposed method. We discuss the experimental results and show the findings in Sect. 4 and finally in Sect. 5, we draw the conclusions of this work.

2   Method

2.1   Subregion Generation

Given a video sequence containing a subject, our goal is to find the RoI that is optimal for pulse-rate measurement. In fact, the pulse-signal can only be extracted from *skin pixels* in the video, so the non-skin pixels (e.g., background) should be discarded for deriving a clean and accurate pulse-signal. Therefore, we propose to initialise a RoI that only contains the skin pixels. Since no prior knowledge/assumption is posed on the location of skin pixels, we first segment the whole video into subregions, where the pulse-signals can be locally measured without inference. Afterwards, the subregions that present clean pulse-signals are automatically selected to initialise the RoI for pulse-rate monitoring.

Since pulse-signals need to be measured in the time-domain, the subregions generated in each video frame have to be concatenated in a time interval (e.g., the interval defined for pulse-signal extraction). Considering the influence of subject motions, the subregions formulated in the time domain may shift the location and lose the temporal consistency. To solve this problem, we propose to segment the video based on the feature corners, which are relatively easy to be tracked between consecutive frames. A spatial mesh, consisting of multiple triangles, is constructed to segment the whole image into subregions by connecting the tracked feature corners.

In this step, we employ the Harris corner detection [9] to find the salient feature points for tracking, which is only ran on the first frame of the pre-defined interval. Subsequently, the detected feature corners are temporally tracked across the remaining frames within the interval using the pyramid optical flow developed by Lucas & Kanade [10]. The outliers, i.e., the corners that disappear or exhibit large tracking errors, are rejected in a forward-backward validation procedure for maintaining the tracking coherence [11]. Finally, we rely on the Delaunay triangulation [12] to connect the tracked feature corners for segmenting the whole image into subregions spatially. The reason for using Delaunay triangulation is that it attempts to maximise the minimum angle of all the angles of the triangles during the triangulation, and thus leads to a more compact image partition, i.e., skinny triangles are avoided.

An example of this step is shown in Fig. 1: (1) shows the tracked feature corners in a predefined time interval. In textured regions like the hand, more feature corners can be detected than the plain regions (e.g., background wall); and (2) shows that the whole image is segmented into multiple triangular subregions, in which some are skin regions (e.g., subregion A) that contain pulse-signals while others are background scenes (e.g., subregion B, C and D). The following steps aim to distinguish the skin regions from non-skin regions (background) by feature extraction and clustering.

2.2   Feature Extraction

Spectral Features. Following the region tracking, we use the chrominance-based method (CHROM) proposed by de Haan & Jeanne [6] to extract the pulse-signals from subregions in the time interval. CHROM uses a combination of two orthogonal chrominance signals $X = R-G$ and $Y = 0.5R-0.5G-B$ and is therefore capable of eliminating the motion-induced specular reflection changes in a white light luminance condition. In order to enable the correct functioning with colour light source, skin-tone standardisation is applied by adapting the equation as:

$$\begin{aligned} X_n &= 3R_n - 2G_n \\ Y_n &= 1.5R_n + G_n - 1.5B_n \end{aligned} \tag{1}$$

where R_n, G_n and B_n are colour channels normalised to their mean over the interval, which make the pulse-signal independent of the brightness of the light

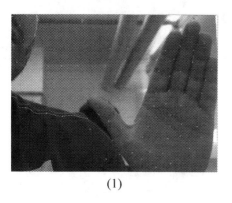
(1)

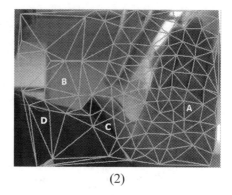
(2)

Fig. 1. An example of subregion generation in the video containing a subject hand. (1) the feature corners in the whole image are detected and tracked in the predefined time interval, (2) Delaunay triangulation is performed to connect the feature corners as an oversegmented mesh, where each triangle is a subregion for local independent pulse extraction.

source. The pulse-signal is defined as:

$$S = X_n - \alpha Y_n, \tag{2}$$

with

$$\alpha = \frac{\sigma(X_n)}{\sigma(Y_n)}, \tag{3}$$

where $\sigma(\cdot)$ denotes the standard deviation operation. Given the fact that the pulse-rate of a healthy subject is within the frequency-band $[40, 220]$BMP, we band-pass filter the S to eliminate out-of-band frequency noise. Figure 2 shows an example of normalised RGB traces and corresponding pulse-signals extracted from skin and non-skin regions respectively.

Obtaining the pulse-signal for each subregion, we can extract the spectral features (e.g., frequency and phase) to summarise the unique characteristics of skin regions. A commonly used term to describe a pulse-signal is the "periodicity", so spectral analysis is performed on S using the Discrete Fourier Transfer (DFT). To retrieve an acceptable spectral resolution, the length of the interval for deriving a pulse-signal is set to $N = 128$ (6.4 s duration in 20 FPS camera) while its frequency spectrum is interpolated by 512 FFT bins. In practice, subject motions and camera noise may degrade the pulse-signals' quality. To reduce the influence of such noise, the pulse-signals are auto-correlated before extracting the periodicity features as:

$$A[\tau] = \sum_{n=0}^{N-1} S[n]S[n - \tau], \quad -\frac{1}{2}N < \tau < \frac{1}{2}N. \tag{4}$$

Figure 2 shows the spectrum magnitude of the pulse-frequency and its auto-correlation. Apparently, the spectrum derived from skin region shows a clear frequency-peak within the pulse-rate frequency-band, whereas the spectrum

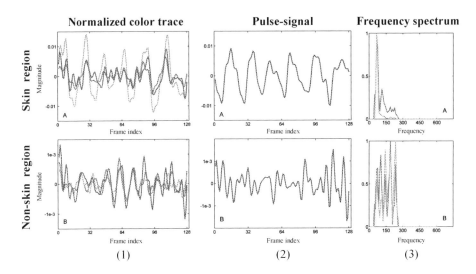

Fig. 2. The signals in the first row are derived from a skin region while the signals in the second row are derived from a non-skin region. (1) the normalised colour channels R_n (red), G_n (green) and B_n (blue); (2) the extracted pulse-signals S using CHROM; and (3) the normalised frequency spectrums (red) and corresponding auto-correlation A (green) of the pulse-signal (Color figure online).

derived from non-skin region does not show a pattern (random noise). In addition, it shows that auto-correlation can significantly suppress the frequency noise in the spectrum.

The frequency F corresponding to the highest peak in the spectrum is considered to be the pulse frequency, which is calculated as:

$$F = \hat{k} \cdot \frac{f_s}{N}, \tag{5}$$

with

$$\hat{k} = \arg\max_{k} \left\{ |\hat{M}[k]| \mid k \in [0, (N-1)/2] \right\} \tag{6}$$

where f_s represents the sampling frequency; N is the parameter controlling the resolution of frequency spectrum; $\hat{M}$ denotes the spectrum magnitude. Furthermore, the pulse-signals extracted from different subjects mostly undergo a different phase, so another commonly used term "spectrum phase" is employed as a discriminate feature, which is calculated as:

$$\theta = \arctan\left(\frac{\mathrm{Im}(S[\hat{k}])}{\mathrm{Re}(S[\hat{k}])}\right) \tag{7}$$

where $\theta \in [-\pi, \pi]$. At this point, the extracted pulse-signal is summarised by two spectral features: frequency (F) and phase (θ), which is sufficient to distinguish between a pulse-signal and a noise signal.

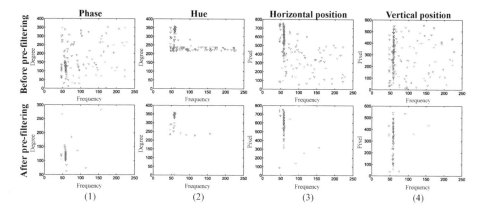

Fig. 3. An example of subregions' distribution in multi-dimensional feature space, where each point represents a subregion. The scatter plots in the first row represent all the oversegmented subregions without pre-filtering; the scatter plots in the second row represent the remaining subregions after the pre-filtering step, where the subregions that explicitly belong to background are filtered out.

Spatial Features. Inspired by the colour-based skin segmentation, we exploit the chromaticity property of skin for skin/non-skin regions classification. In spatial domain (each single frame), image pixels are transformed from RGB colorspace to Hue-Saturation-Value (HSV) colorspace, where the intensity is separated from the intrinsic information of object chromaticity. After that, we only adopt H (hue) component as the spatial feature. Note that in this study, the skin chromaticity feature is exploited in an unsupervised way, which is different from earlier works that used prior knowledge or pre-assumed threshold for skin segmentation. There are two observations behind this step: (1) the skin pixels belonging to the same subject are homogeneous in hue, (2) eliminating S and V components can enhance the robustness towards illumination changes and shadows.

Moreover, the skin regions corresponding to a particular body part are closely located from each other between adjacent frames. The position of the geometric centre $\overline{P}$ of each subregion is calculated as:

$$\overline{P} = \frac{1}{N} \sum_{i=0}^{N-1} P_i, \tag{8}$$

where N is the number of tracked feature corners belonging to a specific subregion; P_i is the spatial position of ith feature corner of this subregion.

To this end, a multi-dimensional feature vector, $\overrightarrow{v} = [F, \theta, H, \overline{P}]$ is constructed as the representation of a subregion, which takes the advantage of pulse and chromaticity properties of skin. Figure 3 shows the scatter plots of subregions' distribution in the feature space, where each point represents a subregion. As can be seen, there is an area has higher density than others, which implies that a group of feature vectors show similar patterns in the feature space.

2.3   Skin-Region Clustering

Pre-filtering. To reduce the obvious non-skin subregions before the skin-region clustering, a pre-filtering step is performed to identify the explicit background regions using certain characteristics of the pulse-signal. Here we introduce several essential conditions for classifying a subregion into non-skin category in a prior stage.

Due to the optical absorptions of blood pulse, the amplitudes of normalised colour traces from skin regions are significant larger than that from non-skin regions, as can been seen in Fig. 2. So the colour variations induced by pulse and noise can be differentiated by the maximum amplitude of normalised colour traces as:

$$\delta_{C_n,max} = \max\{C_n[i] \mid i \in [0, N-1]\} \tag{9}$$

where N is the interval length; $C_n \in \{R_n, G_n, B_n\}$. In our method, the regions whose maximum amplitudes are not in the range $[0.005, 0.15]$ are identified as background.

Essentially, pulse-signals extracted from skin regions exhibit different amplitudes but remain similar phase due to the different blood absorption rates of the light spectrum. Assuming x and y to be two traces (e.g., R_n and G_n) belonging to different colour traces in a subregion, the normalised Sum of Absolute Differences (SAD) is calculated to assess the amplitude differences between the colour traces as:

$$\delta_{SAD} = \frac{\sum_i | x[i] - y[i] |}{\sum_i | x[i] |}, \tag{10}$$

where the subregions with δ_{SAD} outside the range $[0.05, 0.5]$ are identified as background. Afterwards, the Normalised Correlation Coefficient (NCC) between x and y is calculated to find the phase similarities between the colour traces as:

$$\delta_{NCC} = \frac{\sum_i x[i] \cdot y[i]}{\sqrt{\sum_i x[i]^2 \cdot \sum_i y[i]^2}}, \tag{11}$$

where the subregions with δ_{NCC} outside the range $[0.5, 0.99]$ are identified as background. Given that fact that the subregion with real pulse-signal presents a strong frequency-peak in the frequency domain, the ratio between the two largest peaks in the frequency spectrum is calculated to determine the presence of dominant frequency of the pulse as:

$$PR = \frac{\hat{M}_{max2}}{\hat{M}_{max1}}, \tag{12}$$

where $\hat{M}$ represents the amplitude-normalised frequency spectrum. For $\hat{M} \in [0, 1]$, the upper-bound suggests that the amplitudes of the two largest peaks are identical, while the lower-bound indicates that the spectrum is dominated by one frequency. Therefore, the regions where $PR > 0.6$ are identified as background. Moreover, the amplitude of frequency-peak is calculated as:

$$\delta_{\hat{M},max} = \max\{\hat{M}[i] \mid i \in [0, N-1]\}, \tag{13}$$

where the regions with low frequency energy ($\delta_{\hat{M},max} < 0.005$) are identified as background. Note that parameters discussed in this section are specified *empirically*. Figure 3 shows an example of subregions' distribution in feature space after pre-filtering step. The pre-filtering step can effectively prune the subregions that are obviously not belonging to the skin category.

DBSCAN Clustering. After eliminating the explicit background regions in the pre-filtering step, the remaining regions will be clustered into different groups based on extracted feature vectors $\overrightarrow{v}$. In the scatter plots of the feature space of Fig. 3, we can clearly recognise an area where the points' density is considerably higher than other areas. The points from the dense areas are considered to be skin (inliers with similar patterns), whereas the points located outside the dense areas are regarded as background noise (outliers). Based on such observation, we employ a clustering method called Density-Based Spatial Clustering of Applications with Noise (DBSCAN) [13], the clustering method that typically relies on the density notion of data, to separate the skin/non-skin clusters in the feature space.

DBSCAN separates the data into three different types (e.g., cluster core point, border point and noise point) using a density threshold with two parameters: ε specifies a range and $minPts$ denotes a minimal number of sample in a cluster. In our method, data for clustering are the feature vectors in multi-dimensional space. The range of each element in $\overrightarrow{v}$ is controlled by a scaling factor. To determine whether the data samples are located between each other, a boolean factor is calculated as:

$$B_{i,j} = \begin{cases} \text{true} & -\overrightarrow{v}_i - \overrightarrow{v}_j - \leq \epsilon \\ \text{false} & \text{elsewhere} \end{cases} \qquad (14)$$

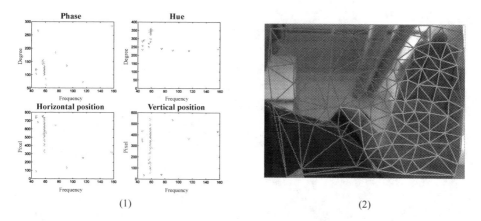

Fig. 4. An example of DBSCAN clustering on feature vectors. (1) scatter plots of the feature vectors along different directions (e.g., phase, hue, x and y) after clustering. The green points represent the feature vectors classified to a skin cluster; (2) the corresponding clustering result in the image (Color figure online).

where $\vec{v}_i$ and $\vec{v}_j$ are two feature vectors; the range $\epsilon = [3, 10, 10, 100]$ is empirically determined. Note that the DBSCAN algorithm exploits the inner distribution/structure of data (e.g., density) for clustering, which does not require a predefined cluster number in advance. Figure 4(1) shows an example of clustering result using the specified parameters, where the feature vectors further to the dense area are classified as background. Figure 4(2) shows the corresponding result in a image, in which most of the subregions representing the skin are clustered correctly.

Cluster Growing. We observe that in some cases, a small amount of skin regions may be falsely identified as the background in the pre-filtering step. Although these regions lack the essential conditions for pre-filtering, they still show a pulse-signal which is sufficient to be clustered into skin group using DBSCAN. In order to recover these misclassified skin regions, a cluster growing procedure is performed. In this process, the average pulse-signal of a cluster is compared to the pulse-signals from its surrounding subregions using Eq. (11). When the pulse-signal of a surrounding subregion is highly correlated with the averaged pulse-signal from the skin cluster ($\gamma_{NCC} > 0.5$), the subregion is added to the cluster. Since the amplitudes of pulse-signals extracted from subregions inside the skin cluster are mostly different, the auto-correlated pulse-signals are normalised as:

$$Sn_i = \frac{S_i}{\sigma(S_i)} \tag{15}$$

where $\sigma(\cdot)$ corresponds to the standard deviation operator. To minimise the impact of noise when combining multiple pulse-signals into a single averaged pulse-signal, we perform the *alpha-trimmed mean* to reject the outliers with extreme values, i.e., α is set to 0.5 in our method.

3   Experiments

3.1   Experimental Setup

The experimental setup consists of a standard 768×572 pixels, 8bit, global shutter CCD camera (type USB UI-2230SE-C of IDS GmbH) operating at 20 frames per second and focused at the subject's face using a flexible C-mount lens (Tamron 21VM412ASIR). The duration of each recording is set to approximately 90 s. The recorded videos are stored uncompressed. We recruit 6 healthy subjects with an equal number of male and female in 3 different skin-types according to the Fitzpatrick scale [14]: *Skin-type II*, *Skin-type III* and *Skin-type V*, as shown in the snapshot of Fig. 5. To mimic the real use-case scenarios for recording, each subject is asked to perform 6 types of head motion: *stationary*, *translation*, *scaling*, *non-rigid* (talking), *rotation* and *mixed motion* (a mixture of all those movements). All recordings are made in a controlled environment using a single illumination source located in front of the subject. In parallel to the video, we synchronously record the raw pulse-oximeter data (PPG signal) from a transmissive pulse-oximeter finger clip of Contec Medical Systems (model CMS50E) using the USB protocol on the device. Both the rPPG and PPG signals are

Skin-type II male Skin-type II female Skin-type III male Skin-type III female Skin-type V male Skin-type V female

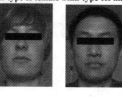

Fig. 5. A snapshot of six subjects with three different skin-types in our recorded video sequences.

treated equally in the post-processing like band-pass filtering. To calculate the pulse-rates, we detect the bin index of the frequency-peak using a 512 point FFT on the Hanning windowed pulse-signals.

3.2 Evaluation Metric

In line with [6], the long-term pulse-signal across the complete video sequence is generated by overlap-adding the signal intervals with a hanning window. The quality of the extracted pulse-signal is measured in Signal-to-Noise Ratio (SNR), which is defined as the ratio between the pulse in-band spectrum energy (1st and 2nd harmonics of the spectrum) and the remaining spectrum energy as:

$$\text{SNR} = 10\log_{10}\left(\frac{\sum_{f=40}^{240}(U_t(f)\hat{M}(f))^2}{\sum_{f=40}^{240}((1-U_t(f))\hat{M}(f))^2}\right), \tag{16}$$

where $\hat{M}$ is the spectrum amplitude of the pulse-signal S; f is the frequency range in Beats Per Minute (BPM); $U_t(f)$ a binary template function made by selecting the first and second harmonics according to the reference PPG-signal. In order to allow for pulse-rate variability, $U_t(f)$ is defined in such a way that 22 bins in the 512 bin spectrum centred around the PPG pulse-rate (9 bins around the first harmonic and 13 bins around the second harmonic) are passed as in-band pulse-signal.

In addition, the pulse-rate accuracy, the percentage of alignment of the pulse-rates given by rPPG and reference PPG sensor, is measured for comparing the rPPG methods' performance. Note that the pulse-rate is calculated by detecting the frequency of the spectrum peak in a pre-defined time interval.

3.3 Compared Method

To benchmark the proposed method, the pulse-signal obtained by our method is compared to that obtained by the rPPG method using automatic face detection [4]. In the compared method, a Viola-Jones face detector [15] is used to locate the face with a bounding box. The middle 60 % width and 100 % height of the bounding box is used as the RoI. The averaged RGB values of pixels inside the RoI are calculated for deriving the pulse-signal using CHROM. In cases that the face (e.g., side-view faces) cannot be detected, the pulse-signal is interpolated with zeroes to ensure the continuity.

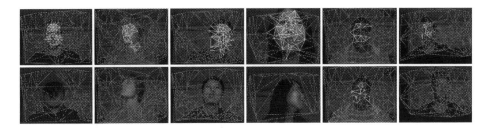

Fig. 6. Examples of detected RoI in recorded video sequences using our method. The first row indicates the cases where skin-regions are correctly found. The second row indicates the cases where our method fails to handle.

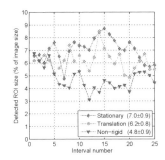

Fig. 7. The consistency of the found RoI is measured in videos of Skin-type II male with stationary, translation and non-rigid motion, where the subject skin captured by camera is assumed to be constant.

4   Results and Discussion

In Fig. 6, it shows the examples of the found RoI for the video sequences using the implementation of the proposed method. As can be seen in the first row of Fig. 6, most of the skin regions are correctly found for the Skin-type II and Skin-type III subjects, whereas the number of skin regions can be found for Skin-type V subjects is significantly reduced. The reason is related to the compositions of skin tissues: the melanin content is much higher in dark skin (e.g., Skin-type V), which absorbs part of the diffuse light reflections that carry the pulse-signal. Thus it leads to a reduced fraction of the diffusely reflected light from the skin [6]. Particularly for the Skin-type V male, our method fails to find the RoI for the complete video sequences using the fixed parameters. Thus the minimum distance between tracked feature corners is increased to 40 pixels while the minimum number of regions in a cluster is decreased to 3 for this specific subject. However, the assumption that the content (e.g., parameters) of a given region is fixed may not be valid anymore. At the same time, decreasing the cluster size could lead to more clusters, and thus increase the risk of false positive detection. As can be seen in the second row of Fig. 6, our method fails to find the skin regions when the subjects' motions are vigorous. The performance degradation

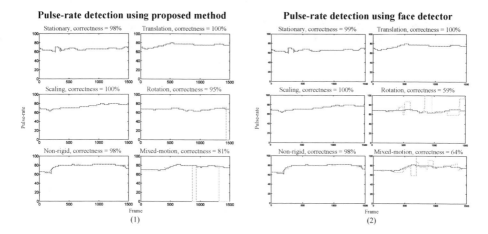

Fig. 8. Example of pulse-rates estimated from Skin-type II male with 6 types of motion. (1) pulse-rates obtained by the proposed method, (2) pulse-rates obtained by the method using Viola-Jones face detector. The x-axis depicts the frame number where 1800 frames correspond to 80 seconds in the video (20 FPS video recording).

is caused by the failure of feature corners tracking. The drifting of feature corners during the large pixel displacement results in the shape change of triangles, which interrupts the temporal consistency for pulse-signal measurement.

To investigate the performance consistency of the proposed method, we analyse the number of skin pixels in the found RoI for the video sequence of Skin-type II male with 3 types of motion: stationary, translation and non-rigid, where the amount of skin pixels captured by camera is assumed to be constant. In Fig. 7, the consistency of the found RoI is denoted as a percentage between the RoI size and image size (e.g., 768×572 pixels). As expected, the proposed method gains the best consistency in videos with stationary subject. In contrast, the results obtained in videos with translation and non-rigid motions are less consistent, which is due to the fact that some skin pixels or regions appear/disappear during the head movements. In general, the proposed method remains fairly consistent performance (with a maximum standard deviation of $\sigma = 1.0$) in dealing with subject motions like translation and non-rigid motion.

In Fig. 8, the pulse-rate obtained by the proposed method is compared to that obtained by the rPPG method with Viola-Jones face detector. The comparison is performed on the videos of Skin-type II male with 6 types of motion. As can be seen that for videos without subject rotation, the pulse-rates obtained by our method remain 98 % of the time consistency with the reference pulse-rates (difference smaller than 3 BPM). For the videos including subject rotation, the accuracy of the pulse-rate obtained by our method decreases to 81 % as the result of the failure in feature corners tracking. The decrease in pulse-rate accuracy is also significant for the face detector (59 %) as the Viola-Jones detector trained with frontal face samples is unable to find the side-view faces. Nevertheless, in situations where the actual pulse-rate is missing, our method outputs a momentary pulse-rate of 0 BPM, whereas the face detector still produces a significant

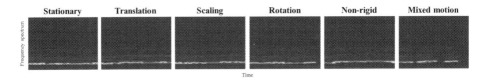

Fig. 9. The frequency spectrum of the pulse-signals obtained from Skin-type II male with 6 types of motion.

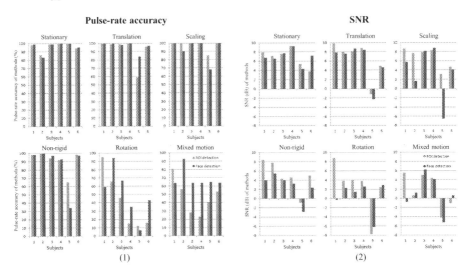

Fig. 10. The pulse-rate accuracy and SNR obtained by the compared methods (the proposed method and the method using Viola-Jones face detector) on 36 video sequences (6 subjects performing 6 types of motion). (1) pulse-rate accuracy, (2) SNR quality.

frequency-peak in the spectrum that does not correspond to the pulse-rate. The reason is that the stronger frequency components are introduced into the pulse-signal due to the repeated and abrupt detection failure of the face detector. Figure 9 provides an additional view of the frequency spectrum of the pulse-signals extracted from 6 types of motion of Skin-type II male, which shows clear pulse-rate frequencies except for the mixed motion.

In Fig. 10, the pulse-rate accuracy and SNR obtained by both methods (the proposed method and the method with Viola-Jones face detector) are given for 36 video sequences (6 subjects with 6 types of motion). Figure 10(1) shows that in the videos without subject rotation (e.g., stationary, translation, scaling, and non-rigid motion), our method achieves an average pulse-rate accuracy of 94 %, which is comparable to that obtained by the rPPG method with face detector (92 %). By comparing the SNR for those video sequences, our method significantly improves the method using face detector from 5.8 dB to 8.6 dB, which is clearly better. The improvement in SNR is probably due to the reason that our method adopts only the skin pixels that present strong pulse-frequency feature for combination, whereas the method with face detector use all the pixels within RoI (including non-skin pixels like nostril and eyebrows) for averaging. This is further demonstrated in the comparison of videos with stationary and non-rigid motion: both methods achieve high pulse-rate accuracy and SNR in videos with stationary subjects, where no significant motion noise is introduced. However, in the videos with non-rigid motion (talking), the performance of the proposed method is apparently better, because it excludes the noisy regions around the subjects' moving mouth/lips where the pulse-signals are distorted by motion. However, both methods suffer from dramatic performance degradation in videos with vigorous head motions such like rotation and mixed motion. As explained before, the major reason causes the failure of the proposed method is due to the problem of feature corners tracking.

5   Conclusions

In this paper, an automatic RoI detection method for camera-based pulse-rate measurements is presented. The proposed method consists of three main steps, namely subregion tracking, feature extraction and skin region clustering. In the first step, subregions are established using triangulation on the tracked feature corners during a pre-defined interval in the video. In the second step, pulse-signals are extracted from the subregions and the spatiotemporal features (e.g., pulse-signal and skin chromaticity) are formulated as discriminative representation. In the third step, subregions containing similar features are classified into the same clusters using a density-based clustering method.

The proposed method is experimentally verified on 36 video sequences consisting of 6 subjects with 3 skin-types and 6 motion-categories. For the videos of subjects with brighter skin and modest motions (without rotation), the proposed method obtains the comparable pulse-rate accuracy but better SNR as compared to the rPPG method using the face detector, which demonstrates the effectiveness of our strategy of only exploiting the skin pixels containing pulse-signals for pulse-rate measurements. However for videos of subjects with darker skin-types or vigorous motions, the pulse-rate accuracy and SNR obtained by both methods decline. In general, compared to the rPPG method using face detection, the novel automatic RoI detection method developed in this study, which takes the advantage of pulse-signals, demonstrates a favourable performance on subjects with brighter skin-tone, yet it is robust to modest body motions.

References

1. Hertzman, A.B.: Photoelectric plethysmography of the fingers and toes in man. Exp. Biol. Med. **37**, 529–534 (1937)
2. Takano, C., Ohta, Y.: Heart rate measurement based on a time-lapse image. Med. Eng. Phys. **29**, 853–857 (2007)
3. Huelsbusch, M., Blazek, V.: Contactless mapping of rhythmical phenomena in tissue perfusion using PPGI. In: Proceedings of the SPIE, vol. 4683, pp. 110–117 (2002)

4. Poh, M.Z., McDuff, D.J., Picard, R.W.: Non-contact, automated cardiac pulse measurements using video imaging and blind source separation. Opt. Express **18**, 10762–10774 (2010)
5. Schmitz, G.: Video camera based photoplethysmography using ambient light. M.S. thesis dissertation, Electrical Engineering. Technische Universiteit Eindhoven, The Netherlands (2011)
6. de Haan, G., Jeanne, V.: Robust pulse rate from chrominance-based RPPG. IEEE Trans. Biomed. Eng. **60**, 2878–2886 (2013)
7. Lewandowska, M., Ruminski, J., Kocejko, T., Nowak, J.: Measuring pulse rate with a webcam - a non-contact method for evaluating cardiac activity. In: 2011 Federated Conference on Computer Science and Information Systems (FedCSIS), pp. 405–410 (2011)
8. Xu, Z., Zhu, M.: Color-based skin detection: survey and evaluation. In: 2006 12th International Multi-Media Modelling Conference Proceedings, p. 10 (2006)
9. Harris, C., Stephens, M.: A combined corner and edge detector. In: Proceedings of the Fourth Alvey Vision Conference, pp. 147–151 (1988)
10. Lucas, B.D., Kanade, T.: An iterative image registration technique with an application to stereo vision. In: Proceedings of the 7th International Joint Conference on Artificial Intelligence, IJCAI 1981, vol. 2, pp. 674–679. Morgan Kaufmann Publishers Inc., San Francisco (1981)
11. Kalal, Z., Mikolajczyk, K., Matas, J.: Forward-backward error: automatic detection of tracking failures. In: 2010 20th International Conference on Pattern Recognition (ICPR), pp. 2756–2759 (2010)
12. Delaunay, B.: Sur la sphère vide. a la mémoire de georges voronoï. Bulletin de l'Académie des Sciences de l'URSS. Classe des sciences mathématiques et na. **6**, 793–800 (1934)
13. Ester, M., Kriegel, H.P., Sander, J., Xu, X.: A density-based algorithm for discovering clusters in large spatial databases with noise. In: KDD, pp. 226–231 (1996)
14. Fitzpatrick, T.: The validity and practicality of sun-reactive skin types i through vi. Arch. Dermatol. **124**, 869–871 (1988)
15. Viola, P., Jones, M.: Rapid object detection using a boosted cascade of simple features. In: Proceedings of the 2001 IEEE Computer Society Conference on Computer Vision and Pattern Recognition, CVPR 2001. vol. 1, pp. I-511–I-518 (2001)

Sparse Optimization for Motion Segmentation

Michael Ying Yang$^{(\boxtimes)}$, Sitong Feng, and Bodo Rosenhahn

Institute for Information Processing (TNT), Leibniz University Hannover,
Hannover, Germany
yang@tnt.uni-hannover.de

Abstract. In this paper, we propose a new framework for segmenting feature-based multiple moving objects with subspace models in affine views. Since the feature data is high-dimensional and complex in the real video sequences, most traditional approaches for motion segmentation use the conventional PCA to obtain a low-dimensional representation, while our proposed framework applies the sparse PCA (SPCA) to obtain a projected subspace, which is a low-dimensional global subspace on a Stiefel manifold with sparse entries. Then, the local subspace separation is achieved via automatically selecting the sparse nearest neighbours. By combining two sparse techniques, the proposed framework segments different motions through a simple spectral clustering on an affinity matrix built with the principal angles. To the best of our knowledge, our framework is the first one to apply the sparse optimization for optimizing the global and local subspace simultaneously. We test our method extensively and compare its performance to several state-of-art motion segmentation methods with experiments on the Hopkins 155 dataset. Our results are comparable with these results, and in many cases exceed them both in terms of segmentation accuracy and computational speed.

1 Introduction

Motion segmentation aims to decompose a video sequence into different moving objects that move throughout the sequence. In the recent years, the tracked features based motion segmentation problem has motivated amount of people to find a fast and high accuracy method. Particularly, different with the traditional motion segmentation method based on the pixel-wise model such as [1,2], which is focused on segmenting the foreground moving objects from their background in an unannotated video, the segmentation of only the few number of tracked features on each moving object can not only solve the occlusions in the video, but also save computation time w.r.t. the pixel-wise methods. Motion segmentation approaches which based on the tracked features of a moving body focus on clustering the sparse or dense feature points into different regions and each region represents a moving object in the sequence.

The goal of a general feature-based motion segmentation is to cluster the union of different point trajectories with different labels and different labels represent the different motions, as shown in Fig. 1. We address the feature trajectories segmentation as a subspace clustering problem under the affine camera

© Springer International Publishing Switzerland 2015
C.V. Jawahar and S. Shan (Eds.): ACCV 2014 Workshops, Part II, LNCS 9009, pp. 375–389, 2015.
DOI: 10.1007/978-3-319-16631-5_28

model. Under the model of subspace, the trajectories which can be extracted during a preprocessing step, such as KLT [3], SIFT [4] or SURF [5], are embedded in a union of local subspaces and each trajectory is represented with a local subspace. Then, the problem of segmenting different feature trajectories changes to cluster the union of different local subspaces.

Fig. 1. Examples of our segmentation result on the real sequences *cars9* and *kananchi1* from the Hopkins155 dataset. (The first row: *from left to right* are the results for frame 1, 40 and 50; the second row: *from left to right* are the results for frame 1, 10 and 20.

Contributions. Our main contribution is an efficient framework for motion segmentation which based on subspace clustering with sparse optimization. We combine two state-of-art sparse representations to optimize both the global and local estimation. *Sparse Principle Components Analysis*(SPCA) is applied for the global optimization, in the same time, we seek a sparse representation for the closest neighbours for the local subspace separation with computing principal angles. As illustrated in Fig. 1, our method can clearly label the moving objects that tracked throughout a video. To the best of our knowledge, our framework is the first one to apply the sparse optimization for optimizing the global and local subspace simultaneously.

The following sections are organized as follows. The related works are discussed in Sect. 2. Section 3 introduces the subspace models for motion segmentation. Our proposed approach described in detail in Sect. 4. In Sect. 5, experimental results are presented. Finally, this work is concluded and future work is discussed in Sect. 6.

2   Related Work

There are a large amount of works on motion segmentation. In general, the methods for motion segmentation can be divided into two classes: affinity-based and subspace-based methods. The affinity-based approach [6] is based on computing the affinity between a pair of trajectories. Our approach is concentrated on the subspace methods, which focuses on segmentation of the different motions with finding the membership from a union of subspaces. The subspace-based methods can divided into: iterative methods, algebraic solution, compressive sensing and subspace estimation. One of the iterative methods is RANSAC algorithm [7], which can deal with outliers and noise. But the number of subspaces needed to be known as prior knowledge, and they require a good initial estimation and parameter selection. The most popular method based on the algebraic solution is the *Generalized Principal Component Analysis* (GPCA) [8]. While GPCA gives a good performance for the subspaces with different dimensions, but GPCA is not robust to the noise and outliers. Agglomerate Lossy Compression (ALC) [9] is a method using the compressive sensing on subspace model. The ALC method is robust to noise and outliers without knowing the subspace dimension and the number of the subspace, but it is highly time-consuming. The other application of compressive sensing on subspace models is based on the sparse representation. One of the most popular methods is the Sparse Subspace Clustering (SSC) [10]. SSC has a goodt performance on the motion segmentation with a part of missing data. But the computation time is quite large.

Our work is most related to Local Subspace Affinity (LSA) [11], which belongs to the subspace estimation method. The overall procedure of LSA is that after the data projection of the feature points that lie in a global low-dimensional subspace. The estimation of the local subspaces can be obtained by computing its nearest neighbours(NNs) and SVD. After the local subspaces have been achieved, LSA builds the affinity matrix by using the principle angles between each local subspace. In the end the subspace segmentation is accomplished by applying the spectral clustering on the affinity matrix. The rank estimation for the global and local subspace is achieved by a model selection(MS) method. The drawback of LSA is that the number of nearest neighbours to estimate the local subspace may lead to the overestimation problem. It means that the nearest neighbours may not belong to the same subspace. This situation is more likely to happen particular with the non-rigid or degenerated motions. The second limitation is that the rank estimation by model selection is based on the parameter k which has to be set depending on the noise and the number of motions. Thus the model selection needs to know the amount of noise as the prior knowledge as well.

3   Subspace Models for Motion Segmentation

In this section, we first introduce the basic idea about using the subspace models for motion segmentation. Subsequently, under the affine camera model we analyse the subspace methods for motion segmentation and show that under the subspace models it is equivalent to clustering multiple low-dimensional linear subspace in a high-dimensional space.

3.1   Subspace Clustering

In practice, in order to clustering the high-dimensional data from the real-world video sequences, one needs to first look for a low-dimensional representation of the high-dimensional data. After projection of the high-dimensional data, the obtained low-dimensional subspace is embedded in a union of multiple subspaces. If we consider the union of low-dimensional projection space as a global subspace, the underlying multiple subspaces can be regarded as the local subspaces. One of the main tasks of subspace clustering is to find out the number of different local subspaces, the other is the separation of the multiple local subspaces which means that one needs to cluster the data according to different subspace, the data from the same subspace should be classified together. It means that the task of cluster the feature data according to different motions changes to cluster the data into different subspaces.

3.2   Multi-body Motion Segmentation with Affine Camera Model

Recently, most popular algorithms for performing the motion segmentation are assuming an affine camera model, which is useful to weak and paraperspective camera models. In this paper, the affine camera model is also used for the moving object in the video sequence. Affine camera model can transform the tracked feature points on the moving object from 3-D coordinates to 2-D position, which is formulated as [12]

$$x_{fp} = A_f \begin{bmatrix} X_F \\ 1 \end{bmatrix}, \tag{1}$$

where the $X_F = \begin{pmatrix} X \\ Y \\ Z \end{pmatrix}$, represent the world coordinate, $A_f = [R_{2f \times 3} | T_{2f \times 1}$ is

the 2×4 affine transformation matrix in the f frame, $\{x_{fp} \in R^2\}_{p=1,\ldots,P}^{f=1,\ldots,F}$ denote the 2-D location of tracked feature trajectory at frame f.

A general input for the motion segmentation under the affine camera model can be formulated as a data matrix containing all of the 2-D positions of tracked features, so-called the trajectory matrix

$$\begin{bmatrix} x_{11} \cdots x_{1P} \\ \vdots \\ x_{F1} \cdots x_{FP} \end{bmatrix}_{2F \times P} = \begin{bmatrix} A_1 \\ \vdots \\ A_F \end{bmatrix}_{2F \times 4} \begin{bmatrix} X_1 \cdots X_P \\ 1 \cdots 1 \end{bmatrix}_{4 \times N}, \tag{2}$$

One can rewrite this as

$$\begin{bmatrix} x_{11} \cdots x_{1P} \\ \vdots \\ x_{F1} \cdots x_{FP} \end{bmatrix} = W_{2F \times P}, \tag{3}$$

$$W_{2F \times P} = M_{2F \times 4} S_{N \times 4}^T, \tag{4}$$

where we call the M as the motion matrix, whereas the S is the structure matrix. The rank of the W trajectory matrix is no more than 4. As a result, the rank of the general trajectory matrix for rigid motion is at most 4.

4   Proposed Framework

Our proposed framework extends the framework of LSA [11] in both the global and the local parts, as shown in Fig. 2. Transformation of the trajectory matrix with the Sparse Principal Component Analysis (SPCA) is used for the global subspace estimation. For the local subspace estimation, instead of the fixed number of local nearest neighbours policy, we adopt a sparse manifold optimization from [13] to automatically extract each local low-dimensional subspaces. In Sect. 4.1, the SPCA is presented. The sparse manifold optimization technique for local subspace estimation is presented in Sect. 4.2.

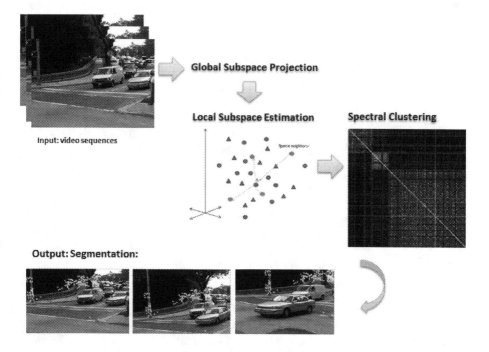

Fig. 2. Proposed framework overview.

4.1   Global Subspace Transformation

As described in Sect. 3.2, the dimensional of a general input trajectory data matrix for subspaces methods for motion segmentation is $2F$, where F denotes the number of the frames. However, the maximum rank of the subspace for one rigid motion is 4. Therefore most of the rank of the trajectory data is redundant.

This has motivated a lot of algorithms to discover a low-dimensional projection for the high-dimensional trajectory data matrix.

Assume that the trajectory matrix $W_{2F}^P \in R^{2F}$ as Eq. 2 is given as input, from Sect. 3.2 the rank of the general motion segmentation for one rigid motion is bounded by 4. This constraints has enforced a customary dimensionality reduction procedure for trajectory matrix W_{2F}^P. Most of the other subspace methods choose to reduce the dimension of W_{2F}^P to $m = 4n$, where n is the number of motions. Following classical PCA we can use the singular value decomposition(SVD) to the matrix W, which decompose the W as follows: $U_{2F \times 2F} D_{2F \times P} V_{P \times P}^T = W_{2F}^P$. A global data transformation is then obtained by considering only the first m columns of V. In the case of SPCA we need to have a sparse construction in the matrix V, for illustration we can choose a l_0-penalty for block sparse PCA through a generalized power method proposed by [14]. After solving a sparse optimization problem on the high-dimensional motion data, we can obtain a new representation matrix denoted by Z^* of W_{2F}^P on the Stiefel manifold S_m^P. Each column of Z^* is a sparse vector $z_i^*, i = 1, \cdots, m$ represents the transformed data.

In order to enforce the sparse entries on the principal components, [15] firstly proposed the direct formulation with *Lasso* to produce sparse principal components. Given the data matrix W_{2F}^P and $\Sigma = W^T W$ is the covariance matrix of W, the classical PCA can be formed as follows,

$$z^* = \max_{z^T z \preceq 1} z^T \Sigma z, \tag{5}$$

The solutions z^* are the principal components of the data matrix W. In [15], they consider a direct reformulation to penalize the nonzero entries of the solutions z,

$$z^* = \max_{z \in B^n} z^T \Sigma z - \gamma \|z\|_0, \tag{6}$$

with the sparsity-controlling parameter $\gamma > 0$, when $\gamma = 0$, the Eq. 6 is relative to the classical PCA problem. B^n refers to a unit Euclidean ball in R^n. Whereas, the author of [14] consider a fixed data $k \in B^n$ which ensure that $(x_i^T k)^2 - \gamma > 0$, where the vector $\{x_i \in W_{2F}^P, i = 1, ..., P\}$, and the Eq. 6 is changed to,

$$z^* = \max_{k \in B^p} \max_{z \in B^n} (k^T W z)^2 - \gamma \|z\|_0, \tag{7}$$

In order to obtain a accurate projected m-dimensional subspace with orthogonal vectors, we can use the block sparse PCA on a Stiefel manifold S_m^P, because the sparse principal components z^* of SPCA are not forced to be orthogonal and cannot be used to the following local subspace separation. To enforce the orthogonal principal components, the author of [14] choose the block form for PCA and solve a block SPCA based on the l_0-Penalty. Following the Eq. 6, when the $\gamma = 0$, we come to the classical PCA situation,

$$z^* = \max_{z \in B^n} z^T \Sigma z, \tag{8}$$

Then the author of [14] extended the Eq. 8 to the block form with a *trace function* as the following reformulation,

$$Z^* = \max_{K \in S_m^P} \max_{Z \in S_m^n} Trace(diag(K^T W Z N)^2) - \sum_{j=1}^{m} \gamma_j \|z_j\|_0, \tag{9}$$

where the m related to the needed transformed dimensional, $m \leq rank(W)$, m-dimensional vector $\gamma = [\gamma_1, ..., \gamma_m]^T$ is positive. The solutions Z^* of Eq. 9 which span the dominant m-dimensional invariant subspace of the matrix W on a size fixed Stiefel manifold S_m^P. The matrix $N = Diag(\mu_1, \mu_2, ..., \mu_m)$ with distinct positive diagonal elements enforces the Eq. 9 to have isolated maximizers. It has been proved in the work of [14] that distinct elements on the diagonal of N enforce the loading vectors of the principal components of sparse PCA that are more orthogonal. Subsequently, Eq. 9 is completely decoupled in the columns of Z^* as follows,

$$Z^* = \max_{K \in S_m^P} \sum_{j=1}^{m} \max_{z_j \in S^n} (\mu_j k_j^T W z_j)^2 - \gamma_j \|z_j\|_0, \tag{10}$$

Equation 10 can be reformulated as a convex object function on the Stiefel manifold S_m^P,

$$Z^* = \max_{K \in S_m^P} \sum_{j=1}^{m} \sum_{i=1}^{n} [\mu_j \mathbf{x}_i^T k_j)^2 - \gamma_j]_+ \tag{11}$$

where the parameters should be given under the condition,

$$(\mu_j a_i^T x_j^*)^2 > \gamma_j \tag{12}$$

As we want a global projection of trajectory matrix W_{2F}^P that is preserved with the sparse loading entries as the matrix $Z^* \in S_m^P$, we use the block SPCA via l_0-Penalty to obtain a sparse low-dimensional representation. In the work of [14] they have already perform an efficient solution to solve the convex objective function in Eq. 11. In the end we can achieve a m-dimensional sparse matrix $Z^* \in S_m^P$ as the projected m-dimensional global subspace. It is equivalent to perform the segmentation of the multiple embedded affine low-dimensional subspaces (local subspaces) on the new global manifold.

4.2   Local Subspace Estimation

A sparse optimization technique *sparse manifold clustering and embedding* (SMCE) [13] can simultaneously estimate the neighbours of each data point from the same manifold and clustering the multiple embedded manifolds. In this work, we adopt the essential idea from SMCE for estimating the neighbours of the local subspace generated by each trajectory from the same low-dimensional local subspace.

SMCE assume that given a data point $x_i \in R^D$ draw from a manifold M_l with dimension d_l, there exists the relative set of points $N_i = \{x_j\}_{j \neq i}$ in M_l that contains only a few number of non-zero elements comes from the same affine subspace that passes near x_i. We call the neighbor set N_i *sparse neighbours*. This assumption can be defined as Eq. 13, which illustrates the optimization of sparse neighbours: among all of the points $\{x_j\}_{j \neq i}$, the ones that are neighbours of x_i in the same manifold span a d_l-dimensional affine subspace that passes through x_i,

$$\|[x_1 - x_i, ..., x_N - x_i]\|_2 \leq \epsilon \quad and \quad 1^T c_i = 1 \tag{13}$$

This assumption can be solved by a simple weighted sparse optimization program under the affine constraint,

$$\min \|Q_i c_i\|_1$$
$$s.t \ \|X_i c_i\|_2 \leqslant \sigma, 1^T c_i = 1 \tag{14}$$

where the Q_i has the diagonal elements $\frac{\|x_{fj} - x_{fi}\|_2}{\sum_{t \neq i} \|x_{ft} - x_{fi}\|_2} \in (0, 1]$ and we can call it a proximity inducing matrix that encourage finding the close neighbours. And X_i denote the normalized new vectors, which is

$$X_i = [\frac{x_{f1} - x_{fi}}{\|x_{f1} - x_{fi}\|_2} \cdots \frac{x_{fP} - x_{fi}}{\|x_{fP} - x_{fi}\|_2}] \in R^{2F \times P - 1} \tag{15}$$

The results $c_i^T = [c_{i1}, \ldots, c_{iP}]$ have only a few non-zero entries which ideally indicate the sparse neighbours of data point x_i from the same subspace.

In our proposed framework, after a global transformation using SPCA and normalizing the projected data, we obtain a sparse global subspace $\widehat{W_m^P}$ on the Stiefel manifold. Following the assumption of SMCE, most points and their sparse neighbours should lie on the same underlying low-dimensional subspace. We can adopt this assumption for searching the sparse representation of the closest neighbourhood of each data point in a global projected subspace. Opposite to the previous works such as LSA [11], who uses the fixed k-nearest neighbours technique, we choose to automatically estimate the sparse neighbours in an adequately large space.

A proper choice of the size for fixed k-nearest neighbours is sometimes critical and has important influence for the results of segmentation. Especially, when there are intersections between different local subspaces, the nearest neighbours can also belong to two different subspaces which could lead to the misclassification. As shown in Fig. 3, if we set the sample x_i as the observed point, Fig. 3(a) is the result of k-nearest neighbour searching. There are two triangles which should belong to the other subspace are found as the nearest neighbours of the circles, which will lead to a misclassification in the final clustering. Whereas Fig. 3(b) illustrates the sparse neighbour searching. It is clearly that if we search the nearby neighbours of each data point in the global subspace instead of looking for only the k-nearest neighbours, the estimation of the local subspace will be more accurate with avoiding the intersection of two different underlying subspace.

In order to conform the subspace geometric property constraint of each point, we choose the distance measure between two different subspaces with principal

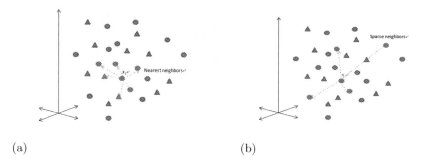

(a)   (b)

Fig. 3. The selection of neighbours of k-nearest neighbours and sparse neighbours in high dimensional space. The circles and triangles represent two different subspace local samples. (a) k-nearest neighbours: the green color denote the searched nearest neighbours of observed data x_i;(b) sparse neighbours: the green color denote the searched sparse neighbours of observed data x_i (Color figure online).

Algorithm 1. Local Subspace Estimation

Require: Projected Normalized Data Matrix $\widetilde{W}_{m \times P} = [x'_1, x'_2, ..., x'_P]_{m \times P}$
Ensure: Set of estimated local subspaces $S_1, S_2, ..., S_P$
1: **for all** $i, j = 1, ..., P$ **do**
2:     $Pr(\theta)_j = acos(x'^T_i x'_j)$
3: **end for**
4: $Pr(\Theta) = [Pr(\theta)_1, Pr(\theta)_2, ..., Pr(\theta)_P]_{P \times P}$

    Compute inducing proximity matrix $Q_i = diag(\dfrac{\theta_i}{\sum_{t \neq i} \theta_t}) \in R^{P-1 \times P-1}, i = 1, 2, ..., P$

5: **for** $i = 1, ..., P$ **do**
6:     solve sparse optimization with parameter λ_1, λ_2
        $C^T_i = [c_{i1}, ..., c_{iP}]^T \leftarrow \min \frac{1}{2} \|Pr(\theta)_i c_i\|^2_2 + \lambda_1 \|Q_i c_i\|_1 - \lambda_2 1^T c_i$
7: **end for**
8: **for** i,j=1,2,...,p **do**
9:     $c^T_i = [c_{i1}, ..., c_{iP}]^T$
10:     $l_i \leftarrow nonzero(c^T_i)$
11: **end for**
12: sparse data matrix $\widetilde{C} = [C^T_1, C^T_2, ..., C^T_P]_{P \times P}$
    estimated sparse neighbours set $L = \{l_i, i = 1, ..., P\}$
13: **for** $i = 1, 2, ..., P$ **do**
14:     local subspace $S_i = \widetilde{W}_{m \times P}(:, C^T_i(1 : l_i, i))$
15: **end for**
16: Estimated local subspaces set $S_1, S_2, ..., S_P$

---

angles. The principal angles between two subspaces S_i and S_j can be defined recursively with $0 \leq \theta_1 \leq, \cdots, \leq \theta_l \leq (\pi/2)$,

$$l = \min(\dim(S_i), \dim(S_j))$$

$$\cos(\theta_k) = \max_{u \in S_j, v \in S_i} u^T v = u_k^T v_k, k = 1, \cdots, l \tag{16}$$

$$s.t. \|u\| = \|v\| = 1, u^T u_q = 0, v^T v_q = 0, q = 1, \cdots, k - 1$$

we can use the distance measure of two subspaces S_i and S_j with the affinity of principal angles of S_i and S_j,

$$A(i, j) = e^{-\sum_{q=1,\cdots,l} \sin^2(\theta_q)} \tag{17}$$

Thus we modify the SMCE with principal angles $0 \leq \theta_1 \leq, \cdots, \leq \theta_l \leq (\pi/2)$,

$$X_i = \theta_i \tag{18}$$

where the $\cos(\theta_i) = max_{u \in S_j, v \in S_i} u^T v = u_k^T v_k, k = 1, \cdots, l$, S_j is the subspace generated by the data point x_j and S_i is generated by x_i, $j \neq i$.

Subsequently we can modify the optimization program in Eq. 14 and define the weight proximity inducing weight matrix and solving the Eq. 14 by the argumented the Lagrange multipliers [16], which is

$$\min \frac{1}{2} \|\theta_i c_i\|_2^2 + \lambda \|Q_i c_i\| \tag{19}$$

$$s.t. \ 1^T c_i = 1,$$

where $Q_i = diag(\frac{\theta_i}{\sum_{t \neq i} \theta_t}) \in R^{N-1 \times N-1}, \theta_i = acos(max_{u \in S_j, v \in S_i} u^T v) = acos$

$(u_i^T v_i), \theta_t = acos(max_{u \in S_t, v \in S_i} u^T v) = acos(u_t^T v_t), t \neq j$. The procedures of the local subspace estimation are summarized in Algorithm 1.

With solving the problem in Eq. 19, the number of the sparse neighbours is obtained from the sparse solutions C_i, which indicates the neighbours of point x_i from the same subspace. The estimation of the local subspace can be achieved by simply extracting the neighbours for each data point. Subsequently, we can use the affinity measure to compute the similarity between each pair of estimated local subspaces and build the symmetric affinity matrix. There are amount of measure techniques for different subspaces like subspace euclidean distances or principal angles. In the end we can easily perform a simple spectral clustering on the built affinity matrix.

4.3   Complete Procedure

Let $W_{2F \times P} = [x_1, x_2, ..., x_P]$, where each $x_i = [x_{1i}, ..., x_{Fi}]^T$ represent a tracked feature trajectory throughout F frames. All of the trajectories are $\{x_i \in R^D\}$ drawn from n different subspaces with dimensions $\{d_j \in R^D, j = 1, ..., n\}$. In general, our proposed subspace-based motion segmentation method can be written with the 3 main steps for clustering the feature data from a union of multiple affine subspaces: (1) Transform the input trajectories into a sparse global subspace with Sparse PCA; (2) Estimate the local subspace with the number of sparse neighbours and the principal angles; (3) Build the similarity matrix with principal angles and clustering the matrix with k-means. The final output is the labelled moving objects. The overall motion segmentation algorithm is summarized in Algorithm 2.

Algorithm 2. Motion segmentation using sparse optimization

Require: Data Matrix $W_{2F \times P} = [x_1, x_2, ..., x_P]_{2F \times P}$, number of motions n
Ensure: A Set of Labels $[1, 2, ..., n]$

    Step 1. Global Projection: $m = 4n, W'_{m \times P} \Leftarrow W_{2F \times P}$
    Sparsity-controlling parameter $\gamma = [\gamma_1, \gamma_2, ..., \gamma_m]^T \geq 0, N = [\mu_1, ..., \nu_m] \geq 0$
1: **for all** $i = 1, ..., P$ **do**
2:    $x'_i \leftarrow SPCA(x_i, \gamma, N)$
3: **end for**
4: $W'_{m \times P} = [x'_1, x'_2, ..., x'_P]_{m \times P}$
5: Normalization $\widetilde{W}_{m \times P} = normalize(W'_{m \times P})$

    Step 2. Local Subspace Estimation:
6: **for all** $i, j = 1, ..., P$ **do**
7:    run Alg.1 to estimate the local subspaces $\mathbf{S} = \{S_1, S_2, ..., S_P\}$ for each projected
      trajectory x'_i
8: **end for**

    Step 3. Perform Spectral Clustering
9: **for** $i, j = 1, 2, ..., P$ **do**
10:    $A(i, j) = Pr(B_i, B_j)$
11: **end for**
12: Perform spectral clustering on A to obtain a set of n Labels

---

5   Experimental Results

In this section we evaluate our method on a standard real-world video sequence benchmark, the Hopkins 155 dataset [17]. We have compared with other state-of-art motion segmentation approaches. We have assumed for all the methods that the number of the motions has already given.

We test all the algorithms on the full original Hopkins 155 dataset [17] with no missing trajectories. The database from the Hopkins 155 composed of 120 sequences with 2 motions and 35 sequences with 3 motions. There are 3 different kinds of motions in the Hopkins 155 dataset: traffic, checkerboard and articulated. As a pre-processing step, the feature points that tracked through all over the sequences have already been obtained. Moreover, all of the errors in tracking were already corrected for each sequence. Hence, this experiment exists

Fig. 4. Comparison of Our approach with ground truth (*the last row*) and the other approaches *the first row*: LSA [11]; *Second row*: MSMC [6], *Third row*: Our Method on the frames 1, 8, 15 and 20 of the *cars02-07* sequence from the Hopkins 155 dataset [17].

no missing entries in the feature trajectories. Most people use the Hopkins 155 dataset for testing the performance of accuracy and computation times.

We have tested SSC [10], LSA [11], RANSAC [7], GPCA [8], MSMC [6], LLMC [18] and our method on the checkboard, traffic and articulated sequences, because of MSMC based on the affinity we have not compare it to the others in the checkboard sequences. The parameter k in LSA has been set to 10^{-6} and the number of nearest neighbours is fixed with 6. The threshold for subspace fitting needed for RANSAC method has chosen to be 0.00002. The sparsity controls parameter λ for SMCE [13] in our method is set to be 20. For all of the methods, the data has been projected into the $4n$ global subspace according to the number of motions. We have computed the average and median misclassification error for comparison, as shown in Tables 1, 2, and 3. These numbers show that our segmentation results are comparable with other state-of-art motion segmentation results, and in many cases exceed them in terms of segmentation accuracy. Figure 4 shows the

Table 1. Mean and median of the missclassification (%) on the Hopkins155 database with 2 motions.

Method	GPCA	ALC	SSC	MSMC	LLMC	LSA	Our method
Articulated 11 sequences							
Mean	2.88	10.70	0.62	2.38	5.23	4.10	0.55
Median	0.00	0.95	0.00	0.00	1.30	0.00	0.00
Traffic: 31 sequences							
Mean	1.41	1.59	0.02	0.06	3.65	5.43	0.59
Median	0.00	1.17	0.00	0.00	0.33	1.48	0.00
Checkerboard 78 sequences							
Mean	6.09	1.55	1.12	NaN	4.65	2.57	1.42
Median	1.03	0.29	0.00	NaN	0.11	0.27	0.27
All 120 sequences							
Mean	4.59	2.40	0.82	NaN	4.44	3.45	1.11
Median	0.38	0.43	0.00	NaN	0.24	0.59	0.00

Table 2. Mean and median of the missclassification (%) on the Hopkins155 database with 3 motions.

Method	GPCA	ALC	SSC	MSMC	LLMC	LSA	Our method
Articulated 2 sequences							
Mean	16.85	21.08	1.91	1.42	9.38	7.25	5.32
Median	16.85	21.08	1.91	1.42	9.38	7.25	5.32
Traffic: 7 sequences							
Mean	19.83	7.75	0.58	0.16	7.79	25.07	4.74
Median	19.55	0.49	0.00	0.00	5.47	23.79	4.04
Checkerboard 26 sequences							
Mean	31.95	5.20	2.97	NaN	12.01	5.80	3.05
Median	32.93	0.67	0.27	NaN	9.22	1.77	0.77
All 35 sequences							
Mean	28.66	6.69	2.45	NaN	11.02	9.73	3.49
Median	28.66	0.67	0.20	NaN	6.81	2.33	1.11

qualitative segmentation examples from the Hopkins 155 dataset. We can infer that our method segments the foreground moving object successfully in comparing with the ground truth and other algorithms.

We also present the run-time of our method, SSC [10], and ALC [9] in Table 4. Comparing with the SSC [10], the performance of our method is better than SSC

Table 3. Mean and median of the missclassification (%) on the whole Hopkins155 database with both 2 and 3 motions.

Method	GPCA	ALC	SSC	LLMC	LSA	Our method
All 155 sequences						
Mean	10.34	3.56	1.24	5.93	4.94	1.65
Median	2.54	0.50	0.00	0.63	0.90	0.32

Table 4. Run-Time (%) for the whole Hopkins155 database with both 2 and 3 motions.

Method	ALC	SSC	Our method
Run-time [s]	88831	14500	**14021**

particularly on the articulated sequence of 2 motions. With only a little loss of accuracy on the other sequence, we save the computation time w.r.t. the SSC [10].

6   Conclusions

We have proposed a feature-based framework for the problem of segmenting different types of the moving objects from a video sequence with combining two sparse subspace optimization methods SPCA [19] and SMCE [13]. The SPCA performs a data projection from a high-dimensional subspace to a low-dimensional global manifold with sparse entries, which ensures the interpretability and accuracy. Simultaneously, we adopt the idea of SMCE that search the sparse closest neighborhood set for each local embedded subspace generated by each trajectory, which efficiently solve the intersection or overestimation problem in LSA framework [11]. The experiments demonstrate that the low misclassification error of our approach on the Hopkins 155 dataset [17], outperforming most of the popular approaches. The limitation of our work is that the number of motions is considered as a prior knowledge. In the future work, our goal is to perform the estimation of the number of motions in the framework as well. Furthermore, we will derive a robust optimization method that can deal with the corrupted trajectory and missing data.

Acknowledgments. The work is funded by the ERC-Starting Grant ('DYNAMIC MINVIP'). The authors gratefully acknowledge the support.

References

1. Lee, Y.J., Kim, J., Grauman, K.: Key-segments for video object segmentation. In: 2011 IEEE International Conference on Computer Vision (ICCV), IEEE, pp. 1995–2002 (2011)
2. Shi, J., Malik, J.: Motion segmentation and tracking using normalized cuts. In: 1998 Sixth International Conference on Computer Vision, IEEE, pp. 1154–1160 (1998)

3. Tomasi, C., Kanade, T.: Detection and Tracking of Point Features. School of Computer Science, Carnegie Mellon University, Pittsburgh (1991)
4. Zhou, H., Yuan, Y., Shi, C.: Object tracking using sift features and mean shift. Comput. Vis. Image Underst. **113**, 345–352 (2009)
5. Bay, H., Ess, A., Tuytelaars, T., Van Gool, L.: Speeded-up robust features (surf). Comput. Vis. Image Underst. **110**, 346–359 (2008)
6. Dragon, R., Rosenhahn, B., Ostermann, J.: Multi-scale clustering of frame-to-frame correspondences for motion segmentation. In: Fitzgibbon, A., Lazebnik, S., Perona, P., Sato, Y., Schmid, C. (eds.) ECCV 2012, Part II. LNCS, vol. 7573, pp. 445–458. Springer, Heidelberg (2012)
7. Fischler, M.A., Bolles, R.C.: Random sample consensus: a paradigm for model fitting with applications to image analysis and automated cartography. Commun. ACM **24**, 381–395 (1981)
8. Vidal, R., Ma, Y., Sastry, S.: Generalized principal component analysis (GPCA). IEEE Trans. Pattern Anal. Mach. Intell. **27**, 1945–1959 (2005)
9. Ma, Y., Derksen, H., Hong, W., Wright, J.: Segmentation of multivariate mixed data via lossy data coding and compression. IEEE Trans. Pattern Anal. Mach. Intell. **29**, 1546–1562 (2007)
10. Elhamifar, E., Vidal, R.: Sparse subspace clustering. In: 2009 IEEE Conference on Computer Vision and Pattern Recognition, CVPR 2009, IEEE, pp. 2790–2797 (2009)
11. Yan, J., Pollefeys, M.: A general framework for motion segmentation: independent, articulated, rigid, non-rigid, degenerate and non-degenerate. In: Leonardis, A., Bischof, H., Pinz, A. (eds.) ECCV 2006. LNCS, vol. 3954, pp. 94–106. Springer, Heidelberg (2006)
12. Hartley, R., Zisserman, A.: Multiple View Geometry in Computer Vision, 2nd edn. Cambridge University Press, New York (2003)
13. Elhamifar, E., Vidal, R.: Sparse manifold clustering and embedding. In: NIPS, pp. 55–63 (2011)
14. Journée, M., Nesterov, Y., Richtárik, P., Sepulchre, R.: Generalized power method for sparse principal component analysis. J. Mach. Learn. Res. **11**, 517–553 (2010)
15. d'Aspremont, A., El Ghaoui, L., Jordan, M.I., Lanckriet, G.R.: A direct formulation for sparse pca using semidefinite programming. NIPS **16**, 41–48 (2004)
16. Tibshirani, R.: Regression shrinkage and selection via the lasso. J. Roy. Stat. Soc. Ser. B (Methodological) **58**, 267–288 (1996)
17. Tron, R., Vidal, R.: A benchmark for the comparison of 3-d motion segmentation algorithms. In: 2007 IEEE Conference on Computer Vision and Pattern Recognition, CVPR 2007, IEEE, pp. 1–8 (2007)
18. Goh, A., Vidal, R.: Segmenting motions of different types by unsupervised manifold clustering. In: 2007 IEEE Conference on Computer Vision and Pattern Recognition, CVPR 2007, IEEE, pp. 1–6 (2007)
19. Zou, H., Hastie, T., Tibshirani, R.: Sparse principal component analysis. J. Comput. Graph. Stat. **15**, 265–286 (2006)

Adaptive Foreground Extraction for Crowd Analytics Surveillance on Unconstrained Environments

Mohamed Abul Hassan[(⊠)], Aamir Saeed Malik, Walter Nicolas,
and Ibrahima Faye

Centre for Intelligent Signal and Imaging Research (CISIR),
Universiti Teknologi PETRONAS, Tronoh, Malaysia
logicbird@yahoo.com

Abstract. Background modeling is one of the key steps in any visual surveillance system. A good background modeling algorithm should be able to detect objects/targets under any environmental condition. The influence of illumination variance has been a major challenge in many background modeling algorithms. These algorithms produce poor object segmentation or consume substantial amount of computational time, which makes them not implementable at real time. In this paper we propose a novel background modeling method based on Gaussian Mixture Method (GMM). The proposed method uses Phase Congruency (PC) edge features to overcome the effect of illumination variance, while preserving efficient background/foreground segmentation. Moreover, our method uses a combination of pixel information of GMM and the Phase texture information of PC, to construct a foreground invariant of the illumination variance.

1 Introduction

Visual surveillance systems are essentially becoming the most attractive research areas in the field of computer vision. Their importance with respect to security and safety in public places is amongst the main reasons of growing attention in this field [1]. The accessibility to inexpensive devices and processors is an additional motivation for the promotion on investigation about visual surveillance systems. Considering that majority of visual surveillance, nevertheless, depends upon human intervention to monitor through video clips [2,3], which is a tiresome and tedious task; keeping track of interesting events that will rarely take place. The large amount of data makes it humanly impossible to analyze and requires computer vision based solutions to automate the process.

Computer aided behavioral understanding of moving objects in a video is often quite challenging task. This requires extraction of related visual data, appropriate representation of information, as well as an interpretation of this visual information with respect to behavior learning and recognition [4]. In automated visual surveillance systems, one of the many key steps in video based human-activity recognition is usually to model the background, which often requires large expanse of processing time of the system.

© Springer International Publishing Switzerland 2015
C.V. Jawahar and S. Shan (Eds.): ACCV 2014 Workshops, Part II, LNCS 9009, pp. 390–400, 2015.
DOI: 10.1007/978-3-319-16631-5_29

Background modeling is comprised of foreground/background segmentation which yields information for various automated behavior understanding applications such as tracking, counting, direction estimation and velocity estimation of objects. Nevertheless background modeling at dynamic environments is a great challenge due to the variation in the nature of backgrounds. The motion of the sun along with the clouds in the sky is the greatest variation on the nature of the background which causes the influence of gradual and sudden illumination variance. The effect of illumination variance on the background often yields detection of false foreground mask due to shadows of the objects and false foreground detection due to sudden illumination variance [5,6]. This provides misinterpreted information regarding the actual scene which would provide poor interpretation of the visual information for behavior learning and recognition applications.

The issue of background modeling for dynamic environments has been addressed by researches, who have proposed many background modeling algorithms based on statistical information [7,8], fuzzy running average, clustering methods [9] and using neural network [10,11]. However, most of these algorithms are based on specific environmental conditions, such as a specific time, place, or activity scenario and carries greater implementation complexity. Considering the implementation simplicity, Gaussian Mixture Model (GMM) is used as the background modeling algorithm in this paper. GMM is one of the most used and implemented background modeling algorithms at present due to its good compromise between computational efficiency and accuracy. However, this algorithm also suffers at above mentioned effects of illumination variance.

2 Related Work

Researches have extended the original GMM algorithm [12] to address the issue by proposing various methods such as, using multiple Gaussians by variable K Gaussian distribution [13,14]. The adaptive background model for compensating sudden illumination variance uses variable assignment of K mean distribution or online variation of K mean distribution of the GMM model. Here the value of K varies between 3 to 5, the multiple Gaussian models, generated could distinguish between the illumination effected areas and eliminate false extraction of foreground mask. This modification dose improve the accuracy of the model but still the model has not been evaluated for different scenarios and is subjected to computational time inefficiency under the effect of sudden illumination and complex scenes.

Another method using phase texture information in place of pixel information [15]. The main advantage is that phase texture of an image is invariant to illumination. The phase texture was obtained using a Gabor filter and a phase based background subtraction was proposed by modelling the phase features independently using the MOG model and the foreground is extracted using distance transform applied at the binary image which is transformed in to a distance map by segmenting and thresholding the distance map to obtain the foreground.

The model proposed was able to compensate sudden illumination efficiently but was evaluated only for static backgrounds.

Manuel et al. [16] proposed Mixture of Merged Gaussian Algorithm (MMGA) using a combination of GMM and Real-Time Dynamic Ellipsoidal Neural Networks (RTDENN), as an updating mechanism. This model uses the conventional GMM to perform the background modeling using different color spaces (i.e. RGB or HSV). The modelled background pixels are stored for every frame, and is used for the RTDENN updating mechanism. The Gaussians are managed based on the comparison to the previous pixel information, where the existence of the Gaussians are decided by merging or creating Gaussian distributions. Eventually the updating mechanism using RTDENN functions based on the excitement of the neurons, which is the inverse of the square root of the Mahalanobis distance between the mean of the sample vectors which excite the neuron and each neuron.

Zezhi, C. and Ellis, T. proposed an illumination compensation background model by self-adaptive background modeling method [17]. This background modeling method followed [18] the, recursive GMM using multidimensional Gaussian kernel density transform. The author of this proposed method [17] used the recursive GMM along with a spatial temporal filter, which suppressed noise and compensated the illumination using median of quotient. This model was evaluated for various surveillance application scenarios alongside with crowd related datasets. However from the evaluated sequences it was observed that the model wasn't evaluated on more divers' environmental conditions.

These proposed methods have been developed in the recent past to effectively solve the issue of sudden illumination variance. However we would implement these methods along with our proposed method in Sect. 4, and evaluate them for various scenarios. In this paper we propose a method to address the issue of illumination compensation for GMM with PC edge features to model the background. The motivation leading to formulate the background modeling method utilizing GMM and Phase congruency edge detector is due to the fact that phase texture of an image is invariant to illumination variance. Therefore, we developed the model using GMM to determine the foreground of the scene using the pixel intensity information while using phase congruency edge detector to extract the phase edge and corner information of the scene. The method utilizes the pixel intensity information and phase edge information of foreground/background to achieve efficient object segmentation, shadow removal and sudden illumination compensation.

3   Proposed Method

Our proposed method for dynamic background/foreground modeling uses a combination of Gaussian Mixture Method (GMM) [12] and Phase Congruency (PC) [19,20] edge features. These are widely used computational tool in detecting background/foreground. However, we use a novel approach on using these methods as features/characteristics for accurate object detection and segregation under dynamic environmental conditions. For each pixel of a frame X at time t in an image sequence, the pixel characteristic is determined based on the intensity of the monochromatic color space. Then, the multidimensional background is modeled based on weighted sum of the probability of observing current pixel values for K Gaussian distributions is given by,

$$P(X_t) = \sum_{i=1}^{K=3} \omega_{i,t}.\eta\,(X_t, \mu_{i,t}, \Sigma_{i,t}) \tag{1}$$

where η is the probability density function, K is the number of Gaussian distributions, $\omega_{i,t}$ is a weight associated with i^{th} Gaussian at the time t with a mean $\mu_{i,t}$ and standard deviation $\Sigma_{i,t}$. The weighted sum of the probability distribution is initialized and at instance of time t the current pixel value X_t is verified if it's in range of the standard deviation of the Gaussian distribution. The pixel in range would be classified as matching to one of the K Gaussian distributions. In this case when a Gaussian distribution matches a pixel X_t, the parameters of the pixels K distributions will be updated by two scenarios. Scenario 1: for unmatched Gaussians distributions mean $\mu_{i,t}$ and standard deviation $\Sigma_{i,t}$ will be unchanged. Scenario 2: for matched Gaussians distributions $\mu_{i,t}$ and $\Sigma_{i,t}$ will be updated as shown in (Eqs. 2–4)

$$mu_{i,t} = (1-\rho)\,\mu_{i,t-1} + \rho.X_t \tag{2}$$

$$\Sigma_{i,t} = (1-\rho)\,\Sigma_{i,t-1} + \rho\,\big|(X_t - \mu_{i,t})^T(X_t - \mu_{i,t})\big|_{diag} \tag{3}$$

$$\rho = \alpha.\eta\,(X_t\,|\mu_{i,t-1}, \Sigma_{i,t-1}|) \tag{4}$$

Once the parameters are initialized, the first foreground detection is performed and the above parameters are updated for time $t+1$ using a criterion ratio, $r_i = \frac{\omega_i}{\sigma_i}$ and the order of Gaussians following the ratio. This ordering depends upon the background pixels which corresponds to a high weight and a weak variance and foreground which corresponds to a low weight and a high variance. The foreground is considered for incoming new frame at instance $t+1$, a match test is performed, to match the incoming pixel to the Gaussian distribution based on the Mahalanobis distance.

$$sqrt\left((X_{t+1} - \mu_{i,t})^T . \sum_{i,t}^{-1}(X_{t+1} - \mu_{i,t})\right) \leq k\sigma_{i,t} \tag{5}$$

where k is a constant established by experimentation whose value equals to 2.5. At this step the binary mask of the foreground frame $F_g\,(X_{t+1})$ is extracted based on two conditions as shown in Fig. 1;

Condition 1: Pixel Matches with one of the K Gaussians. In this case, if the Gaussian distribution is identified as a background, the pixel is classified as background or else classified as foreground.

Condition 2: No match with any of the K Gaussians. In this case, the pixel is classified as foreground. Furthermore the extracted binary mask is filtered using a median filter for reducing noise.

Meanwhile the Phase texture features of the frame X at the time instance $t + 1$ is extracted using Phase Congruency Edge detector [19,20] for every incoming frame. The phase information is extracted via banks of Gabor wavelets tuned to various spatial frequencies, instead of Fourier transform, as given by,

$$PC(X_{t+1}) = \frac{\sum_n W(X_{t+1}) \left[A_n(X(t+1)\alpha_1 - |\alpha_2| - T_{PC} \right]}{\sum_n A_n(X_{t+1}) + \epsilon} \qquad (6)$$

where α_1 denotes, $cos\left(\phi_n\left(X_{t+1}\right)\bar{\phi}\left(t+1\right)\right)$ the term is a factor that weights for frequency spread (congruency over many frequencies is more significant than congruency over a few frequencies) and α_2 denotes, $sin\left(\phi_n\left(X_{t+1}\right)\bar{\phi}\left(t+1\right)\right)$. A small constant ϵ is included to prevent division by zero. The energy values which surpass T_{PC}, the estimated noise effect, are included to the result. The notations within $'[]'$ highlights that this enclosed quantity remains the same while its value is positive, and zero otherwise. For accurate foreground object extraction, we take the intersection of each pixel for the binary mask of the matrices, $F_g\left(X_{t+1}\right)$ and $PC(X_{t+1})$.

$$F = \sum_{i=1}^{H} \sum_{j=1}^{W} F_g\left(X_{t+1}\right) \bigcap PC(X_{t+1}) \qquad (7)$$

where H and W denotes the height and the width of the image frame

Fig. 1. Process of extracting the binary silhouette of the foreground objects. (a) represent the binary silhouette of the foreground for GMM. (b) the foreground after noise suppression. (c), Phase Congruency edge texture image. (d) extracted foreground using intersection of each pixel for the binary mask.

Finally, a set of morphological operations are performed on the extracted foreground object silhouette. Here we use these operations to enhance the boundary connectivity of the silhouette and to region fill the enclosed boundary of the objects. The process is initiated with a morphological opening operation which performs erosion followed by dilation using a 2 by 2 structured matrix as derived in (Eq. 8). Following this step the edges of the silhouette gets connected, this results on an accurate reconstruction of the boundaries of the silhouette.

Then the object silhouettes is completed by performing region filling operation as shown in (Eq. 9), where the results on each of these steps are illustrated on Fig. 2.

$$\gamma_B\left(F\right) = \bigcup_f \{B_x | B_x \subseteq F\} \tag{8}$$

$$F_k = (X_{k-1} \oplus B_x) \bigcap A^c k = 1, 2, 3 \tag{9}$$

where B_x is the structured matrix and A^c is the matrix which contains the filled set and its boundary.

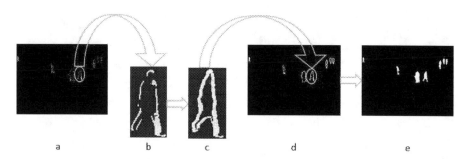

a   b   c   d   e

Fig. 2. Steps of the morphological operations performed on the extracted foreground object silhouette. (a) represent the original foreground silhouette extracted using the proposed method; (b) is the zoomed view of the silhouette of the object; (c) represent the boundary reconstruction of the 2 by 2 structure matrix; (d) is the full view of the boundary reconstructed image; (e) the extracted foreground silhouette after region filling.

4   Experiment Results

The experiments and analysis were carried out qualitatively and quantitatively, where the proposed model along with 5 other model in literature (See Table. 1). The binary silhouettes of the foreground extraction were taken as a final result to determine the efficiency of the models. The results were evaluated using precision recall criteria, this access the ability, to extract the true positive and eliminate the true negative detection of the implemented models.

The developed models were implemented for five different sequences from two most popular crowd surveillance databases for dynamic environments, i.e. PETS2010 [21] and OTCBVS [22]. The description of these sequences are tabulated in Table. 2. The significance of these data sets apart from testing the models on different crowd behaviors, is to challenge the accuracy of segmentation, i.e. detecting small objects from pixel range of 10×25 to 20×75, which challenges the accuracy of segmentation. The non-synchronized nature of the data set adds the effect of sudden and gradual illumination variances continuously after every

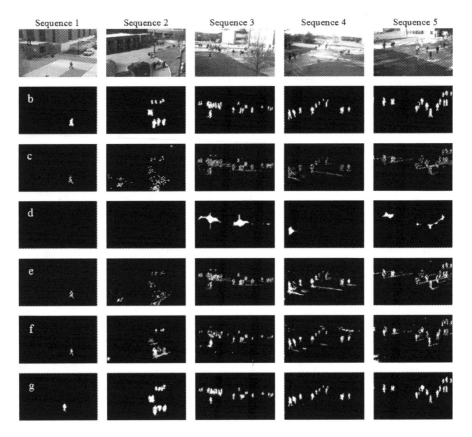

Fig. 3. The sample results of the binary silhouettes of the extracted foreground, from the different background modeling methods. The first row (a) shows the original image, (b) shows the ground truth of the extracted binary silhouettes of the original image, (c) shows the extracted binary silhouettes of GMM1, (d) shows the extracted binary silhouettes of GMM2, (e) shows the extracted binary silhouettes of GMM3 and (f) shows the extracted binary silhouettes of GMM4 followed by (g) which shows the extracted binary silhouettes of proposed method.

30 to 40 frames. To summarize, the selected sequences are to yield the challenges such as, bootstrapping (BS), sudden illumination (SI), gradual illumination (GI), camouflage (CF), waving trees (WT) and shadows (S).

The sample results of the binary silhouettes are shown in Fig. 3. Here the results of each model is arranged in row wise manner with respect to each sequence. The performance of each model was evaluated with respect to the ground truth which is shown in the second row. Most of the models were able to provide an accurate object detection and segregation. However GMM2 model which was based on phase texture information and distance transform failed to detect and segregate the silhouettes of the objects in each sequence. This was due to the small object size in crowd surveillance data and the parameters used

Table 1. Description of the implemented models in Fig. 3.

Model	Methodology	Related publication
GMM1	GMM using multiple Gaussians by variable K Gaussian distribution	[13,14]
GMM2	GMM using phase texture information and distance transform	[15]
GMM3	GMM using RTDENN	[16]
GMM4	Using recursive GMM along with a spatial temporal filter	[17]
Proposed Method	Using a combination of GMM and PC edge features as temporal filter for each pixel	

Table 2. Description of the datasets.

Image sequence	Database	No. of frames	Size	Challenges	Description
Sequence 1	OTCBVS	1507	320 × 240	BS,SI,GI	Dataset 03: OSU Color-Thermal Database
Sequence 2	OTCBVS	1054	320 × 240	GI,BS,S	Dataset 03: OSU Color-Thermal Database
Sequence 3	PETS2010	841	768 × 576	BS,GI,CF	Data set of S0, City Center, view 4
Sequence 4	PETS2010	841	768 × 576	SI,GI, WT, S	Data set of S0, City Center, view 2
Sequence 5	PETS2010	841	768 × 576	SI,GI, WT, S	Data set of S0, City Center, view 1

in the distance transform measures. Meanwhile the other developed models were able to eliminate challenges such as BS, WT, CF and GI.

The effect of shadows were compensated by GMM4 and proposed method, while GMM1 and GMM3 suffered heavily by extracting false foreground in sequence 4 and 5. The effect of SI was the greatest challenge to overcome, where all the models apart from GMM2 and Proposed method, suffered heavily by extracting false foreground mask due to the sudden intensity variation. The GMM2 and Proposed method were clearly able to overcome this issue since these models used phase texture features, which are invariant to illumination changes. However our proposed method was clearly able to overcome all of the above mention issues, and was able to accurately segment the foreground objects. Furthermore, the Proposed model outclassed all the other developed models and resulted in a higher precision recall criteria (See Figs. 4 and 5).

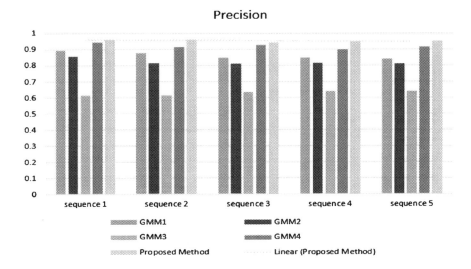

Fig. 4. Quantified Precision results obtained for five different algorithms and five sequences tested.

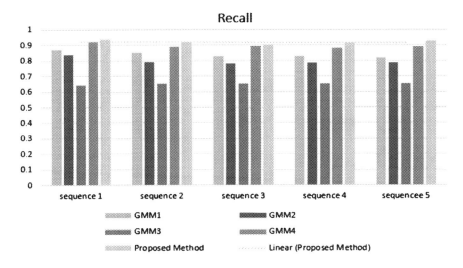

Fig. 5. Quantified Recall results obtained for five different algorithms and five sequences tested.

5   Conclusion

This paper proposed a background modelisation incorporating GMM and PC to adapt to unconstrained environment conditions. The main motivation of this work was to develop a robust background modeling method to extract foreground objects with accurate segregation while retaining the object silhouette. The proposed method was evaluated with four other background modeling methods.

The performance of the models was assessed qualitatively and quantitatively using precision and recall criteria. The results presented demonstrate the superiority of the proposed method in terms of accuracy for the background/foreground extraction and was able to efficiently segregate the individual objects in crowds. Moreover, the model efficiently compensated the challenging effect of sudden illumination and presence of shadows by incorporating the phase texture information along with the pixel gradient information. Future works will focus on using foreground silhouette characteristic of the objects for improving behavior learning crowd analytic algorithms such as Kalman filter and Optical flow.

References

1. Yilmaz, O.J.A., Shah, M.: Object tracking: a survey. ACM Comput. Surv **38**, 45 (2006)
2. Wang, X.: Intelligent multi-camera video surveillance: a review. Pattern Recogn. Lett. **34**, 3–19 (2013)
3. Yasir, S., Malik, A.S.: Comparison of stochastic filtering methods for 3d tracking. Pattern Recognit. **44**, 2711–2737 (2011)
4. Morris, B.T., Trivedi, M.M.: A survey of vision-based trajectory learning and analysis for surveillance. IEEE Trans. Circuits Syst. Video Technol. **18**, 1114–1127 (2008)
5. Pilet, J., Strecha, C., Fua, P.: Making background subtraction robust to sudden illumination changes. In: Forsyth, D., Torr, P., Zisserman, A. (eds.) ECCV 2008, Part IV. LNCS, vol. 5305, pp. 567–580. Springer, Heidelberg (2008)
6. Hassan, A., Aamir, S.M., Nicolas, W., Faye, I.: Mixture of gaussian based background modelling for crowd tracking using multiple cameras. In: International Conference on Intelligent and Advanced Systems vol. 5, pp. 1–4 (2014)
7. Horng-Horng, L.: Regularized background adaptation: a novel learning rate control scheme for gaussian mixture modeling. IEEE Trans. Image Process. **20**, 822–836 (2011)
8. Elgammal, A., Harwood, D., Davis, L.: Non-parametric model for background subtraction. In: Vernon, D. (ed.) ECCV 2000. LNCS, vol. 1843, pp. 751–767. Springer, Heidelberg (2000)
9. Malathi, T., Bhuyan, M.K.: Multiple camera-based codebooks for object detection under sudden illumination change. Int. Conf. Commun. Signal Process. (ICCSP) **20**, 310–314 (2013)
10. Bouwmans, T.: Recent advanced statistical background modeling for foreground detection - a systematic survey (2011)
11. Cuevas, C., Garcia, N.: Versatile bayesian classifier for moving object detection by non-parametric background-foreground modeling. In: 19th IEEE International Conference on Image Processing (ICIP), pp. 313–316 (2012)
12. Stauffer, C., Grimson, W.E.L.: Adaptive background mixture models for real-time tracking. Int. Conf. Comput. Vis. Pattern Recognit. **2**, 252 (1999)
13. Dawei, L., Goodman, E.: Online background learning for illumination-robust foreground detection. In: International Conference on Control Automation Robotics and Vision (ICARCV), vol. 11, pp. 1093–1100 (2010)
14. Huang, T., Fang, X., Qiu, J., Ikenaga, T.: Adaptively adjusted gaussian mixture models for surveillance applications. In: Boll, S., Tian, Q., Zhang, L., Zhang, Z., Phoebe Chen, Y.-P. (eds.) Advances in Multimedia Modeling. LNCS, vol. 5916, pp. 689–694. Springer, Heidelberg (2010)

15. Gengjian, X., Li, S.: Background subtraction based on phase feature and distance transform. In: IEEE 17th International Conference on Image Processing, vol. 17, pp. 3465–3469 (2012)
16. Alvar, M., Rodriguez-Calvo, A., Sanchez-Miralles, A., Arranz, A.: Mixture of merged gaussian algorithm using rtdenn. Mach. Vis. Appl. **25**, 1133–1144 (2014)
17. Chen, Z.: A self-adaptive gaussian mixture model. Comput. Vis. Image Underst. **122**, 35–46 (2013)
18. Zivkovic, Z., van der Heijden, F.: Recursive unsupervised learning of fnite mixture models. IEEE PAMI **5**, 651–656 (2004)
19. Kovesi, P.: Phase congruency detects corners and edges. In: Proceedings of VIIth Digital Image Computing: Techniques and Applications 8, 10–12 (2013)
20. Hassan, A., Aamir, S.M., Nicolas, W., Faye, I.: Foreground extraction for real-time crowd analytics in surveillance system. In: 2014 IEEE 18th International Symposium on Consumer Electronics (ISCE 2014), vol. 18, pp. 1–2 (2014)
21. Ferryman, J., Ellis, A.: Pets2010: dataset and challenge. In: Seventh IEEE International Conference on Advanced Video and Signal Based Surveillance (AVSS), vol. 7, pp. 143–150 (2010)
22. Dataset: O.: http://www.cse.ohio-state.edu/otcbvs-bench/ (2012)

My Car Has Eyes: Intelligent Vehicle with Vision Technology

Driver Assistance System Providing an Intuitive Perspective View of Vehicle Surrounding

Yen-Ting Yeh[1], Chun-Kang Peng[2], Kuan-Wen Chen[3],
Yong-Sheng Chen[2(✉)], and Yi-Ping Hung[1,4]

[1] Graduate Institute of Networking and Multimedia,
National Taiwan University, Taipei, Taiwan
[2] Department of Computer Science, National Chiao Tung University,
Hsinchu, Taiwan
yschen@cs.nctu.edu.tw
[3] Intel-NTU Connected Context Computing Center,
National Taiwan University, Taipei, Taiwan
[4] Department of Computer Science and Information Engineering,
National Taiwan University, Taipei, Taiwan

Abstract. Driver assistance systems can help drivers to avoid car accidents by providing warning signals or visual cues of surrounding situations. Instead of the fixed bird's-eye view monitoring proposed in many previous works, we developed a real-time vehicle surrounding monitoring system that can assist drivers to perceive the vehicle surrounding situations in third-person viewpoints. Four fisheye cameras were mounted around the vehicle in our system. We developed a simple and accurate fisheye camera calibration method to dewarp the captured images into perspective projection ones. Next, we estimated the intrinsic parameters of each undistorted virtual camera by using planar calibration patterns and then obtain the extrinsic camera parameters by using the global patterns on a ground plane. A new method was proposed to tackle the brightness uniformity problem caused by the various lighting conditions of cameras. Finally, we projected the undistorted images onto a 3D hybrid projection model, stitched these images together, and then rendered the images from a third-person viewpoint selected by the driver. The proposed hybrid projection model is composed of a paraboloid model and a columnar model and can achieve rendering results with less distortion. Compared to conventional around-vehicle monitoring systems, our system can provide adaptive, integrated, and intuitive views of vehicle surroundings in a more realistic way.

1 Introduction

Blind spot of the vehicle is one of the major reasons causing car accidents. Helping drivers to perceive the situation around the vehicle while driving is the main goal of intelligent driving assistance systems. Toward this goal, many kinds of driving assistance systems have been developed. For example, parking sensors measure the distances of the obstacles from the vehicles and warn the drivers by beepers.

© Springer International Publishing Switzerland 2015
C.V. Jawahar and S. Shan (Eds.): ACCV 2014 Workshops, Part II, LNCS 9009, pp. 403–417, 2015.
DOI: 10.1007/978-3-319-16631-5_30

Automatic braking system can stop or slow down the vehicle to prevent collisions. Rear-view cameras capture the views behind the vehicles and can help drivers to drive backward. Because the prices of cameras keep dropping, it becomes practical to mount multiple cameras around the car to provide drivers with vehicle surrounding views for better perception of situations around the vehicles.

Ehlgen and Pajdla [1] proposed to use four omnidirectional cameras around a truck. They developed several ways to split the overlap area and provided a bird's-eye view image. But their integrated results are discontinuous on the seam between different cameras. In [2], Liu et al. mounted six fisheye cameras around the vehicle and stitched all six images together to provide a bird's-eye view of the vehicle surrounding environments. In their method, they applied one-dimensional stitching to improve the discontinuity issue on the seam at the expense of large computational costs. Recently, Nissan [3] developed an assistant system, Around View Monitor, which can render the surrounding view of the vehicle by using four wide-angle cameras. This system can provide drivers with both the aerial view and the original images captured by the four cameras. The Eagle Eye system from Luxgen [4] and Multi-View Camera System from Honda [5] used similar approaches to provide drivers better visual perception of surroundings. These two systems project the captured and undistorted images onto a ground plane and leave four black seams between adjacent cameras. Delphi Automotive also proposed a parking system, 360° Surround View System with Parking Guidance [6]. They applied a blending algorithm to provide smooth visual results from the aerial view. In driving situations like lane changing and passing, the visual information from aerial view may be inadequate for the drivers. Fujitsu also released a system, 360° warp-around video imaging system [7], which can project the acquired images onto a 3D curved plane and can provide drivers with a third-person view.

In this paper we propose a monitoring system for driving assistance, as its system flowchart illustrated in Fig. 1. First, we placed the fisheye camera on a rotating device controlled by a stepping motor, as shown in Fig. 4(a), to automatically calibrate the fisheye camera. Next, we mounted four of this kind of fisheye cameras around the vehicle and acquired distorted images from these cameras. Then we applied Zhang's method [8] to estimate the intrinsic parameters of each camera. Once the intrinsic parameters were obtained, these distorted images were dewarpped into perspective ones and then mapped to the ground plane by using homography transformation. Extrinsic parameters of each camera were then calculated by the homography matrix and intrinsic parameters of this camera. Finally, we projected the undistorted images onto a proposed hybrid model and render the images from a driver-selected viewpoint. We used a look-up table approach and can finish the whole image mapping process in real time.

2   Correction of Image Distortion and Estimation of Intrinsic Parameters

This section describes the proposed methods of camera calibration and image dewarping. Because the field of view of the fisheye lens we adopted is so large

Undistorted image

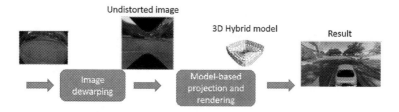

3D Hybrid model

Result

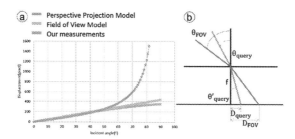

Fig. 1. Flowchart of the proposed system.

Fig. 2. (a) Relationship between incident angle and displacement of image point for perspective model, FOV model, and real fisheye camera. (b) Concept of the FOV model.

(183°), the widely-used FOV (field of view) fisheye lens model can not fit well. Instead of the full calibration procedure to determine the full set of camera parameters, therefore, we developed a simple method to dewarp fisheye camera images according to the relationship between the incident angle and the image formation distance of a point. Radial distortion of fisheye lens causes an inward or outward displacement of a given image point from its ideal location. There are many existing models for the calibration of fisheye cameras [9–12]. Devernay and Faugeras [13] used FOV model to calibrate the fisheye cameras. They used the idea that the distance between the ideal image point and the principal point is roughly proportional to the angle of incidence, as shown in Fig. 2(b). However, the FOV model is not suitable for all fisheye cameras. Figure 2(a) shows the difference between the perspective model, FOV model, and our measurement. As described in Sect. 2.3, we conducted a simple experiment to figure out how the radial distortion changes when the incident angle increases. In Sect. 2.4, we describe how to apply Zhang's method [8] to obtain the intrinsic parameters of the virtual camera, which is the undistorted counterpart of the fisheye camera. Combining with the homography matrix, we can estimate the extrinsic parameters of the virtual camera, as described in Sect. 3.

2.1 Aspect Ratio of Pixel

Because of the non-isometric pixel of fisheye camera, the circular shape of the fisheye lens border may form an elliptic shape in the captured image. The first step of camera calibration is to correct the aspect ratio of pixels. As shown in

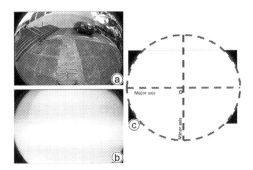

Fig. 3. (a) An image captured by a fisheye camera. (b) The image of a white wall captured by a fisheye camera. (c) Ellipse fitting result.

Fig. 3, we performed the ellipse fitting [14] process to obtain the major and minor axes and center point of the ellipse. Then we rescaled the images along the minor axis so that the length of minor axis is equal to that of the major one. Once we obtained the circular images, we assume that the lens distortion is identical along any axis crossing the image center.

2.2   Feature Point Acquisition

In order to dewarp the fisheye image, we have to estimate the relationship between the incident angle and distortion measure. First, the fisheye camera was mounted on the center of a rotating table with its optic axis perpendicular to the normal vector of the rotating table. The rotating table was controlled by a stepping motor, as shown in Fig. 4(a). Next, we designated a feature point, which was a red circle pattern on a wall and adjusted the rotating table to locate this point at the image center of the camera, as shown in Fig. 4(b). In our experiments, the distance between the wall and the camera was 5 m and we took a sequence of images while rotating the table, one image per 0.9 degrees. Furthermore, when the slanted fisheye camera mounted on the table was rotating, feature point on the image moved along an oblique line, as shown in Fig. 4(c). Consider the aspect ratio of pixels, α, the distance between the ellipse center (x_C, y_C) and the feature point (x_F, y_F) is:

$$D((x_C, y_C), (x_F, y_F)) = \sqrt{(x_C - x_F)^2 + (\alpha(y_C - y_F))^2} \ . \tag{1}$$

As a result, we can construct a table recording the mapping relationship between the rotation angle, which is equal to incident angle, and the distance from the feature point to the ellipse center. Here we denote this mapping relation as:

$$T(D_{f,i}) = \Theta_i \ , \tag{2}$$

where $D_{f,i}$ is the distance from the point i on fisheye image to the ellipse center and Θ_i is the corresponding incident angle.

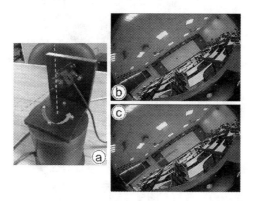

Fig. 4. (a) The rotating table controlled by a stepping motor. (b) The feature point on a wall five meters away from the camera. (c) Locus of feature points.

2.3   Mapping Rule

According to the perspective projection model, incident angle Θ_i and the distance $D_{p,i}$ from the projection point on the projection plane to the center of the plane are related by:

$$f \tan(\Theta_i) = D_{p,i} \ , \tag{3}$$

where f is the focal length of the perspective projection camera model. FOV is the maximum of incident angle which can be projected on the plane. D_{FOV} is the $D_{p,i}$ when point i is projected with incident angle FOV, which is considered as a normalization term. We can then eliminate f by substitution and obtain the following equation:

$$\frac{D_{p,i}}{D_{FOV}} = \frac{\tan(\Theta_i))}{\tan(FOV))} \ . \tag{4}$$

By substituting (2) into (4), the relationship between $D_{f,i}$ and $D_{p,i}$ is derived as:

$$D_{p,i} = \frac{D_{FOV}}{\tan(FOV)} \tan(T^{-1}(D_{f,i})) \ . \tag{5}$$

We can use this relationship to undistort fisheye image to a perspective projection one. Two examples of undistorted fisheye images are shown in Fig. 5.

2.4   Intrinsic Parameter

In order to obtain the relationship between world coordinate system and image coordinate system, we must find out the intrinsic and extrinsic parameters of our fisheye cameras. As mentioned before, the field of view of the fisheye lens we adopted is too large to use FOV model for calibration. Therefore, we estimate the parameters of the virtual camera, which is the undistorted counterpart of the fisheye camera. Following the standard procedure of Zhang's calibration method [8], we made a chess board as a calibration pattern and took multiple

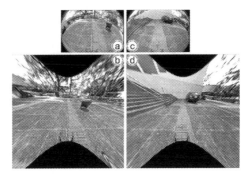

Fig. 5. (a)(c) The images captured by a fisheye camera. (b)(d) The corresponding undistorted results.

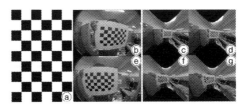

Fig. 6. (a) Chess board calibration pattern, (b)(e) the original fisheye images, (c)(f) undistorted images, and (d)(e) images in the calibration procedure.

pictures with the chess board pattern in different positions and poses. Next, we undistorted the fisheye images by using the mapping rule described above. Finally we obtained camera intrinsic parameters by using the OpenCV procedure of Zhang's method [8], as shown in Fig. 6.

3   Homography Transform and Extrinsic Parameters

3.1   Integration of Camera Views by Homography

After fisheye camera calibration, the next step is to stitch four camera images to construct a monitoring image for vehicle surrounding view. We mounted four fisheye cameras around the vehicle for capturing images toward front, back, left, and right directions, as shown in Fig. 7(a). To stitch four images, we use the homography transformation to map the dewarped images onto a reference ground plane. In this work, the reference ground plane is a bird's eye view image of a plaza captured on the roof of a building, as shown in Fig. 8(b). Forty-eight red circle patterns placed on the plaza were used as the features for the registration of dewarped fisheye images and the reference ground plane image, as shown in Fig. 7(a). At least four pairs of corresponding feature points were identified on both the ground plane image and the undistorted fisheye image. The homography relationship and the coordinate transformation between these two

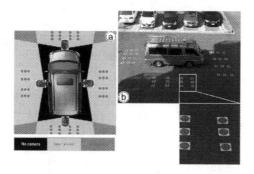

Fig. 7. (a) Four fisheye cameras mounted around the vehicle and 48 circle patterns for calibration. (b) Red circles for the estimation of homography transformation.

images were then estimated from this corresponding relationship. RANSAC [15] (RANdom SAmple Consensus) was applied for better estimation results.

3.2   Extrinsic Parameters

If we obtain the homography transformation matrix, we can know the relationship between each fisheye image and ground plane image. The extrinsic parameters, which denote the coordinate system transformation from fisheye camera coordinate system to the world coordinate system, are calculated by the following equations [8].

$$M_{ext} = \begin{bmatrix} R_1 \ R_2 \ R_3 \ T \end{bmatrix} \tag{6}$$

$$H = \begin{bmatrix} h_1 \ h_2 \ h_3 \end{bmatrix} = s \cdot M_{int} \cdot \begin{bmatrix} R_1 \ R_2 \ T \end{bmatrix} \tag{7}$$

$$R_1 = \frac{1}{s} \cdot M_{int}^{-1} \cdot h_1 \tag{8}$$

$$R_2 = \frac{1}{s} \cdot M_{int}^{-1} \cdot h_2 \tag{9}$$

$$T = \frac{1}{s} \cdot M_{int}^{-1} \cdot h_3 \tag{10}$$

and

$$R_3 = R_1 \times R_2 \ , \tag{11}$$

where h_1, h_2, h_3 are the column of homography matrix H, s is a scaling factor, M_{int} is the intrinsic matrix and M_{ext} is the extrinsic matrix which can be composed to the rotation matrix R_1, R_2, R_3 and translation matrix T.

4   Brightness Uniformity and Image Blending

4.1   Brightness Uniformity

Once the transformation matrix H is obtained, we use it to transform and map the fisheye image onto the ground plane image, as shown in Fig. 8(d). By repeating this process for each of the four fisheye cameras we can stitch all of the fisheye

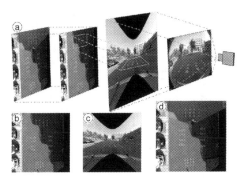

Fig. 8. (a) Homography mapping. (b) The ground reference plane. (c) Undistorted image captured by front fisheye camera. (d) Stitched images of (c) and (b) using homography transform.

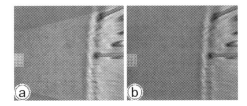

Fig. 9. (a) Before and (b) after brightness uniformity correction.

images into a panoramic one. But the various brightness of different cameras view may cause the obvious edges in the overlapping region, as shown in Fig. 9(a). In this work, we applied the following algorithm to tackle this problem.

Step 1: Separating the overlapping region from stitching image.

Step 2: Transforming from RGB to YUV color space and using Y value as luminance.

Step 3: Finding the sub-region gain of each pixel in overlapping region by the following equation:

$$G_{i,x,y} = \frac{L_{i,x,y} + L_{j,x,y}}{2 \times L_{i,x,y}} \;\;, G_{j,x,y} = \frac{L_{i,x,y} + L_{j,x,y}}{2 \times L_{j,x,y}} \;\;, \tag{12}$$

where i, j represent the indices of the cameras and $L_{i,x,y}$ represent the mean luminance of sub-region in camera i whose centroid coordinate is (x, y). The size of each sub-region is 25×25.

Step 4: Smoothing the available gain with Gaussian kernel of size 25×25.

Step 5: Polynomial fitting to obtain the remaining gains which are not in the overlapping region along horizontal or vertical scan line of stitching image. In order to maintain the automatic gain control mechanism of the used cameras, we set a center value close to unity in order to maintain the original information. As the polynomial fitting shown in Fig. 10, B and F are the two interpolation

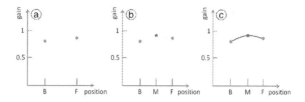

Fig. 10. (a) For each scan line, we first obtain the two boundary points named B and F. The green points indicate the two gain value of B and F. (b) We add a middle position gain named M whose gain value is calculated by averaging equations of the gains of B and F and unity. (c) Polynomial fitting to obtain the remaining gains.

boundary called beginning and ending positions and M is the middle position of the scan line. The gain value on the position M is calculated by using the following averaging equation:

$$G_M = \frac{2 + G_B + G_F}{4} \ .$$

(13)

4.2   Linear Blending

Because of the wide FOV angle of fisheye camera, each corner of the stitching image is within region overlapped to the view of other cameras. We applied linear blending process for the stitching smoothness. As shown below, linear blending is a linear combination process of overlapped pixels:

$$I^{linear}(x, y) = \frac{\sum_{i=1}^{n} I^i(x, y) W^i(x, y)}{\sum_{i=1}^{n} W^i(x, y)} \ ,$$

(14)

where $I^i(x, y)$ is the image intensity of pixel in coordinate (x, y) of image i and $W^i(x, y)$ is the weight in coordinate (x, y) of image i. The shorter distance between overlapped pixel and image center, the larger the weight is. After the blending process, we can obtain the smooth stitched images, as shown in Fig. 11. Figure 12 shows an overall result of dewarping and stitching images captured from four fisheye cameras.

5   Hybrid Model Projection

To build a system that can allow driver to select different third-person view and can synthesize the result image from the four images acquired by fisheye cameras, an intuitive way is to perform view interpolation proposed by Chen [16]. However, in the vehicle surrounding monitoring application, computational cost is large to apply the algorithm of view interpolation when moving objects exist. Therefore, we design a 3D hybrid model and project four undistorted images onto it. Then we use intrinsic and extrinsic parameters of the selected viewpoint to synthesize the results. The concept of view interpolation is shown in Fig. 13(a).

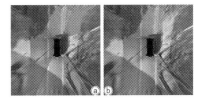

Fig. 11. (a) Stitched image without blending. (b) Stitched image with linear blending.

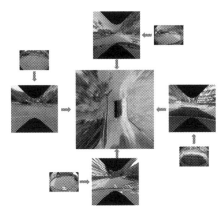

Fig. 12. The stitched result of four images from fisheye cameras.

5.1 View Interpolation Using Hybrid Model

As other vehicle surrounding systems mentioned at Sect. 1, we can stitch four dewarped images into ground plane to see the objects around vehicle. However, there are two major problems in this way. First, distortion in the image is inevitable. Second, image textures above the vanishing point would be projected to an infinite position. In this way, the driver will not be able to see the scene above the vanishing line from the viewpoint. To solve these problems, we calculate the camera extrinsic parameters from homography matrix and camera intrinsic parameters and then project the acquired images to 3D model. We use back-projection to find the texture of 3D model. The projection equation is shown in (15):

$$\begin{bmatrix} x \ y \ 1 \end{bmatrix}^{T} = \frac{1}{w} \cdot M_{int} \cdot M_{ext} \cdot \begin{bmatrix} X \ Y \ Z \ 1 \end{bmatrix}^{T} , \qquad (15)$$

where $\begin{bmatrix} x \ y \ 1 \end{bmatrix}^{T}$ is the coordinate in dewarped image, w is the homogeneous factor, M_{int} and M_{ext} are the intrinsic and extrinsic parameters of fisheye camera, respectively, and $\begin{bmatrix} X \ Y \ Z \ 1 \end{bmatrix}^{T}$ is the coordinate in 3D model.

We use four parameters to define the viewpoint, as shown in Fig. 13(b). Pan is from $-\pi$ to π, describing the main direction specified by the driver. $Tilt$ is set

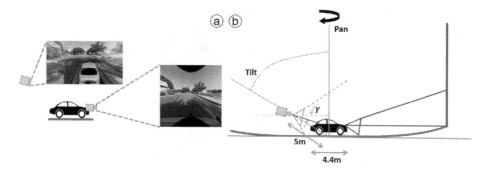

Fig. 13. (a) View interpolation. (b) Variable viewpoint.

as $\frac{\pi}{3}$, which is the angle between the virtual camera and the normal vector of the ground. The elevation angle, γ, is set as $\frac{\pi}{12}$. The distance from the virtual camera to the centroid of curved surface is 5 m.

5.2   Model Comparison

We compared the result images by projecting the acquired images into four different 3D models. The equation of these models are:

Model 1: Ground plane and cylinder surface:

$$\begin{cases} z = 0, & \text{while } x^2 + y^2 < d^2 \\ z \geq 0, & \text{while } x^2 + y^2 = d^2 \end{cases} \tag{16}$$

Model 2: Second degree paraboloid:

$$z = \frac{(x^2 + y^2)}{a^2} \tag{17}$$

Model 3: Fourth degree paraboloid:

$$z = \frac{(x^4 + y^4)}{a^4} \tag{18}$$

Model 4: Hybrid model:

$$\begin{cases} z = \frac{(x^4+y^4)}{a^4}, & \text{while } x^4 + y^4 < c^4 \\ z \geq \frac{(x^4+y^4)}{a^4}, & \text{while } x^4 + y^4 = c^4 \end{cases} \tag{19}$$

where a, c and d are the coefficients of the paraboloid, columnar and cylinder. Figure 14 shows an example of comparison results. After image undistorted, we project the images into ground plane by using homography matrix, as shown in Fig. 14(b). Distortion is large for tall objects like trees and buildings. Another problem is that the scene above the vanishing point does not appear on the

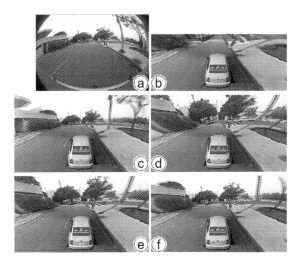

Fig. 14. The results using different 3D models. (a) Acquired image from fisheye camera. (b) After homography transformation. (c) Use ground plane and cylinder model. (d) Use second degree paraboloid. (e) Use fourth degree paraboloid. (f) Use hybrid model.

ground plane. Therefore we project the acquired images onto a 3D surface to reduce the distortions. In Model 1 (Fig. 14(c)), because the ground plane and cylinder is perpendicular, objects bend severely in the image. Although scene looks smooth in Models 2 and 3 (Figs. 14(d)(e)), there is expansion effect between surroundings.

There are two reasons that we choose the hybrid model as our 3D model. First, the columnar surface makes the view more realistic in four directions. Second, the smoothness between fourth degree paraboloid and columnar is better. As shown in Fig. 14(f), better result with less distortion is achieved by using the hybrid model.

5.3 Lookup Table

Instead of warping and stitching the whole images for each frame, we use a lookup table approach to decrease the computational cost. According to the look-up table, we can get the pixel value of output image from the original fisheye camera images by at most seven add operations and six multiplication operations.

Each entry of the lookup table contains three data values and the information of each pixel can be retrieved from single camera frame. One is the camera ID and the other two are the coordinates (x, y) in that image. For those pixels in the overlapping area, there will be seven data values in the lookup table. They are two coordinates of two different images and one linear blending weighting factor. The structure of the lookup table is shown in Fig. 15.

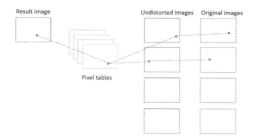

Fig. 15. Structure of the lookup table.

Fig. 16. The results of different viewpoints. (a) Front view. (b) Left view. (c) Top view. (d) Right view. (e) Rear view.

6 Experimental Results and Discussion

Figure 16 shows the result of our system in different viewpoint. We developed our system on a PC with an Intel Xeon 3.3 GHz CPU and 16 G RAM. The software was developed with Visual Studio 2010, Release mode with full code optimization. In our experiments, four fisheye cameras were mounted on a car and we tested our system in a campus and recorded about 300 s. The computation time of each frame was 0.7 s. Drivers can check every side of the vehicle carefully. By using this system, it will be safer to drive through a small lane, change lane in the highway, turn right or left in the intersection, and backward in a narrow area.

The major problem in the Fujitsu system is that the object may disappear in one view and show up again in another view because this system shrink the overlapping area between two neighbouring cameras. To tackle this problem, we choose

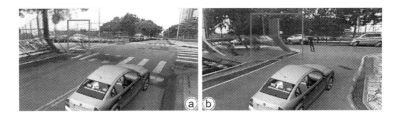

Fig. 17. Ghost effect and image distortion. The color line shows the seam between paraboloid model and columnar model, and the blue line shows the overlapping area. (a) Ghost effect. (b) Image distortion.

the wide angle fisheye camera to enlarge the overlapping area. Inside the overlapping area, the images of objects appear in two different camera image simultaneously. Due to the unknown depth of objects, the same person in different cameras may be projected to different positions and cause the ghost effect in our system, as shown in Fig. 17(a). Because we use one static 3D model for all vehicle surrounding scenes, no matter for the passing pedestrians or the trees in 15 m away, we can project them onto the same 3D model and may cause distortions as shown in Fig. 17(b). We need a 3D model which can adjust its shape according to the 3D positions of surroundings to solve these problems in the future.

7 Conclusions

In this work, we have developed a real-time vehicle surrounding monitoring system. It can provide a realistic and intuitive surrounding scene for drivers. To construct our system, a simple and precise method for fisheye camera calibration and image undistortion are proposed. Furthermore, a novel hyper projection model, which contains a paraboloid surface and a columnar surface, is used. It makes the final rendered view look more realistic. With this monitoring system, driver can use the third-person viewpoint to watch the surroundings of her vehicle. The proposed system can provide the drivers with good comprehension of the surrounding situations and reduce the risk of car accidents.

Acknowledgement. This work was supported in part by Industrial Technology Research Institute, Taiwan, with Grants 102-S-C20 and D301AR3R30.

References

1. Ehlgen, T., Pajdla, T.: Monitoring surrounding areas of truck-trailer combinations. In: Proceedings of the 5th International Conference on Computer Vision Systems (2007)
2. Liu, Y.-C., Lin, K.-Y., Chen, Y.-S.: Bird's-eye view vision system for vehicle surrounding monitoring. In: Sommer, G., Klette, R. (eds.) RobVis 2008. LNCS, vol. 4931, pp. 207–218. Springer, Heidelberg (2008)

3. NISSAN: Around View Monitor TECHNOLOGICAL DEVELOPMENT ACTIV-
ITIES.   http://www.nissan-global.com/EN/TECHNOLOGY/OVERVIEW/avm.
html

4. LUXGEN: Eagle View+ 360 surround imaging system. http://www.luxgen-motors.
vn/module-news-page-tech-cat_id-5-techid-14-lang-en.html

5. HONDA: Honda Develops New Multi-View Camera System to Provide View of
Surrounding Areas to Support Comfortable and Safe Driving. http://world.honda.
com/news/2008/4080918Multi-View-Camera-System/

6. Yu, M., Ma, G.: 360 surround view system with parking guidance. Technical report,
SAE Technical paper (2014)

7. Fujitsu: 360 Wrap-Around Video Imaging Technology. http://www.fujitsu.com/us/
semiconductors/gdc/products/omni.html

8. Zhang, Z.: A flexible new technique for camera calibration. IEEE Trans. Pattern
Anal. Mach. Intell. **22**, 1330–1334 (2000)

9. Weng, J., Cohen, P., Herniou, M.: Camera calibration with distortion models and
accuracy evaluation. IEEE Trans. Pattern Anal. Mach. Intell. **14**, 965–980 (1992)

10. Slama, C.C., Theurer, C., Henriksen, S.W., et al.: Manual of Photogrammetry, 4th
edn. American Society of Photogrammetry, Falls Church (1980)

11. Duane, C.B.: Close-range camera calibration. Photogram. Eng. **37**, 855–866 (1971)

12. Faig, W.: Calibration of close-range photogrammetric systems: mathematical formulation. Photogram. Eng. Remote Sens. **41**, 1479–1486 (1975)

13. Devernay, F., Faugeras, O.: Straight lines have to be straight. Mach. Vis. Appl. **13**,
14–24 (2001)

14. Fitzgibbon, A.W., Fisher, R.B., et al.: A buyer's guide to conic fitting. DAI Research
paper (1996)

15. Fischler, M.A., Bolles, R.C.: Random sample consensus: a paradigm for model
fitting with applications to image analysis and automated cartography. Commun.
ACM **24**, 381–395 (1981)

16. Chen, S.E., Williams, L.: View interpolation for image synthesis. In: Proceedings
of the 20th Annual Conference on Computer Graphics and Interactive Techniques,
pp. 279–288. ACM (1993)

Part-Based RDF for Direction Classification of Pedestrians, and a Benchmark

Junli Tao[(⊠)] and Reinhard Klette

The .enpeda.. Project,
Tamaki Campus, The University of Auckland, Auckland, New Zealand
jtao076@aucklanduni.ac.nz

Abstract. This paper proposes a new benchmark dataset for pedestrian body-direction classification, proposes a new framework for intra-class classification by directly aiming at pedestrian body-direction classification, shows that the proposed framework outperforms a state-of-the-art method,and it also proposes the use of DCT-HOG features (by combining a discrete cosine transform with the histogram of oriented gradients) as a novel approach for defining a random decision forest.

1 Introduction

Human beings are the most important objects for image sequence analysis for *advanced diver-assistance systems* (ADAS) or surveillance applications. Their study in video data attracted extensive research [19]. The appearance of a human in a single frame contains information about body pose, head pose, head direction, body direction, and so forth. Algorithms for pose-estimation tasks typically require high-resolution images as input. Low-resolution cameras, or humans recorded far away from the camera, still support the estimation of global information about the direction of a person expressed by the recorded pose.

Information about body direction helps to improve path predictions in sequences; see [12]. In the ADAS area, pedestrians are the most vulnerable road users. A pedestrian may change a walking path abruptly; motion information acquired in previous frames does not necessarily define an accurate prediction of the future path. For example, Fig. 1 shows a sample from the pedestrian path-prediction benchmark dataset proposed in [22]; the dataset is available on [13]. In video surveillance, a person's direction offers clues for solving specific tasks such as behavior recognition, group detection, or interaction analysis; see [3,16]. In [16], the head direction is noted for analysing the interaction between two persons in a film clip.

Our first contribution in this paper is a proposal of a new *Pedestrian Direction Classification* (PDC) dataset; thus responding to an obvious demand in this area for more benchmark data. The task of discrete body-direction classification is not yet studied as intensively as a generic pedestrian-detection task, and the lack of benchmark datasets might be one reason for this situation. The proposed PDC dataset has been generated based on the *Daimler Mono Pedestrian Classification*

© Springer International Publishing Switzerland 2015
C.V. Jawahar and S. Shan (Eds.): ACCV 2014 Workshops, Part II, LNCS 9009, pp. 418–432, 2015.
DOI: 10.1007/978-3-319-16631-5_31

Benchmark. There are already two existing datasets which provide ground truth on pedestrian directions, the *TUD Multiple View Pedestrian* (TUD) dataset proposed by [1], and the *Human Orientation Classification* (HOC) dataset introduced by [7]. We show that the newly introduced PDC dataset outperforms both the TUD and the HOC dataset with respect to defined criteria. A detailed comparison of the TUD, HOC, and the proposed PDC dataset is given in Sects. 4 and 5.

The second contribution in this paper is the proposal of an efficient framework (PRDF) for pedestrian direction classification. In order to deal with the multiple intra-class (i.e. different body direction classes in one pedestrian class) classification task, see [1,3,14,18,26], previously proposed methods learn multiple classifiers, one for each direction. This excludes sharing of information among classes in the training or classification process, and is (thus) more time consuming for both processes. A random decision forest (RDF) is adopted in [2,24] for direction classification. In both publications, features of a bounding box are used as input for the classifier. Each splitting node in a tree of the RDF, however, selects only one component in the used high-dimensional feature vector. In this paper we propose an PRDF for automatically learning the discrimination of selected body parts for classifying body directions. Experimental results show that the proposed PRDF framework performs better and faster than state-of-the-art methods.

As a third contribution we are proposing a novel feature for direction classification. Features of the histogram of oriented gradients (HOG) are extensively used for pedestrian detection [6]. As reported in [5], complex splitting nodes lead to over-fitting issues. In [24], one or two feature elements are adopted for defining a splitting function. In this paper we propose to perform a discrete cosine transform (DCT) over the HOG feature vector to obtain a more global descriptor before selecting feature elements. Thus, each element contains global frequency information instead of just some local gradient-orientation information.

To summarise the contributions in this paper, we (1) propose a new pedestrian body-direction classification benchmark dataset (PDC), (2) propose a new framework (PRDF) for intra-class classification by directly aiming at pedestrian

Fig. 1. Frames of a body-bending sequence of the Daimler pedestrian path prediction benchmark dataset [13]

body-direction classification, (3) show that the proposed framework outperforms a state-of-the-art method, and we (4) propose the use of DCT-HOG features.

2 Related Work

We briefly review body-direction classification algorithms and classifiers based on random decision forests, given a bounding box containing a pedestrian. For solving the body-direction classification task, as an intra-class classification problem, researchers train multiple two-class classifiers [1,3,14,18,26] or adopt a multi-class classifier [7,24]. Both approaches use features and classifiers as previously known for pedestrian detection. For example, HOG features and support vector machine (SVM) classifiers are extensively employed for body-direction classification [1,3,14].

The authors of [7] propose a weighted array of covariances (WARCO) for deriving features for classifying body- and head-direction. The values of 13 combined feature channels are taken as defining a manifold in feature space; those 13 channels are composed of eight difference-of-offset-Gaussian filter channels, three color channels, gradient magnitude, and a gradient-direction channel. A method based on silhouettes is presented in [18]; used shape descriptors limit the range of body directions to the interval $[0°, 180°]$. The estimation of pedestrian direction is performed in [23] more robustly by selecting a recognition result based on multiple still images, rather than by using just a single image. Multiple random-tree classifiers are trained in [2] and compared with trained SVM classifiers; outputs are integrated using a *mixture of approximated wrapped Gaussians* (MAWG). Using calculated probabilities of multiple outputs obtained from all participating classifiers, the final direction is obtained by maximising the mixed probability of the WAWG.

Direction recognition is difficult as head pose, torso, and body might point into different directions; but their poses are interrelated to each other. For example, body direction is estimated in [3,4] by considering location and head pose, and assuming that tracks are available.

There are also several methods proposed for classifying pedestrians against a background together with their body directions [7,14,24]. The authors of [24] propose to modify the objective function of each split node in the RDF for simultaneously handling both tasks, pedestrian detection and direction classification, a single, or two HOG elements are compared against a randomly generated threshold, and results are selected for optimising a combined objective function. [14] presents a three stage process; Stages 1 and 2 adopt different HoG where blocks are either overlapping or not, to reject non-pedestrian boxes; at Stage 3, four SVM classifiers are trained for the four directions separately using pedestrian samples only. A unified Bayesian model is used in [9], based on shape and motion cues; the proposed method classifies pedestrians with recognizing one of four possible directions.

RDFs are extensively applied for many detection or categorisation subjects, including object detection [11,15], action recognition [25], image labelling [17], or edge detection [8]. The structure of trees in an RDF depends on the training process, which may vary for different subjects. In [15], an RDF is structured for doing pedestrian detection. Instead of using simple algebraic splitting functions, a two-class SVM is adopted for each splitting node. The authors of [20] propose alternating decision forests; instead of independently learning each tree in a forest, a gradient boosting theory is introduced for concurrently training a forest. For each depth level, a significance distribution of training samples is updated based on the performance of the current forest. Miss-classified samples receive more attention when training split nodes at the next depth level. The authors of [21] propose corresponding alternating regression forests.

3   Proposed Algorithm

We detail the proposed algorithm. We start with introducing the used notation following a general RDF framework.

3.1   Random Decision Forest

An RDF acts for a given categorisation problem as a (strong) classifier, defined by a set of trees, each acting as a weak classifier. Let T_t, for $t \in \{1, \dots, N\}$, be a set of randomly trained decision trees which defines an RDF. In each tree, a classification problem is splitted by answering subsequently "simple questions" defined by split functions. In other words, such a decision tree consists of a set of split functions hierarchically arranged into a tree structure.

A decision tree has internal (or split) and terminal (or leaf) nodes. We assign a split function to each split node which has two out-edges connected to two nodes, being either split or leaf nodes. The assigned split function $h_\phi(\cdot)$ decides which of the two nodes comes next. Let $\mathcal{I}$ denote the set of inputs and

$$\mathcal{I}_L(\phi) = \{I \in \mathcal{I} | h_\phi(I) = 0\} \quad \text{and} \quad \mathcal{I}_R(\phi) = \{I \in \mathcal{I} | h_\phi(I) = 1\} \qquad (1)$$

Later we specify split function $h_\phi(\cdot)$ and its parameters ϕ.

A set $\mathcal{I}^{tr} \subset \mathcal{I}$ of labelled pedestrian are used in the training process for expanding the trees of an RDF. Samples $I \in \mathcal{I}^{tr}$ split along internal nodes and end up in leaf nodes of trained trees.

A decision tree is trained by growing subsequently internal nodes, starting at a root node. Suitable functions $h_\phi(\cdot)$ are selected with respect to a predefined target function. Trees of an RDF grow randomly and independently to each-other. Randomness when training a tree is important to ensure some variety in the forest (i.e. trees need to be uncorrelated to ensure that the forest can investigate samples from "different perspectives"). For the assembled forest we intentionally avoid to grow "similar" trees.

A stop criterion defines when a leaf node L is created. The distribution of classes in a leaf node is obtained with respect to those samples $I \in \mathcal{I}^{tr}$ which reach this leaf node. According to this the distribution, the leaf node assigns probabilities $p(d|L_t)$, for $d \in \{N, E, S, W\}$, where N, E, S, W denote body directions.

For testing of an RDF we use a set of input bounding boxes denoted by $\mathcal{I}^{ts} \subset \mathcal{I}$. Any sample $I^{ts} \in \mathcal{I}^{ts}$ is passed through the N trees T_t of the trained forest. Sample I^{ts} ends up in a leaf node L_t in tree T_t. This way we assign N distributions to one test box. The simple rule

$$d^* = \arg\max_d \sum_{t=1}^{N} p(d|L_t) \tag{2}$$

defines a maximum-likelihood decision for classifying the direction of a pedestrian.

3.2   Split and Objective Function

Let V denote a feature vector, and i be the index of a feature element. A split function is then defined as follows:

$$h_\phi(I) = \begin{cases} 0 & \text{if} \quad V(i) > \tau \\ 1 & \text{otherwise} \end{cases} \tag{3}$$

The goal is to split the training samples uniformly for maximising the information gain at each internal node. More specific, for each internal node, parameters $\phi = \{i, \tau\}$ are learned with respect to maximizing a predefined objective function. It is important to choose an appropriate objective function for obtaining "good" split functions during the training process. This is supported by Shannon's entropy-based objective function. Let $E_d(\mathcal{I})$ denote the entropy of direction classes. We use

$$o_d(\phi, \mathcal{I}) = E_d(\mathcal{I}) - \sum_{k \in \{L,R\}} \omega_k E_d(\mathcal{I}_k(\phi)) \tag{4}$$

$$\text{with} \quad E_d(\mathcal{I}) = -\sum_d p(d|\mathcal{I}) \log(p(d|\mathcal{I}))$$

$$\text{and} \quad \omega_k = |\mathcal{I}_k(\phi)|/|\mathcal{I}(\phi)|$$

The ω-values are the weights for balancing the bias caused by varying numbers of samples, going either to the left or right child node. By allowing different objective functions, split nodes can be generated individually.

3.3   DCT-HOG Feature

The HOG was introduced in [6] for pedestrian detection. The steps of HOG calculation can be summarised as follows:

-   An input gray-level image is partitioned into cells of equal sizes (e.g. of 8×8 pixels).

- For each cell, the gradient magnitude at each pixel in the cell votes to "its" discrete phase bin (e.g. nine bins, $[0°, 20°]$, $[20°, 40°]$,...,$[160°, 180°]$). Thus, each cell contains nine elements, where each element corresponds to the sum of gradient magnitudes in those discrete phase ranges.
- To obtain a feature vector, blocks of identical size (e.g. each 2×2 cells) slide through the cell matrix. The cell elements are normalized within the block and combined into one vector (either column or row wise).
- The vectors from all those blocks are now augmented to generate the *HOG feature vector* V_{hog}.

HOG feature vectors are extensively used in object detection because of their positive performance compared to other features (e.g. local binary patterns, or histogram of optical flow). For splitting training samples at a reached node, one feature element is adopted at a time. One feature element in HOG contains specifically local information for one phase bin of one cell. In order to employ more global information in a split node, we propose the DCT-HOG feature $V_{dct-hog}$. It is obtained by applying the discrete cosine transform (DCT) first over the HOG feature vector V_{hog} taken as a 1-dimensional discrete signal:

$$V_{dct-hog} = C^{(|V_{hog}|)} \cdot V_{hog} \tag{5}$$

$$c_{jk}^{(|V_{hog}|)} = \sqrt{\alpha_j/|V_{hog}|} \cdot \cos(\frac{\pi(2k+1)j}{2|V_{hog}|}) \tag{6}$$

where $|V|$ denotes the number of elements in a vector V, $c_{jk}^{(|V_{hog}|)}$ is an element in the orthogonal matrix $C^{(|V_{hog}|)}$ of dimension $|V_{hog}| \times |V_{hog}|$, and $\alpha_0 = 1$, $\alpha_j = 2$. The elements in DCT-HOG contains global frequency information. Thus, global information is adopted in a split node, when splitting based on the DCT-HOG feature elements.

3.4   Part Based Random Decision Forest

A conventional RDF learning procedure is based on feature vectors of the whole object - in our case, of a person. Following the classifier's internal structure, introduced in Sect. 3.2, a randomly selected element from HOG is not a discriminative representation of the sample. Thus, we propose to apply the forest to deduct discriminative information from image patches, i.e. from parts of the human body.

Instead of mixing randomly selected local patches from random locations,we use the same location and size of a patch in each training sample as a tree-training set. In this way, the location of body parts is kind of "encoded" for training. The tree-training procedure focuses on the appearance of the body parts.

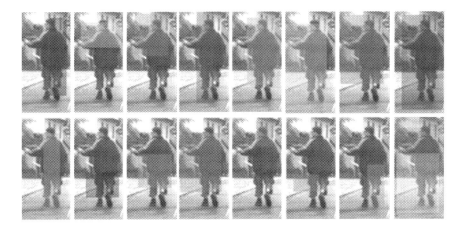

Fig. 2. Selected parts for several trees. *Bottom, right*: All the selected parts for a forest

Algorithm 1. (Training)
Input: All training samples $\mathcal{I}$
Output: Trained trees T_t, for $t = 1, 2, \ldots, N$

```
 1: randomly select the location and size for an image patch, identified by top-left
    coordinates (row_t, col_t) and patch size (width_t, height_t).
 2: calculate feature vector V of each training patch; for different experiments the V
    stands either for V_hog, V_dct−hog, or V_comb.
 3: let T_t = ∅, num = |I|, dep = 0, stop criterion t_num = 20, t_dep = 15, temporal data
    store variables temp_od1 = 0, temp_od2 = 0.
 4: if num < t_num || dep > t_dep then
 5:     calculate p(d|L) with I, according to Equ. (8);
 6:     add leaf L to the tree: T_t = T_t ∪ L
 7:     return T_t.
 8: else
 9:     dep = dep + 1;
10:     for s = 1, . . . , 1000 do
11:         randomly select a feature element index i_s;
12:         find range [τ_min, τ_max] of V_{i_s} with current node samples;
13:         for h = 1, . . . , 10 do
14:             randomly select τ_h ∈ [τ_min, τ_max];
15:             split I into I_{Lh}, I_{Rh} according to Equ. (3);
16:             calculate o_d({i_s, τ_h}, I) with Equ. (4);
17:             if o_d({i_s, τ_h}, I) > temp_od2 then
18:                 temp_od2 = o_d(φ_s, I);
19:                 τ_s = τ_h, φ_s = {i_s, τ_s};
20:             end if
21:         end for
22:         if temp_od2 > temp_od1 then
23:             φ* = φ_s;
24:         end if
25:     end for
26:     expand tree by new split node: T_t = T_t ∪ φ*;
27:     split I into I_L and I_R;
28:     num = |I_L|, I = I_L, and go to Line 4;
29:     num = |I_R|, I = I_R, and go to Line 4;
30: end if
```

Using such localized regions, simple split functions yield better performance compared to an application of the whole bounding box for tree training; see results in Sect. 5. Figure 2 illustrates selected discriminative parts for a tree and a forest. For the considered intra-class classification task, the global appearance of a person is similar to some degree for the whole pedestrian class; the distinctive information among classes actually lies in local body parts.

3.5 Implementation

For the used training and testing algorithms, see Algorithms 1 and 2, respectively. We apply the whole training set when training a tree of the RDF. As reported in [5], randomness is significant for the performance of a forest. Instead of bagging, we introduce randomness by randomly selecting patches and feature elements. The stop criteria parameters, including the tree's depth and the minimum number of samples, is set according to [24].

Because of the different cardinalities of training samples for the different classes, a sample-bias compensation is necessary for calculating probabilities at a leaf node. This is achieved by using a balancing factor r_d, defined as follows:

$$p(d|L) = |\mathcal{I}_d^L| \cdot r_d / \sum_d (|\mathcal{I}_d^L| \cdot r_d) \tag{7}$$

$$\text{with} \quad r_d = |\mathcal{I}^{tr}|/|\mathcal{I}_d^{tr}|$$

where $|\mathcal{I}^{tr}|$ denotes the cardinality of the training samples for each tree; set $\mathcal{I}_d^{tr} \subset \mathcal{I}^{tr}$ contains the training samples for direction d in $\mathcal{I}^{tr}$, and set $\mathcal{I}^L \subset \mathcal{I}^{tr}$ contains all the samples arriving at leaf node L. In our experiments, we set $N = 120$.

During testing, for each tree T_t, the corresponding patch I_{patch}, specified by location (row_t, col_t) and size $(width_t, height_t)$, from a test image I is adopted to calculate the feature vector V, and then passed through the tree. See Algorithm 2 for details.

Algorithm 2. (Testing)
Input: Test bounding box I, trained trees T_t, with $t = 1, 2, \ldots, N$.
Output: Class label d^*.

1: **for** $t = 1, \ldots, N$ **do**
2:    extract corresponding patch location (row_t, col_t) and size $(width_t, height_t)$ from tree t.
3:    calculate feature vector V for the image patch I_{patch}.
4:    pass V through T_t until reaching a leaf node L_t, obtain distribution $p(d|L_t)$.
5: **end for**
6: obtain d^* with Equ. (2);
7: return d^*.

4   Proposed PDC Benchmark Dataset

We introduce our *pedestrian direction classification* (PDC) benchmark dataset.[1]
We compare our dataset with two other available datasets, TUD and HOC.
Sample images are shown in Fig. 3, top.

Fig. 3. *Top row:* PDC samples illustrating ambiguity between *N*, *NE*, and *E*. *Middle and bottom rows:* Sample images from the HOC (*middle*) and TUD (*bottom*) datasets

Besides very advanced research in the driver-assistance area, there is not yet
any publicly available pedestrian body-direction classification dataset available
in a driving context. Thus, we use one popular pedestrian-classification dataset
from Daimler, and manually classified the 12,000 pedestrian bounding boxes
(sized 48×96) into 8 directions (namely *N*, *NE*, *E*, *SE*, *S*, *SW*, *W*, or *NW*). Even
for human beings, it is difficult to classify which direction a pedestrian should
be assigned, e.g. among *N*, *NE*, and *E*, *E*, *SE*, and *S*, *S*, *SW*, and *W*, and *W*,
NW, and *N*. Due to the ambiguity, we classified a person, for example, to *NE*
only if the person is facing into diagonal direction.

As the PDC pedestrians are shown in various scenes, an often cluttered background enables that the PDC dataset gives more general information when used

---

[1] See ccv.wordpress.fos.auckland.ac.nz/data/object-detection/.

for training, and it is more challenging when used for testing; see experimental results in Sect. 5.

The HOC dataset [7] contains 11,881 bounding boxes, including 6,860 training samples and 5,021 test samples. The original image size is 62×132. The bounding boxes are extracted from ETHZ data, see [10]. The sequences are taken from two cameras mounted on top of a trolley. Four directions, N, S, E, W, are labelled for each sample. The TUD dataset [1] contains 5,183 bounding boxes, including 4,935 training samples and 248 test samples. This dataset labels pedestrians for 8 directions, as in our PDC dataset. Sample images from the HOC and TUD datasets are shown in Fig. 3, bottom. Information for all three datasets is summarized in Table 1.

Table 1. PDC, HOC, and TUD datasets summary

	Number of samples	Image channel	Directions
HOC	11,881	Color	N, S, E, W
TUD	5,183	Color	N, NE, E, SE, S, SW, W, NW
PDC	12,000	Gray	N, NE, E, SE, S, SW, W, NW

5 Experiments

We report about several sets of experiments for illustrating the performance of the proposed PRDF framework for all datasets, the merits of the new proposed PDC dataset for both training and testing tasks, the performance of the proposed feature DCT-HOG, and when augmenting DCT-HOG with a HOG feature. Finally, the proposed algorithm is compared with state-of-the-art methods as reported in [1–3, 7, 24]. We also provide the mean processing time for testing one image.

5.1 RDF vs PRDF

To compare the proposed *part-based random decision forest* (PRDF) with the conventional RDF, two algorithms are trained with HOG feature on the three datasets TUD, HOC, and PDC respectively. As the testing set of the TUD dataset only contains 248 images, it is not quite sufficient to evaluate the performance; we use PDC as the test set when using TUD for training, and, correspondingly, TUD as the testing set after training on PDC. The confusion matrices are given in Tables 2 and 3. The average error, depending from the number of applied trees, is shown in Fig. 4.

Figure 4 shows curves for the results of RDFs using different colors for different training sets. For results of PRDFs we use squares in the colors of the used training set. Obviously, the PRDFs outperform the corresponding RDFs (e.g. PRDF-HOC versus RDF-HOC) for all datasets. Thus, for later experiments, PRDF framework is adopted. The confusion matrices quantify how the PRDFs improve the classification performance for body directions.

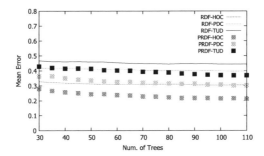

Fig. 4. Mean errors of RDFs and PRDFs for used numbers of trees in the forest

Table 2. RDF confusion matrices. *Left* to *right*: TUD, HOC, PDC

	N	E	S	W		N	E	S	W		N	E	S	W
N	0.67	0.04	0.26	0.04	N	0.77	0.07	0.07	0.10	N	0.63	0.13	0.20	0.04
E	0.17	0.46	0.24	0.13	E	0.10	0.67	0.13	0.10	E	0.06	0.81	0.03	0.09
S	0.45	0.07	0.43	0.04	S	0.09	0.05	0.78	0.07	S	0.33	0.18	0.37	0.12
W	0.15	0.14	0.22	0.49	W	0.11	0.10	0.20	0.60	W	0.08	0.15	0.03	0.74

Table 3. PRDF confusion matrices. *Left to right*: TUD, HOC, PDC

	N	E	S	W		N	E	S	W		N	E	S	W
N	0.76	0.02	0.2	0.02	N	0.84	0.03	0.07	0.06	N	0.70	0.07	0.19	0.04
E	0.16	0.55	0.19	0.09	E	0.05	0.76	0.10	0.09	E	0.05	0.90	0.01	0.04
S	0.41	0.04	0.52	0.02	S	0.05	0.02	0.89	0.04	S	0.27	0.13	0.52	0.08
W	0.16	0.10	0.16	0.57	W	0.07	0.06	0.18	0.69	W	0.11	0.09	0.02	0.78

5.2   Dataset Comparisons

In order to compare the datasets, the PRDFs trained with HOG feature on one of the three datasets are subsequently tested on the other two training sets respectively. For example, a TUD trained PRDF is tested on HOC and PDC training sets. The test results are summarised in Table 4. The mean error is shown in Fig. 5.

In Fig. 5, the TUD-HOC means PRDF trained with TUD training set, and HOC training set is adopted as test set. Both TUD and PDC trained PRDFs perform better than an HOC trained PRDF when tested on TUD and PDC training sets. The PDC trained PRDF significantly outperforms a TUD trained PRDF over HOC, and the HOC trained PRDF over TUD. Using the HOC training set as testing set leads to the largest mean error. The TUD training set is the easiest one according to this experiment. Thus, we conclude that the proposed PDC dataset offers better generalized information for training, and also offers challenges for testing in general.

The confusion matrices show that a TUD-trained PRDF appears to be totally confused with respect to testing on the HOC data, but performs reasonable on

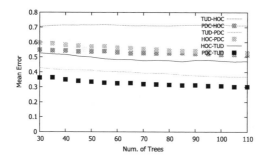

Fig. 5. Mean errors of cross-tested datasets, e.g. TUD-HOC is trained on the TUD training set, but tested on the HOC training set

Table 4. Confusion matrices. *Top, left to right:* TUD-HOC, TUD-PDC, HOC-TUD. *Bottom, left to right:* HOC-PDC, PDC-TUD, PDC-HOC

	N	E	S	W		N	E	S	W		N	E	S	W
N	0.10	0.23	0.36	0.31	N	0.76	0.02	0.2	0.02	N	0.62	0.07	0.22	0.10
E	0.12	0.34	0.23	0.31	E	0.16	0.55	0.19	0.09	E	0.04	0.46	0.28	0.22
S	0.10	0.19	0.37	0.34	S	0.41	0.04	0.52	0.02	S	0.30	0.11	0.47	0.12
W	0.11	0.28	0.27	0.34	W	0.16	0.10	0.16	0.57	W	0.08	0.24	0.18	0.50

	N	E	S	W		N	E	S	W		N	E	S	W
N	0.53	0.13	0.19	0.16	N	0.70	0.07	0.19	0.04	N	0.75	0.09	0.12	0.04
E	0.07	0.29	0.29	0.35	E	0.05	0.90	0.01	0.04	E	0.36	0.42	0.12	0.10
S	0.22	0.12	0.52	0.13	S	0.27	0.13	0.52	0.08	S	0.31	0.16	0.44	0.09
W	0.09	0.19	0.22	0.50	W	0.11	0.09	0.02	0.78	W	0.39	0.17	0.13	0.31

the PDC training set. An HOC-trained PRDF performs worse for classifier E direction on both the TUD and PDC training set, while a PDC-trained PRDF performs best for classifying E.

5.3 Feature Comparison

In this section, the PRDF is trained with three different feature settings, HOG, DCT-HOG, or a combined HOG and DCT-HOG, for the three datasets. The performance of nine PRDFs is shown in Fig. 6. The confusion matrices are illustrated in Tables 3, 5, and 6.

Figure 6 illustrates that the HOG feature performs better than DCT-HOG feature in the mean-error sense. But the confusion matrices show that the N direction classification is improved by using the DCT-HOG feature when training on the TUD and HOC datasets. The combined HOG and DCT-HOG feature slightly improves the overall performance. The confusion matrices tell us that the combined feature improves performance for all direction classifications except for direction S.

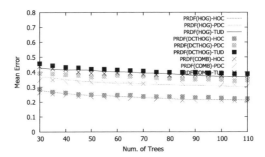

Fig. 6. Mean error of PRDFs for different feature selections. PRDF(HOG)-TUD means trained on the TUD training set using HOG features. PRDF(DCTHOG)-TUD means trained on the TUD training set using DCT-HOG features. PRDF(COMB)-TUD means trained on the TUD training set using combined HOG and DCT-HOG features.

Table 5. PRDF(DCTHOG) confusion matrices. *Left to right*: TUD, HOC, PDC

	N	E	S	W		N	E	S	W		N	E	S	W
N	0.82	0.02	0.15	0.02	N	0.83	0.03	0.08	0.06	N	0.73	0.07	0.16	0.03
E	0.20	0.54	0.12	0.13	E	0.06	0.73	0.11	0.10	E	0.06	0.90	0.01	0.03
S	0.59	0.05	0.32	0.04	S	0.06	0.03	0.86	0.05	S	0.43	0.15	0.39	0.04
W	0.21	0.12	0.10	0.57	W	0.08	0.07	0.17	0.67	W	0.12	0.17	0.02	0.69

Table 6. PRDF(COMB) confusion matrices. *Left to right*: TUD, HOC, PDC

	N	E	S	W		N	E	S	W		N	E	S	W
N	0.79	0.01	0.19	0.01	N	0.84	0.02	0.07	0.07	N	0.72	0.07	0.16	0.04
E	0.17	0.54	0.16	0.13	E	0.05	0.78	0.11	0.07	E	0.05	0.92	0.01	0.02
S	0.51	0.04	0.41	0.04	S	0.03	0.03	0.90	0.04	S	0.36	0.16	0.44	0.04
W	0.18	0.10	0.12	0.60	W	0.06	0.07	0.18	0.69	W	0.08	0.13	0.01	0.78

5.4   Comparison with State-of-the-Art Algorithms

The proposed PRDF(COMB) is compared with two methods proposed in [7], called FEOB, and CBH1. To ensure a fair comparison with the results in [7], PRDF(COMB) is trained with the HOC training set. Three confusion matrices are shown in Table 7. The proposed PRDF(COMG) performs best on classifying N, E, S, and achieved the highest average accuracy (0.79).

For the sake of completeness of experiments, the results of PRDF, trained on the TUD set, are illustrated in Table 8, along with results reported in [1–3,24]. Note that the test set contains 248 images only. We do not consider this as being a sufficient test set for drawing general conclusions.

Processing Time. The mean processing time for one test input over 120 trees is 0.28 s. As trees are independent classifiers, parallel processing could be applied. Thus, the process time could be reduced to 2–3 ms. The given processing time

Table 7. Confusion matrices of FEOB [23], CBH1 [23], and PRDF(COMB) on HOC dataset. *Left*: FEOB (average accuracy 0.78189). *Middle*: CBH1 (average accuracy 0.78692). *Right*: PRDF(COMB) (average accuracy 0.79031)

	N	E	S	W		N	E	S	W		N	E	S	W
N	0.76	0.11	0.04	0.09	N	0.77	0.10	0.04	0.09	N	0.84	0.02	0.07	0.07
E	0.03	0.76	0.15	0.06	E	0.03	0.77	0.14	0.06	E	0.05	0.78	0.11	0.07
S	0.00	0.04	0.88	0.08	S	0.00	0.04	0.88	0.08	S	0.03	0.03	0.90	0.04
W	0.04	0.08	0.15	0.73	W	0.04	0.08	0.15	0.73	W	0.06	0.07	0.18	0.69

Table 8. Test results for 248 test images from TUD

	N	E	S	W
PRDF	0.85	0.82	0.26	0.71
[24]	0.91	0.85	0.37	0.69
[2]	0.76	0.95	0.64	0.86
[3]	0.71	0.65	0.41	0.70
[1]	0.46	0.54	0.4	0.38

was measured for a standard desktop PC, with 3.4 GHz CPU, and 8 GB RAM. All the methods are coded with C++, and compiled with Visual Studio 2010.

6   Conclusions

This paper proposed a new pedestrian-direction classification benchmark dataset and a new framework for solving the pedestrian direction classification task. Experimental results prove that the proposed benchmark dataset outperforms the existing two datasets based on defined criteria. The proposed framework performs better than a state-of-the-art method on the HOC dataset. Results support future work on applying derived direction information while tracking pedestrians in sequences. Pedestrians are assumed to be located in the middle of the given bounding boxes. Localization errors also need to be considered when applying the proposed method. The PRDF framework may be tested with additional features, e.g. PCA-HOG.

References

1. Andriluka, M., Roth, S., Schiele, B.: Monocular 3D pose estimation and tracking by detection. In: CVPR, pp. 623–630 (2010)
2. Baltieri, D., Vezzani, R., Cucchiara, R.: People orientation recognition by mixtures of wrapped distributions on random trees. In: Fitzgibbon, A., Lazebnik, S., Perona, P., Sato, Y., Schmid, C. (eds.) ECCV 2012, Part V. LNCS, vol. 7576, pp. 270–283. Springer, Heidelberg (2012)
3. Chen, C., Heili, A., Odobez, J.-M.: Combined estimation of location and body pose in surveillance video. In: AVSS, pp. 5–10 (2011)

4. Chen, C., Heili, A., Odobez, J.-M.: A joint estimation of head and body orientation cues in surveillance video. In: ICCV Workshops, pp. 860–867 (2011)

5. Criminisi, A., Shotton, J., Konukoglu, E.: Decision forests: a unified framework for classification, regression, density estimation, manifold learning and semi-supervised learning. Found. Trends Comput. Graph. Vis. **7**, 81–227 (2011)

6. Dalal, N., Triggs, B.: Histograms of oriented gradients for human detection. In: CVPR, pp. 886–893 (2005)

7. Diego, T., Spera, M., Cristani, M., Murino, V.: Characterizing humans on riemannian manifolds. IEEE Trans. Pattern Anal. Mach. Intell. **35**(8), 1972–1984 (2013)

8. Dollar, P., Zitnick, L.C.: Structured forests for fast edge detection. In: ICCV (2013)

9. Enzweiler, M., Gavrila, D.M.: Integrated pedestrian classification and orientation estimation. In: CVPR, pp. 982–989 (2010)

10. Ess, A., Leibe, B., Gool, L.: Depth and appearance for mobile scene analysis (2014). http://www.vision.ee.ethz.ch/aess/iccv2007/

11. Gall, J., Lempitsky, V.: Class-specific hough forests for object detection. In: CVPR, pp. 1022–1029 (2009)

12. Gandhi, T., Trivedi, M.M.: Image based estimation of pedestrian orientation for improving path prediction. In: IEEE IV, pp. 506–511 (2008)

13. Gavrila, D.M.: (2014). www.gavrila.net/Datasets

14. Goto, K., Kidono, K., Kimura, Y., Naito, T.: Pedestrian detection and direction estimation by cascade detector with multi-classifiers utilizing feature interaction descriptor. In: IEEE IV, pp. 224–229 (2011)

15. Marin, J., Vazquez, D., Lopez, A.M., Amores, J., Leibe, B.: Random forests of local experts for pedestrian detection. In: ICCV, pp. 2592–2599 (2013)

16. Marin-Jimenez, M.J., Zisserman, A., Eichner, M., Ferrari, V.: Detecting people looking at each other in videos. Int. J. Comput. Vis. **106**(3), 282–296 (2014)

17. Peter, K., Bulo, S.R., Bischof, H., Pelillo, M.: Structured class-labels in random forests for semantic image labelling. In: ICCV, pp. 2190–2197 (2011)

18. Piérard, S., Van Droogenbroeck, M.: Estimation of human orientation based on silhouettes and machine learning principles. In: ICPRAM (2012)

19. Rosenhahn, B., Klette, R., Metaxas, D. (eds.): Human Motion. Springer, Dordrecht (2008)

20. Samuel, S., Wohlhart, P., Leistner, C., Saffari, A., Roth, P.M., Bischof, H.: Alternating decision forests. In: CVPR, pp. 508–515 (2013)

21. Samuel, S., Leistner, C., Wohlhart, P., Roth, P.M., Bischof, H.: Alternating regression forests for object detection and pose estimation. In: ICCV, pp. 417–424 (2013)

22. Schneider, N., Gavrila, D.M.: Pedestrian path prediction with recursive bayesian filters: a comparative study. In: Weickert, J., Hein, M., Schiele, B. (eds.) GCPR 2013. LNCS, vol. 8142, pp. 174–183. Springer, Heidelberg (2013)

23. Shimizu, H., Poggio, T.: Direction estimation of pedestrian from multiple still images. In: IEEE IV, pp. 596–600 (2004)

24. Tao, J., Klette, R.: Integrated pedestrian and direction classification using a random decision forest. In: ICCV Workshop, pp. 230–237 (2013)

25. Yao, A., Gall, J., Gool, L.V.: A Hough transform-based voting framework for action recognition. In: CVPR, pp. 2061–2068 (2010)

26. Zhao, G., Takafumi, M., Shoji, K., Kenji, M.: Video based estimation of pedestrian walking direction for pedestrian protection system. J. Electron. (China) **29**, 72–81 (2012)

Path Planning for Unmanned Vehicle Motion Based on Road Detection Using Online Road Map and Satellite Image

Van-Dung Hoang, Danilo Caceres Hernandez, Alexander Filonenko,
and Kang-Hyun Jo[✉]

School of Electrical Engineering, University of Ulsan, Ulsan, Korea
{hvzung,danilo,alexander}@islab.ulsan.ac.kr, acejo@ulsan.ac.kr

Abstract. This article presents a new methodology for detecting road network and planning the path for vehicle motion using road map and satellite/aerial images. The method estimates road regions from based on network models, which are created from road maps and satellite images on the basis of using image-processing techniques such color filters, difference of Gaussian, and Radon transform. In the case of using the road map images, this method can estimate not only a shape but also a direction of road network, which would not be estimated by the use of the satellite images. However, there are some road segments that branch from the main road are not annotated in road map services. Therefore, it is necessary to detect roads on the satellite image, which is utilized to construct a full path for motion. The scheme of method includes several stages. First, a road network is detected using the road map images, which are collected from online maps services. Second, the detected road network is used to learn a model for road detection in the satellite images. The road network using the satellite images is estimated based on filter models and geometry road structures. Third, the road regions are converted into a Mercator coordinate system and a heuristic based on Dijkstra technique is used to provide the shortest path for vehicle motion. This methodology is tested on the large scene of outdoor areas and the results are documented.

1 Introduction

Nowadays, automatic navigation systems have been developed and applied in many research areas on robotics, autonomous vehicle, intelligent transportation systems, and other industry applications. Motion path planning, localization, and mapping become important research areas in various applications of autonomous navigation. There have been several groups of researchers focusing on autonomous vehicle/robot, especially intelligent transportation in outdoor environments, such as in [1–3]. In automatic navigation of mobile systems, first, they require to provide a global path network for robot/vehicle motion. Therefore, path planning is an important part in every autonomous vehicle system. Recently, path planning methods have been achieved widespread successes

© Springer International Publishing Switzerland 2015
C.V. Jawahar and S. Shan (Eds.): ACCV 2014 Workshops, Part II, LNCS 9009, pp. 433–447, 2015.
DOI: 10.1007/978-3-319-16631-5_32

in several industries as well in academic disciplines, which include unmanned ground mobile robot/vehicle, and aerospace applications. The rapid advance in recent years indicates that the more autonomous applications maybe on the horizon. There are many studies on the planning algorithms and implementations [4]. So far, there have been proposed methods for the path detection and planning [5,6]. In general, there are several approaches to solve the problem of the path planning, but they are usually separated into the global and the local path planning approaches. The global path planning is concerned with the high-level of paths, which is a topological structure and the whole path for motion from the source to the destination of a travel itinerary. It deals with the navigation around the global regions. The local path planning is related to the low-level of paths in detail. It is only a segment of the global path, for obstacle avoidance, dealing with local motion navigation, e.g., the angles of turn, appropriate velocities.

The objective of this paper is to develop a complete, relevant, and efficient application for constructing the shortest path, which provides a real trajectory for autonomous vehicle motion in outdoor environments. In general, user can buy data of the paths for vehicle motion from commercial services. There are several business services, which support a constructed path on real traffic scene of transportation systems. However, in that case, they are expensive and the system becomes dependent on quality, and limitation of the services. This paper focuses on planning the global path, which is self-constructed based on road map and satellite images. The proposed method consists of several steps as follows. First, a road network is estimated by using the road map images, which are retrieved from online map services. Second, road regions on the satellite images are detected. This task is used to solve the problem of road segments, which are not annotated in the road map images. The corresponding road regions, which are detected at the first stage, are used to learn a road color model for detecting the rest of road segments in the satellite images. The road regions are estimated based on color filter, difference of Gaussian (DOG), and Radon transform techniques. The color model of the road is used to filter candidate of road regions. The DOG filter is also used to enhance candidate road borders and roadbeds. The dominant values of the Radon transform are used to detect road regions. The result of road detection is refined based on road joint structure. Third, the shortest path for motion is estimated by using path planning algorithms, e.g., Dijkstra, best-first graph search algorithm (BFS), Rapidly-exploring Randomized Tree (RRT). In this stage, a road network in image pixel coordinates is converted into the Global coordinates, which provides suitable information for the task of online vehicle navigation.

2   Related Work

In recent years, some of the most convincing experimental results have been obtained using promising methods for motion planning [7–12]. A global path planning method based on the modification of rapidly exploring random tree algorithm was presented in [13]. The method was constructed for providing effective partial motion and achieving the global objective. Another group of researchers in [14] presented a motion planning method using guided cluster sampling. That paper developed a point-based Partially-Observable Markov Decision Process (POMDP) approach, which takes into account all the motion errors, sensing errors, and imperfect environment map for robots active sensing capabilities. The experimental results show that the approach contributes an efficient method to balancing sensing and acting to accomplish the given tasks in various uncertain conditions. However, the method requires high computational cost to find an optimal solution [15]. To adapt to changing and uncertain conditions, Toit and Burdick [16] presented a method for motion planning based on integral individual components of dynamic and uncertain environments in planning, prediction, and estimation. In outdoor environments, traffic law guiders are used to estimate the expected behaviors of the dynamic interaction system to predict their future trajectory, and constrain the future location of moving objects in more uncertain environments. In the case of the global path planning for motion under certain maps, computational cost of that method becomes high cost when it is applied for high-level motion planning. Furthermore, authors in [17] focused on developing an interpolation method for optimal cost-path-motion function based on well-known Dijkstra and A^* algorithms. Taking advantages of each of these algorithms, authors exploited to provide an effective method, which estimates the shortest path based on respondent information. The computational cost is significantly reduced by implementing an A^*-like heuristic.

In the field of outdoor path planning, there are several groups of researchers [5,6,18,19], who have been focused on road detection and planning a trajectory for vehicle motion by using satellite/aerial images. Typically, the authors in [6] used a neural network to detect roads on high-resolution aerial images. In that paper, the authors analyzed to learn roads using a road surface context for reducing misdetection, e.g. roofs of buildings are similar to road surfaces without context of surrounding scenes. Chai et al. [5] presented a method to estimate a road network based on the Monte Carlo mechanism using sampling junction-points input images. The network extraction method is focused on investigating shape and extracting structures from nature textures. However, those methods could not overcome the problem of roads fully obscured by high buildings, tunnels, trees.

On the contrary, in this paper, instead of focusing on path detection using only the satellite images, the proposed method interests in both high-level of the road map and the satellite images for detecting a road network to plan the shortest path in outdoor environments. The road map and the satellite images are provided free of charge by online services, such as Google Maps, OpenStreetMap, Bing Maps. The proposed method takes advantages of prior knowledge of the road map images, which provided by map developers, to simplify road detection with high accuracy and low computational cost. This approach does not only construct the path network but also estimate a directed road network. For simplicity, this method believes the prior knowledge of the maps service. Some road segments are not annotated in the road map services, they would be supplemented by detection in the satellite images.

3   Problem Formulation

Regarding the autonomous of robot/vehicle navigation in outdoor environments, global path planning plays an important role in the optimal motion planning applications. Although there are, at present, several different specialized commercial services, which provide complete real-world traffic applications, they are expensive and applied into just several limited applications. Further online road networks are insufficiently and not frequently updated or users should pay extra charge for map updates, as well the required precision and correctness of the trajectory cannot be assured users. In the case of open source projects, free and editable bitmap map layers are provided (satellite, road, boundaries, elevation, etc.). In both cases, the main problem of road map is that they are not fully annotated, especially in areas such as countryside, towns, as depicted in Fig. 1. On the other hand, the problems of road detection based on satellite images are low-resolution, having variations in spectral properties of road surfaces, e.g., vehicle presence and occlusion by buildings, tunnels, overpass, trees, as depicted in Fig. 2. Therefore, in order to address the challenges of global path planning, the authors present a method based on the advantages of multilayer for both road detection and the shortest path estimation applying to autonomous navigation.

For appropriately detected road using road map and satellite images, the characteristics of road can be described as follows:

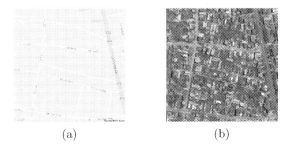

(a)   (b)

Fig. 1. Some road segments are not annotated by map service, (a) road map image, (b) corresponding satellite image

(a)   (b)

Fig. 2. Road segments are occluded by trees: (a) Road markers are appeared in the road map image, (b) Road segments are fully obscured by trees in satellite image.

- Road surfaces can be paved or unpaved. In the first case, the most common type of material are asphalt, concrete, brick. In the second type, roads are designed using gravel or stones materials. Therefore, spectral characteristic of roads are not uniform, particularly in the case of unpaved road. This characteristic is the cause required learns the color model in local region area detection in surroundings areas of the detected road regions based on result of road map.
- Due to the material of road surfaces, roads can be confused with building roofs, grounds, especially in low quality of images. This problem can cause high rate of false detection based on spectrum filter. Therefore, road network structure is useful for reducing false detection.
- The width of road is almost constant. The ratio of length/width of road is usually larger than that of building roofs. Roads are incorporated constructing a road network. It is different to building roof, which is isolated with other parts.
- The detection results of road regions are sometime discontinuous in short distance due to the environment occlusion.

4   Road Network Detection

Recently, there have been several methods for road detection using satellite images were developed [5,6,18]. In contrast to the former methods, this article presents a simple and efficient method to detect roads using both of the satellite and road map images. The scheme of road network detection consists of several following steps.

4.1   Map Images Based Road Network Detection

As prior knowledge, this work believes in annotation of road map services. In order to filter out road regions, the statistic of color channels is used. The representative colors of road annotation in map images are separated into several classes with regard to the number of hierarchy of road maps. The representative colors have specific color characteristics. To investigate the color features, we built our own database for training, which gives the following probability density functions (PDF) of the red, green, and blue channels in Fig. 3. The road candidate regions are estimated by using Gaussian probabilities based on color channels by the following formulation:

$$P(r|x) = \prod_{ch \in C} P(r|x_{ch}) \tag{1}$$

where x is pixel image, C is color channels (red, green, blue), and r is road candidate.

To highlight differences with previous methods [5,6], the road map images are retrieved from the map service with low-resolution image in this paper. The road candidates are disconnected as result of noise and other annotations of the

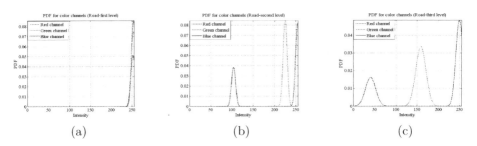

Fig. 3. Probability distribution color channels of road regions, (a) PDF of the first level of road, (b) PDF of the second level of road, (c) PDF of the third level of road.

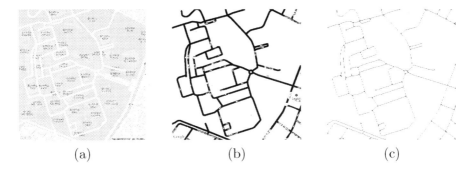

Fig. 4. Path network detection using the road map image, (a) road map image is retrieved from Google Maps, (b) road candidates are estimated by color filter and segmentation, and (c) post process to connect the discontinuous road regions and extract the path network.

road map, see also Fig. 4. It should be noted that some world maps do not allow for removing the annotations in some locations in the maps because of several special purposes related to the map services. To deal with this problem, a rolling ball method is used for connection the discontinuous roads.

Taking advantage of the map images, the road direction is estimated based on arrow signals. The final road network is presented by the directed graph.

4.2   Satellite Images Based Road Detection

In the case of some road segments, which are not annotated by map services, they are detected based on the satellite images. The road regions resulting from previous subsection are used to construct a training dataset from the corresponding regions of the satellite images. Let I_M be a road map image and I_S be a corresponding satellite image. All detected road pixels on I_M are mapped into I_S to construct a dataset for learning spectrum color model, as depicted in Fig. 5(a). The probability density functions (PDF) of the red, green, and blue channels of road colors are shown in Fig. 5(b). This color model is also used to filter out road regions in the satellite images by (1). The result of candidate road

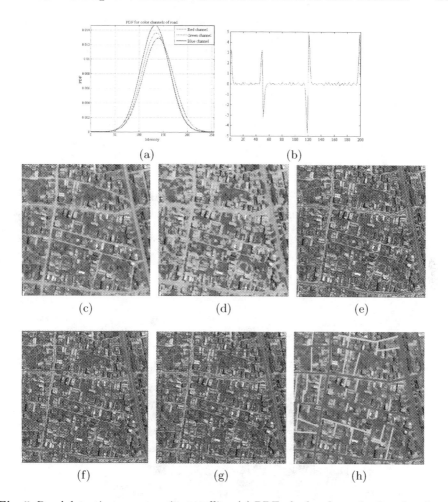

Fig. 5. Road detection process using satellite: (a) PDF of color channels of road regions, (b) DOG filter is used to emphasize object borders, (c) Road regions in satellite image corresponds to road detection results from map image (light- pink) is used for training color model, (d) candidate road regions (light- pink) using color model filter, (e) candidate road borders (dark-cyan) and roadbeds inside of long-edges, (f–g) two grid subregions of DOG image are used for computing the Radon transform to detect local candidate road segments, (h) road detection result based on combine of road map image (red) and satellite image (yellow) (Colour figure online).

regions is shown in Fig. 5(c). In this step, there are many false negative and false positive results due to some properties of roads aforementioned in Sect. 3.

The DOG filter is used to enhance the boundaries of roads. The filter image results are obtained by convolving the grayscale satellite image with difference of two kernels of Gaussian with standard deviations σ_1 and σ_2.

$$F(I, \sigma_1, \sigma_2) = I \otimes \left(\frac{1}{2\pi\sigma_1} e^{-(x^2+y^2)/(2\sigma_1)} - \frac{1}{2\pi\sigma_2} e^{-(x^2+y^2)/(2\sigma_2)} \right) \qquad (2)$$

Image filter using DOG preserves spatial information that lies between the ranges of frequencies that are preserved in the two smoothed images by Gaussian filters. It also removes high-frequency noise while emphasizing edges between regions of different intensity of gray, see also Fig. 5(d) for example. By setting the threshold, border of regions are obtained to construct a binary image, called I_{DOG}. The result is superimposed on the original satellite image in Fig. 5(e).

The roadbeds and road borders are filtered by using Radon transform combining with candidate road regions, which are estimated by color filter in the previous section. The I_{DOG} is divided into grid subregions, as depicted in Fig. 5(f, g), and then the Radon transform is applied for each sub-region to estimate road segments. Two-dimension Radon transform $R(x', \theta)$ of an image $f(x, y)$, is defined in [19] as follows:

$$R(x', \theta) = \iint_D f(x, y)\delta(x \cos \theta + y \sin \theta - x')dxdy \qquad (3)$$

where D is image domain, $f(x, y)$ is binary DOG filter image, $\delta(.)$ is the Dirac function, $\theta \in [0, \pi)$ is a rotation angle from x-axis to the normal direction of x'.

The Radon transform values of the binary DOG filter image are shown in Fig. 6(b). Each triple of adjacent local extreme values is used to predict the

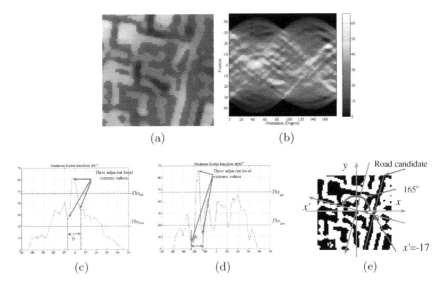

(a)   (b)

(c)   (d)   (e)

Fig. 6. Radon transform for road detection: (a) Binary image of DOG filter result is superimposed on the satellite image, (b) Radon transform results, (c) three adjacent local extreme Radon transform values indicate the non-candidate road region, (d) three adjacent local extreme Radon transform values indicate the candidate road region, (e) Radon transform for detection road borders and roadbeds.

candidate road segments, as depicted in Fig. 6(c–d). Each part of x' is a candidate road segments if three sequence extreme values are alternating lay out two sides of Ths_{up} and Ths_{low}, and the width b is limited by w_m value (the maximal width of roads). The results of candidate road segments from sub-regions of two grids are projected on whole image to discard the road candidates in short distance and only maintain that of long distance. The result is integrated with the result of the color filter to discard the false detections, e.g., rivers, roof of buildings. Finally, the geometry of road structure in [5] is used to post-process for improving the accuracy of road detection. The result is shown in Fig. 5(h).

5   Estimation the Shortest Path for Motion

To make online vehicle navigation more convenient, the road network result in pixel image is converted to the global coordinates (Mercator coordinate system). The details method for converting from pixel image into global coordinate is referred to [20,21] for details. Generally, global image services, e.g. Google Maps, Bing Maps, use similar organization of the world map. The world map can be represented by two-dimensional map, which likes a rectangle of 360 degrees wide and 180 degrees high. The world map is represented by a pyramid of tiles. The origin of a tile is located at the Northwest corner. The top level (zoom level $= 0$) has 256×256 points, next level 512×512 points. For each next level of the tile pyramid, the point space is expanded by doubling of size in both directions x and y. Therefore, the image pixel at zoom level ξ is converted into the Mercator coordinate by follows:

$$Y = Y_0 - \left(y - \frac{h}{2}\right) \times \frac{\tau}{2^\xi} \tag{4}$$

$$X = X_0 + \left(x - \frac{w}{2}\right) \times \frac{\tau}{2^\xi} \tag{5}$$

where (w, h) is the size of image, (x, y) is a location of the point in image, (X_0, Y_0) is the located center of image in the Mercator coordinate. The initial resolution of tile size τ is $156,543.034m$ (the circumference of the Earth in meters $40,075,016.679$ m divide 256 points). The part of equation $(y - h/2) \times (\tau/2^\xi)$ is used to convert image pixel to meter unit in the global coordinate.

A point at the location (x, y) in the Mercator coordinate is converted into the GWS84 coordinate system by following equation [21], with ϕ and λ are latitude and longitude in the GWS84 coordinate, σ is the radius of the Earth.

$$\lambda = \frac{360}{2\pi} \frac{X}{\sigma} \tag{6}$$

$$\phi = \frac{180}{\pi} \left[2\tan^{-1}(e^{Y/\sigma}) - \frac{\pi}{2}\right] \tag{7}$$

This section presents a method to estimate a path for vehicle motion with the minimum cost of feasible trajectory based on the road network configuration. There are many methods for estimating the optimal path [4], e.g. Dijkstra, BFS,

RRT. The shortest path problem in this paper is considered in two-dimensional Euclidean spaces. We construct a discrete directed graph $G(V, E)$. The set of vertex $V = \{v_i, i = i..n\}$ is defined as the set of intersections and ending points of the road network. The set of edges $E = \{e_i, i = 1...m\}$ is defined as the set of road segments between pairs of adjacent intersections or the ending points. A road segment, which connects intersection point to adjacent another one or the ending point, is represented by two edges in opposite direction. In the case of one-way road, it is represented by single directed edge. The Euclidean distance is used to compute the cost of each edge. Given the source position s and the destination position d, the path planning problem is estimation of a feasible trajectory T with the lowest cost for vehicle motion. The cost-function of trajectory is a non-negative cost, which is defined by $c : V \rightarrow R_{\geq 0}$.

The objective of this task is finding the shortest path from the source location to the destination location under assumption that there is no obstacle (the problem of obstacle avoidance will be dealt with in partial motion planning). This paper uses Dijkstra algorithm combining with heuristic based on greedy BFS for fairly flexible and potential searching in a huge area of the map. It is particularly desirable when applying heuristic search techniques in large graphs, which are typically required by a robot operation in outdoor environments, to restrict the point-to-point searching to examine only relevant areas of the input graph [22].

6   Experiment

This section presents evaluated results of the proposed method for automatic extracting the shortest path for vehicle motion in outdoor environments and comparison our method with state of the art methods. The dataset for road network detecting in road map images is manually collected based on annotations of road regions. In general, there are three kinds of color patterns for representing road regions in the road map images. The training dataset for road detecting in satellite images is automatically extracted using road results in the road map images. The color channel distributions are presented in Fig. 5(b). This method is proposed for real application while other methods based on only aerial/satellite image are limited to special conditions. They can not deal with the case of roads fully obscured by high buildings, trees, tunnels in the satellite images, see Fig. 3. In contrast, detection method based on the road map images is dependent on prior knowledge of road marking, for example some road segments are not annotated in the road map services, see Fig. 1. Advantage of the road map images based method is that it does not require high-resolution images, consume low computational time for detection roads due to a simple algorithm. It is suitable to implement the road detection in real applications for autonomous vehicle. The summary of comparison is presented in Table 1. In this experiment, we evaluated our proposed method on road map and satellite images, and compare with the method [5] (denoted by JPP method) and [6] (denoted by neural network method). All most methods result in low accuracy when apply into

Table 1. Comparison of methods using the road map image and the aerial image

	Road map	Satellite image
Require high-resolution images	No	Yes
Overcome occlusion confident	Yes	No
Depend on update of aerial images	No	Yes
Depend on prior knowledge of road annotation	Yes	No
Computational time (second)	<10	>100
Accuracy (%)	>95.0	<67.5

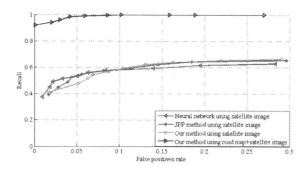

Fig. 7. Comparison results of our proposed method with previous methods on low-resolution satellite image and road map image

low-resolution satellite images. Our method result is slightly better than other methods because it is learned the model based on local spectral color, which accommodates the color model with variety road spectra (source from variety material of road surfaces). Our method is successful when supplement with the result of road detection using road map images.

The image dataset for experiment was automatically retrieved from the Google Maps service. The input parameters of the center location of regions are manually located on map service. In the case of towns and villages, there are many road segments, especially branch roads (byroad), were not frequently updated, while in the cities, almost all road parts were annotated by map services. The experiments were evaluated under configuration of 640×640-pixel resolution images and the zoom level of 15, 16, 17, and 18. The images at the zoom level of 15, 16, 17, and 18 cover areas of about $1,222.99 \times 1,222.99\,\mathrm{m}^2, 611.49 \times 611.49\,\mathrm{m}^2, 305.75 \times 305.75\,\mathrm{m}^2, 152.88 \times 152.88\,\mathrm{m}^2$ respectively. Figure 7 presents the comparison results of our method with other methods. Our method was evaluated on both situations of only using satellite image and using road map and satellite image. Figure 8 shows typical results of images and trajectories of path detection based on our method using both road map and satellite images. The intersection and ending points of roads are ordinally numbered. The experimental results are demonstrated that our method can detect many additional road

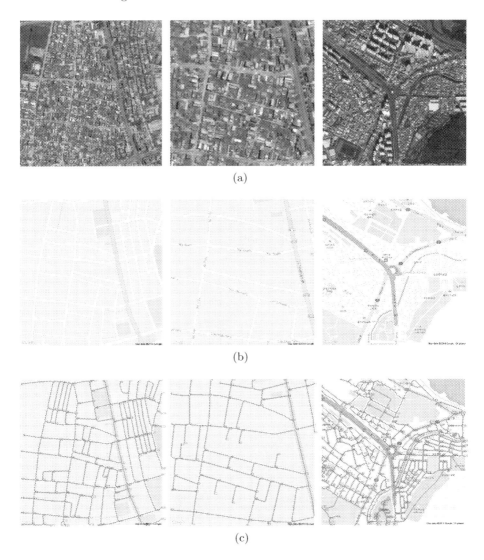

Fig. 8. Typical road detection: (a) Satellite images, (b) road map images, (c) road network estimation results with many additional detected road segments by our proposed method.

segments, which do not available in road maps, as shown in two first columns of Fig. 8(b, c).

The evaluation results are shows in Table 2. The markers on road map service and additionally manual annotation by authors are considered as ground truth data for evaluation and comparison. The sensitivity and precision criteria are used for evaluation of the method. The sensitivity (Recall, True positive rate-TPR) is computed by #TPR = #True positive/(#True positive + #False

Table 2. Comparison of methods using the road map image and the aerial image

Region	Zoom level	Road segments		Intersections		Consuming time
		TPR	Precision	TPR	Precision	
Downtown rotary	15	0.989	0.989	0.997	0.984	5.07
	16	0.995	0.980	100 %	0.974	4.55
	17	100 %	100 %	100 %	100 %	3.90
	18	100 %	100 %	100 %	100 %	3.65
University campus	15	0.994	0.991	0.994	0.988	5.09
	16	100 %	100 %	100 %	100 %	3.92
	17	100 %	100 %	100 %	100 %	3.47
Small town	16	100 %	100 %	0.987	100 %	4.50
	17	100 %	100 %	100 %	100 %	3.78

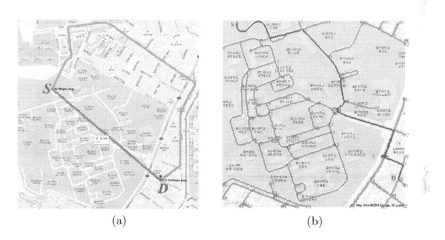

(a)   (b)

Fig. 9. The path planning for motion: (a) The resulted path from Google service with two option gray and turquoise path,(b) Our detected road network (green) and the shortest path for motion (red) (Colour figure online)

negative). False positive rate (FPR) is computed by #FPR = #False positive/ (#False positive + #True negative). The precision is computed by #Precision = #True positive/(#True positive + #False positive). The experimental results show that the road detection result at the higher the zoom level is more precision and vice versa. The road detection is perfect at the zoom level 17 and higher.

The path planning results are presented in Fig. 9. Google service results incorrect path for travel in local areas or the case of unpopular regions, as shown in Fig. 9(a). This problem is solve by our proposed method, as presented in Fig. 9(b). In this experiment, the algorithm (1) is applied to estimate the shortest path for vehicle motion using the images at zoom level 16. The trajectory in red color represents for the shortest path from the source location s to the destination location d with the cost of motion is 1,364 m.

7   Conclusion

This paper presents the method to enhance the efficiency in constructing the path using both road map and satellite images for autonomous vehicle motion in outdoor environments. The method focuses on estimation of paths in the global coordinate for motion without using expensive commercial services. It consists of several parts. First, a road network is estimated using the road map images, which are retrieved from online free charge map services. Second, a road network is also estimated using the satellite images based on prior knowledge of the first stage for learning the color model and using image processing techniques such color filters, difference of Gaussian, and Radon transform for detection road segments. The final road network is constructed and refined based on the geometry structures of road system. Third, the shortest path is estimated using the path planning Dijkstra algorithm combining with heuristic based on BFS technique. The trajectory result of the path network is processed in the global coordinate for convenience in online vehicle navigation when combines with the GPS. By the use of road map images, it takes advantages of maps annotation to provide high confidence of the shortest path for vehicle navigation. One disadvantage of using the road map is that it depends on the update of road information. To compensate this problem, the satellite images are used to detect the lack of annotated road segments for constructing full road network. The experimental results demonstrate the effectiveness of this method under the large scene of the outdoor environments.

Acknowledgement. This work was supported by the National Research Foundation of Korea (NRF) Grant funded by the Korean Government (MOE) (2013R1A1A2009984).

References

1. Zhang, H., Geiger, A., Urtasun, R.: Understanding high-level semantics by modeling traffic patterns. In: 2013 IEEE International Conference on Computer Vision (ICCV), pp. 3056–3063 (2013)
2. Murillo, A.C., Singh, G., Kosecka, J., Guerrero, J.J.: Localization in urban environments using a panoramic gist descriptor. IEEE Trans. Robot. **29**, 146–160 (2013)
3. Hoang, V.D., Hernandez, D.C., Le, M.H., Jo, K.H.: 3d motion estimation based on pitch and azimuth from respective camera and laser rangefinder sensing. In: IEEE/RSJ International Conference on Intelligent Robots and Systems (IROS), pp. 735–740. IEEE (2013)
4. LaValle, S.M.: Planning Algorithms. Cambridge University Press, Cambridge (2006)
5. Chai, D., Forstner, W., Lafarge, F.: Recovering line-networks in images by junction-point processes. In: IEEE Conference on Computer Vision and Pattern Recognition (CVPR), pp. 1894–1901. IEEE (2013)

6. Mnih, V., Hinton, G.E.: Learning to detect roads in high-resolution aerial images. In: Daniilidis, K., Maragos, P., Paragios, N. (eds.) ECCV 2010, Part VI. LNCS, vol. 6316, pp. 210–223. Springer, Heidelberg (2010)

7. Cossell, S., Guivant, J.: Concurrent dynamic programming for grid-based problems and its application for real-time path planning. Robot. Auton. Syst. **62**, 737–751 (2014)

8. Roberge, V., Tarbouchi, M., Labonte, G.: Comparison of parallel genetic algorithm and particle swarm optimization for real-time uav path planning. IEEE Trans. Ind. Inf. **9**, 132–141 (2013)

9. Jaillet, L., Porta, J.M.: Path planning under kinematic constraints by rapidly exploring manifolds. IEEE Trans. Robot. **29**, 105–117 (2013)

10. Achtelik, M.W., Weiss, S., Chli, M., Siegwart, R.: Path planning for motion dependent state estimation on micro aerial vehicles. In: IEEE International Conference on Robotics and Automation (ICRA), pp. 3926–3932. IEEE (2013)

11. Valero-Gomez, A., Gomez, J.V., Garrido, S., Moreno, L.: The path to efficiency: fast marching method for safer, more efficient mobile robot trajectories. IEEE Robot. Autom. Mag. **20**, 111–120 (2013)

12. Xu, B., Stilwell, D.J., Kurdila, A.J.: Fast path re-planning based on fast marching and level sets. J. Intell. Robot. Syst. **71**, 303–317 (2013)

13. Vonasek, V., Saska, M., Kosnar, K., Preucil, L.: Global motion planning for modular robots with local motion primitives. In: IEEE International Conference on Robotics and Automation (ICRA), pp. 2465–2470. IEEE (2013)

14. Kurniawati, H., Bandyopadhyay, T., Patrikalakis, N.M.: Global motion planning under uncertain motion, sensing, and environment map. Auton. Robot. **33**, 255–272 (2012)

15. Grady, D., Moll, M., Kavraki, L.E.: Automated model approximation for robotic navigation with POMDPs. In: IEEE International Conference on Robotics and Automation (ICRA), pp. 78–84. IEEE (2013)

16. Du Toit, N.E., Burdick, J.W.: Robot motion planning in dynamic, uncertain environments. IEEE Trans. Robot. **28**, 101–115 (2012)

17. Yershov, D.S., LaValle, S.M.: Simplicial dijkstra and a* algorithms for optimal feedback planning. In: IEEE/RSJ International Conference on Intelligent Robots and Systems (IROS), pp. 3862–3867. IEEE (2011)

18. Sun, W., Messinger, D.W.: Knowledge-based automated road network extraction system using multispectral images. Opt. Eng. **52**, 047203–047203 (2013)

19. Seo, J.S., Haitsma, J., Kalker, T., Yoo, C.D.: A robust image fingerprinting system using the radon transform. Signal Proc. Image Commun. **19**, 325–339 (2004)

20. Sample, J.T., Ioup, E.: Tile-Based Geospatial Information Systems: Principles and Practices. Springer, New York (2010)

21. Karney, C.F.: Transverse mercator with an accuracy of a few nanometers. J. Geodesy **85**, 475–485 (2011)

22. Murphy, L., Newman, P.: Risky planning on probabilistic costmaps for path planning in outdoor environments. IEEE Trans. Robot. **29**, 445–457 (2013)

Detection and Recognition of Road Markings in Panoramic Images

Cheng Li[1,2], Ivo Creusen[1,2($\boxtimes$)], Lykele Hazelhoff[1,2], and Peter H.N. de With[1,2]

[1] Cyclomedia Technology, Zaltbommel, The Netherlands
[2] Eindhoven University of Technology, Eindhoven, The Netherlands
ivocreusen@gmail.com

Abstract. The detection of road lane markings has many practical applications, such as advanced driver assistance systems and road maintenance. In this paper we propose an algorithm to detect and recognize road lane markings from panoramic images. Our algorithm consists of four steps. First, an inverse perspective mapping is applied to the image, and the potential road markings are segmented based on their intensity difference compared to the surrounding pixels. Second, we extract the distance between the center and the boundary at regular angular steps of each considered potential road marking segment into a feature vector. Third, each segment is classified using a Support Vector Machine (SVM). Finally, by modeling the lane markings, previous false positive detected segments can be rejected based on their orientation and position relative to the lane markings. Our experiments show that the system is capable of recognizing 93 %, 95 % and 91 % of striped line segments, blocks and arrows respectively, as well as 94 % of the lane markings.

1 Introduction

The government is responsible for the maintenance of many objects in the public space, such as traffic signs, street lights, roads and also road markings. Over time, road markings deteriorate due to the constant flow of cars, which can lead to a decreased road traffic safety. Ideally, the condition of road markings should be periodically monitored. Performing this monitoring task manually is too labor intensive and too costly, therefore an automatic or semi-automatic solution would be favorable. Because of the availability of large-scale collections of densely captured street-level panoramic images, this monitoring task can be performed much more efficiently using computer vision algorithms. Moreover, by comparing the results of images captured over different years, the deterioration can be tracked over time. In this paper, we describe the first stage of such a system, which is the detection and recognition of road markings in single panoramic images.

Although detection of road markings is an easy task for humans, we have found that creating an automated solution is not so simple, due to several reasons. The presence of shadows and illumination differences can complicate the road/line segmentation, while other vehicles on the road often (partially) occlude road

© Springer International Publishing Switzerland 2015
C.V. Jawahar and S. Shan (Eds.): ACCV 2014 Workshops, Part II, LNCS 9009, pp. 448–458, 2015.
DOI: 10.1007/978-3-319-16631-5_33

markings. Besides this, roads can also contain regions that have a similar color as road markings. Additionally, the road surface occasionally changes abruptly in appearance and the markings themselves become deteriorated over time.

Most previous work in road marking detection has been performed in the context of Advanced Driver Assistance Systems (ADAS) and Autonomous Vehicles. In those fields, somewhat different constraints and goals apply than for our application. For example, we are interested in all type of road markings including lane markings, while ADAS typically focuses only on lane markings. Besides, there is more focus on real-time performance, the cameras have typically a lower resolution and offer lower quality, based on a smaller field-of-view (and therefore less background clutter) and they operate in more challenging weather conditions. Additionally, since the source data is typically video-based, many of the techniques described in the corresponding literature cannot be directly applied to our panoramic images. An overview of lane tracking techniques can be found in [1]. For instance, a detection and recognition method based on the Hough transform is described in [2] and [3]. A disadvantage of these methods is that they are only suitable for straight lines. In [4], the author proposes an algorithm of extracting lane markings using frequency-domain features, however this methods only supports lane markers and arrows, it is difficult to extend to other shapes and is much more complicated than our proposed method. In [5], a method of road marking detection by image moments and a Bayes classifier is proposed, however, a prior probability is required for Bayes classification. Besides this work, contour orientation [6] and a Histogram of Oriented Gradient (HOG) features [7] are also used for road marking detection. More recently, in [8], the author proposes a practical system based on matching a detected region with template images. However, their matching algorithm is quadratic in the number of features, so that with more features in the image, it becomes rather slow. A method which is capable of crosswalks and arrows recognitions, based on comparing extracted features with known models, is proposed in [9]. However, the accuracy of the comparison is constrained by scale.

There are some important differences between our system, and the previously described literature. Our system should work on high-resolution panoramic images with a very wide field-of-view (and consequently, a lot of background clutter), instead of low-resolution video. We aim to detect many different types of road markings, and it should be easy to extend support for new types of road markings. Compared to the approaches in literature we use a novel shape descriptor, and an innovative method for the modeling of lane markings.

The purpose of the work in this paper is to design a robust detection algorithm for road markings that is sufficiently reliable to handle faded markings and painted road signs. In Sect. 2, we describe the details of the processing stages of the system. In Sect. 3, the experimental results are discussed for lane markings, together with road markings such as line segments, blocks and arrows, which are mainly applied as indications for highways.

2   Road Marking Detection and Recognition

Our proposed system for road marking detection consists of three main steps. First, Inverse Perspective Mapping (IPM) is performed to generate a top-view image. Second, road marking elements are extracted by a segmentation algorithm. Finally and third, from each segment a feature vector is extracted, and a Support Vector Machine (SVM) classifier is employed to distinguish segments based on their geometric features. Since some background segments can have similar shapes as road markings, we model the lanes that appear in the image, using RANSAC and a Catmull-Rom spline [10] with the lane marking candidates classified by SVM. Many non-road marking segments can be rejected based on the lane positions and the orientation of the segments (the orientation should be consistent with the orientation of the lane markings). A block diagram of the entire system is shown in Fig. 1.

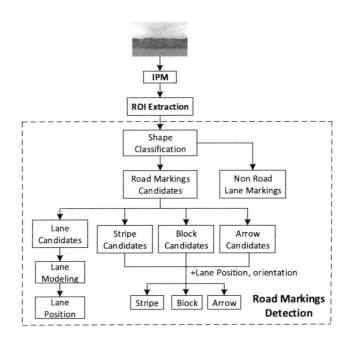

Fig. 1. Overview of the system

2.1   Inverse Perspective Mapping

The original spherical panoramic images are stored in an Equirectangular format of which an example of a panoramic image is shown in Fig. 2a. We transform the panoramic image to a top-view by applying an Inverse Perspective Mapping, similar to [2,8]. The mapping transformation between spherical panoramic images and the top-view image only depends on the camera height of the ground

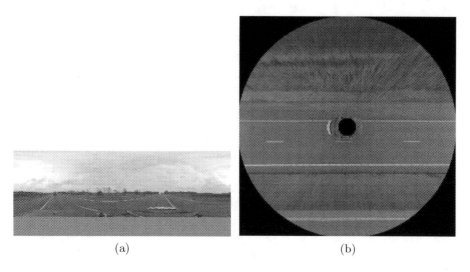

(a)   (b)

Fig. 2. (a) Example of a panoramic image. (b) Top view of the original panoramic image

plane and the azimuth angle of the camera with respect to the ground plane. Since our panoramic images are well calibrated for 360° panoramics, and the orientation of the car in the image is captured by the car positioning system, we simply remap to a top-view, according to the cars position. Figure 2b shows an example of a reconstructed top-view image. The relation between the original panoramic image and the top-view image is specified by the following equations

$$y = \left(m - \frac{\arctan(d/h)}{2\pi} \times n \right) \mod m, \tag{1}$$

$$x = \left(x_{car} + \frac{\arctan(y_o/x_o)}{2\pi} \times n \right) \mod n. \tag{2}$$

In the Eq. (1), x_{car} denotes the x-coordinate of the front of the car in the panoramic image. Parameters (x, y) and (x_o, y_o) denote the pixel position of the original panoramic image and top-view image, respectively, and parameter h is the height of the camera with respect to the ground plane, d the distance between the pixel (x_o, y_o) and the center of the car in the top-view, and (m, n) is the resolution in pixels of the panoramic image.

2.2   Segmentation Algorithm

Road markings are designed to be clearly noticed and are therefore made reflective, so they can be easily observed by drivers and other road users. We design our system to work under various lighting conditions. Usually, pixels of road markings are brighter than neighboring pixels. Based on this observation, we divide our segmentation algorithm into two steps.

1. First, bright pixels are obtained by comparing the value of a pixel with the average of the surroundings. For instance, Fig. 3 shows a sketch of an image with pixel P. We choose a window centered at pixel P and make a subtraction between P and the average pixel value within the window, as in Eq. (3), where g_p is the pixel value of P and (v, w) is the pixel size of the window. If g_p' is larger than a manually defined threshold, then P is considered to be a candidate pixel of road markings. The size of the window is determined empirically, where it was found that a value of a 105×105 pixels provides the best performance for our data. The subtraction of the window average is specified by

$$g_p' = g_p - \frac{1}{vw} \sum_{i=1}^{v} \sum_{j=1}^{w} g_{ij}. \tag{3}$$

2. The high-intensity pixels usually form connected segments. The segments that are too small to be considered as road marking regions are removed. Figure 4b shows an example of the areas of interest extracted with our segmentation algorithm. This example image has highly variable lighting, and it can be seen that the road markings are segmented correctly without being affected by the lighting conditions.

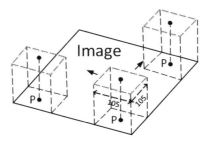

Fig. 3. Brightness normalization

2.3   Shape Classification

Each of the detected regions shown in Fig. 4b is classified into two categories based on their size: long segments and short segments. Long segments whose thickness correspond to the width of a real lane marking are assumed to be lane marking candidates, and split into smaller segments for further classification. Prior to extracting the feature vector of each segment, the centroid location, scaling and rotation should be normalized. The center of each segment is the mean of positions (x, y) of all pixels within the segment in image coordinates, hence

$$\overline{x} = \frac{1}{N} \sum_{n=1}^{N} x_n, \quad \overline{y} = \frac{1}{N} \sum_{n=1}^{N} y_n. \tag{4}$$

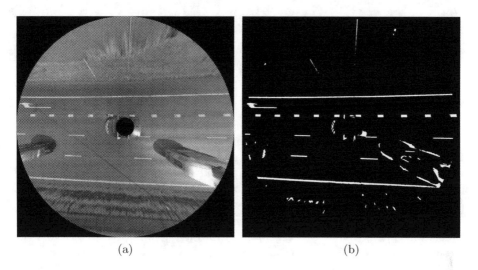

(a)   (b)

Fig. 4. (a) Top-view image. (b) Segmented image of (a) obtained by segmentation algorithm

Therefore each segment can be aligned by subtracting the means from the coordinates,

$$x' = x - \overline{x}, \text{ and } y' = y - \overline{y}. \tag{5}$$

After translation normalization, a scale factor for each segment is defined based on the position of the pixel that is farthest from the center of the segment, which involves the distance computed by

$$a = \sqrt{x'^2_{max} + y'^2_{max}}. \tag{6}$$

Then scale normalization of each segment can be achieved by dividing the coordinates of all pixels in the segment by the computed distance, so that

$$x'' = \frac{x'}{a}, \text{ and } y'' = \frac{y'}{a}. \tag{7}$$

To implement rotational normalization, Principal Component Analysis (PCA) is employed to estimate the orientation angle θ of each segment relative to the horizontal axis, then the segment is rotated using the rotation matrix in Eq. (8)

$$R = \begin{bmatrix} \cos\theta & -\sin\theta \\ \sin\theta & \cos\theta \end{bmatrix}. \tag{8}$$

After translation, scale and rotation are normalized, the features are then extracted by calculating the distance d_i from the center to the boundary of each segment at certain angles. In our experiment, we have chosen these angles from $0°$ to $360°$ with a step size of $30°$. The value of this step size was chosen empirically and based on numerous experiments. Each segment can then be described by the vector $\boldsymbol{v} = (d_1, d_2, ..., d_{12})$. To distinguish the road markings from other regions, we apply a Support Vector Machine (SVM) [11] algorithm to classify the shapes. In our experiment, a non-linear SVM with a radial basis function kernel is used.

2.4 Lane Modeling

Non-road marking segments that have a similar shape as road markings can be misclassified. In order to decrease the number of false detections, we utilize our prior knowledge that the road is bounded by solid lane markings, and any segments found outside cannot belong to road markings. For this purpose, we model lanes, based on lane marking candidates classified by the SVM. Since lanes can appear both straight and curved in the top-view image, two models are applied consecutively: RANSAC to model straight lanes and Catmull-Rom spline to interpolate curved lanes [10]. First, by default we search to fit a straight line using RANSAC. If this method does not converge after a fixed amount of iterations, the lane marker is assumed to be curved, and as a second approach, the Catmull-Rom spline method is employed instead. This method is an efficient way to model both straight and curved lines in a reliable and straightforward manner.

Since the position of the lane in image coordinates can be interpolated by the lane model, road markings can be rejected based on their position relative to the lane markings. In our system, only road marking candidates which are confined by lane markings are preserved. In addition to the location of the segments, their rotation with respect to the lane markings is also used to reject false detections.

3 Experiments and Evaluation

Several experiments are performed to test the performance of our system. The performance is individually measured for the different kinds of road markings, such as lane markings, stripes, blocks and arrows.

A. Dataset
The dataset that we have used to evaluate the performance of our algorithm is captured by a camera mounted on the roof of a driving vehicle, and a panoramic image is captured every 5 meters under very good weather condition. A panoramic image contains a road scene with a resolution of $4,800 \times 2,400$ pixels. We use 910 highway panoramic images under various lighting conditions, and we have manually annotated all road markings that are completely visible on the top-view images of the 910 highway panoramic images. The road markings in our dataset are not damaged or faded, and the images are taken during the day-time in fair weather conditions. This is due to the fact that the panoramic images are only captured during the daytime and when the weather is good, so for this application there is no need to consider these cases. An example of an annotated top-view image is shown in Fig. 5, where different types of road markings are annotated with different colors.

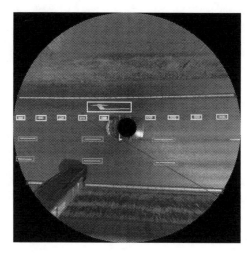

Fig. 5. Example of an annotated top-view image

B. Evaluation metrics

We divide the found detections into two categories, road markings such as line segments, blocks and arrows as well as lane markings, as shown in Fig. 5. Detections of road markings and lane markings are compared with the corresponding annotated top-view image. A detection of a road marking is considered as True Positive (TP), if at least 90 % of the pixels of the detected segment are located within the corresponding bounding box in the annotated top-view image. If a detection of road markings is not annotated in the top-view ground-truth image, it is counted as a False Positive (FP). Since lane markings are long and usually have a variable length in each image, lane markings annotated in the ground-truth image are first split into small blocks. Similar to detections of road markings, if 90 % of the pixels of a detection of a lane marking block overlap with a corresponding annotated lane marking block in the ground-truth image, it is counted as a TP, otherwise it is counted as a FP. If 80 % of the pixels of the lane marking blocks belonging to one lane marking are detected correctly, the lane marking is considered to be correctly detected. The performance is evaluated with a Precision-Recall curve. Precision is defined as $\frac{TP}{TP+FP}$, and recall is defined as $\frac{TP}{TP+FN}$.

C. Experiments

We have tested our algorithm with the previously described dataset. In our experiment, we have selected 50 panoramic images as a training set for the Support Vector Machine, which contain road markings of line segments, blocks, arrows and lane markings. A sliding threshold is applied to the output value of the SVM to obtain a Precision-Recall curve, as shown in Fig. 6. Our algorithm achieves a recall of 93.1 %, 95 % and 91.1 % for line segments, blocks and arrows, respectively, with a corresponding precision of 90.4 %, 95.2 % and 91.7 %. Additionally, it achieves a recall of 96.4 % for lane marking blocks at a precision

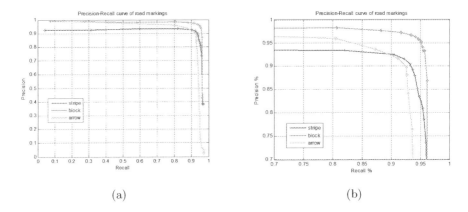

(a)   (b)

Fig. 6. (a) Precision-Recall curve of road markings. (b) Zoomed-in section of (a)

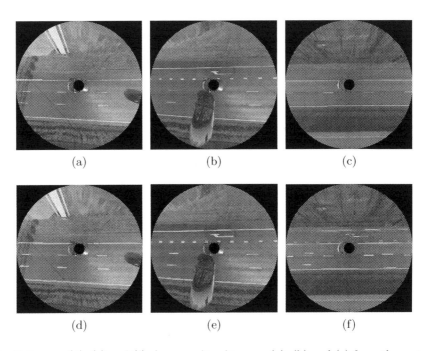

(a)   (b)   (c)

(d)   (e)   (f)

Fig. 7. Images (d), (e) and (f) show results of images (a), (b) and (c) from the system.

of 96.2 %. Considering whole lane markings, it achieves a recall of 94.3 % at a precision of 94.6 %. The performance for blocks and lane markings is thus quite promising. Figure 7 shows some visual example results from our algorithm. After evaluating the results, we have found that most false detections are caused by vehicles on the road that contain visually similar regions to road markings in the top-down image, as well as imperfections in the road surface, of which some examples are shown in Fig. 8.

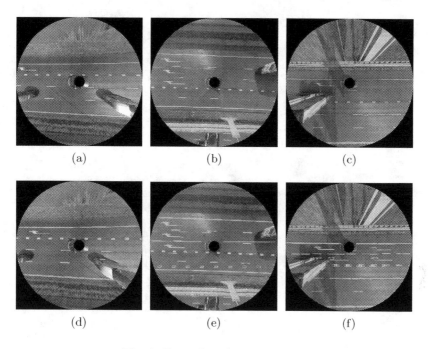

(a)   (b)   (c)

(d)   (e)   (f)

Fig. 8. Examples of false positives

4   Conclusions

We have introduced a system for automatic detection and recognition of road markings in panoramic images. First, an inverse perspective transformation is applied to the image to remap it to a top-down view to modify curved lines into straight lines. Next, regions that are brighter than their surroundings are segmented. Long segments are broken up into smaller pieces, and the shape of each piece is classified using a Support Vector Machine. The shape is described with a feature vector based on boundary distance measurements taken at regular angular steps. In order to decrease the number of false positives, the lane markings are modeled using RANSAC as well as a Catmull-Rom spline for straight and curved lanes, respectively. Finally, road markings are further filtered based on the positions of candidates relative to lane markings, and their orientations. The experimental result shows that 94.3 % of lane markings as well as 93.1 %, 95 %, 91.1 % of dashed line segments, blocks and arrows, considered as road markings are correctly detected. The main cause of the false positives is due to cars on the road, as well as imperfections in the road surface. The false positives can be further reduced by combining the results from nearby panoramic images, since panoramic images are taken every 5 meters. By doing so, we expect that most of the false positives that are due to cars can be rejected. In future work, we will also extend our techniques towards additional road markings.

References

1. McCall, J., Trivedi, M.: Video-based lane estimation and tracking for driver assistance: survey, system, and evaluation. IEEE Trans. Intell. Transp. Syst. **7**, 20–37 (2006)
2. Rebut, J., Bensrhair, A., Toulminet, G.: Image segmentation and pattern recognition for road marking analysis. In: IEEE International Symposium on Industrial Electronics, vol. 1, pp. 727–732 (2004)
3. Maeda, T., Hu, Z., Wang, C., Uchimura, K.: High-speed lane detection for road geometry estimation and vehicle localization. In: SICE Annual Conference, 2008, pp. 860–865 (2008)
4. Kreucher, C., Lakshmanan, S.: Lana: a lane extraction algorithm that uses frequency domain features. IEEE Trans. Rob. Autom. **15**, 343–350 (1999)
5. Li, Y., He, K., Jia, P.: Road markers recognition based on shape information. In: 2007 IEEE Intelligent Vehicles Symposium, pp. 117–122 (2007)
6. Kheyrollahi, A., Breckon, T.: Automatic real-time road marking recognition using a feature driven approach. Mach. Vis. Appl. **23**, 123–133 (2012)
7. Noda, M., Takahashi, T., Deguchi, D., Ide, I., Murase, H., Kojima, Y., Naito, T.: Recognition of road markings from in-vehicle camera images by a generative learning method. In: Proceedings of the IAPR Conference on Machine Vision Applications (2009)
8. Wu, T., Ranganathan, A.: A practical system for road marking detection and recognition. In: Intelligent Vehicles Symposium (IV), pp. 25–30. IEEE (2012)
9. Foucher, P., Sebsadji, Y., Tarel, J.P., Charbonnier, P., Nicolle, P.: Detection and recognition of urban road markings using images. In: 2011 14th International IEEE Conference on Intelligent Transportation Systems (ITSC), pp. 1747–1752 (2011)
10. Wang, Y., Shen, D., Teoh, E.K.: Lane detection using spline model. Pattern Recogn. Lett. **21**, 677–689 (2000)
11. Vapnik, V.N.: The Nature of Statistical Learning Theory. Springer, New York (1995)

A Two Phase Approach for Pedestrian Detection

Soonmin Hwang, Tae-Hyun Oh, and In So Kweon[✉]

Robotics and Computer Vision Laboratory, KAIST, Daejeon, Korea
iskweon77@kaist.ac.kr

Abstract. Most of current pedestrian detectors have pursued high detection rate without carefully considering sample distributions. In this paper, we argue that the following characteristics must be considered; (1) large intra-class variation of pedestrians (multi-modality), and (2) data imbalance between positives and negatives. Pedestrian detection can be regarded as one of *finding needles in a haystack* problems (rare class detection). Inspired by a rare class detection technique, we propose a two-phase classifier integrating an existing baseline detector and a hard negative expert by separately conquering recall and precision. Main idea behind the hard negative expert is to reduce sample space to be learned, so that informative decision boundaries can be effectively learned. The multi-modality problem is dealt with a simple variant of a LDA based random forests as the hard negative expert. We optimally integrate two models by learned integration rules. By virtue of the two-phase structure, our method achieve competitive performance with only little additional computation. Our approach achieves 38.44 % mean miss-rate for the reasonable setting of *Caltech Pedestrian Benchmark*.

1 Introduction

Pedestrian (or Human) detection has been an open research problem in computer vision community for more than decades due to the complexities of human variations and environment. The state-of-the-art approaches still show very high mean miss rate which limits the practical usage [1]. In recent years, pedestrian detection has impressively progressed in terms of feature representations [2–5], learning model [6–15], efficiency [10,12,13].

A challenge mainly comes from the large intra-class variations of human like pose and illumination changes. In addition, a lack of positive (human) samples comparing to negative (non-human) causes high asymmetricity in classification problem. These factors are on data characteristics. We are aware that there are very limited works comprehensibly considering the characteristics.

We argue that by considering the characteristics, one can develop a new effective model from a existing method. Based on our analysis of the data characteristics for pedestrian detection, pedestrian detection can be regarded as a *finding needles in a haystack* problem (rare class detection) [16,17], which is one of generic concepts

---

Electronic supplementary material The online version of this chapter (doi:10. 1007/978-3-319-16631-5_34) contains supplementary material, which is available to authorized users.

© Springer International Publishing Switzerland 2015
C.V. Jawahar and S. Shan (Eds.): ACCV 2014 Workshops, Part II, LNCS 9009, pp. 459–474, 2015.
DOI: 10.1007/978-3-319-16631-5_34

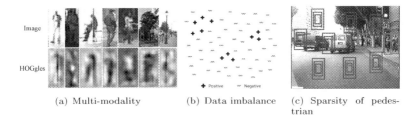

(a) Multi-modality   (b) Data imbalance   (c) Sparsity of pedestrian

Fig. 1. Characteristics of pedestrian detection problem. For HOGgles representation, refer to Vondrick *et al.* [20].

of data mining. Inspired by one of the rare class detection approach [16], we propose a two-phase classifier for pedestrian detection. The proposed two-phase classifier consists of a baseline detector and hard negative expert. We exploit modern successful methods as the first-phase baseline method to reduce sample space to be learned for the second-phase. By virtue of the two-phase approach, we can improve the overall performance with little additional computation without re-computing features. Particularly for the expert model, we extend Random Forest (RF) [18] model to more discriminative one based on the criterion of Fisher's Linear Discriminant Analysis (LDA) [19]. Its purpose is to deal with multi-modality of data automatically and discriminatively, which is not covered by the first-phase. We propose a conjunction rule to effectively fuse the responses of the baseline and expert. As addendum, we present three learning schemes for the expert model to improve discriminative power.

We validate our two-phase model on the challenging *Caltech Pedestrian Benchmark*, and our method achieves the competitive performance against the state-of-the-art methods, although we only use a single feature instead of other rich representations. For reasonable subset, our method achieves at most 38.44 % mean miss rate over the baseline. This achievement is based on the following analyses of pedestrian data.

Analysis of Pedestrian Detection Data. We concentrate on two aspects which make the pedestrian detection problem challenging: (1) multi-modality among intra-class samples (*i.e.* intra-class variations), and (2) data imbalance of positive/negative samples. To achieve more accurate detection, this kinds of characteristics should be seriously considered and reflected to the designed detector. The following analyses go for other single object detection problems such as face detection.

The multi-modality of pedestrians is formed by high intra-class variations due to pose deformation, view points, appearance, resolution, camera hardware, illumination change, background clutters, skin color, and so forth (Some examples of modalities on HOG domain are shown in Fig. 1-(a)). Based on this fact, we believe that positive samples would conform multi-modality rather than uni-modality (see Fig. 1-(b)). It requires a complex learning model.

The data imbalance of pedestrian detection comes from natural statistics. Only pedestrians are considered as positive class and all the others are regarded as negative. In a image pyramid for multi-scale detection, there are only few pedestrians among millions of sliding windows even in crowd scene as illustrated

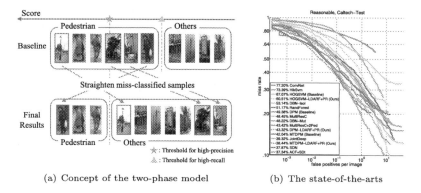

(a) Concept of the two-phase model   (b) The state-of-the-arts

Fig. 2. Illustrations of motivation. (a) We propose a two-phase model which starts from the detection results of existing one and straighten miss-classified samples to achieve low miss-rate and low false-alarms. (b) Existing methods already achieve low miss-rate at high false positives per image, but they did not achieve low false-alarms still (In case of MTDPM [6], we use executable code provided by the authors without context model and get 42.04 % log-average miss rate.).

in Fig. 1-(c). We naturally get a tremendous number of negative samples incomparable to the positives. The imbalance can affect the overall performance of the designed detector through both learning and detection steps. The imbalance on the learning step could cause bias of decision boundary to be negative-oriented, or could induce less informative decision boundaries because some random decision boundary might be misread as work well; *e.g.* For the ensemble model, the imbalance can induce high sub-optimality on the selection of simple weak classifiers of AdaBoost or randomized forests, because they misread their capacity due to easily achieved high recall. Also, most of detection algorithms have a trade-off between false positive and false negative rates. The imbalance on detection step disturbs finding a good trade-off. For this class imbalance problem, a special treatment may be necessary as rare class detection problems did in [16].

2   Related Works

As pedestrian detection is one of attention-getting topics in computer vision, it has long history and many related works. In this section, we focus on the relevant works to our method. One can refer the thorough review on pedestrian detection approaches to [1, 21].

Many works notice that a main challenge of pedestrian detection comes from multi-modality (intra-class variations) of data. Some researches develop robust and distinctive feature representation such as HOG [2], CSS [4], LBP [3], integral channels [5], and temporal feature [22] which invariant to some modalities like illumination changes or color variances.

On the top of rich feature representation, many learning models are also applied to improve accuracy. Most of works try to deal with some specific variations of pedestrians by advanced learning models. Popular Deformable Part Model (DPM) [7] combines a static root detector and part detectors by latent SVM approach.

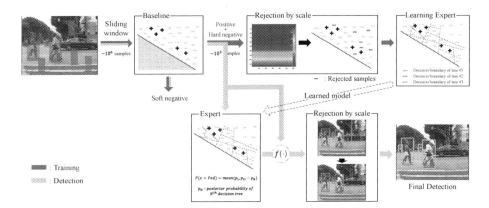

Fig. 3. The proposed two-phase framework. [**Blue line**] In training step, the baseline detector and rejection scheme help gathering positive and hard negative samples for learning the expert discriminatively. [**Yellow line**] In detection step, the baseline discards many data as soft negative. This rejection in the baseline makes the expert check only a small subset of samples. So, a limited amount of additional computation is required to improve performance (Color figure online).

It allows flexibility to handle deformation and partial occlusions by latent variables. Also other works [3,14,15] have been proposed to handle deformation and occlusions. Ouyang *et al.* [23] learn a background diversity for a limited case, two pedestrians being together. Recently, Park *et al.* [24] and Yan *et al.* [6] argue that low and high resolution pedestrians share commonness, but different characteristics should be considered. All these methods provide improved accuracy and higher robustness, but each method focuses on one or two specific intra-class variations. Particularly, the existing linear models are not enough to deal with many kinds of intra-class variations due to its limited parameterizations; *e.g.* The latent variable in DPM is the only parameter to deal with pose variations.

To allow flexibility, ensemble based classifiers are presented for pedestrian detection. Many of them show fast and efficient approach with satisfactory performance. Among the ensemble models, Boosting [8–13] and RF [14,15] based classifiers are popularly applied for pedestrian detection. They learn non-linear decision boundaries with many weak-classifiers, which share a single nature. Our expert model is developed as an extension of RF. For more relationships with other RFs, we will further discuss in Sect. 3.2. Although these approaches allow to learn multi-modality of data, it would not be enough to handle many different kinds of multi-modality (*e.g.* In DPM, latent variables are only parameters to handle deformable parts. The ensemble models dump the flexibility on uni-nature weak classifiers). Our method utilizes two heterogeneous models, and encourages to capture complementary information during learning time.

We are aware that many methods pass over some traditional data mining rules. Among the contexts of data mining, we found that *finding needles in a haystack* problem [16,17] is very relevant to pedestrian detection problem, which detects rarely occurring phenomena in the data. They define a class that has very rare occurrence due to its nature as 'rare class'. The rare class detection problem is especially challenging. High recall can be easily achieved due to class imbalance in the rare class case. Conventional learning models try to achieve high recall and high precision simultaneously. They are prone to find low precision decision boundary, because it easily achieves high recall. This induces performance degradation. They point out that conventional sequential techniques are inadequate for the rare class problem. Joshi *et al.* [16] propose the two-phase rule induction. In the two-phase induction, recall and precision are separately optimized, rather than simultaneously optimizing two measures as most of learning models did. Their experimental result indicate that the two-phase model outperforms AdaBoost for rare class, and consistently produces competitive performance for generic cases. More general concept of two-phase induction model can be found in [16,17]. Our method is built on the top of these philosophy.

3   Two-Phase Classifier Model Under *PNrule*

One of effective approaches for the rare class problem is to separately conquer high recall first and high precision next, which is called as *PNrule* [16]. Inspired by *PNrule*, we propose a two-phase classifier model which minimizes miss rate first, then optimizes our detector to minimize false positives. In the first phase, a detector classifies pedestrians allowing many false positives. Then, the second phase classifier (called as expert) straightens the miss-classifications to achieve high precision. This procedure is illustrated in Fig. 2-(a). By this way, we can achieve low miss-rate at low false positives per image (FPPI) by *PNrule*.

The proposed two-phase classifier model can be viewed as a variant of cascade classifier structures. The conventional approaches learn weak learners which have same properties. Rather than cascading uniform weak learners, exploiting heterogeneous classifiers is more helpful for achieving different objectives (in our case, recall and precision). Even when the same training set is given, heterogeneous classifier models bring out different characteristics in decision boundary or classification results, as well as commonness (intersection regions on feature space among different classifiers). We expect that combining heterogeneous classifiers learns complementary information even from same data, when we carefully choose the classifiers by their properties and data's characteristics.

Our proposed two-phase detector consists of a baseline and a hard negative expert detector as illustrated in Fig. 3. The baseline initially rejects soft negatives which are easily classified with high confidence, and measures how likely positive. Then, the remaining negatives (hard negatives) and positives are passed to the next phase expert. The expert classifies into hard negative or positive on the reduced sample space. By combining results from the baseline and expert, the mis-classifications by the baseline are straighten. If the baseline classifies non-pedestrian as high score, the final results are corrected to have low value. This procedure works like Re-ranking approach [25]. Even though the baseline allows many false positives, the number of samples that have to be checked at the expert detector are surprisingly reduced. Thus, our method only require a limited amount of additional computation, while enhancing overall accuracy.

3.1   First Phase: Baseline Detector

Baseline detector filters out soft negatives, and leaves hard negatives and positives. The main objective of the baseline detector in the two-phase classifier is to minimize miss rate of pedestrians (high recall), while minimizing the number of hard negatives is a subject class designated by the baseline as positive. The minimized number of hard negatives by the baseline can help to reduce the feature space to be learned by the expert. Since the hard negatives guide the learning decision boundaries of the expert, a well designed baseline detector is preferred to grasp informative hard negatives.

We found that many existing detectors already achieve low miss-rate at high FPPI as shown in Fig. 2-(b). At the first phase, one of the existing algorithms can be used for the baseline. Among publicly available methods, we test linear detectors, HOGSVM [2], DPM [7], and MT-DPM [6], as our baseline detector. Our expert adopts an ensemble model, so that linear detector which has different characteristics would be a reasonable choice. These could be a guideline for selecting baseline, but not strict restriction. Users can select the baseline by their own criteria. We observe that improvements are achieved for all of our experiments among the several baselines. We set the rejection threshold of the baseline lower than usual settings suggested by the authors to allow high recall.

3.2   Second Phase: Hard-negative Expert Detector

We build a expert classifier which classifies hard negatives and positives at the second phase. We can improve the performance by applying the second phase detector once again to the detection results of the first phase. Rather than using two models independently with training by the same sample set, we train the expert with the baseline's results, where hard negatives are designated. It learns complementary information from the reduced learning space, where the union of positives and hard negatives. Also, since the baseline eliminates soft negatives, the data imbalance on both detection and learning steps are partially relaxed.

We try to handle multi-modality of features in the expert model, so we exploit a RF model [18][1] which can better handle multi-modality than linear models. Nevertheless, discriminating the non-pedestrians from the hard negatives is still challenging, because it was failed in the baseline once. There are recent variants of discriminative RF [15,27–29]. A fundamental difference is how to learn a linear decision boundaries of each node to obtain more discriminative power ([29]: Ridge regression, [15,28]: SVM, [27]: LDA). According to [18], the standard weak learners in [15,27–29] select few dimensions of data vectors randomly before finding split function to construct forests with less-correlation between trees. The feature selection function $\phi(\cdot)$ maps a high dimension vector to a low dimension vector, and can be represented in a matrix form as: $\mathbf{Y} = \phi(\mathbf{X}) = \mathbf{W} \cdot \mathbf{X}$, where $\mathbf{Y} \in \mathbf{R}^{M \times 1}, \mathbf{X} \in \mathbf{R}^{N \times 1}, \mathbf{W} \in \{W | \{0,1\}^{M \times N}, \mathbf{W} \cdot \mathbf{1} = \mathbf{1} \in 1^{N \times 1}\}$ is a binary selection matrix, and $M \ll N$. Marín et al. [15] extend it to the random sub-group

---

[1] Since RF model is a general concept of AdaBoost, it has more flexibility for complex decision boundary than AdaBoost [26].

feature selection for encouraging part configuration. Our hypothesis is that feature selection may reduce the probability to find good discriminative boundaries, because high dimension feature is more preferable to find a linear separable space than low dimension one in general. Thus, we instead generalize the feature selection in a soft manner, which can automatically select informative features and $\mathbf{W}$ to be $\mathbf{R}^{M \times N}$.

We build our expert model with LDA criterion, because it has an analytical closed-form solution, while SVM based methods cannot avoid expensive numerical optimization. However, the LDA based RF [27] deterministically decides splitting criteria for each node under the strict assumption that each class has exactly same covariance, but the assumption is not satisfied the estimated split function is no longer optimal. By proposing an alternative threshold estimation, we relax the problem. We will explain it later.

Given the hard negative and positive samples, our method learns important and discriminative regions of the pedestrian template based on LDA for each node. By the general definition of Criminisi et al. [18], the split function of the j-th node is defined as $h(\mathbf{s}; \boldsymbol{\psi}_j, \tau_j) = \mathbf{1}[\boldsymbol{\psi}_j^\top \cdot \mathbf{s} < \tau_j]$, where $\mathbf{s}$ denotes a data sample, $\mathbf{1}[\cdot] \in \{0, 1\}$ denotes the indicator function, $\boldsymbol{\psi}$ denotes a transformation that maps the data to a separable space, and τ is a threshold for classification. For a sample $\mathbf{s}$, if $h(\mathbf{s}; \boldsymbol{\psi}_j, \tau_j) = 0$, $\mathbf{s}$ is passed to left (or right), and otherwise vise versa. We obtain the parameter $\boldsymbol{\psi}_j$ from LDA. For each node, we use a maximally separable axis $\boldsymbol{\psi}_j$ computed by the following equation.

$$\boldsymbol{\psi}_j = \Sigma_{W,j}^{-1}(\mu_{j,y=\mathcal{P}} - \mu_{j,y=\mathcal{N}}), \tag{1}$$

where $\mu_{j,y=\{\mathcal{P},\mathcal{N}\}}$ represents the mean vector of the positive and negative data of the node j respectively, and $\Sigma_{W,j}$ represents the within-class scatter matrices of the node j for 2-class case [30].

Also we can easily obtain an optimal decision threshold by $\tau_j = \frac{1}{2}\boldsymbol{\psi}_j^\top (\mu_{y=\mathcal{P}} + \mu_{y=\mathcal{N}})$ with the assumption that two groups have the same covariance matrices. However, the same covariance assumption would be too strict. We propose an alternative threshold computation as the following equation:

$$\tau_j(\alpha) = \boldsymbol{\psi}_j^\top (\alpha \cdot \mu_{y=\mathcal{P}} + (1 - \alpha) \cdot \mu_{y=\mathcal{N}}), \quad \alpha \in [0, 1]. \tag{2}$$

Instead of finding τ, we apply brute-force search on the sampled $\alpha \in [0, 1]$ maximizing the following information gain:

$$I(\mathcal{S}, \Theta) = H(\mathcal{S}) - \sum_{i \in \{L,R\}} \frac{|\mathcal{S}_i|}{|\mathcal{S}|} H(\mathcal{S}_i), \tag{3}$$

where Θ is the set of parameters defining the split function $\boldsymbol{\psi}_j$ and τ_j, $H(S)$ is Shannon's entropy defined as $-\sum_{c \in \mathcal{C}} p(c)\log(p(c))$, and $\mathcal{C}$ is possible class sets (in our case, positive $\mathcal{P}$ and negative $\mathcal{N}$). When each covariance of positive and negative at a node is different, the optimal decision threshold are likely to be between the two means $\mu_{y=\{\mathcal{P},\mathcal{N}\}}$ by Bayes decision rule [30]. This α parameterization gives the bounded sample range, while the range of τ should be estimated from data.

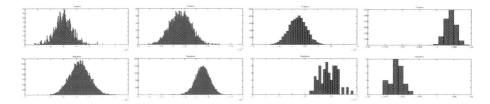

Fig. 4. Distributions of *Caltech* data in four sampled nodes of our proposed LDA based RF. Data samples of each node are projected on LDA axis of the node. (**Top**) Positive distributions. (**Bottom**) Negative distributions. Each column indicates the distributions of the same node. This shows that the data approximately conforms Gaussian distribution in each local partition.

In implementation, we reuse already extracted features by the baseline for all the forests to avoid re-computation. We now discuss the remaining issues of the proposed LDARF.

Discussion of the Proposed LDARF. LDA is optimal in the sense of Bayes error under the assumptions that multivariate normality and equal covariances are satisfied by class data [31]. While we relax the equal covariance assumption by introducing α parameter, we still assume Gaussian distribution of samples.

It is true that the assumed model may not be supported by the given data. The general PDF can be approximated by a non-parametric PDF estimation, but Devijver and Kittler *et al.* [32] claim that the errors on nonparametric PDF estimate may significantly exceed those of simple parametric models, such as Gaussian, when the sample size is limited. Also, parametric models could be better in terms of both accuracy and simplicity under the situation. When learning RF, the data is partitioned as growing trees, so that the sample size of a node is exponentially decreased. Thus, Gaussian approximation would valid by the statement of [32]. Figure 4 show the distribution of the dimension reduced features of *Caltech* data at intermediate nodes. It shows validity of our local Gaussian assumption of each node.

Moreover, the Gaussian assumption is beneficial when determining split functions. In the standard RF [18], Eq. (3) is utilized to measure the goodness of the split function, but Shannon's entropy is defined on discrete distribution which is constructed resultantly when a split function is given. Rather than, suppose that samples follows a parametric model like Gaussian distribution in a small partition. A parametric model has generative property on continuous space and locally spans its feature space, so it is helpful for learning a generalized boundary.

3.3   Integrating the Baseline and Expert

As depicted in Fig. 3, a single final score should be resulted from two scores of the baseline and expert. A well-designed combining rule could straighten miss-classified data from the baseline due to complementary characteristics of two detectors. However, hand-crafted combination rules may not be desirable, so we learn a

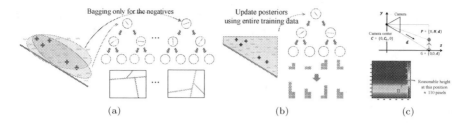

Fig. 5. Illustrations of learning schemes. (a) Soft bagging scheme keeping all the positives to make less-correlated decision trees. (b) Updating the distribution of leaf nodes with whole training set. (c) The estimated reasonable scale map for *Caltech* dataset.

score integration function with the two scores and labels. For simplicity, we model the score integration function with a linear model as $r(\mathbf{x}) = \mathbf{f}^\top \mathbf{x}$, where $\mathbf{f}$ is a weight vector to be learned, and $\mathbf{x}$ is a elementary score vector of which entries come from the baseline and expert scores. To encode several basis rules, we construct the vector $\mathbf{x}$ of each sample as $\mathbf{x} = [s_b|s_e|s_b \cdot s_e|s_b^2|s_e^2]$, where s_b and s_e are the scores obtained by the baseline and expert respectively. This construction can be regarded as kernelization that maps low-dimensional features into a higher dimensional non-linear space.

Our goal is to learn the score function that satisfy the relative score order $r(\mathbf{x}_p) > r(\mathbf{x}_n)$ for all the positive samples $\mathbf{x}_p$ and the negative samples $\mathbf{x}_n$. More explicitly, the pairwise order constraint can be represented as:

$$\mathbf{f}^\top \mathbf{x}_p > \mathbf{f}^\top \mathbf{x}_n \quad \Rightarrow \quad \mathbf{f}^\top (\mathbf{x}_p - \mathbf{x}_n) > 0, \quad \forall p, n. \qquad (4)$$

Finding $\mathbf{f}$ satisfying all the constraints could not be possible due to the presence of outliers or insufficient sorts of the basic rules. Instead of the hard constraint, we encourage the constraints in a soft manner with maximizing margin similar to SVM classification. Then, with adding a regularization term for f, the learning problem leads the following optimization problem as:

$$\arg\min_{\mathbf{f}} \|\mathbf{f}\|_p + \frac{C}{|P| \cdot |N|} \sum_{i \in P} \sum_{j \in N} \max\left(0, 1 - \mathbf{f}^\top (\mathbf{x}_i - \mathbf{x}_j)\right)^2. \qquad (5)$$

where P, N are the positive and negative sample sets of the training set. This formulation shares the similar spirit of the learning to rank technique [33]. Equation (5) encourage the order constraints by the squared hinge loss, and can be regarded as a l_p regularized SVM. The optimization is effectively solved by the off-the-shelf l_p regularized SVM solvers[2] in the l_1 and l_2 cases. For $p \geq 1$, Eq. (5) is convex formulation, so we can obtain a global optimum solution.

We found the optimal parameter $\mathbf{f}$ for both l_1 and l_2 cases, and empirically observed that the l_1 formulation produces slightly better results. Since l_1 is known to have a sparse selection property, it can be possible that only few informative entries of $\mathbf{f}$ have non-zeros, while discarding unhelpful basic rules in $\mathbf{x}$. Thus, we use

---

[2] We use the G-SVM package used in [34].

l_1-norm for all our experiments. Again, we would like to notice that, although we simply model $r(\cdot)$ with a linear function, $\mathbf{x}$ is constructed by kernelizing the baseline and expert scores, so nonlinear integration rules can be considered directly.

4 Learning Schemes for the Expert Model

Additionally, we describe three more simple learning schemes for expert under our analysis in Sect. 1. The described approaches are simple, but improve the performance of our detector.

Bagging with Preserving Positive Distribution. We apply the conventional bagging scheme only for the negative samples with keeping all the positives. It reduces the gap between the number of positive and negative, while positive distribution is preserved. We expect that diverse decision boundaries are around positive distributions. It helps the expert model to well learn decision boundaries and to be generalized by making trees uncorrelated despite the rare positives.

We learn each tree of the hard negative expert ensemble from differently sampled sets by the above bagging scheme. Each tree can share some commonness by sharing the same positives set, while takes different characteristics from different negative subsets.

We simply apply random sampling of negative samples, and it shows plausible results.

Updating Leaf Distribution of LDARF. In RF, each leaf node stores the positive and negative posterior distributions of the training samples arrived at the node. The posterior of a tree $p_t(c|\mathbf{v})$ is defined by the posterior of the leaf node that the sample $\mathbf{v}$ reaches. Given this, the final decision for $\mathbf{v}$ is determined by $c^* = \arg\max_{c \in \{\mathcal{P}, \mathcal{N}\}} p(c|\mathbf{v})$, where $p(c|\mathbf{v}) = \frac{1}{T} \sum_{t=1}^{T} p_t(c|\mathbf{v})$ is the average for every tree by aggregation rule [18].

Constructing accurate posterior distributions is as important as finding good split functions. As we use bagging scheme, each tree is built only with sampled training data, and initially constructed posteriors do not reflect entire training data in our framework. We update the positive and negative posteriors in leaf nodes by expectation with the remaining training data after bagging. Thus, the estimate $p(c|l)$ at a leaf l is calculate as

$$n_l = \frac{1}{n(\mathcal{P})} \cdot n_{p,l} + \frac{1}{n(\mathcal{N})} \cdot n_{n,l}, \quad n_{c,l} = \sum_{\mathbf{v} \in l} \mathbf{1}\left[\mathbf{v} \in c\right], \quad (6)$$

$$p(c|l) = \frac{1}{n(c)} \cdot \frac{n_{c,l}}{n_l}, \quad c \in \{\mathcal{P}, \mathcal{N}\}, \quad (7)$$

where $n(x)$ is the cardinality of a set x, $\mathbf{1}\left[\cdot\right]$ is indicator function. $p(c|l)$ is weighted posterior for relaxing data imbalance between positive and negative.

Learning by Perspective Aware Rejection. Ground plane information is effectively utilized in Hoeim *et al.* [35] and Park *et al.* [24], and is shown to be beneficial for validating the detected locations and scales. Although the perspective information in detection step has been commonly utilized in some cases like the driving scenario, we extent its usage in the learning step.

It is possible that the training sets given by the baseline include unreasonable samples with respect to scale and location in the perspective world. Since samples from different resolution have different natures as argued in [6,24], the hard negative samples with unreasonable scale could introduce unnatural artificial features in the sample space to be learned.

To reject these unreasonable scale samples, we estimate reasonable height map using intrinsic camera parameters given in [1] and geometric relation between a camera and pedestrians. As you can see in Fig. 5-(c), we estimate rough heights of pedestrians in pixels utilizing following relation; $f : d = h : H \Rightarrow h = \frac{fH}{d}$ (f : focal length, d : depth, h : height in image, and H : height in real world)

We allow margins for rejecting samples to alleviate error of the estimated height map. Thus, we reject the candidate boxes which is taller than 1.5 times or shorter than 0.5 times of reasonable height. It could reduce the variance of negative samples in feature space. This seems to be very simple, but it improves the performance of the final detector (see Sect. 5.3).

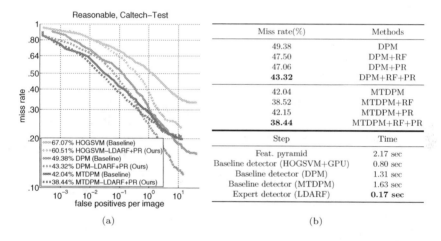

Miss rate(%)	Methods
49.38	DPM
47.50	DPM+RF
47.06	DPM+PR
43.32	**DPM+RF+PR**
42.04	MTDPM
38.52	MTDPM+RF
42.15	MTDPM+PR
38.44	**MTDPM+RF+PR**

Step	Time
Feat. pyramid	2.17 sec
Baseline detector (HOGSVM+GPU)	0.80 sec
Baseline detector (DPM)	1.31 sec
Baseline detector (MTDPM)	1.63 sec
Expert detector (LDARF)	**0.17 sec**

(a)   (b)

Fig. 6. Quantitative evaluations of the proposed two-phase classifier on the reasonable subset of *Caltech*. (a) Miss-rate comparisons between the baseline and the proposed two-phase approach. (b)-[**Top**] Performance comparisons according to combinations of the proposed modules. (b)-[**Bottom**] Computation times.

5   Experimental Results

To focus on the effects of each module, we fixed the used feature with HOG [2]. In this section, the comparisons between our method and other approaches based on HOG feature are only shown to easily compare the effects of the proposed method. Comparisons with more than other recent approaches can be found in the supplementary material.

Miss rate	Methods
49.38	DPM (Baseline)
45.72	DPM+oRF-LDA [29]+PR
44.71	DPM+StdRF [26]+PR
44.39	DPM+SVMRF [15]+PR
43.32	**DPM+LDARF(Ours)+PR**
47.50	DPM+RF
47.06	DPM+PRD
47.77	DPM+RF+PRL
43.50	DPM+RF+PRD
43.32	**DPM+RF+PRL+PRD**
43.85	DPM+RF+PR (+)
43.98	DPM+RF+PR (x)
43.32	**DPM+RF+PR (Opt.)**

(a) Examples of the improved results   (b) Effect of each module

Fig. 7. Effects of the expert model. (a) Sampled results of Straightened detection at 1 FPPI ([**Top**] DPM [7], [**Bottom**] Ours (DPM+LDARF+PR)). We denote the false positives and true positives as *Yellow* and *Magenta* respectively. (b) Comparisons according to the expert types (**Top**, Sect. 3.2), the perspective-aware rejection scheme in the learning and detection step (**Middle**, Sect. 4), and integration rules (**Bottom**, Sect. 3.3)

We evaluated on *Caltech* benchmark [1] which is the challenging and latest pedestrian benchmark. For training both the baselines and expert in our framework, we used both set00-set05 in *Caltech* and training set in *INRIA* dataset [2] due to the lack of information from high-resolution pedestrians in *Caltech*.

To compare performances, we followed full image evaluation and miss rate against FPPI (False Positives Per Image) plot by varying the threshold on the detection confidence as in [1]. For testing, we used set06-set10 in *Caltech*. We repeated the whole training and evaluation process 5 times and report averaged values.

Since the existing detectors find more than 80 % of pedestrians at 10 FPPI as shown in Fig. 2-(b), we set the rejection threshold of the baseline detectors to a value corresponding to 10 FPPI. Total 72 different scale images are used for all the experiments, as in the default setting of [7].

The computation times shown in the bottom of Fig. 6-(b) are measured on a PC with 3.40 GHz i7-4770 CPU and 32 Gb RAM. The computation time of the proposed framework depends on the choice of baseline detector. For **DPM** [7] and **MT-DPM** [6], it takes 3.48 and 3.8 s respectively (feature pyramid construction + applying DPM or MTDPM). **HOGSVM+GPU** [36] takes 0.80 s including the feature pyramid construction and detection times. Our proposed two-phase classifier model only takes additional 0.17 s per an image on MAT-LAB+MEX code to improve the performance. It can easily speed up by parallelization with modern GPU techniques due to the independent structure of trees. Also, the score integration takes 2.3 ms in average.

5.1 Evaluations of Two-Phase Classifier Model

We compare the performances of existing methods with the boosted performances by our two-phase model. We apply our approach to **HOGSVM** [2],

DPM [7] and **MT-DPM** [6] as a baseline, of which the cores are the linear classifier model based on HOG feature.

As depicted in Fig. 3, our expert detector is trained from the detection results of the baseline detector. Thus, a training set for expert depends on the baseline detectors, so each expert is learned differently according to the baseline. As shown in Fig. 6-(a), our model improves the accuracy of the baseline detector from 3.60 % to 6.56 %.

Notice that, for **MT-DPM**, we use executable code provided by the authors without context model, so that the baseline detector trained by only *Caltech benchmark*. Also, the expert is trained with fewer sampled data, because the executable only provides sampled results by non-maximum suppression (NMS) [1], while other experts for **HOGSVM** and **DPM** are learned from samples without NMS. Despite this handicap, the proposed two-phase model still improves the performance of **MT-DPM** as 3.60 %. On *Caltech* benchmark, most of methods exploiting only HOG feature did not achieve below 40 % miss rate except MT-DPM+Context [6][3] which utilizes another vehicle detector to utilize a high level contextual relationship between vehicles and pedestrians.

The proposed two-phase model mines HOG feature for more information to improve performance without additional advanced features and show the consistent improvement which allow better performance than the previous HOG-based algorithms.

5.2 Evaluations of Discriminative Random Forests

We compare expert detectors with the axis-aligned model (**StdRF**) [18], Marín *et al.* (**SVMRF**) [15], the ridge regression based LDA model (**oRF-LDA**) in [29] and the proposed LDA based discriminative RF (**LDARF**). Contrary to Sect. 5.1, in order to maintain training data to be same for all the expert detectors, we fixed the baseline with DPM [7]. For fair comparison, parameters such as the depth of tree and the number of trees were set to 6 and 100 respectively for all the experts.

As shown in the top of Fig. 7-(b), our **LDARF** shows better performance compared to other RF as a expert. This implies that LDARF was learned more discriminatively. Figure 7-(a) shows that mis-classified instances by the baseline are well straightened by our expert.

5.3 Influences According to Each Components

Figure 7-(b) shows the influence according to each component of our framework, such as the choice of the expert types, the perspective-aware rejection (PR), and the integration rule.

As mentioned in Sect. 4, we use PR at both detection and learning step. It means that we intend to completely ignore particular instances that are not matched with reasonable height according to their position during both learning and detection time. As shown in the middle of Fig. 7-(b), applying both **PRL** (PR in Learning

---

[3] Codes for MT-DPM+Context was not provided by author when this paper is submitted.

step) and **PRD** (PR in Detection step) (denoted by **DPM+RF+PRL+PRD**) improves the performance compared to the single usage of PRL or PRD. We notice that **DPM+RF+PRL** would produce more false positives than DPM+RF, because in learning step, **DPM+RF+PRL** ignores the samples with unreasonable heights and supposes that the given candidates for the expert are already filtered by PR. In this case, **DPM+RF+PRL** has not been learned for other resolution candidates, so that could not distinguish un-reasonable height instances from positive. When both PRL and PRD are applied, we can expect that the sample space to be learned get reduced and focused to distinguish the positive and hard negatives with excluding effects of un-reasonable resolutions of instances.

In the bottom of Fig. 7-(b), we try to find a good integration rule of two scores from the baseline and the expert. We compare a simple addition rule $(a \cdot s_B + b \cdot s_E)$, multiplication rule $((a \cdot s_B + b) \cdot s_E)$, and the proposed optimal integration rule in Sect. 3.3. For the addition and multiplication rule, parameter sweeping for a and b is applied to empirically find the best combination. Our optimized integration rule shows better performance. Although the differences of the performances are marginal, the proposed method can suggest more plausible rules than the heuristically found best rules with high probability due to stable performance from the convex formulation. If one adds more complex rules by aggregating to the rule vector, better integration could be automatically found.

5.4   Analysis for Performance Improvement

In the performance measure suggested by [1], miss rates at sampled FPPIs are calculated by varying threshold. This implies that only relative scores (*i.e.* ranks) between true positives and false positives are important. As many true positives get higher ranks, less miss rate can be achieved. Therefore, to analyze why and how our approach improve the performance, we count the number of instances for each rank. We count rank of true positives in each image and summarize them for test images. The proposed two-phase model and the integration of two scores operate as a re-ranking process. The average rank of true positives is reduced from 2.7391 to 2.3072 by applying the proposed method (Fig. 8-(a)). As shown in Fig. 8-(b,c), the re-ranking caused by the proposed method allows higher threshold (less false positives) while true positives are kept.

Rank in an image	Average rank	1	2	3	4	5	6	7	8
DPM	**2.7391**	429	190	75	39	34	24	6	5
Proposed	**2.3072**	457	200	80	37	24	14	20	11

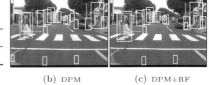

(a) Statistics for rank of true positives   (b) DPM   (c) DPM+RF

Fig. 8. (a) Summary of true positive ranks. Each entry of the table denotes the number of true positives with *i*-rank. (b,c) Sample results by DPM and the proposed two-phase model without PR (DPM+RF). True positives are denoted by *Margenta* color with their rank value (Color figure online).

6   Conclusion

We present a two-phase framework for pedestrian detection inspired by data mining philosophy, especially from the rare class detection. The baseline and expert detectors, which have different characteristics, optimally integrated by max-margin criteria without any heuristics. We validate our method on the systematical experiments and analyze re-ranking effects. We believe that the consistent improvements by our method are mainly comes from the samples space reduction to be learned. The beauty of our approach is that it can be easily adopted to an existing detector as an add-on module, and can improve the performance with little additional computation.

Acknowledgement. This work was supported by the Development of Autonomous Emergency Braking System for Pedestrian Protection project funded by the Ministry of Trade, Industry and Energy of Korea. (MOTIE)(No.10044775)

References

1. Dollár, P., Wojek, C., Schiele, B., Perona, P.: Pedestrian detection: an evaluation of the state of the art. IEEE Trans. PAMI **34**, 743–761 (2012)
2. Dalal, N., Triggs, B.: Histogram of oriented gradient for human detection. In: CVPR (2005)
3. Wang, X., Han, T.X., Yan, S.: An hog-lbp human detector with partial occlusion handling. In: ICCV (2009)
4. Walk, S., Majer, N., Schindler, K., Schiele, B.: New features and insights for pedestrian detection. In: CVPR (2010)
5. Dollár, P., Tu, Z., Perona, P., Belonggie, S.: Integral channel features. In: BMVC (2009)
6. Yan, J., Zhang, X., Lei, Z., Liao, S., Li, S.Z.: Robust multi-resolution pedestrian detection in traffic scenes. In: CVPR (2013)
7. Felzenszwalb, P.F., Girshick, R.B., McAllester, D., Ramanan, D.: Object detection with discriminatively trained part based models. IEEE Trans. PAMI **32**, 1627–1645 (2010)
8. Bourdev, L., Brandt, J.: Robust object detection via soft cascade. In: CVPR (2005)
9. Zhang, C., Viola, P.A.: Multiple-instance pruning for learning efficient cascade detectors. In: NIPS (2007)
10. Dollár, P., Appel, R., Kienzle, W.: Crosstalk cascades for frame-rate pedestrian detection. In: Fitzgibbon, A., Lazebnik, S., Perona, P., Sato, Y., Schmid, C. (eds.) ECCV 2012, Part II. LNCS, vol. 7573, pp. 645–659. Springer, Heidelberg (2012)
11. Viola, P., Jones, M.J.: Robust real-time face detection. IJCV **52**, 137–154 (2004)
12. Benenson, R., Mathias, M., Timofte, R., Van Gool, L.: Pedestrian detection at 100 frames per second. In: CVPR (2012)
13. Dollár, P., Appel, R., Belongie, S., Perona, P.: Fast feature pyramids for object detection. IEEE Trans. PAMI **36**, 1532–1545 (2014)
14. Gall, J., Lempitsky, V.: Class-specific hough forests for object detection. In: CVPR (2009)
15. Marín, J., Vazquez, D., Lopez, A.M., Amores, J., Leibe, B.: Random forests of local experts for pedestrian detection. In: ICCV (2013)

16. Joshi, M.V., Agarwal, R.C., Kumar, V.: Mining needles in a haystack: classifying rare classes via two-phase rule induction. In: ACM SIGMOD, pp. 91–102 (2001)

17. Weiss, G.M.: Mining with rarity: a unifying framework. In: ACM SIGKDD (2004)

18. Criminisi, A., Shotton, J., Konukoglu, E.: Decision forests: a unified framework for classification, regression, density estimation, manifold learning and semi-supervised learning. Found. Trends in Comput. Graph. Vis. **7**, 81–227 (2011)

19. Fisher, R.A.: The use of multiple measurements in taxonomic problems. Ann. Eugen. **7**, 179–188 (1936)

20. Vondrick, C., Khosla, A., Malisiewicz, T., Torralba, A.: Hoggles: Visualizing object detection features. In: ICCV, IEEE (2013)

21. Geronimo, D., Lopez, A.M., Sappa, A.D., Graf, T.: Survey of pedestrian detection for advanced driver assistance systems. IEEE Trans. PAMI **32**, 1239–1258 (2010)

22. Park, D., Zitnick, C.L., Ramanan, D., Dollar, P.: Exploring weak stabilization for motion feature extraction. In: CVPR (2013)

23. Ouyang, W., Wang, X.: Single-pedestrian detection aided by multi-pedestrian detection. In: CVPR (2013)

24. Park, D., Ramanan, D., Fowlkes, C.: Multiresolution models for object detection. In: Daniilidis, K., Maragos, P., Paragios, N. (eds.) ECCV 2010, Part IV. LNCS, vol. 6314, pp. 241–254. Springer, Heidelberg (2010)

25. Hsu, W.H., Kennedy, L.S., Chang, S.F.: Reranking methods for visual search. IEEE MultiMed. **14**, 14–22 (2007)

26. Breiman, L.: Random forests. Mach. Learn. **45**, 5–32 (2001)

27. Lemmond, T.D., Chen, B.Y., Hatch, A.O., Hanley, W.G.: An extended study of the discriminant random forest. Data Mining **8**, 123–146 (2010)

28. Yao, B., Khosla, A., Fei-Fei, L.: Combining randomization and discrimination for fine-grained image categorization. In: CVPR (2011)

29. Menze, B.H., Kelm, B.M., Splitthoff, D.N., Koethe, U., Hamprecht, F.A.: On oblique random forests. In: Gunopulos, D., Hofmann, T., Malerba, D., Vazirgiannis, M. (eds.) ECML PKDD 2011, Part II. LNCS, vol. 6912, pp. 453–469. Springer, Heidelberg (2011)

30. Duda, R., Hart, P., Stork, D.: Pattern Classification. Wiley-Interscience, Hoboken (2001)

31. Hamsici, O.C., Martinez, A.M.: Bayes optimality in linear discriminant analysis. IEEE Trans. PAMI **30**, 647–657 (2008)

32. Devijver, P.A., Kittler, J.: Pattern recognition: a statistical approach. Prentice-Hall, London (1982)

33. Joachims, T.: Optimizing search engines using clickthrough data. In: ACM SIGKDD (2002)

34. Flamary, R., Jrad, N., Phlypo, R., Congedo, M., Rakotomamonjy, A.: Mixed-norm regularization for brain decoding. Comput. Math. Methods Med. **2014**, 1–13 (2014)

35. Hoiem, D., Efros, A., Hebert, M.: Putting objects in perspective. IJCV **80**, 3–15 (2008)

36. Opencv 3.0. http://opencv.org/

Uncertainty Estimation for KLT Tracking

Sameer Sheorey[1]([✉]), Shalini Keshavamurthy[1], Huili Yu[1], Hieu Nguyen[1],
and Clark N. Taylor[2]

[1] UtopiaCompression Corporation, Los Angeles, CA 90064, USA
sameer@utopiacompression.com
[2] Air Force Research Laboratory, Columbus, OH, USA

Abstract. The Kanade-Lucas-Tomasi tracker (KLT) is commonly used for tracking feature points due to its excellent speed and reasonable accuracy. It is a standard algorithm in applications such as video stabilization, image mosaicing, egomotion estimation, structure from motion and Simultaneous Localization and Mapping (SLAM). However, our understanding of errors in the output of KLT tracking is incomplete. In this paper, we perform a theoretical error analysis of KLT tracking. We first focus our analysis on the standard KLT tracker and then extend it to the pyramidal KLT tracker and multiple frame tracking. We show that a simple local covariance estimate is insufficient for error analysis and a Gaussian Mixture Model is required to model the multiple local minima in KLT tracking. We perform Monte Carlo simulations to verify the accuracy of the uncertainty estimates.

1 Introduction

The Kanade-Lucas-Tomasi feature tracker (KLT), developed in [1–3], is the most commonly used approach to feature point tracking in image sequences. KLT searches for the location of a given feature point in the next few images by matching the local image patch intensity. Hierarchical search using image pyramids improves the tracking range (Bouguet [4]). The KLT tracker's excellent speed and reasonable accuracy make it popular in many applications such as video stabilization, egomotion estimation, image mosaicing, 3D reconstruction, visual odometry, and Simultaneously Localization and Mapping (SLAM). There exist many other extensions of the standard KLT algorithm that aim to increase accuracy and efficiency of computations. Baker and Matthews [5] give an overview of the extensions.

While the standard KLT algorithm and its extensions are successful in performing feature tracking, they simply output the estimated displacement without any indication of its accuracy. KLT displacement estimates are noisy due to image intensity noise and corresponding errors in the original feature detection. Complex local image structure is also a major source of error. An error model of the KLT tracker will be useful in downstream applications that aggregate tracking results over many points and frames to produce their output. For example, bundle adjustment for structure and motion estimation naturally uses feature point location covariances. More accurate modelling of the likelihood function

© Springer International Publishing Switzerland 2015
C.V. Jawahar and S. Shan (Eds.): ACCV 2014 Workshops, Part II, LNCS 9009, pp. 475–487, 2015.
DOI: 10.1007/978-3-319-16631-5_35

using these covariances results in more accurate structure and motion estimation (Triggs [6]).

The objective of this paper is to characterize the output uncertainty of the KLT tracker as a function of the uncertainty in its input feature location. (e.g. uncertainty in corner detection.) We first analyse the standard single level KLT tracker in an error propagation framework using least squares estimation theory. We build upon the local covariance representation in Kanazawa and Kanatani [7] and Nickels and Hutchinson [8] and generalize it. Error propagation analysis allows us to extend uncertainty estimation to pyramidal KLT tracking as well as multi-frame KLT tracking. Due to the existence of local minima in KLT tracking, a local single Gaussian covariance representation is insufficient to model the error. To address this issue, our error analysis approach represents the error using a Gaussian Mixture Model (GMM). The GMM model quantifies the probability that KLT tracking will get stuck in different local minima. This GMM error model is the main novel contribution of this work. Further, we approximate the GMM by a single covariance matrix that accounts for multiple local minima for use in downstream applications such as bundle adjustment.

The rest of the paper is organized as follows. We start with a review of related work in uncertainty estimation in computer vision in Sect. 2. Section 3 describes the KLT tracker briefly and Sect. 4 presents an uncertainty analysis of the single level and pyramidal versions. We show experiments to validate our results in Sect. 5. Section 6 concludes the paper.

2   Related Work

The structure from motion (SfM) and ego-motion estimation pipelines consists of feature detection, feature tracking (for image sequences) or matching (for unordered image collections), structure and motion initialization and finally bundle adjustment. Since the final bundle adjustment stage is improved by error estimates of its input, previous work has targeted error analysis on the earlier stages. Some of these include location uncertainty estimation of Harris corner detection (Orgunner and Gustafsson [9]) and SIFT point features (Zeisl *et al.* [10]). Approximate error estimation for feature detection also include Brooks *et al.* [11] and Kanazawa and Kanatani [7]. Nickels and Hutchinson [8] have also used feature tracking error covariance for simple tracking and physical measurements. These approaches only evaluate local covariance matrices and do not consider the common scenarios of pyramidal and multi-frame tracking. We consider full error propagation in the SfM pipeline, including input error from the feature detector and output error to the downstream stages. Our theoretical analysis also shows that local covariance matrices are insufficient uncertainty estimates due to multiple local minima. A full Gaussian Mixture Model (GMM) error representation is required.

Recently, Pfeiffer, Gehrig and Schneider [12] have shown that using confidence information can improve stereo computation. We hope that our work will lead to similar improvements in ego-motion estimation / SLAM and Structure from Motion.

3   The KLT Feature Tracker

We will first give an overview of the KLT tracking algorithm. We will start its error analysis with the basic algorithm in the noise free case and progressively do a more realistic error analysis including image noise and finally conclude with pyramidal and multi-frame tracking.

The KLT tracker starts with a set of sparse image features in the current frame I and attempts to find their locations in the next frame J by matching an image patch around a feature to the corresponding image patch in the next frame. The *brightness constancy* assumption implies that the patch intensities will not change substantially in the next image. Using a patch allows distinguishing between neighboring points of similar intensity. A window function $w(\mathbf{x})$, usually a Gaussian function, is used to emphasize the pixels near the feature point more than those far away. This accounts for the fact that points closer to the feature point are more likely to have similar motion than those that are farther away. The window is scaled so that $\sum_W w(\mathbf{x}) = |W|$ (the number of pixels in W). Some implementations such as Bouguet [4] use a simpler square window with uniform weights. Matching proceeds by calculating the error function

$$\epsilon(\mathbf{d}) = \sum_{W(\mathbf{x}_0)} [J(\mathbf{x} + \mathbf{d}) - I(\mathbf{x})]^2 w \left(\frac{\mathbf{x} - \mathbf{x}_0}{\sigma_w} \right) \tag{1}$$

over the support $W(\mathbf{x}_0)$ of the window function centered around the feature point $\mathbf{x}_0$. Minimizing this weighted (or generalized) nonlinear least squares (weighted NLS) expression yields the estimate for the displacement $\mathbf{d}$ for the feature point located at $\mathbf{x}_0$ in the image I.

This error function is minimized with a Newton-Raphson style algorithm that iteratively linearizes the next image intensity function J using Taylor series about the current feature location estimate. Shi and Tomasi [3] proposed allowing affine deformations of the image patch to check feature tracks extending for more than 5-10 frames, while the simpler displacement model suffices for tracking between consecutive frames. Some implementations such as in the OpenCV library prefer to use the full affine model even when tracking between consecutive frames, though that can cause some slowdown in tracking. We do not analyse the affine extension here.

Standard KLT is unable to track features successfully for large inter frame motion. Bouguet [4] solves this problem by using a pyramidal implementation. A Gaussian image pyramid is created and for each feature, the search starts from the coarsest level. The minimum found at a level is propagated to the next finer level as an initialization. The final feature location is found at the finest (base) level. We will extend our analysis to pyramidal KLT as well.

3.1   Unmodeled Sources of Error

KLT search accounts for sources of error such as image noise and viewpoint change (using affine transformation). Since KLT is a local search, it can get stuck in local minima and miss the actual (global) minimum. Pyramidal KLT greatly helps to reduce this error. Some extensions even allow for lighting and contrast changes. These extensions are more complex to analyse and here we will limit ourselves to image noise and displacements. There are also other sources of error that are not modelled at all. These include specular highlights, defocus and motion blur. Further, if the feature point is at a depth discontinuity, a viewpoint change can completely alter the appearance of its local image patch causing KLT tracking to be erroneous. Similarly, if the point being tracked is on a moving object, the results will cause errors in downstream applications even if KLT tracking succeeds. The last two issues highlight the need for robust estimation in downstream applications. All these unmodeled errors will limit the accuracy of our error estimates.

4   Uncertainty Estimation

KLT tracking is a weighted least squares estimation (WLS). Assume that both images have i.i.d. Gaussian noise with zero mean and σ^2 variance in pixel intensities. We assume that the image J is a shifted version of I, with independent variably distributed $(i.v.d)$ Gaussian noise added.

$$J(\mathbf{x}+\mathbf{d}) = I(\mathbf{x})+e(\mathbf{x}) \text{ with } e(\mathbf{x}) \sim \mathcal{N}(0, 2\sigma^2\text{diag}(w(\mathbf{x}-\mathbf{x}_o)^{-1})) \text{ for } \mathbf{x} \in W(\mathbf{x}_0) \tag{2}$$

For a Gaussian window, the assumed Gaussian noise covariance increases rapidly as we move away from the center of the patch and indicates reduced confidence in the used motion model. The covariance is $2\sigma^2$ at $\mathbf{x} = \mathbf{x}_0$, since it corresponds to the difference of the i.i.d. Gaussian noise from the two images. Farther away pixels are not very likely to have the same displacement as our feature point and their displacement is assumed to be almost uniformly distributed (infinite variance). The simpler box model assumes $i.i.d$ Gaussian noise inside the window and no constraints on the displacements outside it.

4.1   Single Level KLT

We now analyze the error of the KLT tracking. Due to the noise in the first image, the initial location of the feature point $\mathbf{x}_{t-1}$ is uncertain. Figure 1 shows the resulting error function $\epsilon(\mathbf{d})$, where the red ellipse indicates the variance of the initial location of the feature point.

   We first analyze the error for the case where there is no image noise. The KLT tracker essentially performs a local minimization on the error function starting from the feature point location in the current frame. Thus, it will converge to the local minimum $\mathbf{x}_t = \hat{\mathbf{b}}$ corresponding to the basin of attraction of the starting point. In Fig. 1, the starting point is in B_1 and hence it will converge to $\hat{\mathbf{b}} = \mathbf{b}_1$. We can now state the probability of KLT converging to different basin minima as the probability of the starting point being in that basin. This also allows us

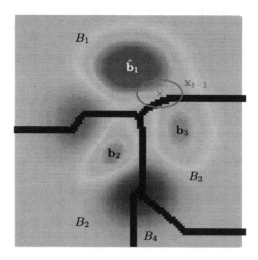

Fig. 1. KLT error surface $\epsilon(\mathbf{d})$. Uncertain detection in previous frame is represented by the red ellipse. Basins of attraction (B_i) on error surface are delineated by the thick black contours and their minima are shown in blue $(\mathbf{b}_i)$. The basin of the initial feature determines the convergence point (Color figure online).

to calculate the mean and variance. With

$$p_i := \Pr(\hat{\mathbf{b}} = \mathbf{b}_i) = \Pr(\mathbf{x}_{t-1} \in B_i) \tag{3}$$

$$\text{mean } \bar{\mathbf{x}}_t = \sum_i p_i \mathbf{b}_i \tag{4}$$

$$\text{covariance } \Sigma = \sum_i p_i(\bar{\mathbf{x}}_t - \mathbf{b}_i)(\bar{\mathbf{x}}_t - \mathbf{b}_i)^T \tag{5}$$

Assuming that the global minimum $(\hat{\mathbf{b}})$ corresponds to the actual feature point location, we then compute the bias

$$\text{bias } = \bar{\mathbf{x}}_t - \hat{\mathbf{b}} \tag{6}$$

We now analyze the uncertainty for the case with image noise. Let us assume that i.i.d. Gaussian noise with zero mean and variance σ^2 is added to each pixel intensity of the images I and J. This is equivalent to adding Gaussian noise with variance $2\sigma^2$ to the (shifted) difference image. We will analyze this problem with multiple local minima by partitioning the parameter space into the basins of attraction of the local minima. We then have a different NLS problem for each sub-domain.

$$P_i : \quad \min_{B_i} \epsilon(\mathbf{d}) \tag{7}$$

The advantage is that each problem is well posed with a unique global minimum and can be analyzed by standard NLS techniques. Finally, the starting point $\mathbf{x}_{t-1}$ is randomly selected, with probability p_i given by Eq. 3, with which these NLS problems will be solved. We will start the analysis with results on NLS for the ith problem (Seber and Wald [13, Sections 2.1.4, 2.8.8]) The displacement parameter $\mathbf{d}$ is now constrained to lie in $\mathbf{B}_i$. Let $\mathbf{H}_i$ be the Hessian matrix of the error function evaluated at the (now global) minimum $\mathbf{b}_i$. If we decompose the error function $\epsilon(\mathbf{d})$ into a sum of its terms $\epsilon_j(\mathbf{d}, \mathbf{x})$ as

$$\epsilon(\mathbf{d}) = \sum_j w\left(\frac{\mathbf{x}_j - \mathbf{x}_0}{\sigma_w}\right)\epsilon_j^2 \text{ with } \epsilon_j := J(\mathbf{x}_j + \mathbf{d}) - I(\mathbf{x}_j) \tag{8}$$

We have

$$\mathbf{H}_i = \mathbf{F}(\mathbf{b}_i)^T \operatorname{diag}(w(\mathbf{x}))\mathbf{F}(\mathbf{b}_i), \tag{9}$$

where $\mathbf{F}(\mathbf{d}) := [F_j(\mathbf{d})]$ with $F_j(\mathbf{d}) := \dfrac{\partial \epsilon_j(\mathbf{d})}{\partial \mathbf{d}}$ is the image gradient. (10)

Here $\operatorname{diag}(w(\mathbf{x}))$ is a diagonal matrix with $w(\mathbf{x})$ as the diagonal. If there are n pixels (n regressors) in the image patch to be compared, $\mathbf{F}(\mathbf{b})$ is an $n \times 2$ gradient matrix. $\mathbf{F}_j$ is a row of $\mathbf{F}$ and contains the gradient of the current image with respect to the shift for each pixel in the patch. Weighted least squares estimation theory tells us that the estimate $\hat{\mathbf{b}}_i$ is asymptotically normally distributed according to

$$\Pr(\hat{\mathbf{b}}_i) \sim \mathcal{N}(\mathbf{b}_i, 2\sigma^2 \mathbf{H}_i^{-1}). \tag{11}$$

Now let us consider the original problem with the full domain. The starting point selects the ith problem for solution with probability p_i, which results in an estimate that is asymptotically normally distributed according to Eq. 11. Consequently, the final estimate $\hat{b} = \mathbf{x}_t$ is distributed according to a Gaussian Mixture Model and we have

$$\Pr(\mathbf{x}_t) = \sum_i p_i g(\mathbf{b}_i, 2\sigma^2 \mathbf{H}_i^{-1}) \tag{12}$$

$$\text{covariance } \Sigma = \sum_i p_i(\mathbf{b}_i \mathbf{b}_i^T + 2\sigma^2 \mathbf{H}_i^{-1}) - \bar{\mathbf{x}}_t \bar{\mathbf{x}}_t^T \tag{13}$$

Mean and bias are given by Eqs. 4 and 6. The function $g(\mu, \Sigma)$ is the Gaussian probability density function with mean μ and covariance matrix Σ. We now have the basic theoretical framework for error analysis of the KLT tracker.

Estimating image noise: We can use weighted least squares theory to estimate the Gaussian noise variance σ present in the image from the error residue at the global minimum as

$$2\hat{\sigma}^2 = \epsilon(\hat{\mathbf{b}}), \tag{14}$$

given that the weights are scaled such that $\sum_W w(\mathbf{x}) = 1$. We will use this value of σ in Eqs. 12 and 13. Unfortunately, this noise estimate is very sensitive to model fidelity — if the image transformation cannot be accurately represented as a shift in the window around the feature point, this is likely to be a gross overestimate. Hence assuming that the entire frame has the same noise, we calculate this as the minimum over all points tracked in a frame.

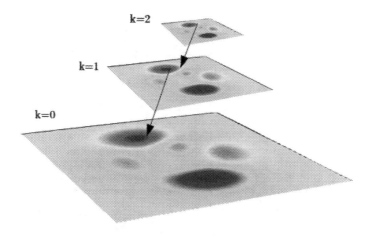

Fig. 2. Error Propagation in Pyramidal KLT Tracking.

4.2   Pyramidal KLT and Multi-frame Tracking

The Pyramidal KLT tracker [4] constructs a Gaussian pyramid of each image in the sequence by low pass filtering and downsampling. The next coarser level of the pyramid is constructed by filtering with a Gaussian of standard deviation 1 (usually approximated by the low pass filter $[1\ 4\ 6\ 4\ 1]/16$) and downsampling by 2. Features are first tracked at the coarsest level. The tracks are then propagated to the next finer level and tracking is repeated using the coarse level initialization. The finest level (original image) tracking results are used as the final tracking results. The tracking process is shown in Fig. 2. Pyramidal KLT offers improved tracking of features with large displacements. Since smoothing and downsampling reduce local minima, the coarse level tracking is more successful. Smoothing also reduces image noise, and consequently the tracking error. The lower levels further refine the displacement. We can propagate errors from the coarsest level L to the finest level 0 (original images) of the pyramid. Since each new level halves the image noise, the noise standard deviation at a level k is $\sigma^{(k)} = 2^{-k}\sigma$.

We will use the residue at the finest scale to estimate σ, i.e. $2\hat{\sigma}^2 = \epsilon^{(0)}(\hat{\mathbf{b}})$, since the translation model is most faithful at this scale. The error distribution at level k is then

$$\Pr(\mathbf{x}_t^{(k)}) = \sum_i p_i^{(k)} g(\mathbf{b}_i^{(k)}, 2^{1-2k}\sigma^2 \mathbf{H}_i^{(k)^{-1}}), \text{ with} \tag{15}$$

$$p_i^{(k)} := \Pr(\hat{\mathbf{b}}^{(k+1)} \in B_i^{(k)}) \tag{16}$$

$p_i^{(k)}$ is the probability that the next coarser level $(k + 1)$ KLT converges to a point $\hat{\mathbf{b}}^{(k+1)}$ that lies inside the level k error function basin $B_i^{(k)}$. The matrix $H_i^{(k)}$ is the Hessian matrix at level k for basin i. Equation 15 can be iterated to

Original Image I displaying feature points

Fig. 3. Street scene image used for KLT error evaluation and the 25 tracked feature points.

propagate the error distribution from the coarsest level k to the finest level 0. The final error is distributed according to a Gaussian mixture model and we can compute the net bias and covariance using Eqs. 6 and 13.

A very similar error propagation analysis can be done for KLT tracking across multiple frames.

5   Evaluation

We conduct simulations to evaluate the performance of the uncertainty estimation method for the KLT tracker. To evaluate consistency, we conduct Monte Carlo simulations on the KLT and use the Average Normalized Estimation Error Squared (ANEES) as the evaluation metric. The ANEES is a standard metric to evaluate the consistency of an estimator [14], and it is defined by

$$\text{ANEES} = \frac{1}{nN} \sum_{i=1}^{N} \epsilon_i, \tag{17}$$

where n is dimension of the parameter vector, N is the total number of Monte Carlo runs, and ϵ_i is the NEES in the ith Monte Carlo run, which is given by

$$\epsilon_i = (\bar{\theta}_i - \hat{\theta}_i)^\top \mathbf{P}^{-1}(\bar{\theta}_i - \hat{\theta}_i), \tag{18}$$

where $\bar{\theta}_i$ is the true parameter vector, $\hat{\theta}_i$ is the estimated parameter vector returned by the ith Monte Carlo run, and $\mathbf{P}$ is the estimator-provided error covariance, which is computed by Eq. 13. The ANEES value of a consistent estimator should be close to 1. Since ANEES is an average ratio, it is best observed on a log scale.

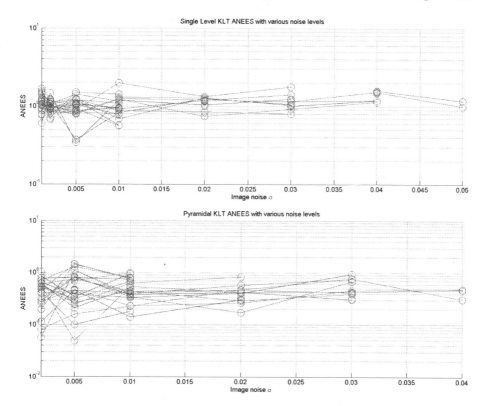

Fig. 4. ANEES values for single level (top) and 2 level pyramidal (bottom) KLT tracker with increasing noise levels. A log plot is used since ANEES is the average of ratios, so equal upwards and downwards deviations from the ANEES=1 line correspond to equal estimation errors. Each trajectory corresponds to the tracking of a single point. The decreasing number of points for larger noise levels corresponds to the fact that tracking fails more often for high noise. Image pixel values are in the range [0,1].

5.1   Simulations Using a Sequence of Shifted Images

In this section, we evaluate the performance of the proposed KLT uncertainty estimation method using a sequence of shifted images. We create a test image sequence by shifting an image I by known values. Points features (such as Harris corners or minimum eigenvalue features) are detected in I and tracked through the image sequence by the KLT tracker. The theoretical error is computed by calculating the KLT error surface and its local minima. The watershed transform (Meyer and Beucher [15]) is used to calculate the basins of attraction of the minima near the initial point. Since the watershed transform does not assign boundary pixels to basins, this computation is done at a higher resolution to prevent ambiguities at the basin borders. For pyramidal KLT, the computation is done by upsampling the higher level image back to the finest scale to maintain accuracy of the minima locations. The error propagation starts with the feature

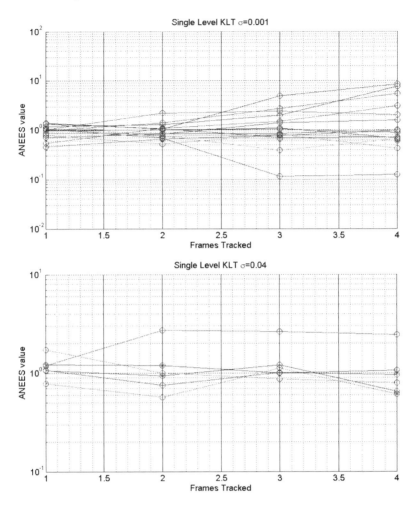

Fig. 5. ANEES for multi-frame single level KLT tracking at two different noise levels. All points out of 25 for which the KLT converged are shown. Image pixel values are in the range [0,1].

point detection error and is then propagated through pyramidal KLT levels as well as different image frames as long as experimental KLT tracking converges.

Next, Monte Carlo simulations are performed by adding Gaussian noise to each image before tracking. We calculate ANEES by aggregating the experimental KLT results from 25 Monte Carlo iterations and using the theoretical error covariance given by Eq. 17. We discard the KLT tracks that do not converge due to numerical issues.

We detect 25 minimum eigenvalue feature points in a test image, as shown in Fig. 3. Our first experiment evaluates the error estimates for a pair of frames for both single level and pyramidal KLT trackers. The graphs in Fig. 4 plot the ANEES value versus different added noise levels for the tracked feature points.

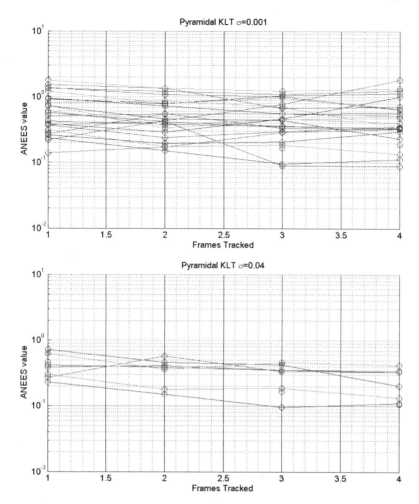

Fig. 6. ANEES for multi-frame pyramidal KLT tracking at two different noise levels. All points out of 25 for which the KLT converged are shown. Image pixel values are in the range [0,1].

The ANEES values are close to 1 for most points even as noise levels increase. The deviation from 1 reflects the limitation of the Hessian approximation for error estimation in non-linear least squares.

The next experiment evaluates the estimates over a sequence of 5 frames (numbered 0–4) for both single level and pyramidal KLT tracking. These ANEES plots are shown in Fig. 5 for the single level KLT and in Fig. 6 for the 2 level pyramidal KLT.

6   Conclusion and Future Work

We have presented a novel comprehensive error analysis of the KLT tracker. We show that the error of the single level, pyramidal as well as the multi-frame

KLT tracker is given by a Gaussian Mixture Model. The components of the mixture model correspond to the local minima that can trap the KLT tracker. Our Monte Carlo simulations show that our uncertainty estimates are accurate for these common use cases.

Our next steps will be to show improvements in real world applications by using the KLT error estimates. This will require improving the speed with which the error estimates are calculated. Since computing watershed transforms for each tracked point can make real time tracking difficult, further work is necessary before the error estimates can be used in practical systems. We would also like to extend the error analysis to allow affine image patch deformations during tracking. Another research direction is exploring new ways to modify the KLT error function to reduce the prevalence of local minima, without compromising its speed.

Acknowledgement. This research was supported by Air Force Research Laboratory (AFRL) under contract FA8650-13-M-1701 with UtopiaCompression Corporation.

References

1. Lucas, B.D., Kanade, T.: An iterative image registration technique with an application to stereo vision. In: Proceedings of the 7th international joint conference on Artificial intelligence - Volume 2. IJCAI 1981, pp. 674–679. Morgan Kaufmann Publishers Inc., San Francisco (1981)
2. Tomasi, C., Kanade, T.: Detection and tracking of point features. School of Computer Science, Carnegie Mellon Univ, Technical report (1991)
3. Shi, J., Tomasi, C.: Good features to track. In: 1994 IEEE Computer Society Conference on Computer Vision and Pattern Recognition, 1994. Proceedings CVPR 1994, pp. 593–600 (1994)
4. Bouguet, J.Y.: Pyramidal implementation of the affine lucas kanade feature tracker description of the algorithm. Intel Corporation 5 (2001)
5. Baker, S., Matthews, I.: Lucas-Kanade 20 Years On: A Unifying Framework. Int. J. Comput. Vis. **56**, 221–255 (2004)
6. Triggs, B., McLauchlan, P.F., Hartley, R.I., Fitzgibbon, A.W.: Bundle adjustment – a modern synthesis. In: Triggs, B., Zisserman, A., Szeliski, R. (eds.) ICCV-WS 1999. LNCS, vol. 1883, pp. 298–372. Springer, Heidelberg (2000)
7. Kanazawa, Y., Kanatani, K.i.: Do we really have to consider covariance matrices for image features? In: Proceedings of the Eighth IEEE International Conference on Computer Vision, 2001. ICCV 2001. vol. 2, pp. 301–306 (2001)
8. Nickels, K., Hutchinson, S.: Estimating uncertainty in SSD-based feature tracking. Image Vis. Comput. **20**, 47–58 (2002)
9. Orguner, U., Gustafsson, F.: Statistical characteristics of harris corner detector. In: IEEE/SP 14th Workshop on Statistical Signal Processing, 2007. SSP 2007, pp. 571–575 (2007)
10. Zeisl, B., Georgel, P.F., Schweiger, F., Steinbach, E.G., Navab, N., Munich, G.E.R.: Estimation of location uncertainty for scale invariant features points. In: BMVC, pp. 1–12 (2009)

11. Brooks, M.J., Chojnacki, W., Gawley, D., Van Den Hengel, A.: What value covariance information in estimating vision parameters? In: Proceedings of the Eighth IEEE International Conference on Computer Vision, 2001. ICCV 2001. vol. 1, pp. 302–308. IEEE (2001)

12. Pfeiffer, D., Gehrig, S., Schneider, N.: Exploiting the power of stereo confidences. In: 2013 IEEE Conference on Computer Vision and Pattern Recognition (CVPR), pp. 297–304. IEEE (2013)

13. Seber, G.A.F., Wild, C.J.: Nonlinear Regression. Wiley, New York (1989)

14. Li, X.R., Zhao, Z., Jilkov, V.P.: Practical measures and test for credibility of an estimator. In: Proceedings of Workshop on Estimation, Tracking, and Fusion-A Tribute to Yaakov Bar-Shalom, Citeseer, pp. 481–495 (2001)

15. Meyer, F., Beucher, S.: Morphological segmentation. J. Vis. Commun. Image Representation **1**, 21–46 (1990)

Third ACCV Workshop on E-Heritage

Combined Hapto-visual and Auditory Rendering of Cultural Heritage Objects

Praseedha Krishnan Aniyath$^{(\boxtimes)}$, Sreeni Kamalalayam Gopalan,
Priyadarshini Kumari, and Subhasis Chaudhuri

Vision and Image Processing Lab, Department of Electrical Engineering,
IIT Bombay, Mumbai, India
praseedhakrishnan@gmail.com

Abstract. In this work, we develop a multi-modal rendering framework
comprising of hapto-visual and auditory data. The prime focus is to haptically render point cloud data representing virtual 3-D models of cultural
significance and also to handle their affine transformations. Cultural heritage objects could potentially be very large and one may be required
to render the object at various scales of details. Further, surface effects
such as texture and friction are incorporated in order to provide a realistic haptic perception to the users. Moreover, the proposed framework
includes an appropriate sound synthesis to bring out the acoustic properties of the object. It also includes a graphical user interface with varied
options such as choosing the desired orientation of 3-D objects and selecting the desired level of spatial resolution adaptively at runtime. A fast,
point proxy-based haptic rendering technique is proposed with proxy
update loop running 100 times faster than the required haptic update
frequency of 1 kHz. The surface properties are integrated in the system
by applying a bilateral filter on the depth data of the virtual 3-D models.
Position dependent sound synthesis is incorporated with the incorporation of appropriate audio clips.

1 Introduction

We interact with our physical world mostly through visual, auditory and tactile sensations. Haptics is an emerging research field which tries to emulate the
latter sensation in the virtual world. Incorporation of haptics in the virtual environment provides a better immersive experience to the users, especially to the
visually impaired persons. Further, a combined hapto-visual rendering enhances
the realism of the haptic interaction even for the sighted users. Many authors
have also suggested the inclusion of surface properties like texture and friction
in the haptic domain for realistic haptic perception. However, in most cases of
prior art, the rendering of surface properties such as texture and friction is not
realistic enough to provide sufficient immersion to the users. This is more so as
common haptic interface devices like Novint Falcon and SensAble Phantom are
kinaesthetic and not tactile.

Haptic rendering methods like god object rendering algorithm [17] work well
in the case of polygon based representation of 3-D models. On the other hand,

© Springer International Publishing Switzerland 2015
C.V. Jawahar and S. Shan (Eds.): ACCV 2014 Workshops, Part II, LNCS 9009, pp. 491–506, 2015.
DOI: 10.1007/978-3-319-16631-5_36

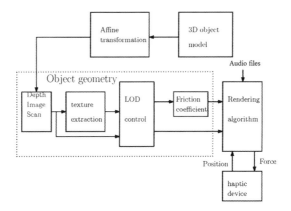

Fig. 1. Block diagram illustrating the proposed system. LoD stands for level of details.

it fails to properly render 3-D objects represented using point cloud data. More-over, 3-D objects may appear at various different scales and orientations, and the user needs to experience the object at different levels of details. A polygon-based rendering scheme is not suitable to implement variations in scale and orientation. This is because recomputing the mesh structure during the inter-action is not feasible as it is time consuming. The authors in [18] did speed up the multi-level hapto-visual rendering using a Monge surface. However, their focus is mainly on fast rendering of a single valued function at different levels of details and cannot handle any rotation of the space and does not provide any surface friction. This paper aims at enhancing the virtual immersion during haptic interaction with point cloud 3-D models by incorporating surface prop-erties such as texture and friction. The coefficient of friction is computed from the texture component extracted from the scanned depth data. But for these properties, the virtual environment would often feel slippery since the direction of the force vector is always perpendicular to the surface and thus would not be sufficiently realistic. The new rendering framework can also handle affine trans-formations such as rotation, translation, expansion and geometric contraction of 3-D objects as illustrated in Fig. 1. Initially, we generate depth data at each point of the model at any desired orientation by reading the contents of depthbuffer in OpenGL using the inbuilt GLUT command `glReadPixels()` and create a Monge surface from it. This surface is then hapto-visually rendered at different scales adaptively at run time. We propose the use of a bilateral filter to extract the local surface texture and compute the dynamic friction as a function of the local texture. We show that the user's experience can be enriched by allowing the user to interact with the object at multiple resolutions. Audio rendering is also incorporated in the proposed system for enhanced virtual experience by playing appropriate audio files based on the position of the haptic probe inside the virtual environment. We also implement a graphical user interface for easier accessibility.

The organization of the paper is as follows. Section 2 provides a review of the related literature. Section 3 explains the proposed method. Haptic and graphic rendering techniques and additional functionalities are discussed in Sect. 4. Section 5 illustrates the results. The conclusions are summarized in Sect. 6.

2   Literature Review

The basic haptic rendering technique is polygon based in which virtual objects are represented using polygonal meshes. In polygon based rendering, each time the haptic interface point (HIP) penetrates the object, the rendering algorithm finds the closest point on the mesh defined surface and computes the penetration depth of HIP inside the object. If $\mathbf{x}$ is the vector representing the penetration depth, the reaction force is calculated as $\mathbf{F} = -k\mathbf{x}$, where k is the stiffness constant of the surface. This method has problems in determining the appropriate force direction while rendering thin objects. Authors in [16,23] independently proposed the concept of god-object algorithm and proxy algorithm, respectively, to solve this problem. The god-object rendering algorithm includes another point in addition to the HIP, called "god-object or proxy". In free space the proxy and the HIP are collocated. However, as the HIP penetrates the virtual object, the proxy is constrained to lie on the surface of the virtual object [7]. But in the case of virtual objects represented using point cloud data, the proxy would slip into the object. It is called the "fall-through" problem of the god-object. This can be avoided by using a spherical proxy instead of a point. However, an increase in radius of the proxy impairs the haptic interaction by smoothing out the surface texture.

Many authors have also developed algorithms to render point cloud data directly. Lee *et al.* have proposed a rendering method with point cloud data which estimates the distance from HIP to the closest point on the moving least square surface defined by the given point set [9]. El-Far *et al.* used axis aligned bounding boxes to fill the voids in the point cloud and then rendered with a god object rendering technique [12]. Leeper *et al.* described a constraint based approach of rendering point cloud based data where the points are replaced by spheres or surface patches of approximate size [10]. Another proxy based technique of rendering a dense 3-D point cloud data was proposed in [19], where the surface normal is estimated locally from the point cloud. Most of the methods are unable to handle scale changes during rendering and variable density of point cloud.

In order to enrich the experience of virtual world, many authors have incorporated surface properties like texture and friction in the haptic domain. Adi *et al.* [1] introduced the technique of rendering the tactile cues from visual information using wavelet transforms. It was more realistic than primitive haptic texture rendering methods implemented using sine waves [2] and Fourier series [21]. But this technique was found to be less stable than the existing methods. Some authors, as in [15], have proposed data driven approaches to realistic haptic texture rendering. However, these methods are computationally expensive for real-time applications. Similarly, Richard *et al.* presented a friction rendering model in haptics using modified Karnopp model [14]. Hayward *et al.* [6] developed a discrete implementation of friction exhibiting four friction regions: sticking, creeping, oscillating, and sliding. Harwin *et al.* [5] proposed the friction cone algorithm for providing friction in haptic environments which also has a few shortcomings as suggested in [11]. Hence, still much needs to be explored in this domain in order to augment the user's virtual experience.

Research in ecological acoustics imply that auditory feedback can effectively convey information about a number of object attributes such as its shape, size and material [3]. The effectiveness of auditory sensations was studied by Lederman *et al.* [8] who have showed that sound plays a dominant role when a probe is used to interact with a surface as compared to the case of direct contact with the bare fingers. We make use of this fact in order to improve the virtual immersion of the user interacting with the virtual environment using a haptic probe. Simultaneous audio rendering is required in our application where we try to render ancient cultural monuments where the supporting pillars have very interesting acoustic (read musical) properties.

3   Proposed Proxy Updation Algorithm

The proposed system incorporates a proxy based algorithm to render a point cloud data. It is to be noted that the proxy is a point and not a sphere as it is popular in point cloud rendering techniques, thus allowing us to perceive surface textures. Let us first assume that one has the depth buffer data corresponding to 3-D models available. Issues related to transformation of the haptic space and scaling will be discussed in Sect. 4.

In order to render the object in the haptic domain, we need to find the collision of HIP with the bounding surface and hence the penetration depth of HIP into the surface. The proposed algorithm tries to translate the proxy over the object surface in short steps during the haptic interaction such that each time it finds the most suitable proxy position which would provide the minimum distance between HIP and proxy and simultaneously applies the reaction force

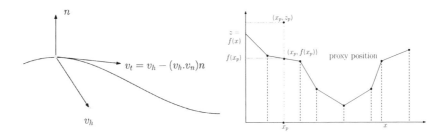

Fig. 2. (a) Illustration of tangent vector evaluation using the current proxy and HIP positions. (b) Illustration of surface approximation from depth values (Color figure online).

normal to the surface at the point of contact. Figure 2(a) illustrates the proxy movement during collision with an arbitrary object represented by a Monge surface. The Monge surface corresponding to a 3-D object is shown as a curve in $1 - D$. HIP is shown with a yellow circle penetrated inside the object and the proxy is also shown with a yellow dot constrained to the surface. The vector $\mathbf{v}_n$ represents the normal at the initial proxy position with $\mathbf{n} = \frac{\mathbf{v}_n}{|\mathbf{v}_n|}$ representing the unit normal and $\mathbf{v}_h$ is the vector from proxy to the HIP and is given by $\mathbf{v}_h = \mathbf{X}_h - \mathbf{X}_p$, where $\mathbf{X}_h$ and $\mathbf{X}_p$ represent the position vectors of HIP and proxy, respectively. The tangential vector in the plane of $\mathbf{v}_n$ and $\mathbf{v}_h$ is evaluated which provides a fast approximation of the direction for the proxy to move so that the distance between proxy and HIP can be minimized. When the proxy is moved continuously along the tangential direction on the curve, proxy will finally come to rest at a point where the angle between $\mathbf{v}_n$ and $\mathbf{v}_h$ is 180 degrees. The tangent vector $\mathbf{v}_t$ can be computed from $\mathbf{n}$ and $\mathbf{v}_h$ using Eq. 1.

$$\mathbf{v}_t = \mathbf{v}_h - (\mathbf{v}_n.\mathbf{v}_h)\mathbf{n}. \tag{1}$$

We use the following proxy update equation to translate the proxy along the tangent plane.

$$\mathbf{X}_p^{(k+1)} = \mathbf{X}_p^{(k)} + \rho\mathbf{v}_t^k. \tag{2}$$

The parameter $\rho < 1$ and is arbitrarily chosen. As the value of ρ increases, the proxy quickly converges, but it does not move close to the surface during the convergence. On the other hand if ρ is very small, the proxy point moves close to the object surface, but needs a little more time to converge. As a matter of fact, the value of ρ relates to the frictional force on the surface and should depend on the material property as explained in Sect. 3.2. A small value of ρ signifies larger surface friction. Hence the proposed method provides an easy way of including dynamic friction during rendering.

Once the proxy moves in the direction of tangential vector, it may deviate from the boundary of the object. This deviation of the proxy from the boundary is avoided by projecting the proxy along $\mathbf{n}$ onto the surface before updating its position along the tangential direction. The proxy position update is performed within 1 ms of time, so that the user's interaction with the object through the haptic device is unhindered and is carried out at 1 kHz. Since the updated proxy location need not be on the chosen lattice for depth representation, the surface needs to be locally interpolated. In case of $2 - D$ depth data, we project the proxy onto the X-Y plane and the corresponding depth value is obtained by interpolating the neighbourhood depth values to form a continuous function $z = f(x, y)$ as illustrated in Fig. 2(b) for a $1 - D$ function $z = f(x)$. Since the available points are sampled quite densely, bilinear interpolation is sufficient to find the bounding surface as shown in Fig. 2(b). In order to check the collision of HIP with the surface, we compare z_p with the depth at the projected point $z = f(x, y)$ for a given proxy position (x_p, z_p). If $f(x_p) > z_p$, the proxy has touched the surface, otherwise it is free to move towards the HIP.

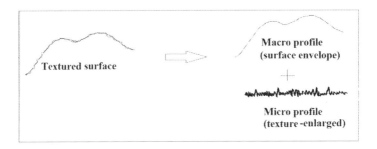

Fig. 3. Envelope subtraction from a textured surface.

3.1 Recovery of Surface Texture

Any real or virtual surface can be represented as a sum of a general shape or envelope of the surface and the minute surface variations called texture. The latter provides a realistic feel to an otherwise slippery envelope. Hence the texture component of the surface can be extracted by subtracting the envelope (low frequency component) from the surface as illustrated in Fig. 3. The figure shows how a textured Monge surface can be represented as a combination of a macro profile (general geometry or shape) and micro profile (texture). In this work, a bilateral filter is used for the purpose of envelope subtraction. Bilateral filtering [20] provides simple and non-iterative edge-preserving data smoothing. The bilateral filter takes a weighted sum of the depth map at a local neighbourhood at the lattice; the weights depend on both the spatial distance and similarity in depth values. In this way, edges are preserved well while "noise" is averaged out. The bilateral filtered output $f_b(x,y)$ of the depth data $f(x,y)$ is obtained from pixels $(\tilde{x}, \tilde{y})$ in the neighbourhood as shown in the following equation.

Fig. 4. Illustration of texture retrieval from depth data using a bilateral filter: (a) Depth image corresponding to 3-D model of Buddha (b) Extracted texture details (Data Courtesy: www.archibaseplanet.com).

$$f_b(x,y) = \frac{1}{W(x,y)} \sum_{\tilde{x}} \sum_{\tilde{y}} G_{\sigma_S}(x - \tilde{x}, y - \tilde{y}) G_{\sigma_R}(f(x,y) - f(\tilde{x}, \tilde{y})) f(\tilde{x}, \tilde{y}) \quad (3)$$

where $G(x, y)$ is a Gaussian kernel and σ_R and σ_S represent the spread in amplitude values and spatial distances, respectively. The term $W(x, y)$ is a normalization factor. Although this is a non-linear filter, computationally efficient algorithms exist to obtain the filtered output [20]. The bilateral filtered output $f_b(x, y)$ of the 3-D object is subtracted from the depth data $f(x, y)$ so as to obtain the texture component alone. Hence

$$h(x, y) = f(x, y) - f_b(x, y). \quad (4)$$

Figure 4(a) shows the depth image of a 3-D model and Fig. 4(b) shows the extracted texture from the depth image using a bilateral filter. As explained earlier, bilateral filter provides edge-preserving smoothing and hence its output is the smoothed depth data without texture details.

3.2   Incorporation of Surface Friction

Haptic rendering techniques that do not consider the surface friction induce the feeling of a very smooth and slippery surface which is not the case in practice. Adding friction along with the texture details to the model provides a more realistic feeling of the surface. Let $\mathbf{f}$ be the reaction force on the haptic device as shown in Fig. 5. The component of the applied force normal to the surface $\mathbf{f}_n$ is given by $|\mathbf{f}| \cos \beta$. Similarly the tangential force on the surface is $|\mathbf{f}| \sin \beta$. The magnitude of the retarding force on the proxy is proportional to the normal force $\mathbf{f}_n$. If μ_s denotes the static friction coefficient, proxy is in static contact with the surface as long as $|\mathbf{f}_t| < \mu_s |\mathbf{f}_n|$. During haptic interaction, dynamic friction exerts a retarding force on the proxy while moving on the surface of the object. μ_d which denotes the coefficient of dynamic friction is made proportional to the resultant curvature of texture component at each point of the rendered surface since friction depends on the surface property of the material. Since the curvature encodes the unevenness of a surface very well, we use μ_d proportional to the curvature. Now, the magnitude of the retarding force is given by $\mu_d |\mathbf{f}_n|$

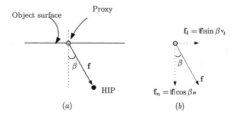

(a)   (b)

Fig. 5. Illustration of calculation of the resultant vector for the proxy movement while incorporating surface friction.

and is in a direction opposite to $\mathbf{f}_t$. The curvature is computed as the resultant of mean and Gaussian curvatures whose magnitudes are given by the following equations. The physical significance of mean curvature(H) is the first variation of the surface area and Gaussian curvature(K) represents the local convexity. Hence the use of resultant curvature as a representative of μ_d can be justified as it represents the amount of "bending" at each point on the surface [22].

$$H = \frac{h_{xx}(1 + h_y^2) + h_{yy}(1 + h_x^2) - 2h_{xy}h_xh_y}{2(1 + h_x^2 + h_y^2)^{\frac{3}{2}}} \tag{5}$$

$$K = \frac{h_{xx}h_{yy} - h_{xy}^2}{(1 + h_x^2 + h_y^2)^2} \tag{6}$$

where the parameters h_x, h_y are first partial derivatives of the texture component $h(x, y)$ of the surface w.r.t x and y axis. Similarly h_{xx}, h_{yy} are second partial derivatives of $h(x, y)$ w.r.t x and y. The dynamic friction coefficient μ_d is computed as follows:

$$\mu_d = \frac{1}{R\sqrt{H^2 + K^2}} \tag{7}$$

where R is the radius of the largest inscribable sphere within the haptic work space. Since R is much larger than the radius of curvature in the micro texture, usually $\mu_d < 1$. The resultant force $\mathbf{f}_r$ on the proxy is given by $\mathbf{f}_t - \mu_d\mathbf{f}_n$.

$$\mathbf{f}_r = \mathbf{f}_t(1 - \mu_d \cot \beta) \quad if \quad |\mathbf{f}_t| \geq \mu_s|\mathbf{f}_n|$$
$$= 0 \quad otherwise \tag{8}$$

Equation 8 provides a direct description of the frictional force. For proxy based haptic rendering, such an equation can easily be incorporated while defining the proxy movement. Comparing Eq. (8) with (2), we observe that $\mathbf{f}_r$ is proportional to $\mathbf{v}_t$. Hence the parameter ρ with friction is given by $\rho_f = \rho(1 - \mu_d \cot \beta)$. Then the proxy update equation is given by:

$$\mathbf{X}_p^{(k+1)} = \mathbf{X}_p^{(k)} + \rho(1 - \mu_d \cot \beta^k)\mathbf{v}_t^{(k)} \tag{9}$$

where k denotes the iteration number.

4   Multimodal Rendering

Haptic rendering involves generating force feedback in order to provide the sensation of touch to the users. Any haptic rendering algorithm would include the following two steps:

1. detection of collision of the HIP with the object.
2. computation of force feedback if a collision is detected.

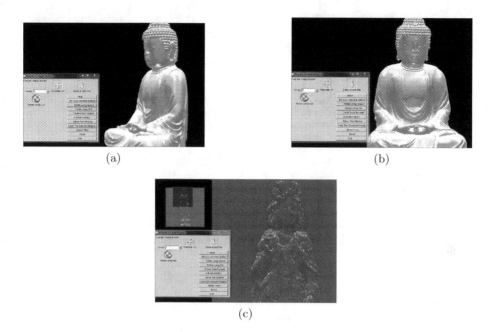

Fig. 6. Illustration of user defined selection of orientation (before depth scan): 3-D model of Buddha (a) before rotation (b) after rotation. (c) Illustration of user defined selection of level of details (at run-time) for another model (Data Courtesy: www. archibaseplanet.com).

If $z_p < f(x_p, y_p)$ in Fig. 2(b) then the proxy has touched the object and a force needs to be fed back to the user through the haptic device. Subsequently as explained in Sect. 2, the reaction force is computed as $\mathbf{F} = -k\mathbf{x}$ where k is the Hooke's constant, and $\mathbf{x}$ is the current penetration depth given by $\mathbf{x} = \mathbf{X}_h - \mathbf{X}_p$, where $\mathbf{X}_h$ is the HIP position and $\mathbf{X}_p$ is the proxy position.

For a combined hapto-visual rendering, the 3-D object surface is displayed as a simple quad mesh formed out of the depth values in OpenGL. We have opted for the mesh-based graphical display in order to give a better perception to the viewer since using point cloud data for graphic display would result in gaps in the visually rendered image. We have used the stereoscopic display technique for creating the effect of depth in the image by presenting two offset images on the screen corresponding to two different points of projection. This provides a 3-D perception to the user. Anaglyphic glasses can be used to view such displays.

4.1   Rendering at Different Scales and Orientations

In practice, 3-D objects come at various physical scales and orientations. In a virtual environment, one should be able to experience objects of all sizes at different scales to get a sense of overall structure to finer details from the same data set. Hence, we have implemented adaptive scaling in both graphical and haptic domains. In order to scale the surface we resize depth data of resolution $N \times N$ depending on the level user selects, with $N \times N$ as the finest level. If we load the level $N \times N$ into the haptic space the full object can be rendered visually as well as haptically. Users can select the level as well as the region of interest at run time either using buttons in the haptic device or using keyboard functions. Additionally, we have developed a graphical user interface for easy accessibility. Figure 6 illustrates the selection of scale and orientation by the user. Figure 6(b) shows the rotated version of Fig. 6(a). Figure 6(c) illustrates how user can select different levels of details and the corresponding textural component is shown.

The user can select the desired orientation before the depth scan using either the mouse controls or using the graphical user interface. Further, depending on the scale selected by the user, the corresponding depth data is dynamically loaded into the active haptic space and an appropriate haptic force is rendered. As only a limited subset of data is loaded, the rendering is very fast. In general, at higher levels of resolution, the user should be able to view finer details. The haptic force also varies accordingly. Hence in order to incorporate realistic haptic and graphic perception, we need to appropriately scale the depth values at each level of depth map. Further, trying to map a large physical dimension over a small haptic work space (typically about 4 in. cube of active space) leads to a lot of unwanted vibrations (something similar in concept to aliasing) during rendering. Hence the depth values need to be smoothed before being down-sampled and mapped into the haptic work space. Multi-scale data generation for procuring different levels of details is performed based on [4].

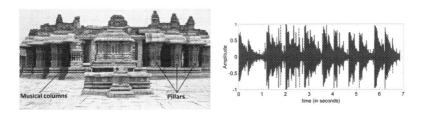

Fig. 7. (a) Front view of Vitthala temple, Hampi (courtesy: [13]). (b) Plot of an audio file recorded from the musical pillar. (courtesy: [13]).

4.2   Audio Rendering

Sounds are incorporated in the rendering framework by playing appropriate audio files based on the position of haptic probe inside the virtual environment. A demonstration of this technique is given in the attached video using the 3-D model of musical pillars at Vitthala temple, Hampi which is an early 14th century world heritage site. The pillars in this temple have musical columns which produces distinct sounds when struck. The temple consists of 56 pillars which are monolithic sculptures each having granite stone columns of height 10 ft as shown in Fig. 7(a). Each major pillar is surrounded by 7 minor pillars that can reverberate at 7 primary notes of Indian classical music. We have used the recorded audio files of these pillars (data courtesy: www.daiict.com) in our work. A sample audio file is shown in Fig. 7(b). A single wavelength of the sound file was extracted from the above audio file and was played back whenever the haptic probe touched the virtual pillar. During audio rendering, one could synthesize various types of sounds synthetically. However, we provide the real data from the actual heritage site to provide a real feel of the musical pillars. During haptic rendering whenever a collision takes place for the first time with a specific pillar, the corresponding note is played, the volume of which is made proportional to the rendered force.

5   Results

The proposed method was implemented in visual C++ in a Windows XP platform with an Intel i5 CPU @ 2.66 GHZ with 2 GB RAM. For obtaining texture details, the depth data obtained from OpenGL depthbuffer is fed to a bilateral filter using OpenCV inbuilt functions and the output from the filter is subtracted from the original depth image. We have experimented with various models of 3-D objects and a few of them are displayed below. Figure 8 shows the model of Ganesh, visually rendered in OpenGL. For haptic rendering we use HAPI library. The blue ball represents the position of the proxy constrained to lie on the surface. The discrete position in the model is displayed in a fixed 200×200 haptic space. The size and spatial resolution of the model depend on two factors: the active space of the haptic device used to render the model, and the resolution at which the model should be displayed. We use a 3-DOF haptic device, NOVINT FALCON with a 4 in. cube of active space. While interacting with the object haptically, the average proxy update time was found to be 0.0056 ms which is much faster than the required upper bound of 1 ms, and hence the user has a very smooth haptic experience. The average time required for dynamic data generation and loading it into the haptic space depends on the resolution of input depth data and it was observed to be around 0.5 s and 0.02 s, respectively, for depth data with a resolution of 800×800. We also carry out the rendering at finer levels of details by successively zooming into the object.In above cases, each figure consists of two parts where the left part is the reference for the users to select the part of the object they wish to explore haptically as shown in Figs. 6 and 8. The right part of the figure corresponds to the selected region at the appropriate resolution for haptic rendering. Figure 8(b) shows the scaled up version of Fig. 8(a). It is quite clear from Fig. 8(c) that the users are able to feel even minute details of the sculpture and have visual perception of closeness in depth. Hence they can have a more realistic experience. Furthermore, the audio-playback feature with respect to the musical pillars augmented the user experience. However, we observe that there is a small time lag between the haptic and audio rendering due to data access time to open the stored audio files. For a small number of musical notes, these audio clips can be stored in RAM to circumvent this problem.

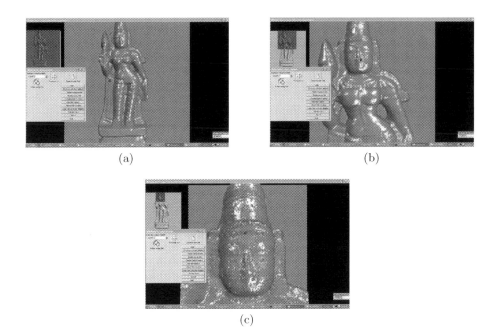

(a)   (b)

(c)

Fig. 8. 3-D model of a statue with texture: at (a) lowest level of details (b) at double the resolution and (c) at the finest resolution. (Data Courtesy: www.archibaseplanet. com).

5.1 Validation of Proposed Method

Validation of result is not quite easy during haptic rendering. Authors in [18] propose a validation technique using a standard 3-D sphere model. But, since rendering of friction in our work involves computation of resultant curvature, the use of a spherical model (having a fixed curvature) is not justified. We demonstrate a validation technique using a known sinusoidal surface. Figure 9 shows the z-component of proxy as a function of time while haptically interacting with the depth data. The green line and the blue line in the figure show the z-component of the proxy point with and without texture and friction, respectively. We observe that the proxy position converges, but with a time-lag in the former case, as expected due to surface effects of texture and friction. The shapes of both the curves are quite similar except for the time lag.

Figure 10(a) shows the reaction force versus time during haptic interaction with a known sinusoidal surface without incorporating the surface effects. In free space the HIP and proxy positions are almost the same and hence the reaction force on the haptic device is zero for initial few seconds as illustrated in figure. As the HIP penetrates the object, the proxy stays on the surface according to the iteration method discussed in Sect. 3. The proxy point moves continuously during interaction, whenever there is a change in HIP position

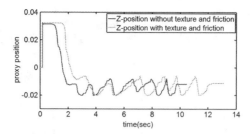

Fig. 9. Plot of z-position of proxy as a function of time to illustrate the effect of surface friction (Color figure online).

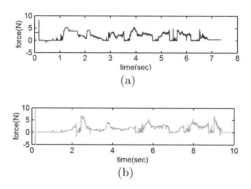

Fig. 10. Force-vs-time graph for a particular interaction with the depth data (a) without surface effects and (b) with surface effects.

and appropriate reaction force is fed back to the haptic device. The force-time graph in Fig. 10(b) illustrates how the net reaction force fed back to the haptic device is effectively delayed by constraining the proxy movement which gives the perception of friction to the user.

In Fig. 11, we show the actual set up of our rendering framework. A user wearing the anaglyphic glasses watches the stereoscopic visual rendering of an object and at the same time haptically interacts with the object through the Falcon device. This provides an excellent hapto-visual immersion of the subject into the virtual object. However, for the visually impaired users, the selection of scale and the location for rendering cannot be based on the small navigation window on the screen. For such subjects, we use the buttons available on the haptic device for the user to explore the object at different scales and locations. We also conducted a user survey on the proposed virtual set-up in order to understand how realistically the users can perceive the shape and surface properties of the virtual models. All subjects were made to sit at a desk with the virtual environment displayed in front of them. Then they were asked to explore the virtual system by grasping the haptic device. After an exploration

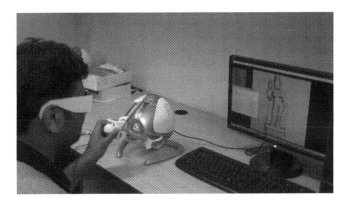

Fig. 11. Illustration of hapto-visual immersion of a subject for a 3-D object. On the left, the user wearing anaglyphic glasses is holding the FALCON haptic device while interacting with the virtual 3-D object displayed on the screen.

time of 1 min, each user was asked to rate the realism of the virtual surface in terms points from 1–10 (with 10 being the highest rating). A total of 12 subjects (8 males and 4 females) volunteered for user study. All the users were in the age group of 18 to 40. The survey resulted in an average user rating of 7.23 (out of 10) for the realism of virtual environment. We propose to conduct richer user study in future including distribution and statistical meaning of user ratings.

6   Conclusions

In this work we have proposed a new technique for multi-modal rendering of 3-D objects represented as a dense depth map data. Rendering of surface properties like texture and friction is found to enrich the user's experience in the virtual world. We include scalability, rotation, translation and stereoscopic display of 3-D models as additional features to enhance the realism in experience. Presently, we have integrated audio rendering with respect to a single spatially segmented 3-D object i.e., the musical pillars at Hampi, India. In future, we propose to include continuous audio rendering in a more generalized framework by annotating the acoustic property at each point in the 3-D model. We conducted experiments with several 3-D models. We also conducted an user survey on a few subjects and observed that hapto-visual and auditory rendering of virtual 3-D models using the proposed method greatly augmented the user's experience.

Acknowledgement. The authors would like to thank DST for the grant provided on the Indian Digital Heritage Project and MCIT for the grant on perception engineering. The authors would also like to thank Prof. Manjunath Joshi and his team for providing us with the audio signals of musical pillars at Hampi which is the input to our proposed rendering framework.

References

1. Adi, W., Sulaiman, S.: Haptic texture rendering based on visual texture information: a study to achieve realistic haptic texture rendering. In: Zaman, H.B., Robinson, P., Maria, M.P., Olivier, P., Schröder, H., Shih, T.K. (eds.) IVIC 2009. LNCS, vol. 5857, pp. 279–287. Springer, Heidelberg (2009)
2. Choi, S., Tan, H.Z.: An analysis of perceptual instability during haptic texture rendering. In: Proceedings of the 10th Symposium on Haptic Interfaces for Virtual Environment and Teleoperator Systems, Orlando, Florida, U.S.A., pp. 129–136 (2002)
3. Gaver, W.W.: What in the world do we hear? an ecological approach to auditory event perception. Ecol. Psychol. J. 5(1), 1–29 (1993)
4. Gluckman, J.: Scale variant image pyramids. In: Proceedings of IEEE Computer Society Conference on Computer Vision and Pattern Recognition, New York, U.S.A., pp. 1069–1075 (2006)
5. Harwin, W.S., Melder, N.: Improved haptic rendering for multi-finger manipulation using friction cone. In: Proceedings of the 2nd Eurohaptics Conference, Edinburgh, UK (2002)
6. Hayward, V., Armstrong, B.: A new computational model of friction applied to haptic rendering. In: Corke, P., Trevelyan, J. (eds.) Experimental Robotics VI. LNICST, vol. 250, pp. 403–412. Springer, Heidelberg (2000)
7. Laycock, S.D., Day, A.M.: A survey of haptic rendering techniques. Comput. Graph. Forum 26, 50–65 (2007). Blackwell Publishing
8. Lederman, S.J., Morgan, T., Hamilton, C., Klatzky, R.L.: Integrating multimodal information about surface texture via a probe: relative contributions of haptic and touch-produced sound sources. In: Proceedings of the 10th Symposium on Haptic Interfaces for Virtual Environment and Teleoperator Systems, Washington, DC, U.S.A., pp. 97–105 (2002)
9. Lee, J.K., Kim, Y.J.: Haptic rendering of point set surfaces. In: World Haptics Conference, Tsukuba, Japan, pp. 513–518 (2007)
10. Leeper, A., Chan, S., Salisbury, K.: Constraint based 3-DoF haptic rendering of arbitrary point cloud data. In: RSS Workshop on RGB-D Cameras, University of Southern California, Los Angeles, U.S.A., June 2011
11. Melder, N., Harwin, W.S.: Extending the friction cone algorithm for arbitrary polygon based haptic objects. In: Proceedings of 12th International Conference on Haptic Interfaces for Virtual Environment and Teleoperator Systems, Washington, DC, U.S.A., pp. 234–241 (2004)
12. El-Far, N.R., Georganas, N.D., Saddik, A.E.: An algorithm for haptically rendering objects described by point clouds. In: Proceedings of the 21st Canadian Conference on Electrical and Computer Engineering, Ontario, Canada (2008)
13. Patil, H.A., Gajbhar, S.S.: Acoustical analysis of musical pillar of great stage of Vitthala temple at Hampi, India. In: International Conference on Signal Processing and Communications (SPCOM), pp. 1–5 (2012)
14. Richard, C., Cutkosky, M.R.: Friction modeling, display in haptic applications involving user performance. In: IEEE International Conference on Robotics and Automation, pp. 605–611 (2002)
15. Romano, J.M., Kuchenbecker, K.J.: Creating realistic virtual textures from contact acceleration data. IEEE Trans. Haptics 5, 109–119 (2012)
16. Ruspini, D.C., Kolarov, K., Khatib, O.: The haptic display of complex graphical environments. In: Proceedings of ACM SIGGRAPH, Los Angeles, U.S.A., pp. 345–352 (1997)

17. Salisbury, K., Conti, F., Barbagli, F.: Haptic rendering: introductory concepts. IEEE Comput. Graph. Appl. Mag., special issue on Haptic Rendering **24**, 24–32 (2004)

18. Sreeni, K.G., Priyadarshini, K., Praseedha, A.K., Chaudhuri, S.: Haptic rendering of cultural heritage objects at different scales. In: Proceedings of the Eurohaptics, Finland, pp. 505–516 (2012)

19. Sreeni, K.G., Chaudhuri, S.: Haptic rendering of dense 3-D point cloud data. In: IEEE Haptics Symposium, Vancouver, BC, Canada, 4–7 March 2012

20. Tomasi, C., Manduchi, R.: Bilateral filtering for gray and color images. In: Proceedings of the Sixth International Conference on Computer Vision, Mumbai, India, pp. 839–846 (1998)

21. Wall, S.: An investigation of temporal and spatial limitation of haptic interfaces. Ph.D. Thesis, Department of Cybernetics, University of Reading (2004)

22. Weisstein, E.: Mean curvature, June 2013. http://mathworld.wolfram.com/MeanCurvature.html

23. Zilles, C.B., Salisbury, J.K.: A constraint-based god-object method for haptic display. In: IEEE/RSJ International Conference on Intelligent Robots and Systems, Pittsburgh, U.S.A., vol. 3, pp. 3146–3151 (1995)

Mesh Denoising Using Multi-scale Curvature-Based Saliency

Somnath Dutta[1], Sumandeep Banerjee[1]([✉]), Prabir K. Biswas[1], and Partha Bhowmick[2]

[1] Department of Electronics and Electrical Communication Engineering,
Indian Institute of Technology, Kharagpur, India
{somdats,sumandeep.banerjee}@gmail.com, pkb@ece.iitkgp.ernet.in
[2] Department of Computer Science and Engineering,
Indian Institute of Technology, Kharagpur, India
pb@cse.iitkgp.ernet.in

Abstract. 3D mesh data acquisition is often afflicted by undesirable measurement noise. Such noise has an aversive impact to further processing and also to human perception, and hence plays a pivotal role in mesh processing. We present here a fast saliency-based algorithm that can reduce the noise while preserving the finer details of the original object. In order to capture the object features at multiple scales, our mesh denoising algorithm estimates the mesh saliency from Gaussian weighted curvatures for vertices at fine and coarse scales. The proposed algorithm finds wide application in digitization of archaeological artifacts, such as statues and sculptures, where it is of paramount importance to capture the 3D surface with all its details as accurately as possible. We have tested the algorithm on several datasets, and the results exhibit its speed and efficiency.

1 Introduction

Mesh denoising is an imperative preprocessing technique for improving meshes containing noise that creeps in during the process of data acquisition and subsequent digitization process. It aims at improving the quality of the reconstructed surface by producing a mesh with better perceptual features. So, it involves removal of noise while retaining most of the original features present in the object, and hence should be robust in nature. An important aspect of a denoising algorithm is to adjust vertex positions without any feature disintegration. The time taken to accomplish the entire process is also one of the important considerations, especially if real-time interactive mesh processing is required. The entire process is iterative, where the number of iterations actually depend on the amount of smoothness required with minimum degradation of the actual content. Low-level human visual perception plays an indispensable role in objective evaluation of geometric processing like denoising. Incorporating perceptuality in mesh denoising analogous to image filtering can improve the processing of the meshes. The requirement of quality graphical mesh and their usage in multidisciplinary applications such as digital heritage, explain the essence of employing

© Springer International Publishing Switzerland 2015

C.V. Jawahar and S. Shan (Eds.): ACCV 2014 Workshops, Part II, LNCS 9009, pp. 507–516, 2015.
DOI: 10.1007/978-3-319-16631-5_37

the concept of human perception into mesh processing. Mesh saliency, as introduced in [1], gives a measure of regional importance, especially for 3D mesh. It can be integrated into graphics applications, such as mesh denoising, mesh simplification, shape matching, and segmentation. Saliency captures features of an object at multiple scales, since what seems interesting at one scale may not remain the same at other scales. It consequently reveals the difference between the vertex and its surrounding context. Various mesh processing methods can be modified to accommodate saliency into the process so that visually salient features can be preserved into the mesh.

1.1   Existing Work

The classical Laplacian smoothing method [2,3] is the simplest surface smoothing method for noise removal. However, it over-smooths the mesh and also causes surface shrinkage. Kernel based Laplacian smoothing method proposed in [4] tries to overcome the problem of classical Laplacian approach. There are various anisotropic filtering approaches which vary on the technique used to preserve prominent features. Some algorithms are geometric diffusion based anisotropic method [5–7]. The other class of algorithms comprise both normal update and vertex update for the purpose of denoising [8–10]. Various methods traditionally used for image denoising have been extended to point cloud denoising as well as mesh denoising. Bilateral filter [11] is one such prominent method that has proved to be an effective edge-preserving filter. The earlier work by Fleishman et al. [12] and Jones et al. [13] focused on modifying the vertex position by a suitable weighted function based on spatial difference as well as normal difference. Yagou et al. proposed mesh denoising based on alpha trimming [14] along with mean, median method [9]. Sun et al. [15] proposed an iterative algorithm of filtering noisy normals and then updating each vertex position based on this modified normal following least square criterion. The recent work of Zheng et al. [16] emphasizes on both local iterative scheme and global non-iterative scheme of mesh denoising. Unlike previous methods of processing normals, it considers normals as a surface signal defined over the original mesh. It also presents comparative analysis of such scheme under different constraints like runtime as well as robustness. Denoising can also be accomplished by considering the point cloud without any mesh representation. The central idea behind such methods is to denoise the points and then triangulate those denoised points to obtain a reconstructed denoised surface. References [17,18] are some of the techniques on point cloud denoising. A different approach based on L_0 minimization is undertaken in [19] for mesh denoising. L_0 norm is used to preserve sharp features and smooth the remaining surface.

The concept of saliency has been studied for images to determine salient image location in [20]. Itti et al. [21] computed a saliency map from the information based on center-surround mechanism. Saliency based methods using the concept of [21] have been used in [22] for computing the saliency map of a 3D dynamic scene. The idea described in [23] has been used to find 3D surface [24] by smoothing noisy data. A user study that compares the previous mesh saliency

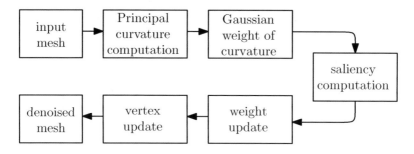

Fig. 1. Proposed denoising process based on saliency.

approach with human eye movements is discussed by Kim et al. [25]. The experimental result discussed in [25] describes the fact that mesh saliency can model human eye movements comparatively better than what can be expected purely by chance. The idea of using only local features or locally prominent salient regions for computing the saliency of a 3D mesh surface is outperformed in [26]. It incorporates the methodology of not only local contrast but even global rarity by defining global saliency on each vertex depending on its contrast with all other vertices.

1.2 Our Work

Our proposed method for mesh denoising is similar to the one followed for denoising in [27] with some modifications. In [27] saliency [1] is combined with contextual discontinuities [10]. In [10], an adaptive smoothing is used to denoise a 2D image, incorporating both inhomogeneity and spatial gradient. Inhomogeneity reveals the incoherence between a pixel and its surrounding pixels. The proposed method also takes mesh saliency into consideration by using curvature as a geometric feature of the object. The saliency value for a vertex is high if it is a salient point and vice versa. The Gaussian weighted average of the principal curvature estimated at each point (vertex) around a neighborhood is considered as a suitable weight function with the weight being amplified depending on certain algorithmic criteria. Curvature of each point v on the surface is computed, and the largest absolute value of the principal curvature is considered for the process, as follows.

$$\kappa(v) = max(|\kappa_{\max}(v)|, |\kappa_{\min}(v)|) \tag{1}$$

where, $\kappa_{\max}(v)$ and $\kappa_{\min}(v)$ denote peak convex and peak concave curvatures respectively.

2 Proposed Method

Figure 1 presents a block diagram of the proposed method on mesh denoising. Its various stages are briefly explained in this section.

2.1   Curvature Estimation

The method of curvature estimation is based on [28]. It is one of the efficient and widely used techniques for estimating curvature of a 3D dataset. Initially, for all points in the dataset, normals are computed. Then, by principal component analysis (PCA) of these normals, the maximum and the minimum principal curvatures for all data points are obtained.

2.2   Neighborhood Determination and Weight Computation

We consider a distance based threshold as opposed to ring neighborhoods used by [27]. We determine the neighborhood of each vertex (point) of the mesh using kd-Tree based ANN algorithm provided by the Point Cloud Library (PCL) [29]. Let the neighborhood $N_2(v, \delta)$ for a vertex v be the set of vertices within a distance δ, measured in L^2 norm. That is, $N_2(v, \delta) = \{x : \|x - v\| < \delta\}$, where x denotes a mesh vertex. The Gaussian-weighted average of the principal curvature is given by

$$G\left(\kappa(v), \delta\right) = \frac{\sum\limits_{x \in N_2(v, 2\delta)} \kappa(x) \exp\left(\frac{-\|x - v\|^2}{2\delta^2}\right)}{\sum\limits_{x \in N_2(v, 2\delta)} \exp\left(\frac{-\|x - v\|^2}{2\delta^2}\right)} \tag{2}$$

where, $\kappa(x)$ is the absolute value of principal curvature, and $N_2(v, 2\delta)$ denotes the neighborhood of a vertex v within a distance 2δ. The value δ is the standard deviation of the Gaussian filter. To incorporate the saliency feature, different values of δ are considered to incorporate the idea of multiple resolutions, which in turn, captures the important features of the object at all perceptually meaningful scales.

2.3   Saliency Computation

The mesh saliency that tends to capture the most prominent features at multiple scales is estimated as

$$S_k(v) = |G(\kappa(v), \delta_k) - G(\kappa(v), \delta_{k+1})| \tag{3}$$

where, $S_k(v)$ is the saliency value of a vertex v at a scale δ_k w.r.t. the next scale δ_{k+1}. The difference between the two captures the importance of saliency of a vertex v at two successive scales. The scales used in our method are from $\{k\varepsilon : 1 \leq k \leq 6\}$, where $\varepsilon = \chi \|\bar{e}\|$ is considered as a multiple of the average edge length $\|\bar{e}\|$ of the mesh.

The *coarseness factor* χ multiplied with $\|\bar{e}\|$ to obtain ε is a parameter supplied manually, depending on the object to be denoised, and we have found values in the range 1–4 to be most effective. For a high value of χ, the finer details are lost, whereas a low value of χ can capture the finer details. The saliency of a vertex is finally estimated as the average of its saliency values computed at different scales, in accordance with the following equation.

$$S(v) = \frac{1}{n_s - 1} \sum_{k=1}^{n_s - 1} S_k(v) \tag{4}$$

where, n_s is the number of scales used (6 in our experiments).

2.4 Weight Update

The maximum saliency, $S_{\max}(v)$, and the minimum saliency, $S_{\min}(v)$, are computed for all mesh vertices, and then the saliency value of each vertex $S(v)$ is normalized for adaptive smoothing. The normalized saliency value is given by

$$\tilde{S}(v) = \frac{S(v) - S_{\min}(v)}{S_{\max}(v) - S_{\min}(v)}. \tag{5}$$

A high saliency value usually corresponds to a surface feature. In other words, vertices with higher saliency values maintain their sharp features during denoising. Hence, we choose a value β such that all saliency values greater than β are amplified, while the rest do retain the original. We choose β in the saliency interval of 60–80th percentile; in most the cases, β is considered as the 80th percentile of the saliency value, while the amplifying factor λ is selected in the range 4–10. As a result, the normalized saliency obtained in Eq. 5 is finally modified as per the following equation.

$$\tilde{S}(v) = \begin{cases} \lambda \tilde{S}(v) & \text{if } \tilde{S}(v) \geq \beta \\ \tilde{S}(v) & \text{otherwise} \end{cases} \tag{6}$$

2.5 Vertex Update

To denoise the mesh, we recompute the position of each vertex in the mesh. This step involves causing a vector displacement to each vertex. We begin by considering a neighborhood within a distance δ around the vertex, and determine the centroid of the concerned neighborhood. We then compute the centroid normal C and the point normal P (mapped vector difference between the centroid and the respective point). The mapped normal overcomes the anomaly of using the true point normal as described in paper [17]. The vertex is replaced by the weighted average computed around the neighborhood of that vertex. The weight function is Gaussian in nature with $P_j \cdot C$ as the variable, and the saliency value from Eq. 6 as the scale.

3 Results and Discussion

The algorithm is tested on various datasets acquired by our system as well as data available from standard databases [30, 31]. The user-defined parameters λ

(a)   (b)   (c)

Fig. 2. (a) `Meduse` original surface (b) Noisy (Gaussian) surface (c) Denoised surface

is used for amplifying the saliency value so that features are preserved during smoothing. In majority of the dataset, only one iteration of denoising is used to ensure that the object does not get over-smoothed and blurry. Figure 2 illustrates the effect of denoising on a surface corrupted by Gaussian noise. The parameters used is $\lambda = 4$. The object shown in Fig. 4 is of a column of a temple. The result obtained for this object shows another instance of the effectiveness of the algorithm in denoising an object while preserving its features. The user-defined parameters in this case is $\lambda = 5$. The `Buddha` dataset obtained using our own scanner without any synthetically added noise also responds effectively to denoising process, as shown in Fig. 3. The parameters used for denoising in this case are the same used for the result shown in Fig. 2.

3.1   Comparative Analysis

The quantitative evaluation as well as the objective evaluation based on perceptual metric can be used for comparing the results, albeit the focus is on quantitatively evaluating the outputs of different denoising method. The quantitative evaluation is based on root mean square (RMS) error method. It takes into consideration the correspondence among the vertices of the two objects under comparison, and hence it is limited to the comparison between two meshes

Table 1. Performance analysis

Algorithms	Data (vertices)	Error (E_v)	Time (in secs)
Laplacian smoothing	`Column` (480932)	2.28×10^1	18
	`Meduse` (358904)	6.66×10^{-3}	14
Bilateral filter (Fleishman et al.)	`Column` (480932)	2.26×10^1	13752
	`Meduse` (358904)	6.68×10^{-3}	11356
Normal filtering (Sun et al.)	`Column` (480932)	2.24×10^1	385
	`Meduse` (358904)	6.51×10^{-3}	215
Saliency based denoising	`Column` (480932)	2.21×10^1	261
	`Meduse` (358904)	6.32×10^{-3}	203

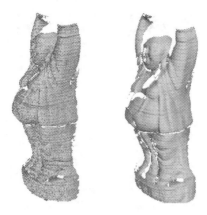

Fig. 3. (a) Buddha laser scanned surface (b) Denoised Buddha surface

(a)　　　　　(b)　　　　　(c)

Fig. 4. (a) Column surface (b) Noisy (Gaussian) surface (c) Denoised surface

sharing the same connectivity. The quantitative error is estimated as

$$Error(A, B) = \frac{1}{n} \sum_{i=1}^{n} \left\| v_i^A - v_i^B \right\|^{1/2} \tag{7}$$

where, n is the number of vertices of the mesh, v_i^B is the vertex of denoised mesh B, and v_i^A is its corresponding vertex in the original mesh A.

A comparison of the proposed method with some of the existing mesh-denoising algorithms is presented in Fig. 5. The parameters of the existing methods have been tuned to generate the best possible outputs. The Laplacian method, as shown in Fig. 5(b), almost smooths the surface, hence degrading its features. Fleishman bilateral filter extends the concept of 2D bilateral filter to 3D mesh denoising. The result shown in Fig. 5(c) is able to preserve details to some extent but takes considerable amount of time to denoise the mesh.

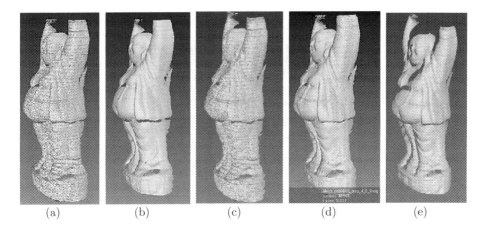

Fig. 5. (a) Noisy surface (b) Laplacian (c) Fleishman bilateral Filter (d) Normal filtering (e) Denoised (Saliency-based) surface

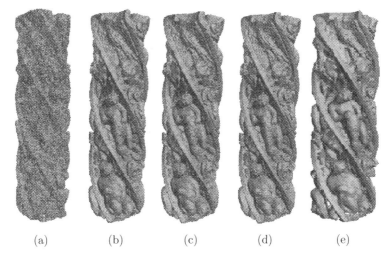

Fig. 6. Various denoised (`Column` dataset) output (a) Noisy surface (b) Laplacian surface (c) Bilateral filtered surface (d) Normal filtered surface and (e) Saliency surface (proposedmethod)

The Normal filtering technique (Sun et al.) requires a large number of iterations initially for the task of normal update and also for vertex update, hence increasing the runtime for denoising. However, the details of the object remain intact to a large extent. The output shown in Fig. 5(d) required ten normal iterations and fifteen vertex iterations to complete the denoising process. The result in Fig. 6 shows the comparison on `Column` data with some existing methods. Table 1 presents a comparison of the proposed method with some of the existing algorithms in terms of quantitative error and also with respect to the execution time of the denoising process.

4   Conclusion

The proposed technique based on mesh saliency is suitable for denoising due to its simplicity, execution speed, and ability to retain the originality of the object while removing the noise. It comes up with a significantly low error rate on different datasets, although its execution is quite fast compared to some of the existing methods. Its fast execution time merits its readiness to larger datasets with archaeological artifacts of historical importance, such as statues, sculptures, monuments, and temple structures.

References

1. Lee, C.H., Varshney, A., Jacobs, D.W.: Mesh saliency. ACM Trans. Graph. **24**, 659–666 (2005)
2. Field, D.A.: Laplacian smoothing and delaunay triangulations. Commun. Appl. Numer. Methods **4**, 709–712 (1988)
3. Vollmer, J., Mencl, R., Muller, H.: Improved laplacian smoothing of noisy surface meshes. Comput. Graph. Forum **18**, 131–138 (1999)
4. Badri, H., El Hassouni, M., Aboutajdine, D.: Kernel-based laplacian smoothing method for 3D mesh denoising. In: Elmoataz, A., Mammass, D., Lezoray, O., Nouboud, F., Aboutajdine, D. (eds.) ICISP 2012. LNCS, vol. 7340, pp. 77–84. Springer, Heidelberg (2012)
5. Desbrun, M., Meyer, M., Schrder, P., Barr, A.H.: Anisotropic feature-preserving denoising of height fields and bivariate data. In: Graphics Interface, pp. 145–152 (2000)
6. Tasdizen, T., Whitaker, R., Burchard, P., Osher, S.: Geometric surface smoothing via anisotropic diffusion of normals. In: Proceedings of the Conference on Visualization 2002, VIS 2002, Washington, DC, pp. 125–132. IEEE Computer Society (2002)
7. Hildebrandt, K., Polthier, K.: Anisotropic filtering of non-linear surface features. Comput. Graph. Forum. **23**, 391–400 (2004)
8. G.Taubin: Linear anisotropic mesh filtering. IBM Research Report RC22213(W0110–051) (2001)
9. Yagou, H., Ohtake, Y., Belyaev, A.: Mesh smoothing via mean and median filtering applied to face normals. In: Proceedings of Geometric Modeling and Processing, pp. 124–131 (2002)
10. Chen, K.: Adaptive smoothing via contextual and local discontinuities. IEEE Trans. Pattern Anal. Mach. Intell. **27**, 1552–1567 (2005)
11. Tomasi, C., Manduchi, R.: Bilateral filtering for gray and color images. In: Sixth International Conference on Computer Vision, pp. 839–846 (1998)
12. Fleishman, S., Drori, I., Cohen-or, D.: Bilateral mesh denosing. ACM Trans. Graph. **22**, 950–953 (2003)
13. Jones, T., Durand, F., Desbrun, M.: Non-iterative, feature preserving mesh smoothing. ACM Trans. Graph. **22**, 943–949 (2003)
14. Yagou, H., Ohtake, Y., Belyaev, A.: Mesh denoising via iterative alpha-trimming and nonlinear diffusion of normals with automatic thresholding. In: Proceedings of Computer Graphics International, pp. 28–33 (2003)
15. Sun, X., Rosin, P., Martin, R., Langbein, F.: Fast and effective feature-preserving mesh denoising. IEEE Trans. Visual. Comput. Graph. **13**, 925–938 (2007)

16. Zheng, Y., Fu, H., Au, O.K.-C., Tai, C.-L.: Bilateral normal filtering for mesh denoising. IEEE Trans. Visual. Comput. Graph. **17**(10), 1521–1530 (2011)
17. Miropolsky, A., Fischer, A.: Reconstruction with 3d geometric bilateral filter. In: Proceedings of the Ninth ACM Symposium on Solid Modeling and Applications, SM 2004, Aire-la-Ville, Switzerland, pp. 225–229. Eurographics Association (2004)
18. Deschaud, J.E., Goulette., F.: Point cloud non local denoising using local surface descriptor similarity. ISPRS Technical Commission III Symposium, PCV 2010 - Photogrammetric Computer Vision and Image Analysis XXXVIII - Part 3A (2010)
19. He, L., Schaefer, S.: Mesh denoising via l0 minimization. ACM Trans. Graph. **32**, 64:1–64:8 (2013)
20. Koch, C., Ullman, S.: Shifts in selective visual attention: towards the underlying neural circuitry. In: Vaina, L. (ed.) Matters of Intelligence. Synthese Library, pp. 115–141. Springer, Heidelberg (1987)
21. Itti, L., Koch, C., Niebur, E.: A model of saliency-based visual attention for rapid scene analysis. IEEE Trans. Pattern Anal. Mach. Intell. **20**, 1254–1259 (1998)
22. Yee, H., Pattanaik, S., Greenberg, D.P.: Spatiotemporal sensitivity and visual attention for efficient rendering of dynamic environments. ACM Trans. Graph. **20**, 39–65 (2001)
23. Guy, G., Medioni, G.: Inferring global perpetual contours from local features. Int. J. Comput. Vis. **20**, 113–133 (1996)
24. Guy, G., Medioni, G.: Inference of surfaces, 3d curves, and junctions from sparse, noisy, 3d data. IEEE Trans. Pattern Anal. Mach. Intell. **19**, 1265–1277 (1997)
25. Kim, Y., Varshney, A., Jacobs, D.W., Guimbretière, F.: Mesh saliency and human eye fixations. ACM Trans. Appl. Percept. **7**, 12:1–12:13 (2010)
26. Wu, J., Shen, X., Zhu, W., Liu, L.: Mesh saliency with global rarity. Graph. Models **75**, 255–264 (2013)
27. Mao, Z.H., Ma, L.Z., Zhao, M.X., Li, Z.: Feature-preserving mesh denoising based on contextual discontinuities. J. Zhejiang Univ. Sci. A **7**(9), 1603–1608 (2006)
28. Taubin, G.: Estimating the tensor of curvature of a surface from a polyhedral approximation. In: Proceedings of Fifth International Conference on Computer Vision, pp. 902–907 (1995)
29. Rusu, R., Cousins, S.: 3d is here: Point cloud library (pcl). In: 2011 IEEE International Conference on Robotics and Automation (ICRA), pp. 1–4 (2011)
30. Epfl, computer graphics and geometry laboratory. http://lgg.epfl.ch/statues.php?p=dataset/
31. Aim@shape repository. http://shapes.aim-at-shape.net/

A Performance Evaluation of Feature Descriptors for Image Stitching in Architectural Images

Prashanth Balasubramanian[✉], Vinay Kumar Verma, and Anurag Mittal

Computer Vision Lab, Indian Institute of Technology Madras,
Chennai 600036, India
{bprash,vkverma,amittal}@cse.iitm.ac.in

Abstract. We present a performance comparison of 4 feature descriptors for the task of feature matching in Panorama Stitching on images taken from architectural scenes and archaeological sites. Such scenes are generally characterized by structured objects that vary in their depth and large homogeneous regions. We test SIFT, LIOP, HRI and HRI-CSLTP on 4 different categories of images: well-structured with some depth variations, partially homogeneous with large depth variations, nearly homogeneous with a little amount of structural details and illumination-variant. These challenges test the distinctiveness and the intensity normalization schemes adopted by these descriptors. HRI-CSLTP and SIFT perform on par with each other and are better than the others on many of the test scenarios while LIOP performs well when the intensity changes are complex. The results of LIOP also show that the order computations of the pixels have to be made in a noise-resilient manner, especially in homogeneous regions.

1 Introduction

Identification of point-correspondences between images is an important problem that finds application in many tasks such as Registration, Stitching, Disparity Matching, 3-D Reconstruction, Tracking, Object Identification and Classification. As the transformations between the images are seldom known a priori, the practice is to localize on distinctive regions of images (called as *keypoints*) and match them under different transformations. Matching of keypoints across 2 images is done by building *feature descriptors* that express the visual characteristics of the regions around the keypoints, and correspond them using a suitable distance metric. The descriptors are expected to be sufficiently distinctive so as to represent the keypoint and be robust to geometric transformations, illumination variations, different blurs, artifacts due to sampling and compression.

Many interesting attempts have been made to design descriptors which satisfy these said characteristics. Early work used the raw pixels of the regions around the keypoints and studied their correlation measure. As correlation measures do not consider geometric information, such measures cannot tolerate localization

© Springer International Publishing Switzerland 2015
C.V. Jawahar and S. Shan (Eds.): ACCV 2014 Workshops, Part II, LNCS 9009, pp. 517–528, 2015.
DOI: 10.1007/978-3-319-16631-5_38

errors of keypoints, and so are good when the regions are exactly registered. Further, these measures can only handle linear changes in intensities while it is well-known that non-linear variations in illuminations are commonplace occurrences, especially in the under-saturation and over-saturation regions.

Gradient-based methods have proposed effective strategies to handle many of these challenges. The popular SIFT [1] algorithm captures the local gradient distributions around the keypoints. Bay et al. [2] propose a faster variant of SIFT called as *SURF*, by computing Haar-wavelet responses using integral images. It is also compact (64 dimensions) and uses the sign of the Laplacian to perform faster indexing. The *GLOH* descriptor [3] improves the robustness and distinctiveness of *SIFT*. It divides the region into a log-polar network of 17 spatial bins, on each of which is a 16 dimensional orientation histogram built. PCA is used to reduce the 272 dimensions to 128 which are used in matching. Ke and Sukthankar [4] propose a dimensionally reduced descriptor *PCA-SIFT* by vectorizing the x and y gradients of the pixels of the normalized patch and linearly projecting the vectors onto a much lower-dimensional ($\sim$30) eigen-space. They argue that an eigen-projection is sufficient to model the variations in the 3D-scene and viewpoints, although the evaluation in [3] shows other descriptors to perform better. Shape Context [5] is another method that bins the orientations of pixels into a log-polar grid. Although the authors applied it only for edge point locations and not orientations, it can be used as a region descriptor as well [3]. Apart from these, there are also other modifications of gradient histograms such as those in [6–8].

Order-based descriptors that are constructed based on the sorting of pixels are an alternative strategy to gradient-based descriptors. Zabih and Woodfill [9] proposed two techniques - *rank* and *census transforms* - that are based on the order of intensities of neighbors of a pixel and the count of flipped point-pairs. Such order-based methods are inherently invariant to monotonic changes in illumination. However, they fail in the presence of pixel noise as a single salt-and-pepper flip can change the counts, which is alleviated to a certain extent by Bhat and Nayar [10]. Mittal and Ramesh [11] improve the latter by penalizing an order-flip in proportion to the change in the intensities of the pixels that underwent the flip. This helps to prevent the movement of pixels due to Gaussian noise. Tang et al. [12] propose the *OSID* descriptor that builds a histogram of orders computed on the entire patch. Though invariant to monotonic illumination variations, it can fail on a patch having many pixels of similar intensities as these tend to shift under Gaussian noise. Gupta and Mittal [13] alleviate this problem by designing a histogram of relative intensities whose bins are adaptively designed for the saturated and the non-saturated regions. Wang et al. [14] improve upon this in their *LIOP* descriptor by inducing rotation invariance to it. The motivation is based on their study [15] that identifies estimation of keypoint orientation as a major source of localization error.

There are other variants of order-based descriptors that are bit-strings of comparisons of pixels. These are attractive because of their minimal storage requirements and their ability to be compared fast. *Local Binary Patterns (LBP)*

[16], first applied for face recognition and texture classification, are formed by the comparison of a pixel with its neighbors and constructing a histogram of these patterns. Since these patterns are rather high-dimensional, variants such as [13,17] compare only certain pixels in the neighborhood without sacrificing the discriminative ability of the *LBP* patterns. Calonder et al. [18] propose the *BRIEF* descriptor that randomly samples 128 or 256 pixel-pairs from the smoothed patch and forms a bit-string based on the outputs of their comparisons. The bit-string turns out to be, surprisingly, discriminative. Because of the manner in which it is constructed, *BRIEF* is not rotation-invariant and Rublee et al. [19] propose the *ORB* descriptor that makes *BRIEF* rotation invariant. Leutenegger et al. [20] design a variant of *BRIEF* called as *BRISK* [20] that is formed by the comparisons of pixels placed uniformly on concentric circles. The region is rotation-normalized according to the orientation estimated from the pixels on the circles. To avoid aliasing while sampling points from the circles, each point is smoothed by a Gaussian window of width that is sufficient to not distort the information content of close-by points. They also propose a fast keypoint detector. The *FREAK* descriptor by Alahi et al. [21], is another binary descriptor that compares intensities of pixels sampled in a pattern as observed in the human retinal system. They also outline the reason behind why such comparison-based binary descriptors work, based on studies of the human visual system.

Mikolajczyk and Schmid [3] provide an extensive comparison of many keypoint descriptors including *SIFT, SURF, Shape-Context, SIFT-PCA, GLOH, Cross-correlation and Steerable Filters* and observe that, although *SIFT* performs well in many scenarios, there is no one particular descriptor which works for all cases. A comparison of the modern descriptors has been made independently by Miksik and Mikolajczyk [22] and Heinly et al. [23].

In this paper, we aim to study the performance of 4 descriptors - *SIFT, LIOP, HRI* and *HRI-CSLTP* for matching keypoints in the applications of Stitching of images of architectural scenes. Such images are characterized by well-structured and textured monuments that can be varying in depth, may have large areas of homogeneous regions especially when shot for a panoramic mosaic and can have varying illumination levels. Accordingly, we test these descriptors on 4 kinds of images from a dataset of archaeological sites and historical monuments: (1) well-structured with sufficient depth variation (2) partly structured and partly homogeneous (3) nearly homogeneous with a few structured regions and (4) illumination change on a dataset. We aim to study the scope of application of these descriptors by testing them on the said challenges. To that end, we plot their response graphs for matching, compare their performance and draw conclusions therefrom.

The paper is organized as follows. Section 2 briefly discusses the challenges that are usually posed by architectural scenes with visual examples. An overview of the descriptors that are tested in the paper is given in Sect. 3. Section 4 discusses the dataset, the groundtruthing technique, the evaluation methodology adopted to test the descriptors. The experimental results are presented along with their analyses in Sect. 5. Section 6 concludes the paper with the lessons drawn from the experiments.

2   Architectural Scenes and Their Challenges

Figure 1 shows some images from a typical dataset of archaeological sites and historical monuments. Such monuments are usually structured[1] with repeated occurrences of textured regions (col. 1 of Fig. 1) at varying levels of depths (cols. 2 and 3 of Fig. 1). These images may also include large homogeneous regions, especially when shot for a panoramic mosaic or 3-D reconstruction, with a vacant landscape in the front or sky in the back (col. 4 of Fig. 1). Homogeneous regions are poor conveyors of distinctive visual information. So, when large areas of images are covered by homogeneous regions, it becomes important to match the available keypoints from the non-homogeneous regions in a reliable and correct manner, and discard as many pseudo-matches as possible. The descriptors have to be highly distinctive to suit this requirement. Further, the lighting conditions and the time of the day when the images are shot govern the intensities of the pixels and can make them vary in a non-linear way (col. 5 of Fig. 1), especially in under-exposed or over-exposed regions (for instance, interior structures that are poorly lit). The descriptors need to be resilient to these changes in intensities by adopting a generic normalization technique.

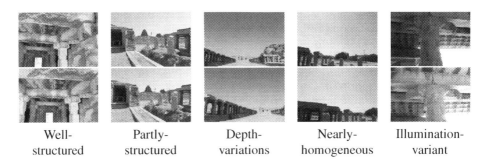

| Well-
structured | Partly-
structured | Depth-
variations | Nearly-
homogeneous | Illumination-
variant |

Fig. 1. Challenges that usually beset a feature matcher.

In the next section, we present a brief overview of the 4 descriptors - *SIFT, HRI, HRI-CSLTP, LIOP* - that are tested in these challenges. While *SIFT* [1] is well-known, *HRI-CSLTP* [13] and *LIOP* [14] are recent order-based descriptors that have performed well on the standard datasets [13, 14, 22].

---

[1] A region of image is well-structured when it is characterized by regular occurrences of homogeneous or textured patches that are flanked by well-defined object gradients. A typical example is that of a building, as opposed to an image of a scenery.

3 Overview of the Descriptors

SIFT descriptors [1] capture the local distribution of the gradients in the patches around the keypoints by tri-linearly binning the gradient magnitudes of the pixels into 8 orientation bins. To make the descriptor robust to small pixel-movements, a patch is divided into 4×4 spatial-grids over which the orientation histograms are built which are then concatenated to form the descriptor of the patch. Robustness to spikes in gradient-magnitudes is handled by capping the gradient-magnitudes to be a maximum of 0.2 and $l2$ normalization of the descriptors make brings resilience to linear changes in illumination. Each patch yields a 128 dimensional, real-valued descriptor.

HRI descriptors [13] capture the relative orders of the pixels of the patch based on their intensities. Orders have the natural ability to be invariant to monotonic changes in illumination. In a *HRI* descriptor, pixels bin their intensities into intervals that are designed based on the intensity distribution of the overall patch. Linear normalization of intensities yields illumination invariance, wherein the min and the max points of the normalization are adaptively chosen for the saturated and the non-saturated regions[2]. Gaussian pixel noise is handled by a uniform distribution of the intensities into the intervals, and trilinear interpolation and spatial-division of the patch into grids handle small pixel movements. It is to be noted that gradient information is not used, in contrast to SIFT [1].

CSLTP descriptors [13] look at the intensity differences of the diagonal neighbors of each pixel and encodes them using 3 categories based on a threshold parameter, T; two of the categories identify differences of opposing contrast, $|i_1 - i_2| > T$, while the third identifies pixels of nearly equal intensities, $|i_1 - i_2| \leq T$. T helps to choose a certain amount of separation between the diagonal pairs. With 2 diagonal pairs, each being encoded with 3 patterns, there are totally 9 different neighborhood patterns which can be treated as the 9 bins of the *CSLTP* histogram. Based on its pattern, each pixel contributes a weighted vote to one of the 9 bins. The weight is designed to eliminate a pixel if it has nearly homogeneous neighbors and, thereby, prevent its movement. The patch is divided into 4×4 grids to counter small spatial errors and the *CSLTP* histograms of the grid are concatenated to yield the *CSLTP* descriptor of the patch.

LIOP descriptors [14] are designed to be rotation and monotonic-illumination invariant by using the order of the intensities of the pixels. The local intensity order pattern of a pixel is a weighted vector that encodes the ranking of its 4 neighbors. The neighbors are sampled from a circular neighborhood in a rotation-invariant manner to avoid the errors in estimation of keypoint orientation [15]. Gaussian noise is handled by giving more weights to the patterns that result from neighbors differing in their intensities by a certain threshold. In addition to the local patterns, the patch is intensity-thresholded using multiple values to yield regions of similar intensities, called as ordinal bins. The *LIOP* pattern of an ordinal bin is the weighted summation of those of its pixels; these *LIOP* patterns

---

[2] A region is saturated if its pixels have intensities either below 10 or above 245.

are concatenated in the order of the ordinal bins resulting in a rotation-invariant *LIOP* descriptor of the patch.

4   Dataset and Evalution Criterion

We evaluate the descriptors on an architectural dataset which contains images of many archaeological monuments and historical sites. The images, ∼50K in all, have been shot in two resolutions (1280 × 960 and 3648 × 2736) and are categorized according to varying details of the structures of the sites and thus, made suitable, for different tasks such as panorama stitching and 3D-reconstruction.

For testing the descriptors on image registration for Mosaicking, images shot with the panoramic constraints[3] have been chosen. Following are the challenges based on the nature of the scene that have been used to test the descriptors: (1) well-structured with sufficient depth variations (2) partly structured and partly homogeneous (3) nearly homogeneous with a few structured regions and (4) illumination changes. Estimation of homography for a pair of images is done with the manual input of 4 point correspondences.

We use the evaluation criterion proposed by Mikolajczyk and Schmid [3] that identifies the correct and the false descriptor matches using ground truth correspondences at a particular region overlap error (50 % in our experiments), as defined by Mikolajczyk et al. [24]. The descriptor matches are obtained using the ratio-test proposed by Lowe [1], the threshold for which is varied to obtain the points on the Precision-Recall response graphs. The correspondences of the regions for a particular overlap error (50 %) and the validation of the descriptor matches have been computed using the code available at the Affine Covariant Features page[4].

DoG keypoints [1] are detected using the co-variant feature detector routine in the *VL-FEAT* library [25]. The minimum absolute value of the cornerness measure is empirically set to 3 for all the experiments. For the *SIFT* and the *LIOP* descriptors, the implementations in the *VL-FEAT* library are used. *HRI* and *HRI-CSLTP* have been implemented by us.

5   Performance Evaluation

5.1   Images with Illumination Variations

Images taken in an uncontrolled environment such as archaeological sites exhibit wide variety of intensity ranges depending on the ambient light which need not illumine the objects in the scene uniformly, especially the interior parts of structures and can thus, result in under-saturated or over-saturated regions. Such variations in the intensities are usually non-linear and hence, the descriptors

---

[3] A set of images is suitable for panoramic stitching if all of them depict a planar scene or are shot with the camera center being fixed.

[4] http://www.robots.ox.ac.uk/~vgg/research/affine/desc_evaluation.html#code.

have to deal with an appropriate normalization scheme. Figure 2 shows the performance of the descriptors on images that vary in their illumination patterns.
These are usually indoors where the natural light doesn't reach all portions of the
scene uniformly. The recall rate is generally low as it is 30 % when the precision
is ∼30 % for the best performer(s), except in Fig. 2(b) which might be due to
the good matches from the well-lit outdoor structures. *SIFT* seems to be doing
consistently well, although *LIOP* is not far behind. Though *HRI-CSLTP* and
HRI use adaptive binning, the changes in these images might be very non-linear
for these methods to perform well.

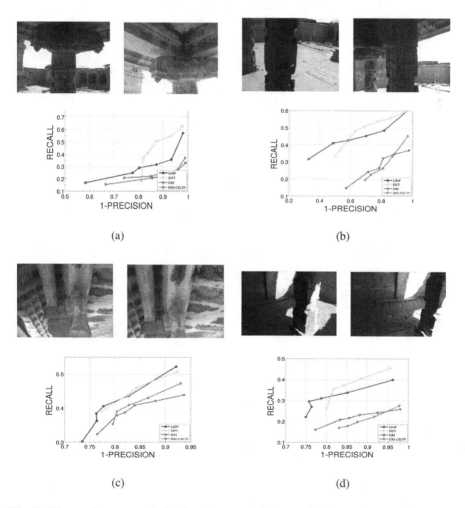

Fig. 2. The performance of the descriptors on images with intensity variations. The
ranges of the plots have been set different for the sake of clarity.

5.2   Structured Images

Figure 3 shows the performance of the descriptors for images that are well-structured with some depth variations and nearly well-lit light conditions. The aim here is to study if the descriptors can match the keypoints output by the detector when they vary in their texture content due to depth and viewpoint changes. *SIFT* and *HRI-CSLTP* perform consistently well in all the 4 cases. The additional edge direction information in *HRI-CSLTP* definitely helps it score better than *HRI*, although the marginal differences in their performances might suggest that *CSLTP* may have to be combined with other descriptors as it captures directional information only in 4 orientations.

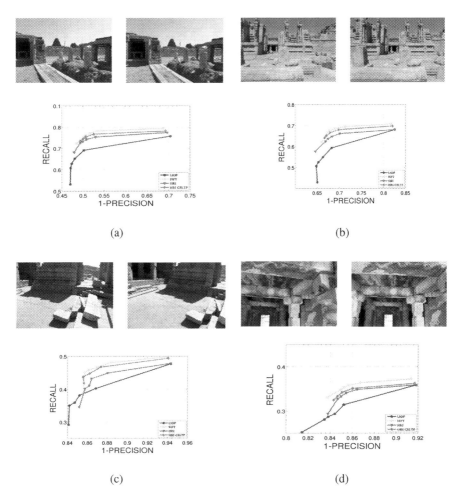

Fig. 3. The performance of the descriptors on well-structured images with some depth variations. The ranges of the plots have been set different for the sake of clarity.

5.3   Partially Homogeneous Images

Figure 4 shows the performance of the descriptors for images that are partially homogeneous containing large depth variations. Such images are usually captured to get a profile of the entire scene when it contains objects that vary significantly in their depths (e.g. a long wall flanked by a bare landscape on its side). For matching, the descriptors have to rely on the keypoints generated from the structured regions of the images. We find that *SIFT* and *HRI-CSLTP* perform well with the differences being very marginal in both the test cases. The orders of the pixels considered in *LIOP* can become noisy in homogeneous regions and that may explain the nature of its performance in these cases.

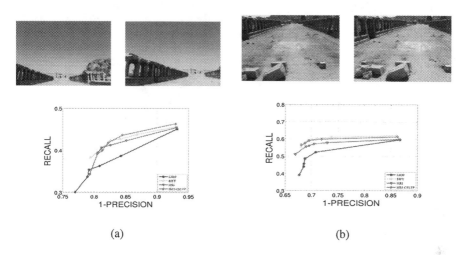

Fig. 4. The performance of the descriptors on partially homogeneous images with significant depth variations. The ranges of the plots have been set different for the sake of clarity.

5.4   Nearly Homogeneous Images

Figure 5 shows the performance of the descriptors for images that are nearly homogeneous with very little amount of structures in them. Such images are usually captured in a panoramic shot of an architectural monument that has a nearly empty landscape in the front. The low ranges of precision in Fig. 5 can be explained by the fact that nearly homogeneous regions tend to result in large number of false matches. The trend exhibited by the descriptors is the same as in the previous 2 challenges. Though the order patterns used in *LIOP* are weighted, the results suggest that the weighting might not be sufficient when there are large areas of homogeneous regions.

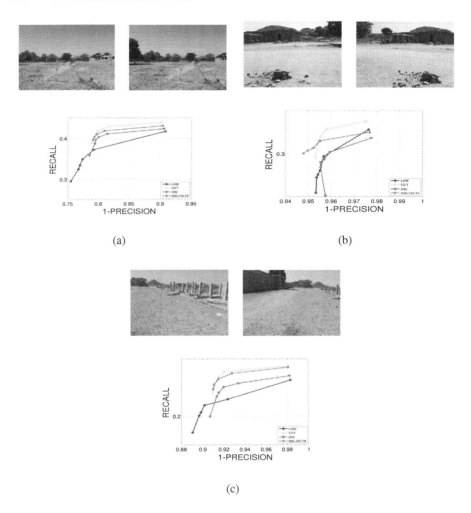

(a)   (b)

(c)

Fig. 5. The performance of the descriptors on nearly homogeneous images with very little structures. The ranges of the plots have been set different for the sake of clarity.

6   Conclusions

We presented a performance evaluation of 4 feature descriptors for the task of feature matching in Image Stitching when the images are of archaeological scenes and architectural sites. As these images are characterized by structures that vary in their textural content and depth and homogeneous regions, we categorized the dataset into 4 classes and tested the descriptors on them. *SIFT* and *HRI-CSLTP* perform better than the others in many of the test cases highlighting their distinctiveness in representing the keypoint regions. *LIOP* performs well when the intensity variations are complex. Also, the results of *LIOP* show that the order computations have to be done in a noise-resilient manner, especially

when homogeneous regions are involved. This performance evaluation can be
extended to other applications like 3-D reconstruction to understand the scope
of applicability osf these descriptors.

References

1. Lowe, D.G.: Distinctive image features from scale-invariant keypoints. Int. J. Comput. Vision **60**, 91–110 (2004)
2. Bay, H., Tuytelaars, T., Van Gool, L.: SURF: speeded up robust features. In: Leonardis, A., Bischof, H., Pinz, A. (eds.) ECCV 2006, Part I. LNCS, vol. 3951, pp. 404–417. Springer, Heidelberg (2006)
3. Mikolajczyk, K., Schmid, C.: A performance evaluation of local descriptors. IEEE Trans. Pattern Anal. Mach. Intell. **27**, 1615–1630 (2005)
4. Ke, Y., Sukthankar, R.: PCA-SIFT: a more distinctive representation for local image descriptors. In: The Proceedings of the IEEE Conference on Computer Vision and Pattern Recognition, vol. 2, pp. II-506–II-513 (2004)
5. Mori, G., Belongie, S., Malik, J.: Efficient shape matching using shape contexts. IEEE Trans. Pattern Anal. Mach. Intell. **27**, 1832–1837 (2005)
6. Lazebnik, S., Schmid, C., Ponce, J.: A sparse texture representation using local affine regions. IEEE Trans. Pattern Anal. Mach. Intell. **27**, 1265–1278 (2005)
7. Mikolajczyk, K., Matas, J.: Improving descriptors for fast tree matching by optimal linear projection. In: The Proceedings of the Eleventh IEEE International Conference on Computer Vision, pp. 1–8 (2007)
8. Freeman, W., Adelson, E.: The design and use of steerable filters. IEEE Trans. Pattern Anal. Mach. Intell. **13**, 891–906 (1991)
9. Zabih, R., Woodfill, J.: Non-parametric local transforms for computing visual correspondence. In: Eklundh, J.O. (ed.) ECCV 1994. Lecture Notes in Computer Science, vol. 801, pp. 151–158. Springer, Berlin Heidelberg (1994)
10. Bhat, D.N., Nayar, S.K.: Ordinal measures for image correspondence. IEEE Trans. Pattern Anal. Mach. Intell. **20**, 415–423 (1998)
11. Mittal, A., Ramesh, V.: An intensity-augmented ordinal measure for visual correspondence. In: The Proceedings of the IEEE Conference on Computer Vision and Pattern Recognition, vol. 1, pp. 849–856 (2006)
12. Tang, F., Lim, S.H., Chang, N., Tao, H.: A novel feature descriptor invariant to complex brightness changes. In: The Proceedings of the IEEE Conference on Computer Vision and Pattern Recognition, pp. 2631–2638 (2009)
13. Gupta, R., Patil, H., Mittal, A.: Robust order-based methods for feature description. In: The Proceedings of the IEEE Conference on Computer Vision and Pattern Recognition, pp. 334–351 (2010)
14. Wang, Z., Fan, B., Wu, F.: Local intensity order pattern for feature description. In: The Proceedings of the Thirteenth IEEE International Conference on Computer Vision, pp. 603–610 (2011)
15. Fan, B., Wu, F., Hu, Z.: Rotationally invariant descriptors using intensity order pooling. IEEE Trans. Pattern Anal. Mach. Intell. **34**, 2031–2045 (2012)
16. Ojala, T., Pietikainen, M., Maenpaa, T.: Multiresolution gray-scale and rotation invariant texture classification with local binary patterns. IEEE Trans. Pattern Anal. Mach. Intell. **24**, 971–987 (2002)
17. Heikkilä, M., Pietikäinen, M., Schmid, C.: Description of interest regions with local binary patterns. Pattern Recogn. **42**, 425–436 (2009)

18. Calonder, M., Lepetit, V., Ozuysal, M., Trzcinski, T., Strecha, C., Fua, P.: Brief: computing a local binary descriptor very fast. IEEE Trans. Pattern Anal. Mach. Intell. **34**, 1281–1298 (2012)
19. Rublee, E., Rabaud, V., Konolige, K., Bradski, G.: ORB: an efficient alternative to SIFT or SURF. In: The Proceedings of the Thirteenth IEEE International Conference on Computer Vision, pp. 2564–2571. IEEE (2011)
20. Leutenegger, S., Chli, M., Siegwart, R.: BRISK: binary robust invariant scalable keypoints. In: The Proceedings of the Thirteenth IEEE International Conference on Computer Vision (2011)
21. Alahi, A., Ortiz, R., Vandergheynst, P.: Freak: fast retina keypoint. In: The Proceedings of the IEEE Conference on Computer Vision and Pattern Recognition, pp. 510–517 (2012)
22. Miksik, O., Mikolajczyk, K.: Evaluation of local detectors and descriptors for fast feature matching. In: The 21st International Conference on Pattern Recognition, pp. 2681–2684 (2012)
23. Heinly, J., Dunn, E., Frahm, J.-M.: Comparative evaluation of binary features. In: Fitzgibbon, A., Lazebnik, S., Perona, P., Sato, Y., Schmid, C. (eds.) ECCV 2012, Part II. LNCS, vol. 7573, pp. 759–773. Springer, Heidelberg (2012)
24. Mikolajczyk, K., Tuytelaars, T., Schmid, C., Zisserman, A., Matas, J., Schaffalitzky, F., Kadir, T., Gool, L.: A comparison of affine region detectors. Int. J. Comput. Vision **65**, 43–72 (2005)
25. Vedaldi, A., Fulkerson, B.: VLFeat : an open and portable library of computer vision algorithms (2008). http://www.vlfeat.org/

Enhancement and Retrieval of Historic Inscription Images

S. Indu[1]([✉]), Ayush Tomar[1], Aman Raj[1], and Santanu Chaudhury[2]

[1] Delhi Technological University, New Delhi, India
s.indu@dce.ac.in, ayushtomar@gmail.com, amanraj9992@hotmail.com
[2] Indian Institute of Technology, New Delhi, India
schaudhury@gmail.com

Abstract. In this paper we have presented a technique for enhancement and retrieval of historic inscription images. Inscription images in general have no distinction between the text layer and background layer due to absence of color difference and possess highly correlated signals and noise; pertaining to which retrieval of such images using search based on feature matching returns inaccurate results. Hence, there is a need to first enhance the readability and then binarize the images to create a digital database for retrieval. Our technique provides a suitable method for the same, by separating the text layer from the non-text layer using the proposed cumulants based Blind Source Extraction(BSE) method, and store them in a digital library with their corresponding historic information. These images are retrieved from database using image search based on Bag-of-Words(BoW) method.

1 Introduction

Enhancement, binarization and retrieval of historic inscription images present one of the many challenging issues in image analysis and preservation of digital heritage. Historic inscriptions are an outlook of the past and an essential part of social, economical and scientific studies. However, pertaining to various factors such as environmental change and human intervention, the quality of such inscriptions degrade with the passage of time. Therefore, there is a need to digitize these inscriptions in form of images, which should be free of any noise and unwanted background information and further, store such images in a digital library so that their retrieval with related information could be performed efficiently in real time.

Significant amount of work has been done in the field of digitization of text document images [1,2]. Many of these works are dependent on background light intensity normalization [3] and exploitation of edge information [4]. Generally, the inscriptions are found engraved or projected out of stones or any durable material. Complexity of the background, uncontrolled illumination and minimal difference between foreground (text) and background in camera-held images of such inscriptions pose a challenging problem for text extraction. Direct use of binarization techniques such as Otsu's method do not give good qualitative

© Springer International Publishing Switzerland 2015
C.V. Jawahar and S. Shan (Eds.): ACCV 2014 Workshops, Part II, LNCS 9009, pp. 529–541, 2015.
DOI: 10.1007/978-3-319-16631-5_39

results due to highly correlated noise present in the images which is not eliminated.

The use of Blind Source Extraction (BSE) and Independent Component Analysis (ICA) [5] is prevalent in the field of digital signal processing; very few applications of the same are found in the field of preservation of our heritage digitally. There are a few ICA-based techniques that have been used for enhancement and digitization of historical inscription images by maximizing text layer information. In the approach as suggested in [6], use of Natural Gradient based flexible ICA (NGFICA) is proposed which minimizes the dependency among the source signals to compute the independent components by considering the slope at each point in the source signals. The authors of [7] suggest a Fast-ICA based method, which aims to enhance the OCR readability of camera-held inscription images.

Image retrieval has been an active field of research. Various techniques involving comparison using local features such as color [8], shape [9] and various other low level features [10] have been proposed but these techniques fail to provide good results under addition of correlated noise or change in scale. There are certain techniques involving comparison of scale invariant features such as SIFT [11] but directly comparing such features is computationally expensive and not realizable in a real time retrieval system as the size of dataset grows. Bag-of-Words (BoW) representation of a images for comparison has been a popular method and has shown excellent results in retrieval of word images [12] and videos [13]. BoW model has been very much suitable for image retrieval from a large database in real time [12].

In our proposed technique, firstly we calculate three independent components i.e. foreground (text-layer), background and middle layer from a linear mixture of unknown sources using cumulant based Blind Source Extraction. After the extraction of three layers, the text information contained in the foreground layer is extracted by further processing which involves local thresholding through Otsu's method [14], morphological operations and median filter for smoothening purpose and finally a binary image is obtained which is free from any noise or unwanted background information and contain maximum text content. Secondly, these binary images are stored in a database and BoW descriptors based on SIFT features [11] of each image are computed and stored, which are used later for image retrieval from the database that are same or similar to the query image.

2   Methodology

2.1   Blind Source Extraction (BSE)

The ICA based techniques as proposed in [15,16] has the potential to become computationally very expensive when the number of source signals is at large, let's say an order of 150 or more. Simultaneous Blind Source Extraction (BSE), overcomes this problem by providing a provision and flexibility to extract the desired number of independent components from a set of linear mixtures of large

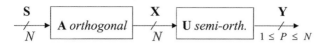

Fig. 1. Signal model of BSE taken from [18]

number of statistically independent source signals. The approach is to use a contrast function to handle the third and fourth order cumulants simultaneously to reduce the computational time overhead.

BSE [15,17] can be explained in a simplest possible way as follows. Let's consider some random sources $s_1, s_2, ..., s_N$ generated by a random process forms the source matrix $\mathbf{S} = [s_1, s_2, s_3, ..., s_N]^T$ as shown in Fig. 1. It is assumed that sources are non-Gaussian with zero mean and have statistical independency. These sources are linearly mixed in a memory-less system represented by a mixing matrix $\mathbf{A}[N\ X\ N]$ such that:

$$\mathbf{X} = \mathbf{AS} \qquad (1)$$

where $\mathbf{X} = [x_1, x_2, x_3, ..., x_N]^T$ are the obtained linear mixtures. It is considered that the mixing matrix $\mathbf{A}$ is orthogonal and non-singular without much loss of generality. In order to extract $P\,(where\,1 \leq P \leq X)$ number of sources from this mixture matrix $\mathbf{X}$, the observations are processed in a linear and memory-less system which is described by a semi-orthogonal de-mixing matrix $\mathbf{U}[P\ X\ N]$, such that output of system gives:

$$\mathbf{Y} = \mathbf{UX} \qquad (2)$$

where $\mathbf{Y} = [y_1, y_2, y_3, ..., y_P]^T$, is the matrix containing the extracted independent components $y_1, y_2, y_3, ..., y_P$ as specified by the user. The semi-orthogonality of matrix $\mathbf{U}$ is important consideration for having a spatial decorrelation in the outputs as in [17].

2.2 Proposed Method: Enhancement and Binarization of Inscription Images Using Cumulants Based BSE

The proposed approach uses a contrast functional that captures higher order cumulants, which is maximized by the blind source extraction procedure to calculate the independent components (ICs). The final binarized image is obtained on further analysis of the ICs.

2.2.1 Mixture Acquisition from Various Source Observations

We consider the text containing inscription image as mixture of sources i.e. text layer, semi-text layer and non-text layer. Our goal is to separate the text-layer from the non-text parts using the suggested Simultaneous Blind Source Extraction method based on higher order cumulants. We have refined our proposed technique by using HSV color space rather than the RGB (as used in NGFICA [6]

Fig. 2. Inscription image taken from mudgal fort gateway Karnatka

Fig. 3. Hue (H) component of original image

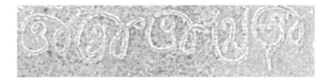

Fig. 4. Saturation (S) component of original image

and Fast-ICA [7] based papers) to obtain three samples of observation from the original image.It is considered that the color of each individual source is uniform in the inscription image. The obtained HSV components of Fig. 2 are shown in Figs. 3, 4 and 5, which constitute $\mathbf{X}$ as per Eq. (2). As proposed in [19], HSV without filtering is better than RGB for ICA algorithms. Data distribution pattern in a HSV model possess strong deviation of the data which makes it an excellent candidate for independent component analysis and the observed results are more comprehensive than RGB model.

2.2.2   Extraction of Independent Components

To find the independent components, Huber [20] suggested finding a contrast functional $\psi(.)$ that maps each p.d.f. of a normalized random variable Y_i to real index $\psi(Y_i)$, where $\mathbf{Y}$ is the vector of the output of estimated sources. Proper optimization of $\psi(Y_i)$ will lead to extraction of one of the independent components.

In general, the blind source separation of the whole set of sources is obtained by maximizing the function below:

$$\max_{U} \sum_{i=1}^{N} \psi(Y_i) \text{ subjected to } Cov(\mathbf{Y}) = I_N \qquad (3)$$

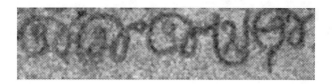

Fig. 5. Value (V) component of original image

I_N is identity matrix with order N (number of mixed sources) and **Y** is the vector of estimated sources [17].

In this paper, independent sources are calculated subject to a cumulants based index, which is a measure of the degree on non-gaussianity or the amount of structure present in the outputs. These indexes have their maxima at the extraction of one of the independent sources. The contrast function is a mathematical function which simultaneously handles these high order cumulants [21]. Thus blind extraction of one of the sources is obtained after solving the following maximization problem:

$$\max_{U} \ \psi(Y_1) \ subjected \ to \ Cov(Y_1) = 1 \tag{4}$$

One generalized form of cumulant index for a random variable is given by [17]:

$$\psi_C um(Y_1) \ = \sum_{r>2} \omega_r^\backslash \centerdot \mid C_{Y_1}^r \mid^{\alpha_r} \tag{5}$$

where r is order of cumulants and $\alpha_r \geq 1$, $\mid C_{Y_1}^r \mid$ denotes modulo of *rth* order auto-cumulant, $Cum(Y_1 \ X \ r)$ and $\omega_r^\backslash = \frac{\omega_r}{r\alpha_r}$ are scaled or normalized as non-negative weighting factors. The low order cumulants having $r=1$ and $r=2$ are excluded from indexing due to normalization constraint.

In case of blind source extraction, we express the Eq. (3) in terms of cumulant index to calculate P out of N total sources, the corresponding cumulant contrast function with largest cumulant index is given by:

$$\psi_C um(\mathbf{Y}) \ = \sum_{i=1}^{P} \sum_{r>2} \omega_r^\backslash \centerdot \mid C_{Y_1}^r \mid^{\alpha_r} \ subject \ to \ Cov(\mathbf{Y}) = I_P \tag{6}$$

The global maxima of this contrast function correspond to the extraction of first P sources from the mixture. Here, in our case $P=N=3$. This way, based on the cumulants index, three estimated Sources (ICs) are calculated from the observed mixtures of the sources (H,S,V components of the original inscription image) as shown in Figs. 6, 7 and 8.

2.2.3   Enhancement of the Text Layer

The estimated sources(ICs) are labeled text, semi-text and non-text layer as per their resemblance with the original text on the inscription image. The text

Fig. 6. Text layer after execution of algorithm

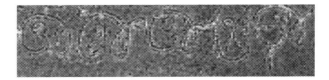

Fig. 7. Semi-text layer after execution of algorithm

Fig. 8. Non-text layer after execution of algorithm

Fig. 9. Final digitized image

layer is selected and is used for further processing. This layer is then binarized calculating a suitable local threshold level as per Otsu's method proposed in [14]. Further the post-processing morphological operations dilation and erosion are used to enhance the readability of text in the text layer followed by a suitable median filter. The final digitized binary image is shown in Fig. 9. Our method provides better qualitative results than as proposed in existing fast-ICA [7] based technique in which thresholding is done after combining all the ICs which leads to adding up of noise also.

2.2.4 Storage of Digitized Image and its Related Information in a Database

We apply the suggested technique for enhancement and binarization on every inscription image taken from a site and obtain the final image similar to as shown in Fig. 9. These final images are in binary form and contain maximum text content. All the non-text information such as noise and background is removed so that final images contain maximum text information, in this way retrieval of such images from a digital database is more accurate due to ease in matching of only important features. As the images contain maximum text information in form of a binary image, these can be used for an OCR, if available. Any information related to the binary image such as history, era, location, civilization etc. are also attached with the image and stored in the digital database.

Creating a digital database using binary images having maximum text content has several advantages as only the features having text information remain in the images, which provide much better results while matching and retrieval giving a better matching score as unwanted information such as noise and background are removed. OCR readability of such images is more than an unprocessed inscription image. Moreover, having a digital database of binary images has less storage space overhead than unprocessed historic images.

2.3 Extraction and Preparation of Bag-of-Words Database

Bag-of-Words was originally used in text classification and retrieval which has been to extended to use in retrieval of images and videos [12,13] and has shown excellent results. An image is a set of unordered visual features. A bag of words is a sparse vector which contains occurrence counts of local image features; in other words, BoW is a sparse histogram over a vocabulary where vocabulary is a set of discrete visual features such as Scale Invariant Feature Transform (SIFT) [11] descriptors. SIFT features have a continuous space and are clustered for a fixed number of bags which are trained with clustered feature descriptors using K-means algorithm. Finally an image is represented as a normalized histogram of the clustered SIFT features having vector length k, where k is the vocabulary size; this histogram is BoW descriptor which can be used to match or classify the image. Figure 10 describes the working of BoW.

2.3.1 Obtaining a Set of Bag-of-Words

We take the set of digitized images from the database and obtain SIFT [11] descriptors based on SIFT feature points that are extracted from each of the image in the database. Further, these feature descriptors are clustered using K-means algorithm into a defined number of bags and are trained, thus descretizing the descriptor space [12,13]. These clustered descriptors are bag of features and functions as a vocabulary in later stages of retrieval.

2.3.2 Obtaining BoW Descriptor of Images in the Database

We take all the digitized images from the database and again find their corresponding local SIFT features. SIFT features are chosen being invariant to image scaling, translation, rotation and illumination changes [11]. Based on these feature points we obtain SIFT descriptors. The feature descriptors so obtained are matched with the vocabulary as proposed in Subsect. 2.3.1 to quantize the extracted features by assigning these features to the closest clustered centroid [12]. The images are finally represented as a normalized histogram of quantized

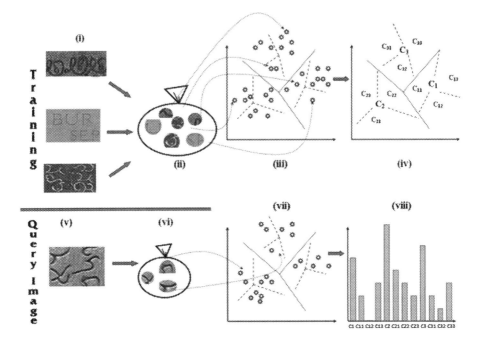

Fig. 10. A Description of working of bag-of-words

SIFT features, which is a count of each word (local visual features) belonging to its bag (clustered features). This histogram is a BoW descriptor, which is computed for each digitized image in the database and stored in a descriptor dictionary (.yml file) on a stable media.

2.4   Retrieval of Image and its Corresponding Information Using BoW Descriptors

In the final step of retrieval, we take a query image which is an image of a historic inscription. The query image is processed using the method as proposed in Subsects. 2.2.1–2.2.3. On the obtained digitized image, BoW descriptors are computed based on SIFT features [11]. BoW descriptor of this image is matched with existing BoW descriptors which are stored in a descriptor dictionary as proposed in the previous subsection. This matching is performed using FLANN (which gives approximate euclidean distance between the descriptors) to compute a dissimilarity score with all the images in the database based on number of "good matches" found. Good matches depend on the euclidean distance between nearest neighbors and represented in terms of dissimilarity score between images. The best image match based on the dissimilarity score was returned and all related information of the matched image in the database was also displayed. Similar images had a dissimilarity score very close to zero.

Table 1. Statistics of result

Query image	Median dissimilarity score	% of Correct Matches
Unprocessed (Raw)	0.231	58 %
Binarized text	0.046	92 %

Query image	Retrived Image	Dissimilarity Score
	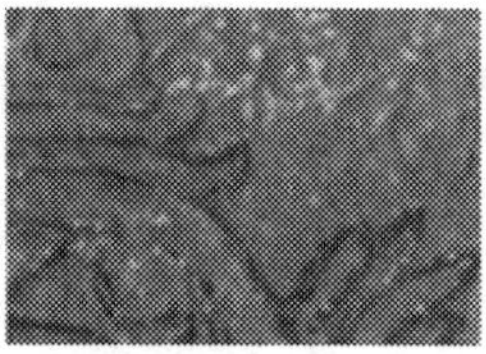	0.27

Fig. 11. Result of an unprocessed image from a unprocessed image database

Query image	Retrived Image	Dissimilarity Score
		0.0192

Fig. 12. Result of a binarized image from a processed image database

3   Results

We implemented the proposed algorithm using the OpenCV library for C++ on Windows platform. The results were obtained using a set of 200 historical inscription images based on a number of query images taking k to be 100, where k is the vector length of clustered SIFT features. The statistics of first fifty query images belonging to each category i.e. unprocessed and binarized text using a raw and processed database respectively are shown in Table 1 with their outputs in Figs. 11 and 12. The system is designed such that it returns a best possible match with input query image based on the least dissimilarity score. The results show the robustness and accuracy of the proposed method in retrieval of corresponding similar images with their attached historic information from the digital library. The results are shown in Figs. 13, 14 and 15, which are accurate even in case of distorted or rotated query images. Figure 16 shows the result when query image is not present in the database, in this case image having the lowest dissimilarity score is retrieved.

Query image	Binarized image	Retrieved image
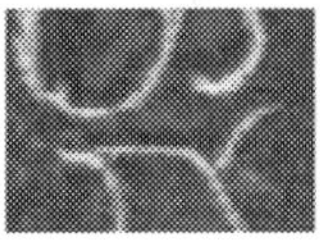		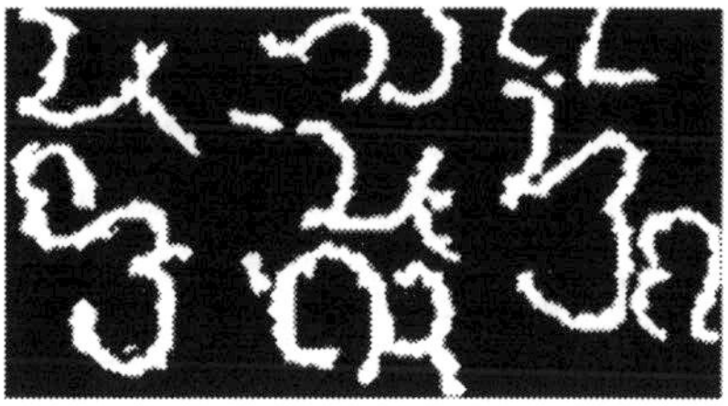
Dissimilarity score		Retrieved information
0.078		Image taken from Badami caves of Karnataka

Fig. 13. Results on query image - I

Query image	Binarized image	Retrieved image
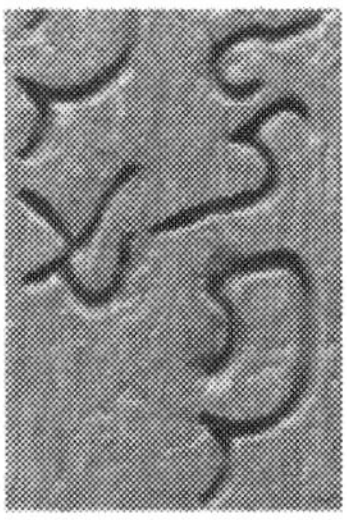	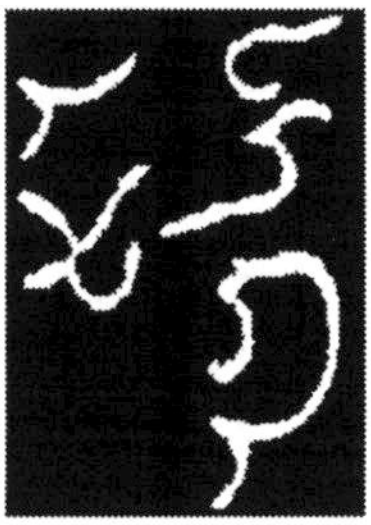	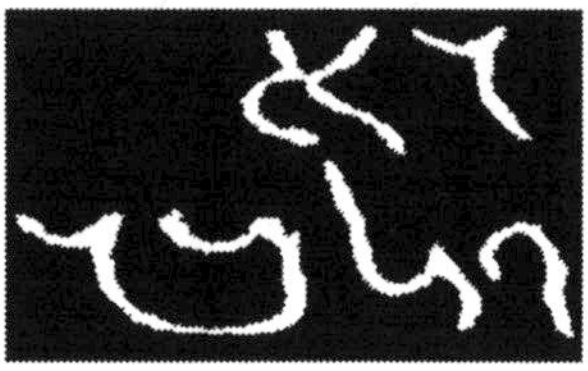
Dissimilarity score		Retrieved information
0.067		Image taken from Edakal caves of Kerela

Fig. 14. Results on query image - II

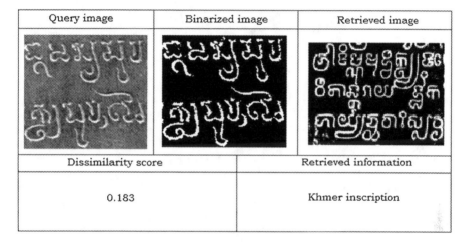

Query image	Binarized image	Retrieved image

Dissimilarity score	Retrieved information
0.038	Image taken from Mudgal fort gateway, Karnataka

Fig. 15. Results on query image - III

Query image	Binarized image	Retrieved image

Dissimilarity score	Retrieved information
0.183	Khmer inscription

Fig. 16. Results on query image - IV

4   Conclusion

In this paper a new robust and efficient technique to enhance, binarize and retrieve historic inscription images and their related information from a database is put forward. The proposed method of enhancement and binarization is found excellent for removal of correlated noise and unwanted background from historic inscription images to extract maximum text content. Further, using a digital library of these images, a suitable method was proposed for retrieval of corresponding inscription image and its related information of historical importance using Bag-of-Words method; the technique was found to be excellent in terms of accuracy of matching based on the obtained results.

5   Future Work

An extension of this technique by using SURF and ORB-Features over SIFT features to check the accuracy and measure the computational overhead is planned for the future work on a larger database. Detailed comparison of proposed enhancement and retrieval techniques with existing techniques is also planned. Further, a web based tool of the technique will be conceptualized in which researchers can add their images and its related information to the existing database to enhance the matching performance of the retrieval process.

References

1. Doermann, D., Liang, J., Li, H.: Progress in camera-based document image analysis. In: Seventh International Conference on Document Analysis and Recognition, Proceedings, pp. 606–616. IEEE (2003)
2. Wolf, C., Jolion, J., Chassaing, F.: Text localization, enhancement and binarization in multimedia documents. In: 16th International Conference on Pattern Recognition, Proceedings, vol. 2, pp. pp. 1037–1040. IEEE (2002)
3. Shi, Z., Govindaraju, V.: Historical document image enhancement using background light intensity normalization. In: Proceedings of the 17th International Conference on Pattern Recognition, ICPR 2004, vol. 1, pp. 473–476. IEEE (2004)
4. Epshtein, B., Ofek, E., Wexler, Y.: Detecting text in natural scenes with stroke width transform. In: 2010 IEEE Conference on Computer Vision and Pattern Recognition (CVPR), pp. 2963–2970. IEEE (2010)
5. Hyvärinen, A., Karhunen, J., Oja, E.: Independent Component Analysis, vol. 46. Wiley, Hoboken (2004)
6. Sreedevi, I., Pandey, R., Jayanthi, N., Bhola, G., Chaudhury, S.: Ngfica based digitization of historic inscription images. Int. Sch. Res. Not. **2013** (2013)
7. Garain, U., Jain, A., Maity, A., Chanda, B.: Machine reading of camera-held low quality text images: an ica-based image enhancement approach for improving ocr accuracy. In: 19th International Conference on Pattern Recognition, ICPR 2008, pp. 1–4. IEEE (2008)
8. Swain, M.J., Ballard, D.H.: Color indexing. Int. J. Comput. Vis. **7**, 11–32 (1991)
9. Rui, Y., She, A.C., Huang, T.S.: Modified fourier descriptors for shape representation-a practical approach. In: Proceedings of First International Workshop on Image Databases and Multi Media Search, pp. 22–23 (1996)
10. Rui, Y., Huang, T.S., Chang, S.F.: Image retrieval: Current techniques, promising directions, and open issues. J. Vis. Commun. Image Represent. **10**, 39–62 (1999)
11. Lowe, D.G.: Distinctive image features from scale-invariant keypoints. Int. J. Comput. Vis. **60**, 91–110 (2004)
12. Shekhar, R., Jawahar, C.: Word image retrieval using bag of visual words. In: 2012 10th IAPR International Workshop on Document Analysis Systems (DAS), pp. 297–301. IEEE (2012)
13. Sivic, J., Zisserman, A.: Video google: a text retrieval approach to object matching in videos. In: Ninth IEEE International Conference on Computer Vision, Proceedings, pp. 1470–1477. IEEE (2003)
14. Otsu, N.: A threshold selection method from gray-level histograms. Automatica **11**, 23–27 (1975)

15. Tonazzini, A., Bedini, L., Salerno, E.: Independent component analysis for document restoration. Doc. Anal. Recognit. **7**, 17–27 (2004)
16. Cichocki, A., Amari, S.I.: Adaptive Blind Signal and Image Processing: Learning Algorithms and Applications, vol. 1. Wiley, Hoboken (2002)
17. Cruces-Alvarez, S.A., Cichocki, A., Amari, S.I.: From blind signal extraction to blind instantaneous signal separation: criteria, algorithms, and stability. IEEE Trans. Neural Netw. **15**, 859–873 (2004)
18. Cruces-Alvarez, S.A., Cichocki, A., Amari, S.I.: On a new blind signal extraction algorithm: different criteria and stability analysis. IEEE Signal Process. Lett. **9**, 233–236 (2002)
19. Katsumata, N., Matsuyama, Y.: Database retrieval for similar images using ica and pca bases. Eng. Appl. Artif. Intell. **18**, 705–717 (2005)
20. Huber, P.J.: Projection pursuit. Ann. Stat. **13**, 435–475 (1985)
21. Blaschke, T., Wiskott, L.: Cubica: Independent component analysis by simultaneous third-and fourth-order cumulant diagonalization. IEEE Trans. Signal Process. **52**, 1250–1256 (2004)

A BRDF Representing Method Based on Gaussian Process

Jianying Hao, Yue Liu, and Dongdong Weng[✉]

School of Optoeletronics, Beijing Institute of Technology,
5 South Zhongguancun Street, Haidian District, Beijing, China
crgj@bit.edu.cn

Abstract. In recent years, digital reconstruction of cultural heritage provides an effective way of protecting historical relics, in which the modeling of surface reflection of historical heritage with high fidelity places a very important role. In this paper Gaussian process (GP) regression based approach is proposed to model the reflection properties of real materials, in which the simulation data generated by the existing model are both used as the training data and the proof that Gaussian process model can be used to describe the material reflection. Matusik's MERL database is also adopted to perform training and inference and obtain the reflection model of the real material. Simulation results show that the proposed GP regression approach can achieve a good fitting of the reflection properties of certain materials, greatly reduce the BRDF measurement time and ensure high realistic rendering at the same time.

1 Introduction

Cultural heritage provide very important physical treasure for studying ancient history, art and development of science and technology. On one hand, the old history cultural heritage is experiencing considerable damage with the passage of time, and needs digital protection, so establishing a digital model of cultural heritage is of great essence. On the other hand, realistic digital display technology is needed to spread the cultural relic's value throughout the world. Precisely relics' digital model should include the original 3D information and correct surface texture information. At present, we can use 3D scanners, three-dimensional modeling software (such as 3DMax, Multi Creator) and many other methods to construct three-dimensional geometric models. But the complexity of reflection phenomenon and the high dimension of reflection data makes the technology of material reflection property acquisition and modeling much more difficult and has been an important research content for a long time.

The light properties on the surface of materials are determined by the interaction between light and objects. The interaction between light and surfaces can be described by a function of 12 dimensions [1]. In practice, for uniform opaque materials, ignoring the space and time changing characteristics, the reflectance function can be simplified as a 4 dimensional bidirectional reflectance distribution function (BRDF) [1]. BRDF describes the appearance of a material by its

ⓒ Springer International Publishing Switzerland 2015
C.V. Jawahar and S. Shan (Eds.): ACCV 2014 Workshops, Part II, LNCS 9009, pp. 542–553, 2015.
DOI: 10.1007/978-3-319-16631-5_40

interaction with light at a surface point and is a function of incident and observation direction vector. The BRDF is denoted symbolically as f, as shown in (1).

$$f(\theta_{in}, \varphi_{in}, \theta_{out}, \varphi_{out}) = \frac{dL_r(\theta_{in}, \varphi_{in}; \theta_{out}, \varphi_{out}; E_{in})}{dE_i(\theta_{in}, \varphi_{out})(sr^{-1})}[sr^{-1}] \qquad (1)$$

A classical BRDF measurement device is gonioreflectometer [1], which samples the angular dependency sequentially by positioning a light source and a detector at various directions. Matusik et al. [2] took images from curved sample to measure isotropic BRDFs and reduced the measuring times significantly. In order to reduce the damage to the cultural relic caused by contacting and being exposed to bright light, measurement times should be reduced. Some researchers also use little measured data to fit the analytical model to represent BRDFs [4,5,7,8,12,13]. But for complex phenomena of reflection, the performance of analysis models with a single lobe is not very good [11]. It is found that there are many extraneous information between BRDFs of different incident and output angles. In this paper, we put forward the method of using Gaussian process, which is based on Bayesian inference method, to learn the relationship between BRDFs of different angles and predict BRDFs of new angle in the whole space. Through the establishment of Gaussian model we can reduce the measurement times dramatically and get more promising rendering performance.

2   Related Work

In order to establish material reflection model and reconstruct the cultural relics with high sense of reality, we studied different existing methods of BRDF modeling. Analytical BRDF models used for rendering can be divided into three categories: empirical (i.e., phenomenological) models, physically-based models and data-driven models. Empirical models such as Phong model [3], the Ward model [4] and anisotropic Phong model [5] etc. focus on using a specific formula to match the surface reflection effect and do not consider the physical mechanism of the light-material interaction explicitly, which makes them concise and the performance becomes idealistic. Torrance and Sparrow [6] supposed that there are many small triangle micro-facets on the material surface and used the micro-facets to describe roughness of the surface. Later, Cook [7] and Blinn [8] improved the Torrance-Sparrow model, and put forward the Cook-Torrance and Blinn model respectively.

Based on optical and electromagnetic wave theory, these physical-based models consider Fresnel effect and the micro-scale geometry of a surface so that they can describe the roughness of real material surface more effectively. Although physically-based BRDF models have a stronger theoretical basis than empirical models, fitting physically-based model parameters to measured BRDF data is not necessarily easier. Also, the surface approximate assumption can't represent the object surface reflection mechanism of different materials exactly, because not all materials meet this relatively simple hypothesis. Analytical models use the sparse sampling of real material acquired by cameras or other optical devices, and fit the mathematical model by using the method of non-linear optimization. Addy Ngan et al. [17] used an existing high-resolution data set of a hundred isotropic materials and computed the best approximation for a variety of analytical model and got the conclusion that in the analysis of the BRDF modeling method, the whole optimization computation is very big, and the calculation result is not very stable.

Another method is to use the measured BRDF data directly in the rendering process. Data-driven method by Matusik and a sampling method by Lawrence used this approach [18, 22]. Since these BRDFs come from the measured data directly, they can preserve the subtleties of the data that may be lost in an analytical model and get very realistic results. The main disadvantage of Matusik et al.'s data-driven model is that each BRDF is stored separately as a tabulated data structure which requires about 17 MB memory. There are also some models combined empirical model or data based analytical model with the surface structure of materials. Marschner [9] and Sadeghi [10] combined the measured BRDF data with the surface structure to build a BRDF model to describe cloth or finished wood. These methods can simulate the appearance of complex highlights and color shifts which cannot be fully handled by pure analytical models. However, the main disadvantage of these models is the need to express the structure of the materials which leads to the description of a specific kind of materials. All the analytical methods use the exact above procedure to represent the material reflection properties and in the process part of the reflection information gets lost.

In recent years, BRDF modeling based machine learning also appeared. Dong Yue [14] proposed manifold bootstrapping for obtaining high-resolution reflectance from sparse captured data to build the BRDF model. Gargan and Neelamkavil [24] presented a model which uses neural networks for approximating reflectance functions. Kurt and Cinsdikici [15] introduced a new BRDF model which uses SOMs and MANs to represent measured BRDF data.

The Gaussian process in machine learning is the generalization of a probability distribution (which describes a finite-dimensional random variables) to functions [16, 20]. It does not give a definite function, but a combination of functions with different weights which can describe more complex information [16]. In addition, computations of Gaussian process required for inference and learning is relatively easy. Over the last decade, theoretical and practical developments have made Gaussian processes a serious competitor for real supervised learning, especially for high dimension and non-linear data. However, we haven't seen anyone apply Gaussian process into material reflection properties modeling in our research. So, in this paper, we introduce Gaussian process model and do the experiments based on Matusik et al.'s measured BRDFs. By using Gaussian process to predict, measurement times of BRDF data can be reduced which is time consuming and causes harm to cultural antiques. Also it can get a good rendering result with small difference compare to the real scene.

3   Representing BRDFs

3.1   Gaussian Process Model

Gaussian process [16] defines a probability distribution (which describes a finite-dimensional random variables) to different functions. The Gaussian process model of BRDF is completely defined by the mean function and covariance function, as shown in formula (2).

$$f(\mathbf{x}) \sim GP\big(m(\mathbf{x}), k(\mathbf{x}, \mathbf{x}')\big) \tag{2}$$

The mean function $m(\mathbf{x})$ and covariance function $k(\mathbf{x}, \mathbf{x}')$ of a process $f(\mathbf{x})$ is defined as:

$$m(\mathbf{x}) = E[f(\mathbf{x})] \tag{3}$$

$$k(\mathbf{x}, \mathbf{x}') = E\big[(f(\mathbf{x}) - m(\mathbf{x}))(f(\mathbf{x}') - m(\mathbf{x}'))\big] \tag{4}$$

Covariance function specifies the covariance between pairs of outputs $f(\mathbf{x}_i)$ and $f(\mathbf{x}_j)$,

$$cov(f(\mathbf{x}_i), f(\mathbf{x}_j)) = k(\mathbf{x}_i, \mathbf{x}_j) \tag{5}$$

The experiments were carried out using different covariance function [16] (see Sect. 4) where relatively simple rational quadratic (RQ) function is chosen as the covariance function. RQ covariance function can be seen as a scale mixture of squared exponential covariance functions (SE) with different length-scales, as shown in Eqs. (6) and (7).

$$k_{SE}(\mathbf{x}_i, \mathbf{x}_j) = \sigma_{sexp}^2 exp\left(-\frac{1}{2}\sum_{k=1}^{d} \frac{(x_{i,k} - x_{j,k})^2}{l_k^2}\right) \tag{6}$$

where l is the characteristic length-scale, σ_{sexp}^2 is signal variance and k is the dimension of input vector.

$$k_{RQ}(\mathbf{x}_i, \mathbf{x}_j) = \left(1 + \frac{1}{2\alpha}\sum_{k=1}^{d} \frac{(x_{i,k} - x_{j,k})^2}{l_k^2}\right)^{-\alpha} \tag{7}$$

with $\alpha, l > 0$ can be seen as a scale mixture of SE covariance functions with different characteristic length scales.

The problem of learning in Gaussian process is exactly the problem of finding suitable properties i.e. parameters for the RQ covariance function. This problem is described in the next two sections of the paper.

3.2   BRDF Prediction

Our task is to map from the input light and observation conditions (given by the angle $\theta_{in}, \varphi_{in}, \theta_{out}, \varphi_{out}$) to the BRDF values. We denote the vector input as $\mathbf{x} = (\theta_{in}, \varphi_{in}, \theta_{out}, \varphi_{out})$ and continuous output (target) as $\mathbf{y}$. Given a training dataset D of n observations, $D = \{X, \mathbf{y}\} = \{(\mathbf{x}_i, \mathbf{y}_i) | i = 1, \ldots, n\}$, we want to make predictions for new input X_* that in the whole space. For a new input points $\mathbf{x}_*$ the output BRDF f_* with the covariance matrix $k(\mathbf{x}_*, \mathbf{x}_*)$ is,

$$f_* \sim N(0, k(\mathbf{x}_*, \mathbf{x}_*)) \tag{8}$$

if there are n training points and n_* test points, $K(X, X_*)$ denotes the $n \times n_*$ matrix of the covariances evaluated at all pairs of training and test data. It is similarly for other matrixes $K(X_*, X)$, $K(X, X)$ and $K(X_*, X_*)$.

In the process of actual measurement the observer cannot access the precise BRDF value because of the measuring conditions, but only noisy observations thereof $\mathbf{y} = f(\theta_{in}, \theta_{out}, \varphi_{in}, \varphi_{out}) + \varepsilon = f(\mathbf{x}) + \varepsilon$. Suppose noise follows an independent, identically distributed Gaussian distribution with zeros mean and variance $\varepsilon \sim N(0, \sigma_n^2)$. So the prior on the BRDF noisy observations is

$$cov(\mathbf{y}) = K(X, X) + \sigma_n^2 I \tag{9}$$

Then the joint distribution of observed BRDF value and function value is

$$\begin{bmatrix} \mathbf{y} \\ f_* \end{bmatrix} = N\left(0, \begin{bmatrix} K(X, X) + \sigma_n^2 I, K(X, X_*) \\ K(X_*, X), K(X_*, X_*) \end{bmatrix}\right) \tag{10}$$

By conditioning the joint Gaussian prior distribution on the training data (observations), the joint prior distribution can be restricted to contain only those functions agree with the observed data points [16].

$$f_* \mid X, \mathbf{y}, X_* \sim N\left(\overline{f}_*, cov(f_*)\right) \tag{11}$$

where,

$$\overline{f}_* \triangleq E[f_* \mid X, \mathbf{y}, X_*] = K(X_*, X)[K(X, X) + \sigma_n^2 I]^{-1} \mathbf{y} \tag{12}$$

$$cov(f_*) = K(X_*, K_*) - K(X_*, X)[K(X, X) + \sigma_n^2 I]^{-1} K(X, X_*) \tag{13}$$

In the formula (11), prediction mean $\overline{f}_*$ is the output of Gaussian regress process i.e. the predicted BRDF value.

3.3   Parameters Training

Rusinkiewicz [19] proposed a BRDF parameterization method which changed the data from the traditional axis $\beta = \beta(\theta_{in}, \theta_{in}, \varphi_{out}, \varphi_{out})$ into new axis $\beta = \beta(\theta_h, \varphi_h, \theta_d, \varphi_d)$. In the new system, isotropic BRDFs are independent of φ_h. The input training data reduced to a three-dimensional vector $\mathbf{x} = (\theta_h, \theta_d, \varphi_d)$ and target $\mathbf{y}$ is the BRDFs associated with the input $\mathbf{x}$. We resampled the MERL dataset as Gaussian training data and did Gaussian process training and inference in new coordinate system.

We use the method of maximizing the marginal likelihood function to determine the hyper-parameters in the covariance, mean function and likelihood function. The marginal likelihood based on output BRDFs value $\mathbf{y}$ is $p(\mathbf{y} \mid X)$.

$$p(\mathbf{y} \mid X) = \int p(\mathbf{y} \mid f, X) p(f \mid X) df \tag{14}$$

In Gaussian process the prior is in line with the Gaussian distribution $f \mid X \sim N(0, K)$, θ is the hyper-parameters, therefore the logarithms of $p(f \mid X, \theta)$ is

$$\log p(f \mid X, \theta) = -\frac{1}{2} f^T (K)^{-1} f - \frac{1}{2} \log |K| - \frac{n}{2} \log(2\pi) \tag{15}$$

And the likelihood is a factorized Gaussian $\mathbf{y} \mid f \sim N(f, \sigma_n^2 I)$, so

$$\log p(\mathbf{y} \mid X, \theta) = -\frac{1}{2} \mathbf{y}^T (K + \sigma_n^2 I)^{-1} \mathbf{y} - \frac{1}{2} \log |K + \sigma_n^2 I| - \frac{n}{2} \log(2\pi) \quad (16)$$

The partial derivatives of the marginal likelihood is

$$\begin{aligned}
\frac{\partial}{\partial \theta_j} \log p(\mathbf{y} \mid X, \theta) &= -\frac{1}{2} \mathbf{y}^T K^{-1} \frac{\partial K}{\partial \theta_j} K^{-1} \mathbf{y} - \frac{1}{2} tr\left(K^{-1} \frac{\partial K}{\partial \theta_j}\right) \\
&= \frac{1}{2} tr\left((\alpha \alpha^T - K^{-1}) \frac{\partial K}{\partial \theta_j}\right)
\end{aligned} \quad (17)$$

where $\alpha = K^{-1}\mathbf{y}$, and θ is the hyper-parameters in the covariance, i.e. length scale l, α and the noise σ_n^2. With maximizing the marginal likelihood function of hyper-parameters, the optimal length scale l, α and the noise σ_n^2. are acquired. After getting the hyper-parameters, the properties of the covariance are determined and the BRDF's predicted value and its variance $\widehat{\sigma_{f_*}}^2$ can be obtained using formula (9).

4 Experiment and Result Evaluation

In order to verify the feasibility of Gaussian process in BRDF prediction, we use the current light model to generate discrete BRDF data, and use GP to fit these examples. Figure 1 shows the BRDF distribution of five different models (empirical model: Phong [3], Blinn Phong [8], anisotropic model Bank Phong [23] and physically-based model: Cook Torrance [7], Blinn [8]) under the fixed light direction and different observation directions. The top row is the ground truth (generated by current models) and the bottom row is the predicted results by GP with the training data shown by red stars.

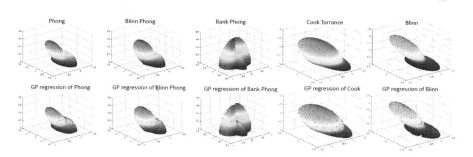

Fig. 1. Comparison between existing models and the GP regression

To evaluate the BRDF value predicted by Gaussian process quantificationally, we use mean absolute error (MAE) and mean-square error (MSE) to measure the difference between predicted BRDF value and the ground-truth,

$$\sigma_{MAE} = \frac{1}{N} \sum_{i=1}^{N} |\overline{\mathbf{y}}(i) - \mathbf{y}(i)| \quad (18)$$

$$\sigma_{MSE} = \frac{1}{N} \sum_{i=1}^{N} \left(\overline{\mathbf{y}}(i) - \mathbf{y}(i) \right)^2 \tag{19}$$

Table 1. Gaussian process prediction error of current models

	Phong	Bank Phong	Blinn Phong	Ward	Blinn
MAE	1.72E-04	0.0021	8.99E-04	1.18E-04	4.31E-04
MSE	0.0017	0.0167	0.0034	0.0037	0.0038

We can see from Fig. 1 and Table 1 that GP can get good fitting and prediction for the current model. Hence, a conclusion can be made that Gaussian process can be used for better BRDF prediction of real materials.

For real materials, we test Gaussian process on MERL [18] material database to test the prediction performance of GP. The measurements of Matusik et al. [18] provide a dense (90*90*180 for values) sampling of many isotropic BRDFs. Every material is described by 1458000 BRDFs in tabular form. It acquires good rendering results compared favorably with real materials, but the main drawback of these representations is that their size is too large. We also made experiments on the sampling rate and error relationships using part of BRDF data to do inference, as shown in Fig. 2. With the increase of sampling rate, the prediction error reduces greatly. Also the corresponding training time will increase significantly. So for each material, we choose 2.58 % (37616 for values) of BRDF data as the training data, use Gaussian process to learn the relationship between BRDFs of different angles, and infer the GP model.

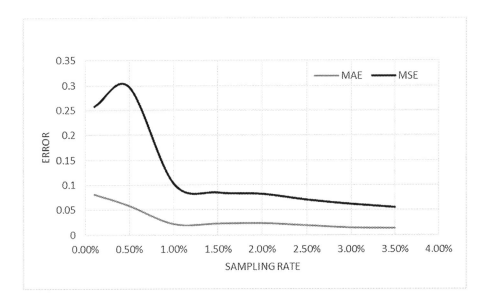

Fig. 2. The prediction error reduces as the increase of sampling rate.

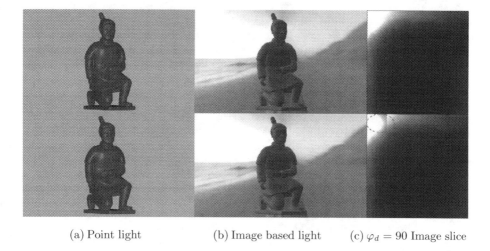

(a) Point light   (b) Image based light   (c) $\varphi_d = 90$ Image slice

Fig. 3. The MERL and Gaussian process prediction rendering results of "Cherry-235". (By using only 2.58 % of the MERL datasets, the proposed work attains almost the same rendering performance.)

The ground truth (rendered using MERL BRDF data directly) and the GP model result of the material "Cherry-235" is shown by Disney's BRDF explorer [21] in Fig. 3. A Terra Cotta Warriors model are lighted by a point light (a) and an environment map (b). The top row is the MERL rendering result and the bottom row shows the result of the proposed work. From left to right, the first column is the rendering results by point lighting, the second column is rendering results based on image lighting and the third is the image slices [21].

$\varphi_d = 90$ image slice is a method to visualize the BRDF features proposed by Brent Burley et al. [21]. All of the interesting features of materials are visible in the $\varphi_d = 90$ image. By comparing the BRDF image slices, we can see the difference between predicted and true BRDF value intuitively. Figure 4 shows the schematic view of the image slice and six image slices of materials. The materials include painting, rubber, plastic, fabric, etc. and use RQ as the Gaussian process's covariance.

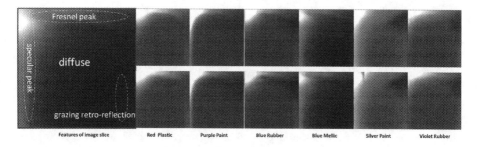

Fig. 4. The BRDF image slices. All of the interesting features of materials such as diffuse, specular Fresnel peak and retro-reflection is visible in the $\varphi_d = 90$ image. The figure shows the different between ground truth and predicted values of six materials.

Figures 3(c) and 4 shows that usually the upper left corner of the slice is the difference between the ground truth and the predicted results, which is near to the specular and Fresnel peak. One reason for such a case is that, BRDF changes dramatically near the specular and Fresnel peak but in the proposed work we choose the same sampling interval for different angles. Increasing the density of sampling near the angles of specular and Fresnel can modify this problem, but as a drawback this will increase the computational complexity of the covariance matrix.

The rendering results of the proposed work are shown in the Fig. 5, where the Gaussian process can get good results under both point lighting and image lighting. It is difficult to distinguish the difference between ground truth and GP regression results with the naked eye. So we calculate the pixel difference value between them by using $1-b/a$ (a, b are the pixel values of two point light image, respectively). The difference of Fig. 5(a) and (b) is much smaller than the difference of point light image, we did not show them) and show the distribution in Fig. 5(e). As Fig. 5(e) shows, the difference is under 5 % of true BRDF rendering value.

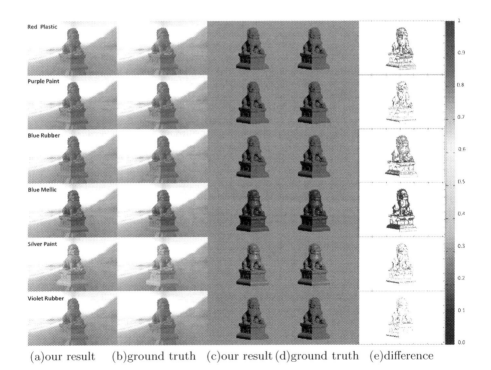

(a)our result   (b)ground truth   (c)our result (d)ground truth   (e)difference

Fig. 5. GP regression results of 6 different materials (red plastic, purple paint, blue rubber, blue melic, silver paint, violet rubber) under the image based light (a) (b) and point light (c) (d). (e) Pixel difference value between ground truth and proposed work (Color figure online).

As the visual effect of the rendering results depends on people's psychological perception, there is still no unified standard to measure. Since PSNR is an objective criterion for evaluation of image, we use peak signal to noise ratio (PSNR) to evaluate the image of object model rendered by predicted BRDF and the ground truth.

$$PSNR = 10 * \log \left(\frac{(2^n - 1)^2}{MSE} \right) \tag{20}$$

where, MSE is the mean square error between the original image and process images.

Table 2 gives the different error analysis of these 6 materials. The representations error of these 6 materials with RQ covariance is very small and our PSNR is much higher than other methods (Phong: 32.09, Blinn-Phong: 30.97, Cook-Torrance: 32.98, Murat Kurt: 52.12) [15].

Covariance functions are used to describe the relationship between different outputs. Also we experimented using different covariance function and their combination [16] (formula 6, 7 and 21, the combination results are shown in Table 3), this is the reason that we chose the rational quadratic (RQ) function with little error to be the covariance function in Sect. 3.1.

Table 2. Gaussian process prediction error of different materials

	Red plastic	Purple paint	Blue rubber	Blue melic	Silver paint	Violet rubber
MAE	0.0144	0.0048	0.0049	0.0028	0.0124	0.0054
MSE	0.0375	0.0248	0.0259	0.0212	0.0345	0.0253
$PSNR$	62.3924	64.1940	63.9993	64.8764	62.7547	64.0961

Matern class function (Matern)

$$k_\nu(\mathbf{x}_i, \mathbf{x}_j) = \sigma_m^2 \frac{2^{1-\nu}}{\Gamma(\nu)} (\sqrt{2\nu}r)^\nu K_\nu(\sqrt{2\nu}r) \tag{21}$$

where, $r = \left(\sum_{k=1}^d \frac{(x_{i,k} - x_{j,k})^2}{l_k^2} \right)^{\frac{1}{2}}$. ν is used to control the roughness of process, and K_ν is the modified Bessel function.

Table 3. Gaussian process prediction error with different covariance

$Error$	Gaussian process prediction error					
	RQ	SE	Matern	Matern+SE	Matern+RQ	RQ+SE
σ_{MAE}	0.0134	0.0156	0.0122	0.0800	0.0187	0.0205
σ_{MSE}	0.0830	0.0875	0.0837	0.4527	0.0992	0.1078
$PSNR$	**54.4767**	50.2725	52.6951	36.1594	**57.2624**	52.9558

5   Future Work

In the proposed work, Gaussian process is used to predict real BRDFs. The experiments show that the error of prediction is small, the peak signal to noise ratio of the rendering image is high and the results can satisfy the demand of practical application.

Using Gaussian process based on Bayesian method can not only achieve accurate prediction but also reduce the complexity and time consumption of BRDF measurement procedure. However, a significant problem with Gaussian process used in BRDF prediction is that for large data processing problems both storing the Gram matrix and solving the associated linear systems are prohibitive on modern workstations. Also, the training process of GP is time-consuming and the learning process is done offline now. With the ascension of computer hardware technology and improvement of the algorithm, real-time calculation is promising to realize.

In this paper, we only considered the isotropic homogeneous materials whose reflection properties is not supposed to change with spatial location. As a four dimensional function, although BRDF describes the illumination properties of different view and light direction on the materials surface, it can only describes the reflection law of homogeneous material. For some uneven material surface of cultural relics, there will be self-occlusion, self-shadow, occlusion and other complicated visual effects. BTF (Bidirectional texture function) can not only capture the changes in light properties along with the light and view direction but also capture the sampling location. In future, we will focus on the research of non-homogeneous relic's material and BTF representations.

Acknowledgement. This work was supported by the National Key Technology Support Program under Grant 2012BAH64F01 and 2013BAH48F01 and National Natural Science Foundation of China 61370134. The authors would like to thank Wojciech Matusik et al. [2] for using their measured BRDF data.

References

1. Nicodemus, F.E.: Geometrical considerations and nomenclature for reflectance. US Department of Commerce, National Bureau of Standards, Washington, D.C. (1977)
2. Matusik, W., et al.: Efficient isotropic BRDF measurement. In: Proceedings of the 14th Eurographics Workshop on Rendering. Eurographics Association (2003)
3. Phong, B.T.: Illumination for computer generated pictures. Commun. ACM **18**, 311–317 (1975)
4. Ward, G.J.: Measuring and modeling anisotropic reflection. ACM SIGGRAPH Comput. Graph. **26**, 265–272 (1992)
5. Ashikhmin, M., Shirley, P.: An anisotropic phong BRDF model. J. Graph. Tool. **5**, 25–32 (2000)
6. Torrance, K.E., Sparrow, E.M.: Theory for off-specular reflection from roughened surfaces. JOSA **57**, 1105–1112 (1967)
7. Cook, R.L., Torrance, K.E.: A reflectance model for computer graphics. ACM Trans. Graph. (TOG) **1**, 7–24 (1982)

8. Blinn, J.F.: Models of light reflection for computer synthesized pictures. ACM SIGGRAPH Comput. Graph. **11**, 192–198 (1977)
9. Marschner, S.R., et al.: Measuring and modeling the appearance of finished wood. ACM Trans. Graph. (TOG) **24**, 727–734 (2005)
10. Sadeghi, I., et al.: A practical microcylinder appearance model for cloth rendering. ACM Trans. Graph. (TOG) **32**, 14 (2013)
11. Ngan, A., Durand, F., Matusik, W.: Experimental analysis of BRDF models. In: Proceedings of the Sixteenth Eurographics Conference on Rendering Techniques. Eurographics Association (2005)
12. Lafortune, E.P.F., et al.: Non-linear approximation of reflectance functions. In: Proceedings of the Sixteenth Eurographics Conference on Rendering Techniques. Eurographics Association (1997)
13. He, X.D., et al.: A comprehensive physical model for light reflection. ACM SIGGRAPH Comput. Graph. **25**, 175–186 (1991)
14. Dong, Y., et al.: Manifold bootstrapping for SVBRDF capture. ACM Tran. Graph. (TOG) **29**, 98 (2010)
15. Kurt, M., Cinsdikici, M.G.: Representing BRDFs using SOMs and MANs. ACM SIGGRAPH Comput. Graph. **42**, 2 (2008)
16. Rasmussen, C.E.: Gaussian Processes for Machine Learning. The MIT Press, Cambridge (2006)
17. Ngan, A., Durand, F., Matusik, W.: Experimental analysis of BRDF models. In: Proceedings of the Sixteenth Eurographics Conference on Rendering Techniques. Eurographics Association (2005)
18. Matusik, W.: A Data-Driven Reflectance Model. University of North Carolina, Chapel Hill (2003)
19. Rusinkiewicz, S.M.: A new change of variables for efficient BRDF representation. In: Drettakis, G., Max, N. (eds.) Rendering Techniques 1998, pp. 11–22. Springer, Vienna (1998)
20. Vanhatalo, J., et al.: GPstuff: Bayesian modeling with Gaussian processes. J. Mach. Learn. Res. **14**, 1175–1179 (2013)
21. Burley, B., Studios, W.D.A.: Physically-Based Shading at Disney. Practical Physically-Based Shading in Film and Game Production, SIGGRAPH (2012)
22. Lawrence, J., Rusinkiewicz, S., Ramamoorthi, R.: Efficient BRDF importance sampling using a factored representation. ACM Trans. Graph. (TOG) **23**, 496–505 (2004)
23. Nicodemus, F.E.: Directional reflectance and emissivity of an opaque surface. Appl. Opt. **4**, 767–773 (1965)
24. Gargan, D., Neelamkavil, F.: Approximating reflectance functions using neural networks. In: Drettakis, G., Max, N. (eds.) Rendering Techniques 1998, pp. 23–34. Springer, Vienna (1998)

Realistic Walkthrough of Cultural Heritage Sites-Hampi

Uma Mudenagudi, Syed Altaf Ganihar[✉], Shreyas Joshi, Shankar Setty,
G. Rahul, Somashekhar Dhotrad, Meera Natampally, and Prem Kalra

NIAS Bangalore, IIT-Delhi, B. V. Bhoomaraddi College of Engineering
and Technology-Hubli, Hubli, India
altafganihar@gmail.com

Abstract. In this paper we discuss the framework for a realistic walk-through of cultural heritage sites. The framework includes 3D data acquisition, different data processing steps, coarse to fine 3D reconstruction and rendering to generate realistic walkthrough. Digital preservation of cultural heritage sites is an important area of research since the accessibility of state of the art techniques in computer vision and graphics. We propose a coarse to fine 3D reconstruction of heritage sites using different 3D data acquisition techniques. We have developed geometry based data processing algorithms for 3D data super resolution and hole filling using Riemannian metric tensor and Christoffel symbols as a novel set of features. We generate a walkthrough of the cultural heritage sites using the coarse to fine 3D reconstructed models. We demonstrate the proposed framework using a walkthrough generated for the Vittala Temple at Hampi.

1 Introduction

In this paper we describe the framework for the generation of realistic digital walkthrough of cultural heritage sites. The advent of digital technology has resulted in a great surge in interest to digitally restore heritage sites [1,2]. A large number of cultural heritage sites are deteriorating or being destroyed over a period of time due to natural weathering, natural disasters and wars. The heritage sites at Hampi, India are largely composed of rock structures which are in a grievous situation as can be seen in Fig. 1 and this necessitates the digital preservation of the sites at Hampi. Digital preservation of the heritage sites can be accomplished using modern techniques in computer vision and graphics.

Digital restoration of cultural heritage sites has been in the purview of computer graphics and vision research since a long time. The notable works reported in the literature are Modeling from Reality [3], The Great Buddha Project [2], Stanford University's Michelangelo Project [1], IBM's Pieta Project [4] and Columbia University's French cathedral project [5] to mention a few. Modeling from Reality [3] discusses the modeling of cultural heritage sites in a precise manner using laser range scanners. The Great Buddha Project [2] describes the pipeline for the digital preservation and restoration of Great Buddhas using a pipeline, consisting of acquiring data, aligning data, aligning multiple range

© Springer International Publishing Switzerland 2015
C.V. Jawahar and S. Shan (Eds.): ACCV 2014 Workshops, Part II, LNCS 9009, pp. 554–566, 2015.
DOI: 10.1007/978-3-319-16631-5_41

images and merging of range images. The Stanford University's Michelangelo Project [1] describes a hardware and software system for digitizing the shape and color of large fragile objects under non-laboratory conditions. Columbia University's French cathedral project [5] describes building of a system which can automatically acquire 3D range scans and 2D images to build 3D models of urban environments.

The acquisition of the 3D data is an integral step in the digital preservation of the cultural heritage sites. The classic 3D modeling tools are often derisory to accurately portray the complex shape of sculptures found at cultural heritage sites. The advent of inexpensive 3D scanning devices like Microsoft Kinect and ToF (Time of Flight) cameras have simplified the 3D data acquisition process. The state of the art 3D laser scanning devices generate very accurate 3D data of the objects. However the scanning of large outdoor objects at the cultural heritage sites invite a lot of tribulations due to the generation of partial meshes. The image based methods like SFM (Structure from Motion) [6] and PMVS (Patch based Multi-View Stereo) [7] consolidate the 3D data acquisition process but do not generate high resolution 3D data to accurately depict the art work at the heritage sites. The occlusions during the scanning process result in the occurrence of missing regions in the 3D data (holes) and generation of partial meshes. This warrants the need for efficient data processing techniques for the digital preservation of the cultural heritage sites.

Fig. 1. The ruins at the Vittala Temple - Hampi, India.

Figure 2 shows the comparison of the rendered scene of the Vittala Temple at Hampi with the original image of the scene. Our framework generates a realizable digital walkthrough of the cultural heritage sites using a coarse to fine 3D reconstruction of the cultural heritages sites. We put into service several data processing algorithms like noise filtering, 3D super resolution, 3D hole filling and texture mapping for the fine level 3D reconstruction of the objects. The fine level 3D reconstructed models at the cultural heritage sites are registered with the coarse level models to generate a coarse to fine 3D reconstructed model. The coarse to fine 3D reconstructed models are subsequently rendered to obtain a

Fig. 2. Comparison of the rendered scene and the original image of the Stone Chariot at Vittala Temple - Hampi, India: Left half is the rendered image and the right half is the original image.

digitally realizable walkthrough of the heritage site. Towards this we make the following contributions:

1. We propose a framework for the generation of realistic walkthrough of cultural heritage sites with coarse to fine 3D reconstruction.
2. We propose 3D super resolution and hole filling algorithms for efficient 3D data processing using concepts of Riemannian geometry with metric tensor and Christoffel symbols as a novel set of features.
3. We demonstrate the proposed framework for Vittala Temple at Hampi, India.

   The rest of the paper is organized as follows. In Sect. 2 we describe the 3D data acquisition techniques employed for the generation of the digital walk-through. In Sect. 3 we discuss the 3D super resolution and hole filling algorithms. In Sect. 4 we explain the coarse to fine 3D reconstruction and rendering of the heritage sites. In Sect. 5 we demonstrate the results of the proposed framework and provide the conclusion in Sect. 6.

2   Proposed Framework and Data Acquisition

The proposed framework of coarse to fine 3D reconstruction is as shown in Fig. 3. The data acquisition step includes acquisition of 3D data for different modalities like CAD model, Single-view model, Kinect model and Multi-view model. The 3D point cloud data generated during acquisition is fed to data processing. Hole filling and 3D super resolution is performed to refine the point cloud data in the data processing step. In the rendering step, the refined data is fed to coarse to fine 3D reconstruction stage. Finally, rendered view is generated using rendering engine.

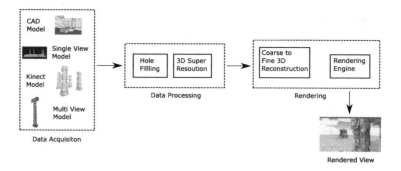

Fig. 3. Framework of coarse to fine 3D reconstruction and rendering to generate realistic walkthrough.

The 3D data acquisition of the cultural heritage sites, is the process of capturing 3D models from the on-site real world objects and is an important part in the digital restoration process. The coarse level models are obtained either using CAD modeling tools or using single view reconstruction. The CAD models obtained do not accurately depict the geometry of the artwork at the cultural heritage sites. The CAD models while modeling are recreated or restored in order to incorporate some of the missing, withered or prophesied part of the cultural heritage site. The CAD models or the single view reconstructed models do not accurately portray the artworks at the cultural heritage sites. The fine level models are hence required to precisely represent the artworks. We acquire the fine level models at the cultural heritage sites in the following ways depending upon the location, size and feasibility of the method.

1. The Microsoft Kinect 3D sensor consisting of a depth and a RGB camera is employed to scan the 3D models. Under appropriate lighting conditions, scanning is done on a 3D model and we use the Kinect Fusion (KinFu) [8] to generate a dense point-cloud or a mesh of the scanned model.
2. A set of images of an object to be reconstructed are captured under appropriate lighting conditions. The images are then fed to dense reconstruction algorithms like SFM [6] or PMVS [7] to generate point cloud models.

3   Data Processing

The data processing algorithms are a vital component in the digital restoration of cultural heritage sites. The obtained data is in the form of a point-cloud which is filtered using Statistical Outliers filter in order to eliminate any noisy data acquired during the scanning process. The data acquired using scanners like laser scanners, Microsoft Kinect or image based methods comprise of certain missing regions (holes), partial meshes or is of low resolution. To address these issues we propose geometry based data processing algorithms for 3D data super resolution and hole filling. The pipeline for the generation of fine level models is shown in Fig. 4.

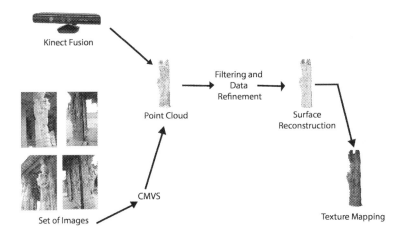

Fig. 4. Overview of the data acquisition and data refinement pipeline for detailed reconstruction.

3.1 Super Resolution

The point-cloud data obtained from the low-resolution 3D scanner like the Microsoft Kinect or from sparse reconstruction algorithms usually fail to capture the accurate geometric properties and detailed structure of the 3D object either due to the presence of occlusions during the scanning process, non-feasibility of the sparse reconstruction algorithm or adverse scanning environment. As a result, these techniques fail to portray all the details in a model's surface resulting in a low-resolution point-cloud data. The generation of high resolution 3D data is important for the realistic rendering of cultural heritage sites. Hence there is an immense requirement to produce a high-resolution point-cloud data from a given low-resolution point-cloud data. Authors in [9] proposed decision framework for super resolution. The decision framework facilitates to obtain the comparatively best fit interpolation curve based on the voting parameters obtained from the point cloud thus producing super-resolved point cloud. However, we propose a learning based super resolution. The overview of the proposed learning based super resolution framework is shown in Fig. 5. Given 3D model is modeled as a set of Riemannian manifolds [10–13] in continuous and discretized space. A Kernel based SVM learning framework [14] is employed to decompose a given 3D model into basic shapes viz., sphere, cone and cylinder using metric tensor and Christoffel symbols as a set of novel geometric features. The decomposed models are then independently super-resolved using selective interpolation techniques for example the spherically decomposed model is super resolved using spherical surface interpolation technique. The independently super resolved algorithms are merged to obtain the final super resolved model.

The metric tensor [10–13] $g_{\mu\nu}$ is a symmetric tensor and in 3-dimensions consists of 6 independent components. The metric tensor gives the quantitative measure

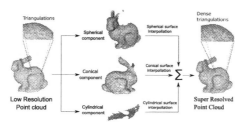

Fig. 5. Proposed learning based super resolution framework.

for the deviation in the manifold from the Euclidean space. The Christoffel symbols [10–13] give a measure of the deviation of the metric tensor as a function of position. The Christoffel symbols in 3-dimensions consists of 18 independent components.

The features used for the decomposition consist of 24 independent components which are in turn dependent on the geometrical position of the point over which the features are calculated. The decomposition of 3D model into basic shapes is carried out using a SVM [14] framework. The training data consists of unit sphere, cone and cylinder which are learned in the SVM framework. The spherical decomposed part of the 3D model is interpolated using spherical surface interpolation method. Similarly the conical and the cylindrical decomposed parts are interpolated using conical surface interpolation and cylindrical surface interpolation method respectively. The interpolated decomposed parts are then fused to generate a super-resolved point-cloud of the 3D model. The algorithm achieves better result than reported in the literature.

3.2   Hole Filling

The 3D data acquired using the proposed techniques consists of missing regions or holes due to occlusions in the surface to be scanned. To address this issue we propose a hole filling algorithm using metric tensor and Christoffel symbols as features. The holes are identified by using the boundary detection algorithm used in [15]. The neighborhood of the hole is decomposed into basic shapes using a kernel based SVM learning framework with metric tensor and Christoffel symbols as features. The overview of the proposed hole filling algorithm is shown in Fig. 6. The decomposed regions in the neighborhood of the hole are interpolated using selective surface interpolation techniques. The centroid of the hole region is computed and the selective surface interpolation is carried out along the directional vector.

The point-cloud is surface reconstructed using Poisson surface reconstruction [16] or Ball-pivoting surface reconstruction algorithm [17]. The surface reconstructed model is texture mapped using image alignment with mutual information [18] and parameterization of the registered rasters for the surface reconstructed model.

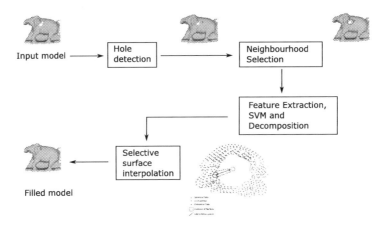

Fig. 6. Proposed hole filling algorithm.

4   Coarse to Fine 3D Reconstruction and Rendering

In this section we present the coarse to fine 3D reconstruction and the rendering of the reconstructed models for the generation of digital walkthrough. We carry out coarse level 3D reconstruction using methods such as single-view 3D reconstruction [19] or from modeling tools. The models generated using modeling tools and single view reconstruction do not accurately portray the geometrical complexities of the artwork at the cultural heritage sites. However, the fine level 3D reconstruction of large scale outdoor objects is not feasible using the techniques discussed

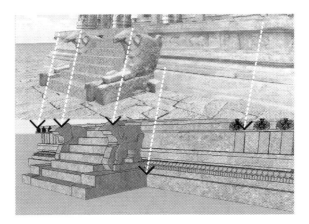

Fig. 7. Coarse to fine reconstruction of 3D objects using ICP registration with corresponding points in the coarse model and the fine model. Upper part of the image shows the fine reconstruction model and lower part of the image shows the coarse reconstruction model.

in the previous section. To resolve this issue we propose a coarse to fine level 3D reconstruction of the cultural heritage sites. The coarse to fine level 3D reconstruction is achieved by registering the coarse level 3D models with the fine level 3D models. The fine level 3D models are superimposed on the coarse level 3D models by interactively selecting the correspondence points in the model. The coarse and fine level 3D models are subsequently registered using the ICP (Iterative Closest Point) algorithm [20] for the corresponding points as shown in Fig. 7.

The coarse to fine 3D reconstructed models are rendered for the generation of the digital walkthrough. The rendering of the reconstructed models is carried out using either a rendering engine like OGRE 3D or a gaming engine like Unity 3D.

5   Results and Discussion

We demonstrate the proposed framework for Vittala Temple at Hampi, India. The data processing algorithms are implemented on Intel(R) Xeon(R) CPU E5-2665 0 @2.40 GHz (16 CPU's) and 64 GB RAM with NVIDIA Quadro K5000 graphics, 4 GB DDR3 graphics memory.

5.1   3D Reconstruction

The coarse level 3D models at the cultural heritage site are obtained either using single view reconstruction or using modeling tools. The CAD model for the Vittala Temple and the single view reconstruction of the Kalyan Mantap at Vittala Temple is as shown in Fig. 8. The fine level models are obtained using 3D scanning devices like laser scanner, Microsoft Kinect and image based methods like SFM and PMVS as shown in Fig. 9.

5.2   3D Super Resolution and Hole Filling

The fine level 3D models are processed using the proposed 3D super resolution and 3D hole filling algorithm. The processed 3D models are then surface reconstructed using the Poisson surface reconstruction algorithm with the following parameter values *Octree depth* = 12, *Solver divide* = 10, *Samples per*

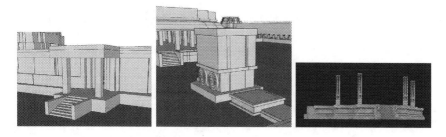

Fig. 8. Coarse level models obtained using CAD and single view reconstruction for Maha Mantapa, Stone Chariot and kalyan mantap at Vittala Temple - Hampi

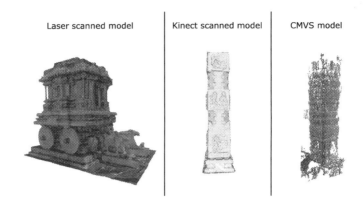

Fig. 9. Figure shows the 3D reconstructed models from Laser scanner for stone chariot, Kinect model for a pillar at main mantapa and CMVS model for a pillar at kalyan mantapa.

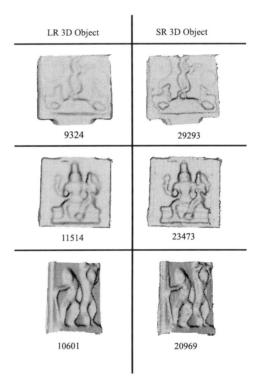

Fig. 10. Results for the proposed super resolution algorithm. Left column shows the 3D Objects of Low Resolution (LR) point cloud data. Right column shows the 3D Objects of Super Resolved (SR) point cloud data.

$node = 2$, $Surface\ offsetting = 1$. The surface reconstructed models are subsequently textured mapped using image alignment mutual information and registration of rasters. Figure 10 shows super resolution models generated for different artifacts of one of the pillers at Main Mantap - Hampi with magnification factor of approx 2. Figure 11 (a) shows hole filling for a part of the Stone Chariot at Vittala Temple - Hampi and Fig. 11 (b) shows hole filling for a part of the artifact of one of the piller's at Main Mantap - Hampi.

5.3   Coarse to Fine 3D Reconstruction

The coarse level models and fine level models are registered using ICP algorithm [20]. The coarse to fine level reconstruction of Kalyan Mantapa, Vittala Temple is shown in Fig. 12. The pillars at Kalyan Mantap can be classified into five different types. The fine level models for the five variants of the pillars are obtained using the proposed pipeline and are as shown in Fig. 12. The fine level models of the pillars comprise of roughly $\approx 300,000$ vertices and $\approx 600,000$ triangles.

5.4   Realistic Rendering

The coarse to fine level reconstructed models are rendered using OGRE 3D rendering engine and Unity 3D gaming engine and the rendered views are shown

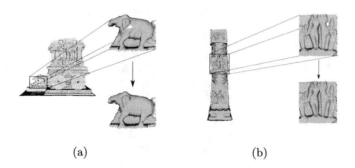

(a)   (b)

Fig. 11. Results for the proposed hole filling algorithm.

Fig. 12. Coarse to Fine 3D reconstructed model of Kalyan Mantapa along with Reconstructed models of the five variants of the pillars at Kalyan Mantapa, Vittala Temple - Hampi.

Fig. 13. Rendered views of Kalyan mantap and stone chariot using Unity 3D gaming engine and OGRE 3D rendering engine.

Fig. 14. Comparison of rendered views and the original images at Vittala Temple. Left: Rendered scene, Right: Original image.

in Fig. 13. The closeups of the rendered scene and the original images at the Vittala Temple are shown in Fig. 14.

6   Conclusion

In this paper we have proposed a framework for the realization of digital walk-through of cultural heritage sites. Digital restoration and preservation of cultural heritage sites is an important area of research due to the availability of techniques in data acquisition, data processing and rendering. The main goal of the paper is to create a framework for the generation of digital walkthrough of cultural heritage sites. To accomplish this we have proposed a framework for coarse to fine level 3D reconstruction using coarse level 3D reconstruction of the cultural heritage sites and fine level 3D reconstruction of the artworks at the cultural heritage sites. We also proposed data processing algorithms like 3D super resolution and 3D hole filling using concepts of Riemannian geometry with metric tensor and Christoffel symbols as a novel set of features. We have demonstrated the proposed framework for Vittala Temple - Hampi, India.

Acknowledgement. This work is supported by the Department of science and technology (DST), India, under grant NRDMS/11/201/Phase-III/ as a part of India Digital Heritage project. We thank DST, NIAS Banglore and IIT Delhi for the support.

References

1. Levoy, M., Pulli, K., Curless, B., Rusinkiewicz, S., Koller, D., Pereira, L., Ginzton, M., Anderson, S., Davis, J., Ginsberg, J., Shade, J., Fulk, D.: The digital michelangelo project: 3d scanning of large statues. In: Proceedings of the 27th Annual Conference on Computer Graphics and Interactive Techniques. SIGGRAPH 2000, pp. 131–144. ACM Press/Addison-Wesley Publishing Co., New York, NY, USA (2000)
2. Ikeuchi, K., Oishi, T., Takamatsu, J., Sagawa, R., Nakazawa, A., Kurazume, R., Nishino, K., Kamakura, M., Okamoto, Y.: The great buddha project: digitally archiving, restoring, and analyzing cultural heritage objects. Int. J. Comput. Vis. **75**, 189–208 (2007)
3. Ikeuchi, K., Sato, Y.: Modeling from Reality. Kulwer Academic Press, Boston (2001)
4. Wasserman, J.: Michelangelo's Florence Peita. Princeton University Press, Princeton (2003)
5. Stamos, I., Allen, P.K.: Automatic registration of 2-d with 3-d imagery in urban environments. In: ICCV, pp. 731–737 (2001)
6. Snavely, N., Seitz, S.M., Szeliski, R.: Photo tourism: exploring photo collections in 3d. ACM Trans. Graph. **25**, 835–846 (2006)
7. Furukawa, Y., Ponce, J.: Accurate, dense, and robust multi-view stereopsis. IEEE Trans. Pattern Anal. Mach. Intell. **32**, 1362–1376 (2010)
8. Izadi, S., Kim, D., Hilliges, O., Molyneaux, D., Newcombe, R., Kohli, P., Shotton, J., Hodges, S., Freeman, D., Davison, A., Fitzgibbon, A.: Kinectfusion: real-time 3d reconstruction and interaction using a moving depth camera. In: Proceedings of the 24th Annual ACM Symposium on User Interface Software and Technology. UIST 2011, pp. 559–568. ACM, New York, NY, USA (2011)
9. Ganihar, S., Joshi, S., Patil, N., Mudenagudi, U., Okade, M.: Voting-based decision framework for optimum selection of interpolation technique for 3d rendering applications. In: Students' Technology Symposium (TechSym), pp. 270–275. IEEE (2014)
10. Jost, J.: Riemannian Geometry and Geometric Analysis. Springer Universitat texts, New York (2005)
11. Kumaresan, S.: A Course in Differential Geometry and Lie Groups. Texts and Readings in Mathematics. Hindustan Book Agency, Cambridge (2002)

12. Weinberg, S.: Gravitation and Cosmology: Principles and Applications of the General Theory of Relativity. Wiley, New York (1972)
13. Misner, C., Thorne, K., Wheeler, J.: Gravitation. W.H. Freeman and Company, San Francisco (1973)
14. Burges, C.J.: A tutorial on support vector machines for pattern recognition. Data Min. Knowl. Disc. **2**, 121–167 (1998)
15. Liepa, P.: Filling holes in meshes. In: Proceedings of the 2003 Eurographics/ACM SIGGRAPH Symposium on Geometry Processing. SGP 2003, pp. 200–205. Eurographics Association, Aire-la-Ville, Switzerland (2003)
16. Kazhdan, M., Bolitho, M., Hoppe, H.: Poisson surface reconstruction. In: Proceedings of the Fourth Eurographics Symposium on Geometry Processing. SGP 2006, pp. 61–70. Eurographics Association, Aire-la-Ville, Switzerland (2006)
17. Bernardini, F., Mittleman, J., Rushmeier, H., Silva, C., Taubin, G.: The ball-pivoting algorithm for surface reconstruction. IEEE Trans. Vis. Comput. Graph. **5**, 349–359 (1999)
18. Corsini, M., Dellepiane, M., Ponchio, F., Scopigno, R.: Image-to-geometry registration: a mutual information method exploiting illumination-related geometric properties. Comput. Graph. Forum **28**, 1755–1764 (2009)
19. Koutsourakis, P., Simon, L., Teboul, O., Tziritas, G., Paragios, N.: Single view reconstruction using shape grammars for urban environments. In: 2009 IEEE 12th International Conference on Computer Vision, pp. 1795–1802 (2009)
20. Rusinkiewicz, S., Levoy, M.: Efficient variants of the ICP algorithm. In: Third International Conference on 3D Digital Imaging and Modeling (3DIM) (2001)

Categorization of Aztec Potsherds Using 3D Local Descriptors

Edgar Roman-Rangel[1]([✉]), Diego Jimenez-Badillo[2],
and Estibaliz Aguayo-Ortiz[2]

[1] CVMLab, University of Geneva, Geneva, Switzerland
edgar.romanrangel@unige.ch
[2] National Anthropology and History Institute of Mexico (INAH),
Mexico City, Mexico

Abstract. We introduce the Tepalcatl project, an ongoing bi-disciplinary effort conducted by archaeologists and computer vision researchers, which focuses on developing statistical methods for the automatic categorization of potsherds; more precisely, potsherds from ancient Mexico including the Teotihuacan and Aztec civilizations. We captured 3D models of several potsherds, and annotated them using seven taxonomic criteria appropriate for categorization. Our first task consisted in exploiting the descriptive power of two state-of-the-art 3D descriptors. Then, we evaluated their retrieval and classification performance. Finally, we investigated the effects of dimensionality reduction for categorization of our data. Our results are promising and demonstrate the potential of computer vision techniques for archaeological classification of potsherds.

1 Introduction

The application of computer vision technologies for the preservation, management, and analysis of cultural heritage artifacts has witnessed a rapid growth during the last decade [1]. This is especially true with regard to the creation and use of digital 3D models, which enable capabilities that would not be available using the original artifacts, such as automatic and semi-automatic content analysis [2,3], virtual reconstructions [4,5], more efficient archiving [6,7], sharing documentation online [1,7], training of novel scholars, etc.

An area of especial interest is the statistical analysis of features observed on 3D models of potsherds with the purpose of categorizing archaeological pottery [3], which is one of the most important jobs in archaeology [2,4]. In this work, we use the term *categorization* with a slight different connotation than that of *classification*. More precisely, categorization defines a series of decisions that have to be made about a sherd of interest, which is different from the traditional classification scenarios where a single class label is assigned to a query instance. For example, besides of classifying a sherd in function of the type of vessel it comes from, we might also be interested in knowing the type of rim it has. These distinctions are of high relevance, as knowing certain of such characteristics helps archaeologist to infer the final class of the sherd.

ⓒ Springer International Publishing Switzerland 2015
C.V. Jawahar and S. Shan (Eds.): ACCV 2014 Workshops, Part II, LNCS 9009, pp. 567–582, 2015.
DOI: 10.1007/978-3-319-16631-5_42

This paper introduces the Tepalcatl project, an ongoing effort that brings together archaeologist from the National Anthropology and History Institute of Mexico (INAH) and researchers in computer vision. This project is oriented to developing a system for the automatic categorization of ceramic sherds. Figure 1, shows examples of Aztec potsherds used as study case in this work.

(a) Cajete.   (b) Cajete Tripode.   (c) Cup.

Fig. 1. Pictures provided by the project urban archaeology of the INAH directed by the archaeologist Raul Barrera. These artifacts are from excavations conducted near the Templo Mayor in Mexico City.

The contributions of this paper are as follows.

1. We design a recognition system for potsherd. First, we acquired a set of 3D surfaces of sherds belonging to specific vessel shapes. These model instances are then used for semi-supervised training of the recognition system.
2. We reformulate the archaeological task of sherd categorization into a computer vision problem, namely 3D image classification and retrieval. For this purpose, we defined seven different taxonomic criteria for which the recognition system is able to perform sherds categorization. For example, it is able to distinguish between several types of rim; or using another criterion, it is able to recognize the type of curvature of the fragment. Note that the training of the system happens only once, but it works well with the seven different criteria nevertheless.
3. We exploit the discriminative power of two state-of-the-art local descriptors, namely the Scale Invariant Feature Transform (SIFT) [8,9] and the Spin Images descriptor [9,10]. These descriptors are invariant to rotation and scale transformations, and therefore avoid the need of a pre-processing step to normalize the scale and orientation of the sherds, as required by other methods [3,11]. Our approach proved to be effective for potsherd categorization.

Our approach allows examining different combinations of local descriptor, dictionary sizes and taxonomic criteria in order to find out which one achieves the best performance. To the best of our knowledge, this approach has not been tried before in archaeological classification.

The rest of this paper is organized as follows. Section 2 describes the details of the Tepalcatl project. Section 3 discusses detailed work in areas of digital preservation of pottery and their automatic analysis. Section 4 summarizes the statistical descriptors that we used. Section 5 explains the data that we used and our experimental protocol. Section 6 discusses our findings. Finally, Sect. 7 presents our conclusions and future directions for our project.

2   Tepalcatl Project

The analysis and classification of pottery sherds constitutes one of the most important jobs in archaeology. It provides cultural information of past societies at multiple scales, from the identification of human activity areas to the determination of site chronology and/or the elucidation of regional economic systems.

Unfortunately, a detailed analysis of ceramic fragments is also one of the most cumbersome and time-consuming tasks for archaeologists. This is due not only to the long learning curve involved in mastering a ceramic classification system but also to the vast quantity of sherds normally recovered from the field. A typical excavation in Central Mexico, for example, produces tens of thousands of fragments and a single site may contain ceramics dating from a very long period, encompassing occupations of the Teotihuacan, Toltec and/or Aztec civilizations, that is, sherds dating from approximately 100 B.C.E. to 1521 C.E. Figure 2 shows the settlement region of the Teotihuacan, Toltec and Aztec civilizations.

(a) Settlement region of Teotihua-can, Toltec, and Aztec cultures.

(b) Purple circle: Teotihuacan; blue: Toltec; red: Aztec.

Fig. 2. Settlement region of Teotihuacan, Toltec (city of Tula), and Aztec (city of Tenochtitlan) cultures. (a) Localization within current Mexico. (b) Details of (a) indicating the capital cities.

2.1   Sherd Categorization

A very relevant property in ceramic analysis is shape [3]. Archaeologists are expected to reconstruct the profile of a whole vessel by examining the form of a surviving fragment. A traditional method to accomplish this task, has been a visual comparison of a sherd with a manual drawing of the silhouette and diameter of vessels already known. However, this approach is time consuming and requires the knowledge of highly trained people. A common practice to reduce the complexity of the task is selecting only fragments that are considered diagnostic to identify important ceramic types. This includes certain body parts like rims, legs, handles, etc., which are then used in a more detailed typological analysis to deduce chronology, cultural style, etc. Figure 3 shows examples of potential sherds from the type of ceramic we use.

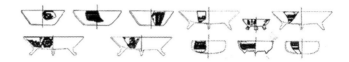

Fig. 3. Potential sherds from the type of pottery pieces in our project.

During the past decade, some professionals have been experimenting with digital technologies oriented to reducing learning curves and improving the quality of the classification process. One of the most interesting proposals has been the acquisition of 3D models of sherds with the purpose of applying mathematical, computer vision and/or machine learning techniques that perform classifications more accurately in an automatic or semi-automatic way [2,3]. The extraction of 3D digital data has also brought extra benefits, such as the possibility to undertake new types of content analyses, as well as an easier sharing of information among professionals, the design of better ceramic documentation and archiving systems [6,7], and the performance of virtual reconstruction of vessels [4,5].

2.2   Goals of the Tepalcatl Project

To help in this categorization endeavor, we decided to experiment with new techniques, hoping that a better and more efficient classification method will emerge from successive iterations that benefit from the join efforts of archaeologists and computer vision specialists.

Namely, the goals that our project pursuits are:

1. **Design and compilation of a 3D dataset.** We will digitize a large collection of potsherds from the Aztec and Teotihuacan cultures that ranks in the order of tens of thousands. This task will generate a new digital dataset that poses several challenges in terms of visual description and automatic categorization. This dataset will be rich in terms of labels, as its instances will be annotated using different taxonomic criteria.
2. **Advance the state-of-the-art in 3D classification.** We will develop new methods for better description and representation of 3D models. These new models will overcome limitations potentially found in current approaches. Also, we will explore more efficient methods for classification.
3. **Improve the archaeological process of ceramic classification.** Using different taxonomic criteria (explained in Sect. 5.1), we will conduct several analyses of similarity that could lead us to design efficient methods for accurate categorization of potsherds.

4. **Further assessment.** We will assess the potential that our and other methods that could have to deal with more archaeological tasks, such as retrieval of complete vessels from potsherds, and virtual reconstruction of vessels.

2.3   Work in Data Collection

Our project contemplates the study of potsherds belonging to the ancient Teotihuacan (100–600 C.E.) and Aztec civilizations (1325–1521 C.E.). Teotihuacan was the first urban center in central Mexico and produced one the most influential cultures in the ancient Americas, whereas the Aztecs were the dominant civilization in the same region by the time of the European arrival. Remains of both cultures, including ceramic sherds, are often found in different strata of the same archaeological sites. Figure 2 shows the settlement location of these cultures.

Aztec III Black on Orange Ceramics. The potsherds used in this work belong to a ceramic style known as "Aztec III Black on Orange", which is one of the most important wares in Mesoamerican archaeology. Along with the so-called Aztec I, II and IV it is part of the sequence that helps understand the cultural evolution in Central Mexico during Postclassic times [13].

The geographic distribution of Aztec III Black on Orange covers the entire Basin of Mexico. Vessels of this type were part of the utilitarian assemblage common in households during the late Aztec period (1350–1520 C.E.). For this reason, it is considered a diagnostic ware to infer the presence of Aztec settlements as opposed to those controlled by previous ethnic groups. Another reason for its importance is that the emergence of this style coincides with the formation and development of the Aztec empire.

Vessels belonging to this ceramic complex show black painted designs drawn on a polished orange surface. The designs are formed by thin brush strokes that follow parallel and curved lines combined with other motifs, such as birds, fishes, plants, geometric figures etc. The designs are found both on the interior and exterior surfaces of the vessels. Figure 1 shows examples of the Aztec III Black on Orange ceramics. A feature of this ceramic complex, and a major challenge for the classification of sherds, is the abundance of shapes given to the vessels. The most common are cajetes with three legs (Fig. 1b), comales (a flat circular surface to cook corn tortillas), bowls, jars, pots, plates and the so-called apaxtles.

A great quantity of this type of ceramics was collected on the surface during the 1970's by the Valley of Mexico Survey Projects [14–17]. During the 1990's, paste composition and stylistic analyses of those materials was performed, comparing their geographic distribution with historically-documented polities [18,19]. They found that a number of decorative motifs are exclusive to particular paste groups and managed to identify three main production areas in the Basin of Mexico. This in turn allowed to hypothesize about the exchange systems of the Aztec empire. Thanks to these studies we know in greater detail the economic relations between dependent communities and the Aztec capital.

Most recently, Mexican researchers have uncovered new material inside the boundaries of the Sacred Precinct of Tenochtitlan, the main religious center of the Aztec capital (whose remains lay in close proximity to the main square in Mexico City). The findings have drawn the attention of many specialists, mainly because it proceeds from controlled excavations as opposed to surface surveys. However, the great quantity of findings has made it difficult to classify and publish the reports.

3   Related Work

Computer-aided classification and reconstruction of ceramics has been a subject of research for at least 20 years, with pioneering work by Hall and Lafling [20], who created a software package called NEWTS for drawing, archiving, and editing ceramic profiles using B-spline functions.

Since then, the focus has continued to be the acquisition of 2D profiles, either from sherds or from complete vessels, which can then be classified by archaeologists using their curvature functions [3]. A notable example of these efforts is the creation of the so-called profilograph (http://www.dolmazon.de/profilograph_e.htm), a device to draw profiles from 3D objects (i.e. sherds).

Thanks to the growing accessibility of 3D laser scanners, some researchers have shown the advantages of acquiring 3D models of ceramics with the purpose of extracting automatically the 2D profiles [4–6,21–25]. They based their work on the assumption that the majority of vessels studied by archaeologists are axially symmetric. For perfectly symmetric vessels the profile corresponds to a cross section in the direction of the rotational axis. Thus, by calculating the rotational axis, different methods are able to draw the profile of the vessels with different levels of success. Unfortunately, archaeological vessels are not perfectly symmetric, especially if they were hand-crafted as opposed to thrown-wheel manufactured. Therefore, extracting the rotational axis possesses serious challenges [3]. Perhaps the best method to address this tasks, is the one proposed by Karasik and Smilansky [3] because it takes into account the many possible deviations in the axis of symmetry.

The calculation of the axis of symmetry also requires a pre-processing stage of orientation and alignment of the sherds [26], which is normally a time consuming task [21]. Our analytic approach is invariant to rotation and scale and therefore avoids the need of normalizing orientation and scale.

Another drawback of the current approaches is that they require a whole model of the sherd. This involves taking several scans of the sherd, which are then merged together to complete the model [3]. Our method does not rely on capturing a whole model. Instead it uses only surface information from the back and front views of the sherd, which expedites the process significantly.

As for the comparison of sherds for classification, some methods compare sherds using a point-to-point matching process [27]. Our method compares histograms (i.e. bag-of-visual-words) that represent the statistical frequency of local descriptors (i.e. SIFT and spin images).

Perhaps the most complete and sophisticated approaches under the paradigm of profile extraction that include classification of sherds are due to Gilboa [28] and to Hörr [26]. Hörr [26] proposed a series of mathematical algorithms for profile segmentation, feature extraction and, more importantly, clustering of sherd descriptions using algebraic functions, which in turn allows query and hierarchical classification. The method, however, was applied to complete or semi-complete vessels and results for sherds are not available in the original paper.

4   Descriptors

This section describes the two local descriptors that we used in our work. Namely, the Local Invariant Feature Transform (SIFT) [8] and the Spin Images [10], that were originally proposed to deal with 2D images, and whose 3D version consists in scale invariant points of interest in a 3D space [9].

4.1   Spin Images

Spin images are well known local descriptors developed to match mesh-surfaces by the individual point-to-point matching of its vertices [10]. However, it might as well be implemented under a bag-of-visual-words approach [9,12].

In the spin images methodology, each vertex v_i is characterized by

$$v_i = (x_i, y_i, z_i, \theta_i, r_i, h_i), \tag{1}$$

where, (x_i, y_i, z_i) defines the three coordinates position of the vertex; θ_i denotes its local orientation, which corresponds to the orientation of the plane touching the closest points that are connected to v_i; and r_i and h_i are respectively the radius and height of a cylindrical support volume centered at (x_i, y_i, z_i).

The base idea of this method consists in *scanning* the 3D surface by *spinning* a sheet around the axis defined by the point orientation θ_i while counting the amount of nearby points [10]. More formally, the cylindrical support volume is divided into R radial segments (*rings*) and L vertical segments (*layers*), thus generating $K = R x L$ spatial bins. Finally, the counting of the number of nearby points $\#\{p_j\}$ falling within each resulting bin $b_k, k = 1, \ldots, K$, is arranged as a K-dimensional feature vector s termed spin image [29],

$$s(k) = \#\{p_j : p_j \in b_k\}. \tag{2}$$

The above mentioned methodology can be applied to each vertex in the mesh [10]. However, it might also be applied only to certain vertices selected randomly or by other approaches. One of such approaches is presented in Sect. 4.3.

4.2   SIFT

The Scale-Invariant Feature Transform (SIFT) is a state-of-the-art and one of the most popular image descriptor for gray-scale images. It detects points of interest as points that maximize the difference of Gaussian (DoG) response in a Gaussian scale space, where such a maximization is evaluated both in location and scale spaces. The resulting points are invariant to scale and rotation variations [8].

Recently, the Local-Depth SIFT (LD-SIFT) variant was proposed to deal with 3D images [9], representing the vicinity of an interest point as a depth map. In this approach, the 3D mesh around vertex v_i^s is locally projected onto its dominant plane, where the dominant plane is defined by it local orientation θ_i, and s denotes the characteristic scale (or locality span) of the vertex of interest, which can be computed as explained in Sect. 4.3.

A 2D patch (image) is obtained after projecting the 3D mesh onto the dominant plane, for which a traditional SIFT descriptor is computed [8].

4.3   Scale-Invariant 3D Local Descriptors

One of the approaches for selecting a subset of vertices, proposes to detect a set of so-called points of interest [8,9], where a vertex is considered as a point of interest if it maximizes the local variation of a function of the image of 3D model, both in location and scale spaces. In the case of 3D models, this is the straightforward extension of the Gaussian scale space used by Lowe to select points of interest in 2D images [8], where the Gaussian scale space is constructed by cascading a Gaussian filter and a smoothing of the image, which create a set of filtered images called octaves.

However, an important difference between applying Gaussian filters to 2D images and 3D meshes is that points on a mesh are not necessarily uniformly spaced in the depth axis [9]. This characteristic, might harm the repeatability rate and the scale-invariant capabilities of the selected points [9,30]. To cope with this issue, several approaches have been proposed [9,27,31,32], among which, the use of local filters with uniform weights to build the mesh octaves outperforms the others methods in terms of repeatability rate and scale invariance [9].

In this approach [9], the smoothing step required to build the Gaussian scale space is given by the estimation of each vertex v_i^{s+1} at scale $s+1$ as,

$$v_i^{s+1} = \frac{1}{|V_i^s|} \sum_{v_j^s \in V_i^s} v_j^s, \tag{3}$$

where, V_i^s is the set of first order neighbors of v_i^s, $|\cdot|$ denotes the cardinality operator, and the summation of the vertices is performed component-wise on the x, y, and z components of v.

Once the smoothing of the vertex is performed, the difference of Gaussians (DoG) is computed by,

$$d_i^s = \frac{1}{(s \cdot \sigma_{i,0}^2)} \left(v_i^s - v_i^{s+1} \right), \tag{4}$$

where, σ_0 is the initial variance of the integration parameter [8] and it is independently estimated for each vertex as,

$$\sigma_{i,0} = \frac{1}{|V_i^s|} \sum_{v_j^s \in V_i^s} abs \left(v_i^s - v_j^s \right), \tag{5}$$

where, $abs(\cdot)$ denotes the absolute value operator.

Using this formulation, it is possible to select a subset of points of interest for which a local descriptor (SIFT or spin image) will be computed, i.e., those points whose d_i^s is maximal both in location and scale spaces. Section 6 will show that this scheme to detect points of interest and select their characteristic scale achieves good retrieval and classification performance when it is combined with Spin images and SIFT descriptors.

5 Experiments

This section introduces the dataset generated by our project during its initial phase, as well as the set of experiments that we have performed.

5.1 Data

Our data consists of 149 3D surfaces that were manually scanned from 16 sherds of Aztec ceramics, and annotated under different taxonomic criteria by experts in archaeology. More precisely, surfaces obtained from different views were scanned from the internal and external sides of the sherds, and the following taxonomic criteria were used to annotate them according to their visual information,

- *type*: defines the type of ceramic where the sherd comes from. At this stage of the project, three types of ceramics were available to us: Apaxtle, Cajete, and Cajete with handler (see Fig. 1).
- *side*: the side of the sherd with two possible values: internal and external.
- *form*: this label indicates the form of the ceramic where the sherds comes from, with two possible values: open and close.
- *wall*: corresponds to the type of wall of the original ceramic, with three possible values: simple wall, divergent, and straight.
- *curvature*: this label indicates whether the sherd corresponds to a flat or curved section of the ceramic. It has two possible values: NONE and YES.
- *border*: gives information about the type of border of the ceramic. It has three possible values: NONE when no border is visible in the 3D surface, straight, and round.
- *base*: this label indicates whether the sherd contains part of the base of the ceramic, and it has two possible values: NONE and YES.

Figure 4a shows the amount of 3D surfaces scanned from each of the 16 sherds. Note that the dataset tends to be balanced in the amount of internal and external views that were scanned, despite the fact that only external views were available for sherd 11. Also, Fig. 4b shows the distribution of labels on the 149 3D surfaces. Note that only 139 3D surfaces are used for the label 'border', because 10 models have borders different from the three possible values (NONE, straight, round), but they have only one or two instances, which makes difficult to define relevant sets, and therefore, these instances were excluded for this taxonomy.

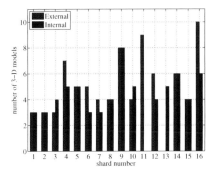
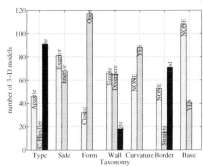

(a) Amount of 3D surfaces per sherd.   (b) Amount of 3D surfaces per label.

Fig. 4. (a) Amount of 3D surfaces scanned per sherd: blue bars correspond to external views, whereas red bars indicate internal views. (b) Distribution of labels assigned to each 3D surface.

Although the available data after scanning the sherds contains both geometry (mesh defined by points and faces) and texture (surface photos), we started our analysis relying solely on the geometric information. For this purpose, the SIFT descriptor was tailored to work with orientation information instead of intensities, the same way as the work of Darom and Keller [9].

5.2   Experimental Protocol

The experimental protocol that we followed consists of:

1. Compute the sets of local descriptors (SIFT [8] or spin images [10]) for each of the 149 3D surfaces.
2. Estimate a visual vocabulary using a subset of 700 local descriptors for each of the 16 sherds, thus 11200 local descriptors. We computed visual vocabularies of different sizes, i.e., 100, 250, 500, 1000, and 2000 visual words.
3. Build a bag-of-words representation (BoW) of each 3D surface.
4. Perform retrieval experiments by sorting the 3D surfaces in terms of visual similarity, where such similarity is computed using the BoW. Note that these retrieval experiments were performed using a leave-one-out full-cross-validation approach. We evaluated the retrieval performance using the different taxonomic criteria to define the relevant set for each 3D query surface.
5. Using PCA [33], we found the most informative principal components of the BoW's, and repeated the retrieval experiments using only them.
6. Finally, we performed a series of classification experiments using the kNN approach for the seven taxonomic criteria.

This protocol was repeated independently for each of the local descriptors and each visual vocabulary. We used a Matlab implementation [9] to compute both SIFT and spin images local descriptors. In Sect. 6, we report our results in terms of mean average precision computed for the first 25 3D surfaces retrieved

($mAP@25$) and the standard recall vs average precision curves. We also report the average classification rate and show the most interesting confusion matrices.

6   Results

Table 1 shows the $mAP@25$ obtained with both the SIFT [8] and the spin images [10] local descriptors, and using different vocabulary sizes and the different taxonomic criteria described in Sect. 5.1.

The first observation available from Table 1, is that small vocabularies are adequate for description of the 3D surfaces of potsherds for most of the taxonomic criteria, i.e., 100 and 200 visual words in most cases. Two exceptions to this observation were found using the Spin images along with the Curvature and Border information, where respectively, larger vocabularies of 2000 and 1000 visual words achieved the best retrieval performance.

Table 1. $mAP@25$ obtained with the SIFT and spin images descriptors using different vocabulary sizes and taxonomies. Best result per taxonomy is in bold.

Vocabulary	Taxonomy						
SIFT	Type	Side	Form	Wall	Curvature	Border	Base
100	**0.688**	**0.749**	0.802	**0.651**	**0.704**	**0.643**	**0.767**
200	0.677	0.734	**0.816**	0.637	0.694	0.635	0.744
500	0.649	0.721	0.812	0.605	0.687	0.633	0.736
1000	0.618	0.709	0.807	0.592	0.687	0.640	0.727
2000	0.613	0.687	0.810	0.574	0.670	0.627	0.701
Spin	Type	Side	Form	Wall	Curvature	Border	Base
100	**0.579**	0.610	**0.753**	0.526	0.619	0.533	**0.660**
200	0.577	**0.618**	0.751	**0.532**	0.617	0.540	0.655
500	0.573	0.617	0.752	0.527	0.620	0.544	0.638
1000	0.569	0.616	0.751	0.521	0.625	**0.549**	0.634
2000	0.564	0.622	0.747	0.522	**0.626**	0.544	0.626

Another observation from Table 1 is that, good retrieval performance is obtained by using the Form information to build the relevant sets, i.e., 0.816 with the SIFT descriptors and 0.753 with the spin images, which suggests that it is relatively easy to differentiate between open and close forms. However, this result must be read with caution, since the dataset is not well balanced under such a taxonomic criterion as shown in Fig. 4b, where the set of models with closed forms is about a third of the set with open forms.

The construction of the relevant set using the taxonomic criteria Side, Curvature, and Base also provided with acceptable retrieval performance, i.e., above 0.7 using the SIFT descriptors and above 0.6 with the spin images. Note that the bars in Fig. 4b indicate that all these three taxonomies define a binary classification problem, with balanced relevant sets for the Side and Curvature cases. Thus suggesting that the use of such taxonomies can provide with good hints for automatic categorization. The remaining three taxonomic criteria Type, Wall, and Border, all divide the dataset into three classes, and all obtained the poorer retrieval performance for either of the descriptors.

Finally, note that the SIFT descriptors are more appropriate than the spin images for the statistical visual description of 3D surfaces of potsherds, as they obtained better retrieval performance in all cases, i.e., about 10% better on average, which confirms observations from previous comparisons [9,29].

Using Principal Component Analysis (PCA) [33], we performed dimensionality reduction on the BoW's and evaluated the retrieval performance. We noticed that the retrieval precision increased in all cases, specially when large vocabularies are used, and with the SIFT descriptors. Table 2 shows the best retrieval results, which on average were obtained using 1000 visual words.

Table 2. $mAP@25$ obtained using 1000 words and principal component analysis. The number of principal components used with each taxonomy is also shown.

Vocabulary	Taxonomy						
SIFT	Type	Side	Form	Wall	Curvature	Border	Base
1000	0.718	0.788	0.821	0.678	0.695	0.658	0.776
Num. Comp.	5	9	5	9	20	20	10
Spin	Type	Side	Form	Wall	Curvature	Border	Base
1000	0.590	0.627	0.734	0.527	0.634	0.560	0.678
Num Comp	9	18	9	20	5	5	6

Looking at the Type column in Table 2, we can see that by using only 5 principal components from a 1000 vocabulary, higher retrieval precision is obtained, i.e., 3% higher than using 100 words, and 10% higher than using the full 1000 words vocabulary. Due to space constraints, we only present here the results with 1000 words, as they represent the case of the larger increment in retrieval precision. However, this behavior resulted true for most of our experiments.

Figures 5a and 5b show, respectively for SIFT and Spin images, the average precision vs standard recall curves obtained with the visual vocabulary of 1000 words. This is, the average drop of the retrieval precision as more relevant instances are included within the vector of sorted elements. Note that the drop of the retrieval precision is smooth in all cases. Also note that these curves are consistent with the results shown in Tables 1 and 2 for both descriptors, i.e., Form information provides the best retrieval performance (red-dotted line with diamond markers), whereas the Border and Wall taxonomies obtained the lowest retrieval performance (olive-dotted line with triangle markers and cyan-solid line with asterisk markers respectively).

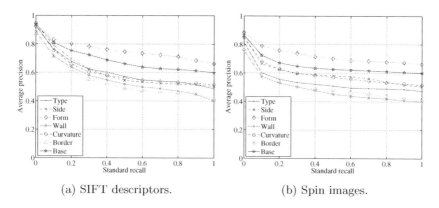

(a) SIFT descriptors.   (b) Spin images.

Fig. 5. Average precision vs standard recall obtained with the different taxonomic criteria. These result were obtained using 1000 visual words.

Table 3. Average classification rate per taxonomic criteria. These results were obtained using the principal components of the 1000 words vocabularies, as indicated in Table 2.

	Type	Side	Form	Wall	Curvature	Border	Base
SIFT	0.70 ± 0.1	0.88 ± 0.0	0.77 ± 0.2	0.67 ± 0.3	0.78 ± 0.1	0.64 ± 0.3	0.74 ± 0.2
spin	0.43 ± 0.2	0.58 ± 0.0	0.57 ± 0.4	0.42 ± 0.2	0.58 ± 0.1	0.44 ± 0.1	0.56 ± 0.3

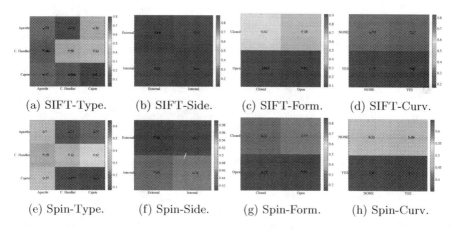

(a) SIFT-Type.   (b) SIFT-Side.   (c) SIFT-Form.   (d) SIFT-Curv.

(e) Spin-Type.   (f) Spin-Side.   (g) Spin-Form.   (h) Spin-Curv.

Fig. 6. Confusion matrices for type, side, form, and curvature information, computed using PCA and the number of components indicated in Table 2. Best viewed as pdf.

Finally, we performed a series of classification experiments using the kNN (k=1). For these experiments, we also used the principal components of the 1000 words vocabularies, as indicated in Table 2.

Table 3 shows the average classification rate achieved by both the SIFT descriptor and the spin images. Note that, similarly to the retrieval results, classifying potsherds using the Side, Form, and Curvature information gives the highest accuracy. However, using the Type information as labels also gives acceptable classification rates. Also note that, the use of he Wall and Border information gives the poorest results, which contain high levels of standard deviation.

A more detailed explanation of the classification performance is shown in the confusion matrices of Fig. 6. Due to space constraints, we only show the most interesting confusion matrices.

7   Conclusions and Future Work

We presented a novel project towards the integration of computer vision techniques in the analysis of potsherds from the ancient Teotihuacan and Aztec cultures. Our interdisciplinary approach is rich and genuine, as it addresses needs and open problems in archaeology, i.e., those of helping the efficient analysis and categorization of potsherds.

Different from classical works on analysis of 3D potsherds, ours approaches the categorization problem under 3 assumptions: we use rotation and scale invariant local descriptors, which avoid the need of normalizing the 3D sherds under such transformations; we avoid the need of manual registration of different views of the sherds by relying on 3D surfaces; and instead of performing point-to-point matching of the 3D models, we use statistical representations that are much more efficient.

In this work, we designed a recognition system, which after a single unsupervised training, is able to categorize 3D surfaces of potsherds under different taxonomic criteria that in archaeology are used to infer the final class of sherds.

Within the context of our project, we started the collection of a new and challenging dataset of 3D surfaces of potsherds. The initial strategy is constrained to the use of Aztec III ceramics, due to the availability of this material. However, we plan to scan a much larger collection of sherds from the Teotihuacan culture, which ranks in the order of tens of thousands, and that is well documented. When this dataset is completely scanned, it will be also annotated using the different taxonomic criteria.

We presented a study of the descriptive potential of two state-of-the-art 3D descriptors for the pieces of our datasets, and combined them with the efficient bag-of-visual-words representation, and dimensionality reduction techniques based on PCA. Overall, our methodology is promising, as the results show good categorization performance using the different taxonomic criteria. More precisely, achieving an average accuracy rate of 70 %, while leaving room to conduct more research in the topic. For instances, collecting more data and designing better descriptors.

Acknowledgement. This work was funded by the Swiss NSF through the project Tepalcatl-P2ELP2-152166, and supported by the project Urban Archaeology of the National Anthropology and History Institute of Mexico (INAH).

References

1. Scopigno, R., Callieri, M., Cignoni, P., Corsini, M., Dellepiane, M., Ponchio, F., Ranzuglia, G.: 3D models for cultural heritage: beyond plain visualization. Computer **44**, 48–55 (2011)
2. Kampel, M., Sablatnig, R., Costa, E.: Classification of archaeological fragments using profile primitives. In: Proceedings of Computer Vision, Computer Graphics and Photogrammetry - a Common Viewpoint, Workshop of the Austrian Association for Pattern Recognition (2001)
3. Karasik, A., Smilanski, U.: 3D scanning technology as a standard archaeological tool for pottery analysis: practice and theory. J. Archaeol. Sci. **35**, 1148–1168 (2008)
4. Leymarie, F., Cooper, D., Joukowsky, M., Kimia, B., Laidlaw, D.H., Mumford, D., Vote, E.: The SHAPE lab - new technology and software for archaeologists. In: Proceedings of Computer Applications in Archaeology, BAR International Series 931 (2000)
5. Kampel, M., Sablatnig, R.: Virtual reconstruction of broken and unbroken pottery. In: Proceedings of the International Conference on 3-D Digital Imaging and Modeling (2003)
6. Razdan, A., Liu, D., Bae, M., Zhu, M., Farin, G.: Using geometric modeling for archiving and searching 3D archaeological vessels. In: Proceedings of the International Conference on Imaging Science, Systems, and Technology CISST (2001)
7. Larue, F., Di-Benedetto, M., Dellepiane, M., Scopigno, R.: From the digitization of cultural artifacts to the web publishing of digital 3D collections: an automatic pipeline for knowledge sharing. J. Multimedia **7**, 132–144 (2012)
8. Lowe, D.G.: Distinctive image features from scale-invariant keypoints. Int. J. Comput. Vis. **60**, 91–110 (2004)
9. Darom, T., Keller, Y.: Scale-invariant features for 3-D mesh models. IEEE Trans. Image Process. **21**, 2758–2769 (2012)
10. Johnson, A.E., Hebert, M.: Using spin images for efficient object recognition in cluttered 3D scenes. IEEE Trans. Pattern Anal. Mach. Intell. **21**, 433–449 (1999)
11. Maiza, C., Gaildrat, V.: Automatic classification of archaeological potsherds. In: Proceedings of the International Conference on Computer Graphics and Artificial Intelligence (2005)
12. Quelhas, P., Monay, F., Odobez, J., Gatica-Perez, D., Tuytelaars, T.: A thousand words in a scene. IEEE Trans. Pattern Anal. Mach. Intell. **29**, 1575–1589 (2007)
13. Cervantes Rosado, J., Fournier, P., Carballal, M.: La cerámica del Posclásico en la Cuenca de México. In: La producción Alfarera en el México Antiguo. Volume V of Coleccin Cientfica. Insituto Nacional de Antropología e Historia (2007)
14. Parsons, J.R.: Prehistoric Settlement Patterns in the Texcoco Region, Mexico. Memoirs of the Museum of Anthropology, vol. 3. University of Michigan, Ann Arbor (1971)
15. Blanton, R.E.: Prehispanic settlement patterns of the ixtapalapa region, Mexico, vol. 6, The Pennsylvania State University Occasional Papers in Anthropology, University Park, Pennsylvania (1972)
16. Sanders, W.T., Parsons, J., Santley, R.S.: The Basin of Mexico: Ecological Processes in the Evolution of a Civilization. Academic Press, New York (1979)
17. Parsons, J.R., Brumfiel, E.S., Parsons, M.H., Wilson, D.J.: Prehistoric settlement patterns in the southern valley of mexico: the chalco-xochimilco regions. In: Memoirs of the Museum of Anthropology. vol. 14. University of Michigan (1982)

18. Hodge, M.G., Minc, L.D.: The spatial patterning of aztec ceramics: implications for prehispanic exchange systems in the valley of Mexico. J. Field Archaeol. **17**, 415–437 (1990)

19. Hodge, M.G., Neff, H., Blackman, M.J., Minc, L.D.: Black-on-orange ceramic production in the aztec empire's heartland. Lat. Am. Antiq. **4**, 130–157 (1993)

20. Hall, N., Laflin, S.: A computer aided design technique for pottery profiles. In: Computer Applications in Archaeology, Centre for Computing and Computer Science. University of Birmingham (1984)

21. Adler, K., Kampel, M., Kastler, R., Penz, M., Sablatnig, R., Schindler, K., Tosovic, S.: Computer aided classification of ceramics-achievements and problems. In: Proceedings International Workshop on Archaeology and Computers (2001)

22. Kampel, M., Sablatnig, R.: Rule based system for archaeological pottery classification. Pattern Recogn. Lett. **28**, 740–747 (2007)

23. Melero, F.J., Len, A.J., Contreras, F., Torres, J.C.: A New System for Interactive Vessel Reconstruction and Drawing. Bar International Series, vol. 1227. Archaeopress, Oxford (2004)

24. Sablatnig, R., Menard, C.: Computer based acquisition of archaeological finds: the first step towards automatic classification. In: Proceedings of the Third International Symposium on Computing and Archaeology (1996)

25. Schurmans, U., Razdan, A., Simon, A., McCartney, P., Marzke, M., Alfen, D.V., Jones, G., Rowe, J., Farin, G., Collins, D., Zhu, M., Liu, D., Bae, M.: Advances in geometric modeling and feature extraction on pots, rocks and bones for representation and Query via the internet. In: Computer Applications and Quantitative Methods in Archaeology (CAA) (2001)

26. Hörr, C., Brunner, D., Brunnett, G.: Feature extraction on axially symmetric pottery for hierarchical classification. Comput. Aided Des. Appl. **4**, 375–384 (2007)

27. Castellani, U., Cristani, M., Fantoni, S., Murino, V.: Sparse points matching by combining 3D mesh saliency with statistical descriptors. Comput. Graph. Forum **27**, 643–652 (2008)

28. Gilboa, A., Karasik, A., Sharon, I., Smilansky, U.: Towards computerized typology and classification of ceramics. J. Archaeol. Sci. **31**, 681–694 (2004)

29. Frome, A., Huber, D., Kolluri, R., Bülow, T., Malik, J.: Recognizing objects in range data using regional point descriptors. In: Pajdla, T., Matas, J.G. (eds.) ECCV 2004. LNCS, vol. 3023, pp. 224–237. Springer, Heidelberg (2004)

30. Boyer, E., Bronstein, A.M., Bronstein, M.M., Bustos, B., Darom, T., Horaud, R., Hotz, I., Keller, Y., Keustermans, J., Kovnatsky, A., Litman, R., Reininghaus, J., Sipiran, I., Smeets, D., Suetens, P., Vandermeulen, D., Zaharescu, A., Zobel, V.: SHREC 2011: robust feature detection and description benchmark. In: Proceedings of the 4th Eurographics Conference on 3D Object Retrieval (2011)

31. Zaharescu, A., Boyer, E., Varanasi, K., Horaud, R.: Surface feature detection and description with applications to mesh matching. In: IEEE Conference on Computer Vision and Pattern Recognition (2009)

32. Mian, A., Bennamoun, M., Owens, R.: On the repeatability and quality of keypoints for local feature-based 3D object retrieval from cluttered scenes. Int. J. Comput. Vision **89**, 348–361 (2010)

33. Jolliffe, I.T.: Principal Component Analysis, 2nd edn. Springer, New York (2002)

Image Parallax Based Modeling of Depth-Layer Architecture

Yong Hu[1,2], Bei Chu[1], and Yue Qi[1(✉)]

[1] State Key Laboratory of Virtual Reality Technology and Systems,
Beihang University, Beijing, China
`qy@buaa.edu.cn`
[2] School of New Media Art and Design, Beihang University, Beijing, China

Abstract. We present a method to generate a textured 3D model of architecture with a structure of multiple floors and depth layers from image collections. Images are usually used to reconstruct 3D point cloud or analyze facade structure. However, it is still a challenging problem to deal with architecture with depth-layer structure. For example, planar walls and curved roofs appear alternately, front and back layers occlude each other with different depth values, similar materials, and irregular boundaries. A statistic-based top-bottom segmentation algorithm is proposed to divide the 3D point cloud generated by structure-from-motion (SFM) method into different floors. And for each floor with depth layers, a repetition based depth-layer decomposition algorithm based on parallax-shift is proposed to separate the front and back layers, especially for the irregular boundaries. Finally, architecture components are modeled to construct a textured 3D model utilizing the extracting parameters from the segmentation results. Our system has the distinct advantage of producing realistic 3D architecture models with accurate depth information between front and back layers, which is demonstrated by multiple examples in the paper.

1 Introduction

Realistic and flexible 3D architecture models are very important for many applications including culture heritage protection, games, movies and augmented reality navigation etc. Depth images from 3D scanners or color images from digital cameras are the two most popular data sources to model the architecture. Obviously, digital cameras are more common and inexpensive, and also provide rich color texture information which is very important for realistic modeling. Therefore, we focus our work on the problem of image based architecture modeling.

Many works have been proposed for this problem, e.g. [1] focuses on piecewise planar architecture; [2] utilizes a single image to model symmetry architecture, but it needs a lot of manual interactions; [3] makes use of a rectangular plane or a developable surface to generate buildings. There are also a few commercial tools which all require tedious manual works. However, both of them can not deal with architecture with front and back depth layers and get the accurate depth between these layers automatically. Front-back depth layers are very

© Springer International Publishing Switzerland 2015
C.V. Jawahar and S. Shan (Eds.): ACCV 2014 Workshops, Part II, LNCS 9009, pp. 583–597, 2015.
DOI: 10.1007/978-3-319-16631-5_43

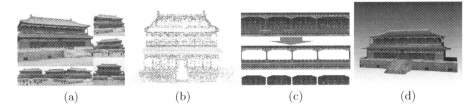

(a)   (b)   (c)   (d)

Fig. 1. Our modeling results. (a) Image collections. (b) Top-bottom segmentation result. (c) Depth-layer decomposition result. (d) The final textured 3D model.

common in architecture, e.g. in Fig. 1(a), the second floor of this building are composed of two layers which are pillars layer and windows layer respectively. This property is particularly worthy of being modeled, [4] proposes a 2D-3D fusion method to decompose these layers to rectangular planar fragments, and a 3D LiDAR scanner is also needed. We handle this problem only with digital images especially for decomposing layers with irregular boundaries (as shown in the corner regions in Fig. 1(c)) as one of our contributions. Another kind of methods are based on multi-view stereos, such as [5,6], etc. The reconstruction results are dense point clouds or meshes and can not be further segmented to generate meaningful architecture components which are more important to extend the range of architecture modeling application.

Given image collections of one building, our goal is to generate a visually compelling 3D model, in which accurate architecture components segmentation is critical. Although high resolution texture information can be acquired from images, but it is hard to segment the components in the image space solely. A common case is that different components which are occluded by each other may have the same material, such as the pillars and their back windows in Fig. 1(a). 3D point cloud can be generated from multiple images by structure-from-motion (SFM) method, but these 3D points are too sparse to be segmented into architecture components directly. The complementary characteristics of these two data sources are combined to handle more complex architecture modeling problems in this paper. We first propose a statistic-based top-bottom segmentation algorithm and divide the sparse 3D point cloud to several horizontal floors vertically along the ground plane normal as in Fig. 1(b). For each horizontal floor with depth layers, we propose a repetition based depth-layer decomposition algorithm to divide its sparse 3D points and images into several repetitive components as in Fig. 1(c). The key observation of our depth-layer decomposition algorithm is the parallax-shift among repetitive structures in a single image or multiple view images. Finally, textured architecture component models are reconstructed from the segmented sparse 3D points and texture parts to make up the 3D architecture model. An example of modeling results is shown in Fig. 1(d), and others are shown in the section Experiments and Discussion, which prove that our work can be used in practical applications.

2   Related Works

Image based architecture modeling have received a lot of research interest, with a large spectrum of modeling systems developed to build realistic 3D models. We classify the up to date and most relevant studies according to the data sources, single view or multiple view images without being exhaustive.

Single View Based Modeling. Reference [7] represents a scene as a layered collection of depth images, but assigns depth values and extracts layers manually. Reference [8] presents a fully automatic method for creating a rough 3D model from a single photograph, the model is composed of several texture-mapped planar billboards. Reference [9] automatically extracts shape grammar rules from facade images for procedural modeling technology whose modeling results are similar as ours, but the depth are also assigned manually. Reference [10] makes use of Manhattan structure for man made building and models the building as a number of vertical walls and a ground plane. Reference [2] calibrates the camera and reconstructs a roughly sparse point clouds from a single image by exploiting symmetry, but user must interactively marks out components on the 2D image to complete the modeling work. Realistic textured 3D models are reconstructed but depth layer decomposition are not handled. Reference [11] proposes a repetition-based dense single-view reconstruction method, but the repetition is necessary, and depths are roughly estimated from the repetition intervals.

Multiple Views Based Modeling. Image collections from different view-points are able to provide more 3D geometric information. Computer vision based multi-view stereo (MVS) algorithms, such as [5,6], generate architecture meshes on dense stereo reconstruction method. References [12,13] develop a real-time video images registration method and focus on the global reconstruction of dense stereo results. Proper modeling of the structure from reconstructed point clouds or meshes has not yet been addressed. Recently, some methods use 3D points from SFM or MVS to guide users for marking out the architecture components interactively and efficiently. Reference [1] uses image collections to assist interactively reconstructing the architecture composed of planar elements. Reference [3] proposes a semi-automatic method to segment the architecture and optimizes a depth value for each component using reconstructed 3D points. The following paper [14] proposes a partition scheme to separate the scene into independent blocks and extends their methods to reconstruct street-side city block. Reference [15] introduces a schematic representation for architecture, in which transport curves and profile curves are extracted from 3D point cloud to generate an architecture model with some swept surfaces. Reference [16] uses baseline and profile, a similar representation of [15], to model the architecture facade in the unwrapped space and get a textured architecture model. But the above methods are not suited for separating the depth layers with irregular boundaries and similar appearance. Reference [4] uses LiDAR data which are manually registered with photos to generate building with multiple layers. However, their layer decomposition method needs denser point cloud than SFM and only deals with rectangular components.

3   Overview

In the real world, most buildings are multistory and also have different depth layers. For the purpose of reconstructing the structure and components of the architecture, we first segment the architecture into isolated floors, and further divide some floors to repetitive or non-repetitive parts, and finally decompose these parts to layers with different depth values. In this paper, floor-segmentation is implemented in a 3D point cloud space generated from multiple view images by SFM algorithm, and we call it top-bottom floors segmentation. Integration of the above 3D points segmentation results and the multiple view images, repetition detection and further segmentation is performed to solve the decomposition problem for the floors with depth-layer structure, and we call it depth-layer decomposition. The pipeline of our architecture modeling method is composed of four major stages.

(1) **3D Point Cloud from SFM.** From the captured images, a sparse 3D point cloud is reconstructed by SFM method, then outliers removing and normals estimation are performed. The reconstruction results are used as our input for the following segmentation.

(2) **Top-Bottom Floors Segmentation.** Manhattan directions [17] are first estimated from the normals of the 3D points. Along the direction which is parallel to the normal of the ground plane, the 3D points are partitioned vertically to different horizontal floors. For some horizontal floors, 3D points are further partitioned to different layers according to their depth values.

(3) **Depth-Layer Decomposition.** For the candidate decomposition floor, 3D points at different layers are projected back to images and help us to detect the horizontal repetition at different layers respectively in the image space. Then, for each repetitive region, per-pixel parallax-shift values are estimated using SIFT-flow method [18], and the region is further decomposed into front and back layers by solving a per-pixel label-assignment optimization problem.

(4) **Architecture components modeling.** Parameters are extracted from the corresponding 3D points clusters and their projection images to generate the geometry of the architecture components. Then the components' textures are repaired from the multiple view segmented images, and a textured 3D architecture model is reconstructed finally.

4   Statistic-Based Top-Bottom Floors Segmentation

Our top-bottom floors segmentation algorithm is based on two intuitive criteria: normal variation that separates components like roofs and walls, and the depth variation that distinguishes layers with different depth values. Firstly, point cloud is generated by SFM algorithm and preprocessed by outliers removing and normals estimation. Secondly, Manhattan-direction is estimated as our segmentation direction. Finally, 3D points are segmented to floors along the ground normal direction according to the points' normal variation and some floors are further segmented to layers with different depth values along the facade direction.

4.1   Image Capture and Point Cloud Preprocessing

About 100 images for each dataset are captured with the positions distributing on an 180-degree arc in front of the architecture. A 3D point cloud is reconstructed using VisualSFM [19] because of its stability and ease of use. Generally, SFM point cloud contains outliers, which can be removed by performing a radius outlier removing method. In addition, point normals can only be inferred from the point cloud dataset directly. The problem of point normal estimation is approximated by the problem of estimating the normal of a plane tangent to the surface, which in turn becomes a least-square plane fitting estimation problem. For each point, an analysis of the eigenvectors and eigenvalues of a covariance matrix created from the nearest neighbors is implemented to estimate its normal. PCL-Point Cloud Library [20] is used to complete our outliers removing and point normal estimation.

4.2   Manhattan Axes Estimation

The 3D point cloud from SFM is in the camera coordinate system of the primary image as shown in the top three images of Fig. 2(a), which can not be directly used to vertically segment the 3D points into horizontal floors. Therefore, the ground plane normal should be estimated firstly. Many existing methods adapt the Manhattan assumption, estimate the Manhattan-axis in the case of piecewise planar architecture, and generate axis-aligned plane segments. We relax this assumption to non-piecewise planar architecture that includes oblique or curved surfaces like roofs. Let X_M, Y_M, Z_M be the Manhattan axes, and in this paper Y_M and Z_M also represent the ground plane normal direction and the facade normal direction of the architecture respectively. Let X_C, Y_C, Z_C be the coordinate axes of the original 3D point cloud. Since the captured images are taken without much yawing and rolling, the facade normal direction Z_M is nearly perpendicular to Y_C. Therefore, a histogram of angles between all the point normals and Y_C is created as shown in Fig. 2(b). The longest column with horizontal coordinate value near to 90 is selected and the corresponding group of point normals are considered as the candidates to estimate Z_M. Another histogram of angles between candidate points' normals and their mean normal is created repeatedly to remove outlier points until all angles between candidate point normals and the mean normal are less than 2 degree. Z_M is assigned as the mean normal of the remaining candidate points after outliers removing. Then, $X_M = Z_M \times Y_C$ and $Y_M = Z_M \times X_M$. Finally, the 3D point cloud is transformed to the Manhattan axes system as shown in the bottom three images of Fig. 2(a).

4.3   Segmentation

Normal Variation Based Segmentation. With respect to the artificial architecture, it is intuitive that point normals in the same floor vary slightly or smoothly. Therefore, we split the 3D point cloud into small slices along Y_M

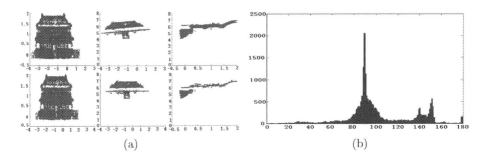

Fig. 2. (a) Top three images are the three orthographic views of the reconstructed 3D point cloud before Manhattan rectification and bottom three images are the rectification results. (b) The histogram of all the 3D point normals distribution. The vertical axis is the 3D point number and the horizontal axis is the angle between the point normal and ground plane normal.

and compute the variance of point normals in each slice. The positions of the local maximum values in the curve composed of the variance values can be considered as the potential split lines to segment the point cloud into different horizontal floors. More specifically, a series of uniform sampling planes which are perpendicular to Y_M are created to divide the 3D point cloud into point slices. In each slice, the variance of the dot product between the point normals and Y_M are computed to form the curve. The interval of these sampling planes are only determined by the ratio of the 3D point cloud height to the real architecture height. Because of the noise points, false split lines will be chosen and result in over-segmentation results. Nonetheless, this over-segmentation will be resolved by the subsequent merging operation. The segmentation result of this step are shown in Fig. 3(a).

Depth-Based Segmentation. In this step, some floors are further divided into separated layers according to the depth values along facade normal direction Z_M. First, roof floors are recognized and removed by the dot products between their mean normals and Y_M. Then, for each remaining floor, a series of uniform sampling planes perpendicular to the axis Z_M are created to divide the current floor into point slices. A statistic curve composed of the numbers of all point slices is built, and the positions of its local maximum values are considered as the split lines along Z_M direction as shown in the middle images of Fig. 3(b). In order to remove the noise, some local maximum values are filtered if the corresponding slices contain less than two percent of all points. The final depth-based segmentation result is shown in the top image of Fig. 3(b).

Merging. Due to the presence of noise points, the 3D point cloud will be over segmented by the above two segmentation steps. To solve this problem, any two segmented parts are traversed to determine whether they are able to merge. Two segmented parts with similar normals will be merged if the two distances between their bounding boxes along Y_M and Z_M directions are both less than the

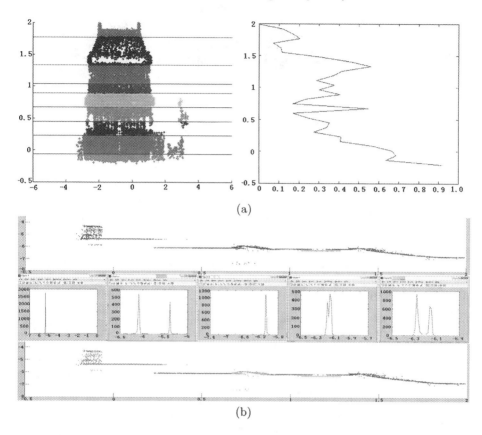

Fig. 3. (a) The result of normal variation based segmentation. Local maximum values of the curve in right image generate the split lines in the left image. (b) The result of depth variation based segmentation. Different segmented parts are represented by different colors in the top image. Statistic curves formed by point numbers of slices for the potential floors to be partitioned are shown in the middle images. Bottom image shows the merging result.

sampling plane interval. The above merging operation is implemented repeatedly until there are no mergeable parts and the result is shown in the bottom image of Fig. 3(b).

5   Repetition Based Depth-Layer Decomposition

In order to reconstruct the layered and textured 3D architecture model, we should decompose the images into different layers according to their corresponding depth values. During the SFM process, some pairs of 2D pixels and 3D points are established, but they are too sparse to do the direct layer assignment for all pixels. Nevertheless, these sparse pairs can provide enough information for accurate repetition detection in the image with depth-layer structure. Because of the perspective projection and the camera position, image deviation exists among

(a)   (b)   (c)   (d)

Fig. 4. (a) Image rectification by [22]. (b) Repetition detection by [22]. (c) Image rectification by our method. (d) Repetition detection by our method.

back layer regions behind the front layer repetitive structures, and we name this deviation as *parallax-shift*. A *parallax-shift* estimation based coarse image segmentation algorithm is proposed to perform the initial depth-layer decomposition. Generally, the boundaries between front and back layers are often irregular, but also have a high edge response. Based on above-mentioned characteristic and coarse decomposition result, we design a per-pixel label-assignment formulation and deploy a graph-cut optimization to refine the depth-layer decomposition.

Our repetition based depth-layer decomposition algorithm decomposes the image into components with different depth values in the following stages: (1) Carry out repetition detection in each image to get rectangle repetitive regions. (2) Perform a coarse depth-layer decomposition based on *parallax-shift* estimation between these repetitive regions. (3) Refine the coarse depth-layer decomposition by per-pixel layer assignment using graph-cut [21] energy minimization. (4) Decompose non-repetitive regions with multiple images.

5.1   Repetition Detection

After vertical segmentation, architecture images are segmented to image strips according to different floors. The repetitive structures appear only along the horizontal direction. We first test the method in [22] to detect rectangle repetitive regions, but there exist two problems in our case. First, large roof area and high frequency repetition of tiles always lead to wrong vanish point detection and result in incorrect image rectification (Fig. 4(a)). Second, repetitive structures at different depth layers always affect the estimation of the symmetry axis location (Fig. 4(b)).

In order to satisfy our subsequent decomposition requirement, we improve Wu's method [22] in three ways to achieve higher accuracy and stability. (1) Some 3D points are selected randomly, and the lines across these points and parallel to the Manhattan coordinate axes are computed and projected on the captured images to estimate the vanish points. This improvement utilizes the 3D point cloud information and results in consistent rectification for multiple view images. Meanwhile, wrong rectification results caused by parallel lines which are not vertical or horizontal such as parallel lines on the roof are avoided. (2) According to the former segmentation in point cloud, we are able to pick out the SIFT points at the front layer to estimate the repetitive region size and the symmetry

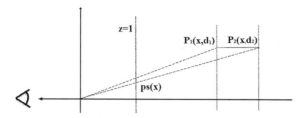

Fig. 5. The principle of parallax-shift between two points which is represented by the red solid line segment $ps(x)$.

axis location. This improvement may remove the interference by the repetition at different layers. (3) Vector quantization is used to estimate the similarity of different repetitive regions. A quad tree is constructed for the image, the root node is the whole image, and the image region belonging to each child node is one quarter of its parent node. The image region size in the leaf node is half size of the smallest repetitive region. SIFT descriptors are computed for all nodes and clustered to 256 categories (In our experiment, 256 is totally enough to detect the repetitive region similarity). Any two candidate regions are confirmed as repetitive regions if their SIFT descriptors belong to the same category. This improvement avoids the sensitive threshold value determination in Wu's method [22]. Our image rectification and repetition detection results are shown in Fig. 4(c) and (d).

5.2 Coarse Depth-Layer Segmentation

For the rectified image, the camera projection plane is parallel to the architecture facade. The parallax-shift between the projection pixels of two points with the same horizontal coordinates but different depth values can be computed according to

$$ps(x) = x \times (d_2 - d_1)/d_1 d_2 = Cx \qquad (1)$$

as in Fig. 5.

Convert to the repetitive regions in the image, repeating points at the back layer project to different pixel locations in respective repeating regions (Fig. 6(a)). More specifically, given a set of repeating regions $\{R_1, R_2, .., R_n\}$ with symmetry axes $\{X_1, X_2, .., X_n\}$, for any point P_i at the front layer and its corresponding point P'_i at the back layer in R_i, its corresponding repetitive point in R_j is denoted as P_j and P'_j, they satisfy the following equations:

$$I_x(P_i) - X_i = I_x(P_j + (t_{ij}, 0)) - X_j, \qquad (2)$$

where $I_x(P_i)$ is the x coordinate of the projection of point P_i on image I, t_{ij} is the distance between R_i and R_j.

$$I_x(P'_i) - X_i = I_x(P'_j + (t_{ij}, 0)) - X_j + ps_{ij}, \qquad (3)$$

where $ps_{ij} = t_{ij} \times C$ is the parallax-shift between R_i and R_j.

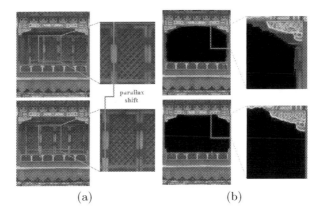

(a)   (b)

Fig. 6. (a) The blue lines are repeating back layer structures in different repeating regions. The parallax-shift is represented by a short red line. (b) Depth-layer decomposition results for one repetitive region. Part of the image is extracted to show details (Color figure online).

Given two repetitive regions, SIFT-Flow method is used to compute the parallax-shift and obtain a flow vector map. The horizontal component of the flow vector is a coarse indicator of the corresponding pixel's layer label, but it is very unfaithful. Therefore, flow vector maps between multi-pair repetitive regions are computed to estimate a consistent confidence map for the repetitive regions. The confidence map is used as the input for the graph-cut optimization method to refine the depth-layer decomposition in the next section. Our confidence map calculation includes two stages. (1) Local confidence map calculation. For each region R_i, the flow vector maps between it and R_{i-2}, R_{i-1}, R_{i+1}, R_{i+2} are computed respectively. These flow vector maps are converted to confidence maps as the following equation.

$$cm_{ij}(s) = \begin{cases} 1 & \text{if } |(h_{i,j}(s)/t_{ij}| < C/5 \\ -1 & \text{else} \end{cases} \tag{4}$$

where $cm_{ij}(s)$ denotes the confidence map contributed from R_j to R_i, $h_{ij}(t)$ denotes the horizontal component of the flow vector map between R_i and R_j. Local confidence map of R_i is calculated by accumulating all the confidence maps from neighboring repetitive regions (Fig. 7(a)). (2) Global confidence map calculation. For a group of repetitive regions, only one uniform confidence map is calculated by summing all the regions' local confidence maps. During the summation of local confidence maps, the axial and translational symmetry are also considered to increase the number of votes and improve the robustness of the global confidence map (Fig. 7(b)).

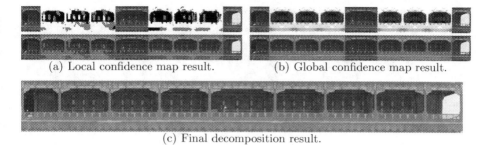

(a) Local confidence map result.   (b) Global confidence map result.

(c) Final decomposition result.

Fig. 7. For (a) and (b), top image shows the confidence value directly. Bottom image shows the segmentation results by blending the confidence value and pixel color value. (c) Final decomposition result including non-repetitive regions.

5.3   Decomposition Refinement

After coarse depth-layer decomposition, a uniform decomposition result for each repetitive region is obtained and shown as in the top image of Fig. 6(b). However, the boundary of the decomposition result is not accurate. In order to refine it, local edge information is utilized to optimize the boundary to the pixels with maximal gradient variation. Inspired by the interactive Graph-Cut method [21], a similar Markov Random Field (MRF) energy function is constructed, global confidence map for each repetitive region is assigned as the data term instead of the interactive constrain in [21], and the pixel edge response is assigned as the smooth term as the following equations.

$$E(L) = \sum E_{data}(L(s), s) + \sum_{L(p)!=L(q)} E_{smooth}(p, q) \tag{5}$$

where L denotes the layer label map of a repetitive region. There are only two kinds of values in L: 1 means front layer and 0 means back layer. s denotes one pixel in a repetitive region, (p, q) denotes one pair of neighboring pixels.

$$E_{data}(L(s), s) = \frac{1}{\sigma} e^{-cm'(L(s),s)} \tag{6}$$

$$E_{smooth}(p, q) = \frac{1}{dist(p, q)} e^{-\frac{edge(p)+edge(q)}{\gamma}} \tag{7}$$

$edge$ is the canny edge response of the region image, $edge(p) = 1$ means p is a edge pixel, or else $edge(p) = 0$, and γ is the smooth factor. The data term in the energy function is constructed by the confidence maps from coarse decomposition as:

$$cm'(L(s), s) = \begin{cases} K & \text{if } L(s) = 1 \& cm(s) > 0 \\ K & \text{if } L(s) = 0 \& cm(s) < 0 \\ -|cm(s)| & \text{else} \end{cases} \tag{8}$$

where K is set to 9 in the case of 2D image due to the constructed eight-connected graph as in method [21]. Equation 5 is optimized by a graph-cut algorithm [23] and the final refined result is shown in the bottom image of Fig. 6(b).

Fig. 8. Left images show the roof modeling parameters and result. Right image shows the depth-layer modeling result.

5.4   Non-repetitive Region Decomposition

Some repetitive regions over a wide distance are very difficult to detected (Left region and right region in Fig. 7(a) and (b)), and there also exist some non-repetitive regions (Center region in Fig. 7(a) and (b)). Fortunately, our depth-layer decomposition algorithm in a single image can be easily extended to multiple view images with approximate camera parameters. The remaining non-repetitive regions are decomposed by utilizing multiple view images and the results are shown in Fig. 7(c).

6   Experiments and Discussion

6.1   Architecture Geometry Modeling

After segmentation, we get architecture components composed of 3D points and texture images. In order to construct a 3D textured architecture model, textured plane models are used to fit planar components, and parametric surface models to fit non-planar components. The boundary of each component's sparse 3D points is projected onto the images to get a coarse texture boundary. Along the texture boundary, image windows with 100-pixel width are created, where pixels with maximum gradient variation are selected to refine the boundary and extract the modeling parameters. For example, roof is modeled by a quadric surface which is fitted by the parameters (*topwidth*, *bottomwidth*, *height* and *depth*) extracted from segmented texture and 3D points as in Fig. 8. The back layer of the architecture is usually occluded by the front layer, textures need to be repaired from multi-view images. For a planar object in 3D space, there exists an affine homography between each view. Image from front-parallel view is chosen as the reference to estimate the affine transforms with other views. The texture holes in this reference image are repaired by warping images from other views with the computed affine transforms. The modeling result of the floor with depth layers is shown in Fig. 8.

6.2   Implementation and Results

We demonstrate the results of our approach with three data sets: the HongYi Ge, the Chairman Mao Memorial Hall as shown in Fig. 9 and the Hall of Central Harmony as shown in Fig. 10.

Fig. 9. Left to right: 3D SFM points, top-bottom segmentation, depth-layer decomposition, final reconstructed model.

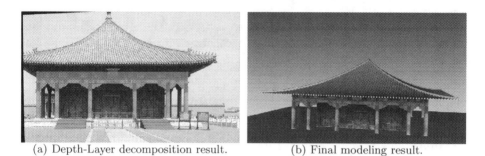

(a) Depth-Layer decomposition result.   (b) Final modeling result.

Fig. 10. Modeling results of Hall of Central Harmony.

We first evaluate our method on the ancient Chinese architecture HongYi Ge (Top row in Fig. 9). The resolution of our photos is about twenty million pixels. By using continuous shooting mode of the digital camera, these photos are taken in a few minutes for each architecture. VisualSFM [19] is used to generate 3D point cloud as shown in the first column. The result of top-bottom floors segmentation is shown in the second column. The building is segmented successfully into curved roofs and planar facades. For the facade floors with depth layers, repetition based depth-layer decomposition is implemented, and the result is shown in the third column. In this data set, the segmentation on the second floor is very difficult due to the same appearance between the front pillars and the back layer. However, by using both the parallax-shift and edge response, we can get excellent segmentation results. The textured 3D model of this building is shown in the fourth column of Fig. 9. Our method also works well on regular modern architectures. In these cases, top-bottom segmentation does not always output meaningful parts, and the output layers are all treated as simple planes as shown in the second row in Fig. 9. We also evaluate our method on the architecture without repetitive regions at the front layer as in Fig. 10. The modeling result proves that our depth-layer decomposition algorithm can also be applied to non-repetitive architecture.

Limitations. As most feature-based methods, SFM result is poor on non-Lambert material and textureless areas. If there exist gaps along the ground plane normal direction, our top-bottom segmentation algorithm will fail. The sparsity of SIFT features may also affect our parallax-shift detection algorithm, which is important in our depth-layer decomposition algorithm based on SIFT-Flow.

7   Conclusion and Future Work

We have presented an image-based modeling approach for architecture with a structure of multiple floors and depth layers. Repetition detection in the image region where repetition interference at different layers exists and irregular architecture components decomposition are handled well in our method.

The possible future work includes several directions. The geometry and texture of the architecture components are reconstructed simply which can be further refined to get geometry details and appearance from zoom-in images and make the 3D architecture model relightable. Procedural rules can be extracted to get an editable procedural 3D architecture model and generate large architecture scene quickly.

Acknowledgement. This work was supported in part by "National Natural Science Foundation of China (NSFC) 61202235, 61272348", "Humanity and Social Science Youth foundation of Ministry of Education of China 11YJCZH064", "Ph.D. Program Foundation of Ministry of Education of China (no. 20111102110018)", 863 Plan "2013AA013803", and "YWF-13-D2-JC-20".

References

1. Sinha, S.N., Steedly, D., Szeliski, R., Agrawala, M., Pollefeys, M.: Interactive 3d architectural modeling from unordered photo collections. In: ACM SIGGRAPH Asia 2008 papers, pp. 159:1–159:10. ACM (2008)
2. Jiang, N., Tan, P., Cheong, L.F.: Symmetric architecture modeling with a single image. In: ACM SIGGRAPH Asia 2009 papers, pp. 113:1–113:8. ACM (2009)
3. Xiao, J., Fang, T., Tan, P., Zhao, P., Ofek, E., Quan, L.: Image-based facade modeling. In: ACM SIGGRAPH Asia 2008 papers, pp. 161:1–161:10. ACM (2008)
4. Li, Y., Zheng, Q., Sharf, A., Cohen-Or, D., Chen, B., Mitra, N.J.: 2d–3d fusion for layer decomposition of urban facades. In: Proceedings of the 2011 International Conference on Computer Vision, pp. 882–889. IEEE Computer Society (2011)
5. Goesele, M., Snavely, N., Curless, B., Hoppe, H., Seitz, S.: Multi-view stereo for community photo collections. In: Proceedings of the 2007 International Conference on Computer Vision, pp. 1–8 (2007)
6. Agarwal, S., Snavely, N., Simon, I., Seitz, S., Szeliski, R.: Building rome in a day. In: Proceedings of the 2009 International Conference on Computer Vision, pp. 72–79 (2009)
7. Oh, B.M., Chen, M., Dorsey, J., Durand, F.: Image-based modeling and photo editing. In: ACM SIGGRAPH 2001 Papers, pp. 433–442. ACM (2001)

8. Hoiem, D., Efros, A.A., Hebert, M.: Automatic photo pop-up. In: ACM SIG-GRAPH 2005 Papers, pp. 577–584. ACM (2005)

9. Müller, P., Zeng, G., Wonka, P., Van Gool, L.: Image-based procedural modeling of facades. In: ACM SIGGRAPH 2007 papers. ACM (2007)

10. Barinova, O., Konushin, V., Yakubenko, A., Lee, K.C., Lim, H., Konushin, A.: Fast automatic single-view 3-d reconstruction of urban scenes. In: Forsyth, D., Torr, P., Zisserman, A. (eds.) ECCV 2008, Part II. LNCS, vol. 5303, pp. 100–113. Springer, Heidelberg (2008)

11. Wu, C., Frahm, J., Pollefeys, M.: Repetition-based dense single-view reconstruction. In: Proceedings of the 2011 IEEE Conference on Computer Vision and Pattern Recognition, pp. 3113–3120. IEEE Computer Society (2011)

12. Pollefeys, M., Nistér, D., Frahm, J.M., Akbarzadeh, A., Mordohai, P., Clipp, B., Engels, C., Gallup, D., Kim, S.J., Merrell, P., Salmi, C., Sinha, S., Talton, B., Wang, L., Yang, Q., Stewénius, H., Yang, R., Welch, G., Towles, H.: Detailed real-time urban 3d reconstruction from video. Int. J. Comput. Vis. **78**, 143–167 (2008)

13. Cornelis, N., Leibe, B., Cornelis, K., Gool, L.: 3d urban scene modeling integrating recognition and reconstruction. Int. J. Comput. Vis. **78**, 121–141 (2008)

14. Xiao, J., Fang, T., Zhao, P., Lhuillier, M., Quan, L.: Image-based street-side city modeling. In: ACM SIGGRAPH Asia 2009 papers, pp. 114:1–114:12. ACM (2009)

15. Wu, C., Agarwal, S., Curless, B., Seitz, S.M.: Schematic surface reconstruction. In: Proceedings of the 2012 IEEE Conference on Computer Vision and Pattern Recognition, pp. 1498–1505. IEEE Computer Society (2012)

16. Fang, T., Wang, Z., Zhang, H., Quan, L.: Image-based modeling of unwrappable facades. IEEE Trans. Vis. Comput. Graph. **19**, 1720–1731 (2013)

17. Coughlan, J.M., Yuille, A.L.: Manhattan world: compass direction from a single image by bayesian inference. In: Proceedings of the 1999 International Conference on Computer Vision, pp. 941–947. IEEE Computer Society (1999)

18. Liu, C., Yuen, J., Torralba, A.: Sift flow: dense correspondence across scenes and its applicationn. Pattern Anal. Mach. Intell. **33**, 978–994 (2011)

19. Wu, C., Agarwal, S., Curless, B., Seitz, S.: Multicore bundle adjustment. In: Proceedings of the 2011 IEEE Conference on Computer Vision and Pattern Recognition, pp. 3057–3064 (2011)

20. Rusu, R., Cousins, S.: 3d is here: point cloud library (pcl). In: 2011 IEEE International Conference on Robotics and Automation (ICRA), pp. 1–4 (2011)

21. Boykov, Y., Jolly, M.P.: Interactive graph cuts for optimal boundary and region segmentation of objects in n-d images. In: Proceedings of the 2001 International Conference on Computer Vision, vol. 1, pp. 105–112 (2001)

22. Wu, C., Frahm, J.-M., Pollefeys, M.: Detecting large repetitive structures with salient boundaries. In: Daniilidis, K., Maragos, P., Paragios, N. (eds.) ECCV 2010, Part II. LNCS, vol. 6312, pp. 142–155. Springer, Heidelberg (2010)

23. Fulkerson, B., Vedaldi, A., Soatto, S.: Class segmentation and object localization with superpixel neighborhoods. In: Proceedings of the 2009 International Conference on Computer Vision (2009)

A Method for Extracting Text from Stone Inscriptions Using Character Spotting

Shashaank M. Aswatha[✉], Ananth Nath Talla, Jayanta Mukhopadhyay,
and Partha Bhowmick

Indian Institute of Technology Kharagpur, Kharagpur, India
shashaankama@gmail.com

Abstract. A novel interactive technique for extraction of text characters from the images of stone inscriptions is introduced in this paper. It is designed particularly for on-site processing of inscription images acquired at various historic palaces, monuments, and temples. Its underlying principle is made of several robust character-analytic elements like HoG features, vowel diacritics, and location-bounded scan lines. Since the process involves character spotting and extraction of the inscribed information to editable text, it would subsequently help the archaeologists for epigraphy, transliteration, and translation of rock inscriptions, particularly for the ones having high degradations, noise, and a variety of styles according to the mason origin and reign. The spotted characters can also be used to create a database for ancient script analysis and related archaeological work. We have tested our method on various stone inscriptions collected from some of the heritage sites of Karnataka, India, and the results are quite promising. An Android application of the proposed work is also developed to aid the epigraphers in the study of inscriptions using a tablet or a mobile phone.

1 Introduction

Many of the stone inscriptions that are found across different regions of the world reveal the details of extravagance, lifestyle, economic condition, culture, and also of the administrative regulations followed by various rulers and dynasties particular to those regions. The information gained from these inscriptions can be corroborated with the information from other sources, in order to provide an insight into world's dynastic history, which otherwise lacks the completeness of contemporary historical records. Epigraphists use this information to identify the graphemes, clarify their meaning, classify their uses according to dates and cultural contexts, and to draw conclusions about the writing and its writers. Texts inscribed on stone are usually put up for public view to exhibit different cultures that prevailed during the period of inscription.

The inscriptions considered in our work are from Indian subcontinent. These Indic inscriptions have a composite mix of characters that evolved during the reign of several dynasties and kingdoms. They are usually found to be engraved on a variety of stones and other durable materials. Conventionally, they are studied

© Springer International Publishing Switzerland 2015
C.V. Jawahar and S. Shan (Eds.): ACCV 2014 Workshops, Part II, LNCS 9009, pp. 598–611, 2015.
DOI: 10.1007/978-3-319-16631-5_44

offline by generating *estampages* of the inscription surface. For this, the surface of the stone inscription is first cleaned with water-soaked brush. Then, the stone surface is carpeted by a large piece of wet paper (or layers of paper), which is gently patted by a dabber made of some soft material. The dabber is smeared with Indian ink to get the impression of the surface. The paper is allowed to dry on the stone surface before taking it off. The ink impression (estampage) comes out as white letters (grooves of the characters) against black background on the paper. Epigraphers usually take several days to few weeks for reading, transliterating, and for translating these estampages.

With elapsing time, these inscriptions are gradually deteriorating to a undecipherable state. Although estampages are taken for many of them, it is very difficult to preserve these estampages, as they fade away very soon. Frequent generation of estampages would cause the inscription to degrade more, since it involves a physical activity on the inscription surface. Hence, with advancing technology, various imaging techniques have emerged for acquiring images with considerable originality and economical viability. However, extraction and processing of information and text from these images is a challenging problem due to various factors, such as uncontrolled illumination, multilingual text, low-contrast distinction between the groove and the surrounding surface, distortions due to perspective projection, and administrative constraints in using imaging devices at heritage sites.

2   Related Work and Our Contribution

There exists a series of research work on historical document processing, which deals mainly with preprocessing, word spotting, classification, and optical character recognition (OCR) [1,2]. However, a little has been done so far to address the problem of extracting text and symbols from age-old inscriptions having historical importance. Moreover, the problem complexity and the related issues are often different in case of recognizing inscribed texts, as evident from the related literature [3–12]. Hence, for binarization of inscription images, an exclusive method is proposed recently in [12]. As the text in an inscription has a chiseled and engraved effect, which has often a degraded form owing to erosion through centuries, high-precision 3D measuring techniques are found to be useful, as shown in [3,7]. Although these 3D techniques help in producing almost exact copies of the original inscriptions with an objective of processing them for better legibility, they are both expensive and time-consuming. To achieve a better processing speed, a GPU-based method for optical character transcription is proposed in [6], which is focused only on inscriptions written in a script that is highly structured in both horizontal and vertical directions. To identify the dating of a stone inscription by identifying its writer, other methodologies can be seen in [4,5,8,9,11]. In [5], enhancement of inscription images for recognizing the text using OCR is performed using *natural gradient based flexible ICA* (NGFICA). This is carried out by grouping the inscriptions by their inscribers [11]. A technique for identifying the consonants of a language from an inscription image by

600   S.M. Aswatha et al.

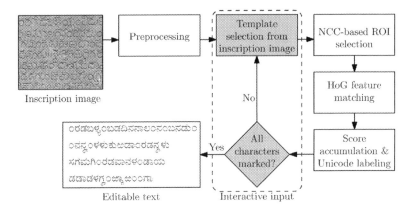

Fig. 1. The proposed character spotting algorithm. Note that the output editable text contains the modern Kannada characters equivalent to the ones found in the inscription image.

finding the feature vectors and classifying the characters based on SVM classifier, is also reported in [10].

This paper proposes an on-site semi-automatic method to extract text from the images of stone inscriptions using a novel technique of character spotting[1]. The proposed technique for character spotting uses *histogram of oriented gradients* (HoG) [13] as feature descriptor, and it does not use any classifier or training dataset. A labeling is made by the user through an interactive process, which is the input to subsequent processes. This requires a knowledgeable user, who is acquainted with the script, to convert the inscription to an editable text document, which can be used for further analysis. Since the marked and the spotted characters exist on the same surface of an inscription, the spotting process is robust to writing style and font, with a reasonable presumption that the entire inscription has a uniformity of script. The editable text can later be used for different analysis purposes like epigraphy, translation, and transliteration. The method accommodates human expertise in the process of delivering quality output.

The rest of the paper is organized as follows. The proposed method is explained in Sect. 3. We have tested our method on various stone inscriptions collected from the heritage sites of Karnataka, India. Some of these test results and our experimental setup are presented in Sect. 4. Finally, we conclude the paper in Sect. 5.

3   The Character Spotting Technique

The character spotting algorithm is shown in Fig. 1. The image of an inscription is first preprocessed for contrast adjustment and denoising. Then the user starts

---

[1] Patent pending: System and method for converting substrate inscription into electronically editable format; patent filing reference no.: 452/KOL/2014, April 2014.

Fig. 2. An inscription image (cropped) with character spotting result (left: highlighted in yellow) corresponding to a template image (middle) having the HoG visualization as shown (right) (Color figure online).

by selecting a character in the inscription by specifying a rectangular box around the character (henceforth referred to as *template*) and selects the corresponding Unicode related to its equivalent modern character as input. Figure 2 shows an example of the *search* image and the *template* image. Iteratively, the first occurrence of each character is marked by the user, which is the spotted throughout the inscription. If any character is missed in the spotting process, it can be marked again in subsequent trials. The template is divided into equally sized N overlapping parts, whose scores of the *normalized cross correlation (NCC)* are accumulated at the center region of the original image portion corresponding to the template image fragments. For a template size equivalent to 100×100, $N = 8$ is observed to yield the desired result for our data.

The resulting correlation surface is thresholded to obtain the regions of interest (ROI) of possible character occurrences in the original image. All spotted characters from each of the iterations are marked using a distinct color. The above process is repeated until all the characters are marked. The process of character spotting involves manual interaction to choose the Unicodes of the characters that are being marked. The spotted characters and their locations are saved in a text file. The last stage of the character spotting process involves reading the spotted character positions and their Unicodes from the text file and generating an editable text document of the inscription in required script.

3.1   Feature Points and Thresholding

The normalized cross correlation is carried out in parts. Let I_s be the stone inscription image and I_t be the character template selected from I_s. Let I_{tp} be one of the overlapping parts of I_t. At each pixel location (x, y), $I_{st}(x, y)$ is an image patch in I_s, centered at (x, y), confined under the character template fragment when I_{tp} is placed at that location. By performing correlation of all the parts, we compute the character centers in I_s, which are similar to the character in the template image I_t. The template image, which is divided into 8 overlapping

Fig. 3. Overlapping parts for the correlation corresponding to the template shown in Fig. 2.

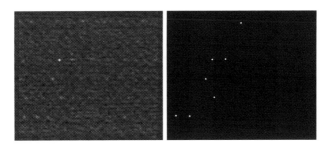

Fig. 4. Correlation result (left) and its thresholded image (right) corresponding to the inscription image and the template shown in Fig. 2.

parts, is shown in Fig. 3. The normalized cross correlation at each of the pixel locations, for each of the parts is given by

$$\gamma(x,y) = \frac{\sum (I_{st}(x,y) - \overline{I_{st}(x,y)})(I_{tp} - \overline{I_{tp}})}{\sqrt{\sum (I_{st}(x,y) - \overline{I_{st}(x,y)})^2 (I_{tp} - \overline{I_{tp}})^2}},$$

where, $\overline{I_{tp}}$ is the constant-intensity image with the mean value of pixel intensities in I_{tp}, and $\overline{I_{st}(x,y)}$ is the similar one corresponding to $I_{st}(x,y)$. The maximum in the window of radius 4 is computed at each pixel location of the correlated image for each template fragment. The correlation values of all the parts are aggregated to get the correlation result for the entire template. The correlation result is then thresholded based on a threshold value, t_c. For the data considered in our work, the value of t_c has been empirically set to 30 % of the number of fragments. After thresholding, we obtain points where the possible occurrences of the template exist. These points are the probable centers of the character in the template image. Figure 4 shows the correlation result and the thresholded correlation image.

3.2   HoG Feature Description and Matching

HoG features capture the local edge information, with some tolerance to deformation, using the distribution of intensity gradients [13]. The local shape information gets captured even without precise information about the location of the edges. The HoG feature is computed by dividing the image region into small non-overlapping regions, called *cells*. A histogram of gradient directions or edge orientations is computed for each cell. Overlapping neighboring cell histograms are combined to form blocks, which are individually normalized to unit magnitude. A combined histogram entry from all the blocks is used as the feature vector describing the image region. HoG features are used in our work accounting to their relative invariance to local geometric and photometric transformations. The key points (character centers) that are identified by correlation results are considered to prune the search space for spotting the marked character. The character spotting is performed by describing the image regions of possible occurrences at each of the key points by patch descriptors. Similar to the correlation process, a neighborhood of the size of ROI is considered, which is described using HoG feature descriptor. To take into account the partial deterioration of the characters in stone inscriptions, the matching is performed by computing the scalar product of corresponding normalized feature vectors of each of the blocks. The mean value of all the block feature projection scores is accumulated to obtain a final score of the matching. If the template image is described by M blocks, then an identical image patch in the search image would have a maximum matching score of M. These mean scores for all the possible regions of occurrences are thresholded by a threshold value, t_f, to spot the selected character. The threshold is empirically set to 80 % of the number of blocks in the template image. Figure 2(right) shows the HoG descriptor for a template image.

For computing HoG features, we have chosen a cell size of 8×8 pixels, the number of orientation bins as 9, and the block size to be of 32×32 pixels (that is, 4×4 cells). The input image is preprocessed using *non-local means (NLM)* denoising process [14]. It is found that application of Gaussian smoothing before applying the derivative mask did not improve the results, but yielded relatively poorer results in matching similar patches. Characters identified by the HoG matching, for each template, are displayed in a distinct color, as shown in Fig. 2. For every spotting result, respective template size, spatial location of the spotted character center, matching score, and Unicode of the character in the template image are written to a text file.

This process of character spotting continues iteratively, until all the characters are spotted. Figure 5 shows an example of the search image (partial), after spotting 26 character templates.

3.3 Working with Vowel Diacritics

A *diacritic* is a glyph added to a letter. It is a modifier that changes the pronunciation of a consonant by associating either a consonant or a vowel to the base character. In general, diacritical marks appear above or below a consonant, or in some other position, such as within the letter or between two letters. However, in Kannada script, the diacritics can appear above, below, before, or after the consonants. Figure 6 shows four typical instances of vowel diacritics of consonants in Kannada.

To resolve the ambiguity in character diacritics, the matching scores and the areas associated with the character templates are considered. When multiple templates are found to match with the same character in the search image,

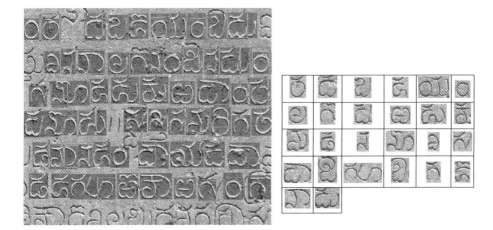

Fig. 5. Partial character spotting result for the inscription image shown in Fig. 2 corresponding to 26 templates.

the areas of the corresponding template images are used to decide the best. If the maximum area is at least 1.5 times the second maximum, then the one associated with the maximum area is considered and the rest is discarded. Otherwise, the template associated with the maximum score is considered. Figure 7 shows an example of spotting a Kannada consonant 'na' (template). The spotted character is a part of consonant's vowel combination, 'nu', shown in blue rectangle. Here, matching scores are used to resolve the diacritic ambiguity.

3.4   Text Generation

The information about the spotted characters, which are obtained from the above process, contains the locations of the characters in the image. During the spotting process, the locations of the spotted characters along with the Unicodes are stored in a text file. After completely spotting all the characters in the image (which is inspected manually), the text file contains all the locations and Unicodes of respective characters in the image. The entries of character centers in the text file are almost in a sorted order, in terms of the coordinates of the character centers in the search image. However, due to the complexity and non-uniformity involved in the script, the characters centers may not always lie exactly in the same line as in the search image. Figure 8 shows an image with the character centers along with their corresponding Unicodes. So, these

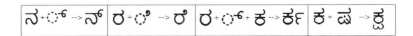

Fig. 6. Four typical cases of vowel diacritic of a consonant.

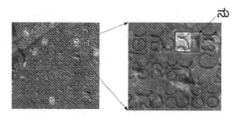

Fig. 7. An example of ambiguity during character spotting (Color figure online).

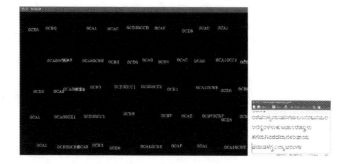

Fig. 8. An image with Unicodes at the locations of their respective characters (left) and the corresponding text image (right).

character centers are sorted in row-major (i.e., left-to-right and top-to-bottom) order. For each character center (Fig. 9), the upper boundary row index and the lower boundary row index, denoted by a and b respectively, are computed from the dimensions of the character. For the top-left character center $p(x, y)$, which serves as the *anchor point*, all the centers whose row index is less than b are considered, and the row indices of these centers are assigned the same index as p. The process is iterated with subsequent anchor points, as shown in Fig. 9. After rearranging the character centers, the Unicodes corresponding to

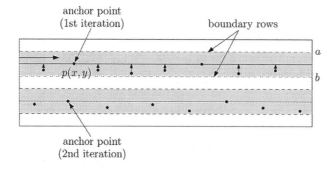

Fig. 9. Location-bounded scan for sorting the lines.

the sorted character centers are updated in the text file. Finally, these Unicodes are parsed to generate the inscription text.

4   Experimental Results

The proposed character spotting method is tested on the images of different stone inscriptions collected from various parts of Karnataka, India. Stone inscriptions made by various dynasties that ruled over Karnataka display unique features in their style with reference to the type of stone, polishing, composition of text, inscribing on stone with color, engraving the text on stone, and also based on the position of erecting the stone in an appropriate place. Many of these inscriptions are deteriorated so badly that it is difficult to identify the meaningful data, particularly when the surface is broken or etched. Due to centuries of deterioration, majority of these ancient texts are in very poor condition, and many text portions are already missing. The damage has occurred to such an extent that either the fragments do not exist, or sections are no longer recognizable and beyond recovery. The performance result of the character spotting process on such images is also reported in this section.

4.1   Experimental Setup

A group of volunteers consisting of Sanskrit scholars, Kannada scholars, regional journalists, and archaeological students were identified to put the tool into use. These volunteers belong to different age groups and sects, and have knowledge of more than three regional dialects. A questionnaire form was given to each of the volunteers to identify the background of the volunteer. Feedbacks of the spotting tool by the volunteers are collected. It is found that the tool is indeed useful for extracting the text from stone inscriptions. Also, the volunteers have agreed that this tool will considerably reduce the effort and interpretation time in deciphering the inscriptions. Some of the inscriptions that were considered in our experiments, procured from various places in Karnataka, India, are mentioned below.

1. *Arasikere Temple, Karnataka:* Stone inscriptions found in Arasikere temples are characterized by artistic writing on huge slabs with small engravings. So, the images of such inscriptions were taken by us in slices. Figure 10 shows the image slice of a stone inscription of size 7098×1114 on which the character spotting method was applied. This image has a non-uniform noise distribution owing to stone texture, which is difficult to separate from the foreground characters. The performance of character spotting on this image is shown in Table 1.

2. *Chitradurga Fort, Karnataka:* The image in Fig. 10 is of an inscription in the fort of Chitradurga, Karnataka. The inscriptions found in this fort are made of black granite—a homogeneous rock that changes its chemical properties according to climatic conditions. The character spotting process is executed in the top segment of the inscription, and the corresponding result is shown in Table 1.

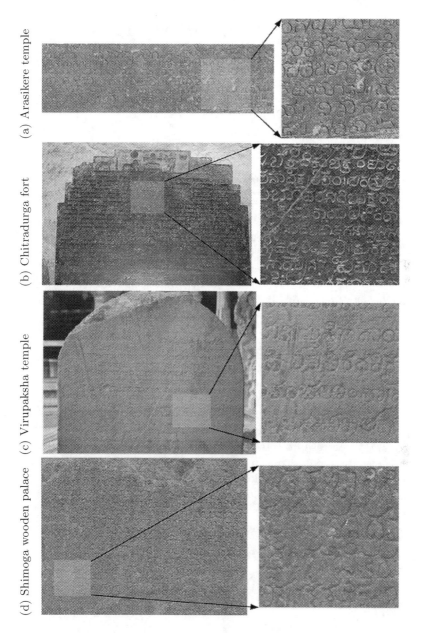

Fig. 10. Some of the test images of stone inscriptions collected by us from different heritage sites of Karnataka, India.

3. *Virupaksha Temple, Hampi, Karnataka:* Virupaksha temple in Hampi is a part of the group of monuments at the sites of Hampi, designated by UNESCO as one of the world heritage sites. Figure 10 shows one of the inscriptions

found at the temple entrance. The performance result on the image of this inscription is shown in Table 1.

4. *Shimoga Wooden Palace, Karnataka:* Shimoga is a city in the central part of the state of Karnataka, India. The inscriptions found in the palace are not clear as there is a little difference between the characters and the background. Many of them are almost faded away due to constant exposure to climatic variations. Figure 10 shows an inscription found in the palace, and the performance of our algorithm on this is shown in Table 1.

4.2   Desktop Implementation

The spotting algorithm is implemented in C++ under Ubuntu 12.04 environment, using OpenCV [15] and Qt standard libraries. In all the experiments, inscriptions in Kannada script were considered. These images were acquired by high-resolution cameras (8 to 18 mega pixels) in daylight under auto settings without any external illumination. Some are sheltered indoor, while majority of them are kept in open grounds. The modern Kannada characters are displayed by using *Baraha* font library.

4.3   Performance Analysis

The performance of the spotting method is analyzed by estimating the measures of *sensitivity, specificity, positive predictive value* (PPV), and *negative predictive value* (NPV). For the spotting result of each character, we calculate the number of *true positives* (TP), *false positives* (FP), *true negatives* (TN), and *false positives* (FN), which correspond to correctly spotted, incorrectly spotted, correctly rejected, and incorrectly rejected characters, respectively. The ground truth for each of the inscriptions is obtained by the experts of Archeological Survey of India (ASI) and manual inspection.

Sensitivity is the ability of the method to correctly spot a character, given a similar character as input. Specificity is defined as the proportion of other characters (excluding the selected character) that are not correctly spotted. Precision is the fraction of the spotted characters that are same as the input character. The mathematical expressions of these parameters are as follows.

$$Sensitivity\ (True\ Positive\ Rate) = \frac{TP}{(TP + FN)}$$

$$Specificity\ (True\ Negative\ Rate) = \frac{TN}{(FP + TN)}$$

$$Precision = \frac{TP}{TP + FP}$$

$$Accuracy = \frac{(TP + TN)}{(TP + FN + FP + TN)}$$

Table 1. Desktop performance (in %) and average CPU time (per spotting per template) for the inscription images shown in Fig. 10.

Performance measure	Fig. 10a	Fig. 10b	Fig. 10c	Fig. 10d
Sensitivity	86.05	73.82	68.27	52.50
Specificity	99.14	84.31	84.12	92.52
Precision	78.88	43.04	30.39	51.59
Accuracy	99.00	83.92	83.01	90.74
CPU time (sec.)	1.10	0.72	0.76	0.88

Arasikere inscription 1.10 s Chitradurga inscription 0.72 s Viroopaksha inscription 0.76 s Shimoga inscription 0.88 s.

It is observed from our experiments that, on varying the thresholds t_c and t_f (Sects. 3.1 and 3.2), the results also vary to a great extent. If t_c and t_f are decreased from the chosen values, then the number of false positives increases. Since the search space also increases with decreasing thresholds, the time taken for execution increases. If t_c and t_f are increased beyond the chosen values, then the number of false negatives increases. In the latter case, since the search space decreases, the execution time also decreases. These thresholds are chosen empirically.

4.4   Android Implementation

The proposed method of character spotting presented and its related implementation with an appropriate interface caters the historical information to an archaeologist, who already has a database of stone inscription images taken by professional photographers. Wide adoption of mobile devices, especially smart phones with the app-store mobile application distribution model, supports an archaeological group to work out many problems on-site. Hence, we have also made an implementation of our method in mobile devices and tablets to initiate the interpretation

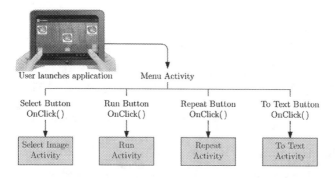

Fig. 11. Android application activity calls.

process on-site, without necessarily requiring off-line workspace. The hierarchy of activity calls in our Android application is shown in Fig. 11.

5   Conclusion

In this paper, we have proposed a useful interactive tool for epigraphers to read and archive ancient inscriptions in a convenient way. It can be used to replace the tedious task of obtaining estampages from stone inscriptions by ink-smeared manual dabbers, as followed in the conventional practice. The Android application developed by us can be used by the epigraphers and historians to analyze and interpret the inscriptions on-site. The character spotting results can be used to create a dataset of various characters of the concerned language, which would be helpful in studying the hierarchy of evolution of the script. This dataset can also be used to train a classifier for recognition process, which is in the purview of our future work.

Acknowledgement. This work is partially sponsored by Department of Science and Technology, Govt. of India through sanction number NRDMS/11/1586/2009.

References

1. Vamvakas, G., Gatos, B., Stamatopoulos, N., Perantonis, S.J.: A complete optical character recognition methodology for historical documents. In: International Workshop on Document Analysis Systems, pp. 525–532 (2008)
2. Kim, S.K., Sin, B.K., Lee, S.W.: Character spotting using image-based stochastic models. In: International Conference on Document Analysis and Recognition, pp. 60–63 (2001)
3. Barmpoutis, A., Bozia, E., Wagman, R.: A novel framework for 3d reconstruction and analysis of ancient inscriptions. Mach. Vis. Appl. **21**, 989–998 (2010)
4. Galanopoulos, G., Papaodysseus, C., Arabadjis, D., Exarhos, M.: Exploiting 3D digital representations of ancient inscriptions to identify their writer. In: Bebis, G., et al. (eds.) ISVC 2012, Part II. LNCS, vol. 7432, pp. 188–198. Springer, Heidelberg (2012)
5. Shridevi, I.: Enhancement of inscription images. In: National Conference on Communications, pp. 1–5 (2013)
6. Mara, H., Hering, J., Kromker, S.: GPU based optical character transcription for ancient inscription recognition. In: International Conference on Virtual Systems and Multimedia, pp. 154–159 (2009)
7. Schmidt, N., Boochs, F., Schütze, R.: Capture and processing of high resolution 3D-data of Sutra inscriptions in China. In: Ioannides, M., Fellner, D., Georgopoulos, A., Hadjimitsis, D.G. (eds.) EuroMed 2010. LNCS, vol. 6436, pp. 125–139. Springer, Heidelberg (2010)
8. Papaodysseus, C., Rousopoulos, P., Arabadjis, D., Panopoulou, F., Panagopoulos, M.: Handwriting automatic classification: application to ancient greek inscriptions. In: International Conference on Autonomous and Intelligent Systems, pp. 1–6 (2010)

9. Papaodysseus, C., Rousopoulos, P., Giannopoulos, F., Zannos, S., Arabadjis, D., Panagopoulos, M., Kalfa, E., Blackwell, C., Tracy, S.: Identifying the writer of ancient inscriptions and Byzantine codices. A novel approach. Comput. Vis. Image Underst. **121**, 57–73 (2014)

10. Rajakumar, S.: Eighth century Tamil consonants recognition from stone inscriptions. In: International Conference on Recent Trends in Information Technology, pp. 40–43 (2012)

11. Rousopoulos, P., Panagopoulos, M., Papaodysseus, C., Panopoulou, F., Arabadjis, D., Tracy, S., Giannopoulos, F., Zannos, S.: A new approach for ancient inscriptions' writer identification. In: International Conference on Digital Signal Processing, pp. 1–6 (2011)

12. Shaus, A., Turkel, E., Piasetzky, E.: Binarization of first temple period inscriptions: performance of existing algorithms and a new registration based scheme. In: International Conference on Frontiers in Handwriting Recognition, pp. 645–650 (2012)

13. Dalal, N., Triggs, B.: Histograms of oriented gradients for human detection. In: International Conference on Computer Vision and Pattern Recognition, pp. 886–893 (2005)

14. Buades, A., Coll, B., Morel, J.M.: A non-local algorithm for image denoising. In: International Conference on Computer Vision and Pattern Recognition, pp. 60–65 (2005)

15. OpenCVDocumentation: (OpenCV documentation). http://docs.opencv.org/modulesimgprocdocobject_detection.html. Accessed 10 Nov 2013

3D Model Automatic Exploration: Smooth and Intelligent Virtual Camera Control

Zaynab Habibi[✉], Guillaume Caron, and El Mustapha Mouaddib

MIS Laboratory, University of Picardie Jules Verne, Amiens, France
zaynab.habibi@u-picardie.fr

Abstract. In a 3D dense point clouds model, virtual tour without assistance is a complex and difficult task discouraging users from doing so. The aim of this work is to achieve a virtual navigation support tool. It will help users to perform virtual tour to explore the 3D model. In particular, the tool will allow to guide the camera automatically. We assume that the user is attracted by rich information areas in the model. This important assumption will be modelled by entropy. Secondly, in order to achieve a realistic automatic navigation we must avoid obstacles, ensure a relevant camera orientation during its motion and regulate the visual movement in the produced image. In this paper, we propose a solution to this problem based on a hierarchical algorithm, which combines the main task to be achieved and the realistic constraints. We validate the system on different complex 3D models: lab, urban environment and a cathedral.

1 Introduction

The last few years, 3D virtual applications such as architectural walk-through and videogames are spreading on and receive much attention from researchers in computer animation. However, existing tools for 3D exploration of virtual environments show some inconveniences, particularly for uninitiated people: the difficulty of finding a relevant point of view and the synthetic appearance of the camera movement. Thus, controlling the virtual camera is a task of primary importance in order to transmit relevant information to the user, i.e. to ensure the appropriate visualization within a 3D model.

The aim of the paper is to address the visual content-based automated and relevant control of a virtual camera. Considering a 3D model, a result that we want to automatically obtain is illustrated in Fig. 1. The virtual camera moves in the 3D scene in order to visualize the most important information.

Existing automatic camera control approaches are either path planning methods or image-based ones.

For path planning methods, preliminarily to the planning stage itself, waypoints must be defined, either manually or automatically, by sampling the camera configuration space. The selection of good viewpoints, that is to visualize the maximum information of the scene, generally exploits information theory-based measures. Indeed, under the hypothesis that users are naturally pulled through

© Springer International Publishing Switzerland 2015
C.V. Jawahar and S. Shan (Eds.): ACCV 2014 Workshops, Part II, LNCS 9009, pp. 612–626, 2015.
DOI: 10.1007/978-3-319-16631-5_45

Fig. 1. Ideal camera path maximizing the relevant information in the image.

areas of high information, the entropy is exploited to characterize the attraction of an area of the 3D model. In a static environment, the viewpoint quality can be evaluated by computing a geometric visual entropy [22]. Polygonal meshes are considered in the latter work. To compute the probability distribution for entropy expression, [22] use the relative area of 3D mesh projected faces over the sphere of directions centered in the viewpoint. The maximum entropy is reached when a certain viewpoint can see all the faces with the same projected area. All candidate views have been manually determined [22] and then validate on the entropy criterion.

In lighting context, illumination entropy has been exploited [11] to find the view that maximize the scene illumination. An adaptive search algorithm uses pixel brightness values (Y tristimulus value of the CIE 1931 standardized color model) to compute the probability distribution of the entropy.

In multi-room 3D model, [1] exploit the geometric visual entropy in order to find the most relevant viewpoints. Then build a path exploring all the cells using a backtracking algorithm.

Indeed, connecting between several viewpoints, whatever the technique to determine them (manually, based or not on geometric [22] or photometric [11] entropy), is the next step for path planning. Other path-planning techniques, as the A-star algorithm [23], the traveling salesman problem [20], roadmaps [16,19], spatial decomposition methods [4,14] and potential field-based approaches [5,13] can be used for connecting viewpoints. And finally, the virtual camera path is built by the set of these viewpoints using an interpolation method (a spline curve, for instance [2]) to determine the intermediate camera positions.

The second family of virtual camera control is image-based control methods. The camera path is, thus, defined by optimizing a cost function, defined using some image properties. An iterative process is performed to update the six camera degrees of freedom. Within this context, Courty et al. [8] proposed a visual servoing based approach. This approach requires adding constraints on the virtual camera to define the positioning task in the environment. This technique allows, on one hand, following a particular object on the screen, and on the other hand, provides assistance to create a camera path using cinematographic primitives.

To have a wider view of camera control methods, one can refer to the detailed state of the art made by Christie et al. [7].

Previously mentioned path-planning methods exploit the visual entropy in a preliminary process to make use of the selected viewpoints later. These approaches can be accompanied by an obstacle avoidance constraint [8] or not. Thus, only waypoints are relevant, not the camera motion, visually or dynamically speaking, to reach them (unknown problem for interactive methods because the user can manage speed). The problem comes from the fact that such virtual camera control proceeds following a set of different and, almost, disconnected stages. On the opposite, the virtual visual servoing framework, while not dealing with visual entropy, elegantly merges, in the same solver, all constraints simultaneously. Thus, we propose to address the entropy-based virtual camera control under the virtual visual servoing framework, with new constraints on the camera configuration, described below.

1.1   General Strategy

We propose to implement the photometric visual entropy, computed from all image intensities, since it depends on the viewpoint but it is independent of the 3D model type (mesh or point cloud). The challenge is, then, to perform the visual servoing-based virtual camera control using the photometric entropy feature, maximizing the amount of perceived information in the image. Indeed, it will allow the determination of a relevant direction of the camera movement from any viewpoint without any planning stage or the need to connect between several viewpoints independently. Furthermore, to mechanically ensure a relevant camera orientation during its motion, five operational degrees of freedom (three translations, then pan and, then, tilt orientations) are considered for the camera, instead of the six Cartesian degrees of freedom, generally considered in the virtual visual servoing.

The virtual camera must avoid obstacles during its movement. A lot of work in camera control has been tackled to deal with the latter issue such as ray tracing methods [3,6,21], bounding volumes methods [8,10] and image-based methods [18]. To avoid obstacles, Courty et al. [8] define a function that reaches an infinite value when the distance between the camera and the obstacle is null. We adapted this method that consists of maximizing the distance between the camera and all the closest 3D points of the 3D model.

The maximization of photometric entropy and the maximization of distances between the camera and obstacles are merged into the same problem as a hybrid control law [12]. This allows avoiding sudden and repetitive motion direction changes as soon as an obstacle is detected.

When the photometric entropy and obstacle avoidance based control law converges, the static relevance of the image is locally ensured. The images got during the camera motion have an increasing static relevance along the achieved camera path, as it gets closer to the optimum. This is ensured by the optimization method implemented by virtual visual servoing [8]. However, the dynamic relevance is not guaranteed as jumps between two successive images during the camera motion might be observed. To overcome this discontinuity issue, particularly in the image motion appearance, we propose the first optical flow-based control method to regulate or confine the camera motion. Indeed, nothing has been found on that topic in the state of the art of image-based methods.

To merge the entropy-obstacle hybrid control with the optical flow regulation, we choose to implement a hierarchical control in which the direction of motion in the five operational camera degrees of freedom space is obtained from the entropy-obstacle hybrid control law and its magnitude by the optical flow regulation law.

Obviously, once the photometric visual entropy reaches a local maximum in the 3D model, no camera motion will ever be observed again. Still in an automated approach, we implemented an alternation of entropy maximization and minimization: first, a photometric entropy maximization process is led in order to bring the camera to a rich information area and second, a photometric entropy minimization process is applied to cause a change of the area. The alternation between these two processes is performed to move away from the local maxima/minima and therefore allows the generation of a long camera path. During the entropy maximization process, since the system is driving the camera through richer and richer visual information, it is not a drawback if the camera speed is slightly slowing down. That is why optical flow confinement-based control law is mentioned above. However, during the entropy minimization process, slow motions are not relevant since the visual information is getting poorer and poorer. Thus, during the entropy minimization process, the optical flow is not confined but regulated in order to have a visual motion of a constant speed, in a mean, as can be observed in cultural heritage documentaries, for instance.

1.2 Contributions

The general strategy is to achieve an automatic navigation carrying out important criteria. The contributions are as follows:

- We use the entropy to enssure a relevant motion direction.
- We add constraints to get realistic movement.
- We present a hierarchical optimization method to build relevant camera path.
- We tested our algorithm in indoor and outdoor navigation on different 3D complex models of colored 3D point clouds.

The remainder of this paper presents, first, a description of the proposed method. Then, we present some experimental results, and a conclusion ends the paper.

2 Proposed Method

To combine the principal task and the constraints, we could intuitively stack all of them in a single matrix and make an overall resolution. However, the result is falling below our expectations because in some particular situations one of the constraints is not fully taken into account. In image based approaches, the redundancy formalism is often used to combine several tasks. This technique is performed by assigning to each task a specific number of degrees of freedom of the camera. Courty et al. [8] have exploited this method to perform the main task of tracking objects, which does not require all the degrees of freedom of the camera, combined with a secondary task (occlusions management or lighting ...). In the redundancy formalism, it is not allowed for a degree of freedom to be operated by several tasks concurrently. This formalism can not be applied to our problem because the principal task that we consider (maximizing photometric entropy) requires all the degrees of freedom of the camera.

We propose a new algorithm to control automatically the camera while achieving the desired constraints. We choose a hierarchical approach done in two steps. As illustrated in Fig. 2, our method takes as input the 3D model and the camera pose $\mathbf{r}$. During the first step, we compute the entropy-obstacle hybrid control law to estimate the motion direction. Then, during the second step, using the optical flow based constraint, we adjust the movement magnitude. A camera path is built in an iterative optimization process taking into account these two steps.

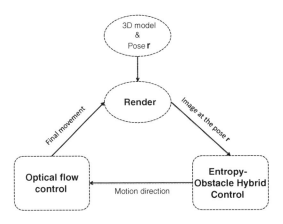

Fig. 2. The hierarchical strategy.

2.1   Task and Constraints Specification

Photometric Entropy: We consider $I(\mathbf{r})$ the image at the pose $\mathbf{r} = (t_X, t_Y, t_Z, \theta u_X, \theta u_Y, \theta u_Z)^T$, where the three translations are $[t_X, t_Y, t_Z]$ and the rotation is represented by an angle θ and an unit vector (axis of rotation) $\mathbf{u} = [u_X, u_Y, u_Z]$. The photometric entropy of this image is given by:

$$E(\mathbf{r}) = -\sum_i P_i(\mathbf{r}) log_2(P_i(\mathbf{r})) \tag{1}$$

where i is in the range of image grey levels of a pixel in the image, $i \in [0; 255]$ and $P_i(\mathbf{r})$ is the probability of the luminance i to exist in image I. For clarity, in the following $P_i(\mathbf{r})$ is replaced by P_i.

The probability distribution is obtained using the normalized histogram of intensity value. Inspired from Dame et al. in [9] for mutual information, in a different context. We approximate the normalized histogram using a second order

B-spline [17] ϕ that makes the photometric entropy twice differentiable, so the second order optimization can be achieved.

$$P_i = \frac{1}{N_{\mathbf{u}}} \sum_{\mathbf{u}} \phi(i - I(\mathbf{u}, \mathbf{r})) \tag{2}$$

where $\mathbf{u}(u, v)$ are the coordinates of a pixel $\mathbf{u}$, $I(\mathbf{u}, \mathbf{r})$ is the intensity at the current pose $\mathbf{r}$ and $N_{\mathbf{u}}$ is the number of image pixels.

Obstacles Avoidance Cost Function: All the vertices within a distance below one meter from the camera will be considered. The used cost function defined in [8], tends to infinity when the distance between the camera and the obstacle is null. Each vertex is considered as an obstacle and we maximize the distance between all the considered vertices and the camera which is equivalent to minimize:

$$O(\mathbf{r}) = \sum_k \frac{1}{2||{}^w\mathbf{p}_c - {}^w\mathbf{p}_{o_k}||^2} \tag{3}$$

Such that k is the index of a vertex. ${}^w\mathbf{p}_c = (x_c, y_c, z_c)^T$ is the camera position and ${}^w\mathbf{p}_{o_k} = (X_{o_k}, Y_{o_k}, Z_{o_k})^T$ is the position of one vertex in the world frame $\Re_w$.

Optical Flow Description: The goal of this constraint is to ensure a smooth movement in the image space. Our approach consists in fixing an optical flow magnitude between images. We choose a magnitude for the optical flow $\triangle\mathbf{u}_{des} = \left[\triangle u_{des}, \triangle v_{des}\right]^T$ and use the intensity conservation equation in order to link $\triangle\mathbf{u}_{des}$ and the camera pose $\mathbf{r}$:

$$I(\mathbf{u}, \mathbf{r}) = I(\mathbf{u} + \triangle\mathbf{u_{des}}, \mathbf{r} + \triangle\mathbf{r}) \tag{4}$$

2.2 Solving Process

The principal task and constraints are nonlinear equations w.r.t the camera pose $\mathbf{r}$. So, we choose an iterative nonlinear optimization scheme. The linearization will be achieved using the Taylor expansion. We present first, the camera control for the six Cartesian camera degrees of freedom and the transformation to five operational camera degrees of freedom will be explained at the end of this section. The iterative scheme consists in estimating the pose increment $\triangle\mathbf{r}$ to modify the current pose $\mathbf{r}_k$ (step k).:

$$\mathbf{r}_{k+1} = \mathbf{r}_k + \triangle\mathbf{r} \tag{5}$$

Photometric Entropy Criterion: To maximize the photometric entropy is equivalent to minimize its opposite $-E(\mathbf{r})$. Applying the first-order Taylor expansion leads to:

$$\frac{\partial(-E(\mathbf{r_0}))}{\partial\mathbf{r}} + \left(\frac{\partial^2(-E(\mathbf{r_0}))}{\partial\mathbf{r}^2}\right)\triangle\mathbf{r} = 0 \tag{6}$$

To solve this system, we need to compute all the derivatives of the photometric entropy. Starting by the first derivative of the entropy and using Eq. (1), we obtain:

$$\frac{\partial}{\partial \mathbf{r}}(-E(I(\mathbf{r_0}))) = \sum_i \frac{\partial P_i}{\partial \mathbf{r}}(1 + log(P_i)) \tag{7}$$

Using the expression of the probability expressed in Eq. (2) leads to:

$$\frac{\partial P_i}{\partial \mathbf{r}} = \frac{1}{N_{\mathbf{u}}} \sum_{\mathbf{u}} \left(-\frac{\partial \phi(i - I(\mathbf{u}, \mathbf{r}))}{\partial (i - I(\mathbf{u}, \mathbf{r}))} \nabla I \, \mathbf{L_u} \right) \tag{8}$$

where $\nabla I = (\nabla_u I, \nabla_v I)$ is the gradient of the image $I(\mathbf{r})$ and $\mathbf{L_u}$ the interaction matrix at $\mathbf{u} = (u, v)$.

$$\mathbf{L_u} = \frac{\partial \mathbf{u}}{\partial \mathbf{x}} L_{\mathbf{x}} \tag{9}$$

$\mathbf{x} = (x, y)$, where:

$$\begin{cases} u = & \alpha_u x + u_0 \\ v & \alpha_v y + v_0 \end{cases} \tag{10}$$

with $(\alpha_u, \alpha_v, u_0, v_0)$ are the intrinsic camera parameters and $L_{\mathbf{u}}$ is the interaction maxtrix defined in [10].

The development below represents the second derivative of the photometric entropy:

$$\frac{\partial^2(-E(\mathbf{r_0}))}{\partial \mathbf{r}^2} = \sum_i \frac{\partial P_i}{\partial \mathbf{r}} \frac{\partial P_i^T}{\partial \mathbf{r}} \frac{1}{P_i} + \frac{\partial^2 P_i}{\partial \mathbf{r}^2}(1 + log(P_i)) \tag{11}$$

Given the equation of $\frac{\partial P_i}{\partial \mathbf{r}}$, the second derivative of the probability is given by:

$$\frac{\partial^2 P_i}{\partial \mathbf{r}^2} = \frac{1}{N_{\mathbf{u}}} \sum_{\mathbf{u}} \left(\frac{\partial^2 \phi(i - I(\mathbf{u}, \mathbf{r}))}{\partial (i - I(\mathbf{u}, \mathbf{r}))^2} \mathbf{L_I^T L_I} - \frac{\partial \phi(i - I(\mathbf{u}, \mathbf{r}))}{\partial (i - I(\mathbf{u}, \mathbf{r}))} \mathbf{H_I} \right) \tag{12}$$

In this last equation, $\mathbf{L_I} = \nabla I \, \mathbf{L_u}$ and $H_I = \mathbf{L_u^T} \nabla^2 I \, \mathbf{L_u} + \nabla_u I \, \mathbf{H_u} + \nabla_v I \, \mathbf{H_v}$ where $\nabla^2 I = \begin{pmatrix} \nabla I_{uu} & \nabla I_{uv} \\ \nabla I_{vu} & \nabla I_{vv} \end{pmatrix}$ is the gradient of image gradient. $\mathbf{H_u}$ and $\mathbf{H_v}$ are the hessian of size 6×6 of the two coordinates of the point $\mathbf{u}$ [15].

Obstacles Avoidance Constraint: The goal is to minimize Eq. (3). Using the Taylor expansion, we have:

$$O(\mathbf{r}) \simeq O(\mathbf{r_0}) + \left(\frac{\partial(O(\mathbf{r_0}))}{\partial \mathbf{r}} \right) \triangle \mathbf{r} \tag{13}$$

As we want to avoid that the camera crosses an obstacle, only the position and not the orientation is considered because we suppose that our camera is a point (optical center). In the camera reference frame $(\Re_c)$ $^c\mathbf{p}_c = (0, 0, 0)^T$, then [8]:

$$\frac{\partial(O(\mathbf{r_0}))}{\partial \mathbf{r}} = \sum_k \frac{1}{||^c P_{o_k}||^2} \left(^c X_{o_k}, ^c Y_{o_k}, ^c Z_{o_k}, 0, 0, 0 \right) \tag{14}$$

Optical Flow Control Constraint: From the Eq. (4), and using again the Taylor expansion, we obtain:

$$I(\mathbf{u} + \triangle_{\mathbf{u}_{des}}, \mathbf{r} + \triangle\mathbf{r}_f) = I(\mathbf{u}, \mathbf{r}) + \frac{\partial I}{\partial \mathbf{u}}\triangle\mathbf{u}_{des} + \frac{\partial I}{\partial \mathbf{r}}\triangle\mathbf{r}_f \tag{15}$$

Substituting Eqs. (4) in (15) gives:

$$\frac{\partial I}{\partial \mathbf{u}}\triangle\mathbf{u}_{des} + \frac{\partial I}{\partial \mathbf{u}}\mathbf{L_u}\triangle\mathbf{r}_f = 0 \tag{16}$$

This last equation will be used later Eq. (20) to compute the final camera displacement.

Hierarchical Algorithm: In order to satisfy the principal task and the two constraints presented in the previous section, we propose a hierarchical strategy. It consists in combining photometric entropy optimization (max/min) with the obstacles avoidance by stacking them into the same matrix to estimate $\triangle\mathbf{r}$. This will give us the motion direction. Then, we estimate the "amplitude" (to regulate the movement) of $\triangle\mathbf{r}$ by keeping the same direction previously calculated. This "amplitude" must respect a globally constant optical flow to have a visual regular velocity. Figure 3a describes the adopted camera control algorithm.

Step 1: Motion direction estimation

– It is obtained directly by stacking Eqs. (6) and (13) and using the pseudo inverse ()$^+$ of this system of seven equations. The estimation of $\triangle\mathbf{r}$ is given by:

$$\triangle\mathbf{r} = -\left(\begin{array}{c}\frac{\partial^2(-E(\mathbf{r}_0))}{\partial \mathbf{r}} \\ \frac{\partial(O(\mathbf{r}_0))}{\partial \mathbf{r}}\end{array}\right)^+ \left(\begin{array}{c}\frac{\partial}{\partial \mathbf{r}}(-E(\mathbf{r})) \\ O(\mathbf{r})\end{array}\right) = -\mathbf{H}^+\mathbf{g} \tag{17}$$

Hence, from the displacement $\triangle\mathbf{r}$, we will use the exponential map of the SE(3) to update the current pose $\mathbf{r}$ and thus obtain the transformation matrix at this pose.

Step 2: Final pose computation. In this second step we regulate the "amplitude" of the movement.

– Compute the theoretical optical flow as in Fig. 3b, where $I(\mathbf{r})$ and $I(\mathbf{r} + \triangle\mathbf{r})$ are the generated synthesized images.
– We will fix a constant value C as a desired amplitude of the global optical flow.

The desired flow is computed as follows:

$$\triangle u_{des} = \frac{C.\triangle u}{\sqrt{\triangle u^2 + \triangle v^2}} \tag{18}$$

$$\triangle v_{des} = \frac{C.\triangle v}{\sqrt{\triangle u^2 + \triangle v^2}} \tag{19}$$

such that C is the constant corresponding to the optical flow mean amplitude and $\triangle u$ and $\triangle v$ are the u and v components of the flow.

From the Eq. (16) we compute the final pose displacement:

$$\triangle \mathbf{r}_f = -\mathbf{L_u^+}\triangle \mathbf{u}_{des} \tag{20}$$

Where $\mathbf{L_u^+}$ is the pseudo inverse of $\mathbf{L_u}$. All matching pixels between the two images are used and $\mathbf{L_u}$ is over-determined.

Managing Rotations: The automatic navigation of the camera in the 3D model provides unrealistic rotations around the optical axis of the camera "Roll". Therefore, we consider that the camera has only five operational degrees of freedom, three translations (t_X, t_Y, t_Z) and two rotations (θ_p, θ_t) corresponding to the "pan" and the "tilt" instead of the six Cartesian degrees of freedom generally considered in the virtual visual servoing.

The camera is simulated as a virtual robot with only five degrees of freedom, where the three translations take place on the cartesian frame (world frame $\Re_w$) and the two rotations on the end-effector (camera frame $\Re_c$). We note by $\mathbf{q}_k$ the position of the robot at the step k.

We start by computing the robot jacobian on the $\Re_c$ frame $^c\mathbf{J}_c$:

$$^c\mathbf{J}_c =^c \mathbf{V}_w^w\mathbf{J}_w \tag{21}$$

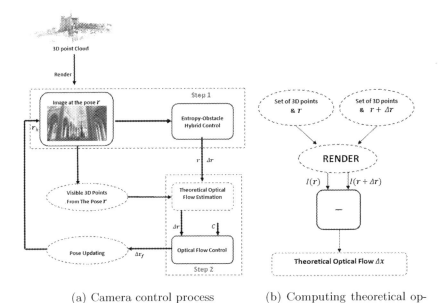

(a) Camera control process

(b) Computing theoretical optical flow

Fig. 3. Overall view of the algorithm.

where, ${}^{w}\mathbf{J}_{w}$ is the robot jacobian expressed in the $\Re_{w}$ frame and ${}^{c}\mathbf{V}_{w}$ a twist transformation matrix from $\Re_{w}$ to $\Re_{c}$.

Thus the Eq. (17) becomes:

$$\dot{\mathbf{q}}_{k} = (\mathbf{H}^{c}\mathbf{J}_{c})^{+}\mathbf{g} \tag{22}$$

The same steps are performed a second time for Eq. (20):

$$\dot{\mathbf{q}}_{kf} = (\mathbf{L}_{\mathbf{u}}^{c}\mathbf{J}_{c})^{+}\triangle_{\mathbf{u}_{des}} \tag{23}$$

$\dot{\mathbf{q}}_{kf}$ represents the five operational degrees of freedom increment. The iterative scheme consists to modify the current pose $\mathbf{q}_{k}$ using:

$$\mathbf{q}_{k+1} = \mathbf{q}_{k} + \dot{\mathbf{q}}_{kf} \tag{24}$$

The transformation matrix is built from the translation vector ${}^{c}\mathbf{t}_{w}$ and the rotation matrix ${}^{c}\mathbf{R}_{w}$, where ${}^{c}\mathbf{R}_{w} = R_{X}(\theta_{t})R_{Z}(\theta_{p})$ and the translation vector ${}^{c}\mathbf{t}_{w} = -{}^{c}\mathbf{R}_{w}^{c}\mathbf{t}_{w}$ such as ${}^{c}\mathbf{t}_{w} = (\mathbf{q}_{k+1}[0], \mathbf{q}_{k+1}[1], \mathbf{q}_{k+1}[2])^{T}$.

3   Experimental Results

We tested our approach on three different 3D models composed from 3D colored point clouds. Our algorithm is designed to run on extremely large data sets. An example of such data, is a set of N merged scans of a cathedral scan and urban environments. Scans are obtained by a FARO laser scanner. Environments have not been scanned totally, this is why we notice some holes, that have a white color on the images. There are now about 100 million of 3D colored points in each model. For the visualization of the 3D model we use the OGRE 3D graphics engine. On the 3D model of the cathedral we perform an indoor and an outdoor navigation. However, on the urban environments 3D models, we just perform an outdoor navigation.

This section demonstrates our results on this type of data. We applied the proposed approach to perform an automatic navigation in these 3D models based on the principal task and the two constraints previously described.

3.1   Photometric Entropy Maximization

This first set of experiment aims to show the behaviour of photometric visual entropy maximization. First, it is applied on a simple and flat 3D point cloud: a lab room with white wall and posters. Applying the entropy maximization algorithm, the path followed by the virtual camera tends to maximize the presence of the posters in the camera field of view Fig. 4. This is a first experimental validation of the entropy maximization to drive the virtual camera to interesting areas. The entropy maximization is applied to a virtual camera that is outside of a cathedral huge point cloud. Starting from a pose from which the cathedral is not filling the camera field of view Fig. 5, the entropy based control law drives it to maximize the amount of information in the image. Figure 6 shows the photometric visual entropy evolution over frames.

Fig. 4. Resulting visual camera path maximizing the relevant information in the image.

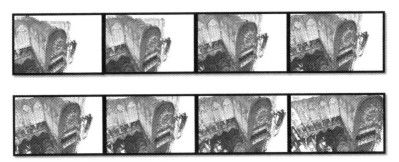

Fig. 5. Automatic navigation with photometric entropy maximization: view from outside.

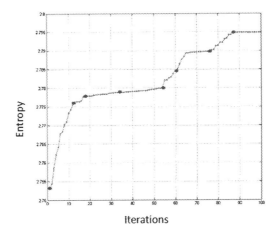

Fig. 6. Photometric entropy evolution. Blue dots correspond to images of Fig. 5 (from left to right and top to bottom) (Colour figure online).

3.2   Obstacles Avoidance

While experiments of Sect. 3.1 did not drive the camera through obstacles, this situation may be encountred inside the 3D point cloud of the cathedral Fig. 7 with chairs, pilars, and so on. Figure 7 demonstrates the behaviour of the camera for the entropy and obstacles hybrid control law Eq. (17). We observe the synthetic images Fig. 7a captured at each camera viewpoint and corresponding images that contain obstacles (in our case vertices) surrounding the viewpoint

(a) Images sequence in indoor navigation.

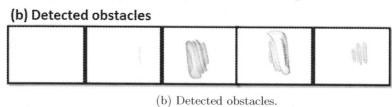

(b) Detected obstacles.

Fig. 7. Obstacles avoidance constraint behaviour.

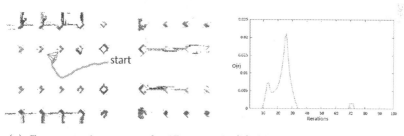

(a) Camera trajectory on the 2D map of   (b) Obstacles cost function evo-
the 3D model of the cathedral   lution

Fig. 8. Camera path.

below one-meter depth Fig. 7b. The equirectangular camera was implemented through GPU shaders programming. We notice that starting from the fourth image the camera moves away from obstacles, since the obstacles (vertices) number on the equirectangular image decreases.

In Fig. 8a, we observe in green the camera trajectory projected on a 2D plane of the 3D model of the cathedral which passes between two pilliars and the Fig. 8b shows the evolution of the cost function during the generated camera movement.

3.3 Optical Flow Control

In this third experiment, the camera is guided by the proposed hierarchical control law.

Starting from the same initial pose of the camera, two images sequences were generated. The first sequence considers the constraint of optical flow control and the second doesn't. The two curves of Fig. 9 show the evolution of the average

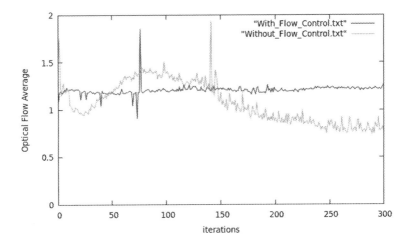

Fig. 9. Difference between the average of the optical flow norm for an images sequence considering the optical flow control constraint (red), and the other not considering it(green) (Colour figure online).

optical flow norm calculated for both sequences. The red curve clearly shows that we keep almost regular flow compared to the green curve where the evolution of the flow is disturbed during the whole sequence. In order to generate a regular camera movements avoiding significant jumps over its trajectory in the 3D model, this constraint is necessary and yields to the desired result. The video accompanying this paper visualy illustrates the difference between the two camera paths obtained before and after the flow control.

3.4 "Realistic" Navigation

In this last experiment, the optical flow control depends on the entropy maximization or minimization. Indeed, maximization and minimization of the entropy is alternating to reach and, then, leave local maxima. The optical flow is confined

Fig. 10. Automatic navigation.

in order to spend more time in rich information parts of the scene. However, during the entropy minimization process, slow motions are not relevant since the visual information is getting poorer and poorer. Thus, during the entropy minimization process, the optical flow is not confined but regulated in order to have a visual motion of a constant speed.

Figure 10 shows some images extracted from the video generated by applying the proposed algorithm. Starting from a small facade, the algorithm moves the camera to include the entire facade of the most interesting building of the 3D model.

4   Conclusion and Future Works

In this paper we presented a new approach to achieve automatic navigation in a 3D dense colored point cloud model. The proposed method generates smooth and realistic camera trajectories. The method implements a hierarchical algorithm performed in two steps. The first step concerns the motion direction estimation by stacking the principal task (photometric entropy maximization) and one of our constraints (obstacles avoidance).We show the benefits of using a photometric entropy to guide the camera to interesting areas. The second step controls the optical flow in the image in order to avoid sudden jumps of the camera. We have demonstrated, particularly in Fig. 10, that from a starting camera viewpoint where the image is poor, the algorithm moves the camera to a rich viewpoint by generating a smooth and realistic movement. The method is general and can integrate other criteria to optimize.

One of our future perspectives is to achieve a framework that introduce interactivity to this automatic method by adding other constraints in order to assist the user during the exploration of the 3D model. Then, we plan to carry out a study based on user opinion to evaluate the system, and maybe even improve it.

References

1. Andújar, C., Vázquez, P., Fairén, M.: Way-finder: guided tours through complex walkthrough models. Comput. Graph. Forum **23**, 499–508 (2004). Wiley Online Library
2. William, B., Scott, M.D., Christina, B., Somying, T.: Virtual 3D camera composition from frame constraints. in: Proceedings of the Eighth ACM International Conference on Multimedia, pp. 177–186. ACM (2000)
3. Bares, W.H., Zettlemoyer, L.S., Rodriguez, D.W., Lester, J.C.: Task-sensitive cinematography interfaces for interactive 3D learning environments. In: Proceedings of the 3rd International Conference on Intelligent User Interfaces, pp. 81–88. ACM (1998)
4. Srikanth, B., Daniel, T.: Space discretization for efficient human navigation. Comput. Graph. Forum **17**(3), 195–206 (1998). Wiley Online Library
5. Beckhaus, S., Ritter, F., Strothotte, T.: Cubicalpath-dynamic potential fields for guided exploration in virtual environments. In: Proceedings in the Eighth Pacific Conference on Computer Graphics and Applications, pp. 387–459. IEEE (2000)

6. Owen, B., Abdul, S., Scott, G.: A constraint-based autonomous 3d camera system. Constraints **13**(1–2), 180–205 (2008). Springer

7. Christie, M., Machap, R., Normand, J.-M., Olivier, P., Pickering, J.H.: Virtual camera planning: a survey. In: Butz, A., Fisher, B., Krüger, A., Olivier, P. (eds.) SG 2005. LNCS, vol. 3638, pp. 40–52. Springer, Heidelberg (2005)

8. Courty, N., Marchand, E.: Computer animation: a new application for image-based visual servoing. In: Proceedings 2001 ICRA. IEEE International Conference on Robotics and Automation, vol. 1, pp. 223–228. IEEE (2001)

9. Amaury, D., Marchand, E.: Entropy-based visual servoing. In: IEEE International Conference Robotics and Automation ICRA 2009, pp. 707–713 (2009)

10. Feddema, J.T., Mitchell, O.R.: Vision-guided servoing with feature-based trajectory generation for robot. IEEE Trans. Robot. Autom. **5**(5), 691–700 (1989)

11. Stefan, G.: Maximum entropy light source placement. In: Visualization VIS. IEEE (2002)

12. Hager, G.D.: A modular system for robust positioning using feedback from stereo vision. IEEE Trans. Robot. Autom. **13**(4), 582–595 (1997)

13. Nicolas, H., Ralf, H., Thomas, S.: A camera engine for computer games: managing the trade-off between constraint satisfaction and frame coherence. Comput. Graph. Forum **20**(3), 174–183 (2001). Wiley Online Library

14. Fabrice, L.: Topoplan: a topological path planner for real time human navigation under floor and ceiling constraints. Comput. Graph. Forum **28**(2), 649–658 (2009). Wiley Online Library

15. Jean-Thierry, L., Youcef, M.: A Hessian approach to visual servoing. IEEE Intell. Robot. Syst. IROS **1**, 998–1003 (2004)

16. Li, T.-Y., Cheng, C.-C.: Real-time camera planning for navigation in virtual environments. In: Butz, A., Fisher, B., Krüger, A., Olivier, P., Christie, M. (eds.) SG 2008. LNCS, vol. 5166, pp. 118–129. Springer, Heidelberg (2008)

17. Frederik, M., Andre, C., Dirk, V., Guy, M., Paul, S.: Multimodality image registration by maximization of mutual information. IEEE Trans. Med. Imaging **16**(2), 187–198 (1997)

18. Marchand, E., Hager, G.D.: Dynamic sensor planning in visual servoing. In: Proceedings International Conference on Robotics and Automation, vol. 3, pp. 1988–1993. IEEE (1998)

19. Salomon, B., Garber, M., Lin M.C., Manocha, D.: Interactive navigation in complex environments using path planning. In: Proceedings of the 2003 Symposium on Interactive 3D Graphics, pp. 41–50. ACM (2003)

20. Ekrem, S., Adali, H.: Serdar Balcisoy Selim: automatic path generation for terrain navigation. Comput. Graph. **36**(8), 1013–1024 (2012). Elsevier

21. Bill, T., Bruce, B., Delphine, N.: Expressive autonomous cinematography for interactive virtual environments. In: Proceedings of the Fourth International Conference on Autonomous Agents, pp. 317–324. ACM (2000)

22. Vázquez, P.-P., Feixas, M., Sbert, M., Heidrich, W.: Viewpoint selection using viewpoint entropy. VMV **1**, 273–280 (2001)

23. Yeh, I., Lin, C.-H., Chien, H.-J., Lee, T.-Y.: Efficient camera path planning algorithm for human motion overview. Comput. Animat. Virtual Worlds **22**(2–3), 239–250 (2011). Wiley Online Library

Workshop on Computer Vision for Affective Computing (CV4AC)

A Robust Learning Framework Using PSM and Ameliorated SVMs for Emotional Recognition

Jinhui Chen[✉], Yosuke Kitano, Yiting Li, Tetsuya Takiguchi, and Yasuo Ariki

Graduate School of System Informatics, Kobe University, Kobe 657-8501, Japan
{ianchen,kitano,liyiting}@me.cs.scitec.kobe-u.ac.jp,
{takigu,ariki}@kobe-u.ac.jp

Abstract. This paper proposes a novel machine-learning framework for facial-expression recognition, which is capable of processing images fast and accurately even without having to rely on a large-scale dataset. The framework is derived from Support Vector Machines (SVMs) but distinguishes itself in three key technical points. First, the measure of the samples normalization is based on the Perturbed Subspace Method (PSM), which is an effective way to improve the robustness of a training system. Second, the framework adopts SURF (Speeded Up Robust Features) as features, which is more suitable for dealing with real-time situations. Third, we use region attributes to revise incorrectly detected visual features (described by invisible image attributes at segmented regions of the image). Combining these approaches, the efficiency of machine learning can be improved. Experiments show that the proposed approach is capable of reducing the number of samples effectively, resulting in an obvious reduction in training time.

1 Introduction

Facial expressions recognition is a typical multi-class classification problem in computer vision. Furthermore, since it is one of the most significant technologies for auto-analyzing human behaviors, and it is able to be widely applied into many domains. Therefore, the need for this kind of technology in various different areas keeps pushing the research forward every year.

As the main detectors, Adaboost and SVMs, etc. are widely used in this field of research. In 1995, Freund and Schapire [1] supplied the AdaBoost algorithm for realizing the learning framework of Boosted Trees, which could be referred to Probably Approximately Correct (PAC) learning proposed by Valiant [2]. Since then great advances have been made based on AdaBoost, especially milestone work by Viola and Jones [3]. But some ideal strong classifiers are usually required a large number of training samples and very time-consuming training experiments. Even recently, many researchers are trying to solve these problems. Li *et al.* [4] proposed a new learning SURF cascade for ameliorating boosting cascade frameworks. It improved the training efficiency, but the need for large-scale data gathering and extensive preparations create a critical bottleneck. On the other hand, similar problems also exist in methods based on SVMs, because

© Springer International Publishing Switzerland 2015
C.V. Jawahar and S. Shan (Eds.): ACCV 2014 Workshops, Part II, LNCS 9009, pp. 629–643, 2015.
DOI: 10.1007/978-3-319-16631-5_46

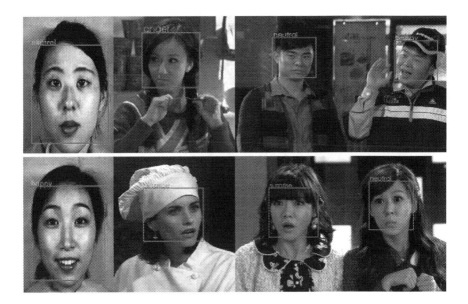

Fig. 1. Examples of Recognition Results

of the limitation of length (they will be not enumerated here). Therefore, collecting many training samples and the long training time lead to considerable work and difficulty for researchers in the field of pattern recognition. Since training is a critical infrastructure for recognition engines, the research on training is significant for learning machines. Hence, there is a great need to solve the problem mentioned above.

This paper brings together new normalization measures, visual features and image attributes to construct a framework for facial expressions recognition. As almost all of our approaches relate to vectors processing, the classifier of our proposed method is built based on SVMs. There are three main approaches with emphasis on reducing training samples and improving the efficiency of learning machines. First, PSM is used to extend the training data, which allows for the generation of ideal strong classifiers without having to collect a large number of training samples. Second, the features are described by local multi-dimensional SURF descriptors [5], which are spatial regions with windows and are good at processing real-time scenes. Moreover, SURF is much faster and more efficient than most of the existing local features algorithms [4], such as SIFT [6], HoG [7] *etc.* Third, the region attributes of images are adopted to revise incorrect detection of classifiers relying on visual features, which are represented by feature vectors in a segmented region. Therefore, the discriminative capability can guarantee the proposed framework will be more robust.

In summary, there are these main contributions in this paper. The first is that the efficiency of machine learning can be improved. Experiments show that the proposed approach is capable of reducing the number of samples effectively, resulting in an obvious reduction in training time. The second is that the recognition accuracy is comparable to the state-of-the-art algorithms. In experiments, the results show that although using a mini-sized database of training samples, our approaches can also construct a robust facial expressions recognition system, which is comparable to the state-of-the-art methods. Some examples of recognition result are shown in Fig. 1. We believe the proposed method is a good try for machine learning, because recognition accuracy plays a very significant role in machine learning, but without doubt, the training efficiency and is also equally important.

The rest of this paper is organized as follows: we will first revisit related works in Sect. 2, then we describe the samples normalization in Sect. 3 and classifying framework in Sect. 4 respectively. Section 5 elaborates on region attributes estimation. Section 6 shows the experiments and conclusions are drawn in Sect. 7.

2 Related Work

Facial-expression recognition is a hot research topic in computer vision due to its many applications, and many researchers attach great importance to this field. For instance, Lyons *et al.* [8] adopted PCA and LDA to analyze facial expressions through closed experiments, and they achieved 92 % accuracy on JAFFE [9,10]; Bartlett *et al.* [11] proposed a Gabor feature based AdaSVM method for expression recognition, obtaining a good performance based on the use of the Cohen-Kanade expression database [12]. However, it has been shown that the processing speed of these approaches is too slow to deal with real-time scenes. More recently, Anderson *et al.* [13] and Chen *et al.* [14] proposed their approaches for real-time expression recognition separately, but their methods required a large amount of data for training.

Our approach enables competition that completes a recent line of papers that use third-party software tools to obtain mirror images of samples for training in their object/facial-expression recognition systems, which we briefly review here. To the best of our knowledge, our approach is the first to apply the proposed method into both of object recognition and facial expression recognition. Meanwhile, it is the first to employ the PSM directly for detectors training, but not use any tool. Experiments show it has the greatest impact on the performance of training efficiency, because time can be saved, which would be spent on collecting vast amounts of data from the Internet or using third-party software to deal with the samples for getting mirror images of these samples. In addition, in our framework, the classifier is based on SVMs, furthermore, its function is ameliorated, which can guarantee the results will be more reliable.

3 PSM for Samples Normalization

PSM is applied to the normalization of samples, but different from previous versions (many-to-one mapping), in this paper, PSM is a one-to-many mapping, we use it to extend the subspace of samples. There are many existing methods based on virtual images, which seem similar with ours, but most of them rested on pixel-level transformation (such as [15] and [16] *etc.*), so that after the calculation, some features data would be damaged easily. Moreover, they required some handwork, and in the training period, the program had to read a large number of virtual image files again, these lead to time waste. We don't have these problems. Therefore, it is an effective way to improve the robustness of the training system.

3.1   Training-Sample Normalization

In order to reduce the noise, the size of the images is unified by $m \times n$ pixels, and the original samples are normalized by the mean value and variance of pixels transformation. Therefore, the images after the normalization can be obtained according to the following equation:

$$I'(x,y) = a\frac{I(x,y) - \mu}{2\sqrt{2}\sigma} + b. \tag{1}$$

Here σ is the standard deviation, and

$$\sigma = \sqrt{\frac{1}{mn}\sum_{x=1}^{m}\sum_{y=1}^{n}(I(x,y) - \mu)^2}. \tag{2}$$

(a, b) is used to adjust the value of pixels (In this paper, we used regular samples in experiments, therefore, a was set as 1, b was set as 0). μ is the mean value of pixels, and it can be computed through image traversal by the equation:

$$\mu = \frac{1}{mn}\sum_{x=1}^{m}\sum_{y=1}^{n}I(x,y). \tag{3}$$

3.2   Changing Facial Orientations

After the calculation of Sect. 3.1, we can thus extend the subspace of samples through changing the facial directions of the images. In this paper, we use the method proposed by Chen *et al.* [14] to reconstruct three-dimensional faces and obtain three-dimensional data, and we indicate it in Algorithm 1.

In Algorithm 1, when E_r is upper a threshold ε, or K landmarks are processed over, the while loop would be stopped, and the three-dimensional data will be output. Here $\beta = (\beta_1, \beta_2, \cdots, \beta_m)^T$ is the shape parameter and m is the dimensionality of the shape parameter, which is used to adjust three-dimensional shape data. S_{3D} is a $3 \times n$ matrix, P is a 2×3 orthographic projection matrix, T is a $3 \times n$ translation matrix consisting of n translation vectors $t = [t_x, t_y, t_z]^T$, and R_θ is a 3×3 rotation matrix where the yaw angle is θ. In this paper, θ is set as $\pm 15°$, $\pm 30°$, $\pm 60°$. Thus, through Algorithm 1, we can obtain the three-dimensional

Algorithm 1. Reconstruct Three-dimensional Face

Require:

Input: two-dimensional shape vector: $S_{2D} \in R^2$

Output: three-dimensional shape vector: $S_{3D} \in R^3$

Initialization: set $\beta_0 = 0$, $i = 0$

while $i < K$ or $E_r \leq \varepsilon$ **do**

   1. Let

$$S_{3D} \Leftarrow s_0 + \sum_{i=1}^{m} \beta_i s_i$$

   2. Alignment: S_{2D} is aligned with the two-dimensional shape, which is obtained by projecting the frontal three-dimensional shape (s_i) onto the $x - y$ plane.

   3. Minimize

$$\|P(R_\theta S_{3D} + T) - S_{2D}\|^2$$

   4. Reconstruct $(S_{3D})_i$ using the shape parameter β_i.

   5. Update R_θ and T with the fixed shape parameter and

$$E_r \Leftarrow \|P(R_\theta S_{3D} + T) - S_{2D}\|^2$$

   6. Let

$$i \Leftarrow i + 1$$

end while

   7. Reconstruct three-dimensional shape using the final shape parameters.

   8. Output S_{3D}.

---

data $X = (x, y, z)^T$ from the original images. Hence, according to the transformation matrix formula:

$$X' = T_z \cdot T_y \cdot T_x \cdot S \cdot R_z \cdot R_y \cdot R_x \cdot X. \tag{4}$$

we can convert the facial directions to extend the subspace of the training samples. Here T and R are the shear mapping transformation matrix and the rotation matrix respectively, and S is represented by the scaling matrix.

3.3 Changing Illumination Attributes

The illuminative change is conducted according to the following equation:

$$V_2^{(n)} = V_1^{(n)} + \sum_{m=1}^{K} w_m \cdot e_m^{(n)}. \tag{5}$$

where V_1 is the changing feature, V_2 is the result after the changes, n is the dimensionality of the feature vector, w is the weight coefficient, and e is the basis of illumination-change-factor vectors.

In this paper, e is obtained through processing the luminance normalized rendering images by principal component analysis (PCA), wherein, m is the principal component ($m = 1, \cdots, 8$). The rendering images are gained by the

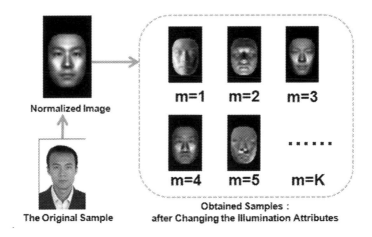

Normalized Image

The Original Sample

Obtained Samples :
after Changing the Illumination Attributes

Fig. 2. Illumination-attribute Changes Used for Extending Training Samples

treatment of three-dimensional images obtained in Sect. 3.2. An example of results is shown in Fig. 2.

4   Classifying Framework

This section will provide the framework used for SVM machine learning through adopting SURF features. Moreover we will also employ the region attributes of image to revise miss-detection of classifier relying on visual features. We will describe them separately in this section.

4.1   Feature Description

SURF is a scale- and rotation-invariant interest point detector and descriptor. It is faster than SIFT [6] and more robust against different image transformations. In this paper, we adopt an 8-bin T2 descriptor to describe the local feature, which is inspired by [4]. Different from [17], we further allow different aspect ratio for each patch (the ratio of width and height), because this can make the speed of image traversal become quicker. Meanwhile we imported diagonal and anti-diagonal filters, this can improve description capability of SURF descriptors.

Given a detection window, we define rectangular local patches within it, each patch with 4 spatial cells and allow the patch size ranging from 12×12 pixels to 40×40 pixels. Each patch is represented by a 32-dimensional SURF descriptor. The descriptor can be computed quickly based on sums of two-dimensional Haar wavelet responses and we can make an efficient use of Integral Images [3]. Suppose d_x as the horizontal gradient image, which can be obtained using the filter kernel $[-1, 0, 1]$, and d_y is the vertical gradient image, which can be obtained using the filter kernel $[-1, 0, 1]^T$; Define d_D as the diagonal image and d_{AD} as the anti-diagonal image, both of which can be computed using two-dimensional filter

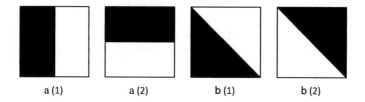

Fig. 3. Haar-type Filter Used for Computing SURF Descriptor

kernels $diag\ (-1, 0, 1)$ and $antidiag\ (-1, 0, 1)$. Therefore, 8-bin T2 is able to be defined as $v = (\sum(|d_x| + d_x), \sum(|d_x| - d_x), \sum(|d_y| + d_y), \sum(|d_y| - d_y), \sum(|d_D| + d_D), \sum(|d_D| - d_D), \sum(|d_{AD}| + d_{AD}), \sum(|d_{AD}| - d_{AD}))$. Here, d_x, d_y, d_D, and d_{AD} can be computed individually by filters shown in Fig. 3 in use of Integral Images, the details about how to compute two-dimensional Haar responses with Integral Images, please refer to [3].

The detection template for SURF is 40×40 with 4 spatial cells, and allow the patch size ranging from 12×12 pixels to 40×40 pixels. We slide the patch over the detection template with 4 pixels forward to ensure enough feature-level difference. We further allow different aspect ratio for each patch (the ratio of width and height). The local candidate region of the features is divided into 4 cells. The descriptor is extracted in each cell. Hence, concatenating features in 4 cells together yields a 32-dimensional feature vector. About feature normalization, in practice, L_2 normalization followed by clipping and renormalization ($L_2 Hys$) [18] is shown working best.

4.2   Classifier Construction

The classifier of our framework is built based on One-Versus-Rest SVMs (OVR-SVMs). OVR strategy consists of constructing one SVM per class, which is trained to distinguish the samples of one class from the samples of all remaining classes. Normally, classification of an unknown object is carried out by adopting the maximum output among all SVMs. The proposed method is based on OVR-SVMs classifier, and implemented by re-developing liblinear SDK [19].

Usually, most of researchers estimate posterior probability by mapping the outputs of each SVM into probability separately. The method was proposed by Platt [20]. It applies an additional sigmoid function:

$$H(\omega_j | f_j(x)) = \frac{1}{1 + exp\ (c_j f_j(x) + d_j)}. \tag{6}$$

$f_j(x)$ denotes the output of the SVM trained to separate the class ω_j from the other classes (total samples are M). Then, for each sigmoid the parameters c_j and d_j are optimized by minimizing the local negative log-likelihood:

$$-\sum_{k=1}^{N} \{p_k log(h_k) + (1 - p_k) log(1 - h_k)\}. \tag{7}$$

Here are N outputs of the sigmoid function, where h_k is the output of the sigmoid function with the probability p_k event. In order to solve this optimization problem, [20] applied a model-trust minimization algorithm based on the Levenberg-Marquardt algorithm. But in [21], Lin *et al.* pointed out that there are some problems in this method, meanwhile they proposed another minimization algorithm based on Newton's method with backtracking line search.

But unfortunately, there is nothing to guarantee that:

$$\sum_{j=1}^{M} H(\omega_j | f_j(x)) = 1. \tag{8}$$

For this reason, it is necessary to normalize the probabilities as following:

$$H(\omega_j | x) = \frac{H(\omega_j | f_j(x))}{\sum_{j'=1}^{M} H(\omega_{j'} | f_{j'}(x))}. \tag{9}$$

Thus, we use another approach to estimate posterior probability, using OVR-SVMs to exploit the outputs of all SVMs to estimate overall probabilities. In order to achieve this goal, we apply the softmax function, regarding it as a generalization of sigmoid function for the multi-SVMs case. Thus, in the spirit of the improved Platt's algorithm [22], this paper applies a parametric form of the softmax function to normalize the probabilities by:

$$H(\omega_j | x) = \frac{exp \ (c_j f_j(x) + d_j)}{\sum_{j'=1}^{M} exp \ (c_{j'} f_{j'}(x) + d_{j'})}. \tag{10}$$

And here the parameters c_j and d_j are optimized by minimizing the global negative log-likelihood

$$-\sum_{k=1}^{N} log(H(\omega_k | x_k)). \tag{11}$$

Optimizing the parameters c_j and d_j are done with intention of obtaining the lowest error rate on testing dataset. The reason of why we use the negative log-likelihood is not only because it can optimize the parameters c_j and d_j, but also because it can be used for comparing the various probability estimates, in other words, it can evaluate the error rate on machine learning and reject some of unsatisfactory candidate expression regions described by SURF features.

5   Region Attributes Estimation

After classifying the OVR-SVMs model, we can obtain the recognition result classified by classifiers based on visual features. However, it is necessary to make further efforts on reducing misrecognition; thus, we use invisible image attributes to realize this purpose.

The detected face region is divided into 9×10 blocks, and the feature vector of each block is computed. We call this region attributes. It can be obtained after normalization by equalizing the value and variance of the luminance, while the norm is set as 1. The region attribute is estimated using the following score equation.

$$d = \left\| X - \bar{X} \right\|^2 - \sum_{i=1}^{N} \frac{\lambda_i}{\lambda_i + \delta^2} (\varphi_i (X - \bar{X}))^2. \tag{12}$$

here φ is eigenvector and λ is eigenvalue, δ^2 is the image noise correct divisor. When $\delta^2 = 0$, it means that the distances of all feature vectors of the current image projecting into subspace are unified, in the other words, the noise is negligible. X is estimated image region attributes, and $\bar{X}$ is the average feature vector of samples. The value of distance is smaller, the score is higher, namely, the probability of miss-detection is lower.

In this paper, the most significant way to ameliorate OVR-SVMs is based on two conflicted criteria, inspired by Boosting cascade: A error rate evaluating threshold e ($e^n = (1 - d)$), its function is similar with false-positive-rate in Boosting cascade [2]. And recognition rate evaluating threshold d, its function is similar with hit-rate in Boosting cascade. They are used for the detection-error tradeoff: $e < 0.5$ the classifying result will be considered as miss-detection and OVR-SVMs classifying model is executed repeatedly until a given Boolean condition $d \leq 0.2$ is met.

6　Experiments

In this section we will show the details of implementation, dataset, and evaluation results. The proposed method is designed for Neutral-, Happy-, Anger- and Surprise-expression recognition. And the recognition result examples are shown in Fig. 1.

We implemented all training and detection programs in C++ on RHEL (Red Hat Enterprise Linux) 6.5 OS. The facial recognition part used the source code of Open CV, which was based on Viola and Jones framework [3]. The expressional recognition part was implemented based on the proposed framework. The experiments were done on the PC with Core i7-2600 3.40 GHz CPU and 8 GB RAM, the training procedure was fully automatic. For SURF extraction, we adopted Integral Image to speedup the computation as described in Sect. 3.1. For machine learning, we built the OVR-SVMs through re-developing liblinear software [19].

6.1　Experimental Dataset

Training Database Set. We used Cohn-Kanade expression database (CK+) [12] as training database, which is a set of front face images posed by 123 posers, but not all of posers posed each type of expressions what we need. Therefore, we also collected some samples by online image search engine, fianlly we obtained 240 initial facial samples for each type of emotion. All of facial samples were normalized to 90×100-pixel patches and processed by histogram equalization, no color information was used.

Testing Database Set. In order to evaluate both of the real life and ideal situations, we used two parts of testing sets. One part was obtained from soap operas, because many public databases were processed by providers in advance, or the other reasons, such as the images cannot represent real-life scene, because they are not continuous images *etc.*, hence, we had to use some video clips of comedy dramas, which had the total of 10 persons whose facial expressions were similar to the training samples. These images of these actors and actresses are on 8 video clips having a length of 120 seconds. We marked this set as Test Set A. The other testing set was the JAFFE database [9, 10], whose facial samples are totally different from CK+ database. 213 images of JAFFE were mixed randomly and one image can be used repeatedly (ensure that there are enough images for different videos making), these images were also made into 8 120-second-long videos, and we marked this set as Test Set B. All of the test videos have speed of 60 FPS (Frames Per Second)

6.2   Experimental Evaluation

Training Experiments. The training database of all methods was mentioned above, but only the proposed method did not adopt any process to obtain plenty of mirror samples. Hence, it reduced a mass of samples and took only 49.8 min to complete the whole process. Besides, the training procedure was fully automatic. The relative data are shown in Table 1.

Table 1. Training Efficiency Evaluation Results

Method	Proposed	K-means [23]	LUT_Ada [14]
Time cost	49.8 min	1,589 min	172.5 min

However, in order to enhance the generalization performance of comparison method [23] and comparison method [14], we had to deal with the images by some transformations (mirror reflection and rotate the images by horizontal and vertical angles $\pm 15°, \pm 30°, \pm 60°$ *etc.*), finally, we obtained each class $30, 960$, total $123, 840$ facial samples for training classifiers. Therefore, they are very time-consuming tasks.

Testing Experiments. In order to be evaluated easily, CSV files were created automatically by the experimental program, and the tested results for each frame image of test videos were stored in these files. After doing test experiments, tested videos were divided into images, the tested results of CSV files were checked one by one with these images. These measures can guarantee the validity of experimental data. All of approaches of this paper were evaluated though testing experiments, the details are indicated as follows:

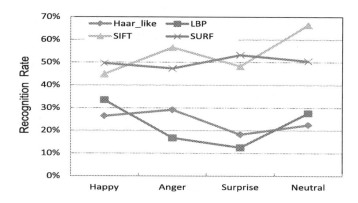

Fig. 4. Green: Recognition rate for OVR-SVMs with SIFT. Purple: Recognition rate for OVR-SVMs with SURF. Blue: Recognition rate for OVR-SVMs with Haar-like. Red: Recognition rate for OVR-SVMs with LBP. Features using SURF and SIFT obtained the more accurate results, but the feature extraction speed of SIFT was low (Colour figure online).

Figure 4 indicates the expression-recognition rate for different feature detectors based on our ameliorated SVMs detector. The aim of this experiment was evaluating the performance of the proposed detector using different methods of feature extraction. Hence, this experiment was done without a PSM model. Feature detectors using SURF and SIFT obtained the more accurate recognition rates, but the average speed of the SIFT detectors version was only 16.8 FPS. In comparison, the speed of the SURFs version reached 39.4 FPS. Theoretically, 16.8 FPS is also too slow to deal with complex scenes, such as real-time scenes. Thus, SURF was selected as the feature detector.

In this paper, We placed 450 local patches on the 40×40-pixel size detection template with 4 spatial cells, and allowed the patch size ranging from 12×12 pixels to 40×40 pixels. We slide the patch over the detection template with 4 pixels forward to ensure enough feature-level difference. Different from [17], we further allow different aspect ratio for each patch (the ratio of width and height). Our framework adopted 8-bin T2 descriptor as descriptor. It obtained similar precision of recognition results to the accuracy of original SURF's version and even SIFT's one, but dominated others on feature extraction speed. In fact, in our experiments, 8-bin T2 descriptors had almost the same accuracy as the original SURF; however, its extraction speed can reach more than 39.4 FPS, while the speed of original SURF version had only about 19 FPS, which was also too slow. Therefore, the feature descriptor based on 8-bin T2 SURF is the best choice for our framework.

In Fig. 5, the component selection of the proposed method was carried out to investigate how each component contributes to the recognition rate. In fact, in our experiments, OVR-SVMs had almost the same convergence speed as the original SVM and non-linear SVMs with the RBF kernel. However, its accuracy was the best one. As we known, the original SVM cannot be applied to the

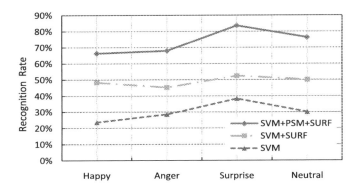

Fig. 5. Top: Recognition rate for proposed method. Middle: Proposed method without PSM; *i.e.*, the OVR-SVMs+SURF model. Bottom: Only OVR-SVMs. OVR-SVMs+PSM+SURF (proposed method) is the most accurate version of our detector.

continuous classification, if it includes more than 2-classe categories. Meanwhile, we also tried the non-linear SVMs using the RBF kernel, but its average precision of 4-type emotional classification was only approximately 33.6 %. As a result, the OVR-SVMs+PSM+SURF model was the most accurate version of our classifier.

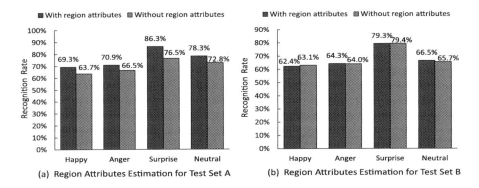

Fig. 6. Evaluation Results for Expressional Region Attributes

Figure 6 shows the results of the evaluation experiments for expressional region attributes. Figure 6(a) shows the results for Test Set A videos, and Fig. 6(b) shows the results for Test Set B. In the experiments, we found that after introducing the region attributes model, the recognition accuracy of Test Set A improved approximately by 7 %. On the other hand, the results of Test Set B were almost unchanged, since the videos in Test Set B consisted of JAFFE images, and these images had been normalized by the supplier [9]. But the videos of Test Set A were used without any normalization. Therefore, this approach is capable of dealing with original images better; *i.e.*, it is good at processing real-life videos.

Table 2. Experimental Results for Test Set A

	Proposed	K-means [23]	LUT_Ada [14]
Happy	69.3 %	52.0 %	61.2 %
Anger	70.9 %	64.5 %	50.9 %
Suprise	86.3 %	42.8 %	68.6 %
Neutral	78.3 %	37.1 %	65.6 %

Tables 2 and 3 indicate the recognition accuracies, and they show the performance of the proposed method compared to other classifiers ([14] is one of the latest methods for facial expressions recognition, and it was based on AdaBoost; [23] is a typical expressions recognition method using K-means). Table 2 shows the recognition rate of evaluation experiments for Test Set A. Since the races and facial expressions of Test Set As people were similar to those of the training samples, the region attributes model was effective for Test Set A in which there are videos from real life. Consequently, its accuracy was quite better than the Test Set B's. The maximum recognition precision of the proposed method was 86.3 %, and the worst result was 69.3 %.

Table 3. Experimental Results for Test Set B

	Proposed	K-means [23]	LUT_Ada [14]
Happy	62.4 %	55.3 %	57.7 %
Anger	64.2 %	59.5 %	48.2 %
Suprise	79.3 %	44.8 %	68.4 %
Neutral	66.5 %	32.6 %	71.6 %

On the other hand, Table 3 shows the recognition accuracies for Test Set B. Due to the variation and complexity of facial expressions across different cultures and races, the region attributes model was not effective for facial recognition. The results for this test set were not better than Test Set A's. But on the whole, the results of both test sets show that the proposed method was the more accurate version of these methods. Note that the proposed method used training samples without any image-mirror process here. Namely, based on the mini-size training set, the proposed method can also obtain a better result; thus, this model allows for generating ideal strong classifiers without the need for large volumes of training samples. Hence, under these experimental conditions, the validity of the proposed methods was proved.

7 Conclusions

This paper brings together new normalization measures, visual features and image attributes to construct a novel framework, which minimizes the training data but improves the training efficiency. It may well have broader application in machine learning.

PSM is an effective approach for alleviating suffering of collecting plenty of training samples. Then, through doing a great many of experiments, we find SURF is the most suitable feature descriptor for our detector, and the region attributes of images can revise some miss-detection caused by visual features. Combining these approaches together, a robust expression recognition framework can be constructed, but due to the variation and complexity of the facial expression across different cultures and human races, using mini-size training set to obtain high recognition precision, there are many difficult challenges have to be overcome.

About future plan, considering a possible implementation in a real scenario, we are inclined to consider these points: (1) Try to use region attributes as binary latent variables, which are incorporated into the SVMs model for inference. (2) Ameliorate approaches on the construction of SVMs to improve accuracy and to make it be qualified for more complex tasks. (3) applying the approach to object recognition such as human detection, car detection, events understanding *etc.*

References

1. Freund, Y., Schapire, R.: A desicion-theoretic generalization of on-line learning and an application to boosting. In: Vitányi, P. (ed.) Computational Learning Theory. LNCS, vol. 904, pp. 23–37. Springer, Heidelberg (1995)
2. Valiant, L.G.: A Theory of the Learnable. Communications of the ACM **27**, 1134–1142 (1984)
3. Viola, P., Jones, M.: Rapid object detection using a boosted cascade of simple features. In: Proceedings of the IEEE Computer Society Conference on Computer Vision and Pattern Recognition (CVPR), vol. 1, pp. I-511-I-518 (2001)
4. Li, J., Zhang, Y.: Learning SURF Cascade for Fast and Accurate Object Detection. In: Proceedings of the IEEE Computer Society Conference on Computer Vision and Pattern Recognition (CVPR), pp. 3468–3475 (2013)
5. Bay, H., Tuytelaars, T., Van Gool, L.: SURF: speeded up robust features. In: Leonardis, A., Bischof, H., Pinz, A. (eds.) ECCV 2006, Part I. LNCS, vol. 3951, pp. 404–417. Springer, Heidelberg (2006)
6. Lowe, D.G.: Object recognition from local scale-invariant features. In: The Proceedings of the Seventh IEEE International Conference on Computer Vision (ICCV)
7. Dalal, N., Triggs, B.: Histograms of oriented gradients for human detection. Proc. IEEE Comput. Soc. Conf. Comput. Vis. Pattern Recognit. (CVPR) **1**, 886–893 (2005)
8. Dailey, M.N., Joyce, C., Lyons, M.J., Kamachi, M., Ishi, H., Gyoba, J., Cottrell, G.W.: Evidence and a computational explanation of cultural differences in facial expression recognition. Emotion **10**, 874–893 (2010)
9. Kamachi, M., Lyons, M., Gyoba, J.: The Japanese Female Facial Expression (JAFFE) Database, vol. 21 (1998). http://www.kasrl.org/jaffe.html
10. Pantic, M., Rothkrantz, L.J.M.: Automatic analysis of facial expressions: the state of the art. IEEE Trans. Pattern Anal. Mach. Intell. **22**, 1424–1445 (2000)

11. Bartlett, M., Littlewort, G., Fasel, I., Movellan, J.: Real time face detection and facial expression recognition: development and applications to human computer interaction. In: Proceedings of the IEEE Computer Society Conference on Computer Vision and Pattern Recognition Workshops (CVPR Workshops), vol. 5, pp. 53–53 (2003)

12. Lucey, P., Cohn, J.F., Kanade, T., Saragih, J., Ambadar, Z., Matthews, I.: The extended Cohn-Kanade dataset (CK+): a complete dataset for action unit and emotion-specified expression. In: Proceedings of the IEEE Computer Society Conference on Computer Vision and Pattern Recognition Workshops (CVPR Workshops), pp. 94–101 (2010)

13. Anderson, K., McOwan, P.W.: A real-time automated system for the recognition of human facial expressions. IEEE Trans. Syst. Man Cybern. Part B: Cybern. **36**, 96–105 (2006)

14. Chen, J., Ariki, Y., Takiguchi, T.: Robust facial expressions recognition using 3d average face and ameliorated adaboost. In: Proceedings of the 21st ACM International Conference on Multimedia (ACM MM), pp. 661–664 (2013)

15. Chu, W.S., Huang, C.R., Chen, C.S.: Gender classification from unaligned facial images using support subspaces. Inf. Sci. **221**, 98–109 (2013)

16. Decoste, D., Schölkopf, B.: Training invariant support vector machines. Mach. Learn. **46**, 161–190 (2002)

17. Li, J., Wang, T., Zhang, Y.: Face detection using surf cascade. In: IEEE International Conference on Computer Vision Workshops (ICCV Workshops), pp. 2183–2190 (2011)

18. Dalal, N., Triggs, B.: Histograms of oriented gradients for human detection. In: Proceedings of the IEEE Computer Society Conference on Computer Vision and Pattern Recognition (CVPR), vol. 1, pp. 886–893 (2005)

19. Fan, R.E., Chang, K.W., Hsieh, C.J., Wang, X.R., Lin, C.J.: LIBLINEAR: a library for large linear classification. J. Mach. Learn. Res. **9**, 1871–1874 (2008)

20. Platt, J.: Probabilities for SV Machines. In: Advances in Large Margin Classifiers (2000)

21. Lin, H.T., Lin, C.J., Weng, R.C.: A note on platts probabilistic outputs for support vector machines. Mach. Learn. **68**, 267–276 (2007)

22. Sun, Z., Ampornpunt, N., Varma, M., Vishwanathan, S.: Multiple kernel learning and the SMO algorithm. In: Advances in Neural Information Processing Systems (NIPS), pp. 2361–2369 (2010)

23. Alldrin, N., Smith, A., Turnbull, D.: Classifying facial expression with radial basis function networks, using gradient descent and K-means. In: CSE253 (2003)

Subtle Expression Recognition Using Optical Strain Weighted Features

Sze-Teng Liong[1(✉)], John See[2], Raphael C.-W. Phan[3], Anh Cat Le Ngo[3], Yee-Hui Oh[3], and KokSheik Wong[1]

[1] Faculty of Computer Science and Information Technology, University of Malaya, Kuala Lumpur, Malaysia
szeteng1206@hotmail.com
[2] Faculty of Computing and Informatics, Multimedia University, Cyberjaya, Malaysia
johnsee@mmu.edu.my
[3] Faculty of Engineering, Multimedia University, Cyberjaya, Malaysia
raphael@mmu.edu.my, {lengoanhcat,yeehui716}@gmail.com

Abstract. Optical strain characterizes the relative amount of displacement by a moving object within a time interval. Its ability to compute any small muscular movements on faces can be advantageous to subtle expression research. This paper proposes a novel optical strain weighted feature extraction scheme for subtle facial micro-expression recognition. Motion information is derived from optical strain magnitudes, which is then pooled spatio-temporally to obtain block-wise weights for the spatial image plane. By simple product with the weights, the resulting feature histograms are intuitively scaled to accommodate the importance of block regions. Experiments conducted on two recent spontaneous micro-expression databases–CASMEII and SMIC, demonstrated promising improvement over the baseline results.

1 Introduction

Facial based emotion recognition attracts research attention both in the computer vision and psychology community. Six basic facial expressions which are commonly considered are happy, surprise, anger, sad, fear and disgust [1]. Contributing to this interest in emotion recognition is the increased research into affective computing, i.e. the ability for software and machines to react to human emotions as they are performing their tasks.

Facial *micro-expressions* were discovered by Ekman [2] in 1969 when he analyzed the interview video of a patient stricken with depression who tried to commit suicide. According to Ekman, micro-expressions cannot be controlled by humans and they are able to reveal concealed emotions. Micro-expressions occur at a high speed (within one twenty-fifth to one fifth of a second) and they are usually involuntary facial expressions [3]. The fact that they occur in a short duration and potentially in only one part of the face makes it hard to detect them with naked eye in real-time conversations.

Work done in project UbeAware funded by TM.

ⓒ Springer International Publishing Switzerland 2015
C.V. Jawahar and S. Shan (Eds.): ACCV 2014 Workshops, Part II, LNCS 9009, pp. 644–657, 2015.
DOI: 10.1007/978-3-319-16631-5_47

There are various applications that support why micro-expressions are important to be analysed, such as clinical diagnosis, national security and interrogation [4–6]. To date, detection of micro-expressions is still a great challenge to researchers in the field of computer vision due to its extremely short duration and low intensity.

Optical strain is the relative amount of deformation of an object [7]. It is able to calculate any small changes on the facial expression, including small muscular movements on the face. In this paper, we propose a new optical strain weighting scheme that utilizes the block-based optical strain magnitudes to extract weighted spatio-temporal features for subtle micro-expression recognition. Firstly, the optical strain map images are computed and normalized from the optical strain magnitudes. Then, the spatial plane (XY) is partitioned into $N \times N$ non-overlapping blocks, where spatio-temporal pooling is applied to obtain a single magnitude for each block. The histograms obtained from the feature extractor are then multiplied with the optical strain weights to form the final feature histogram.

2   Related Work

Optical strain patterns justify its superiorty over the raw image in face recognition as the computation of the magnitudes is based on biomechanics. It is also robust to the lighting condition, heavy make up and under camouflage [8, 9]. In [10], Shreve et al. used the optical strain technique to automatically spot macro- and micro-expressions on facial samples. They could achieve 100 % accuracy in detecting seven micro-expressions in the USF dataset. However the micro-expressions in the database are posed expressions rather than spontaneous ones.

Two years later, [11] an extensive testing was carried on two larger datasets (Canal-9 [12] and found videos [13]), containing a total of 124 micro-expressions by implementing a modified algorithm to spot the micro-expressions. To overcome the noises caused by irrelevant movements on the face, some parts of the face were masked. The face was partitioned into eight regions for the optical strain magnitude to be calculated locally. They extended the work by modifying the algorithms [14]. However, they mentioned that the background and some parts of the face should be masked to avoid the inaccurate optical flow values affect the spotting accuracy.

Block-based method in feature extraction process is widely used in detecting or recognizing micro-expressions, as demonstrated in [15–18]. The face image is partitioned into multiple $N \times N$ non-overlapping or overlapping blocks. The Local Binary Pattern with Three Orthogonal Planes (LBP-TOP) histograms in each block are computed and concatenated into a single histogram. By doing so, the local information of facial expression at its spatial location are taken into account.

Pooling is a method to decrease the number of features (lower dimension) in image recognition. If all the features are extracted, it may result in overfitting. Spatial pooling summarizes the values in the neighbouring locations to achieve better robustness to noise [19]. In [20], Philippe et al. demonstrated several combinations of temporal pooling over a time period and it has been proven to improve the performance of automatic annotation and ranking music audio.

Gaussian filter is one of the effective and adaptive filters to remove Gaussian noises on an image [21]. To track the action units (AUs) on facial expressions using Facial Action Coding System (FACS) [22], a 5×5 Gaussian filter is applied to smooth the images and different sizes of gradient filter are used on different regions of face [23]. In [24], an adaptive Gaussian filter is used to reduce the noises on images in order to compute the illumination change of one person or Expression Ratio Image (ERI) resulted from deformation of the person's face.

To analyze the micro-expressions through a recognition system, it is necessary to have a database to act as a test data set for the researchers to be able to compare the results. There are plenty of facial expression databases available for evaluation [25]. However, there are only a few well-established databases for micro-expressions. This brings to an even bigger obstacle in classifiying the micro-expressions and training the detection algorithms. For example, the micro expressions are posed rather than spontaneous in USF-HD [26] and Polikovsky's database [27]. On the other hand, there are insufficient videos in YorkDDT [28] and SMIC [17] databases.

3   Motion and Feature Extraction

3.1   Optical Flow

Optical flow specifies the velocity of each image pixel's movement between adjacent frames [29]. Computation of differential optical flow is by measuring the spatial and temporal changes of intensity to find a matching pixel in the next frames [30]. As this estimation method is highly sensitive to any changes in brightness, hence it is assumed that all temporal intensity changes are due to motion only. There are three assumptions to measure the optical flow. First, brightness constancy, where the brightness intensity of moving objects between two image frames are assumed to remain constant. Second, spatial coherence, where the pixels in a small image window are assumed to be originating from the same surface and having similar velocity. Third, temporal persistence, where it assumes objects changes gradually over time. The optical flow gradient equation is often expressed as:

$$\nabla I \bullet \boldsymbol{p} + I_t = 0, \tag{1}$$

where $I(x,y,t)$ is the image intensity function at point (x, y) at time t. $\nabla I = (I_x, I_y)$ is the spatial gradients and I_t denotes the temporal gradient of the intensity function. $\boldsymbol{p} = [p = dx/dt, q = dy/dt]^T$ represents the horizontal and vertical motion vector.

3.2   Optical Strain

Using optical strain in identifying deformable results performs better than optical flow [31] as it can well distinguish the time interval of the occuring of micro-expressions. A deformable object can be described in two dimensional space by using a displacement vector $\mathbf{u} = [u, v]^T$. Assuming that the moving object is in small motion, the finite strain tensor can be represented as:

$$\varepsilon = \frac{1}{2}[\nabla\mathbf{u} + (\nabla\mathbf{u})^T] \tag{2}$$

or in an expanded form:

$$\varepsilon = \begin{bmatrix} \varepsilon_{xx} = \frac{\partial u}{\partial x} & \varepsilon_{xy} = \frac{1}{2}(\frac{\partial u}{\partial y} + \frac{\partial v}{\partial x}) \\ \varepsilon_{yx} = \frac{1}{2}(\frac{\partial v}{\partial x} + \frac{\partial u}{\partial y}) & \varepsilon_{yy} = \frac{\partial v}{\partial y} \end{bmatrix} \tag{3}$$

where $(\varepsilon_{xx}, \varepsilon_{yy})$ are normal strain components and $(\varepsilon_{xy}, \varepsilon_{yx})$ are shear strain components.

The magnitude of the optical strain can be computed as follows:

$$\varepsilon = \sqrt{\varepsilon_{xx}{}^2 + \varepsilon_{yy}{}^2 + \varepsilon_{xy}{}^2 + \varepsilon_{yx}{}^2} \tag{4}$$

An *optical strain map* (OSM) provides a visual representation of the motion intensity for each pixel in a video frame. To visualize the OSM, the optical strain magnitudes for each point (x, y) in image space at time t can be normalized to intensity values 0–255. By observing the OSM, we can clearly notice regions in the image frame that contain the most prominent (large values) or least prominent (small values) motion in terms of spatial displacement. To obtain a summed OSM for the entire sequence, all the individual generated OSMs can be summed across the temporal dimension. This accumulates all motion displacements in the whole sequence, a pooling operation that will be discussed later in Subsect. 4.2. Figure 1 shows a sample optical strain map image (for two adjacent frames), and a summed optical strain map image (for all frames, temporal sum pooled) after applying intensity normalization.

3.3 Block-Based LBP-TOP

Block-based LBP-TOP is implemented by partitioning each frame of the video into $N \times N$ non-overlapping blocks then concatenate them into a single histogram. Figure 2 shows the process of extracting the features from three orthogonal plane for one block volume and concatenate them into a histogram. The feature histogram of block-based LBP-TOP [15] can be defined as follows:

$$H_{i,j,c,b} = \sum_{x,y,t} I f_c(x, y, t) = b, \ b = 0, \ldots, n_c - 1; \ c = 0, 1, 2; \ i, j \in 1 \ldots N \tag{5}$$

where n_c is the number of different labels produced by the LBP operator in the cth plane ($c = 0 : XY, 1 : XT$ and $2 : YT$), $f_c(x, y, t)$ is the LBP code of the central pixel (x, y, t) in c-th plane, $x \in \{0, \ldots, X - 1\}, y \in \{0, \ldots, Y - 1\}, t \in \{0, \ldots, T - 1\}$,

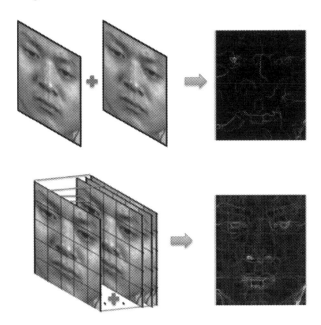

Fig. 1. Example of optical strain map for two image frames *(top row)* and for all the frames in sequence *(bottom row)* for a tense micro-expression

$$I\{A\} = \begin{cases} 1, & \text{if } A \text{ is true;} \\ 0, & \text{otherwise.} \end{cases} \tag{6}$$

The histogram is normalized to get a coherent description:

$$\overline{H}_{i,j,c,b} = \frac{H_{i,j,c,b}}{\sum_{k=0}^{n_c-1} H_{i,j,c,k}} \tag{7}$$

We denote LBP-TOP parameters by $LBP\text{-}TOP_{P_XY,P_XY,P_YT,R_X,R_Y,R_T}$ where the P parameters indicate the number of neighbor points for each of the three orthogonal planes, while the R parameters denote the radii along the X, Y, and T dimensions of the descriptor.

4   Proposed Algorithm

4.1   Block-Wise Optical Strain Magnitudes

The magnitude of optical strain for each pixel is very small. Much of the surrounding pixels that contain very little flow corresponds to very minute values. As such, we hypothesize that using the optical strain map magnitudes directly as features or by extraction of LBP patterns for classification may result in a loss of essential information from its original image intensity values.

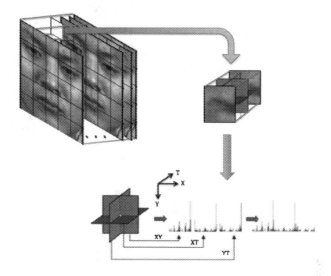

Fig. 2. Block-based LBP-TOP: Feature extraction from three orthogonal planes for one block volume

However, optical strain maps provide valuable motion information between successive frames, more so in the case of subtle expressions that may be difficult to distinguish at the feature level. In this paper, we propose a new technique that uses optical strain information as a weighting function for the LBP-TOP feature extractor. This is because pixels with a large displacement in space (large optical strain magnitude) indicate large motion at that particular location and vice versa. Hence, we can increase (or decrease) the importance of the extracted features by placing more (or less) emphasis through the use of weights.

To obtain the optical strain magnitudes, first, horizontal and vertical optical flow vectors, (p, q) are calculated for each image frames in a video [32]. Then optical strain magnitude, ε of each pixel for each frame in a video will be computed.

4.2   Spatio-Temporal Pooling

Spatial sum pooling is applied on each optical strain image, where each of the strain map image will first be partitioned into $N \times N$ non-overlapping blocks, then all the pixels in that particular block will be summed up. Spatial sum pooling can be computed for each block in an image as follows:

$$s_{i,j} = \sum_{y=(j-1)H+1}^{jH} \sum_{x=(i-1)L+1}^{iL} \varepsilon_{x,y}, \ i,j \in 1 \ldots N \tag{8}$$

where (i, j) and (X, Y) are the block's coordinate and width and height of the frame in (x, y). L and H are equal to X/N and Y/N respectively. Temporal sum

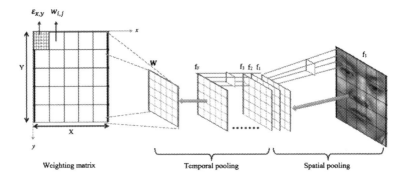

Fig. 3. Spatial-temporal sum pooling of a strain image divided into 5×5 non-overlapping blocks

pooling is then performed by summing up the resultant optical strain magnitudes of each block from the first frame, f_i to the last frame, f_F.

Hence, for each video, a weight matrix $W = \{w_{i,j}\}_{i,j=1}^{N}$ is formed using spatial-temporal sum pooling (process illustrated in Fig. 3) where each block weight value is given by:

$$w_{i,j} = \sum_{t=1}^{F} s_{i,j} = \sum_{t=1}^{F} \sum_{y=(j-1)H+1}^{jH} \sum_{x=(i-1)L+1}^{iL} \varepsilon_{x,y} \tag{9}$$

4.3   Obtaining Block Weights for XY Plane Histogram

Subsequently, all weight matrices are max-normalized to increase the significance of each weighting value. As optical strain magnitudes only describe the expression details in spatial information, the weighting values should be effective on the XY plane only. As such, the resultant histogram is obtained by multiplying the histogram bins of the XY plane with the weighting values, as illustrated in Fig. 4. The new feature histogram is given as:

$$G_{i,j,c,b} = \begin{cases} w_{i,j}\overline{H}_{i,j,c,b}, & \text{if } c = 0 \\ \overline{H}_{i,j,c,b}, & \text{else} \end{cases} \tag{10}$$

5   Experiments

5.1   Subtle Expression Databases

There are only a few known subtle or micro-expression databases available due to numerous difficulties in the creation process; proper elicitation of stimuli and ground-truth labelling. To evaluate our proposed methods, we consider two of the

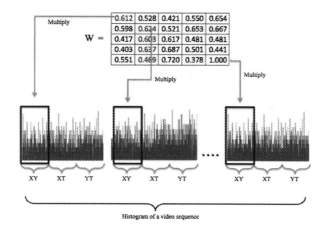

Fig. 4. Multiplication of weighting matrix to X-Y plane of histogram bins

most recent and comprehensive databases: CASMEII [16] and SMIC (Spontaneous Micro-expression Database) [17]. The databases are recorded under constrained lab condition and all the images have been preprocessed with face registration and alignment.

CASMEII consists of 26 candidates (mean age of 22.03 years), containing 247 spontaneous and dynamic micro-expression clips. The videos are recorded using Point Grey GRAS-03K2C camera with a frame rate of 200 fps and a spatial resolution of 280 × 340 pixels. There are 5 micro-expression classes (tense, repression, happiness, disgust and surprise) and selection was done by two coders then marked based on the AUs, participants' self report as well as the content of the clip episodes. Each sample contains the ground-truth of onset and offset frames, emotions labeled and AUs represented. The baseline performance reported in CASMEII for 5-category classification is 63.41%. This was obtained using a block-based LBP-TOP consisting of 5×5 blocks. Support Vector Machine (SVM) was used as classifier with leave-one-out cross validation (LOOCV).

SMIC contains 164 micro-expression samples from 16 participants (mean age of 28.1 years). The camera used to capture participant's face is a high speed camera (PixeLINK PL-B774U) with 100 fps and a resolution of 640 × 480 pixels. There are three classes of micro-expressions: positive (happy), negative (sad, fear, digust) and surprise. The micro-expressions are selected by two coders based on participants' self report and the suggestion by [2] to view the video frame-by-frame with increasing speed. The reported baseline 3-class recognition performance for SMIC is 48.78% using polynomial kernel of degree six in SVM classifier based on leave-one-subject-out cross-validation (LOSOCV) setting. All image frames from each video are first interpolated to ten frames by temporal interpolation model (TIM) [18], while features were extracted using $LBP\text{-}TOP_{4,4,4,1,1,3}$ with block size of 8 × 8.

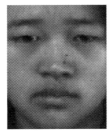 

Fig. 5. Sample image from CASMEII before (left) and after (right) applying Gaussian filter

5.2   Pre-processing

Gaussian Filtering. Since the motions characterized by the subtle facial expressions are very fine and we are using the cropped and resampled frames for both databases, it is likely that the presence of unwarranted noise from the acquisition or down-sampling process might be incorrectly identified as fine facial motions. Thus, as a feasible pre-processing step, all the images are filtered by 5×5 pixel Gaussian filter ($\sigma = 0.5$) to suppress the background noise present in the images. The filter size and standard deviation value are empirically determined. Figure 5 shows the difference of an image before and after filtering.

Noise Block Removal. The two bottom corner blocks (bottom left and bottom right) are removed entirely from consideration in the feature histogram by setting their respective weights to zero, i.e. $\{w_{N,1}, w_{N,N}\} = 0$. This results in only the remaining $N^2 - 2$ weights to be effective on the XY-plane histograms. The reason of removing these 2 blocks from consideration is because there are unexpectedly high optical strain magnitudes that do not correspond to the desired facial movements but are rather unfortunately caused by background/clothing texture noise or wirings from the headset worn by the participants. This problem is consistent across both CASMEII and SMIC datasets, as can be clearly seen in Fig. 6. Analogously, the authors of [11,14] applied masking technique at consistently noisy regions of the face that unnecessarily affect the optical strain, such as eyes (blinking) and mouth (opening/closing) regions.

5.3   Results and Discussions

Experiments were conducted on both CASMEII and SMIC databases based on carefully configured settings in order to validate the effectiveness of our proposed method in improving recognition of subtle facial expressions. In our experiments, we performed classification using SVM with leave-one-out cross-validation (LOOCV) in CASMEII and leave-one-subject-out cross-validation (LOSOCV) in SMIC in order to appropriately compare with the baselines reported in the original CASMEII and SMIC papers. In our work, CASMEII is evaluated using linear and RBF kernel, whereas SMIC uses linear, RBF and polynomial kernel

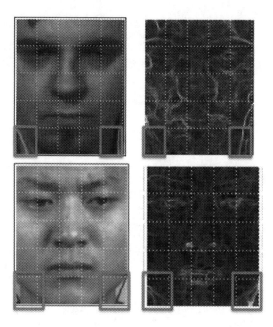

Fig. 6. Top row: sample image from SMIC (left) and its optical strain map image (right). Bottom row: sample image from CASMEII (left) and its optical strain map image (right).

with degree six. There are two ways to calculate the classification performance in LOSOCV approach, which are *macro-* and *micro*-averaging. Macro-averaged results are the average accuracy of per-subject results. Micro-averaged results are the average accuracy across all individual results (per sample) which can be obtained from the confusion table that summarizes the overall performance.

To establish our baseline evaluation, the standard methods employed by the original authors of CASMEII and SMIC [16,18]—LBP-TOP for feature extraction and SVM for classification, were used. For CASMEII, we opt for the best reported configuration, that is $LBP\text{-}TOP_{4,4,4,1,1,4}$. As for SMIC, we used both $LBP\text{-}TOP_{4,4,4,1,1,3}$ and $LBP\text{-}TOP_{4,4,4,1,1,4}$. CASMEII baseline used the block configuration of 5×5 blocks, whereas SMIC used 8×8 blocks.

In our experiments, we evaluated our proposed Optical Strain Weighted (OSW) LBP-TOP method (denoted as **OSW-LBP-TOP** in table of results) against the baseline method of **LBP-TOP**. Apart from that, we also examined the method with pre-processing, which filters all the images using Gaussian filter and removes two specific "noise blocks" that are contributing to surplus motions unrelated to facial expressions. For the basic weighted method, all $(N \times N)$ weight coefficients are multiplied with the respective histogram bins of the XY plane. The tables shows the recognition accuracy of the evaluated methods for both CASMEII and SMIC, using SVM classifier with leave-one-out cross-validation (LOOCV) and leave-one-subject-out cross-validation (LOSOCV) respectively.

Table 1. Accuracy results (%) on CASMEII database based on LOOCV

Methods	SVM kernel	
	RBF	Linear
Baseline: LBP-TOP	63.97	61.94
OSW-LBP-TOP	**65.59**	**62.75**

Table 2. Accuracy results (%) on CASMEII database based on LOOCV with pre-processing (PP)

Methods	SVM kernel	
	RBF	Linear
Baseline: LBP-TOP (with PP)	63.56	63.97
OSW-LBP-TOP (with PP)	**66.40**	62.75

Generally, the recognition capabilities of the LBP-TOP descriptor demonstrated encouraging signs of improvement when the features are weighted using the proposed scheme. The pooled optical strain magnitudes as block weights intuitively increases the classification accuracy. Crucially, more weightage is assigned to blocks that exhibit more movements, and vice versa, so that the significance of each block histogram can be scaled accordingly. The OSW-LBP-TOP method, with pre-processing obtained the best CASMEII result of 66.4 % (RBF kernel), an increase of 2.84 % over the baseline. It managed to achieve 65.59 % (RBF kernel) without pre-processing, an increase of 1.62 % over the baseline. The recognition results of CASMEII are illustrated in Tables 1 and 2.

On the other hand, the OSW-LBP-TOP method is consistently superior in the SMIC database. With the $LBP\text{-}TOP_{4,4,4,1,1,3}$ setting from the original paper [17], we are able to obtain an improvement of 3.6 % (linear and RBF kernel) without pre-processing and 4.49 % (polynomial kernel) of increment with pre-processing, as shown in Tables 3 and 4 respectively. However, we discovered that with parameters $LBP\text{-}TOP_{4,4,4,1,1,4}$, we are able to generate better baselines while the proposed OSW-LBP-TOP method performed even better with pre-processing. An increment of 1.83 % (polynomial kernel) and 5.13 % (RBF kernel) were achieved in cases of without and with pre-processing, as tabulated in Tables 5 and 6 respectively.

The improvement in accuracy is apparent on both databases, albeit the fact that the choice of SVM kernel seems to play an equally important role as well. Notably, the OSW-LBP-TOP methods easily outperform the CASMEII baseline result when the RBF kernel is used for the SVM classifier. In the case of SMIC when the OSW-LBP-TOP methods are used, all the three kernels consistently produced improved results. This is an interesting finding that requires further investigation as to how these weights impact and alter the sample distribution to the advantage of specific linear or nonlinear (RBF in this case) kernel types.

Table 3. Accuracy results (%) on SMIC database using $LBP\text{-}TOP_{4,4,4,1,1,3}$ based on LOSOCV

Methods	SVM kernel					
	Macro			Micro		
	RBF	Linear	Poly	RBF	Linear	Poly
Baseline	43.11	43.11	51.63	43.29	43.29	48.78
OSW-LBP-TOP	**46.71**	**46.71**	**51.70**	**46.34**	**46.34**	**49.39**

Table 4. Accuracy results (%) on SMIC database using $LBP\text{-}TOP_{4,4,4,1,1,3}$ based on LOSOCV with pre-processing (PP)

Methods	SVM kernel					
	Macro			Micro		
	RBF	Linear	Poly	RBF	Linear	Poly
Baseline (with PP)	44.06	44.06	48.94	42.07	42.07	46.34
OSW-LBP-TOP (with PP)	**47.17**	**47.17**	**53.43**	**46.34**	**46.34**	**50.00**

Table 5. Accuracy results (%) on SMIC database using $LBP\text{-}TOP_{4,4,4,1,1,4}$ based on LOSOCV

Methods	SVM kernel					
	Macro			Micro		
	RBF	Linear	Poly	RBF	Linear	Poly
Baseline	55.65	55.65	57.63	51.83	51.83	51.83
OSW-LBP-TOP	**57.34**	**57.34**	**57.71**	**53.05**	**53.05**	**53.66**

Table 6. Accuracy results (%) on SMIC database using $LBP\text{-}TOP_{4,4,4,1,1,4}$ based on LOSOCV with pre-processing (PP)

Methods	SVM kernel					
	Macro			Micro		
	RBF	Linear	Poly	RBF	Linear	Poly
Baseline (with PP)	51.66	51.66	55.04	47.56	47.56	50.00
OSW-LBP-TOP (with PP)	**56.79**	**56.09**	**57.54**	**53.05**	**52.44**	**53.05**

Another observation that is worth highlighting for subtle micro-expression research is that sufficient attention should be given to deal with the impact of noise on the recognition performance. The addition of essential pre-processing steps to suppress image noise and remove the noisy blocks are able to produce better results. This can be attributed to the discarding of the histogram bins (set to zero) or features that belong to those noisy regions of the image.

6   Conclusion

In this paper, we have presented a novel method for recognizing subtle expressions in video sequence. The proposed optical strain weighted feature extraction method for subtle expression recognition is able to achieve 66.4 % accuracy for a five-class classification on CASMEII database and a 57.71 % accuracy for a three-class classification on SMIC database. However, due to the subtlety of facial micro-expressions, the presence of image noise is a challenging problem that requires attention. For future works, the weighting scheme can be extended to the classifier kernel distances to further increase the effectiveness and robustness in the classification stage. Also, noise suppression schemes can also be introduced to reduce the impact of noisy textures.

References

1. Ekman, P., Friesen, W.V.: Constants across cultures in the face and emotion. J. Pers. Soc. Psychol. **17**(2), 124 (1971)
2. Ekman, P.: Lie catching and microexpressions. In: Martin, C. (ed.) The Philosophy of Deception, pp. 118–133. Oxford University Press, New York (2009)
3. Porter, S., ten Brinke, L.: Reading between the lies identifying concealed and falsified emotions in universal facial expressions. Psych. Sci. **19**, 508–514 (2008)
4. Frank, M.G., Herbasz, M., Sinuk, K., Keller, A., Kurylo, A., Nolan, C.: I see how you feel: training laypeople and professionals to recognize fleeting emotions. In: Annual Meeting of the International Communication Association, Sheraton New York, New York City, NY (2009)
5. O'Sullivan, M., Frank, M.G., Hurley, C.M., Tiwana, J.: Police lie detection accuracy: the effect of lie scenario. Law Hum. Behav. **33**(6), 530–538 (2009)
6. Frank, M.G., Maccario, C.J., Govindaraju, V.: Protecting Airline Passengers in the Age of Terrorism. ABC-CLIO, Santa Barbara (2009)
7. D'hooge, J., Heimdal, A., Jamal, F., Kukulski, T., Bijnens, B., Rademakers, F., Hatle, L., Suetens, P., Sutherland, G.R.: Regional strain and strain rate measurements by cardiac ultrasound: principles, implementation and limitations. Eur. J. Echocardiogr. **1**(3), 154–170 (2000)
8. Shreve, M., Manohar, V., Goldgof, D., Sarkar, S.: Face recognition under camouflage and adverse illumination. In: 4th IEEE International Conference on Biometrics: Theory Applications and Systems (BTAS), pp. 1–6 (2010)
9. Manohar, V., Goldgof, D., Sarkar, S.: Facial strain pattern as a soft forensic evidence. In: Applications of Computer Vision (WACV) (2007)
10. Shreve, M., Godavarthy, S., Manohar, V., Goldgof, D., Sarkar, S.: Towards macro- and micro-expression spotting in video using strain patterns. In: Applications of Computer Vision (WACV), pp. 1–6 (2009)
11. Shreve, M., Godavarthy, S., Goldgof, D., Sarkar, S.: Macro-and micro-expression spotting in long videos using spatio- temporal strain. In: Automatic Face, Gesture Recognition and Workshops, pp. 51–56 (2011)
12. Vinciarelli, A., Dielmann, A., Favre, S., Salamin, H.: Canal9: a database of political debates for analysis of social interactions. In: Affective Computing and Intelligent Interaction and Workshops, pp. 1–4 (2009)

13. Ekman, P.: Telling Lies: Clues to Deceit in the Marketplace, Politics, and Marriage. W. W. Norton and Company, New York (2009)
14. Shreve, M., Brizzi, J., Fefilatyev, S., Luguev, T., Goldgof, D., Sarkar, S.: Automatic expression spotting in videos. Image Vis. Comput. **32**(8), 476–486 (2014)
15. Zhao, G., Pietikainen, M.: Dynamic texture recognition using local binary patterns with an application to facial expressions. IEEE Trans. Pattern Anal. Mach. Intell. **20**(6), 915–928 (2007)
16. Yan, W.J., Wang, S.J., Zhao, G., Li, X., Liu, Y.J., Chen, Y.H., Fu, X.: Casme ii: an improved spontaneous micro-expression database and the baseline evaluation. PLoS ONE **9**, e86041 (2014)
17. Pfister, T., Li, X., Zhao, G., Pietikainen, M.: Recognising spontaneous facial micro-expressions. In: Computer Vision (ICCV), pp. 1449–1456 (2011)
18. Li, X., Pfister, T., Huang, X., Zhao, G., Pietikainen, M.: A spontaneous micro-expression database: inducement, collection and baseline. In: Automatic Face and Gesture Recognition, pp. 1–6 (2013)
19. Boureau, Y.L., Ponce, J., LeCun, Y.: A theoretical analysis of feature pooling in visual recognition. In: Machine Learning (ICML 2010), pp. 11–118 (2010)
20. Hamel, P., Lemieux, S., Bengio, Y., Eck, D.: Temporal pooling and multiscale learning for automatic annotation and ranking of music audio. In: International Society for Music Information Retrieval Conference, pp. 729–734 (2011)
21. Forsyth, D.A., Ponce, J.: Computer Vision: A Modern Approach. Prentice Hall, Upper Saddle River (2002)
22. Ekman, P., Friesen, W.V.: Facial Action Coding System. Consulting Psychologists Press, Palo Alto (1978)
23. Lien, J.J.J., Kanade, T., Cohn, J.F., Li, C.C.: Detection, tracking, and classification of action units in facial expression. Rob. Auton. Syst. **31**(3), 131–146 (2000)
24. Liu, Z., Shan, Y., Zhang, Z.: Expressive expression mapping with ratio images. In: Computer Graphics and Interactive Techniques, pp. 271–276 (2001)
25. Anitha, C., Venkatesha, M.K., Adiga, B.S.: A survey on facial expression databases. Int. J. Eng. Sci. Technol. **2**(10), 5158–5174 (2010)
26. Yan, W.J., Wang, S.J., Liu, Y.J., Wu, Q., Fu, X.: For micro-expression recognition: database and suggestions. Neurocomputing **136**, 82–87 (2014)
27. Polikovsky, S., Kameda, Y.,O.Y.: Facial micro-expressions recognition using high speed camera and 3D-gradient descriptor. In: Crime Detection and Prevention
28. Warren, G., Schertler, E., Bull, P.: Detecting deception from emotional and unemotional cues. J. Nonverbal Behav. **33**(1), 59–59 (2009)
29. Barron, J.L., Thacker, N.A.: Tutorial: Computing 2D and 3D optical flow (2005) Imaging Science and Biomedical Engineering Division, Medical School, University of Manchester
30. Jain, R., Kasturi, R., Schunck, B.G.: Machine Vision, vol. 5. McGraw-Hill Education, New York (1995)
31. Godavarthy, S.: Microexpression spotting in video using optical strain. Masters thesis, University of South Florida (2010)
32. Sun, D., Roth, S., Black, M.J.: Secrets of optical flow estimation and their principles. In: Computer Vision and Pattern Recognition, pp. 2432–2439 (2010)

Task-Driven Saliency Detection on Music Video

Shunsuke Numano[(⊠)], Naoko Enami, and Yasuo Ariki

Kobe University, Kobe, Japan
numano@me.cs.scitec.kobe-u.ac.jp

Abstract. We propose a saliency model to estimate the task-driven eye-movement. Human eye movement patterns is affected by observer's task and mental state [1]. However, the existing saliency model are detected from the low-level image features such as bright regions, edges, colors, etc. In this paper, the tasks (e.g., evaluation of a piano performance) are given to the observer who is watching the music videos. Unlike existing visual-based methods, we use musical score features and image features to detect a saliency. We show that our saliency model outperforms existing models that use eye movement patterns.

1 Introduction

Yarbus suggested that human eye-movement patterns are modulated top down by different task demands [1]. After that, many works had showed the relationship between eye-movement patterns and cognitive factor [2–4]. From these analyses, Itti *et al.* [4] proposed two hypotheses which are called the saliency hypothesis with relation to eye-movement. Hypothesis 1 is that the eye-movement patterns is affected by the task such as driving. Hypothesis 2 is that human gazes at the area which has the saliency of color, edge and the intensity when they are not given the task. The method for the estimation of the task such as the documents in reading [5] using eye-movement patterns is proposed based on hypothesis 1. However, no method is proposed that estimates the eye-movement patterns for different tasks from the image. On the other hand, based on hypothesis 2, Itti *et al.* [6] proposed the method for the estimation of the "saliency" area where human gazed in terms of the image feature such as color, intensity and orientation. However, the existing saliency model are not considered the hypothesis 1.

Our goal is to detect the image saliency so that we can estimate the task-driven eye-movement. In this work, we give the observers two tasks when viewing the music videos. Task 1 is to evaluate the piano performance, and task 2 is to grasp the melody of the performance. These tasks are affected by the cognitive factor related to music. We can introduce the cognitive factor to the evaluation of the musical sense that can be obtained in the musical education. Thus, we need to show the relationship between the cognitive factor and the eye-movement patterns related to music. In accordance with above, we should consider our tasks as well as the conventional image feature when constructing the saliency model. In addition, the observers listen to the sound as well as viewing the image in

© Springer International Publishing Switzerland 2015
C.V. Jawahar and S. Shan (Eds.): ACCV 2014 Workshops, Part II, LNCS 9009, pp. 658–671, 2015.
DOI: 10.1007/978-3-319-16631-5_48

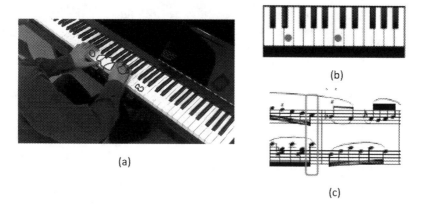

Fig. 1. Our saliency model adding the musical information. (a) Our saliency model (blue line), the ground truth (red line). (b) The striking key on the keyboard. (c) The striking key on the musical score (red box) (Color figure online).

our tasks. So, we expect that the eye-movement is affected by the music video including the sound and the tasks.

Although most of existing saliency models are detected from the low-level image features, the saliency affects the observer's knowledge (e.g., the musical sense in this work) to the observation object and the cognitive factor. In this paper, we propose a novel saliency model that is added the information of the musical score in order to achieve our goal as shown in Fig. 1(a). The musical score has a lot of information necessary for the performance such as the musical note, the dynamics, and so on, and the performance is conducted in accordance with the musical score. We therefore can add more information related to the performance by using the musical score information than the sound information, and we expect to obtain the saliency related to the knowledge of the music. In this paper, we add the information of the musical note (as shown in Fig. 1(b) and (c)) to the conventional saliency map that is proposed by Itti *et al.* [6] who detected the saliency using the image feature and that is the state-of-the-art proposed by Yang *et al.* [7]. Next, we evaluate the proposed saliency model. Our goal is to construct the saliency model for the estimation of the task-driven eye-movement, so we use the task-driven eye-movement as the ground truth in the evaluation. The method where the eye-movement is used for the estimation of the saliency map that is detected from the video is proposed [8]. However, Riche *et al.* [8] used the eye-movement that was not considered the task. Thus, considering the task in this paper, we construct the dataset which is constituted by the eye-movement and the music video of our task. We treat this dataset as the ground truth and show the effectiveness of the proposed method by the measure for the evaluation which is proposed in [8]. The contributions in this paper are as follows. (1) We propose the saliency model for the task-driven eye-movement. (2) We add the information which is not the image feature as well as the image feature to the saliency model. (3) We introduce the musical information in the form of the musical note to the saliency model of the music video. (4) We treat the eye-movement which is observed in our tasks as the ground truth when evaluating the saliency model.

The rest of the paper is organized as follows. Section 2 is described the problem setting. Section 3 is described the proposed saliency model. Section 4 is described the dataset for the evaluation of the saliency model. Section 5 demonstrates experimental results. Section 6 is described the discussion and Sect. 7 is the conclusion.

2   Problems

Our goal is to detect the saliency for the estimation of the task-driven eye-movement. First, we describe the task setting, and then, we describe the saliency model based on the image feature that is baselines of our work.

2.1   Task Setting

Yarbus [1], Henderson [2], Angelusa [3] and Itti [4] gave the observers some tasks when observing a still image, and measured the eye-movement. In this paper, we gave the observers two tasks as follows in order to observe the task-driven eye-movements.

– Task 1: To evaluate the piano performance.
– Task 2: To memorize the music.

We consider that these tasks are affected by the musical discipline of the observers. So, the observers have the experience of learning to play the piano for more than one year in this paper. They are considered to have more chances of developing the musical sense than someone who has never learned the piano. The music videos in our work are related to the piano performance and collected from the video site "You Tube". We perform the previous measurement for our work. In the previous measurement, we measured the gaze behavior in watching the music video without observer's head fixed. The music video included whole body of the performer, and the video was taken from the right side of the performer. The music video also included the sound of the piano performance. The length of the video was 1 min and the part of the tune was used. We found as follows in the previous measurement. First, since the observer's head was not fixed, the change of face direction and the range of movement of the observer's head was not restricted. We therefore consider that the accurate position of the gaze point was not obtained by the eye tracker. Second, most observers paid attention to the head of the performer. Third, observers tended to gaze at the center of the frame or a certain location as the time passed. According to the previous measurement, we selected the videos that meets the following conditions. (1) The music video includes the sound of the piano performance. (2) The music video was taken by the fixed camera. (3) The music video was taken from the position where the keyboard and the performer's hands are seen (Fig. 1(a)). (4) The head of the performer is not seen in the video. (5) The person in the video is only the performer. There are no restrictions to the performer, music, the background of the video and piano. The performer basically play the piano according to the musical score. The observers also have difficulty in recognizing the motion of all of the fingers, so the ambiguity of the musical score and the sound is not considered in this paper. We use the eye-movement of the observers in two tasks as the ground truth to evaluate the saliency model. However, the ground truth is obtained by the eye-movement of some observers as is the case in [2–4].

2.2   The Saliency Map Based on the Image Feature

There are many works where the saliency is detected by the image feature based on the saliency hypothesis 2. These works are divided into two types. One is the saliency model for the estimation of the fixation points from the natural images [6,9–11]. The other is the saliency model for the detection of the salient object in the image [7,12,13]. We use the saliency models proposed by Itti [6] and Yang [7] (the state-of-the-art) as the baselines and add the score feature to these baselines. We also describe each baseline. Itti constructed the three feature maps (intensity, color, orientation), and summed them to obtain the saliency map. Since we use the video, the optical flow is added as the dynamic feature. Yang constructs the graph where each superpixel extracted from the image is used as the node. First, these nodes are compared with the nodes of four sides of the image (the top, bottom, left and right of the image) as labeled background queries, and compute the salient nodes based on their relevances (*i.e.*, ranking score) to those background queries, so that the labeled maps of each side are obtained. Then, these four labeled maps are integrated to generated a saliency map. Second, the labeled foreground nodes are taken as saliency queries, and the saliency of each nodes is computed based on the relevance to foreground queries for the final map.

3   Our Approach

In this section, we describe the proposed saliency of the music video that includes the musical score information.

3.1   The Saliency Adding the Musical Score Feature

In this paper, we consider the image feature extracted from the music videos and the score feature S extracted from the musical score corresponding to the music video frame. We use these features to construct the proposed saliency model of the music videos. We do not intend to match the musical score to the striking keys in the musical video, so we do by hand. The automation of matching the musical score to the striking keys in the video is possible by the digital piano.

The musical score feature: We extract the score feature from the musical score. The method for extraction is as follows. In this paper, we use the musical score as the musical feature. First, we related the notes on the score to the striking keys by using the virtual keyboard. The virtual keyboard has 52 white keys whose size are $75pixels \times 5pixels$ and 36 black keys whose size are $25pixels \times 5pixels$. The position of keying is defined as the center of each key. We transform the position of the notes on the virtual keyboard to the position of the keyboard in the music video. We also generate the normal distribution whose mean is the center of each position of keying. We consider the view angle of the visual field (0–5 degrees from the center of the visual field), so we define the variance as 20 pixels of this normal distribution [14]. In this way, we can obtain the feature map of the musical note.

Our model based on Itti's method: We construct the image feature map (the color feature map C, the orientation feature map O, the intensity feature map I, the dynamic feature map M) from the image as is the case in [6]. Next, we summed these image feature maps and the musical score feature map with the weighted linear combination as following formula,

$$Sal_{Itti+S} = a_1 C + a_2 O + a_3 I + a_4 M + a_5 S, \tag{1}$$

where in $a_i (i = 1, \ldots, 5)$ is the weight of each feature map (as shown in Table 2). Figure 2(a) is an example of the saliency calculated by Itti's method, Fig. 2(b) is an example of the saliency map where the optical flow is added as the dynamic feature, Fig. 2(d) and (e) are examples of the proposed saliency maps where the musical score feature is added.

Our model based on Yang's method: We also calculate the saliency by the Yang's method. In Yang's method, the superpixels are extracted from the image as the node, and the salient region is detected using the graph-based ranking score of each node. We add the musical score feature to the ranking score defined as following formula.

$$\boldsymbol{f}^* = (\boldsymbol{D} - \alpha\boldsymbol{W})^{-1}\boldsymbol{y}, \tag{2}$$

where $\boldsymbol{f}^*$ is the ranking value of each node, $W = [w_{i,j}]_n \times n$ is an affinity matrix of the nodes, $D = diag d_{11}, \ldots, d_{nn}$ is the degree matrix, where $d_{ii} = \sum_j w_i j$, and y_i is whether the $node_i$ is the query. We add the musical score feature to the ranking value. First, we split the musical score map into the superpixels that are the same as that of the frame of the music video. The resolution of the musical score map is the same as that of the music video frame. We also compute the average of each superpixel of the musical score map, so that we obtain the musical value MV for each superpixel. We use the normalized value of MV and compute a new ranking score as follows,

$$\boldsymbol{fnew}_i^* = f_i^* \cdot MV_i. \tag{3}$$

We obtain the four labeled maps (the query of each map is the top, bottom, left and right of the image) and integrate them to obtain the saliency map based on

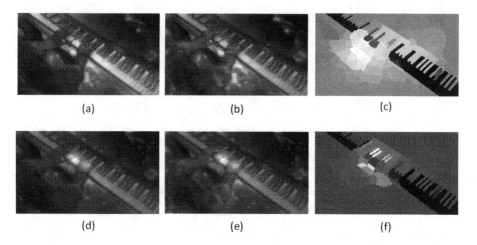

Fig. 2. The saliency map. (a) is the map proposed by Itti, (b) is the proposed map based on (a), (c) is the map which adds feature of the optical flow to (a), (d) is the proposed map based on (d), (e) is the map proposed by Yang, (f) is the proposed map based on (e)

the ranking score $fnew_i^*$. Figure 2(c) is an example of the saliency calculated by baseline of Yang, Fig. 2(f) is an example of the proposed saliency map where the musical score feature is added.

4   DataSet

Our saliency model is task-driven, and the evaluation of our model needs the ground truth of the task driven eye-movements. For overt attention and the detection of eye fixation positions, the ground truth of human attentional behavior can be measured using an eye tracking device. The eye tracker provides the binocular gaze point at a joint sampling frequency of 60 fps. The framerate of the viewing video is 30 fps with a resolution of 640 × 480 pixels. The dataset has been proposed that includes the video and the eye tracking data. However, the task is not given to the observers in the existing dataset, and the video in the dataset does not include the sound. We therefore construct the dataset of the saliency in order to estimate the task-driven eye-movement. The observers consist of 12 observers who have learned the piano and 10 observers without the experience of a piano performance. The observer are 18–32 years old, both males and females. The number of years of learning the piano is 1–21. In addition, a professional artists is not contained in the observer. The time of each music video was about 30 s. We give the observers following two tasks. Task 1 is to evaluate the performance by five levels in four items (e.g., the mellifluence, the strength of sound, the accuracy of keying, the rhythm of the performance). Task 2 is to memorize the tune and select the score of the performed tune from three scores on the display after watching the music video. The observers watched eight videos in task

Table 1. The dataset of the gaze data.

Dataset contents

–The serial data (the frame counter, XY coordinate, data of gazing points, pupil diameter)
–The viewing image with gazing points
–The questionnaire

Subjects

–22 subjects of 12 experienced person and 10 inexperienced person

Video The Subjects Watched

–10 music videos(Each video is 30 seconds)
–Tune list(Number of tune, Name of classical composer, Title of tune, Degree of the visibility of the tunes)
No.1, Beethoven,Fur Elise, 2.425.
No.2,Chopin,Nocturne Op.9-2,1.93.
No.3,Beethoven,Piano Sonata No.17 "Tempest",1.27.
No.4,Rubinstein,Op.44,No1,1.04.
No.5,Schumann = Liszt, Widmung,1.14.
No.6,Chopin, Etudes Op.10-8,1.04.

1 and two videos in task 2. Each video is 30 s. We measured the eye-movement of the observers in each task. The music videos were collected from "You Tube" and met the conditions as described in Sect. 2.1. The detail of the music videos is shown in Table 1. We also asked the visibility of each tune in three levels (ex., unknown tune, known tune, played tune) to the observers. Since the musical sense developed in the musical discipline is considered in particular, we used the tune with high visibility among the observers. Additionally, we describe the generation of the ground truth of the task-driven eye-movements as shown in Fig. 3. The ground truth is the distribution of the gaze data of the observers in watching the music video, and generated per frame. We first overlap the visual fields of all the observers with the music video frame. Humans recognize the object in the viewing area centering on the gaze point. The area where human can visualize the object accurately is restricted because of the structure of the retina. The visual angle of this area is known as 0–5 degrees. However, the method for estimation of the accurate visual fields individually has not been established yet, so many works define the visual fields as the circular area. We also approximate the visual fields as follows,

$$R_{cm} = d \times tan(\frac{\theta}{2} \times \frac{\pi}{180}), \tag{4}$$

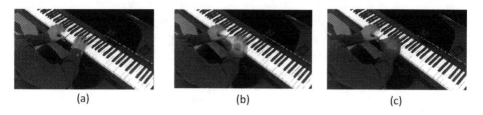

Fig. 3. The ground truth. (a) The gaze position is obtained per frame. (b) The visual fields overlapped on the music video frame. (c) The ground truth obtained by binarization of the map (b).

$$R_{px} = R_{cm} \times \frac{w_{px}}{w_{cm}}, \tag{5}$$

where R_{cm} and R_{px} represent the radius of the visual fields in centimeters and in pixels, θ represents the visual angle of the visual fields, w_{cm} represents the width of the display in centimeters, w_{px} represents the width of the music video frame in pixel. In our work, θ was $5\,degree$, w_{cm} was 59.79 cm, w_{px} was 1920 pixels. From these parameters, R_{px} was 59 pixels. The visual fields of all the subjects were overlapped with one frame and binarized, so that we can obtained the ground truth. The threshold for binarization was 0.7 of the maximum of the overlapped frame [8].

5   Experiments

We evaluate three baselines and three our methods by comparison between the saliency map and the ground truth per frame of the music video. We use the tune with high visibility that is well-known among the observers.

5.1   Evaluation Metric

For the evaluation of the saliency map, we used Normalized Scanpath Saliency (NSS) [15], the Correlation Coefficients (CC) [16], the area under the Receiver Operating Characteristics (AUC-ROC) [17], and precision, recall, F-measure as the evaluation value.

Normalized Scanpath Saliency (NSS). For computing the NSS value, the saliency map was linearly normalized to have zero mean and unit standard deviation. NSS is obtained as following,

$$NSS = \sum_{i=1}^{n} \frac{s(x_i^h, y_i^h) - \mu_s}{\sigma_s}, \tag{6}$$

where $s(x_i^h, y_i^h)$ is the normalized saliency value of the locations of the ground truth, μ_s and σ_s are the mean and variance of the normalized saliency map. A value greater than zero suggest that the saliency map correspond to the eye position of the ground truth, a value of zero indicates no correspondence between the saliency map and the ground truth, and a value less than zero indicate an anti-correspondence.

Correlation Coefficients (CC). CC is obtained as follows,

$$CC(s, h) = \frac{cov(s, h)}{\sigma_s \sigma_h}, \tag{7}$$

where s is the map of the ground truth, h is the saliency map, σ_s and σ_h are the variance of each map. The value close to 1 indicates that the saliency map correspond to the ground truth map, the value close to 0 indicates no correspondence between the saliency map and the ground truth, and the value close to -1 indicates an anti-correspondence.

The area under the Receiver Operating Characteristics (AUC-ROC). ROC curve is the signal detection theory. First, fixation pixels of the ground truth are positive set, and the same number of random pixels are chosen from the saliency map as the negative set. The saliency map is then treated as a binary classifier, and all points above threshold indicates positive samples and all points below threshold indicates negative samples. By plotting true positive rate vs. false positive rate for any particular value of the threshold, an ROC curve can be drawn and the Area Under the Curve (AUC) computed. An ideal score is one while random classification provides 0.5.

Precision, Recall, F-measure. We use precision, recall and F-measure for the evaluation. In order to calculate these values, we determined the adaptive threshold to binarize the saliency map. The threshold is obtained as follows [18],

$$T_\alpha = \frac{\alpha}{W \times H} \sum_{x=0}^{W-1} \sum_{y=0}^{H-1} S(x, y). \tag{8}$$

The adaptive threshold is α times the mean saliency of the music video frame. α is 1 to 5 and we adopted the F-measure value where the mean of F-measure of 6 saliency models is the highest in the level of α. F-measure is described as follows.

$$F_\beta = \frac{(1 + \beta^2) Precision \times Recall}{\beta^2 \times Precision + Recall}. \tag{9}$$

We use $\beta^2 = 0.3$ to weigh precision more than recall [18].

5.2   Result

In this section, we evaluate the baseline models and our proposed models by the evaluation values above stated. As the ground truth, we used the gaze behavior when watching the part of the tune "Fur Elise (Beethoven)" as shown in Table 1 whose visibility was high among the observers. We also set the weight of our saliency models based on Itti's method as shown in Table 2. From the result

Table 2. The weight of each saliency model based on Itti's method

| | Parameter | | | | |
	Intensity	Color	Orientation	Motion	Score
Itti(Baseline)	0.33	0.34	0.33	0	0
Itti+M(Baseline)	0.3	0.3	0.2	0.2	0
Itti+S(w1)	0.3	0.3	0.2	0	0.2
Itti+M+S(w1)	0.25	0.25	0.15	0.2	0.15
Itti+S(w2)	0.25	0.25	0.25	0	0.25
Itti+M+S(w2)	0.2	0.2	0.2	0.2	0.2
Itti+S(w3)	0.2	0.2	0.2	0	0.4
Itti+M+S(w3)	0.15	0.15	0.15	0.15	0.4

shown in Table 3, our models adding the musical score feature outperform the baselines that are constructed from the image feature. From the 7th to 10th row in Table 3 are our saliency models that weights the music score map more than the image feature as shown in Table 2. From the result, the evaluation values increase as the weight of the musical score map is large. Additionally, our saliency model of "Itti+M+S(w3)" outperform the score map in most values. This indicates that we need not only the musical score feature but also the image feature in order to construct the saliency map of the music video. As for the threshold (the formula 8), the F-measure value is the highest when α is 3. We therefore use that threshold for the following evaluation.

Table 3. Result of the evaluation of the saliency maps

	AUCROC	CC	NSS	F-measure	Precision	Recall
Itti(Baseline)	70.3 %	0.105	0.65	6.91 %	6.57 %	9.64 %
Itti+M(Baseline)	73.6 %	0.127	0.779	1.8 %	2.13 %	1.38 %
Yang(Baseline)	84.2 %	0.299	1.85	9.8 %	10.8 %	11.5 %
Itti+S(w1)	84.2 %	0.24	1.49	19.4	21.1 %	17.8 %
Itti+M+S(w1)	86.1 %	0.256	1.59	18.5 %	27 %	10.7 %
Yang+S	76.7 %	0.174	1.03	5.35 %	9.21 %	2.8 %
Itti+S(w2)	84.7 %	0.259	1.61	22.2 %	21.8 %	26.3 %
Itti+M+S(w2)	86.8 %	0.280	1.74	24.1 %	**27.1 %**	20.2 %
Itti+S(w3)	85.8 %	0.321	2.23	24.3 %	24.2 %	40.1 %
Itti+M+S(w3)	**87.7 %**	**0.354**	**2.22**	**26.5 %**	24.3 %	**41.1 %**
Score Map	75.2 %	0.355	2.23	22.0 %	18.6 %	60.0 %

6   Discussion

In this section, we also evaluate the saliency map under some condition, and discuss the result. First, we generate the ground truth from the gaze behavior of the inexperienced subjects. In our work, we consider that everyone has the possibility of having the musical sense. We therefore measured the gaze behavior of the inexperienced subjects in watching the music video, and evaluated the saliency maps using that ground truth. The result is shown in Table 4. Compared to the baselines that are constructed from the image feature, our saliency models that are added the musical score feature give high values. Our saliency model based on Itti's method gives high values when the musical score feature map is weighted more. From this result and Table 3, we consider that the musical score feature other than the image feature can detect the saliency that is close to the observed gaze behavior when watching the music video. The evaluation value of the Table 4 tends to be lower than that of the gaze behavior of the experienced subjects. We consider that the experienced subjects watched the piano performance with attention to the performer's hands and the tune. Regardless of the experience of learning the piano, the recall of the musical score map and the saliency map where the musical score map are weighted is high, which shows that the area where the ground truth correspond to the score feature map is large, but the salient area detected by the image feature does not correspond to the ground truth.

Second, we generated the musical score map for four patterns, where the number of notes are different as follows,

- Case 1: The score feature is extracted from the striking keys and one note after the striking keys.

Table 4. The result of the evaluation of the saliency maps (the ground truth is generated from the gaze behavior of the inexperienced subjects).

	AUCROC	CC	NSS	F-measure	Precision	Recall
Itti(Baseline)	67.7%	0.078	0.47	3.3%	3.3%	4.4%
Itti+M(Baseline)	73.1%	0.11	0.70	0.9%	1.3%	0.6%
Yang(Baseline)	85.2%	0.28	1.82	5.5%	5.3%	9.1%
Itti+S(w1)	82.5%	0.21	1.34	16.8%	18.4%	15.2%
Itti+M+S(w1)	86.1%	0.24	1.56	17.6%	**25.3%**	10.4%
Yang+S	77.3%	0.16	1.00	3.3%	6.3%	1.9%
Itti+S(w2)	82.7%	0.23	1.47	19.1%	18.9%	22.8%
Itti+M+S(w2)	86.5%	0.26	1.69	22.0%	24.5%	19.4%
Itti+S(w3)	83.8%	0.29	1.87	22.9%	21.2%	36.8%
Itti+M+S(w3)	**87.1%**	**0.34**	**2.16**	24.1%	22.2%	**39.7%**
Score Map	74.5%	0.34	2.18	20%	17.0%	58.6%

– Case 2: The score feature is extracted from the striking keys and one note before and after the striking keys.
– Case 3: The score feature is extracted from the striking keys and one and two notes after the striking keys.
– Case 4: The score feature is extracted from the striking keys and one and two notes before and after the striking keys.

In each case, we weigh the striking keys more than other notes. We use these musical features to construct our proposed saliency models. The result of the evaluation is shown in Table 5, where the ground truth is generated by the gaze behavior of the experienced subjects. The evaluation values of Case 1 and Case 2 outperform the result of Table 4. We therefore find that we can obtain more appropriate model to the eye-movements in our tasks by adding the musical notes other than the striking keys as the musical score feature. However, most of the evaluation values in Case 3 and Case 4 are lower than that of Case 1 and Case 2. To add the musical notes as the musical score feature leads to expand the salient area, which increases the area corresponding not to the ground truth.

Table 5. The result of the evaluation of the saliency map. Case 1 is the result that the saliency map includes the musical information of the current tunes and a subsequent tunes. Case 2 is the result that the saliency map includes the musical information of the current tunes, a subsequent tunes and a previous tunes. Case 3 is the result that the saliency map includes the current tunes and 2 subsequent tunes. Case 4 is the result that the saliency map includes the current tunes, 2 subsequent tunes and 2 previous tunes.

Case	Method	AUCROC	CC	NSS	F-measure	Precision	Recall
1	Itti+S(w3)	88.1%	0.33	2.02	24.0%	21.9%	38.5%
	Itti+M+S(w3)	89.6%	0.36	2.23	**24.5%**	**22.2%**	40.9%
	Yang+S	**90.1%**	0.27	1.63	11.0%	9.2%	39.2%
	Score Map	78.0%	**0.38**	**2.32**	21.5%	19.1%	**66.8%**
2	Itti+S(w3)	88.5%	0.33	2.05	24.5%	22.6%	38.1%
	Itti+M+S(w3)	**90.2%**	0.37	2.28	**25.3%**	**23.3%**	39.4%
	Yang+S	88.4%	0.26	1.55	10.9%	9.1%	39.6%
	Score Map	82.5%	**0.38**	**2.39**	23.0%	19.1%	**77.6%**
3	Itti+S(w3)	84.8%	0.28	1.68	19.6%	18.2%	29.0%
	Itti+M+S(w3)	86.7%	**0.31**	**1.87**	20.7%	**19.2%**	31.3%
	Yang+S	**88.6%**	0.27	1.62	13.4%	11.0%	54.2%
	Score Map	72.4%	0.29	1.75	18.6%	15.6%	**56.4%**
4	Itti+S(w3)	85.2%	0.28	1.68	18.4%	17.0%	28.5%
	Itti+M+S(w3)	87.5%	**0.31**	**1.86**	19.2%	**17.6%**	29.9%
	Yang+S	**89.5%**	0.25	1.54	9.9%	8.1%	40.9%
	Score Map	73.9%	0.29	1.76	18.0%	15.0%	**59.0%**

7   Conclusion

We proposed the task-driven saliency model of the music video. We added the musical score map generated from other than the image feature, and constructed the saliency model based on the baselines constructed from the image feature. The proposed saliency models were evaluated by comparison to the ground truth that was generated the gaze behavior when watching the music video, and we showed that the musical score feature as well as the image feature are needed for detecting the saliency of the music video. In this paper, we used the musical notes as the musical information. However, other information related to the musical sense is possibly preferable to detect the task-driven saliency of our tasks, such as the sound intensity, the rhythm, and so on. Additionally, the ground truth was generated by the observer having learned the piano more than one year because we consider that they have relatively many chances of developing the musical sense. However, we should consider other information that is related to the musical information except the experience of learning the piano.

References

1. Yarbus, A.L.: Eye Movements and Vision. Plenum Press, New York (1967)
2. Henderson, J.M., Shinkareva, S.V., Wang, J., Luke, S.G., Olejarczyk, J.: Predicting cognitive state from eye movements. PLoS One **8**(5), e64937 (2013)
3. DeAngelusa, M., Pelza, J.B.: Top-down control of eye movements: yarbus revisited. Visual Cogn. **17**, 790–811 (2009)
4. Borji, A., Itti, L.: Defending Yarbus: eye movements reveal observers' task. J. Vis. **14**(3), 29 (2014)
5. Kunze, K., Utsumi, Y., Shiga, Y., Kise, K., Bulling, A.: I know what you are reading: recognition of document types using mobile eye tracking. In: International Symposium on Wearable Computers (2013)
6. Itti, L., Koch, C., Niebur, E.: A model of saliency-based visual attention for rapid scene analysis. PAMI **20**(11), 1254–1259 (1998)
7. Yang, C., Zhang, L., Lu, H., Ruan, X., Yang, M.H.: Saliency detection via graph-based manifold ranking. In: CVPR, pp. 3166–3173 (2013)
8. Riche, N., Mancas, M., Culibrk, D., Crnojevic, V., Gosselin, B., Dutoit, T.: Dynamic saliency models and human attention: a comparative study on videos. In: Lee, K.M., Matsushita, Y., Rehg, J.M., Hu, Z. (eds.) ACCV 2012, Part III. LNCS, vol. 7726, pp. 586–598. Springer, Heidelberg (2013)
9. Bruce, N., Tsotsos, J.: Saliency based on information maximization. In: NIPS (2005)
10. Harel, J., Koch, C., Perona, P.: Graph-based visual saliency. In: NIPS (2006)
11. Seo, H.J., Milanfar, P.: Static and space-time visual saliency detection by self-resemblance. J. Vis. **9**(15), 1–27 (2009)
12. Wang, L., Xue, J., Zheng, N., Hua, G.: Automatic salient object extraction with contextual cue. In: ICCV (2011)
13. Shi, K., Wang, K., Lu, J., Lin, L.: PISA: pixelwise image saliency by aggregating complementary appearance contrast measures with spatial priors. In: CVPR (2013)
14. Iwatsuki, A., Hirayama, T., Mase, K.: Analysis of soccer coach's eye gaze behavior. In: Proceedings of ASVAI (2013)

15. Peters, R.J., et al.: Components of bottom-up gaze allocation in natural images. Vis. Res. **45**(18), 2397–2416 (2005)
16. Ouerhani, N., von Wartburg, R., Hugli, H., Muri, R.M.: Empirical validation of saliency-based model of visual attention. Electron. Lett. Comput. Vis. Image Anal. **3**(1), 13–24 (2003)
17. Bruce, N., Tsotsos, J.: Saliency based on information maximization. In: Advances in Neural Information Processing Systems (2005)
18. Achanta, R., Hemami, S., Estrada, F., Susstrunk, S.: Frequency-tuned salient region detection. In: CVPR (2009)

Recognition of Facial Action Units with Action Unit Classifiers and an Association Network

Junkai Chen[1]([✉]), Zenghai Chen[1], Zheru Chi[1], and Hong Fu[1,2]

[1] Department of Electronic and Information Engineering,
The Hong Kong Polytechnic University, Hong Kong, China
{Junkai.Chen,Zenghai.Chen}@connect.polyu.hk, chi.zheru@polyu.edu.hk
[2] Department of Computer Science, Chu Hai College of Higher Education,
Hong Kong, China
hongfu@chuhai.edu.hk

Abstract. Most previous work of facial action recognition focused only on verifying whether a certain facial action unit appeared or not on a face image. In this paper, we report our investigation on the semantic relationships of facial action units and introduce a novel method for facial action unit recognition based on action unit classifiers and a Bayes network called Facial Action Unit Association Network (FAUAN). Compared with other methods, the proposed method attempts to identify a set of facial action units of a face image simultaneously. We achieve this goal by three steps. At first, the histogram of oriented gradients (HOG) is extracted as features and after that, a Multi-Layer Perceptron (MLP) is trained for the preliminary detection of each individual facial action unit. At last, FAUAN fuses the responses of all the facial action unit classifiers to determine a best set of facial action units. The proposed method achieves a promising performance on the extended Cohn-Kanade Dataset. Experimental results also show that when the individual unit classifiers are not so good, the performance could improve by nearly 10 % in some cases when FAUAN is used.

1 Introduction

Facial expression is a very powerful and important nonverbal way for people to transmit message in daily life. Facial expression recognition has attracted an increasing attention in the past decade. Facial expressions are caused by facial muscle movements. These facial muscle movements are called facial action units. Ekman et al. [1] developed the Facial Action Coding System which was used for describing facial expressions by action units (AUs). In the 44 AUs defined, 30 AUs are anatomically related to the contractions of specific facial muscles: 12 are for upper face, and 18 are for lower face [2]. Although the number of action units is relative small, more than 7000 different AU combinations have been observed [3]. Facial action units provide an important cue for facial expression recognition.

We have witnessed much progress about facial action unit recognition. Two mainstream approaches, appearance based and geometry based [4], are widely

© Springer International Publishing Switzerland 2015
C.V. Jawahar and S. Shan (Eds.): ACCV 2014 Workshops, Part II, LNCS 9009, pp. 672–683, 2015.
DOI: 10.1007/978-3-319-16631-5_49

employed to handle this problem. Bartlett et al. [5] proposed an automatic spontaneous facial action units recognition system based on Gabor filters, AdaBoost
and Support Vector Machine (SVM) classifiers. They applied the AdaBoost to
select Gabor filters and the outputs of the selected Gabor filters were employed
to train a SVM. Valstar et al. [6] combined SVMs and hidden Markov models
to model facial action temporal dynamics. In their system, a set of carefully
selected geometrical features were used to separate a facial action unit into several temporal phases. They utilized an SVM and the hybrid SVM / Hidden
Markov Model (HMM) as the classifiers, respectively. Senechal et al. [7] computed the Local Gabor Binary Pattern (LGBP) histograms of the neutral and
expressive faces. The differences between the two histograms were used as features to train an SVM with a Histogram Difference Intersection (HDI) kernel.
Simon et al. [8] introduced a segment-based SVM to detect action units. They
explored two widely used models: static modeling, typically evaluated each video
frame independently, and temporal modeling, typically modeling action units
with a variant of dynamic Bayesian networks and integrated the benefits of the
two models. The system beats state-of-the-art static methods for AU detection.
Lucey et al. [9] described an active appearance model (AAM)-based system that
can automatically detect the frames in videos, in which a patient is in pain.
They defined the AU combinations as the pain emotion. The pain emotion was
predicted through detecting the related Action Units. Tong et al. [10] analyzed
the semantic relationships among AUs and proposed a novel method to handle Action Units recognition. A Dynamic Bayes network (DBN) was employed
to model the relationships among different AUs. Experiments illustrated that
the integration of AU relationships and AU dynamics with AU measurements
could improve the performance of AU recognition. Chu et al. [11] considered that
most existing Automatic Facial Action (AFA) unit detection methods neglected
individual differences in target persons. They introduced a transductive learning method called Selective Transfer Machine (STM), to personalize a generic
classifier by attenuating person-specific biases. The STM could learn a classifier
and re-weight the training samples that were most relevant to the test subject.

  However, most previous works mentioned above focused on single facial action
unit recognition. In fact, facial action unit combination is very important for
facial expression analysis. Each of most facial expressions consists of several
facial action units. Compare with the methods which recognize facial expressions
directly, facial action unit combination provides another meaningful way for
facial expression recognition. Instead of verifying a single facial action unit on a
face image, in this paper, we propose a method to identify the facial action unit
combination of a face image. We achieve this goal through three steps, feature
extraction, single AU detection and AU combination recognition. We conduct
the experiments on the extend Cohn-Kanade Dataset [12] and achieve a good
performance.

  The rest of the paper is organized as follows. In Sect. 2, we describe our proposed AU combination recognition system. We report and analyze experimental
results in Sect. 3. The paper is included in Sect. 4.

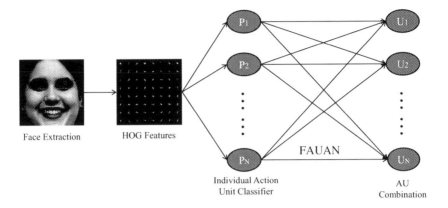

Fig. 1. Our proposed system for AU combination recognition.

2   An AU Combination Recognition System

The system includes three parts. At first, the Viola-Jones face detector [13] is employed to detect the face and Histogram Oriented Gradients (HOG) are used to encode the face. After that, we train a Multi-Layer Perceptron (MLP) for each facial action unit detection. At last, on the basis of the semantic relationships of facial action units, we construct a Bayes network called Facial Action Units Association Network (FAUAN) to combine the responses of all individual facial action unit classifiers. Figure 1 shows our proposed system.

2.1   Feature Extraction

Facial action units are caused by the corresponding facial muscle movements. These movements are subtle and transient. How to capture and represent these muscle movements is a long standing problem in facial expression analysis. Many different features like SIFT [14], Gabor filters [15], Local Binary Pattern (LBP) [16], Histogram Oriented Gradients (HOG) [17] and Local Phase Quantization (LPQ) [18] have been proposed for facial expression analysis. Gabor filters and LBP have been widely used for facial expression analysis. In this paper, we adopt the HOG to represent the facial images. HOG was first introduced by Dalal and Triggs in 2005 [17]. It is very popular in computer vision community and widely used in many object detection applications, especially in pedestrian detection. HOG calculates the occurrences of gradient orientations in a local patch of an image. The distribution of the local gradient intensities and orientations can describe the local object appearance and shape.

HOG is very sensitive to object deformations. After analyzing facial action units, we find that these facial action units could regard as some sorts of deformations. For example, there are several facial action units related to the lip. Such as lip stretcher, lip tightener and lip pressor etc. These facial action units are different distortions of the lip. Compared with the other feature descriptors,

HOG can better characterize these facial actions. In our study, we divide the detected face regions into many overlapped small blocks. Each block includes 2×2 cells. The cell size is set to 8×8. The bin size is set to 9. There are two orientation ranges used, $0°$–$360°$ and $0°$–$180°$. We set the orientation range to $0°$–$180°$ in our study.

2.2   Individual Facial Action Unit Classifiers

In this step, we build a visual classifier for each facial action unit. It is used to detect whether a face image containing a certain action unit.

For computer vision and pattern recognition, SVM [19] and Multilayer Perceptron(MLP) [20] are two commonly adopted classifiers and have been successfully used in many applications. In this paper, MLPs are trained as facial action unit classifiers. An MLP maps inputs to appropriate outputs through hidden layers and transform functions. In general, a supervised learning technique called back propagation [22] is utilized to train the MLP. With the hidden layer and nonlinear activition functions, MLP can discriminate the data which could not be separately linearly.

2.3   Facial Action Unit Association Network

Besides the visual features are related to facial action units, we also find that some facial action units appear together on face images. A study reported in [9] and [12] has shown that some universal expressions like happy, angry, and surprise etc. have specific sets of facial action units. It indicates that some facial action units have strong correlations. These correlations can help to recognize facial action units in groups. In [23], Fu et al. proposed a Concept Association Network (CAN) for image annotation. The CAN utilized the correlations of the concepts. Inspired by their work, we construct a Bayes network called Facial Actin Unit Association Network (FAUAN) for the recognition of facial action units in groups. Suppose that we have facial action units in the FAUAN, the appearances of action units are denoted by

$$F = (f_1, ..., f_i, ..., f_N) \tag{1}$$

where f_i denotes the number of occurrences of facial action unit i, the relationship between each pair of facial action units are defined by

$$\mathbf{W} = \{w_{ij}\}, i, j = 1, ...N \tag{2}$$

where w_{ij} is the number of co-occurrences of action unit i and j appear together on face image. With the Bayes rule, we can obtain the co-occurrence matrix among facial action units as:

$$M = \{m_{ij}, i \neq j\} = \{\frac{w_{ij}}{\sum_{k \neq j} w_{kj}}\}, i, j, k = 1, ...N \tag{3}$$

where m_{ij} denotes the occurrence frequency (an estimation of probability) of action unit i when action unit j appears.

2.4   Facial Action Unit Group Recognition with FAUAN

The FAUAN can combine the responses of individual action unit classifiers to identify which facial action unit group appears on a face image. Given a test face image, the pre-trained action unit classifiers can give a likelihood of the face image including a certain action unit:

$$\mathbf{P} = \{P_1, ...P_t, ...P_N\} \tag{4}$$

where subscript t is the index of the facial action unit, for example, P_1 means the probability of a test face image including action unit (AU)1. Because there are correlations among facial action units, when we consider an action unit in a face image, it is necessary to consider the appearance of the other action units. Based on the outputs of individual action unit classifiers and the FAUAN, we define

$$\mathbf{U} = \{U_1, ...U_k, ...U_N\} \tag{5}$$

$$U_k = P_k + \sum_{j \neq k} P_j m_{jk} \tag{6}$$

where k is the index of action unit, P is the output of individual action unit classifiers and m is the correlation coefficient of each pair of action units, defined in Eq. (3). U is the final output. Using Eq. (6), we can obtain the output for each facial action unit.

$$U_1 = P_1 + \sum_{j \neq 1} P_j m_{1j}$$

$$U_2 = P_2 + \sum_{j \neq 2} P_j m_{2j}$$

$$\vdots \tag{7}$$

$$U_N = P_N + \sum_{j \neq N} P_j m_{Nj}$$

Equation (7) can be rewritten as:

$$\begin{bmatrix} U_1 \\ U_2 \\ \vdots \\ \vdots \\ U_N \end{bmatrix} = \begin{bmatrix} 1 & m_{12} & m_{13} & \cdots & m_{1N} \\ m_{21} & 1 & m_{23} & \cdots & m_{2N} \\ m_{31} & m_{32} & 1 & \cdots & m_{3N} \\ \cdots & \cdots & \cdots & \cdots & \cdots \\ m_{N1} & m_{N2} & \cdots & m_{NN-1} & 1 \end{bmatrix} \begin{bmatrix} P_1 \\ P_2 \\ \vdots \\ \vdots \\ P_N \end{bmatrix} \tag{8}$$

From Eq. (8), given a test face image, we can obtain the response of the image to each facial action unit. Each individual action unit classifier at first computes the likelihood. All the likelihoods go through the FAUAN and the response of each action unit is determined. And the facial action units which have the largest responses will be selected with the face image labeled accordingly.

3   Experimental Results and Discussions

In order to evaluate our method, we conduct the experiments on the extend Cohn-Kanade Dataset [12]. This dataset consists of 123 subjects between the age of 18 to 50 years, of which 69 % female, 81 % Euro-American, 13 % Afro-American, and 6 % other groups. There are 593 image sequences from 123 subjects. The peak frame of each sequence has been labeled by FACS [24]. It means that there are 593 face images including facial action unit labels. For our experiment, we pick out all the peak frames and divide them into two groups. We randomly select 60 face images as the test set and the remaining 533 face images are used for training.

For the feature extraction, we apply the Viola-Jones face detector [13] to detect the face and resize the face to the 64×64 from the original image size of 640×490, the final HOG features for each face is a 1764×1 vector.

Table 1. The indexes and names of the selected 15 AUs.

AU	Name	AU	Name	AU	Name
1	Inner Brow Raiser	7	Lid Tightener	20	Lip Stretcher
2	Outer Brow Raiser	9	Nose Wrinkler	23	Lip Tightener
4	Brow Lowerer	12	Lip Corner Puller	24	Lip Pressor
5	Upper Lip Raiser	15	Lip Corner Depressor	25	Lips Part
6	Cheek Raiser	17	Chin Raiser	27	Mouth Stretch

There are 30 different action units (AUs) in the dataset. We select 15 AUs which are universal and appear in the dataset with a high frequency. The indexes and names of 15 AUs are shown in Table 1. From Table 1, we could find that most action units are related with the lip. This may due to that among all the facial components (nose, eye, lip, brow etc.); the lip is the most flexible. It could generate many different actions. There are three action units (AU1, AU2 and AU4) related to the brow. Each of the other facial components (nose and cheek) has one action unit.

For individual facial action unit classification, we employed a 3-layer MLP. The input is a 1764×1 vector, followed by a hidden layer with 100 nodes and the output with two nodes. The target output $(1, 0)$ represents a positive sample and $(0, 1)$ means a negative sample. During training, the face images which include the specific action unit are positive samples and the other face images are negative samples.

In order to compute the co-occurrence matrix, we count the co-appearances of each action unit pair based on the training set first. The frequency distribution of action unit pairs in the training set is shown in Table 2.

A diagonal value in Table 2 is the number of appearances of an action unit in the training set. From Table 2, we can see some pairs have a large value and

Table 2. The distribution of the action unit pairs in the training set.

AU	1	2	4	5	6	7	9	12	15	17	20	23	24	25	27
1	157	102	64	76	8	16	0	7	34	46	42	7	3	110	62
2	102	102	18	70	1	1	0	3	11	16	17	2	3	87	62
4	64	18	178	23	32	89	44	8	37	116	48	41	34	54	2
5	76	70	23	90	3	6	1	3	2	11	19	6	0	81	54
6	8	1	32	3	113	44	23	74	1	21	16	8	5	78	0
7	16	1	89	6	44	112	44	11	6	64	26	31	30	38	0
9	0	0	44	1	23	44	66	4	4	46	2	9	10	11	0
12	7	3	8	3	74	11	4	120	0	2	9	0	2	84	2
15	34	11	37	2	1	6	4	0	88	86	1	9	8	1	0
17	46	16	116	11	21	64	46	2	86	186	6	44	41	8	0
20	42	17	48	19	16	26	2	9	1	6	72	1	0	69	1
23	7	2	41	6	8	31	9	0	9	44	1	55	32	0	0
24	3	3	34	0	5	30	10	2	8	41	0	32	53	0	0
25	110	87	54	81	78	38	11	84	1	8	69	0	0	285	70
27	62	62	2	54	0	0	0	2	0	0	1	0	0	70	70

some pairs have a small value. It indicates that the pairs with a large value have a strong correlation and the pairs with a small value have a weak correlation. There are some zeros in Table 2, indicating that some action units do not appear together. They have litte correlations. Applying Eq. (3) we can get the co-occurrence matrix as is shown in Table 3.

We determine a set of action units using two methods. One way is to obtain the output from the individual action unit classifiers directly without the FAUAN. It means that the results totally depend on the performance of individual action unit classifiers. Another way is our proposed method. We employ the FAUAN to fine tune the results. The classification rate of each individual action unit classifier is shown in Fig. 2.

In order to compare the performance of the two methods, we fix the number of outputs (AUs) and compute the F1-score, recall and precision respectively. The definitions of the three metrics are given as follows [25]:

$$precision = \frac{TP}{TP + FP}, recall = \frac{TP}{TP + FN}, F1 = \frac{2TP}{2TP + FP + FN} \quad (9)$$

where TP is the true positive, in our experiment, it means that an action unit predicted actually appears in the test face image. FP is the false positive. In our experiment, an FP occurs when an action unit predicted does not appear in the test face image. FN is the false negative, which occurs when an action unit appears in the test face image but it is missed in the prediction. Table 4 shows the F1-score, recall and precision with or without FAUAN.

Table 3. The co-occurrence matrix.

AU	1	2	4	5	6	7	9	12	15	17	20	23	24	25	27
1	1.00	0.26	0.10	0.21	0.03	0.04	0.00	0.03	0.17	0.09	0.16	0.04	0.02	0.16	0.25
2	0.18	1.00	0.03	0.20	0.00	0.00	0.00	0.01	0.06	0.03	0.06	0.01	0.02	0.13	0.25
4	0.11	0.05	1.00	0.06	0.10	0.22	0.22	0.04	0.19	0.23	0.19	0.22	0.20	0.08	0.01
5	0.13	0.18	0.04	1.00	0.01	0.01	0.01	0.01	0.01	0.02	0.07	0.03	0.00	0.12	0.21
6	0.01	0.00	0.05	0.01	1.00	0.11	0.12	0.35	0.01	0.04	0.06	0.04	0.03	0.11	0.00
7	0.03	0.00	0.15	0.02	0.14	1.00	0.22	0.05	0.03	0.13	0.10	0.16	0.18	0.06	0.00
9	0.00	0.00	0.07	0.00	0.07	0.11	1.00	0.02	0.02	0.09	0.01	0.05	0.06	0.02	0.00
12	0.01	0.01	0.01	0.01	0.24	0.03	0.02	1.00	0.00	0.00	0.04	0.00	0.01	0.12	0.01
15	0.06	0.03	0.06	0.01	0.00	0.01	0.02	0.00	1.00	0.17	0.00	0.05	0.05	0.00	0.00
17	0.08	0.04	0.19	0.03	0.07	0.16	0.23	0.01	0.43	1.00	0.02	0.23	0.24	0.01	0.00
20	0.07	0.04	0.07	0.05	0.05	0.06	0.01	0.04	0.01	0.01	1.00	0.01	0.00	0.10	0.40
23	0.01	0.01	0.07	0.02	0.03	0.08	0.05	0.00	0.05	0.09	0.00	1.00	0.19	0.00	0.00
24	0.01	0.01	0.06	0.00	0.02	0.07	0.05	0.01	0.04	0.09	0.00	0.17	1.00	0.00	0.00
25	0.19	0.22	0.09	0.23	0.25	0.09	0.06	0.40	0.01	0.02	0.27	0.00	0.00	1.00	0.28
27	0.12	0.16	0.00	0.15	0.00	0.00	0.00	0.01	0.00	0.00	0.00	0.00	0.00	0.10	1.00

Table 4. The F1-score, recall and precision of the two methods tested (the final classifiers).

Outputs/N	With FAUAN			Without FAUAN		
	F1	Recall	Precision	F1	Recall	Precision
3	0.6361	0.6178	0.6556	0.6146	0.5969	0.6333
4	0.6636	0.7487	0.5958	0.6404	0.7225	0.5750
5	0.6640	0.8534	0.5433	0.6354	0.8168	0.5200
6	0.6352	0.9162	0.4861	0.6134	0.8848	0.4694
7	0.5892	0.9424	0.4286	0.5728	0.9162	0.4167
Avg	0.6376	0.8157	0.5419	0.6153	0.7874	0.5229

We can see that through the FAUAN, the F1-score, recall and precision all become higher, meaning that the performance is improved with the FAUAN. And we also draw the ROC curve and compute the area under the ROC curve. Figure 5(a) shows the ROC of the two methods. The area under curve with FAUAN is 0.9291 and without FAUAN is 0.9160 respectively.

Intuitively, it is easy to foresee that the performance of individual action unit classifiers would influence the final results. We want to explore whether or not the FAUAN could improve the performance when individual action unit classifiers is not so good. We set the number of iterations for training to control the classification rate of each action unit classifier and to evaluate the AU detection performance with and without FAUAN.

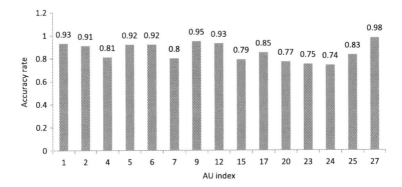

Fig. 2. The classification rate of each individual action unit classifier (the final classifiers).

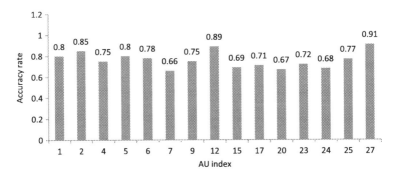

Fig. 3. The classification rate of each individual action unit classifier (the classifiers trained with 5 iterations).

Table 5. The F1-score, recall and precision of the two methods (the classifiers trained with 5 iterations).

Outputs/N	With FAUAN			Without FAUAN		
	F1	Recall	Precision	F1	Recall	Precision
3	0.5984	0.5812	0.6167	0.5013	0.4869	0.5167
4	0.6311	0.7120	0.5667	0.5383	0.6073	0.4833
5	0.6395	0.8220	0.5233	0.5499	0.7068	0.4500
6	0.6098	0.8796	0.4667	0.5554	0.8010	0.4250
7	0.5827	0.9319	0.4238	0.5237	0.8377	0.3810
Avg	0.6123	0.7853	0.5194	0.5337	0.6879	0.4512

Figure 3 shows the classification rate of each individual action unit classifier when the iteration is set to 5. Compared with Figs. 3 and 2, we can see that the classification rates in Fig. 3 are smaller than those in Fig. 2. The F1-score,

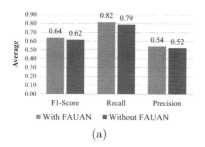
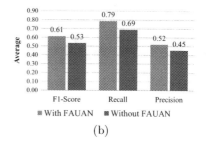

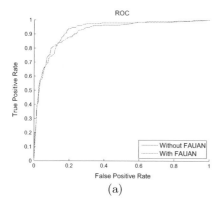
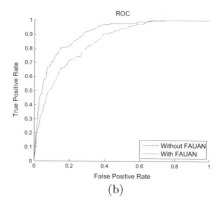

Fig. 4. The average F1-score, recall and precision of the two methods tested. (a) the final classifiers. (b) the classifiers trained with only 5 iterations.

Fig. 5. The ROCs of the two methods tested. In (a), the area under curve with FAUAN is 0.9291 and that without FAUAN is 0.9160. In (b), the area under curve with FAUAN is 0.9038 and that without FAUAN is 0.8452.

recall and precision with or without FAUAN are shown in Table 5. The average F1-score, recall and precision of the two test methods are shown in Fig. 4. From Table 5, we can see that when the classification rate of each individual action unit classifier becomes lower, the improvement with FAUAN is more obvious. The F1-score, recall and precision all improve by about 10 % when the numbers of outputs are 3, 4 and 5, respectively. Figure 5(b) shows the ROC of the two methods. The area under curve with FAUAN is 0.9038 and without FAUAN is 0.8452.

4   Conclusion

Facial expression recognition is still a challenge problem in computer vision. Facial action units provide an important cue to solve this probelm. We can detect the facial action units in groups and then recognize the facial expressions from the facial action units. In this paper, we explore the correlations among facial action units and propose a Facial Action Unit Association Network (FAUAN) for the recognition of facial action units in groups. To evaluate our proposed method, we conduct the experiments on the extended Cohn-Kanade Dataset. Experimental results show that the FAUAN can improve the AU detection results. When the classification rates of individual action unit classifiers are poor, the improvement with FAUAN is more obvious. The future work will include the application of the proposed method to facial expression recognition.

Acknowledgement. This work reported in this paper was partly supported by a research grant from The Hong Kong Polytechnic University (Project Code: G-YL77).

References

1. Ekman, P., Friesen, V.F.: Facial Action Coding System: A Technique for the Measurement of Facial Movement. Consulting Psychologists Press, Palo Alto (1978)
2. Tian, Y., Kanado, T., Cohn, J.F.: Recognizing action units for facial expression analysis. IEEE T PAMI **23**, 97–115 (2001)
3. Scherer, K., Ekman, P.: Handbook of Methods in Nonverbal Behavior Research. Cambridge University Press, Cambridge (1982)
4. Li, S.Z., Jain, A.K.: Handbook of Face Recognition. Springer, New York (2011)
5. Bartlett, M.S., Littlewort, G.C., Frank, M.G., et al.: Automatic recognition of facial actions in spontaneous expressions. J. Multimedia **1**, 22–35 (2006)
6. Valstar, M.F., Pantic, M.: Combined support vector machines and hidden Markov models for modeling facial action temporal dynamics. In: Lew, M., Sebe, N., Huang, T.S., Bakker, E.M. (eds.) Human Computer Interaction. LNCS, vol. 4796, pp. 118–127. Springer, Heidelberg (2007)
7. Senechal, T., Bailly, K., Prevost, L.: Automatic facial action detection using histogram variation between emotional states. In: ICPR, pp. 3752–3755 (2010)
8. Simon, T., Nguyen, M.H., De la Torre, F., Cohn, J.F.: Action unit detection with segment-based SVMs. In: CVPR, pp. 2737–2744 (2010)
9. Lucey, P., Cohn, J.F., Matthews, I., Lucey, S., Sridharan, S., Howlett, J., Prkachin, K.M.: Automatically detecting pain in video through facial action units. IEEE T SMC **41**, 664–674 (2011)
10. Tong, Y., Liao, W., Ji, Q.: Facial action unit recognition by exploiting their dynamic and semantic relationships. IEEE T PAMI **29**, 1683–1699 (2007)
11. Chu, W., Torre, F., Cohn, J.F.: Selective transfer machine for personalized facial action unit detection. In: CVPR, pp. 3515–3522 (2013)
12. Lucey, P., Cohn, J.F., Kanade, T., Saragih, J., Ambadar, Z., Matthews, I.: The extended Cohn-Kanade dataset: a complete dataset for action unit and emotion-specified expression. In: CVPRW, pp. 94–101 (2010)
13. Viola, P., Jones, M.: Robust real-time face detection. IJCV **57**, 137–154 (2004)
14. Lowe, D.G.: Distinctive image features from scale-invariant keypoints. IJCV **60**, 91–110 (2004)
15. Feichtinger, H.G., Strohmer, T.: Gabor Analysis and Algorithms: Theory and Applications. Springer, Berlin (1998)
16. Ojala, T., Pietikainen, M., Maenpaa, T.: Multiresolution gray-scale and rotation invariant texture classification with local binary patterns. IEEE T PAMI **24**, 971–987 (2002)

17. Dalal, N., Triggs, B.: Histograms of oriented gradients for human detection. CVPR **1**, 886–893 (2005)
18. Ojansivu, V., Heikkil, J.: Blur insensitive texture classification using local phase quantization. In: Elmoataz, A., Lezoray, O., Nouboud, F., Mammass, D. (eds.) Image and Signal Processing. LNCS, vol. 5099, pp. 236–243. Springer, Heidelberg (2008)
19. Burges, C.: A tutorial on support vector machines for pattern recognition. Data Min. Knowl. Disc. **2**, 121–167 (1998)
20. Haykin, S.: Neural Networks: A Comprehensive Foundation. Prentice Hall PTR, Upper Saddle River (1994)
21. Rosenblatt, F.: Principles of neurodynamics. perceptrons and the theory of brain mechanisms. In: DTIC Document (1961)
22. Rumelhart, D.E., Hinton, G.E., Williams, R.J.: Learning internal representations by error propagation. In: DTIC Document (1985)
23. Fu, H., Chi, Z., Feng, D.: Recognition of attentive objects with a concept association network for image annotation. Pattern Recognit. **43**, 3539–3547 (2010)
24. Ekman, P., Friesen, W.V., Hager, J.C.: Facial action coding system: the manual on CD ROM. In: A Human Face (2002)
25. Fawcett, T.: An introduction to ROC analysis. Pattern Recogn. Lett. **27**, 861–874 (2006)

A Non-invasive Facial Visual-Infrared Stereo Vision Based Measurement as an Alternative for Physiological Measurement

Mohd Norzali Haji Mohd[1,2][✉], Masayuki Kashima[1],
Kiminori Sato[1], and Mutsumi Watanabe[1]

[1] Department of Information Science and Biomedical Engineering,
Graduate School of Sciences and Engineering, Kagoshima University,
Korimoto 1-21-40, Kagoshima 890-0065, Japan
`mnorzali@gmail.com, norzali@uthm.edu.my`
[2] Department of Computer Engineering, Faculty of Electrical and Electronic
Engineering, University Tun Hussein Onn Malaysia (UTHM),
86400 Parit Raja, Batu Pahat, Johor, Malaysia

Abstract. Our main aim is to propose a vision-based measurement as an alternative to physiological measurement for recognizing mental stress. The development of this emotion recognition system involved three stages: experimental setup for vision and physiological sensing, facial feature extraction in visual-thermal domain, mental stress stimulus experiment and data analysis and classification based on Support Vector Machine. In this research, 3 vision-based measurement and 2 physiological measurement were implemented in the system. Vision based measurement in facial vision domain consists of eyes blinking and in facial thermal domain consists 3 ROI's temperature value and blood vessel volume at Supraorbital area. Two physiological measurement were done to measure the ground truth value which is heart rate and salivary amylase level. We also propose a new calibration chessboard attach with fever plaster to locate calibration point in stereo view. A new method of integration of two different sensors for detecting facial feature in both thermal and visual is also presented by applying nostril mask, which allows one to find facial feature namely nose area in thermal and visual domain. Extraction of thermal-visual feature images was done by using SIFT feature detector and extractor to verify the method of using nostril mask. Based on the experiment conducted, 88.6 % of correct matching was detected. In the eyes blinking experiment, almost 98 % match was detected successfully for without glasses and 89 % with glasses. Graph cut algorithm was applied to remove unwanted ROI. The recognition rate of 3 ROI's was about 90 %–96 %. We also presented new method of automatic detection of blood vessel volume at Supraorbital monitored by LWIR camera. The recognition rate of correctly detected pixel was about 93 %. An experiment to measure mental stress by using the proposed system based on Support Vector Machine classification had been proposed and conducted and showed promising results.

© Springer International Publishing Switzerland 2015
C.V. Jawahar and S. Shan (Eds.): ACCV 2014 Workshops, Part II, LNCS 9009, pp. 684–697, 2015.
DOI: 10.1007/978-3-319-16631-5_50

1   Introduction

Various methods for internal state measurement such as mental stress have been carried out which utilize the changes of physiological through electroencephalography (EEG), blood volume pulse (BVP), heart rate variability (HRV), galvanic skin response (GSR) and electromyography (EMG) measurement [2]. However it requires the individuals to wear or touch electrodes or sensors. On the other hand, physical signals for measuring stress include eye gaze, pupil diameter, voice characteristic and face movement and these quantities are measured invasively by the use of expensive instruments.

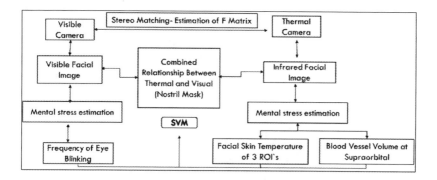

Fig. 1. Overall system.

1.1   Problems and Related Work

Face recognition system based on visual images has reached significant level of maturity with some practical successes. However, the performance of visual face recognition may degrade under poor illumination conditions, especially for subjects of various skin color and the changes in facial expression. The use of infrared in face recognition allows the limitations of visible face recognition to be solved. However, infrared suffers from other limitations like opacity to glasses. Hence, multi modal fusion comes with the promise of combining the best of each modalities and overcoming their limitations [4]. Dvijesh Shantri et al. [10] proposed a novel framework for quantifying physiological stress at a distance via thermal imaging. The method captures stress-induced responses on the perinasal area that manifest as transient perspiration. It is based on morphology and spatial isotropic wavelets. The limitation of this propose perinasal imaging is it depends on an expensive sensor, cannot operate if the perinasal area is covered and requires support from reliable face tracker.

1.2   Key Contribution Area

In this paper we introduce an integrated non-invasive measurement via imaging techniques. Our aim is to propose vision-based measurement as an alternative to physiological measurement. The fusion of physiological vision-measurement from thermal IR and visual is shown in Fig. 1. Our research consists of three vision based physiological measurements which are eye blinking from visual sensor, skin temperature of 3 ROI's and blood vessel volume at supraorbital from thermal IR camera. The primary physical measurement in detecting mental stress is heart rate variability [2] and Salivary amylase level [3] as proof by earlier researcher. The normal heart rate ranges from 60–100 bpm. In other hand, Salivary amylase level increased significantly and is suggested as the better index of mental stress [3]. Salivary amylase with level more than 60 $KU \setminus L$ is considered to have mental stress. Both of the measurement was used as a ground truth measurement in our research. We will also publish the data set consisting the proposed vision measurement.

In this paper, we also propose a new calibration chessboard attach with fever plaster to locate calibration point in stereo Visual-Thermal view. A new method of integration of two different sensors for detecting facial feature in both thermal and visual is also presented by applying nostril mask, which allows one to find facial feature namely nose area in thermal and visual domain (Fig. 2).

Fig. 2. Screening station.

2   Methodology

2.1   Registration of IR and Visible

We propose three ways to compute the relationship between Visible-IR camera as below.

Registration of IR-Visible by Using Fever Plaster Calibration Board. The relative position between the IR and visible cameras is calibrated by using the special calibration board. With some small adjustment and preparation where cold fever plasters are attached to the back of the calibration points on a chessboard, the calibration points can be reliably located in thermal IR and visible domain [15]. One of the common strategies to simplify correspondence problem between IR and visible domain is to exploit epipolar geometry (Fig. 4). The epipolar line is the intersection of an epipolar plane with the image plane. All epipolar lines intersect at the epipole. An epipolar plane intersects left and right image plane in epipolar lines, and defines the correspondence between lines [6]. The relative position between thermal-visible stereo cameras is calculated using calibration board. The origin of the world coordinate system is defined to be on the calibration board (Fig. 3).

Fig. 3. Propose calibration chessboard.

Fig. 4. Detected F matrix and epipolar lines from Th-Vi calibrated images.

The comparison between traditional heated chessboard and the proposed calibration board is cheap in comparing to the other calibration board made from polished copper plate coated with high emissivity paint or calibration rig. Even though heating is required in both method, it is difficult to get an even coverage which the pattern that can last longer than 10 min. We also propose different ways to compute the relationship between Visible-IR camera.

SIFT Feature Matching. Facial feature in both thermal and visible that can be detected in view is nostril part (Fig. 5). SIFT features are extracted from images to help in reliable matching between different views in same object [9]. The extracted features are variant to scale and orientation, and are highly distinctive of the image. They are extracted in four steps. The first steps computes the location of potential interest points in the images by detecting the maxima and minima of a set of Difference of Gaussian (DoG) filters applied at different scales all over the images. Then, these location are refined by discarding points of low contrast. An orientation is then assigned to each key point based on local image features. Finally, a local feature descriptor is computed at each key point. This descriptor is based on the local image gradient, transformed according to the orientation of the key point to provide orientation invariance. Every feature is a vector of dimension 128 distinctively identifying the neighborhood around the key points.

At first, the nostril area in thermal IR is detected using our pair calibrated Thermal-Visible camera [14]. Then, the face part in visible domain is detected and SIFT feature point is calculated in both domain. The feature matching is done and the correct matching is recorded in Table 1.

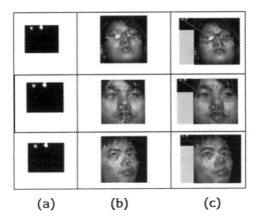

Fig. 5. (a) Feature point (nostril mask in thermal), (b) Feature point (facial in visual) (c) Sift feature matching.

Stereo Matching: Estimation of F Matrix. The correspondence problem remains of central interest in the field of image analysis. In the case of uncalibrated cameras, establishing correspondence between cameras is crucial. In our research, the matching of feature points correspondence between stereo pairs of visible and infrared facial is done as one of the matching strategy. Our focus is only on frontal faces as the stimulus is located in front of correspondent. One of the common strategies to simplify correspondence problem between IR and visible domain is to exploit epipolar geometry. Our system is based on uncalibrated images taken from thermal and visual camera (Fig. 6). The reason of this step is to show the correspondence and location of facial images taken from thermal IR and visible camera. By knowing the point correspondence in two images we can compute and estimate F matrix. SURF feature point is used to extract and match feature points correspondence. The outliers are then removed by using epipolar

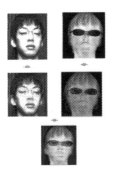

Fig. 6. Estimating F matrix, remove outliers using epipolar constraint and visualize rectified images by composites image.

constrain. Then the images are being rectified on F matrix.

$$m'^T Fm = 0 \tag{1}$$

Lastly, transform points are visualized together using composites image.

2.2 Facial Feature Extraction in Visual and Thermal Domain

Frequency of Eye Blinking from Visible Domain. At first we detect the face using Viola and Jones' boosting algorithm and a set of cascade structure with Haar-like features [16]. The first 30 frames of detected face that shows the location is marked as template. The edge image of the face shows the location of the eye more precisely because of the contour of the eyes area. Eyes regions are detected by calculation of integral image rectangular filter (Fig. 7).

$$S_r = (ii(x, y) + ii(x - W, y - L) - ii(x - W, y) - ii(x, y - L)) \tag{2}$$

S_r sum of the pixel values of the rectangle D is obtained by calculating the difference between the sum of the 4-points. The filter is minimum. When the location of eyes region is detected, the location of the iris is determined where there are the most black pixel in the horizontal axis. Figure 8 below shows the steps in locating iris area.

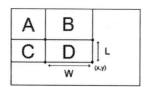

Fig. 7. Calculation of the integral image rectangular filter.

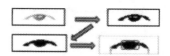

Fig. 8. Steps (binarization, opening and closing to remove noise) in locating iris area.

We locate blink detection by the ratio of the white and black pixel detected mask. When the ratio is bigger than the threshold value, the eyes are considered open and when it is less, they are considered close (Fig. 9).

Facial Skin Temperature of 3 ROI's (Supraorbital, Periorbital and Maxillary) from Thermal Domain. In our experiment, measurement is done at three facial areas of sympathetic importance which is periorbital, supraorbital and maxillary in Fig. 10. Based on the past researchers [7,8,10], we have found out that 3 ROI's are the most affected during mental stress and we focus on the

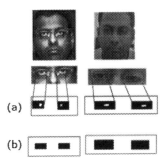

Fig. 9. Detection of frequency of blinking with and without glasses, (a) Ratio more than threshold (b) Ratio less than threshold, eyes close.

temperature on this areas. At first, we detect the face region using Viola and Jones's boosting algorithm [1,13] and a set of Cascade structure with Haar-like features. The 3 ROI's are detected based on the face ration Then, the collaboration of ROI with temperature value is done based on the relationship as below. $Temperature = 0.0819 * GrayLevel + 23.762$. Next, we assume the centroid of the detected face area as nose area and nostril mask is used to recalculate the detected area and map into the facial thermal image in Fig. 10.

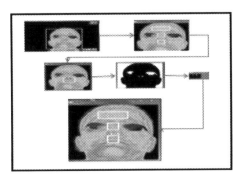

Fig. 10. Proposed automatic thermal face, supraorbital, periorbital, maxillary and nostril detection.

We then analyze the detected thermal face of the 3 ROI's with sympathetic importance for person wearing spectacles and without it. We found out that there are unwanted ROI such as spectacles and hair which can be excluded so that temperature reading can be done precisely as shown in Fig. 11.

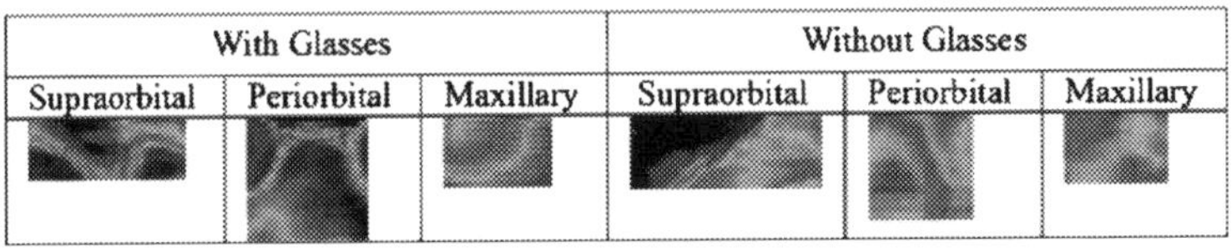

With Glasses			Without Glasses		
Supraorbital	Periorbital	Maxillary	Supraorbital	Periorbital	Maxillary

Fig. 11. Detected thermal face with and without spectacles.

To overcome this problem, unwanted ROI is removed by using Graph Cut method as shown in Fig. 12.

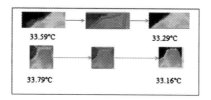

Fig. 12. The unwanted ROI removed using Graph Cut.

Blood Vessel Volume at Supraorbital from Thermal Domain. Based on recent study conducted, [5,10], we have found out that user stress is correlated with the increased blood flow in three facial areas of sympathetic importance which is periorbital, supraorbital and maxillary. This increased blood flow dissipates convective heat which can be monitored through mid-wave infrared (MWIR, 3–5 μm) camera. Our approach is different which is trying to implement it on a long-wave infrared (LWIR, 8–14 μm) camera.

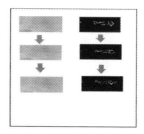

Fig. 13. Detected blood vessel during stress stimulus.

The methodology for detecting blood vessel is shown in the flow chart in Fig. 13. After the detection of supraorbital area in thermal IR, image morphing is applied on the diffused image to extract the blood vessels that are relatively low contrast compared with the surrounding tissue. We employ, top hat segmentation method, which is the combination of the erosion and dilation operations. We are interested in the bright (hot) like structure which correspond to blood vessel. For this reason we employ White Top-hat segmentation (WTH) as in the equation below.

$$I \circ S = (I \ominus S) \oplus S, \, WTH = I - (I \circ S) \tag{3}$$

where I: is the original image, I∘ S: opened image, S: structuring element, $\ominus$: erosion, $\oplus$: dilation. In order to enhance the edge and reduce noise, bilateral filter is employed. Bilateral filter is a nonlinear, edge preserving and noise reducing smoothing filter. The intensity value at each pixel in an image is replaced by a weighted average of intensity values from nearby pixels.

$$I^{filtered} = \Sigma_{x_i \epsilon \Omega} I(x_i) f_r(||I(x_i) - l(x)||) g_s(||x_i - x||) \tag{4}$$

where $I^{filtered}$ is the filtered image; I is the original input image to be filtered; x is the coordinates of the pixel to be filtered; Ω is the window centered in x; f_r is the range kernel for smoothing difference in intensities. This function can be Gaussian function, g_s is the spatial kernel for smoothing differences in coordinates. After applying bilateral filter, based on the RGB pixel value, the red range of pixel which is associated to blood vessel is then segmented (Fig. 13). The ground truth comparison images of blood vessel in MWIR thermal IR and Visual, detected in our system are shown in Fig. 14.

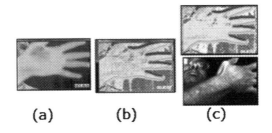

(a)   **(b)**   **(c)**

Fig. 14. Hand blood vessel in (a) thermal-actual, (b) thermal-white top hat segmentation (c) thermal-bilateral filter and visual image.

2.3   Experimental Setup and Result

Experimental Setup. Our screening environment consists of light and sound proof screening station, NEC TH7800 thermal camera (right) and USB CMOS Imaging Source DFK 22AUCO3 (left), Stimulus screening monitor, temperature and humidity sensor for monitoring screening station PICO RH-02, Pulse Oximeter SAT-2200 to monitor pulse rate, Nipro cocoro meter to measure salivary amylase level, responder seat and operator machine. About 20 people age from 18 to 30 both male and female participated in this study. The subject were asked to sit comfortably in the screening station and have him/her rest for about 5–10 min before and after the mental stress stimulus test. Mental Stress experiment adopted is designed by Dr Soren Brage of MRC Epidemiology Unit, Cambridge University [2]. It consists of 4 min of stroop color word test with 30 s of rest in between and 4 min of math test with 30 s of rest in between.

Result of Feature Matching and Detection in Visual and Thermal Camera. The result of matching SIFT feature between nostril area in thermal IR and the face part in visual in our pair calibrated Thermal-visible camera is shown in Table 1 below.

Table 1. Results of matching features in thermal-visible

Frame	Pro. time [ms]	Fea. point (TH)	Fea. point (VI)	Correct match [%]
1	170.759	2	157	100
2	168.443	2	93	100
3	171.866	11	106	36
4	162.89	2	81	100
5	170.921	2	168	50
6	154.105	1	168	100
7	160.873	1	114	100
8	153.269	1	112	100
9	170.35	1	146	100
10	170.972	1	157	100

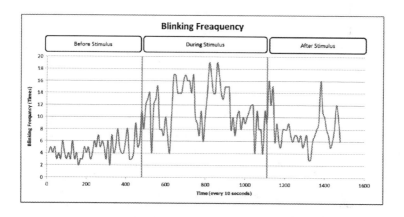

Fig. 15. Frequency of blinking before, during and after stress stimulus.

Experiment showed 86 % of correct matching. To further improve its accuracy, it requires more accurate samples. The development of automatic thermal face, supraorbital, periorbital and maxillary and nostril detection showed 90–96% of correct measurement of ROI and correct temperature was detected. It was calculated based on correctly recognize ROI of 10 correspondent within 4 min of stimulus experiment. For detection of eyes blinking in visual domain, almost 98 % of eyes blinking were detected successfully for without eye glasses and 89 % with glasses. The recognition rate for detected pixel for blood vessel detection was about 93 %.

Result of Stimulus Experiment and Emotion Recognition System. Five experiments were conducted for monitoring before, during and after mental stress stimulus. The first experiment was to measure frequency of eyes blinking in real time (Fig. 15). Before the stimulus, blinking frequency was below

10 times for every second and it doubled to 20 times every second during stimulus and decreased to an average of 13 times afterwards.

The second experiment was to monitor the 3 ROI's (Supraorbital, Periorbital and Maxillary) temperatures before, during and after the stress stimulus experiment. In overall, during pre-stress stimulus which was 1–2 min, temperature in the 3 ROI's was stable in value. During stimulus, which was 2–4 min and 4–6 min, the temperature fluctuated and slight increase tendency can be seen. During rest time which was 4–5 min, there was a slight decrease of temperature at 3 ROI's and it increased gradually after the period. During post-stimulus which was 6–7 min, the temperatures in 3 ROI's were stable and slight temperature decrease tendency can be seen. In most cases, especially in Fig. 17(D) supraorbital area was the most hottest part followed by periorbital and maxillary. In the third experiment, thermal facial image was also monitored for possible blood vessel which was visible through the stress experiment at the supraorbital area. We have managed to segment out the blood vessel automatically and calculate the pixel value before, during and after experiment in real time. In overall, during rest time which was 4–5 min, less blood vessel pixel was detected in comparison with during mental stress experiment (Fig. 16).

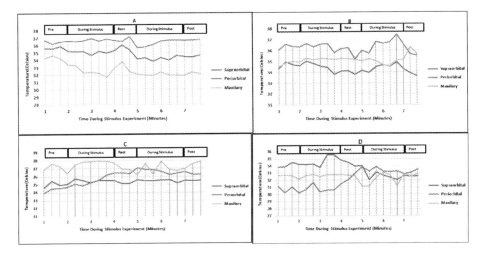

Fig. 16. A-D temperature of 4 subjects for the first time stimulus experiment conducted at 3 ROI's.

In the fourth experiment (ground truth), pulse rate and salivary amylase level was monitored for three correspondent through 2 types of stimulus. In overall, during 10–11 min, there was a slight decrease of pulse rate and salivary amylase level. This physiological measurement of pulse rate and salivary amylase level was consistent and correlate with the other vision based measurement proposed in this research (Fig. 18).

In the fifth experiment, classification experiment was done based on Support Vector Machine [11]. Seven features were used as the data entry which were mean value of number of blink frequency, average temperature of 3 ROI, average pixel blood vessel volume, pulse rate and Salivary amylase level. Number of

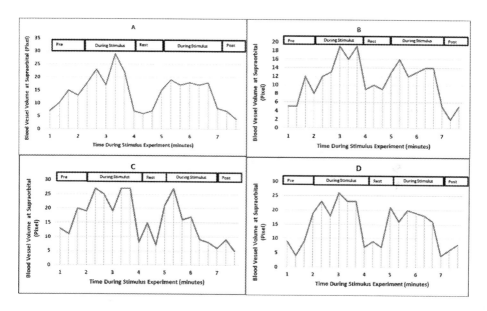

Fig. 17. Blood vessel volume at supraorbital for 4 correspondent.

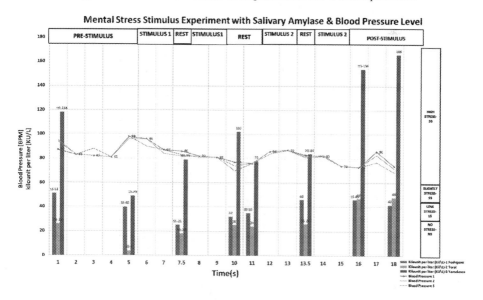

Fig. 18. Blood pressure and salivary amylase level for 3 correspondent.

classes were 3 which were before, during and after stimulus experiment. At this stage only 10 experimental subjects were collected and divided into 210 data entries. 20 samples were used as a test samples and the remaining to train the classifier. To evaluate the classification ability of the system under the different conditions, the total classification accuracy, which was the number of correctly classified samples divided by the number of total samples, was calculated for each classification conditions. The overall accuracy reached was 82.46 %.

2.4   Discussion

In this research, we only consider frontal face in both thermal and infrared case as the stimulus monitor is located in front of the correspondent in our screening environment. However, facial occlusion in stereo visual-infrared is still a challenging issue. In our first experiment, the recognition rate of eyes blinking in glasses case is slightly low from without glasses because of the shadow from the light which appears on the glass surface and the frame of the glasses. However it can be overcome by implementing proper screening environment. From this real time data it shows that blinking frequency is highly correlated with mental stress. More experiment data should be taken in the later stage. In the second experiment, we have also found out that user stress is correlated with the increased blood flow in three facial areas of sympathetic importance which is periorbital, supraorbital and maxillary which somehow increase the temperature during stimulus. From the third and fourth experiment, we can conclude that mental stress is highly correlated with the activation of the corrugator muscle on the forehead. However segmenting the thermal imprints of the supraorbital vessels is challenging because they are fuzzy due to thermal diffusion and exhibit significant inter-individual and intra-individual variation. On the average the diameter of the blood vessels is 10–15 m, which is too small for accurate detection and 0.1 °C warmer than the adjacent skin. In the fifth experiment, the overall SVM classification accuracy is considered high at 82.46 %. C.D. Katsis et al. [12] reported accuracy of only 79.3 % using SVM classification method by features extracted from EMG, ECG, respiration and EDA biosignals. Moreover, the image processing techniques is not used by [12] to extract facial characteristic as the the drivers need to wear a helmet. Physiological measurement of pulse rate and salivary amylase level are considered to be the most reliable measurement of mental stress according to a lot of researchers and they are used as a ground truth value in our research. They also shows correlation with other vision based measurement proposed in this research.

2.5   Conclusion

Our objective is to propose vision-based measurement as an alternative to physiological measurement. In this research three vision-based and two physiological measurement had been proposed and shows promising results. Usually in physiological measurement, combination of measures may be redundant with others and this may cause collection of unnecessarily large volumes of data and unnecessary processing time. This motivates the use of vision-based as it only requires crucial data after pattern recognition and processing of partial image. Our methodology to estimates emotional state from human subjects by extracting facial characteristic shows good performance.

Acknowledgment. This work was partially funded by MEXT/JSPS Kakenhi grant number: 50325768 and University Tun Hussein Onn Malaysia (UTHM). We would like to give special thanks to the laboratory members for their invaluable inputs and assistance.

References

1. Viola, P.A., Jones, M.J.: Rapid object detection using a boosted cascade of simple features. In: IEEE CVPR, pp. 511–518 (2001)
2. Sharma, N., Gedeon, T.: Objective measures, sensors and computational techniques for stress recognition and classification: a survey. Comput. Meth. Prog. Biomed. **3**, 1287–1301 (2012). Elsevier
3. Takai, N., Yamaguchi, M., et al.: Effect of physiology stress on the salivary cortisol and amylase levels in healthy young adults. Arch. Oral Biol. **49**, 963–968 (2004). Elsevier
4. Akhloufi, M., Bendada, A., Batsale, J.C.: State of the art in infrared face recognition. QIRT J., **X-N°X/2008**, 1–24 (2008)
5. Levine, J., Pavlidis, I., Cooper, M.: The face of fear. Lancet **357**, 1757 (2001)
6. Hartley, R., Zisserman, A.: Multiple View Geometry in Computer Vision. Cambridge University Press, Cambridge (2000)
7. Tan, E., Levine, J.: Human behavior: seeing through the face of deception. Nature **415**, 6867 (2002)
8. Merla, D.A., Tsiamyrtzis, P., Pavlidis, I.: Imaging facial signs of neurophysiological responses. IEEE Trans. Biomed. Eng. **56**, 2 (2009)
9. Lowe, D.G.: Distictive image features fromscale-invariant keypoints. Int. J. Comput. Vis. **60**, 91–110 (2004)
10. Shatri, D., Papadakis, M., Tsiamyrtzis, P.: Perinasal imaging of physiological stress and its affective potential. IEEE Trans. Affect. Comput. **3**, 366–378 (2012)
11. Chang, C.C., Lin, C.J.: LIBSVM: a library for support vector machines (2001). http://www.csie.ntu.edu.tw/cjlin/libsvm
12. Katsis, C.D., Katertsidis, N., et al.: Toward emotion recognition in car-racing drivers: a biosignal processing approach. IEEE Trans. Syst. Man Cybern. Part A: Syst. Hum **38**, 502–512 (2008)
13. Majumder, A., Bahera, L., Subramanian, V.K.: Automatic and robust detection of facial features in frontal face images. IEEE Trans. Syst. Man UKsim Int. Conf. Comput. Model Simul **3**, 331–336 (2011)
14. Mohd, M.N.H., Kashima, M., Sato, K., Watanabe, M.: Thermal-visual facial feature extraction based on nostril mask. IAPR-Mach. Vis. Appl. **4–20**, 113–116 (2013)
15. Mohd, M.N.H., Kashima, M., Sato, K., Watanabe, M.: Effective geometric calibration and facial feature extraction using multi sensors. Int. J. Eng. Sci. Innov. Technol. (IJESIT) **1–2**, 170–178 (2012)
16. Wang, J.-G.: Facial feature extraction in an infrared image by proxy with visible face image. IEEE Trans. Affect. Comput. **3–3**, 2057–2066 (2007)

A Delaunay-Based Temporal Coding Model for Micro-expression Recognition

Zhaoyu Lu, Ziqi Luo, Huicheng Zheng[✉], Jikai Chen, and Weihong Li

School of Information Science and Technology, Sun Yat-sen University,
Guangzhou 510006, China
{luzhaoyu,luoziqi,chenjik,liweih3}@mail2.sysu.edu.cn,
zhenghch@mail.sysu.edu.cn

Abstract. Micro-expression recognition has been a challenging problem in computer vision research due to its briefness and subtlety. Previous psychological study shows that even human being can only recognize micro-expressions with low average recognition rates. In this paper, we propose an effective and efficient method to encode the micro-expressions for recognition. The proposed method, referred to as Delaunay-based temporal coding model (DTCM), encodes texture variations corresponding to muscle activities on face due to dynamical micro-expressions. Image sequences of micro-expressions are normalized not only temporally but also spatially based on Delaunay triangulation, so that the influence of personal appearance irrelevant to micro-expressions can be suppressed. Encoding temporal variations at local subregions and selecting spatial salient subregions in the face area escalates the capacity of our method to locate spatiotemporally important features related to the micro-expressions of interest. Extensive experiments on publicly available datasets, including SMIC, CASME, and CASME II, verified the effectiveness of the proposed model.

1 Introduction

Expression recognition has important applications in human computer interaction and psychological study. Facial expressions can be broadly categorized into normal (macro-) and micro-expressions [5]. Usually, normal facial expressions provide rich information about the emotions of human being. But macro-expressions get ineffective when people are hiding their true emotions with faked expressions or do not show macro facial activities at all. On the contrary, micro-expressions often reveal true intents or hidden emotions corresponding to brief and subtle facial activities, which are often hard to be concealed. Ekman noticed the difference between micro-expressions and normal expressions, and started to study micro-expressions in the 1990's [5]. Micro-expressions are rapid, subtle, and involuntary facial expressions which tend to be concealed when people communicate with each other [7]. These micro-expressions reflect the true feelings. But it is hard for humans to notice or recognize such expressions when they take place.

So far, there are few studies on micro-expression recognition and numerous challenges are still to be solved. Firstly, micro-expressions are subtle, which

© Springer International Publishing Switzerland 2015
C.V. Jawahar and S. Shan (Eds.): ACCV 2014 Workshops, Part II, LNCS 9009, pp. 698–711, 2015.
DOI: 10.1007/978-3-319-16631-5_51

means that the facial activities would be minute and sparse, leading to difficulties of extracting features related to micro-expressions. For example, due to the subtleness, the extracted features are often dramatically influenced by the personal variance of appearance and facial movement irrelevant to expressions. Secondly, because micro-expressions often last for no more than 1/5 of a second [17], there are only a limited number of frames which contain the facial movements corresponding to micro-expressions captured in a video. It is challenging to recognize such facial activities, which are subtle and brief whereas complex and rich in emotional information.

In this paper, we propose the Delaunay-based temporal coding model (DTCM) to address the challenges in micro-expression recognition. In the proposed model, the image sequences containing micro-expressions are normalized not only temporally, but also spatially based on the Delaunay triangulation, to remove the influence of personal appearance on micro-expression recognition. In view of the sensitiveness of locations of facial feature points to noise for micro-expressions, we consider salient coding of texture variations in the subregions generated by triangulation for representation of micro-expressions. Instead of considering fixed areas of the face as in [19], we select subregions related to micro-expressions based on magnitudes of local texture fluctuations, which corresponds to spatial salient coding of local features and handles sparse and subtle facial movements well. For classification, we implement random forest (RF) [2] and support vector machine (SVM) [15]. Extensive experimental results on publicly available SMIC, CASME, and CASME II datasets verified the effectiveness of the proposed method.

The rest of this paper is organized as follows. Section 2 reviews representative related work. In Sect. 3, the overview of the proposed model is presented. Section 4 presents the coding space of the proposed model, including temporal and spatial normalization of the image sequences corresponding to micro-expressions, and construction of the coding space based on Delaunay triangulation. In Sect. 5, we explain how we carry out temporal coding at local subregions and select spatially salient subregions. Extensive experimental results are demonstrated in Sect. 6. This paper is concluded by Sect. 7.

2   Related Work

Micro- and macro-expression recognition are closely related. Research under one case often inspires that under the other case. Relatively more extensive research has been done for macro-expression recognition, which mainly focus on selection of emotion-related facial areas or extraction of features related to expressions.

Hamm et al. proposed a system for analysis of human facial movements, referred to as facial action coding system (FACS), which was successfully used in automatic expression recognition [9]. Micro-expressions are not considered in FACS. Moreover, it is not convenient to apply the system for expression recognition in practice, since action units in the system have to be marked manually. In [16,19], regions of interest related to facial expressions are selected from face images for recognition. However, the selected regions are fixed, ignoring distinct distributions among different micro-expressions.

There are mainly two strategies for expression representation, namely static approaches [3,13] and dynamic approaches [10,14,18,20,23]. Static approaches only focus on the peak frame of an image sequence and do not exploit the rich information in the dynamical interactions of facial muscles. On the other hand, dynamic approaches extract expression-related features from the full image sequences. For example, in [10], a method based on nonrigid registration using free-form deformations was proposed to represent the texture variation of action units [9] in the spatial-temporal domain. Wang et al. [20] introduced a temporal Bayesian network to capture complex spatiotemporal relations among facial muscles for expression recognition.

For micro-expression recognition, Pfister et al. [14] proposed a temporal interpolation model (TIM) to generate sufficient frames from an image sequence. They utilize the active shape model to detect the facial feature points, and implement local binary patterns extracted from three orthogonal planes (LBP-TOP) [23] to describe the spatiotemporal local textures. In later research, LBP-TOP is combined with tensor independent color space (TICS) and demonstrates better recognition results [19].

In this paper, we apply Delaunay triangulation and standard deviation analysis to locate facial sub-areas related to micro-expressions, which is very effective due to its adaptability to different micro-expressions. We encode the variations of textures instead of the movements of feature points in traditional methods such as [20], since micro-expressions are subtle and feature points are very likely to be static or sensitive to noises. As verified by extensive experiments, the proposed model provides a very effective representation of micro-expressions and leads to promising accuracies for micro-expression recognition.

3   Overview of the Proposed Model

Figure 1 illustrates the basic steps in the proposed method. First, image sequences containing micro-expressions are normalized in the time domain by using the temporal interpolation model [14]. Then Delaunay triangulation [1] and mapping [6] is implemented according to the landmarks generated with the active appearance model (AAM) [4] fitting. The face area is normalized spatially based on Delaunay triangulation to avoid the disastrous influence of personal appearance variances. By dividing the overall face area into a collection of local subregions, rich discriminative local descriptors may be extracted for micro-expression representation. More specifically, we extract texture features from the generated Delaunay triangles. We analyze the variations of Delaunay triangles and select those potentially related to micro-expressions according to the magnitudes of local variations. The selected Delaunay triangles are encoded in the time domain to preserve temporal information in micro-expression representation. Finally, the micro-expressions are recognized by using RF or SVM classifiers.

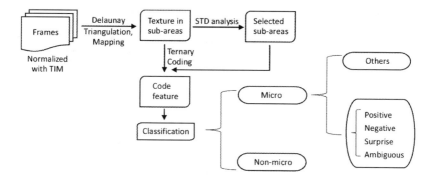

Fig. 1. Overview of the proposed model.

4   Coding Space for DTCM

In this section, we construct the spatiotemporal coding space for the proposed DTCM, which will be used later for representation of micro-expressions in image sequences.

4.1   Temporal Normalization

To remove temporal fluctuations of micro-expressions, which may introduce noise irrelevant to classification of micro-expressions, we apply TIM [14] to normalize image sequences corresponding to micro-expressions temporally. TIM is a manifold-based interpolation method, which builds a low-dimensional embedding of an image sequence, and then interpolates a curve in the low-dimensional space. The interpolated frames are mapped back to the original high-dimensional space to form the temporally normalized image sequence.

4.2   Delaunay Triangulation and Mapping

In order to extract rich description of facial expressions, we define 68 feature points as in [8]. The AAM is implemented to obtain these feature points in a frame with a neutral face from the image sequence. The AAM exploits both texture and shape information to trace facial feature points with a model building phase and a face fitting phase. The fitting process iterates until the Euclidean distance between the model texture vector and the instance texture vector converges. It provides satisfactory feature points for micro-expression recognition based on our method.

Since micro-expressions are revealed by subtle muscle movements in the face, we focus on variations of the image sequence in the facial area. In traditional methods, subregions remain fixed and not adaptive to different faces, which would introduce expression-irrelevant noise. To have a detailed and descriptive representation of the facial expression, we implement Delaunay triangulation

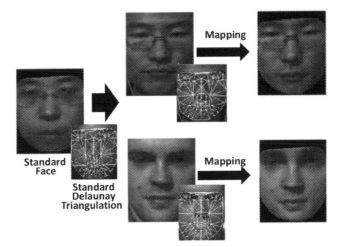

Fig. 2. Delaunay triangulation and mapping. With a standard face, all face images are mapped into the same domain defined by 68 feature points.

based on the point set in the facial area. The Delaunay triangulation is unique for a given set of feature points in general position and there is no other feature point other than the three vertices in the circumcircle of each triangle [12]. The uniqueness with respect to the feature points can be exploited to normalize all faces into a standard space. In traditional methods, subregions remain fixed and not adaptive to different faces, which would introduce expression-irrelevant noise. The triangulation process divides the overall facial area into a number of triangular subregions based on the previously detected feature points. We denote the n sub-areas by $P = \{p_1, p_2, \ldots, p_n\}$. The texture variations in the subregions signal the movements of facial muscles and can be used to characterize micro-expressions.

As micro-expressions are subtle, the landmarks corresponding to feature points rarely change in the image sequence. Based on this observation, we propose to focus on the variations of local textures in the triangles instead of the feature points themselves.

To normalize all faces into a standard space and remove the influence of personal appearances on micro-expression recognition, triangular mapping is applied to all the images in the sequence. As illustrated in Fig. 2, a standard Delaunay triangulation is first determined based on a chosen standard face. With this standard triangulation, other face images with different appearances will be mapped to the same triangulation. In this way, personal appearance difference irrelevant to micro-expressions will be largely removed. After normalization based on the triangular mapping, all face images will have the same amount of pixels in each triangular subregion. The standard face can be a neutral face of any subject in the dataset. In our experiments, we simply choose the neutral face in the first sequence of the first subject for each dataset.

4.3   Construction of the Coding Space

Let $F = \{f_1, f_2, \ldots, f_t\}$ be the collection of t frames in the temporally normalized image sequence and $p_{i,j}$, with $j = 1, 2, \ldots, t$, be the subregion p_i in the j-th frame. We define the differences between neighboring texture signatures corresponding to $p_{i,j-1}$ and $p_{i,j}$ as the local temporal variations (LTVs) $X = \{x_{i,j} | i = 1, 2, \ldots, n; j = 2, 3, \ldots, t\}$. A spatiotemporal coding space $\Gamma = \{F, P, X\}$ is then constructed. In this paper, we simply choose the grayscale values in each subregion $p_{i,j}$ as the texture feature. For micro-expressions, the facial movements are subtle. Our primary purpose is to describe the weak variations of the expression-related subregions and construct strong descriptors out of the weak ones efficiently. The temporal grayscale variation appears to be a good feature for subtle facial variations according to our experiments. More complex textual features may easily contain a lot of expression-irrelevant information such as the appearance, which overshadows real facial movements.

5   Micro-expression Coding Based on DTCM

This section explains how we encode a micro-expression using DTCM based on the previously constructed coding space.

5.1   Extraction of Local Temporal Variations

The LTV $x_{i,j}$ reflects the difference between the texture features of neighboring frames in the subregion p_i. Let $T_{i,j} \in R^m$ be an m-dimensional texture vector extracted from the j-th frame in the i-th subregion $p_{i,j}$. In this paper, we define the LTV as follows,

$$x_{i,j} = \frac{\sum_{k=1}^{m} (T_{i,j,k} - T_{i,j-1,k})}{m} \tag{1}$$

where $T_{i,j,k}$ denotes the k-th element in $T_{i,j}$.

In this paper, $T_{i,j}$ consists of all pixel gray values in the subregion $p_{i,j}$. So m is simply the number of pixels in $p_{i,j}$, and $x_{i,j}$ the difference between mean values of the subregion p_i computed from neighboring frames. The extracted LTVs capture the overall variations in the corresponding subregions between nearby frames. Specifically, a positive $x_{i,j}$ signals the overall lightening in the corresponding subregion and a negative one signals the overall darkening.

5.2   Adaptive Threshold and Ternary Coding

The facial movement amplitude, indicated by the value of $x_{i,j}$, is sensitive to noise. To improve the robustness, we only preserve local temporal variations with significant magnitudes, which corresponds to the idea of salient coding. Let $x_{i,j,r}$ be LTVs of the r-th sequence, where $r = 1, \ldots, R$. We define $x_{i,j,r}^+ = \{x_{i,j,r} | x_{i,j,r} > 0\}$ and $x_{i,j,r}^- = \{x_{i,j,r} | x_{i,j,r} < 0\}$. Then a positive and a negative thresholds are defined for each subregion p_i as follows,

$$\tau_i^+ = \frac{a}{R} \sum_{r=1}^{R} \text{mean}\{x_{i,j,r}^+\} \tag{2}$$

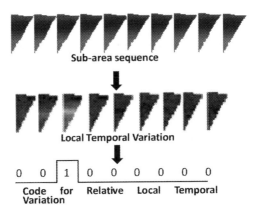

Fig. 3. Coding process for the local temporal variations.

$$\tau_i^- = \frac{a}{R} \sum_{r=1}^{R} \text{mean}\{x_{i,j,r}^-\} \tag{3}$$

where mean($\cdot$) denotes the mean value of the set, and a is a positive real number to be chosen appropriately. The thresholds defined in (2) and (3) are adaptive for the subregions to cope with spatially-varying noise.

We then propose a ternary coding method to encode the LTVs for representation of micro-expressions by defining

$$c_{i,j} = \begin{cases} 1, & x_{i,j} \geqslant \tau_i^+ \\ 0, & \tau_i^- < x_{i,j} < \tau_i^+ \\ -1, & x_{i,j} \leqslant \tau_i^- \end{cases} \tag{4}$$

The code values stand for three different relative variations in texture, namely, lightening, remaining, and darkening, which can be used to describe significant variations in the subregions. With the salient coding method, noise corresponding to illumination variations and inaccurate alignments can be removed. It is also useful for reducing the influence of personal appearances. The coding process is illustrated in Fig. 3.

5.3   Feature Selection

With Delaunay triangulation, abundant subregions will be generated, which leads to a large number of local features. Some of these local features may be noisy and irrelevant to micro-expressions. Therefore, we introduce a feature selection process in our model to determine subregions related to micro-expressions. The approach is based on analysis of the standard deviation (STD) of $x_{i,j}$, which reveals the saliency of a subregion relevant to micro-expressions. Specifically, the proposed STD analysis allows feature selection without a data-dependent pre-training stage, which is prone to overfitting as observed in traditional discriminative methods. Furthermore, the proposed method allows the selected features

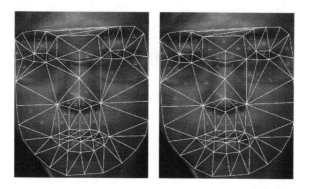

Fig. 4. Top 60 subregions determined as most related to the corresponding micro-expressions based on STD analysis. The small triangles mark the identified subregions. The subregions most related to the corresponding negative and positive micro-expressions are indicated in the left and right pictures, respectively.

to be adaptive to specific input instances. The personal STD is computed in each subregion and for each image sequence individually as

$$
\mathrm{STD}_i = \sqrt{\frac{1}{t-1} \sum_{j=2}^{t} \left(x_{i,j} - \frac{1}{t-1} \sum_{k=2}^{t} x_{i,k} \right)^2}
\tag{5}
$$

The magnitude of STD_i reflects the strength of facial movement in the corresponding facial subregion. We select N subregions with the largest STD values, which are assumed to be mostly related to the micro-expressions. Other subregions with lower STD values are assumed to be irrelevant to micro-expressions and coded as zeros. Figure 4 shows two examples where 60 subregions determined as most related to the corresponding positive or negative micro-expression are identified based on the STD analysis.

We concatenate the code sequences of all subregions in the image sequence into one feature vector as representation of the micro-expression. Then the feature vector is classified by using the RF or SVM classifier, which have demonstrated good performance when dealing with high-dimensional data.

6   Experimental Results

In this section, we test the proposed DTCM on three publicly available micro-expression datasets, including SMIC, CASME, and CASME II to verify its effectiveness for micro-expression recognition. These datasets consist of spontaneous micro-expressions which appear in real life. For each dataset, we first compare the proposed method to state-of-the-art methods in terms of the micro-expression recognition rates. Then we investigate the influence of the parameters of DTCM on the recognition results. Leave-one-subject-out cross validation is used in all experiments.

Table 1. The results of separating micro and non-micro expressions

Method	Accuracy (%)
LBP-TOP+TIM10+RF [14]	74.3
DTCM+RF (Test 1)	79.80
DTCM+RF (Test 2)	85.86
DTCM+RF (Test 3)	**88.89**

Table 2. Classification of positive and negative micro-expressions

Method	Accuracy (%)
LBP-TOP+TIM10+MKL [14]	71.4
DTCM+SVM (SMO) (Test 1)	74.3
DTCM+SVM (SMO) (Test 2)	80
DTCM+SVM (SMO) (Test 3)	**82.86**

6.1 Results on SMIC

We first verify the performance of the proposed method on the SMIC dataset. The image sequences taken with a high speed camera are used in the experiments. This dataset consists of 70 sequences of negative micro-expressions, 51 sequences of positive micro-expressions, and 164 sequences without micro-expressions. All the sequences last for no more than $1/2$ s, which means that the number of frames are less than 50 for a sequence captured with a camera at 100fps [11].

We follow the same experimental protocol as in [14], which combines TIM and LBP-TOP for recognizing micro-expressions on the same dataset. More specifically, we randomly select 77 image sequences without micro-expressions, $18/17$ sequences containing negative/positive micro-expressions from the SMIC dataset. We repeat the random selection process several times and record the recognition results.

Table 1 shows the results of separating micro-expressions from non-micro-expressions, where RF refers to the random forest algorithm. The parameters $a = 1.0$ and $N = 60$ in the proposed method. As we can see, in several random runs of the proposed method, the proposed method obtains better results than that reported in [14]. Table 2 illustrates the results of classifying positive and negative micro-expressions, where SVM (SMO) refers to support vector machine with sequential minimization optimization. Linear SVM is implemented in the experiments. The parameters $a = 1.0$ and $N = 70$ in the proposed method. Again, in several random runs, the proposed method reports higher accuracies than the method based on LBP-TOP+TIM [14].

There are mainly two parameters in the proposed method, namely a for the thresholds τ_i^+ and τ_i^- defined in (2) and (3), and N for the number of selected subregions related to expressions in Sect. 5.3. a controls the thresholds for determining if local temporal variations are sufficiently salient. Small values of a lead to

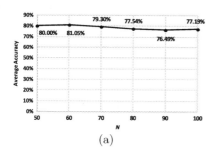 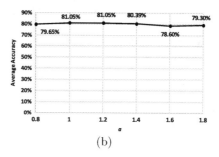

(a)   (b)

Fig. 5. Influence of parameters (a) N and (b) a for separating micro-expressions from non-micro-expressions.

higher sensitivity to temporal noises while large values may suppress useful temporal information. The number of selected local regions N controls the sensitivity to noise in the spatial domain. Small values of N can suppress most spatial noise but could also lose useful information. On the other hand, large values of N would be sensitive to noise. To investigate the influence of these parameters in practice, we used all videos including 51 sequences with positive micro-expressions, 70 sequences with negative micro-expressions, and 164 sequences with non-micro-expressions from the SMIC dataset.

Figure 5a shows the influence of N for separating micro-expressions from non-micro-expressions, where a is fixed at 1.0. As we can see, for N from about 50 to 70, which corresponds to about 45–64 % of the number of local regions in the face area, the proposed model shows reasonably good classification accuracies. Figure 5b illustrates the influence of a in the experiments, where N is fixed at 60. In the tested range of values, the proposed method shows promising results. Generally, a value of a between 0.8 and 1.4 seems to be a good choice.

For the experiments of separating positive and negative micro-expressions, the influence of the parameters N and a are illustrated in Figure. 6a (a is fixed at 1.0) and Figure. 6b (N is fixed at 70), respectively. It seems that the results are sensitive to the choice of N and a. This is probably due to the fact that separating two types of micro-expressions is more challenging than separating micro-expressions from non-micro-expressions. Nevertheless, the proposed method obtained reasonable results in the tested range of N. A choice of around $N = 70$ gives the best results. For the parameter a, the performance is best for a between 0.8 and 1.2, consistent with the conclusion drawn from Figure. 5b.

To verify the effectiveness of Delaunay triangulation, we conduct another experiment for comparison, where Delaunay triangulation is replaced by small-grid-based segmentation as in [14]. Specifically, an image is split into 10×10 square regions and top 20 regions are selected with STD analysis, which correspond to about the same area as 60 salient Delaunay triangles. The accuracy drops to 63.6 % for positive and negative micro-expression classification. This is because the grid-based segmentation is fixed and can not be adapted to different faces. Due to different facial appearances, subregions located at the same coordinate may not correspond to the same semantic region of different people.

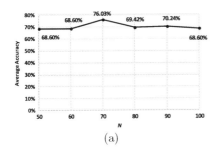 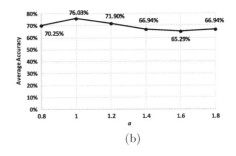

(a)   (b)

Fig. 6. Influence of parameters (a) N and (b) a for separating positive and negative micro-expressions.

Table 3. Results of four-category micro-expression recognition on CASME

Method	Accuracy (%)
LBP-TOP+TIM70+TICS+SVM [19]	61.86
DTCM (Grid segmentation)	52.58
DTCM+RF	**64.95**

6.2   Results on CASME

Experiments in this section are based on sequences in Class B of the CASME dataset, which were recorded by Point Grey GRAS-03K2C camera at 60fps [22]. The dataset contains eight categories of micro-expressions. Since some categories contain very few samples, we follow the experimental settings in [19] and carry out four-category classification on this dataset. The first category is positive, which corresponds to the happiness micro-expression. The second category is negative, which consists of disgust, sadness, and fear micro-expressions. The third category is surprise, and the last category contains ambiguous micro-expressions, including tense and repression. The four categories contain 4, 47, 13, and 33 samples, respectively.

Table 3 shows the results of four-category micro-expression recognition on CASME. The parameters $a = 1.0$ and $N = 40$ in the proposed method. As we can see, under the same experimental protocol, we achieve a higher recognition rate than the method combining LBP-TOP, TIM and TICS [19]. When Delaunay triangulation is replaced by grid-based segmentation (10×10 squares), the accuracy drops by 12.37 %.

Figures 7a and b show the influence of the parameters N and a in the proposed method, where a and N are fixed at 1.0 and 40, respectively. As we can see, a choice of N around 40 gives the best result, which corresponds to about 36 % of the subregions in the face area. For the parameter a, a choice between 0.8 and 1.2 would still be recommended.

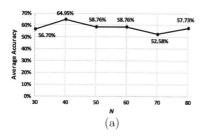 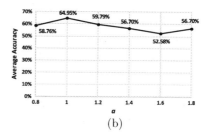

(a)   (b)

Fig. 7. Influence of parameters (a) N and (b) a for four-category micro-expression recognition on CASME.

6.3 Results on CASME II

CASME II [21] contains high-quality sequences captured by a Point Grey GRAS-03K2C camera at 200fps, which consists of five categories of micro-expressions, including happiness, surprise, disgust, repression, and others. There are 32, 25, 64, 27, and 99 samples in these categories, respectively.

Wang et al. [19] reported an accuracy of 58.54 % in the five-category classification. However, since "others" contains a mixture of different facial activities, in our experiments, this category is singled out first by our method, with a success rate of 72.06 %. Our experiments show that the approach is not very sensitive to the number of selected subregions N. For the parameter a, a choice between about 0.8 and 1.2 is still recommended. Setting $a = 0.8$ and $N = 50$ leads to the overall best accuracy 72.06 %. In the experiments of four-category micro-expression recognition, with a fixed at 1.0 and N around 70 (which corresponds to about 63 % of the facial subregions), the best recognition accuracy 64.19 % is obtained. For the parameter a, we observe that when it is around 1.0, a good recognition accuracy can be obtained, consistent with conclusions in the previous experiments.

7   Conclusions

In this paper, we propose a method for micro-expression recognition with Delaunay-based temporal coding (DTCM). An arbitrary sequence of micro-expression is normalized not only temporally, but also spatially based on Delaunay triangulation, which aims to remove the influence of personal appearance on recognition. It is especially meaningful for micro-expressions, which are subtle and brief, and can easily be concealed by appearances irrelevant to the expressions of interest. Furthermore, we propose to consider texture information in the subregions generated by the triangulation instead of the fitted feature points, since for micro-expressions, locations of feature points are not discriminative enough and tend to be dominated by noise. The features extracted from local subregions are then coded temporally base on local variations, which captures the saliency distribution in the time domain. Furthermore, spatial selection of local subregions are carried out to extract spatial saliency distribution information. Extensive experimental results on public datasets including SMIC, CASME, and CASME II demonstrate the effectiveness of the proposed method for micro-expression recognition.

Acknowledgement. This work is supported by National Natural Science Foundation of China (No. 61172141), Key Projects in the National Science &Technology Pillar Program during the 12th Five-Year Plan Period (No. 2012BAK16B06), Science and Technology Program of Guangzhou, China (2014J4100092), and National Undergraduate Scientific and Technological Innovation Project of China.

References

1. Barber, C., Dobkin, D., Venue, H.: The quickhull algorithm for convex hulls. ACM Trans. Math. Softw. **22**, 469–483 (1996)
2. Breiman, L.: Random forests. Mach. Learn. **45**, 5–32 (2001)
3. Chu, W., Torre, F., Cohn, J.: Selective transfer machine for personalized facial action unit detection. In: International Conference on Computer Vision, pp. 3515–3522 (2013)
4. Cootes, T., Edwards, G., Taylor, C.: Active appearance models. IEEE Trans. Pattern Anal. Mach. Intell. **23**, 681–685 (2001)
5. Ekman, P.: Facial expressions of emotion: an old controversy and new findings. Philos. Trans. R. Soc. Lond. Ser. B. Biol. Sci. **335**, 63–69 (1992)
6. Fournier, A., Montuno, D.: Triangulating simple polygons and equivalent problems. ACM Trans. Graph. **3**, 153–174 (1984)
7. Frank, M., Herbasz, M., Sinuk, K., Keller, A., Nolan, C.: I see how you feel: Training laypeople and professionals to recognize fleeting emotions. In: The Annual Meeting of the International Communication Association, pp. 3515–3522, New York (2013)
8. Gross, R., Matthews, I., Cohn, J., Kanade, T., Baker, S.: Multi-PIE. Image Vis. Comput. **28**, 807–813 (2010)
9. Hamm, J., Verma, R., Kohler, C., Gur, R.: Automated facial action coding system for dynamic analysis of facial expressions in neuropsychiatric disorders. J. Neurosci. Methods **200**, 237–256 (2011)
10. Koelstra, S., Pantic, M., Patras, I.: A dynamic texture-based approach to recognition of facial actions and their temporal models. IEEE Trans. Pattern Anal. Mach. Intell. **32**, 1940–1954 (2010)
11. Li, X., Pfister, T., Huang, X., Zhao, G., Pietikainen, M.: A spontaneous micro-expression database: Inducement, collection and baseline. In: IEEE Conference on Automatic Face and Gesture Recognition, pp. 1–6 (2013)
12. Mark, B., Cheong, O., Kreveld, M., Overmars, M.: Computational Geometry: Algorithms and Applications. Springer, Berlin (2008)
13. Park, S., Kim, D.: Spontaneous facial expression classification with facial motion vectors. In: IEEE Conference on Automatic Face and Gesture Recognition, pp. 1–6 (2008)
14. Pfister, T., Li, X., Zhao, G., Pietikainen, M.: Recognising spontaneous facial micro-expressions. In: International Conference on Computer Vision, pp. 1449–1456 (2011)
15. Platt, J.: Sequential minimal optimization: a fast algorithm for training support vector machines. Microsoft Research, MSR-TR-98-14, pp. 1–21 (1998)

16. Polikovsky, S., Kameda, Y., Ohta, Y.: Facial micro-expressions recognition using high speed camera and 3D-gradient descriptor. In: International Conference on Distributed Platforms, pp. 1–6 (2009)
17. Shen, X.B., Wu, Q., Fu, X.L.: Effects of the duration of expressions on the recognition of microexpressions. J. Zhejiang Univ. Sci. B **13**, 221–230 (2012)
18. Shreve, M., Godavarthy, S., Goldgof, D., Sarkar, S.: Macro-and micro-expression spotting in long videos using spatio-temporal strain. In: IEEE Conference on Automatic Face and Gesture Recognition, pp. 51–56 (2011)
19. Wang, S., Yan, W., Li, X., Zhao, G., Fu, X.: Micro-expression recognition using dynamic textures on tensor independent color space. In: International Conference on Pattern Recognition (2014)
20. Wang, Z., Wang, S., Ji, Q.: Capturing complex spatio-temporal relations among facial muscles for facial expression recognition. In: Conference on Computer Vision and Pattern Recognition, pp. 3422–3429 (2013)
21. Yan, W., Li, X., Wang, S., Zhao, G., Liu, Y., Chen, Y., Fu, X.: CASME II: an improved spontaneous micro-expression database and the baseline evaluation. PLoS ONE **9**, 1–8 (2014)
22. Yan, W., Wu, Q., Liu, Y., Wang, S., Fu, X.: CASME database: a dataset of spontaneous micro-expressions collected from neutralized faces. In: IEEE Conference on Automatic Face and Gesture Recognition, pp. 1–7 (2013)
23. Zhao, G., Pietikainen, M.: Dynamic texture recognition using local binary patterns with an application to facial expressions. IEEE Trans. Pattern Anal. Mach. Intell. **29**, 915–928 (2007)

Author Index